ELASTICITY

Theory and Applications

ELASTICITY

Theory and Applications

Second Edition, Revised and Updated

Adel S. Saada, Ing., E.C.P., Ph.D.
Professor of Civil Engineering
Case Western Reserve University

ISBN-13: 978-1-60427-019-8

Printed and bound in the U.S.A. Printed on acid-free paper
10 9 8 7 6 5 4 3 2 1

This J. Ross Publishing edition, first published in 2009, is a revised, updated, and reformatted publication of the works originally published by Krieger Publishing in 1993.

Library of Congress Cataloging-in-Publication Data

Elasticity : theory and applications / by Adel S. Saada. -- 2nd ed., rev. & updated.

p. cm.

Includes bibliographical references and index.

ISBN 978-1-60427-019-8 (pbk. : alk. paper)

1. Elasticity--Textbooks. I. Title.

QA931.S2 2009

531'.382--dc22 2008055164

Direct all inquiries to J. Ross Publishing, Inc., 5765 N. Andrews Way, Fort Lauderdale, FL 33309.

Phone: (954) 727-9333
Fax: (561) 892-0700
Web: www.jrosspub.com

Contents

Part II

THEORY OF STRESS

About the Author

Adel S. Saada (Ing., E.C.P., Ph.D., Princeton University) is presently Professor of Civil Engineering at Case Western Reserve University, Cleveland, Ohio. Dr. Saada received his *Ingénieur des Arts et Manufactures* degree from École Centrale des Arts et Manufactures de Paris, France, and the equivalent of a Master of Science degree from the University of Grenoble, France. Before coming to Princeton University the author was a practicing structural engineer in France. Dr. Saada's teaching activities are in two major areas: the first is that of the mechanics of solids and in particular elasticity; the second is that of mechanics applied to soils and foundations. His research activities are primarily in the area of stress-strain relations and failure of transversely isotropic geomaterials. Much of his research work has been supported by personal grants from the National Science Foundation, the U.S. Army, and the U.S. Air Force. Dr. Saada is a member of several professional societies, a consulting engineer, and the author of numerous papers published in both national and international journals.

Preface

This book is intended to give advanced undergraduate and graduate students sound foundations on which to build advanced courses such as mathematical elasticity, plates and shells, plasticity, fracture, and those branches of mechanics which require the analysis of strain and stress. The book is divided into three parts: *Part I* is concerned with the kinematics of continuous media, *Part II* with the analysis of stress, and *Part III* with the theory of elasticity and its applications to engineering problems.

In *Part I*, the use of the notion of linear transformation of points makes it possible to present the geometry of deformation in a language that is easily understood by the majority of engineering students. It is agreed that tensor calculus is the most elegant tool available to mechanicists, but experience has shown that most engineering students are not ready to accept it without a reasonable amount of preparation. The study of finite and linear strains, using the notion of linear transformation, gradually introduces the tensor concept and removes part of the abstraction commonly associated with it. Orthogonal curvilinear coordinates are examined in detail and the results extensively used throughout the text.

In *Part II*, the study of stress proceeds along the same lines as that of strain, and the similarities between the two are pointed out. All seven chapters of Parts I and II are essential to the understanding of Part III and serve as a common base for all branches of mechanics.

In *Part III*, Chapter 8 covers the three-dimensional theory of linear elasticity and the requirements for the solution of elasticity problems. The method of potentials is presented in Chapter 9. Torsion is discussed in Chapter 10 and topics related to cylinders, disks, and spheres are treated in Chapter 11. Straight and curved beams are analyzed in Chapters 12 and 13 respectively, and the answers of the elementary theories are compared to the more rigorous results of the theory of elasticity. In Chapter 14, the semi-infinite elastic medium and some of its related problems are studied using the results of Chapter 9.

Energy principles and variational methods are presented in Chapter 15 and their application illustrated by a large number of simple examples. Columns and beam-columns are discussed in Chapter 16 and the bending of thin flat plates in Chapter 17. Chapter 18 is more than an introduction to the theory of thin shells. It includes a relatively detailed presentation of the theory of surfaces which is necessary for the full understanding of the analysis of thin shells. In this Chapter, as well as throughout this text, geometry and the relations between strain and displacement are emphasized; since it is my conviction that once geometry is mastered most of the difficulties in studying the mechanics of solids will have disappeared.

Chapter 19 deals with the use of complex variables in elasticity. This tool is at the heart of many elegant solutions in solid mechanics, and in particular in fracture mechanics.

The theories and applications presented in a book of this nature do not change from year to year. What changes is the background of the students in our classrooms, the nature of the questions they ask and the tools that are available to answer those questions. Changes in curricula invariably involve the removal of material, sometimes basic, in favor of newer courses deemed more important. This creates gaps that need to be filled when one is introducing topics related to such material.

In response to questions from instructors and students alike, an addendum to the second edition was written and widely distributed. It contains comments, detailed explanations, new problems and a large number of numerical solved examples, all aimed at helping readers get a better understanding of the topic at hand.

In this edition, the addendum has been bound with the main text and cross references (*) direct the reader to the pages where supplemental information can be found.

Nearly all the problems in this textbook require numerical answers, an approach that the author considers indispensable to engineering education and the development of engineering judgment. The need for computers is minimal, meaning that all problems can be solved using a simple calculator. However, much time can be saved if this calculator is programmable or if one has access to software dealing with operations on matrices, eigenvalues and the solution of sets of linear equations. A solution manual is available to instructors.

I wish to express my gratitude to Dr. T.P. Kicher who reviewed the manuscript of the first edition of this book; to Dr. G.P. Sendeckyj with whom many sections were discussed; and to the late Dr. A. Mendelson for allowing the use of his class notes in preparing Chapter 19 and for reviewing this chapter. Mrs. W. Reeves very ably handled the typing for the first edition and Mrs. K. Ballou accomplished the near impossible task of typing Chapter 19 and the addendum; both deserve praise for their patience and ability.

Last but not least, I wish to acknowledge the encouragement and understanding of my wife Nancy during the various stages of writing this book.

Adel S. Saada

PART I

KINEMATICS OF CONTINUOUS MEDIA

(Displacement, Deformation, Strain)

Chapter 1

INTRODUCTION TO THE KINEMATICS OF CONTINUOUS MEDIA

1.1 Formulation of the Problem

The theory of deformation of continuous media is a purely mathematical one. It is concerned with the study of the intrinsic properties of the deformations independent of their physical causes. It is most conveniently expressed by the notion of transformation, which implies displacement and change in shape. The problem is formulated as follows: Given the positions of the points of a body in its initial state (i.e., before transformation) and in its final state (i.e., after transformation), it is required to determine the change in length and in direction of a line element joining two arbitrary points originally at an infinitesimal distance from one another.

In the following, we shall make use primarily of orthogonal sets of cartesian coordinates. Let x_1, x_2, x_3 be the coordinates of a point M of a body B before transformation. After transformation, this point becomes M^* with coordinates ξ_1, ξ_2, ξ_3 :

$$\begin{aligned} \xi_1 &= x_1 + u_1 \\ \xi_2 &= x_2 + u_2 \\ \xi_3 &= x_3 + u_3, \end{aligned} \tag{1.1.1}$$

where u_1, u_2, u_3 are the projections of $\overline{MM^*}$ on the three axes OX_1, OX_2, OX_3 (Fig. 1.1). We shall assume that u_1, u_2, u_3, as well as their

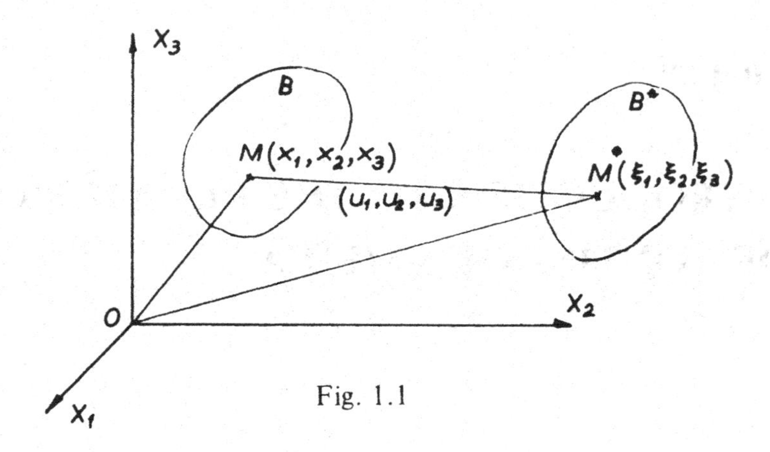

Fig. 1.1

partial derivatives with respect to x_1, x_2, x_3, are continuous functions of x_1, x_2, x_3. Eqs. (1.1.1) can therefore be written as:

$$\begin{aligned}\xi_1 &= x_1 + u_1(x_1, x_2, x_3)\\ \xi_2 &= x_2 + u_2(x_1, x_2, x_3)\\ \xi_3 &= x_3 + u_3(x_1, x_2, x_3).\end{aligned} \tag{1.1.2}$$

Let us consider two points, $M(x_1, x_2, x_3)$ and $N(x_1 + dx_1, x_2 + dx_2, x_3 + dx_3)$, infinitesimally near one another. As a result of the transformation, M is displaced to $M^*(\xi_1, \xi_2, \xi_3)$ and N is displaced to $N^*(\xi_1 + d\xi_1, \xi_2 + d\xi_2, \xi_3 + d\xi_3)$ (Fig. 1.2). The coordinates of N^* are

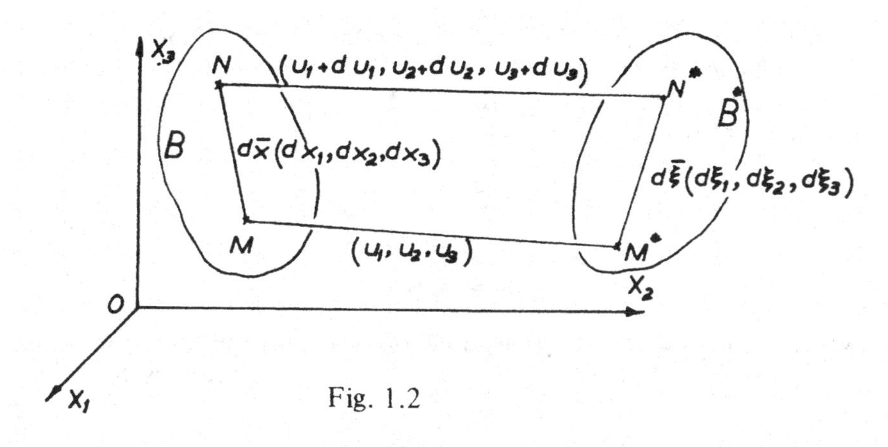

Fig. 1.2

given by:

$$\begin{aligned} \xi_1 + d\xi_1 &= x_1 + dx_1 + u_1 + du_1 \\ \xi_2 + d\xi_2 &= x_2 + dx_2 + u_2 + du_2 \\ \xi_3 + d\xi_3 &= x_3 + dx_3 + u_3 + du_3 . \end{aligned} \tag{1.1.3}$$

Because of the assumptions on u_1, u_2, u_3, we can write the displacement of N under the form of a Taylor series in the neighborhood of M:

$$u_1 + du_1 = (u_1)_M + \left(\frac{\partial u_1}{\partial x_1}\right)_M dx_1 + \left(\frac{\partial u_1}{\partial x_2}\right)_M dx_2 + \left(\frac{\partial u_1}{\partial x_3}\right)_M dx_3 + \ldots$$

$$u_2 + du_2 = (u_2)_M + \left(\frac{\partial u_2}{\partial x_1}\right)_M dx_1 + \left(\frac{\partial u_2}{\partial x_2}\right)_M dx_2 + \left(\frac{\partial u_2}{\partial x_3}\right)_M dx_3 + \ldots$$

$$u_3 + du_3 = (u_3)_M + \left(\frac{\partial u_3}{\partial x_1}\right)_M dx_1 + \left(\frac{\partial u_3}{\partial x_2}\right)_M dx_2 + \left(\frac{\partial u_3}{\partial x_3}\right)_M dx_3 + \ldots$$

(1.1.4)

If we substitute Eqs. (1.1.4) in Eqs. (1.1.3), and subtract Eqs. (1.1.1) from the resulting equations, we obtain:

$$d\xi_1 = \left[1 + \left(\frac{\partial u_1}{\partial x_1}\right)_M\right] dx_1 + \left(\frac{\partial u_1}{\partial x_2}\right)_M dx_2 + \left(\frac{\partial u_1}{\partial x_3}\right)_M dx_3 + \ldots$$

$$d\xi_2 = \left(\frac{\partial u_2}{\partial x_1}\right)_M dx_1 + \left[1 + \left(\frac{\partial u_2}{\partial x_2}\right)_M\right] dx_2 + \left(\frac{\partial u_2}{\partial x_3}\right)_M dx_3 + \ldots \tag{1.1.5}$$

$$d\xi_3 = \left(\frac{\partial u_3}{\partial x_1}\right)_M dx_1 + \left(\frac{\partial u_3}{\partial x_2}\right)_M dx_2 + \left[1 + \left(\frac{\partial u_3}{\partial x_3}\right)_M\right] dx_3 + \ldots ,$$

If, in Eqs. (1.1.5), we neglect the higher-order terms of Taylor's series, the relations between $d\xi_1, d\xi_2, d\xi_3$ and dx_1, dx_2, dx_3 become linear. The system of equations can be looked upon as an operation which transforms a vector $d\bar{x}$ (dx_1, dx_2, dx_3) of length ds to a vector $d\bar{\xi}$ $(d\xi_1, d\xi_2, d\xi_3)$ of length ds^*. This type of transformation is called *linear transformation*. It is the linearization of Eqs. (1.1.5) that allows us to assume that the vector $d\bar{x}$ is transformed to a vector $d\bar{\xi}$ and not to a curve. The properties of linear transformations are discussed in Chapter 3. If we omit the subscript M, Eqs. (1.1.5) are written as:

$$\begin{aligned} d\xi_1 &= \left(1 + \frac{\partial u_1}{\partial x_1}\right)dx_1 + \frac{\partial u_1}{\partial x_2}dx_2 + \frac{\partial u_1}{\partial x_3}dx_3 \\ d\xi_2 &= \frac{\partial u_2}{\partial x_1}dx_1 + \left(1 + \frac{\partial u_2}{\partial x_2}\right)dx_2 + \frac{\partial u_2}{\partial x_3}dx_3 \\ d\xi_3 &= \frac{\partial u_3}{\partial x_1}dx_1 + \frac{\partial u_3}{\partial x_2}dx_2 + \left(1 + \frac{\partial u_3}{\partial x_3}\right)dx_3, \end{aligned} \tag{1.1.6}$$

provided we keep in mind that the partial derivatives of the functions u_1, u_2, u_3 are taken at the point M.

In essentially static problems, while little consideration is given to rigid body displacements, particular attention is given to the changes in length and in orientation of elements like ds. These changes are described by the three components of the relative displacement vector du_1, du_2, du_3 (Fig. 1.3):

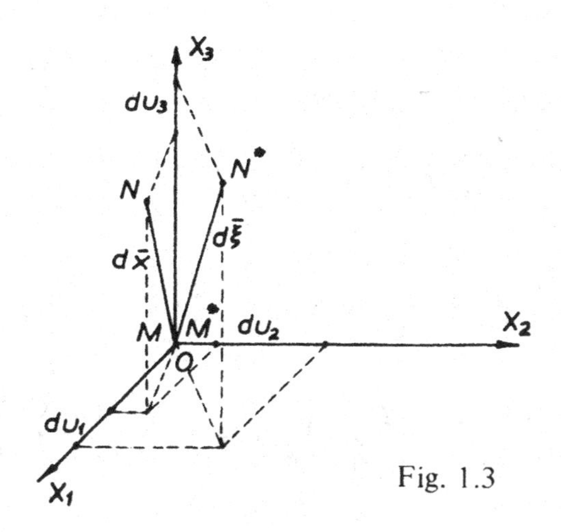

Fig. 1.3

$$
\begin{aligned}
du_1 &= d\xi_1 - dx_1 = \frac{\partial u_1}{\partial x_1} dx_1 + \frac{\partial u_1}{\partial x_2} dx_2 + \frac{\partial u_1}{\partial x_3} dx_3 \\
du_2 &= d\xi_2 - dx_2 = \frac{\partial u_2}{\partial x_1} dx_1 + \frac{\partial u_2}{\partial x_2} dx_2 + \frac{\partial u_2}{\partial x_3} dx_3 \\
du_3 &= d\xi_3 - dx_3 = \frac{\partial u_3}{\partial x_1} dx_1 + \frac{\partial u_3}{\partial x_2} dx_2 + \frac{\partial u_3}{\partial x_3} dx_3 .
\end{aligned}
\tag{1.1.7}
$$

The kinematics of continuous media is centered on the two sets of Eqs. (1.1.6) and (1.1.7). Within the scope of this text, the necessary mathematical tool required to study these equations is the notion of linear transformation. Since matrix algebra was developed primarily to express linear transformations in a concise and lucid manner, it is natural that it should be employed in the formulation and the solution of kinematics problems. A brief review of matrix algebra is given in Chapter 2.

1.2 Notation

The following system of notation will be adhered to throughout this text:

$$
\begin{aligned}
&\frac{\partial u_1}{\partial x_1} = e_{11} \qquad \frac{1}{2}\left(\frac{\partial u_1}{\partial x_2} + \frac{\partial u_2}{\partial x_1}\right) = e_{12} \qquad \frac{1}{2}\left(\frac{\partial u_2}{\partial x_1} - \frac{\partial u_1}{\partial x_2}\right) = \omega_{21} \\
&\frac{\partial u_2}{\partial x_2} = e_{22} \qquad \frac{1}{2}\left(\frac{\partial u_1}{\partial x_3} + \frac{\partial u_3}{\partial x_1}\right) = e_{13} \qquad \frac{1}{2}\left(\frac{\partial u_1}{\partial x_3} - \frac{\partial u_3}{\partial x_1}\right) = \omega_{13} \\
&\frac{\partial u_3}{\partial x_3} = e_{33} \qquad \frac{1}{2}\left(\frac{\partial u_2}{\partial x_3} + \frac{\partial u_3}{\partial x_2}\right) = e_{23} \qquad \frac{1}{2}\left(\frac{\partial u_3}{\partial x_2} - \frac{\partial u_2}{\partial x_3}\right) = \omega_{32}
\end{aligned}
\tag{1.2.1}
$$

In Eqs. (1.2.1), the e's remain unchanged and the ω's change sign when the indices are interchanged. Thus,

$$
\begin{aligned}
e_{12} = e_{21} &= \frac{1}{2}\left(\frac{\partial u_2}{\partial x_1} + \frac{\partial u_1}{\partial x_2}\right) \\
e_{13} = e_{31} &= \frac{1}{2}\left(\frac{\partial u_3}{\partial x_1} + \frac{\partial u_1}{\partial x_3}\right) \\
e_{23} = e_{32} &= \frac{1}{2}\left(\frac{\partial u_3}{\partial x_2} + \frac{\partial u_2}{\partial x_3}\right)
\end{aligned}
\tag{1.2.2}
$$

and

$$
\begin{aligned}
-\omega_{12} = +\omega_{21} &= \frac{1}{2}\left(\frac{\partial u_2}{\partial x_1} - \frac{\partial u_1}{\partial x_2}\right) \\
-\omega_{31} = +\omega_{13} &= \frac{1}{2}\left(\frac{\partial u_1}{\partial x_3} - \frac{\partial u_3}{\partial x_1}\right) \qquad (1.2.3) \\
-\omega_{23} = +\omega_{32} &= \frac{1}{2}\left(\frac{\partial u_3}{\partial x_2} - \frac{\partial u_2}{\partial x_3}\right).
\end{aligned}
$$

With these notations, Eqs. (1.1.6) become:

$$
\begin{aligned}
d\xi_1 &= (1 + e_{11})dx_1 + (e_{12} - \omega_{21})dx_2 + (e_{13} + \omega_{13})dx_3 \\
d\xi_2 &= (e_{12} + \omega_{21})dx_1 + (1 + e_{22})dx_2 + (e_{23} - \omega_{32})dx_3 \qquad (1.2.4) \\
d\xi_3 &= (e_{13} - \omega_{13})dx_1 + (e_{23} + \omega_{32})dx_2 + (1 + e_{33})dx_3 .
\end{aligned}
$$

Eqs. (1.1.7) become:

$$
\begin{aligned}
du_1 &= e_{11}\,dx_1 + (e_{12} - \omega_{21})dx_2 + (e_{13} + \omega_{13})dx_3 \\
du_2 &= (e_{12} + \omega_{21})dx_1 + e_{22}\,dx_2 + (e_{23} - \omega_{32})dx_3 \qquad (1.2.5) \\
du_3 &= (e_{13} - \omega_{13})dx_1 + (e_{23} + \omega_{32})dx_2 + e_{33}\,dx_3 .
\end{aligned}
$$

In all the previous equations, the coordinates of the points of the body in the transformed state are expressed in terms of their coordinates in the initial state. This is known as the Lagrangian Method of describing the transformation of a continuous medium. Another method, the Eulerian Method, expresses the coordinates in the initial state in terms of the coordinates in the final state. Each method has its advantages. It is, however, more convenient in the study of the mechanics of solids to use the Lagrangian approach because the initial state of the body often possesses symmetries which make it susceptible to description in a simple system of coordinates. The Lagrangian Method is exclusively used in this text. (*)

* See pages A-1 and A-2 for additional information and problems.

Chapter 2

REVIEW OF MATRIX ALGEBRA

2.1 Introduction

The use of matrices in mechanics introduces a notation that enables one to see the components of the entities being studied in their totality, while providing great conciseness. In this chapter, the basic definitions and the operations of matrix algebra which will be needed in this text are given.

2.2 Definition of a Matrix. Special Matrices

A matrix is an array of elements arranged in rows and columns. For instance, a matrix of m rows and n columns is written:

$$[a] = \begin{bmatrix} a_{11} & a_{12} & \cdots & a_{1n} \\ a_{21} & a_{22} & \cdots & a_{2n} \\ \cdot & \cdot & \cdots & \cdot \\ a_{m1} & a_{m2} & \cdots & a_{mn} \end{bmatrix}, \qquad (2.2.1)$$

and is called an $(m \times n)$ matrix. The first subscript i of each element a_{ij} represents the number of the row, and the second subscript j represents the number of the column. The a_{ij}'s can be pure numbers, functions, instructions to a computer, or other matrices. In this text, the elements are all real. A square matrix with n rows and columns is said to be of order n.

A symmetric matrix has elements which satisfy the condition $a_{ij} = a_{ji}$. This means that elements symmetrically located with respect to the

main diagonal of the matrix are equal in magnitude and sign.
An antisymmetric or skew symmetric matrix has elements which satisfy the condition $a_{ij} = -a_{ji}$. This means that elements symmetrically located with respect to the main diagonal are equal in magnitude and opposite in sign, and that the elements of the diagonal are equal to zero.
A diagonal matrix is a matrix whose elements a_{ij} vanish except for $i = j$. These non-vanishing elements constitute the main diagonal of the matrix.
A unit matrix is a diagonal matrix whose elements are equal to unity. It is written [1].
A null matrix has all its elements equal to zero. It is written [0].
A column matrix has m rows and one column. it is also called a column vector and is written:

$$\{\bar{a}\} = \begin{bmatrix} a_{11} \\ a_{21} \\ \cdot \\ \cdot \\ a_{m1} \end{bmatrix}.$$

A row matrix is a matrix with one row and n columns. It is also called a row vector and is written:

$$[\bar{a}] = [a_{11} \ldots a_{1n}].$$

The transpose of a matrix $[a]$ is a matrix $[a]'$, whose rows are the same as the columns of $[a]$. Thus, a symmetric matrix is its own transpose and the transpose of a column matrix is a row matrix.
A scalar matrix is a diagonal matrix whose elements are identical.

2.3 Index Notation and Summation Convention

The introduction of numerical subscripts in Chapter 1 to denote the reference axes makes the use of indices in writing the components of vectors quite natural. When writing relations between vectors or other directional quantities (such as tensors), a great deal of space is saved when a shorthand notation is introduced. In this text, the only indices to be used are subscripts and the following conventions will be adhered to:

The range convention: Whenever a subscript is repeated in a term, it is understood to represent a summation over the range 1, 2, 3 unless otherwise stated. Also, an index never appears more than twice in the same term. For example, the expression

$$\xi_i = a_{ij} x_j \tag{2.3.1}$$

contains, in the right-hand term, the index j which is repeated. Therefore, taking the values of $i = 1, 2, 3$ in turn, we obtain the three linear equations:

$$\begin{aligned} \xi_1 &= a_{11} x_1 + a_{12} x_2 + a_{13} x_3 \\ \xi_2 &= a_{21} x_1 + a_{22} x_2 + a_{23} x_3 \\ \xi_3 &= a_{31} x_1 + a_{32} x_2 + a_{33} x_3 . \end{aligned} \tag{2.3.2}$$

i is the *identifying index* and j is the *summation index*. We notice that the summation index can be changed at will and is therefore called a *dummy index*. Thus, Eqs. (2.3.2) can also be written:

$$\xi_i = a_{ik} x_k .$$

The index k is similar to the dummy variable of integration in a definite integral and can be changed freely.

For convenience, it is sometimes useful to introduce the two following symbols:

The Kronecker delta, δ_{ij}, which by definition is such that:

$$\begin{aligned} \delta_{ij} &= 1, \text{ when } i = j \text{ and} \\ \delta_{ij} &= 0, \text{ when } i \neq j. \end{aligned} \tag{2.3.3}$$

The alternating symbol, ε_{ijk}, which by definition is such that:

$$\begin{aligned} \varepsilon_{ijk} &= 0, \text{ when any two of } i, j, k \text{ are equal} \\ \varepsilon_{ijk} &= 1, \text{ when } i, j, k \text{ are different and in cyclic order } (1,2,3, \\ &\quad 1,2,3,\ldots) \\ \varepsilon_{ijk} &= -1, \text{ when } i, j, k \text{ are different and not in cyclic order } (1, \\ &\quad 3,2,1,3,2,\ldots). \end{aligned} \tag{2.3.4}$$

Examples

1). $\delta_{ik} x_k$ for $i = 1$ is equal to:

$$\delta_{11} x_1 + \delta_{12} x_2 + \delta_{13} x_3 = x_1 = x_i.$$

2). $\delta_{ii} = \delta_{11} + \delta_{22} + \delta_{33} = 3.$

3). A vector $\bar{x}$ whose components are x_1, x_2, x_3, has a magnitude $|\bar{x}| = \sqrt{x_1^2 + x_2^2 + x_3^2} = \sqrt{x_i x_i}$. Its direction cosines are given by $\ell_i = x_i / \sqrt{x_j x_j}$.

4). The sum of the diagonal elements of a matrix $[a]$ is called the trace of $[a]$ and is written a_{ii}.

5). The determinant of the matrix $[a]$ is written $\varepsilon_{ijk} a_{1i} a_{2j} a_{3k}$.

2.4 Equality of Matrices. Addition and Subtraction

Let us turn now to the rules governing the manipulation of the arrays of elements forming a matrix. Two matrices $[a]$ and $[b]$ of the same order are said to be equal if, and only if, their corresponding elements are identical; that is, we have:

$$[a] = [b], \tag{2.4.1}$$

provided that

$$a_{ij} = b_{ij} \text{ for all } i \text{ and } j. \tag{2.4.2}$$

If $[a]$ and $[b]$ are matrices of the same order, then the sum of $[a]$ and $[b]$ is defined to be a matrix $[c]$, the typical element of which is $c_{ij} = a_{ij} + b_{ij}$. In other words, by definition:

$$[c] = [a] + [b], \tag{2.4.3}$$

provided

$$c_{ij} = a_{ij} + b_{ij}. \tag{2.4.4}$$

In a similar manner, we have:

$$[d] = [a] - [b], \tag{2.4.5}$$

provided

$$d_{ij} = a_{ij} - b_{ij}. \tag{2.4.6}$$

From the above definitions, it can be shown that the following operations are valid:

$$[a] + [b] = [b] + [a] \tag{2.4.7}$$

$$([a] + [b]) + [c] = [a] + ([b] + [c]). \tag{2.4.8}$$

An important property of square matrices, which follows from the laws of addition and subtraction, is that any square matrix may be given as the sum of a symmetric and of an antisymmetric matrix. Indeed, if $[a]$ is a square matrix, then

$$[a] = \frac{[a] + [a]'}{2} + \frac{[a] - [a]'}{2}. \tag{2.4.9}$$

2.5 Multiplication of Matrices

The product of a matrix $[a]$ by a matrix $[b]$ is defined by the equation

$$[a][b] = [c], \tag{2.5.1}$$

where the elements of $[c]$ are given by:

$$c_{ij} = a_{ik} b_{kj}. \tag{2.5.2}$$

Thus

$$\begin{bmatrix} a_{11} & a_{12} & a_{13} \\ a_{21} & a_{22} & a_{23} \\ a_{31} & a_{32} & a_{33} \end{bmatrix} \begin{bmatrix} b_{11} & b_{12} & b_{13} \\ b_{21} & b_{22} & b_{23} \\ b_{31} & b_{32} & b_{33} \end{bmatrix} = \begin{bmatrix} a_{11} b_{11} + a_{12} b_{21} + a_{13} b_{31} & a_{11} b_{12} + a_{12} b_{22} + a_{13} b_{32} & a_{11} b_{13} + a_{12} b_{23} + a_{13} b_{33} \\ a_{21} b_{11} + a_{22} b_{21} + a_{23} b_{31} & a_{21} b_{12} + a_{22} b_{22} + a_{23} b_{32} & a_{21} b_{13} + a_{22} b_{23} + a_{23} b_{33} \\ a_{31} b_{11} + a_{32} b_{21} + a_{33} b_{31} & a_{31} b_{12} + a_{32} b_{22} + a_{33} b_{32} & a_{31} b_{13} + a_{32} b_{23} + a_{33} b_{33} \end{bmatrix}.$$

Two matrices can be multiplied by each other only if they are *conformable*, which means that the number of the columns of the first is equal to the number of the rows of the second. Thus, if $[a]$ is an $(m \times p)$ matrix and $[b]$ is a $(p \times n)$ matrix, then $[c]$ is an $(m \times n)$ matrix.
Two nonzero matrices can be multiplied by each other and result in a zero matrix. For example,

$$\begin{bmatrix} 1 & 1 & 0 \\ 0 & 0 & 0 \\ 0 & 0 & 0 \end{bmatrix} \begin{bmatrix} 0 & 0 & 0 \\ 0 & 0 & 0 \\ 1 & 0 & 0 \end{bmatrix} = \begin{bmatrix} 0 & 0 & 0 \\ 0 & 0 & 0 \\ 0 & 0 & 0 \end{bmatrix}.$$

A permutation of the matrices will lead to a different result:

$$\begin{bmatrix} 0 & 0 & 0 \\ 0 & 0 & 0 \\ 1 & 0 & 0 \end{bmatrix} \begin{bmatrix} 1 & 1 & 0 \\ 0 & 0 & 0 \\ 0 & 0 & 0 \end{bmatrix} = \begin{bmatrix} 0 & 0 & 0 \\ 0 & 0 & 0 \\ 1 & 1 & 0 \end{bmatrix}.$$

The product $[b]$ $[a]$ is, in general, not equal to $[a]$ $[b]$. Therefore, it is necessary to differentiate between *premultiplication*, as when $[b]$ is premultiplied by $[a]$ to yield the product $[a]$ $[b]$, and *postmultiplication*, as when $[b]$ is postmultiplied by $[a]$ to yield $[b]$ $[a]$. If we have two matrices which are such that

$$[a][b] = [b][a], \tag{2.5.3}$$

these matrices are said to *commute* or to be permutable.

Of particular importance is the *associative law* of continued products,

$$[d] = ([a][b])[c] = [a]([b][c]), \tag{2.5.4}$$

which allows one to dispense with parentheses and to write $[a]$ $[b]$ $[c]$ without ambiguity since the double summation

$$d_{ij} = a_{ik} b_{kl} c_{lj} \tag{2.5.5}$$

can be carried out in either of the orders indicated. It must be noticed that the product of a chain of matrices will have meaning only if the adjacent matrices are conformable.

The product of matrices is *distributive*, that is

$$[a]([b] + [c]) = [a][b] + [a][c]. \tag{2.5.6}$$

The multiplication of a matrix $[a]$ by a scalar k is defined by:

$$k[a] = [b], \tag{2.5.7}$$

where

$$b_{ij} = ka_{ij}.$$

Using the definition of the transpose and the laws of addition and multiplication of matrices, it can be shown that:

$$([a] + [b])' = [a]' + [b]' \tag{2.5.8}$$

$$(k[a])' = k[a]' \tag{2.5.9}$$

$$([a][b])' = [b]'[a]' \text{ (note the order).} \tag{2.5.10}$$

For the case of the unit matrix, we have:

$$[a][1] = [1][a] = [a] \tag{2.5.11}$$

and, if k is a constant,

$$[a]k[1] = k[a][1] = k[a] = k[1][a]. \tag{2.5.12}$$

An important result in the theory of matrices is that the determinant of the product of two square matrices is equal to the product of their determinants. Thus,

$$\|[a][b]\| = \left(\|[a]\|\right)\left(\|[b]\|\right) = \left(\|[b]\|\right)\left(\|[a]\|\right). \tag{2.5.13}$$

Among the special matrices defined in Sec. 2.2, the diagonal matrix plays an important part in operations involving matrices. The premultiplication of a matrix $[a]$ by a diagonal matrix $[d]$ produces a matrix whose rows are those of $[a]$ multiplied by the element in the corresponding row of $[d]$:

$$\begin{bmatrix} d_1 & 0 & 0 \\ 0 & d_2 & 0 \\ 0 & 0 & d_3 \end{bmatrix} \begin{bmatrix} a_{11} & a_{12} & a_{13} \\ a_{21} & a_{22} & a_{23} \\ a_{31} & a_{32} & a_{33} \end{bmatrix} = \begin{bmatrix} d_1 a_{11} & d_1 a_{12} & d_1 a_{13} \\ d_2 a_{21} & d_2 a_{22} & d_2 a_{23} \\ d_3 a_{31} & d_3 a_{32} & d_3 a_{33} \end{bmatrix}. \tag{2.5.14}$$

The postmultiplication of $[a]$ by $[d]$ produces a matrix whose columns are those of [a] multiplied by the element in the corresponding column of $[d]$:

$$\begin{bmatrix} a_{11} & a_{12} & a_{13} \\ a_{21} & a_{22} & a_{23} \\ a_{31} & a_{32} & a_{33} \end{bmatrix} \begin{bmatrix} d_1 & 0 & 0 \\ 0 & d_2 & 0 \\ 0 & 0 & d_3 \end{bmatrix} = \begin{bmatrix} d_1 a_{11} & d_2 a_{12} & d_3 a_{13} \\ d_1 a_{21} & d_2 a_{22} & d_3 a_{23} \\ d_1 a_{31} & d_2 a_{32} & d_3 a_{33} \end{bmatrix}. \tag{2.5.15}$$

The diagonal matrix $[d]$ is therefore a convenient tool for writing groups of equations under the form of one single matrix equation. For example, a whole group of systems of equations, such as

$$[a]\{\bar{x}\} = \lambda_1\{\bar{b}\}, \ [a]\{\bar{y}\} = \lambda_2\{\bar{c}\}, \ [a]\{\bar{z}\} = \lambda_3\{\bar{e}\},$$

can be written:

$$[a][f] = [h][d],$$

where $[f]$ has columns formed by $\{\bar{x}\}$, $\{\bar{y}\}$, $\{\bar{z}\}$; $[h]$ has columns formed by $\{\bar{b}\}$, $\{\bar{c}\}$, $\{\bar{e}\}$; and $[d]$ is a diagonal matrix with elements λ_1, λ_2, and λ_3. If $[a]$ and $[b]$ are diagonal matrices of the same order, they are commutative with each other so that $[a][b] = [b][a]$.

2.6 Matrix Division. The Inverse Matrix

If the determinant $|[a]|$ of a matrix $[a]$ does not vanish, $[a]$ is said to be nonsingular and possesses a reciprocal or inverse matrix $[a]^{-1}$, such that

$$[a][a]^{-1} = [1] = [a]^{-1}[a]. \tag{2.6.1}$$

The cofactor matrix of any square matrix $[a]$ is the matrix obtained by replacing each element of $[a]$ by its cofactor. It will be remembered that the cofactor A_{ij} of any element a_{ij} of a determinant $|[a]|$, is the minor of that element with a sign attached to it determined by the numbers i and j which fix the position of a_{ij} in the determinant. The sign is given by the equation giving the cofactor A_{ij}:

$$A_{ij} = (-1)^{i+j} M_{ij}, \tag{2.6.2}$$

where M_{ij} is the minor of the element a_{ij}. For example, the cofactor A_{23} of the determinant of the matrix

$$[a] = \begin{bmatrix} a_{11} & a_{12} & a_{13} \\ a_{21} & a_{22} & a_{23} \\ a_{31} & a_{32} & a_{33} \end{bmatrix}, \tag{2.6.3}$$

is

$$A_{23} = (-1)^5 \begin{vmatrix} a_{11} & a_{12} \\ a_{31} & a_{32} \end{vmatrix} = a_{12}a_{31} - a_{11}a_{32}; \tag{2.6.4}$$

and the cofactor matrix $[C_o A]$ of the square matrix $[A]$ is:

$$[C_o A] = \begin{bmatrix} A_{11} & A_{12} & A_{13} \\ A_{21} & A_{22} & A_{23} \\ A_{31} & A_{32} & A_{33} \end{bmatrix} \tag{2.6.5}$$

If the determinant $\|[a]\|$ of the matrix $[a]$ is not equal to zero, in other words if $[a]$ is nonsingular, then

$$[a]^{-1} = \frac{[C_o A]'}{\|[a]\|} = \begin{bmatrix} \frac{A_{11}}{\|[a]\|} & \frac{A_{21}}{\|[a]\|} & \frac{A_{31}}{\|[a]\|} \\ \frac{A_{12}}{\|[a]\|} & \frac{A_{22}}{\|[a]\|} & \frac{A_{32}}{\|[a]\|} \\ \frac{A_{13}}{\|[a]\|} & \frac{A_{23}}{\|[a]\|} & \frac{A_{33}}{\|[a]\|} \end{bmatrix}. \tag{2.6.6}$$

The transpose of the cofactor matrix is also called the adjoint of $[a]$. Thus,

$$[a]^{-1} = \frac{\text{adj}[a]}{\|[a]\|} \tag{2.6.7}$$

The previous equation can be verified by direct substitution in Eq. (2.6.1). Various methods are available for the inversion of matrices. The bibliography at the end of this chapter gives detailed information on the subject.

In matrix algebra, multiplication by the inverse of a matrix plays the same role as division in ordinary algebra. That is, if we have:

$$[a][b] = [c][e], \tag{2.6.8}$$

where $[a]$ is a nonsingular matrix, then on premultiplying by $[a]^{-1}$, the inverse of $[a]$, we obtain:

$$[a]^{-1}[a][b] = [a]^{-1}[c][e] \tag{2.6.9}$$

and, because of Eq. (2.6.1),

$$[b] = [a]^{-1}[c][e]. \tag{2.6.10}$$

Using the definition of the transpose and that of the inverse, it can be shown that

$$\{[a]^{-1}\}' = \{[a]'\}^{-1}, \tag{2.6.11}$$

which means that the transpose of the inverse of a matrix is equal to the inverse of its transpose. Also,

$$\{[a][b]\}^{-1} = [b]^{-1}[a]^{-1}, \tag{2.6.12}$$

which means that the inverse of the product of two matrices is equal to the product of the inverse of the second by the inverse of the first. The inverse of a diagonal matrix is a diagonal matrix whose elements are the reciprocals of those of the matrix itself. The inverse of a symmetric matrix is a symmetric matrix.

PROBLEMS

1. Given

$$[a] = \begin{bmatrix} 3 & -2 & 5 \\ 6 & 0 & 3 \\ 1 & 5 & 4 \end{bmatrix} \text{ and } [b] = \begin{bmatrix} 2 & 3 & -1 \\ 4 & 1 & 0 \\ 5 & 2 & -1 \end{bmatrix}.$$

(a) compute $[a] + [b]$ and $[a] - [b]$.
(b) Verify: $[a] + ([b] - [c]) = ([a] + [b]) - [c]$.
(c) Split $[a]$ into its symmetric and its antisymmetric parts.

2. Given

$$[a] = \begin{bmatrix} 1 & -3 & 2 \\ 2 & 1 & -3 \\ 4 & -3 & -1 \end{bmatrix}, [b] = \begin{bmatrix} 1 & 4 & 1 & 0 \\ 2 & 1 & 1 & 1 \\ 1 & -2 & 1 & 2 \end{bmatrix},$$

and

$$[c] = \begin{bmatrix} 2 & 1 & -1 & -2 \\ 3 & -2 & -1 & -1 \\ 2 & -5 & -1 & 0 \end{bmatrix}, \text{ show that } [a][b] = [a][c]$$

in spite of the fact that $[b] \neq [c]$.

3. If $[b] = [a][a]'$, show that $[b] = [b]'$.
4. If $\{\bar{a}\}$ is a column matrix, show that $\{\bar{a}\}\{\bar{a}\}' = [c]$, where $[c]$ is a square matrix with the property that $[c] = [c]'$.
5. Given

$$[a] = \begin{bmatrix} 5 & -2 & 0 \\ -2 & 3 & -1 \\ 0 & -1 & 1 \end{bmatrix} \text{ and } \{\bar{b}\} = \begin{bmatrix} 3 \\ -2 \\ 1 \end{bmatrix},$$

compute the product $[a]\{\bar{b}\}$.

6. If $[a]$ is a square matrix of order 3, show that its determinant is given by $\epsilon_{ijk}\, a_{1i}\, a_{2j}\, a_{3k}\, (i, j, k, = 1, 2, 3)$.
7. Write out in full the following expressions :

a) $a_{ij}\, x_i\, x_j$ b) $\delta_{ij}\, x_i\, x_j$ c) $\sigma_{ni} = \sigma_{ji}\, \ell_j$

d) $\sigma'_{ij} = \ell_{ik}\, \ell_{jm}\, \sigma_{km}$ e) $\sigma_{ij} = 2\mu e_{ij} + \lambda \delta_{ij}\, e_{kk}$

The subscripts i, j, k, and m take the values 1, 2, and 3.

8. Find the inverse of the matrices

$$\begin{bmatrix} 2 & -2 & 4 \\ 2 & 3 & 2 \\ -1 & 1 & -1 \end{bmatrix} \text{ and } \begin{bmatrix} 1 & 2 & 3 \\ 0 & 1 & 5 \\ 4 & 0 & 3 \end{bmatrix}. \text{ (*)}$$

REFERENCES

1. R. A. Frazier, W. J. Duncan, A. R. Collar, *Elementary Matrices*, MacMillan, New York, N. Y., 1947.
2. S. Perlis, *Theory of Matrices*, Addison-Wesley, Reading, Mass., 1952.
3. F. B. Hildebrand, *Methods of Applied Mathematics*, Prentice-Hall, New York, N. Y., 1952.
4. L. Fox, *An Introduction to Numerical Linear Algebra*, Oxford University Press, New York, N. Y., 1965.
5. J. B. Scarborough, *Numerical Mathematical Analysis*, The Johns Hopkins Press, Baltimore, Md., 1966.

* See page A-3 for additional problems.

CHAPTER 3

LINEAR TRANSFORMATION OF POINTS

3.1 Introduction

The importance of linear transformations for the study of kinematics was indicated in Chapter 1. In the present chapter, this kind of transformation is examined in detail, and whenever possible the results are interpreted geometrically. This interpretation is essential if one is to visualize the deformation of continuous media.

In addition to kinematics, topics such as stress, moments of inertia of surfaces and volumes, and curvature of surfaces, to mention only a few, involve linear transformations. This chapter serves, therefore, as a foundation common to a wide variety of subjects in mechanics.

3.2 Definitions and Elementary Operations

In a trirectangular system of coordinates OX_1, OX_2, OX_3 (Fig. 3.1), consider the linear equations giving the coordinates of a point $M^*(\xi_1, \xi_2, \xi_3)$ in terms of those of $M(x_1, x_2, x_3)$:

$$\begin{aligned} \xi_1 &= a_{11}x_1 + a_{12}x_2 + a_{13}x_3 \\ \xi_2 &= a_{21}x_1 + a_{22}x_2 + a_{23}x_3 \\ \xi_3 &= a_{31}x_1 + a_{32}x_2 + a_{33}x_3, \end{aligned} \tag{3.2.1}$$

where the a_{ij}' s are constants. These equations are said to transform the point M to the point M^*. One may choose to consider that Eqs. (3.2.1) transform the vector $\overline{OM}$ to the vector $\overline{OM^*}$. In such a case, however,

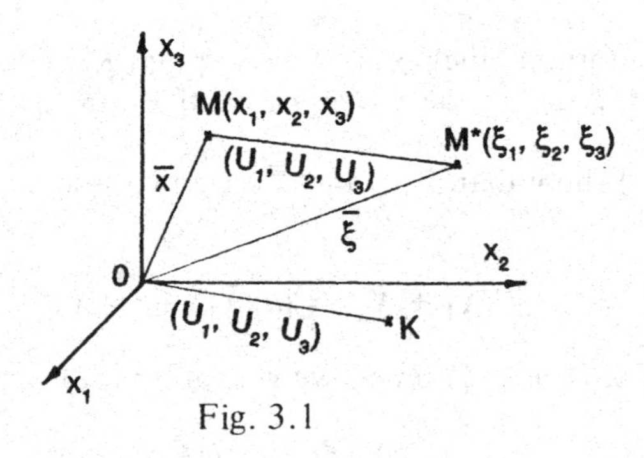

Fig. 3.1

these two vectors are tied to the point O and are not free vectors. The transformation expressed by Eqs. (3.2.1) is called a point-to-point linear transformation or, simply, a linear transformation. It can be written in any of the following forms:

$$\begin{bmatrix} \xi_1 \\ \xi_2 \\ \xi_3 \end{bmatrix} = \begin{bmatrix} a_{11} & a_{12} & a_{13} \\ a_{21} & a_{22} & a_{23} \\ a_{31} & a_{32} & a_{33} \end{bmatrix} \begin{bmatrix} x_1 \\ x_2 \\ x_3 \end{bmatrix},$$

$$\{\overline{OM^*}\} = [a]\{\overline{OM}\},$$

$$\xi_i = a_{ij} x_j.$$

We can look upon the matrix $[a]$ as an operator acting on the column vector $\{\overline{OM}\}$ to give the column vector $\{\overline{OM^*}\}$. The inverse of this transformation gives $\{\overline{OM}\}$ in terms of $\{\overline{OM^*}\}$, provided the matrix $[a]$ is nonsingular:

$$\{\overline{OM}\} = [a]^{-1}\{\overline{OM^*}\}. \tag{3.2.2}$$

The vector $\overline{MM^*}$ is called the displacement of M. Its components are given by:

$$\begin{bmatrix} u_1 \\ u_2 \\ u_3 \end{bmatrix} = \begin{bmatrix} \xi_1 - x_1 \\ \xi_2 - x_2 \\ \xi_3 - x_3 \end{bmatrix} = \begin{bmatrix} a_{11} - 1 & a_{12} & a_{13} \\ a_{21} & a_{22} - 1 & a_{23} \\ a_{31} & a_{32} & a_{33} - 1 \end{bmatrix} \begin{bmatrix} x_1 \\ x_2 \\ x_3 \end{bmatrix}. \tag{3.2.3}$$

If, from the origin O (Fig. 3.1), we draw a vector $\overline{OK}$ parallel to $\overline{MM^*}$ and whose components are u_1, u_2, and u_3, Eqs. (3.2.3) can be looked upon as transforming the point M to the point K. Such a transformation from M to K is called the *hodograph* of the transformation from M to M^*.

Eqs. (3.2.1) show that if $\overline{X}_1$ and $\overline{X}_2$ are two vectors tied to the origin O, then

$$[a]\{\overline{X}_1 + \overline{X}_2\} = [a]\{\overline{X}_1\} + [a]\{\overline{X}_2\}. \tag{3.2.4}$$

Using Eq. (3.2.4) $(n - 1)$ times, we get for a vector $\overline{X}$:

$$[a]\{n\overline{X}\} = n[a]\{\overline{X}\}, \tag{3.2.5}$$

where n is an integer. Eq. (3.2.5) can be generalized for fractional values of n.

Eq. (3.2.5) shows that a linear transformation of points transforms a straight line OAB through the origin to another straight line $O\alpha\beta$ through the origin, and that (Fig. 3.2):

$$\frac{O\beta}{O\alpha} = \frac{OB}{OA}.$$

Eq. (3.2.4) shows that a parallelogram $OACB$ is transformed by Eqs. (3.2.1) to a parallelogram $O\alpha\gamma\beta$ (Fig. 3.3). Thus, a free vector $\overline{AC}$ is transformed to another free vector $\overline{\alpha\gamma}$, a parallelogram to a parallelogram, a plane to a plane, and a parallelepiped to a parallelepiped.

If a matrix $[a]$ transforms a vector $\overline{OP}$ to $\overline{O\Pi}_1$, and a matrix $[b]$ transforms $\overline{OP}$ to $\overline{O\Pi}_2$ — i.e., if

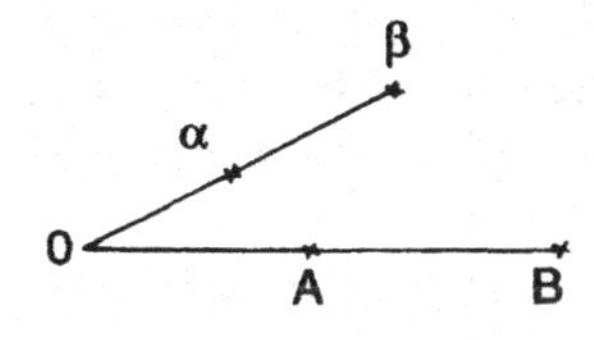

Fig. 3.2

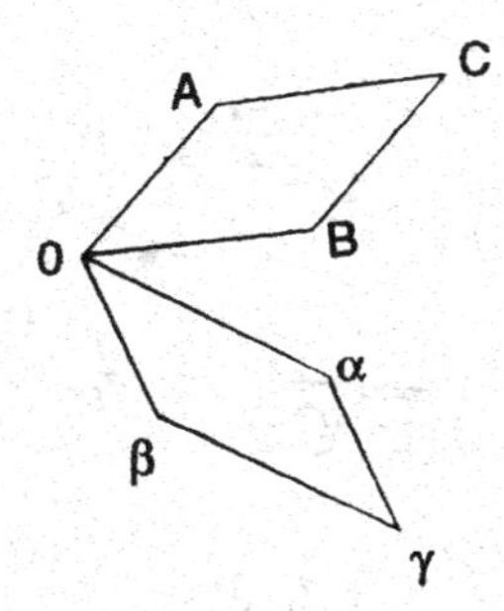

Fig. 3.3

$$[a]\{\overline{OP}\} = \{\overline{O\Pi}_1\} \tag{3.2.6}$$

and

$$[b]\{\overline{OP}\} = \{\overline{O\Pi}_2\}, \tag{3.2.7}$$

then $\overline{O\Pi}_1 + \overline{O\Pi}_2 = \overline{O\Pi}$ is defined as the *sum of the two transformations* (Fig. 3.4). Therefore,

$$[a]\{\overline{OP}\} + [b]\{\overline{OP}\} = [c]\{\overline{OP}\} = \overline{O\Pi}, \tag{3.2.8}$$

where

$$[c] = [a] + [b]. \tag{3.2.9}$$

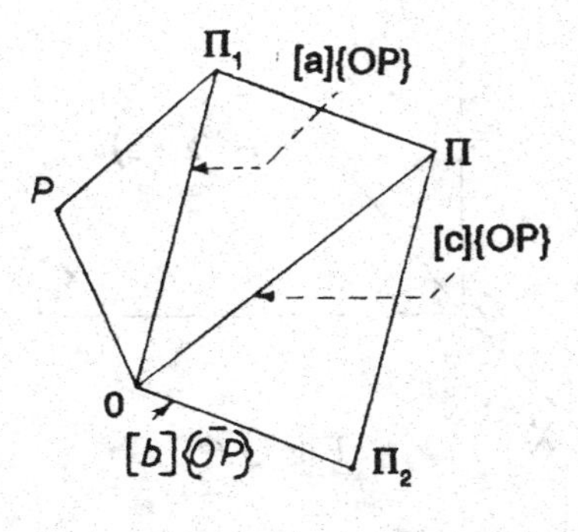

Fig. 3.4

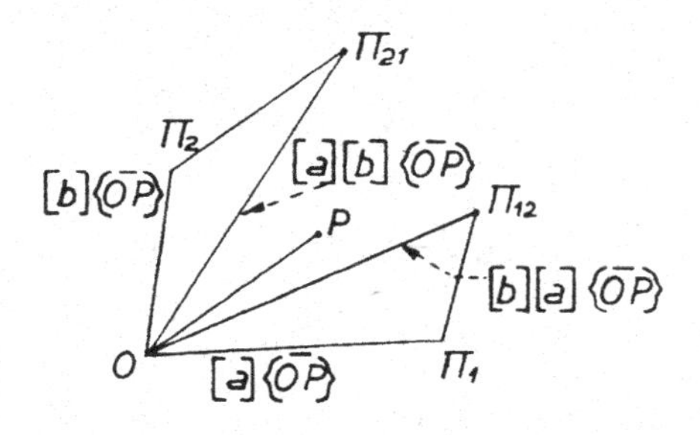

Fig. 3.5

If a matrix $[a]$ transforms a vector $\overline{OP}$ to $\overline{O\Pi}_1$, and another matrix $[b]$ transforms $\overline{O\Pi}_1$ to $\overline{O\Pi}_{12}$, the transformation which brings $\overline{OP}$ to $\overline{O\Pi}_{12}$ is defined as the *product of the two transformations* (Fig. 3.5). Thus,

$$[b][a]\{\overline{OP}\} = [c]\{\overline{OP}\} = \{\overline{O\Pi}_{12}\}. \tag{3.2.10}$$

As shown in the previous chapter, in general,

$$[b][a]\{\overline{OP}\} \neq [a][b]\{\overline{OP}\} = \{\overline{O\Pi}_{21}\}, \tag{3.2.11}$$

and the two points Π_{21} and Π_{12} do not coincide (Fig. 3.5).

A *small transformation* is one whose matrix is nearly equal to the identity or unit matrix. For example, the transformation (Fig. 3.6)

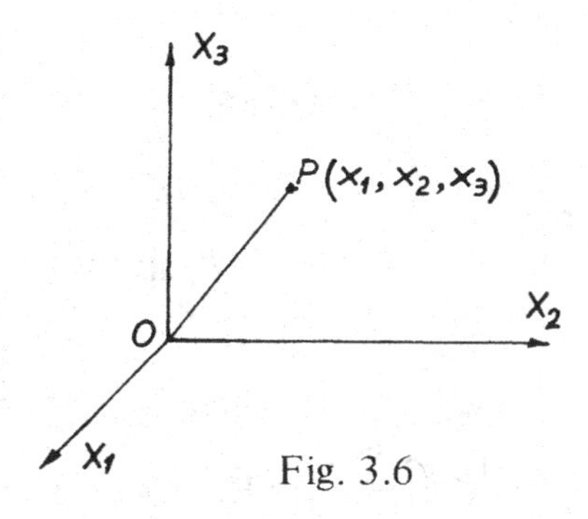

Fig. 3.6

$$[[1] + [a]]\{\overline{OP}\} = \begin{bmatrix} 1 + a_{11} & a_{12} & a_{13} \\ a_{21} & 1 + a_{22} & a_{23} \\ a_{31} & a_{32} & 1 + a_{33} \end{bmatrix} \begin{bmatrix} x_1 \\ x_2 \\ x_3 \end{bmatrix} \qquad (3.2.12)$$

is said to be small when all the a_{ij}' s are small with respect to unity. *The product of two small transformations* gives:

$$[[1] + [a]][[1] + [b]]\{\overline{OP}\} = [[1] + [a] + [b] + [a][b]]\{\overline{OP}\}. \qquad (3.2.13)$$

If the terms of the second order are neglected, then

$$[[1] + [a]][[1] + [b]]\{\overline{OP}\} = [[1] + [a] + [b]]\{\overline{OP}\}. \qquad (3.2.14)$$

We notice that the order no longer intervenes, which means that the operation is commutative.

3.3 Conjugate and Principal Directions and Planes in a Linear *Transformation*

Consider a body B which is transformed to β by a point-to-point linear transformation (Fig. 3.7). A plane P and a straight line D in B are transformed to a plane Π and a straight line Δ in β. By definition, if Δ is normal to Π, P and D are called *conjugate*. If P and D, as well as Δ and Π, are perpendicular to one another, then P and D are called *principal plane* and *principal direction*, respectively.

In other words, a direction is called principal if any right angle involving this direction remains a right angle after transformation.

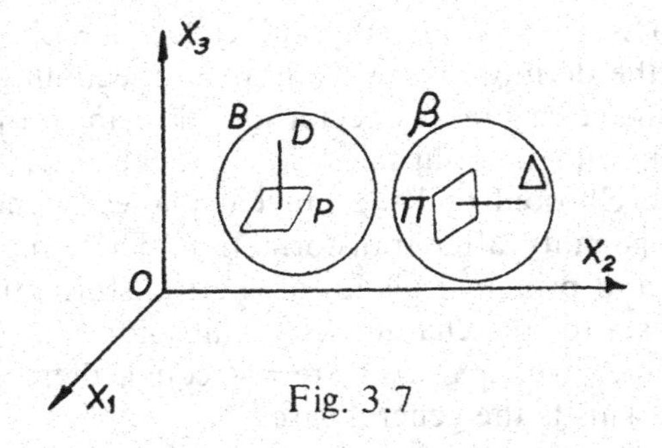

Fig. 3.7

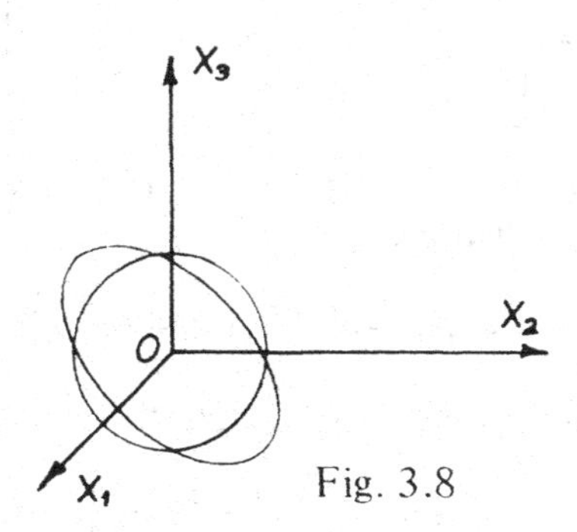

Fig. 3.8

Let us find the shape of the surface whose points after transformation fall on a sphere of radius R and which is centered at the origin. The equation of the sphere is (Fig. 3.8):

$$\xi_1^2 + \xi_2^2 + \xi_3^2 = R^2 \tag{3.3.1}$$

and that of the initial surface is:

$$(a_{11}x_1 + a_{12}x_2 + a_{13}x_3)^2 + (a_{21}x_1 + a_{22}x_2 + a_{23}x_3)^2 + (a_{31}x_1 + a_{32}x_2 + a_{33}x_3)^2 = R^2$$

or

$$\sum_j (a_{ji}x_i)^2 = R^2. \tag{3.3.2}$$

Eq. (3.3.2) is that of an ellipsoid called the characteristic ellipsoid. Recalling the definitions set forth at the beginning of this section, we conclude that every radius vector and the tangent plane at its point of intersection with the ellipsoid are conjugate. Also the three principal axes of the ellipsoid and the tangent planes at their extremities keep their orthogonality after transformation. Therefore, they are principal directions and principal planes of the transformation. There are three possible cases for the characteristic ellipsoid:
1. If the three principal axes are not equal, there exist three principal directions: This is the general case.

2. If two of the three principal axes are equal, the characteristic ellipsoid is an ellipsoid of revolution. If, for instance, OX_1 is the axis of revolution, all the axes of the ellipsoid normal to OX_1 are principal directions and all the planes parallel to OX_1 are principal planes.
3. If the three principal axes of the ellipsoid are equal, it becomes a sphere. All radii and all planes are principal axes and principal planes.(*)

3.4 Orthogonal Transformations

Let us examine what conditions are to be imposed on the matrix $[a]$ in the point-to-point transformation (Fig. 3.9):

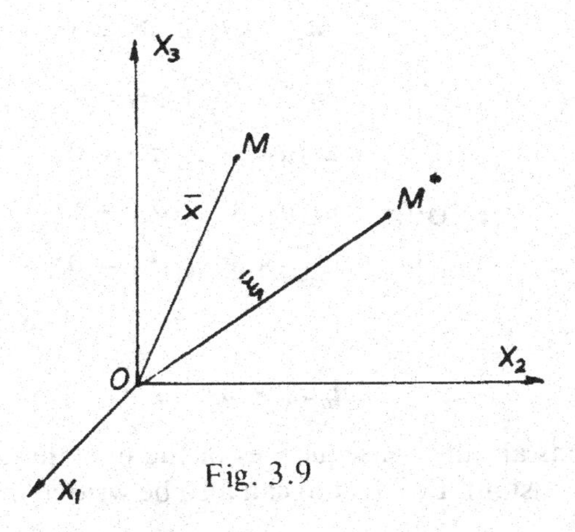

Fig. 3.9

$$\xi_i = a_{ij} x_j \tag{3.4.1}$$

so that the length of the vector $\overline{OM}$ remains unchanged. This obviously would correspond to a rotation or a rotation followed by a reflection. If the length of $\bar{x}$ is to be unchanged, then

$$x_i x_i = \xi_i \xi_i . \tag{3.4.2}$$

Substituting Eq. (3.4.1) in Eq. (3.4.2), we get:

$$x_i x_i = (a_{ij} x_j)(a_{ik} x_k). \tag{3.4.3}$$

* See pages A-4 to A-6 for additional information and a solved example.

where the dummy index in one of the brackets has been changed from j to k to conform with the summation convention. Eq. (3.4.3) can also be written:

$$a_{ij}a_{ik}x_jx_k = \delta_{jk}x_jx_k, \tag{3.4.4}$$

since

$$\delta_{jk}x_jx_k = x_kx_k = x_ix_i. \tag{3.4.5}$$

Equating the coefficients of like products in Eq. (3.4.4), we obtain the following six equations:

$$\begin{aligned} a_{11}^2 + a_{21}^2 + a_{31}^2 &= 1 \\ a_{12}^2 + a_{22}^2 + a_{32}^2 &= 1 \\ a_{13}^2 + a_{23}^2 + a_{33}^2 &= 1 \\ a_{12}a_{13} + a_{22}a_{23} + a_{32}a_{33} &= 0 \\ a_{13}a_{11} + a_{23}a_{21} + a_{33}a_{31} &= 0 \\ a_{11}a_{12} + a_{21}a_{22} + a_{31}a_{32} &= 0 \end{aligned} \tag{3.4.6}$$

or

$$a_{ij}a_{ik} = \delta_{jk}. \tag{3.4.6a}$$

These equations are the consequences of the hypothesis that the length of $\bar{x}$ remains constant. Eqs. (3.4.6) can also be written in matrix notation as follows:

$$[a]'[a] = [1]. \tag{3.4.6b}$$

Since

$$[a]^{-1}[a] = [1].$$

Therefore,

$$[a]^{-1} = [a]'. \tag{3.4.7}$$

From the rules of multiplication of determinants,

$$(|[a]'|)(|[a]|) = 1,$$

and since the value of a determinant does not change when the rows and columns are interchanged, then

$$(|[a]'|)(|[a]|) = |[a]|^2 = 1. \tag{3.4.8}$$

Thus, in a linear transformation in which the length of the vector $\bar{x}$ remains constant, the inverse of the matrix of the transformation is equal to its transpose and the square of its determinant is equal to unity.

A linear transformation $\xi_i = a_{ij} x_j$ in which $a_{ij} a_{ik} = \delta_{jk}$ is called an *orthogonal transformation*. It is called a transformation of rotation when $|[a]| = +1$, and the matrix $[a]$ is referred to as *a proper orthogonal matrix*. It is called a transformation of reflection when $|[a]| = -1$, and the matrix $[a]$ is referred to as *an improper orthogonal matrix*.

Since

$$[a]^{-1}[a] = [a][a]^{-1} = [1],$$

then

$$[a][a]' = [1] \tag{3.4.9}$$

and

$$a_{ji} a_{ki} = \delta_{jk}. \tag{3.4.10}$$

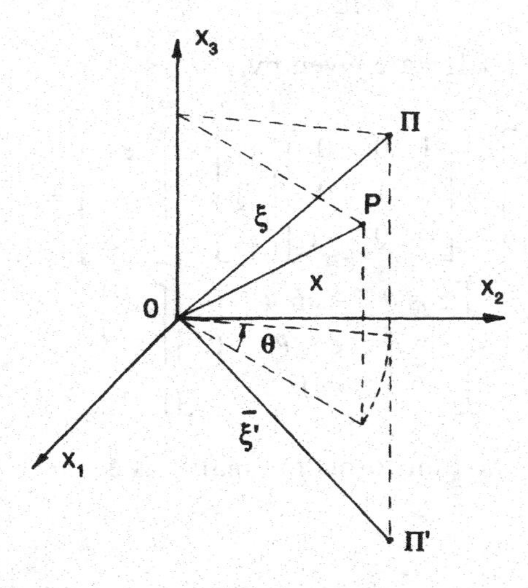

Fig. 3.10

Eqs. (3.4.6) and (3.4.10) show that both the columns and the rows of the orthogonal matrix $[a]$ form a system of trirectangular unit vectors. The transpose of $[a]$, and consequently its inverse, is also an orthogonal matrix.

Example

In Fig. 3.10, consider the vector $\overline{OP}$ which is transformed by rotation around the OX_3 axis to $\overline{O\Pi}$. The equations of transformation are written:

$$\begin{bmatrix} \xi_1 \\ \xi_2 \\ \xi_3 \end{bmatrix} = \begin{bmatrix} \cos\theta & -\sin\theta & 0 \\ \sin\theta & \cos\theta & 0 \\ 0 & 0 & +1 \end{bmatrix} \begin{bmatrix} x_1 \\ x_2 \\ x_3 \end{bmatrix}.$$

The determinant of the matrix of the transformation is equal to $+1$. The vectors formed by its columns are mutually orthogonal. This is also the case for the vectors formed by its rows. If the transformation from $\overline{OP}$ to $\overline{O\Pi}$ is followed by a reflection with respect to the plane $X_1 OX_2$, one obtains the vector $\overline{O\Pi}'$, which is of the same length as $\overline{OP}$. The matrix $[R]$ describing the reflection is written:

$$[R] = \begin{bmatrix} 1 & 0 & 0 \\ 0 & 1 & 0 \\ 0 & 0 & -1 \end{bmatrix},$$

and the coordinates of Π' are given by

$$\begin{bmatrix} \xi'_1 \\ \xi'_2 \\ \xi'_3 \end{bmatrix} = \begin{bmatrix} 1 & 0 & 0 \\ 0 & 1 & 0 \\ 0 & 0 & -1 \end{bmatrix} \begin{bmatrix} \xi_1 \\ \xi_2 \\ \xi_3 \end{bmatrix} = \begin{bmatrix} \xi_1 \\ \xi_2 \\ -\xi_3 \end{bmatrix} =$$

$$\begin{bmatrix} \cos\theta & -\sin\theta & 0 \\ \sin\theta & \cos\theta & 0 \\ 0 & 0 & -1 \end{bmatrix} \begin{bmatrix} x_1 \\ x_2 \\ x_3 \end{bmatrix}.$$

The determinant of the transformation matrix is equal to -1.(*)

* See pages A-7 to A-10 for additional explanations.

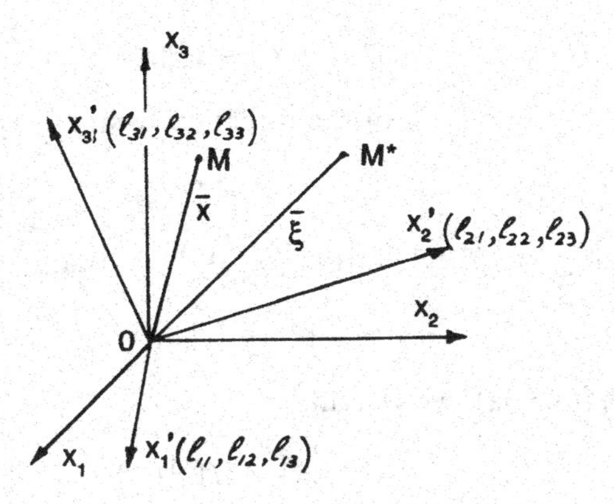

Fig. 3.11

3.5 Changes of Axes in a Linear Transformation

In a trirectangular system of coordinates OX_1, OX_2, OX_3 (Fig. 3.11), let the linear transformation

$$\{\bar{\xi}\} = [a]\{\bar{x}\} \tag{3.5.1}$$

transform the vector $\bar{x}$ to $\bar{\xi}$. This means that the matrix $[a]$ operating on the vector $\bar{x}$ gives the vector $\bar{\xi}$ in the base OX_1, OX_2, OX_3. Let us now consider a second system of axes OX'_1, OX'_2, OX'_3, obtained by means of any rotation of the first system around O, and ask the question: In this second system, what is the form of the matrix which transforms $\bar{x}$ to $\bar{\xi}$? In other words, if in the second system we write:

$$\{\bar{\xi}'\} = [a']\{\bar{x}'\}, \tag{3.5.2}$$

what are the values of the $\bar{a}'_{ij}$'s? ξ'_1, ξ'_2, ξ'_3, and x'_1, x'_2, x'_3 are the components of the two vectors $\bar{\xi}$ and $\bar{x}$ in the second system. The second system of coordinates is often defined by means of the direction cosines of its axes with respect to the first system. If the direction cosines of OX'_1 are (l_{11}, l_{12}, l_{13}), those of OX'_2 are (l_{21}, l_{22}, l_{23}), and those of OX'_3 are (l_{31}, l_{32}, l_{33}); then

$$\begin{aligned} x'_1 &= l_{11}x_1 + l_{12}x_2 + l_{13}x_3 \\ x'_2 &= l_{21}x_1 + l_{22}x_2 + l_{23}x_3 \\ x'_3 &= l_{31}x_1 + l_{32}x_2 + l_{33}x_3 . \end{aligned} \tag{3.5.3}$$

In matrix notation,

$$\{\bar{x}'\} = [\ell]\{\bar{x}\}, \tag{3.5.3a}$$

where $[\ell]$ is the matrix of the direction cosines. Also,

$$\{\bar{\xi}'\} = [\ell]\{\bar{\xi}\}. \tag{3.5.4}$$

Substituting Eq. (3.5.1) in Eq. (3.5.4), we obtain:

$$\{\bar{\xi}'\} = [\ell][a]\{\bar{x}\}. \tag{3.5.5}$$

From Eq. (3.5.3a), we obtain:

$$\{\bar{x}\} = [\ell]^{-1}\{\bar{x}'\}. \tag{3.5.6}$$

If we substitute Eq. (3.5.6) in Eq. (3.5.5), we obtain:

$$\{\bar{\xi}'\} = [\ell][a][\ell]^{-1}\{\bar{x}'\}, \tag{3.5.7}$$

which means that the matrix $[a']$ of Eq. (3.5.2) is given by:

$$[a'] = [\ell][a][\ell]^{-1}. \tag{3.5.8}$$

This answers the question asked at the beginning of this section.

It is customary to set

$$[\ell]^{-1} = [m],$$

so that Eq. (3.5.8) becomes:

$$[a'] = [m]^{-1}[a][m]. \tag{3.5.9}$$

The matrix $[\ell]$ is formed by the direction cosines of the new axes with respect to the old ones. Therefore, the components of $[\ell]$ satisfy the orthogonality relation

$$\ell_{ij}\,\ell_{kj} = \delta_{ik}. \tag{3.5.10}$$

The transpose of $[\ell]$ is equal to its inverse, and both are orthogonal matrices.

In index notation, Eq. (3.5.6) can now be written:

$$x_m = \ell_{nm}\,x'_n. \tag{3.5.6a}$$

Eq. (3.5.7) becomes:

$$\xi'_i = \ell_{ij} a_{jm} \ell_{nm} x'_n . \tag{3.5.7a}$$

Eq. (3.5.8) becomes:

$$a'_{ik} = \ell_{ij} \ell_{km} a_{jm} . \tag{3.5.8a}$$

Eq. (3.5.9) becomes:

$$a'_{ik} = m_{ji} m_{mk} a_{jm} . \tag{3.5.9a}$$

Eq. (3.5.9) is a particular case of a class of transformations called similarity transformations. In general, if there exists a nonsingular matrix $[s]$ such that

$$[s]^{-1}[a][s] = [b] \tag{3.5.11}$$

for any two square matrices $[a]$ and $[b]$ of the same order, then $[a]$ and $[b]$ are called *similar*. The transformation of $[a]$ to $[b]$ is a similarity transformation. Taking the determinant of both sides of Eq. (3.5.11), we obtain:

$$|[s]^{-1}[a][s]| = (|[s]^{-1}|)(|[a]|)(|[s]|) = |[a]| = |[b]| . \tag{3.5.12}$$

Thus the determinants of two similar matrices are equal.

3.6 Characteristic Equations and Eigenvalues

In Sec. 3.2, it was mentioned that in the transformation $\xi_i = a_{ij} x_j$, the matrix $[a]$ could be looked upon as an operator acting on the vector $\bar{x}$ to give the vector $\bar{\xi}$. Let us ask the following question: For a given transformation matrix $[a]$, does there exist a vector (or vectors) $\bar{x}$ which when transformed remains parallel to itself (Fig. 3.12)? The answer can be found by writing that the vector $\bar{\xi}$ is parallel to $\bar{x}$—in other words, is a scalar multiple of $\bar{x}$. Thus,

$$\{\bar{\xi}\} = \lambda\{\bar{x}\} = [a]\{\bar{x}\} = \lambda[1]\{\bar{x}\} \tag{3.6.1}$$

or

$$\xi_i = \lambda x_i = a_{ij} x_j ,$$

and the scalar multiplier λ must be determined.

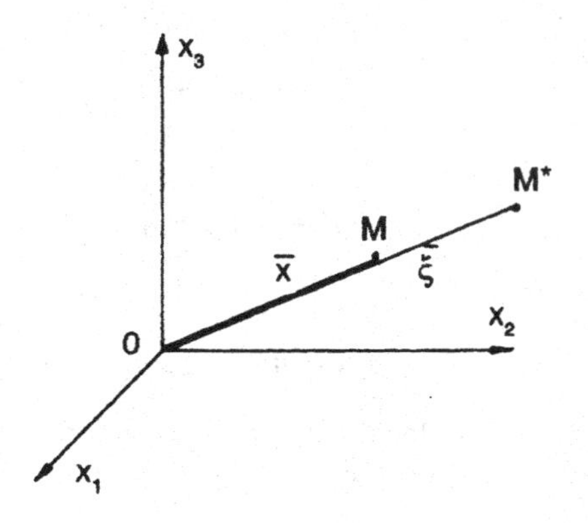

Fig. 3.12

Eq. (3.6.1) is written:

$$[[a] - \lambda[1]]\{\bar{x}\} = 0 \tag{3.6.2}$$

or

$$(a_{ij} - \delta_{ij}\lambda)x_j = 0.$$

This is a set of three linear homogeneous equations with three unknowns, a nontrivial solution of which exists only when the determinant of the matrix $[[a] - \lambda[1]]$ vanishes, or

$$|[[a] - \lambda[1]]| = 0. \tag{3.6.3}$$

By expanding the determinant, we obtain a polynomial in λ, which when equated to zero gives the characteristic equation of the matrix $[a]$:

$$\begin{aligned}\lambda^3 &- (a_{11} + a_{22} + a_{33})\lambda^2 + (a_{11}a_{22} + a_{22}a_{33} + a_{33}a_{11} - a_{12}a_{21} - a_{23}a_{32} \\ &- a_{13}a_{31})\lambda - (a_{11}a_{22}a_{33} + a_{12}a_{23}a_{31} \\ &+ a_{13}a_{21}a_{32} - a_{12}a_{21}a_{33} - a_{11}a_{23}a_{32} - a_{13}a_{22}a_{31}) = 0.\end{aligned} \tag{3.6.4}$$

It is thus apparent that the scalar multiplier λ must be a root of Eq. (3.6.4). There are three such roots, real or imaginary, with possible equal roots. These roots are called eigenvalues or characteristic roots. Each root λ_i, when substituted in Eq. (3.6.2), gives a set of three linear equations which are not all independent. By assuming one value for one of the components of $\bar{x}$ (or more than one value when there are equal roots) and discarding one of the equations, one can solve for the two other components of $\bar{x}$. In other words, the eigenvector $\bar{x}$ (also called characteristic vector) is determined within an arbitrary multiplicative constant. This is due to the homogeneity of the equations: If $\bar{x}$ is a solution of Eq. (3.6.1), then $K\bar{x}$ is also a solution. Therefore, the answer to the question is that three directions exist (of which at least one is real) which remain parallel to themselves after transformation.

Let us assume for the present that the characteristic equation (3.6.4) has three distinct roots. In such a case, there are three distinct eigenvectors $\{\bar{x}\}_i$ that satisfy Eq. (3.6.1); that is,

$$\{\bar{x}\}_1 = \begin{bmatrix} x_{11} \\ x_{21} \\ x_{31} \end{bmatrix}$$

is the eigenvector corresponding to the eigenvalue λ_1.

$$\{\bar{x}\}_2 = \begin{bmatrix} x_{12} \\ x_{22} \\ x_{32} \end{bmatrix}$$

is the eigenvector corresponding to the eigenvalue λ_2.

$$\{\bar{x}\}_3 = \begin{bmatrix} x_{13} \\ x_{23} \\ x_{33} \end{bmatrix}$$

is the eigenvector corresponding to the eigenvalue λ_3. The eigenvectors are generally expressed in normalized form; that is, the elements of the vector $\bar{x}_1$, for example, are chosen in such a way that

$$x_{11}^2 + x_{21}^2 + x_{31}^2 = 1. \tag{3.6.5}$$

This is possible because we have a free choice of one of the components of each eigenvector.

Let a square matrix $[M]$ be constructed from the eigenvector columns $\{\bar{x}\}_i$ in the following manner:

$$[M] = \begin{bmatrix} x_{11} & x_{12} & x_{13} \\ x_{21} & x_{22} & x_{23} \\ x_{31} & x_{32} & x_{33} \end{bmatrix}; \qquad (3.6.6)$$

that is, the columns of $[M]$ are the eigenvectors of $[a]$. The matrix $[M]$ is called the modal matrix of $[a]$. The diagonal matrix formed by the eigenvalues of $[a]$ is called the spectral matrix:

$$[D] = \begin{bmatrix} \lambda_1 & 0 & 0 \\ 0 & \lambda_2 & 0 \\ 0 & 0 & \lambda_3 \end{bmatrix}. \qquad (3.6.7)$$

Recalling the remarks made in Sec. 2.5 regarding the use of diagonal matrices to write groups of equations under the form of one matrix equation, the three groups of three equations represented by Eq. (3.6.1) can be written:

$$[a][M] = [M][D]. \qquad (3.6.8)$$

Premultiplying both sides by $[M]^{-1}$, we obtain:

$$[D] = [M]^{-1}[a][M]. \qquad (3.6.9)$$

This is a similarity transformation of $[a]$ to a diagonal matrix $[D]$ through the use of the modal matrix of $[a]$ and its inverse. The transformation expressed by Eq. (3.6.9) is referred to as the diagonalization of the matrix $[a]$.

An important property of linear transformations is that *the eigenvalues obtained from a matrix similar to* $[a]$ *are equal to those obtained from* $[a]$ *itself.* To prove this property, let us replace $[a]$ in Eq. (3.6.3) by a similar matrix $[s]^{-1}[a][s]$:

$$|[s]^{-1}[a][s] - \lambda[1]| = (|[s]^{-1}|)(|[a] - \lambda[1]|)(|[s]|) = |[a] - \lambda[1]|. \qquad (3.6.10)$$

It follows that the characteristic equation associated with $[s]^{-1}[a][s]$ is the same as the one associated with $[a]$, and hence their roots are identical.

If we now consider the transpose of $[a]$ instead of $[a]$ itself in Eq. (3.6.3), we obtain the same characteristic Eq. (3.6.4) and the same eigenvalues. Let the modal matrix of $[a]'$ be called $[N]$. Its spectral matrix is still $[D]$. Then,

$$[a]'[N] = [N][D]. \qquad (3.6.11)$$

Taking the transpose of Eq. (3.6.11), we obtain:

$$[N]'[a] = [D][N]'. \qquad (3.6.12)$$

If we now premultiply both sides of Eq. (3.6.8) by $[N]'$ and postmultiply both sides of Eq. (3.6.12) by $[M]$, we obtain:

$$[N]'[a][M] = [N]'[M][D] \qquad (3.6.13)$$

and

$$[N]'[a][M] = [D][N]'[M]. \qquad (3.6.14)$$

The comparison of Eqs. (3.6.13) and (3.6.14) shows that the product of the two matrices $[N]'$ and $[M]$ must result in a diagonal matrix. Every one of the column vectors of these two modal matrices can be multiplied by a suitable constant and adjusted so that the resulting diagonal matrix is the unit matrix $[1]$. Thus,

$$[N]'[M] = [1]. \qquad (3.6.15)$$

The vectors of $[N]$ and $[M]$ are said to be normalized. Eq. (3.6.15) can be written in index notation as:

$$N_{ij} M_{ik} = \delta_{jk}. \qquad (3.6.15a)$$

This is the orthogonality relation of the eigenvectors of the matrix $[a]$ and of its transpose $[a]'$. If the matrix $[a]$ is symmetric so that $[a] = [a]'$, then $[N] = [M]$, and Eq. (3.6.15a) becomes:

$$M_{ij} M_{ik} = \delta_{jk}, \qquad (3.6.16)$$

which is the orthogonality condition for the matrix $[M]$. *Therefore the eigenvectors of a symmetric matrix are orthogonal.*

Another *property of real symmetric matrices is that their eigenvalues are always real.* To prove this property, let $\{x\}_1$ be an eigenvector of the matrix $[a]$ associated with the eigenvalue λ_1. Then,

$$[a]\{x\}_1 = \lambda_1 \{x\}_1.$$

If λ_1 is a complex number and the elements of $[a]$ are real, then the elements of $\{x\}_1$ must be complex numbers. Let the complex quantities of the previous equation be replaced by their complex conjugates. Then,

$$[a]\{\underline{\bar{x}}\}_1 = \underline{\lambda}_1\{\underline{\bar{x}}\}_1,$$

where the bar under the complex quantity denotes its conjugate. If we now premultiply the two previous equations by $\{\underline{\bar{x}}\}_1'$ and $\{\bar{x}\}_1'$, respectively, we obtain:

$$\{\underline{\bar{x}}\}_1'[a]\{\bar{x}\}_1 = \{\underline{\bar{x}}\}_1'\lambda_1\{\bar{x}\}_1 \tag{3.6.17}$$

and

$$\{\bar{x}\}_1'[a]\{\underline{\bar{x}}\}_1 = \{\bar{x}\}_1'\underline{\lambda}_1\{\underline{\bar{x}}\}_1. \tag{3.6.18}$$

Let us take the transpose of both members of Eq. (3.6.18):

$$\{\underline{\bar{x}}\}_1'[a]\{\bar{x}\}_1 = \{\underline{\bar{x}}\}_1'\underline{\lambda}_1\{\bar{x}\}_1.$$

Subtracting Eq. (3.6.17) from the previous equation, we obtain:

$$(\underline{\lambda}_1 - \lambda_1)\{\underline{\bar{x}}\}_1'\{\bar{x}\}_1 = 0.$$

Since $\{\underline{\bar{x}}\}_1'\{\bar{x}\}_1$ is the sum of positive quantities and cannot be equal to zero, then

$$\underline{\lambda}_1 = \lambda_1,$$

and therefore λ_1 must be real.

The eigenvalue problem, introduced here through the idea of a vector remaining parallel to itself after transformation, arises in many branches of mechanics. The elements of the matrix $[a]$ can represent, among others, stresses, strains, moments of inertia, and couples. This problem will be encountered several times in the coming sections. Various methods for obtaining eigenvalues and eigenvectors have been devised. They can be found in the bibliography at the end of Chapter 2.

3.7 Invariants of the Transformation Matrix in a Linear Transformation

There are some interesting relations among the characteristic values and the coefficients of λ in Eq. (3.6.4). From the theory of equations, we know that: In an equation in λ of degree n in which the coefficient of λ^n is unity, the sum of the roots equals the negative of the coefficient of λ^{n-1}, the sum of the products of the roots two at a time equals the coefficient of λ^{n-2}, the sum of the products of the roots three at a time equals the negative of the coefficient of λ^{n-3}, etc.; finally the product of the roots equals the constant term or its negative, depending on whether n is even or odd. Thus, if the roots of the characteristic equation are $\lambda_1, \lambda_2, \lambda_3$, we have:

$$\lambda_1 + \lambda_2 + \lambda_3 = a_{11} + a_{22} + a_{33} = a_{ii} = I_1 \tag{3.7.1}$$

$$\begin{aligned}\lambda_1\lambda_2 + \lambda_2\lambda_3 + \lambda_3\lambda_1 = a_{11}a_{22} + a_{22}a_{33} + a_{33}a_{11} - a_{12}a_{21} \\ - a_{23}a_{32} - a_{13}a_{31} = I_2\end{aligned} \tag{3.7.2}$$

$$\begin{aligned}\lambda_1\lambda_2\lambda_3 = a_{11}a_{22}a_{33} + a_{12}a_{23}a_{31} + a_{13}a_{21}a_{32} - a_{12}a_{21}a_{33} \\ - a_{11}a_{23}a_{32} - a_{13}a_{22}a_{31} = \varepsilon_{ijk}a_{1i}a_{2j}a_{3k} = I_3 .\end{aligned} \tag{3.7.3}$$

In Sec. 3.5, it was shown that, in a change of coordinates, the elements of the transformation matrix could be obtained through the similarity transformation of Eq. (3.5.9); in Sec. 3.6, it was shown that characteristic equations associated with similar matrices were the same and led to the same eigenvalues: Thus, whatever be the system of coordinates at the start, the elements of the matrix of a given linear transformation satisfy the three relations (3.7.1), (3.7.2), and (3.7.3). These three combinations of the elements of the matrix are, respectively, called the first, second, and third invariant of the transformation. For convenience, Eq. (3.6.4) is written:

$$\lambda^3 - I_1\lambda^2 + I_2\lambda - I_3 = 0. \tag{3.7.4}$$

3.8 Invariant Directions of a Linear Transformation

A straight line is said to have an invariant direction if it keeps the same orientation under a linear transformation. If the line initially passes through the origin, both the line and its transform coincide (Fig. 3.13). If the line OM, for example, has an invariant direction under a linear transformation, the coordinates of M^* must be given by:

$$\xi_i = \lambda x_i = a_{ij}x_j; \tag{3.8.1}$$

that is,

$$(a_{ij} - \lambda\delta_{ij})x_j = 0. \tag{3.8.2}$$

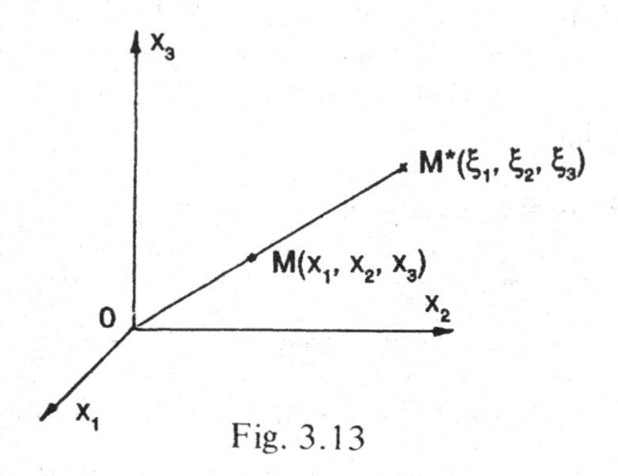

Fig. 3.13

The solution of this set of three linear homogeneous equations in x_1, x_2, x_3 has already been examined in Sec. 3.6. It was used to introduce the eigenvalue problem and has yielded three possible values for λ, each one corresponding to a given eigenvector $\{\bar{x}\}_i$. If the roots of the characteristic equation λ_i are all real, we obtain three invariant directions such that (Fig. 3.14):

$$\frac{\overline{OM^*}_1}{\overline{OM}_1} = \lambda_1, \quad \frac{\overline{OM^*}_2}{\overline{OM}_2} = \lambda_2, \frac{\overline{OM^*}_3}{\overline{OM}_3} = \lambda_3. \tag{3.8.3}$$

The three directions are not necessarily orthogonal. They define three invariant planes—in other words, three planes which do not change

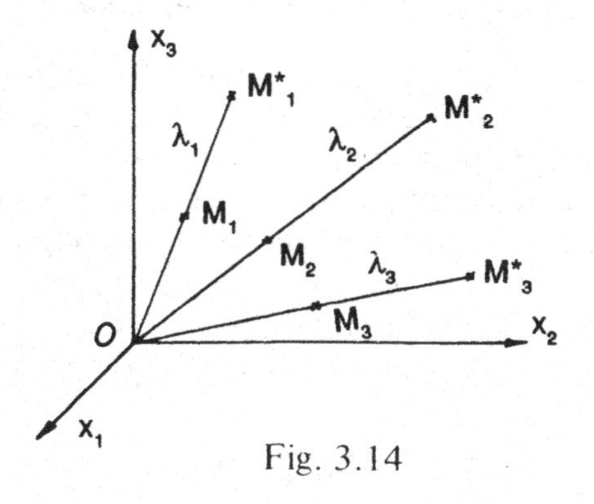

Fig. 3.14

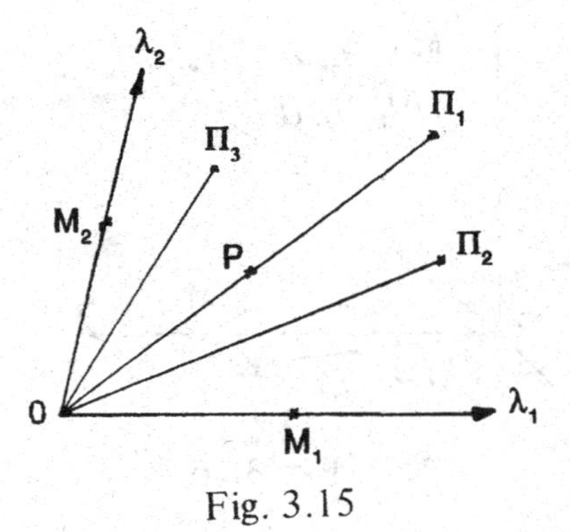

Fig. 3.15

during the transformation. If one of the roots is real and the two other are complex, we obtain only one invariant direction.

It is instructive to examine geometric.lly the previous analysis: In an invariant plane defined by the two invariant directions corresponding to λ_1 and λ_2 (Fig. 3.15), the vector $\overline{OP}$ is transformed to $\overline{O\Pi_i}$ $(i = 1, 2)$. If $\lambda_1 = \lambda_2$, point P moves to Π_1. If $\lambda_1 > \lambda_2$, $\overline{OP}$ rotates toward $\overline{OM_1}$ to $\overline{O\Pi_2}$. If $\lambda_2 > \lambda_1$, $\overline{OP}$ rotates towards $\overline{OM_2}$ to $\overline{O\Pi_3}$. In other words, the rotation occurs towards the line corresponding to the larger λ.

The same invariant directions are obtained if the hodograph of the transformation (3.8.1) is used in place of the transformation itself. In such a case we write:

$$u_i = \xi_i - x_i = (\lambda - 1)x_i = (a_{ij} - \delta_{ij})x_j; \qquad (3.8.4)$$

that is,

$$(a_{ij} - \lambda\delta_{ij})x_j = 0, \qquad (3.8.5)$$

which is the same as Eq. (3.8.2). (*)

3.9 Antisymmetric Linear Transformations

A linear transformation is called antisymmetric or asymmetric when its hodograph is expressed by an antisymmetric matrix. Thus (Fig. 3.16), the transformation of M to M^* is antisymmetric if

---

* See pages A-10 to A-15 for a detailed solved example.

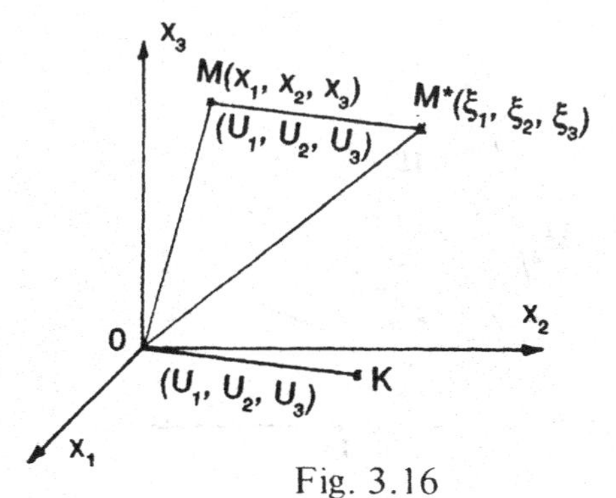

Fig. 3.16

$$\begin{bmatrix} u_1 \\ u_2 \\ u_3 \end{bmatrix} = \begin{bmatrix} 0 & -a_{12} & a_{13} \\ a_{12} & 0 & -a_{23} \\ -a_{13} & a_{23} & 0 \end{bmatrix} \begin{bmatrix} x_1 \\ x_2 \\ x_3 \end{bmatrix}. \tag{3.9.1}$$

It is of interest to find what type of displacement this transformation gives to a point $M(x_1, x_2, x_3)$. Let $\overline{OH}$ be a vector whose components η_1, η_2, η_3 (Fig. 3.17) are:

$$\eta_1 = a_{23}, \quad \eta_2 = a_{13}, \quad \eta_3 = a_{12}. \tag{3.9.2}$$

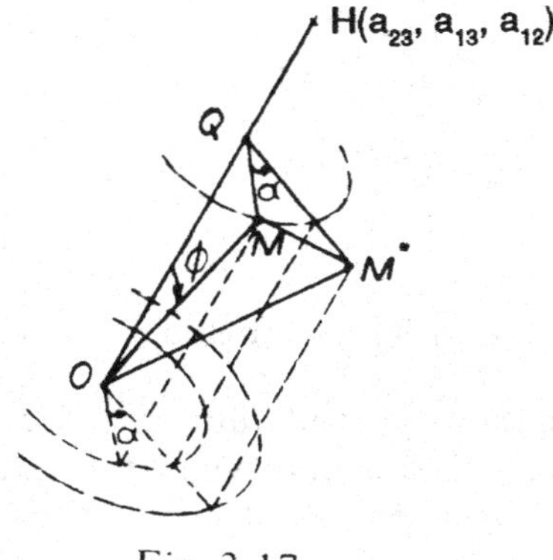

Fig. 3.17

The transformation expressed by Eqs. (3.9.1) when applied to H gives:

$$\begin{bmatrix} u_1 \\ u_2 \\ u_3 \end{bmatrix} = \begin{bmatrix} 0 & -a_{12} & a_{13} \\ a_{12} & 0 & -a_{23} \\ -a_{13} & a_{23} & 0 \end{bmatrix} \begin{bmatrix} a_{23} \\ a_{13} \\ a_{12} \end{bmatrix} = \begin{bmatrix} 0 \\ 0 \\ 0 \end{bmatrix}, \tag{3.9.3}$$

which means that H does not change under the antisymmetric transformation. Therefore, $\overline{OH}$ has an invariant direction and all points on it are fixed during the transformation. When the transformation (3.9.1) is applied to any point M with coordinates x_1, x_2, x_3, we obtain the components of $\overline{MM^*}$; namely:

$$\begin{bmatrix} u_1 \\ u_2 \\ u_3 \end{bmatrix} = \begin{bmatrix} 0 & -a_{12} & a_{13} \\ a_{12} & 0 & -a_{23} \\ -a_{13} & a_{23} & 0 \end{bmatrix} \begin{bmatrix} x_1 \\ x_2 \\ x_3 \end{bmatrix} = \begin{bmatrix} a_{13}x_3 - a_{12}x_2 \\ a_{12}x_1 - a_{23}x_3 \\ a_{23}x_2 - a_{13}x_1 \end{bmatrix}. \tag{3.9.4}$$

The magnitude and direction of $\overline{MM^*}$ with respect to $\overline{OH}$ and $\overline{OM}$ can be found by computing the vector product $\overline{OH} \times \overline{OM}$ and the scalar products $\overline{OM} \cdot \overline{MM^*}$ and $\overline{OH} \cdot \overline{MM^*}$:

$$\begin{aligned} \overline{OH} \times \overline{OM} &= \begin{vmatrix} \bar{i}_1 & \bar{i}_2 & \bar{i}_3 \\ a_{23} & a_{13} & a_{12} \\ x_1 & x_2 & x_3 \end{vmatrix} \\ &= [\bar{i}_1(a_{13}x_3 - a_{12}x_2) + \bar{i}_2(a_{12}x_1 - a_{23}x_3) \\ &\quad + \bar{i}_3(a_{23}x_2 - a_{13}x_1)] \end{aligned} \tag{3.9.5}$$

$$\begin{aligned} \overline{OM} \cdot \overline{MM^*} &= x_1(a_{13}x_3 - a_{12}x_2) + x_2(a_{12}x_1 - a_{23}x_3) \\ &\quad + x_3(a_{23}x_2 - a_{13}x_1) = 0 \end{aligned} \tag{3.9.6}$$

$$\begin{aligned} \overline{OH} \cdot \overline{MM^*} &= a_{23}(a_{13}x_3 - a_{12}x_2) + a_{13}(a_{12}x_1 - a_{23}x_3) \\ &\quad + a_{12}(a_{23}x_2 - a_{13}x_1) = 0. \end{aligned} \tag{3.9.7}$$

Eqs. (3.9.4) to (3.9.7) show that $\overline{MM^*}$ is normal to the plane $(\overline{OH}, \overline{OM})$, and that it is given in magnitude and direction by $\overline{OH} \times \overline{OM}$.

In Fig. 3.17, let MQ be the normal from M on $\overline{OH}$ and α be the angle between $\overline{QM}$ and $\overline{QM^*}$. The following relations deduced from the geometry of Fig. 3.17 can be written:

$$|\overline{MM^*}| = (OH)(OM)\sin\phi = (OH)(OM)\frac{(QM)}{(OM)} = (OH)(QM) \quad (3.9.8).$$

$$\tan\alpha = \frac{(MM^*)}{QM} = OH = \sqrt{(a_{23})^2 + (a_{13})^2 + (a_{12})^2} \quad (3.9.9)$$

$$\cos\alpha = \frac{1}{\sqrt{1 + (a_{23})^2 + (a_{13})^2 + (a_{12})^2}}. \quad (3.9.10)$$

We note that the angle α depends only on the values of the coefficients of the transformation matrix. It is therefore the same for all points like M, and always varies between O and $\Pi/2$ in a direction following the right-hand rule with the thumb in the direction of $\overline{OH}$. Thus, under the transformation, M rotates by an angle α around $\overline{OH}$ in a plane normal to $\overline{OH}$. However, this rotation does not occur alone, but is associated with a radial displacement normal to $\overline{OH}$. The unit value of this radial displacement is given by:

$$\begin{aligned}\varepsilon_r &= \frac{QM^* - QM}{QM} = \frac{1 - \cos\alpha}{\cos\alpha} \\ &= \sqrt{1 + (a_{23})^2 + (a_{13})^2 + (a_{12})^2} - 1.\end{aligned} \quad (3.9.11)$$

In summary, the transformation expressed by an antisymmetric matrix is the product of a rotation α and a cylindrical dilatation around an axis $\overline{OH}$ whose direction ratios are given by the coefficients of the matrix. The product of the rotation and the dilatation is commutative. $\overline{OH}$ is not only invariant but is also a principal direction. Any other direction in the plane $(\overline{QM}, \overline{QM^*})$ normal to $\overline{OH}$ is also a principal direction. If the angle of rotation α is small, so that α^2 is very small compared to unity, ε_r can be neglected since it is of the second order with respect to α. This is shown by:

$$\varepsilon_r = \frac{1 - \cos\alpha}{\cos\alpha} = \frac{1 - 1 + \frac{\alpha^2}{2!} - \frac{\alpha^4}{4!} + \dots}{1 - \frac{\alpha^2}{2!} + \dots} \approx \frac{\alpha^2}{2}.(*) \quad (3.9.12)$$

3.10 Symmetric Transformations. Definitions and General Theorems

A linear transformation is said to be symmetric if the matrix of the transformation is symmetric. Thus,

$$\{\xi\} = [a]\{x\}, \text{ with } [a] = [a]', \quad (3.10.1)$$

* See pages A-16 to A-18 for more information on antisymmetric linear transformations.

represents a linear symmetric transformation. In index notation, we write:

$$\xi_i = a_{ij} x_j, \text{ with } a_{ij} = a_{ji}. \qquad (3.10.2)$$

Linear symmetric transformations occur commonly in the study of stresses and deformations in continuous media. In the following, we shall prove some theorems related to this particular type of linear transformations.

Theorem I: *The symmetric transformations are the only ones to possess the property of reciprocity*. By reciprocity we mean that if a vector $\overline{OM}_1$ is transformed into $\overline{OM^*}_1$ and another vector $\overline{OM}_2$ is transformed into $\overline{OM^*}_2$, then

$$\overline{OM}_1 \cdot \overline{OM^*}_2 = \overline{OM}_2 \cdot \overline{OM^*}_1. \qquad (3.10.3)$$

To prove this theorem, let us consider a general matrix $[a]$ operating on both $\overline{OM}_1(x_1, x_2, x_3)$ and $\overline{OM}_2(y_1, y_2, y_3)$ to give $\overline{OM^*}_1(\xi_1, \xi_2, \xi_3)$ and $\overline{OM^*}_2(\eta_1, \eta_2, \eta_3)$, respectively. In tabular form, this is written:

$\overline{OM}_2$		y_1	y_2	y_3
	$\overline{OM}_1$	x_1	x_2	x_3
η_1	ξ_1	a_{11}	a_{12}	a_{13}
η_2	ξ_2	a_{21}	a_{22}	a_{23}
η_3	ξ_3	a_{31}	a_{32}	a_{33}

(3.10.4)

Using the results of Eq. (3.10.4), we get:

$$\begin{aligned} \overline{OM}_1 \cdot \overline{OM^*}_2 = {} & x_1(a_{11}y_1 + a_{12}y_2 + a_{13}y_3) \\ & + x_2(a_{21}y_1 + a_{22}y_2 + a_{23}y_3) \\ & + x_3(a_{31}y_1 + a_{32}y_2 + a_{33}y_3) \end{aligned} \qquad (3.10.5)$$

and

$$\begin{aligned} \overline{OM}_2 \cdot \overline{OM^*}_1 = {} & y_1(a_{11}x_1 + a_{12}x_2 + a_{13}x_3) \\ & + y_2(a_{21}x_1 + a_{22}x_2 + a_{23}x_3) \\ & + y_3(a_{31}x_1 + a_{32}x_2 + a_{33}x_3). \end{aligned} \qquad (3.10.6)$$

We see that Eq. (3.10.5) cannot be equal to Eq. (3.10.6) unless $a_{12} = a_{21}$, $a_{13} = a_{31}$, $a_{23} = a_{32}$; in other words, unless $[a]$ is a symmetric matrix. The previous analysis is based on scalar products and is

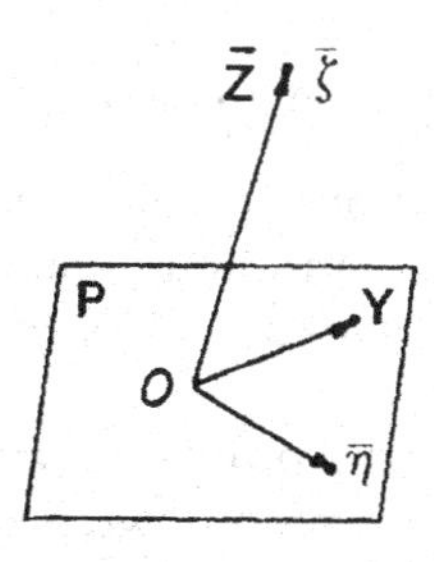

Fig. 3.18

independent of the chosen rectangular system of coordinates.

Theorem II: *The symmetric transformations are non-rotational: Any invariant direction is a principal direction.* Let $\overline{Z}$ be a vector along an invariant direction. A linear symmetric transformation transforms $\overline{Z}$ to a vector $\overline{\zeta}$ along the same direction (Fig. 3.18). Let P be a plane normal to $\overline{Z}$, and $\overline{Y}$ be any vector in it. The same linear symmetric transformation transforms $\overline{Y}$ to $\overline{\eta}$. Since $\overline{Y} \cdot \overline{\zeta} = 0$, then $\overline{\eta} \cdot \overline{Z} = 0$ because of the property of reciprocity. Therefore, $\overline{\eta}$ lies in the plane P. When $\overline{Y}$ sweeps the plane P, $\overline{\eta}$ sweeps the same plane. Therefore, $\overline{Z}$ is a principal direction since, when coupled with any vector like $\overline{Y}$, the angle between them remains a right angle after transformation.

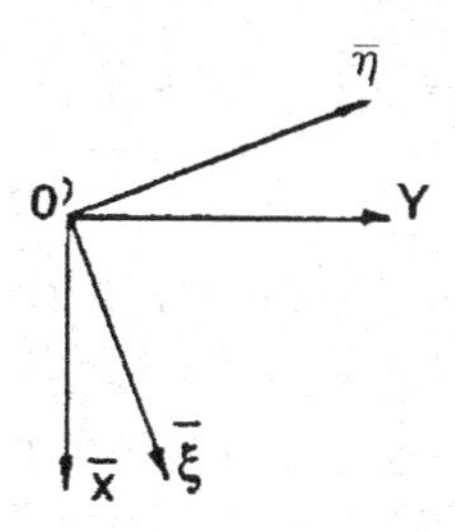

Fig. 3.19

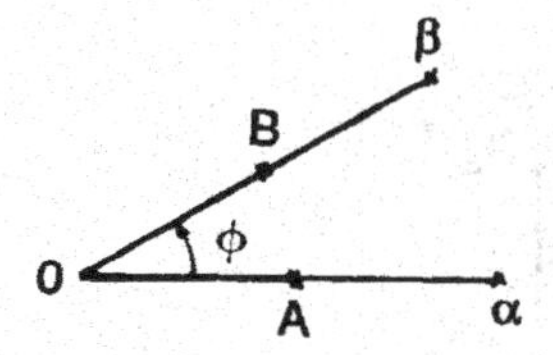

Fig. 3.20

Let us now consider two other vectors $\overline{X}$ and $\overline{Y}$ along the two other principal directions (Fig. 3.19). They cannot rotate to $\overline{\xi}$ and $\overline{\eta}$ in the transformation since the property of reciprocity would not be satisfied: Indeed, $\overline{X} \cdot \overline{\eta}$ is negative and $\overline{Y} \cdot \overline{\xi}$ is positive. Thus the two principal directions $\overline{X}$ and $\overline{Y}$ are also invariant.

Theorem III: *Non-rotational transformations are the only ones to have the property of reciprocity.* Let us consider a general linear transformation—in other words, a transformation which is not necessarily symmetric. If $\overline{OA}$ and $\overline{OB}$ are two vectors (Fig. 3.20) along two invariant directions, points A and B will move to α and β after transformation in a way such that:

$$\overline{O\alpha} = \lambda_1(\overline{OA}) \tag{3.10.7}$$

$$\overline{O\beta} = \lambda_2(\overline{OB}). \tag{3.10.8}$$

Since $\lambda_1 \neq \lambda_2$ and $\cos\phi \neq 0$, then

$$\overline{OA} \cdot \overline{O\beta} \neq \overline{OB} \cdot \overline{O\alpha},$$

which means that the property of reciprocity is not satisfied. If $\cos\phi = 0$, the two invariant directions are also principal, which means that the linear transformation is a symmetric one and consequently possesses the property of reciprocity.

In Fig. 3.21, let the three coordinate axes OX_1, OX_2, OX_3 lie along the principal directions of a linear symmetric transformation. These three directions are invariant and do not rotate. Under the transformation, any vector $\overline{OP}$ will become OQ through a rotation and a change in

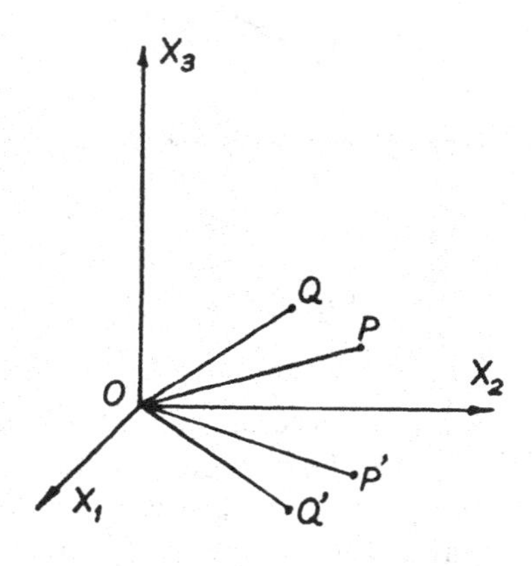

Fig. 3.21

length. A vector $\overline{OP'}$ symmetric of $\overline{OP}$ with respect to any of the three principal planes, will become $\overline{OQ'}$ symmetric of $\overline{OQ}$ with respect to the same plane. Thus, in a symmetric transformation all the directions rotate except the three principal ones. The rotations, however, compensate one another symmetrically with respect to the three orthogonal principal planes. The expression, symmetric transformation, therefore implies the symmetry of the matrix as well as the symmetry of rotation of the various directions. From the previous theorems, we conclude that the terms non-rotational, reciprocal, pure deformation, and symmetric transformation are all equivalent.

3.11 Principal Directions and Principal Unit Displacements of a Symmetric Transformation

In Sec. 3.3, it was shown that in a linear transformation three orthogonal directions exist that remain orthogonal after the transformation. These directions are called the principal directions. In Sec. 3.10, it was shown that for a symmetric transformation these directions were also invariant directions of the transformation. Thus the search for

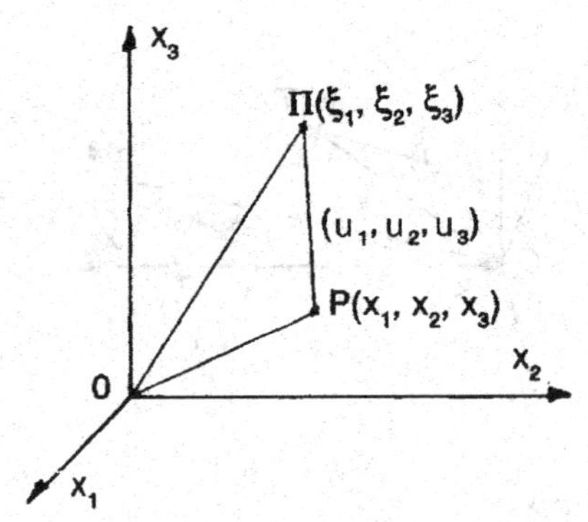

Fig. 3.22

principal directions, in this case, is equivalent to the search for invariant directions. This search was examined in both Secs. 3.6 and 3.8.

For convenience, let us write the transformation which brings P to Π (Fig. 3.22), as follows:

$$\begin{bmatrix} \xi_1 \\ \xi_2 \\ \xi_3 \end{bmatrix} = \begin{bmatrix} 1 + a_{11} & a_{12} & a_{13} \\ a_{12} & 1 + a_{22} & a_{23} \\ a_{13} & a_{23} & 1 + a_{33} \end{bmatrix} \begin{bmatrix} x_1 \\ x_2 \\ x_3 \end{bmatrix} \tag{3.11.1}$$

or

$$\xi_i = (a_{ij} + \delta_{ij})x_j,$$

with $a_{ij} = a_{ji}$. The hodograph of the transformation (in other words, the components of the displacement) is given by:

$$u_i = a_{ij}x_j. \tag{3.11.2}$$

The invariant directions are obtained by writing:

$$\xi_i = (1 + \lambda)x_i, \tag{3.11.3}$$

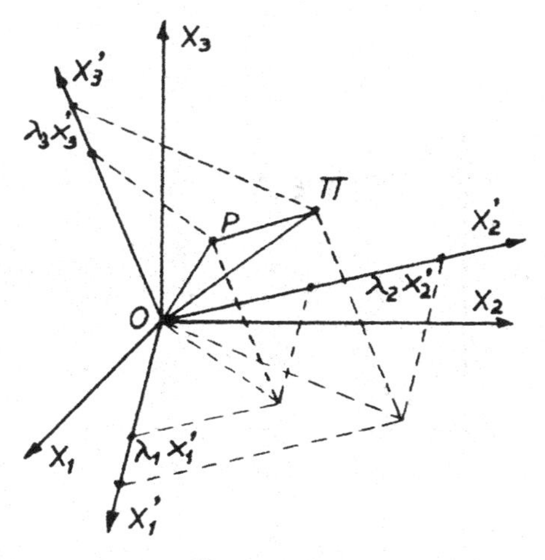

Fig. 3.23

in which it is seen that λ represents the unit displacement along the invariant (also principal) directions. The characteristic equation has three roots, λ_1, λ_2, and λ_3, which are then used to find the three principal directions. The same results are obtained if we use Eq. (3.11.2) in conjunction with

$$u_i = \lambda x_i . \tag{3.11.4}$$

If $\lambda_3 < \lambda_2 < \lambda_1$, λ_3 is called the minor principal unit displacement, λ_2 is called the intermediate principal unit displacement, and λ_1 is called the major principal unit displacement. The principal directions define a trirectangular system of coordinates, which is often very convenient to use. Let the axes of such a system be called OX'_1, OX'_2, OX'_3 (Fig. 3.23), and let the components of $\overline{OP}$, $\overline{P\Pi}$, and $\overline{O\Pi}$ in it be (x'_1, x'_2, x'_3), (u'_1, u'_2, u'_3), and (ξ'_1, ξ'_2, ξ'_3), respectively. Thus,

$$\begin{bmatrix} \xi'_1 \\ \xi'_2 \\ \xi'_3 \end{bmatrix} = \begin{bmatrix} 1+\lambda_1 & 0 & 0 \\ 0 & 1+\lambda_2 & 0 \\ 0 & 0 & 1+\lambda_3 \end{bmatrix} \begin{bmatrix} x'_1 \\ x'_2 \\ x'_3 \end{bmatrix} \tag{3.11.5}$$

and

$$\begin{bmatrix} u'_1 \\ u'_2 \\ u'_3 \end{bmatrix} = \begin{bmatrix} \lambda_1 & 0 & 0 \\ 0 & \lambda_2 & 0 \\ 0 & 0 & \lambda_3 \end{bmatrix} \begin{bmatrix} x'_1 \\ x'_2 \\ x'_3 \end{bmatrix}. \tag{3.11.6}$$

In the principal system of axes, the transformation is seen to be made by a diagonal matrix. The displacement of P is obtained by simple extension or contraction of the projections of $\overline{OP}$ on the principal axes: Indeed, those directions are also invariant in the transformation. The principal directions form three planes called the principal planes of the transformation and, regardless of which system of coordinates we start from, those directions will always be the same.

If the linear symmetric transformation (3.11.1) is such that one of the roots of the characteristic equation is repeated, three different eigenvectors can still be found. Consider, for example, the homogeneous system:

$$\begin{bmatrix} 3-\lambda & 0 & 1 \\ 0 & 2-\lambda & 0 \\ 1 & 0 & 3-\lambda \end{bmatrix} \begin{bmatrix} x_1 \\ x_2 \\ x_3 \end{bmatrix} = 0, \tag{3.11.7}$$

whose characteristic equations is

$$\lambda^3 - 8\lambda^2 + 20\lambda - 16 = 0. \tag{3.11.8}$$

The roots of this equation are $\lambda_1 = \lambda_2 = 2$ and $\lambda_3 = 4$. For $\lambda = 2$, we obtain one equation for determing $\{\overline{x}\}_1$, namely

$$x_{11} + x_{31} = 0. \tag{3.11.9}$$

In Sec. 3.6, it was indicated that in the eigenvalue problem, one of the equations is discarded and replaced by a free choice of a component. As a choice, it is customary to use the normalization condition:

$$\sqrt{x_{11}^2 + x_{21}^2 + x_{31}^2} = 1, \tag{3.11.10}$$

which reduces the components of the eigenvector to direction cosines. In our case, we are entitled to a second free choice of a component since one of the eigenvalues is repeated. We shall choose $x_{11} = 1/\sqrt{2}$. These two choices, together with Eq. (3.11.9), give:

$$\{\bar{x}\}_1 = \begin{bmatrix} \frac{1}{\sqrt{2}} \\ 0 \\ \frac{-1}{\sqrt{2}} \end{bmatrix}. \tag{3.11.11}$$

The other eigenvector corresponding to $\lambda = 2$ must be such that

$$x_{12} + x_{32} = 0 \tag{3.11.12}$$

and, since it is orthogonal to $\{\bar{x}\}_1$, then

$$(x_{11})(x_{12}) + (x_{21})(x_{22}) + (x_{31})(x_{32}) = 0 \tag{3.11.13}$$

or

$$\frac{1}{\sqrt{2}}x_{12} + (0)(x_{22}) - \frac{1}{\sqrt{2}}x_{32} = 0. \tag{3.11.14}$$

Eqs. (3.11.12), (3.11.14), and the normalization condition show that

$$\{\bar{x}\}_2 = \begin{bmatrix} 0 \\ 1 \\ 0 \end{bmatrix}. \tag{3.11.15}$$

For $\lambda = 4$, we have the system of equations

$$\begin{aligned} -x_{13} + x_{33} &= 0 \\ -2x_{23} &= 0, \end{aligned} \tag{3.11.16}$$

added to the normalization condition (3.11.10). The solution of this system is:

$$\{\bar{x}\}_3 = \begin{bmatrix} \frac{1}{\sqrt{2}} \\ 0 \\ \frac{1}{\sqrt{2}} \end{bmatrix}. (*) \tag{3.11.17}$$

* See pages A-18 to A-19 for another example.

3.12 Quadratic Forms

If x and y are two sets of n variables (three in our case), a function which is linear and homogeneous in the variables of each set separately is called a bilinear form. Thus,

$$A(x,y) = [y_1 y_2 y_3] \begin{bmatrix} a_{11} & a_{12} & a_{13} \\ a_{21} & a_{22} & a_{23} \\ a_{31} & a_{32} & a_{33} \end{bmatrix} \begin{bmatrix} x_1 \\ x_2 \\ x_3 \end{bmatrix} \tag{3.12.1}$$

is a bilinear form. When the sets of variables are identical so that $[y] = \{x\}'$, the bilinear form becomes a quadratic form. It was shown in Sec. 2.4 that any square matrix can be given as the sum of a symmetric and of an antisymmetric component. Setting $[y] = \{x\}'$ and decomposing the matrix $[a]$ into its two components, we obtain for $A\ (x,\ x)$ the sum:

$$[x_1 x_2 x_3] \begin{bmatrix} a_{11} & \dfrac{a_{12}+a_{21}}{2} & \dfrac{a_{13}+a_{31}}{2} \\ \dfrac{a_{12}+a_{21}}{2} & a_{22} & \dfrac{a_{23}+a_{32}}{2} \\ \dfrac{a_{13}+a_{31}}{2} & \dfrac{a_{23}+a_{32}}{2} & a_{33} \end{bmatrix} \begin{bmatrix} x_1 \\ x_2 \\ x_3 \end{bmatrix}$$

$$+ [x_1 x_2 x_3] \begin{bmatrix} 0 & \dfrac{a_{12}-a_{21}}{2} & \dfrac{a_{13}-a_{31}}{2} \\ \dfrac{a_{21}-a_{12}}{2} & 0 & \dfrac{a_{23}-a_{32}}{2} \\ \dfrac{a_{31}-a_{13}}{2} & \dfrac{a_{32}-a_{23}}{2} & 0 \end{bmatrix} \begin{bmatrix} x_1 \\ x_2 \\ x_3 \end{bmatrix}. \tag{3.12.2}$$

The second term is equal to zero, while the first one with its symmetric square matrix is found to be equal to $\{\overline{x}\}'[a]\{\overline{x}\}$. Therefore, in a quadratic form, the antisymmetric component has no effect, and a convenient expression for $A(x,x)$ becomes:

$$A(x,x) = \{\overline{x}\}'[a]\{\overline{x}\}, \tag{3.12.3}$$

where $[a]$ is a symmetric matrix. In index notation, the *scalar* $A(x,x)$ is written:

$$A(x,x) = a_{ij} x_i x_j . \tag{3.12.4}$$

If A is a constant, Eq. (3.12.4) represents a quadric surface with its center at the origin. The nature of the quadric depends on the value of the elements a_{ij}. If the determinant of $[a]$ does not vanish, the quadric is either an ellipsoid or a hyperboloid. If the determinant of $[a]$ vanishes, the surface degenerates into a cylinder of the elliptic or hyperbolic type or else into two parallel planes symmetrically situated with respect to the origin.

Let us assume that the quadric surface is an ellipsoid. This ellipsoid will, in general, have three principal axes different in length so that their direction is uniquely determined (Fig. 3.24). A suitable rotation of the

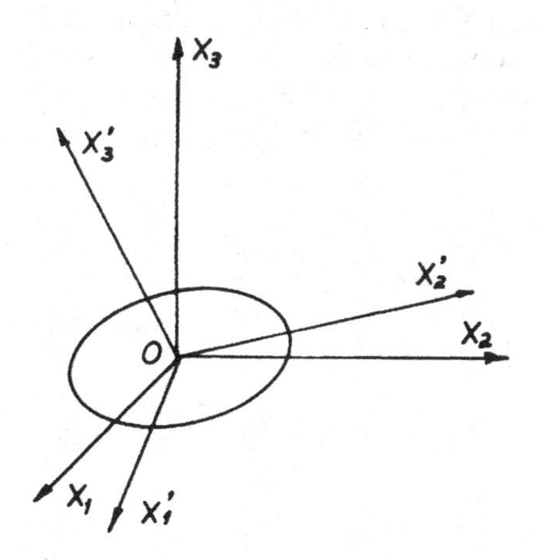

Fig. 3.24

system of reference axes can be made to bring it in coincidence with the principal axes of the ellipsoid. In this new system, the expression for A becomes:

$$A(x', x') = \lambda_1 (x'_1)^2 + \lambda_2 (x'_2)^2 + \lambda_3 (x'_3)^2 \tag{3.12.5}$$

or

$$A(x', x') = \{\bar{x}'\}'[D]\{\bar{x}'\}, \tag{3.12.6}$$

where $[D]$ is a diagonal matrix with elements λ_1, λ_2, and λ_3. The change of axes from the system OX'_1, OX'_2, OX'_3 to OX_1, OX_2, OX_3 can be brought about by (Sec. 3.5):

$$\{\bar{x}\} = [m]\{\bar{x}'\}, \tag{3.12.7}$$

where $[m]$ is a matrix whose columns are the direction cosines of the new system with respect to the old one. Introducing Eq. (3.12.7) into Eq. (3.12.3), we obtain:

$$A(x, x) = \{\bar{x}'\}'[m]'[a][m]\{\bar{x}'\}. \tag{3.12.8}$$

Equating Eq. (3.12.6) to Eq. (3.12.8), we get:

$$\{\bar{x}'\}'[D]\{\bar{x}'\} = \{\bar{x}'\}'[m]'[a][m]\{\bar{x}'\}. \tag{3.12.9}$$

Thus,

$$[D] = [m]'[a][m] = [m]^{-1}[a][m]. \tag{3.12.10}$$

The operation of Eq. (3.12.10) diagonalizes the matrix $[a]$ so that the elements of $[D]$ are the eigenvalues. The columns of $[m]$ are the eigenvectors giving the directions of the new system of axes (Modal Matrix). They are immediately obtained once the λ's are determined from the characteristic equation of $[a]$. If, for instance, we set $A = 1$, the lengths of the principal diameters of the ellipsoid are given by $2/\sqrt{\lambda_1}$, $2/\sqrt{\lambda_2}$, $2/\sqrt{\lambda_3}$.

Thus far we have assumed that the principal axes of the ellipsoid are of unequal length. When two out of the three eigenvalues are equal, the ellipsoid is of revolution around one of its principal axes, and any other axis normal to it is a principal axis. When the three eigenvalues are equal, the ellipsoid degenerates into a sphere: Any three mutually perpendicular axes are principal axes.

Many problems associated with quadratic forms are intimately related to problems associated with sets of linear equations. We may notice that if we write:

$$\xi_i = \frac{1}{2}\frac{\partial A}{\partial x_i}, \tag{3.12.11}$$

we obtain the equations:

$$\begin{aligned}
\xi_1 &= a_{11}x_1 + a_{12}x_2 + a_{13}x_3 \\
\xi_2 &= a_{12}x_1 + a_{22}x_2 + a_{23}x_3 \\
\xi_3 &= a_{13}x_1 + a_{23}x_2 + a_{33}x_3
\end{aligned} \tag{3.12.12}$$

or $\xi_i = a_{ij}x_j$, with $a_{ij} = a_{ji}$. Eqs. (3.12.12) are those of the symmetric transformation of $\{\bar{x}\}$ to $\{\xi\}$. When A is formed by a sum of squares as in Eq. (3.12.5), with no cross-product terms, we say that A is reduced to a canonical form. To reduce a quadratic form to its canonical form, the system of Eqs. (3.12.12) can first be obtained and would correspond to a symmetric transformation. The matrix of the transformation is then diagonalized to give λ_1, λ_2, and λ_3. The example given in Sec. 3.11 can be used to demonstrate the reduction of a quadratic form to its canonical form. Starting from the quadratic form:

$$A(x,x) = 3x_1^2 + 2x_2^2 + 3x_3^2 + 2x_1x_3,$$

Eq. (3.12.11) gives:

$$\begin{aligned} \xi_1 &= 3x_1 + 0x_2 + x_3 \\ \xi_2 &= 0x_1 + 2x_2 + 0x_3 \\ \xi_3 &= x_1 + 0x_2 + 3x_3 \end{aligned} \qquad \text{or} \qquad \begin{bmatrix} \xi_1 \\ \xi_2 \\ \xi_3 \end{bmatrix} = \begin{bmatrix} 3 & 0 & 1 \\ 0 & 2 & 0 \\ 1 & 0 & 3 \end{bmatrix} \begin{bmatrix} x_1 \\ x_2 \\ x_3 \end{bmatrix}.$$

The eigenvalues and eigenvectors are obtained as shown in Sec. 3.11 and the canonical form is:

$$A(x',x') = 2[(x_1')^2 + (x_2')^2] + 4(x_3')^2.(*)$$

3.13 Normal and Tangential Displacements in a Symmetric Transformation. Mohr's Representation

Consider a point $P(x_1, x_2, x_3)$ located on a unit sphere centered at the origin (Fig. 3.25). Under a linear symmetric transformation, P goes to Π. The components of the displacement $\overline{P\Pi}$ are u_1, u_2, and u_3. The equation of the sphere is:

$$x_1^2 + x_2^2 + x_3^2 = 1. \tag{3.13.1}$$

In the plane $OP\Pi$, the vector $\overline{PN} = \bar{n}$ is the projection of $\overline{P\Pi}$ on the line OPN and the vector $\overline{PT} = \bar{t}$ is the projection of $\overline{P\Pi}$ on the tangent plane to the sphere at P. $\bar{n}$ and $\bar{t}$ are respectively called the *normal and the tangential components of the displacement of P*. Since the magnitude of

* See pages A-19 to A-20 for another example.

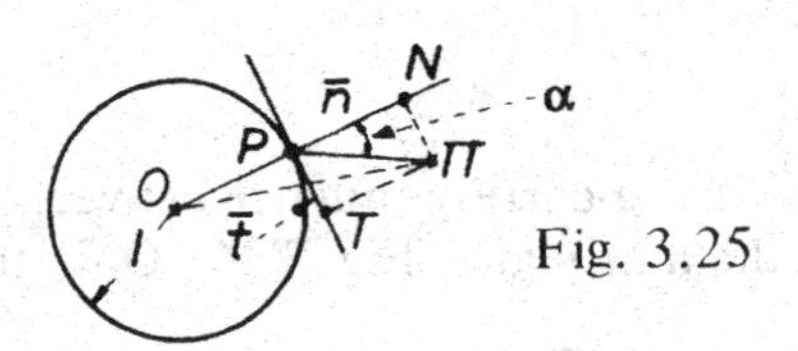

Fig. 3.25

$\overline{OP}$ is unity, the magnitude of $\bar{n}$ is given by the scalar product of $\overline{OP}(x_1, x_2, x_3)$ and $\overline{P\Pi}(u_1, u_2, u_3)$. Using the principal axes as axes of reference, i.e., using Eqs. (3.11.6), we obtain:

$$n = P\Pi \cos \alpha = \lambda_1 x_1^2 + \lambda_2 x_2^2 + \lambda_3 x_3^2. \tag{3.13.2}$$

The magnitude of $\bar{t}$ is given by the magnitude (but not by the direction) of the vector product of $\overline{OP}$ and $\overline{P\Pi}$. The components of this vector product are:

$$(\lambda_3 - \lambda_2)x_2 x_3, \quad (\lambda_1 - \lambda_3)x_3 x_1, \quad (\lambda_2 - \lambda_1)x_1 x_2. \tag{3.13.3}$$

Therefore,

$$\begin{aligned} t^2 = (P\Pi)^2 \sin^2 \alpha &= (\lambda_2 - \lambda_1)^2 x_1^2 x_2^2 + (\lambda_1 - \lambda_3)^2 x_3^2 x_1^2 \\ &\quad + (\lambda_3 - \lambda_2)^2 x_2^2 x_3^2 \end{aligned} \tag{3.13.4}$$

and

$$n^2 + t^2 = \lambda_1^2 x_1^2 + \lambda_2^2 x_2^2 + \lambda_3^2 x_3^2. \tag{3.13.5}$$

Thus, in a given linear symmetric transformation characterized by specific values of λ_1, λ_2, and λ_3, one can compute the normal and tangential displacements of any point $P(x_1, x_2, x_3)$ of the unit sphere. The same operation can be made graphically by means of a construction due to O. Mohr. For that, one has to solve Eqs. (3.13.1), (3.13.2), and (3.13.5), and obtain the expression of x_1, x_2, x_3 in terms of n and t. These are written:

$$x_1^2 = \frac{t^2 + (n - \lambda_2)(n - \lambda_3)}{(\lambda_1 - \lambda_2)(\lambda_1 - \lambda_3)} \tag{3.13.6a}$$

$$x_2^2 = \frac{t^2 + (n - \lambda_3)(n - \lambda_1)}{(\lambda_2 - \lambda_3)(\lambda_2 - \lambda_1)} \tag{3.13.6b}$$

$$x_3^2 = \frac{t^2 + (n - \lambda_1)(n - \lambda_2)}{(\lambda_3 - \lambda_1)(\lambda_3 - \lambda_2)}. \tag{3.13.6c}$$

Mohr's construction establishes a correspondence between points on the unit sphere and points in the n , t plane (Fig. 3.26). Only the squares of

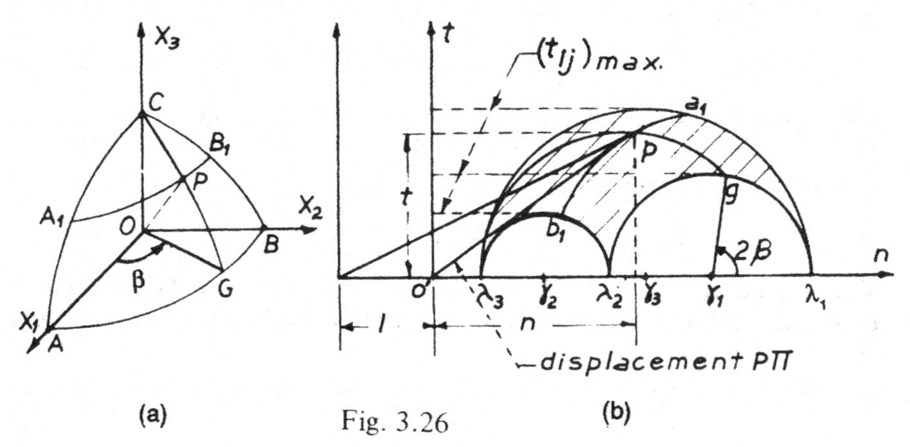

Fig. 3.26

x_1, x_2, x_3, and t appear in Eqs. (3.13.6) so that one-eighth of the sphere and one-half of the n , t plane are required to study the correspondence. Let us assume $\lambda_1 > \lambda_2 > \lambda_3$. In Figs. 3.26a and 3.26b, the curve corresponding to the principal circle AB is obtained by setting $x_3 = 0$ in Eq. (3.13.6c). Therefore,

$$t^2 + (n - \lambda_1)(n - \lambda_2) = 0. \tag{3.13.7}$$

This is the equation of a circle $\lambda_1 \lambda_2$, whose radius is $(\lambda_1 - \lambda_2)/2$. The center is γ_1, which is given by $o'\gamma_1 = (\lambda_1 + \lambda_2)/2$. In the same way, the circle $\lambda_2 \lambda_3$ corresponds to BC, and the circle $\lambda_1 \lambda_3$ corresponds to AC. Thus, all the points on the sphere have in the $(n\ ,\ t)$ plane an image which falls in the hatched region limited by the three circles.

If, on the surface of the sphere, we consider a circle $A_1 B_1$ parallel to the plane OX_1, OX_2 and at a height $x_3 = h$, the corresponding curve in the n , t plane is obtained by setting $x_3 = h$ in Eq. (3.13.6c). This gives:

$$t^2 + (n - \lambda_1)(n - \lambda_2) = h^2(\lambda_3 - \lambda_1)(\lambda_3 - \lambda_2). \tag{3.13.8}$$

This is the equation of a circle of center γ_1 and of radius

$$R = \sqrt{\left(\frac{\lambda_1 - \lambda_2}{2}\right)^2 + h^2(\lambda_3 - \lambda_1)(\lambda_3 - \lambda_2)}\,. \tag{3.13.9}$$

Points on a meridian through C and G satisfy the equation:

$$x_2 = Kx_1 \tag{3.13.10}$$

Eqs. (3.13.6a), (3.13.6b), and (3.13.10) give, in the n , t plane, a circle centered on the $o'n$ axis and passing through g and λ_3 (Fig. 3.26b). We thus have three families of parallels and three families of meridians whose images are circles on Mohr's diagram.

In Fig. 3.26a, let the angle $AOG = \beta$. The coordinates of point G are:

$$x_{1G} = \cos\beta, \quad x_{2G} = \sin\beta, \quad x_{3G} = 0. \tag{3.13.11}$$

Eqs. (3.13.2) and (3.13.4) give for G:

$$n_G = \frac{\lambda_1 + \lambda_2}{2} + \frac{\lambda_1 - \lambda_2}{2}\cos 2\beta \tag{3.13.12}$$

$$\left|t_G\right| = \left|\frac{\lambda_1 - \lambda_2}{2}\sin 2\beta\right|, \tag{3.13.13}$$

in which the sign of t_G is not considered at the present time. Eqs. (3.13.12) and (3.13.13) show that point G defined by β in Fig. 3.26a has for an image the point g defined by 2β in Fig. 3.26b. Thus, when G describes the quarter of a circle AB , g describes half a circle on Mohr's diagram. This property is only true on the three principal circles AB , BC , and CA .

Every point P on the sphere has an image on the (n, t) plane. This image falls in the area limited by the three circles. It is obtained by plotting the image of the parallel and that of the meridian passing through P. This is a simple matter once one has plotted the three principal circles corresponding to AB, BC, and CA . Since the sphere has a unit radius, the coordinates of P , x_1, x_2, x_3 are also its direction cosines ℓ_1, ℓ_2, ℓ_3. Eq. (3.13.9) gives the radius of the circle corresponding to $A_1 B_1$, and β is obtained from

$$\cos\beta = \frac{\ell_1}{\sqrt{1 - \ell_3^2}}. \tag{3.13.14}$$

The point p on the n , t plane corresponding to P on the sphere is the intersection of the circle $a_1 b_1$ centered at γ_1 and of the circle $\lambda_3 g$ whose center falls on the axis $o'n$. The abcissa and the ordinate of P give the normal and tangential displacements of P .

Mohr's construction shows that:

1. Among all the points located on a unit sphere, the one which has the largest normal displacement under a linear symmetric transformation is the point A located on the major principal axis (Fig. 3.26).
2. In each principal plane, the point located on the bisector of the principal axes has the largest tangential displacement. It is equal in magnitude to one-half of the difference between the two principal normal displacements (Fig. 3.26b).
3. The maximum tangential displacement occurs in the plane of the major and minor principal axes.
4. If the same quantity h is added to the three principal unit displacements, Mohr's circles in the n , t plane keep the same diameter and are simply displaced by h on the $o'n$ axis. All the normal displacements n become $n + h$. In particular, if $h = 1$, Mohr's construction gives the transformed vector itself (ξ_1, ξ_2, ξ_3) and not the displacement vector (u_1, u_2, u_3) (Fig. 3.26b).

In all the previous equations, the coordinates x_1, x_2, x_3 of P on the unit sphere can by replaced by the direction cosines ℓ_1, ℓ_2, ℓ_3 of the line OP.(*)

3.14 Spherical Dilatation and Deviation in a Linear Symmetric Transformation

Let us introduce the following notation:

$$\frac{\lambda_1 + \lambda_2 + \lambda_3}{3} = \lambda_m \tag{3.14.1}$$

$$\lambda_1 = \lambda_m + \lambda_1', \quad \lambda_2 = \lambda_m + \lambda_2', \quad \lambda_3 = \lambda_m + \lambda_3'. \tag{3.14.2}$$

We notice that

$$\lambda_1' + \lambda_2' + \lambda_3' = 0. \tag{3.14.3}$$

Using the principal directions as coordinate axes, the substitution of Eqs. (3.14.1) and (3.14.2) in Eqs. (3.11.6) gives:

* See pages A-20 and A-21 for more information on Mohr's circles.

$$\begin{bmatrix} u_1 \\ u_2 \\ u_3 \end{bmatrix} = \begin{bmatrix} \lambda_m & 0 & 0 \\ 0 & \lambda_m & 0 \\ 0 & 0 & \lambda_m \end{bmatrix} \begin{bmatrix} x_1 \\ x_2 \\ x_3 \end{bmatrix} + \begin{bmatrix} \lambda'_1 & 0 & 0 \\ 0 & \lambda'_2 & 0 \\ 0 & 0 & \lambda'_3 \end{bmatrix} \begin{bmatrix} x_1 \\ x_2 \\ x_3 \end{bmatrix}.$$

or

$$\overline{P\Pi} = \overline{P\Pi'} + \overline{P\Pi''}. \tag{3.14.4}$$

Eq. (3.14.4) shows that the displacement of a point P under a linear symmetric transformation is the sum of the two vectors:

(1) *A* vector along OP equal to $\overline{P\Pi'}$

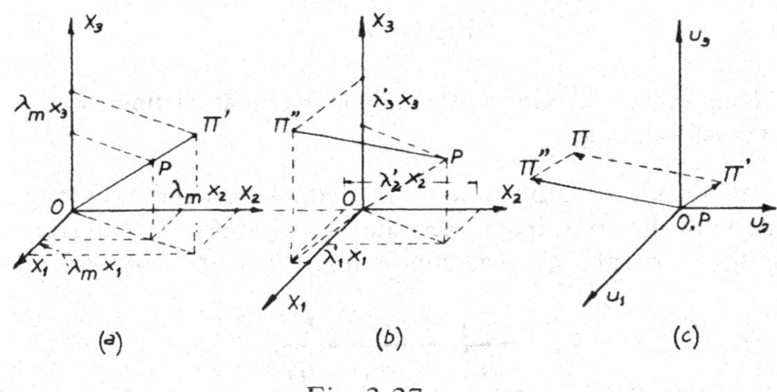

Fig. 3.27

(2) *A* vector $\overline{P\Pi''}$ characterized by a transformation matrix whose trace is equal to zero.

The first vector $\overline{P\Pi'}$ is called the spherical dilatation (or dilation) because it corresponds to an extension or a contraction along the original vector $\overline{OP}$. The second vector is called the deviation. Figs. 3.27a and 3.27b show each one of the components. The two components are added in a displacement space in Fig. 3.27c.

The previous discussion can easily be represented on a Mohr diagram. Knowing λ_1, λ_2, and λ_3, one can plot the three corresponding Mohr circles (Fig. 3.28). The point o'' is taken on $o'n$ such that $o'o'' = \lambda_m$. The diagram with the origin at o' is used to find the normal and tangential unit displacements for any vector $\overline{OP}$ with known direction cosines. The same diagram with origin at o'' allows one to obtain the normal and tangential unit displacements due to the deviation alone. On Mohr's diagram, the change of origin does not affect the tangential components; only the normal ones are affected.

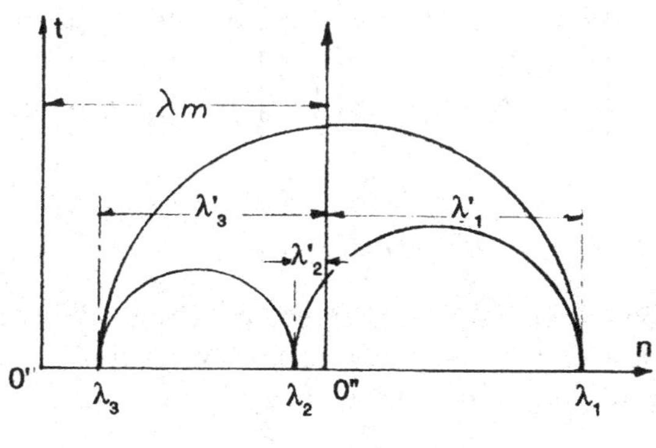

Fig. 3.28

3.15 Geometrical Meaning of the a_{ij}'s in a Linear Symmetric Transformation

Let us consider a unit cube whose edges OA, OB, OC (Fig. 3.29) coincide with the coordinate axes, and let us apply the transformation (3.11.1) to A, B, and C. The components of their displacements are:

	A	B	C
u_1	a_{11}	a_{12}	a_{13}
u_2	a_{12}	a_{22}	a_{23}
u_3	a_{13}	a_{23}	a_{33}

(3.15.1)

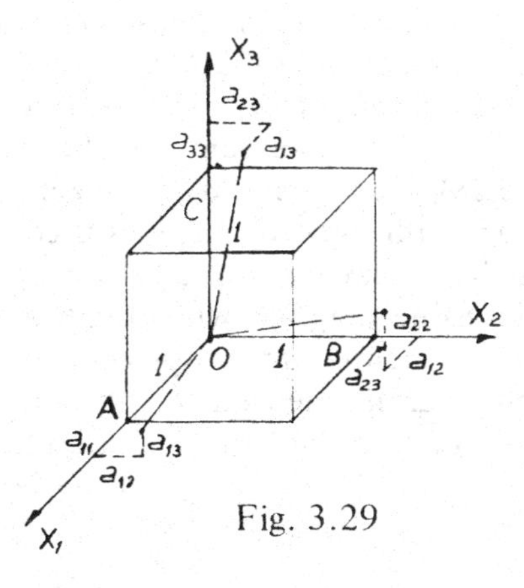

Fig. 3.29

a_{11} is seen to be the normal component of the displacement of A. a_{12} and a_{13} contribute only second-order terms to the change in length of OA. Therefore, for small transformations, a_{11} gives to a good approximation the change in length per unit length of a vector initially parallel to OX_1. Similarly, a_{22} and a_{33} give the change in length per unit length of vectors initially parallel to OX_2 and OX_3, respectively.

a_{12} is seen to be the projection on OX_2 of the tangential component of the displacement of A; it is also the projection on OX_1 of the tangential component of the displacement of B. Fig. 3.29 shows that a_{11}, a_{13}, a_{22}, and a_{23} contribute only second-order terms to the change in the right angle AOB. Therefore, for small transformations, $2a_{12}$ gives the change in the right angle between two vectors initially parallel to OX_1 and OX_2, respectively. Similarly, $2a_{13}$ gives the change in the right angle between two vectors initially parallel to OX_1 and OX_3, and $2a_{23}$ gives the change in the right angle between two vectors initially parallel to OX_2 and OX_3.

3.16 Linear Symmetric Transformation in Two Dimensions

Let Π be a principal plane of a symmetric transformation, and let OX_1 and OX_2 be two reference axes in this plane (Fig. 3.30); OX_3 is the principal direction normal to Π. Since Π is also invariant, the transformation of its points is such that

$$\begin{bmatrix} u_1 \\ u_2 \end{bmatrix} = \begin{bmatrix} a_{11} & a_{12} \\ a_{12} & a_{22} \end{bmatrix} \begin{bmatrix} x_1 \\ x_2 \end{bmatrix}, \tag{3.16.1}$$

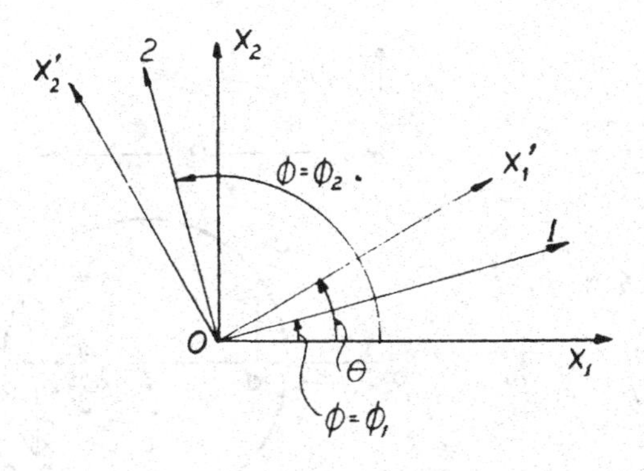

Fig. 3.30

and $u_3 = 0$. If the reference axes are rotated by an angle θ around OX_3, the a_{ij} ' s are transformed according to Eq. (3.5.8a) with

$$\begin{aligned} \ell_{11} &= \ell_{22} = \cos\theta & \ell_{12} &= -\ell_{21} = \sin\theta \\ \ell_{13} &= \ell_{23} = \ell_{31} = \ell_{32} = 0 & \ell_{33} &= 1 \end{aligned} \tag{3.16.2}$$

Thus,

$$\begin{aligned} a'_{11} &= a_{11}\cos^2\theta + a_{22}\sin^2\theta + 2a_{12}\sin\theta\cos\theta \\ &= \frac{a_{11}+a_{22}}{2} + \frac{a_{11}-a_{22}}{2}\cos 2\theta + a_{12}\sin 2\theta \end{aligned} \tag{3.16.3}$$

$$\begin{aligned} a'_{22} &= a_{11}\sin^2\theta + a_{22}\cos^2\theta - 2a_{12}\sin\theta\cos\theta \\ &= \frac{a_{11}+a_{22}}{2} - \frac{a_{11}-a_{22}}{2}\cos 2\theta - a_{12}\sin 2\theta \end{aligned} \tag{3.16.4}$$

$$a'_{12} = -\frac{a_{11}-a_{22}}{2}\sin 2\theta + a_{12}\cos 2\theta. \tag{3.16.5}$$

The eigenvalue problem in the plane yields two principal directions, $O1$ and$O2$ given by (Fig. 3.30):

$$\tan 2\phi = \frac{2a_{12}}{a_{11}-a_{22}}, \tag{3.16.6}$$

and two principal unit displacements, λ_1 and λ_2, given by:

$$\begin{matrix}\lambda_1 \\ \lambda_2\end{matrix} = \frac{a_{11}+a_{22}}{2} \pm \sqrt{\left(\frac{a_{11}-a_{22}}{2}\right)^2 + a_{12}^2}\,. \tag{3.16.7}$$

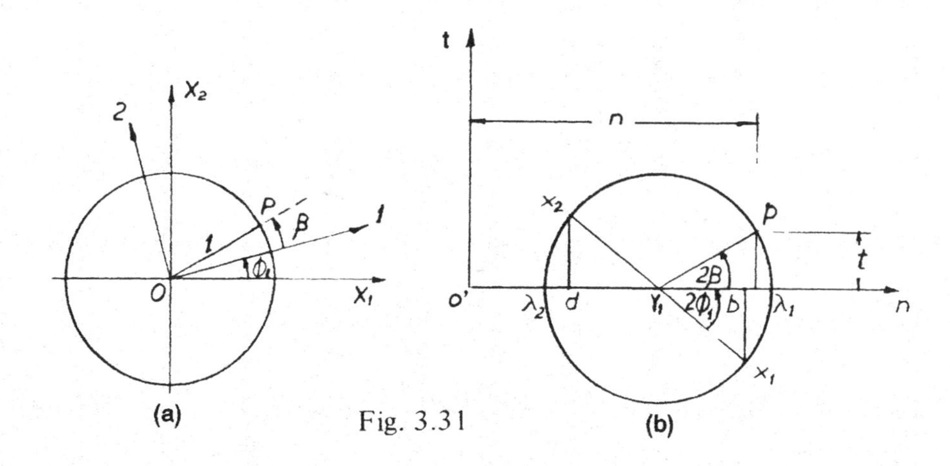

Fig. 3.31

The normal and tangential components of the displacement of any point P (Fig. 3.31a) located on a unit circle centered at O are obtained from Eqs. (3.13.2) and (3.13.4); in these equations, the values of x_1, x_2, and x_3 are set equal to cos β, sin β, and zero, respectively. Thus,

$$n = \lambda_1 \cos^2\beta + \lambda_2 \sin^2\beta$$

$$|t| = \left|\frac{\lambda_1 - \lambda_2}{2} \sin 2\beta\right|.$$

When referred to the system of axes OX_1, OX_2, n and t are given by:

$$n = \frac{a_{11} + a_{22}}{2} + \frac{a_{11} - a_{22}}{2} \cos 2\theta + a_{12} \sin 2\theta$$

$$t = -\frac{a_{11} - a_{22}}{2} \sin 2\theta + a_{12} \cos 2\theta.$$

If λ_1 and λ_2 are known, Mohr's circle for the plane Π can be constructed (Fig. 3.31b). Every point on the unit circle of Fig. 3.31a has an image on Mohr's circle. For example, the image of P is p, where $\gamma_1 p$ makes an angle 2β with $o'n$; the image of a point on the OX_1 axis is x_1, where $\gamma_1 x_1$ makes an angle $2\phi_1$ with $o'n$ in a clockwise direction; the image of a point on the OX_2 axis is diametrically opposite to the previous one. In Fig. 3.31b, one can verify that $o'b = a_{11}$, $o'd = a_{22}$, and $bx_1 = a_{12}$. Thus, knowing the elements of the transformation matrix in a system of coordinates OX_1, OX_2, one can draw Mohr's circle and obtain the principal directions and the principal unit displacements. The convention is that when a_{11} and a_{22} are positive they are plotted to the right of o' on the $o'n$ axis, and when they are negative they are plotted to its left; when a_{12} is positive it is plotted vertically under b (Fig. 3.31b) and when it is negative it is plotted vertically above b. In Sec. 3.15, it was shown that $2a_{12}$ is a measure of the change in the right angle between two vectors initially parallel to the coordinate axes; when a_{12} is positive the angle becomes acute and when it is negative the angle becomes obtuse.

Example

Given a linear transformation whose hodograph is expressed by:

$$\begin{bmatrix} u_1 \\ u_2 \end{bmatrix} = \begin{bmatrix} 0.1 & 0.04 \\ 0.04 & 0.04 \end{bmatrix} \begin{bmatrix} x_1 \\ x_2 \end{bmatrix},$$

draw Mohr's circle and obtain the principal directions and the principal unit displacements. Find the normal and tangential displacements of a

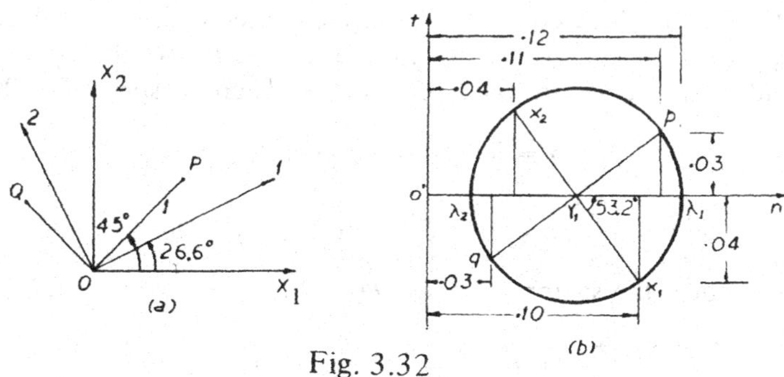

Fig. 3.32

point P whose coordinates are $(1/\sqrt{2}, 1/\sqrt{2})$. What are the coefficients of the transformation matrix in a coordinate system formed by OP and OQ (Fig. 3.32a)?

Fig. 3.32b shows Mohr's circle for the given data. The principal directions make 26.6° and 116.6° with the OX_1 axis. The principal unit displacements are 0.12 and 0.02. The normal and tangential displacements of P are 0.11 and 0.03, respectively. In the coordinate system formed by OP and OQ , the transformation matrix is:

$$\begin{bmatrix} 0.11 & -0.03 \\ -0.03 & 0.03 \end{bmatrix}.$$

Notice that the sign of a_{12} in this new coordinate system is negative, since p is above $o'n$. The right angle between two vectors parallel to OX_1 and OX_2 becomes acute after transformation. On the other hand, the right angle between two vectors parallel to OP and OQ becomes obtuse.

PROBLEMS

1. For $a_{jm} = a_{mj}$, write in full the six equations given in a condensed form by Eq. (3.5.8a).
2. Find the invariant directions of the linear transformation whose matrix is

$$\begin{bmatrix} 2 & -2 & 3 \\ 1 & 1 & 1 \\ 1 & 3 & -1 \end{bmatrix}.$$

What is the angle between those directions?

3. Show that the linear transformation whose matrix is

$$\begin{bmatrix} 1 & -2 \\ 1 & -1 \end{bmatrix}$$

does not possess a real invariant direction.

4. Find the principal directions of the linear transformation whose matrix

$$[a] = \begin{bmatrix} 7 & -2 & 1 \\ -2 & 10 & -2 \\ 1 & -2 & 7 \end{bmatrix}.$$

If $[M]$ and $[D]$ are its modal and spectral matrices, respectively, verify that $[D] = [M]^{-1}[a][M]$.

5. If the axis of rotation $\overline{OH}$ in Sec. 3.9, coincides with the OX_3 axis, find the angle of rotation and the radial extension due to antisymmetric transformation.

6. Given the quadratic form:

$$4x_1^2 + 4x_2^2 + 2x_3^2 - 4x_1 x_2 + 4x_1 x_3 + 4x_2 x_3 = A,$$

determine the associated linear transformation. If $A = 1$, find the magnitudes and directions of the principal axes of the quadric surface.

7. Reduce the quadratic form,

$$25x_1^2 + 34x_2^2 + 41x_3^2 - 24x_2 x_3 = A,$$

to its canonical form.

8. Find the eigenvalues and eigenvectors of the following system:

$$\begin{bmatrix} 3 & 0 & 2 \\ 0 & 5 & 0 \\ 2 & 0 & 3 \end{bmatrix} \begin{bmatrix} x_1 \\ x_2 \\ x_3 \end{bmatrix} = \lambda \begin{bmatrix} x_1 \\ x_2 \\ x_3 \end{bmatrix}.$$

9. In a coordinate system OX_1, OX_2, OX_3, the system

$$\begin{bmatrix} \xi_1 \\ \xi_2 \\ \xi_3 \end{bmatrix} = \begin{bmatrix} 1 & 1 & 1 \\ 1 & 2 & 3 \\ 1 & 3 & 2 \end{bmatrix} \begin{bmatrix} x_1 \\ x_2 \\ x_3 \end{bmatrix}$$

transforms $\{\bar{x}\}$ to $\{\bar{\xi}\}$.

(a) Find the elements of the transformation matrix in a new system of coordinates whose axes have direction cosines (0, 0, 1), $(1/\sqrt{2}, 1/\sqrt{2}, 0)$, $(1/\sqrt{2}, -1/\sqrt{2}, 0)$.

(b) Find the principal directions of the transformation and the principal unit displacements.

(c) Find the normal and tangential displacements of the point whose coordinates are $1/\sqrt{3}$, $1/\sqrt{3}$, $1/\sqrt{3}$. Show the position of this point on Mohr's diagram.

10. Given a linear transformation whose hodograph is

$$\begin{bmatrix} u_1 \\ u_2 \end{bmatrix} = \begin{bmatrix} 0.13 & -0.045 \\ -0.045 & 0.05 \end{bmatrix} \begin{bmatrix} x_1 \\ x_2 \end{bmatrix},$$

find the principal directions and the principal unit displacements. What are the normal and tangential displacements of a point P such that OP is inclined 30° on the OX_1 axis and its length is equal to 5?(*)

---

* See pages A-22 and A-23 for additional problems.

CHAPTER 4

GENERAL ANALYSIS OF STRAIN IN CARTESIAN COORDINATES

4.1 Introduction

In this chapter, the properties of linear transformations are used to study the problem formulated in Chapter 1. The two sets of Eqs. (1.2.4) and (1.2.5) express the change in length and direction of an element MN at M (Fig. 1.3). Every point of the body is associated with two such sets of equations and is the origin of a linear transformation. The e_{ij}' s and ω_{ij}' s are numbers specific to the point, and in general they vary from one point to another. In matrix form, Eqs. (1.2.4) and (1.2.5) are written as follows:

$$\begin{bmatrix} d\xi_1 \\ d\xi_2 \\ d\xi_3 \end{bmatrix} = \begin{bmatrix} 1 + e_{11} & e_{12} - \omega_{21} & e_{13} + \omega_{13} \\ e_{12} + \omega_{21} & 1 + e_{22} & e_{23} - \omega_{32} \\ e_{13} - \omega_{13} & e_{23} + \omega_{32} & 1 + e_{33} \end{bmatrix} \begin{bmatrix} dx_1 \\ dx_2 \\ dx_3 \end{bmatrix}, \qquad (4.1.1)$$

$$\begin{bmatrix} du_1 \\ du_2 \\ du_3 \end{bmatrix} = \begin{bmatrix} e_{11} & e_{12} - \omega_{21} & e_{13} + \omega_{13} \\ e_{12} + \omega_{21} & e_{22} & e_{23} - \omega_{32} \\ e_{13} - \omega_{13} & e_{23} + \omega_{32} & e_{33} \end{bmatrix} \begin{bmatrix} dx_1 \\ dx_2 \\ dx_3 \end{bmatrix}. \qquad (4.1.2)$$

If the element MN is of unit length, dx_1, dx_2, and dx_3 become its direction cosines.

In a body that is subjected to large transformations, a straight element seldom remains straight: A curved element is more likely to

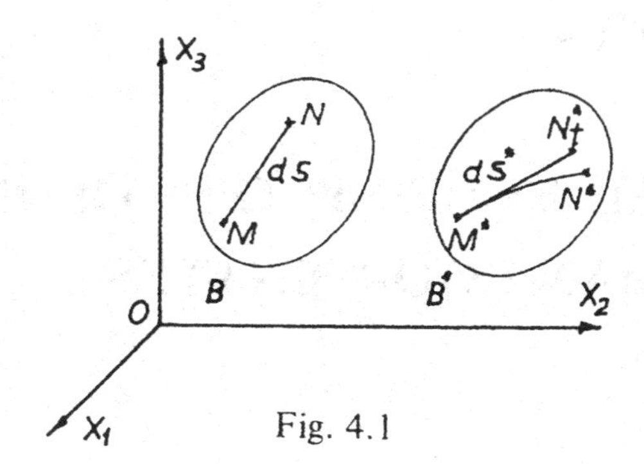

Fig. 4.1

result (Fig. 4.1). The use of the linearized Eqs. (4.1.1) and (4.1.2) to express the transformation of MN amounts to a substitution of the tangent to the curve at M^* for the curve itself. This is why the transformation is sometimes called a *linear tangent transformation*. It is obvious that the smaller the element ds, the better the approximation of M^*N^* by its tangent $M^*N_t^*$. At every point:

1) There are three principal directions and three principal planes. Although these directions and planes can be obtained from the equation of the characteristic ellipsoid, this method is operationally tedious and a more direct one will be given in a later section.

2) There are three invariant directions and three invariant planes. These directions and planes may or may not all be real; they are obtained as shown in Sec. 3.6.

3) The transformation can be split into a symmetric and an antisymmetric one. Eq. (4.1.2) becomes:

$$\begin{bmatrix} du_1 \\ du_2 \\ du_3 \end{bmatrix} = \begin{bmatrix} e_{11} & e_{12} & e_{13} \\ e_{12} & e_{22} & e_{23} \\ e_{13} & e_{23} & e_{33} \end{bmatrix} \begin{bmatrix} dx_1 \\ dx_2 \\ dx_3 \end{bmatrix} + \begin{bmatrix} 0 & -\omega_{21} & \omega_{13} \\ \omega_{21} & 0 & -\omega_{32} \\ -\omega_{13} & \omega_{32} & 0 \end{bmatrix} \begin{bmatrix} dx_1 \\ dx_2 \\ dx_3 \end{bmatrix}. \quad (4.1.3)$$

The displacement of point N (with respect to M) is the sum of the displacements caused by the two transformations; each has been studied in detail in Chapter 3 and their summation is to be made as indicated in Sec. 3.2.

In the following sections, the changes in length and direction of the elements of a body under transformation will be determined, the concept of strain at a point will be introduced, and the various definitions of strain will be discussed. For clarity, cartesian coordinates are solely used in this chapter; analyses in curvilinear coordinates are presented in Chapter 6.

4.2 Changes in Length and Directions of Elements Initially Parallel to the Coordinate Axes

Let us assume that the element MN of Fig. 1.2 is parallel to the OX_1 axis. After a linear transformation, the components of M^*N^* are given by Eq. (4.1.1) as:

$$d\xi_1 = (1 + e_{11})dx_1 \tag{4.2.1}$$

$$d\xi_2 = (e_{12} + \omega_{21})dx_1 \tag{4.2.2}$$

$$d\xi_3 = (e_{13} - \omega_{13})dx_1 . \tag{4.2.3}$$

The elongation of MN per unit length is given by:

$$E_{MN} = \frac{M^*N^* - MN}{MN} = E_{x1} . \tag{4.2.4}$$

The direction of the transformed element M^*N^* is given by its three direction cosines:

$$\ell_1 = \frac{(1 + e_{11})dx_1}{(1 + E_{MN})dx_1} = \frac{1 + e_{11}}{1 + E_{x1}} \tag{4.2.5}$$

$$\ell_2 = \frac{(e_{12} + \omega_{21})dx_1}{(1 + E_{MN})dx_1} = \frac{e_{12} + \omega_{21}}{1 + E_{x1}} \tag{4.2.6}$$

$$\ell_3 = \frac{(e_{13} - \omega_{13})dx_1}{(1 + E_{MN})dx_1} = \frac{e_{13} - \omega_{13}}{1 + E_{x1}} . \tag{4.2.7}$$

Similar equations can be written for elements MP and MQ initially parallel to OX_2 and OX_3. Unless OX_1, OX_2, OX_3 are along principal directions of the transformation, the elements will not remain orthogonal (Fig. 4.2). Their direction cosines are listed in the following table:

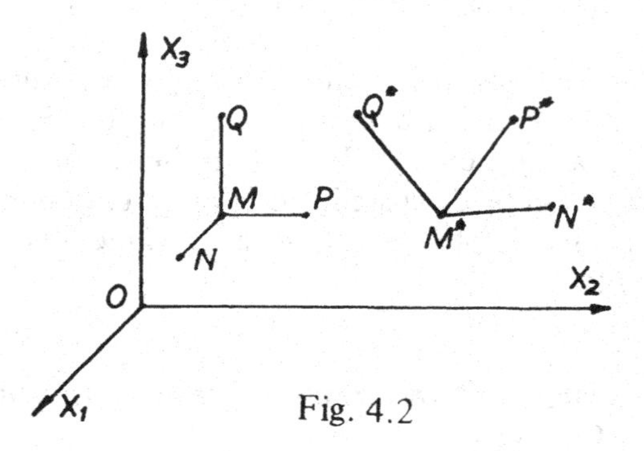

Fig. 4.2

	M^*N^* Initially Along OX_1	M^*P^* Initially Along OX_2	M^*Q^* Initially Along OX_3
l_1	$\dfrac{1+e_{11}}{1+E_{x1}}$	$\dfrac{e_{12}-\omega_{21}}{1+E_{x2}}$	$\dfrac{e_{13}+\omega_{13}}{1+E_{x3}}$
l_2	$\dfrac{e_{12}+\omega_{21}}{1+E_{x1}}$	$\dfrac{1+e_{22}}{1+E_{x2}}$	$\dfrac{e_{23}-\omega_{32}}{1+E_{x3}}$
l_3	$\dfrac{e_{13}-\omega_{13}}{1+E_{x1}}$	$\dfrac{e_{23}+\omega_{32}}{1+E_{x2}}$	$\dfrac{1+e_{33}}{1+E_{x3}}$

(4.2.8)

Since

$$M^*N^* = \sqrt{(d\xi_1)^2 + (d\xi_2)^2 + (d\xi_3)^2}, \tag{4.2.9}$$

then

$$1 + E_{x1} = \sqrt{(1 + e_{11})^2 + (e_{12} + \omega_{21})^2 + (e_{13} - \omega_{13})^2}\ . \tag{4.2.10}$$

Similarly,

$$1 + E_{x2} = \sqrt{(e_{12} - \omega_{21})^2 + (1 + e_{22})^2 + (e_{23} + \omega_{32})^2} \tag{4.2.11}$$

$$1 + E_{x3} = \sqrt{(e_{13} + \omega_{13})^2 + (e_{23} - \omega_{32})^2 + (1 + e_{33})^2}\ . \tag{4.2.12}$$

4.3 Components of the State of Strain at a Point

In Fig. 1.2, let us compute the change in length of the elements ds after a linear transformation:

$$(ds^*)^2 - (ds)^2 = \sum_i (d\xi_i)^2 - \sum_i (dx_i)^2 = \delta_{ij}(d\xi_i d\xi_j - dx_i dx_j). \quad (4.3.1)$$

Introducing Eqs. (1.1.6) in Eq. (4.3.1), we obtain:

$$\begin{aligned}(ds^*)^2 - (ds)^2 = 2[&\varepsilon_{11}(dx_1)^2 + \varepsilon_{22}(dx_2)^2 + \varepsilon_{33}(dx_3)^2 + 2\varepsilon_{12}dx_1 dx_2 \\ &+ 2\varepsilon_{13}dx_1 dx_3 + 2\varepsilon_{23}dx_2 dx_3],\end{aligned} \quad (4.3.2)$$

where the following substitutions of Eqs. (4.3.3) have been made:

$$\begin{aligned}
\varepsilon_{11} &= \frac{\partial u_1}{\partial x_1} + \frac{1}{2}\left[\left(\frac{\partial u_1}{\partial x_1}\right)^2 + \left(\frac{\partial u_2}{\partial x_1}\right)^2 + \left(\frac{\partial u_3}{\partial x_1}\right)^2\right] \\
&= e_{11} + \frac{1}{2}[(e_{11})^2 + (e_{12} + \omega_{21})^2 + (e_{13} - \omega_{13})^2] \\
\varepsilon_{22} &= \frac{\partial u_2}{\partial x_2} + \frac{1}{2}\left[\left(\frac{\partial u_1}{\partial x_2}\right)^2 + \left(\frac{\partial u_2}{\partial x_2}\right)^2 + \left(\frac{\partial u_3}{\partial x_2}\right)^2\right] \\
&= e_{22} + \frac{1}{2}[(e_{12} - \omega_{21})^2 + (e_{22})^2 + (e_{23} + \omega_{32})^2] \\
\varepsilon_{33} &= \frac{\partial u_3}{\partial x_3} + \frac{1}{2}\left[\left(\frac{\partial u_1}{\partial x_3}\right)^2 + \left(\frac{\partial u_2}{\partial x_3}\right)^2 + \left(\frac{\partial u_3}{\partial x_3}\right)^2\right] \\
&= e_{33} + \frac{1}{2}[(e_{13} + \omega_{13})^2 + (e_{23} - \omega_{32})^2 + (e_{33})^2] \\
\varepsilon_{12} &= \frac{1}{2}\left(\frac{\partial u_1}{\partial x_2} + \frac{\partial u_2}{\partial x_1} + \frac{\partial u_1}{\partial x_1}\frac{\partial u_1}{\partial x_2} + \frac{\partial u_2}{\partial x_1}\frac{\partial u_2}{\partial x_2} + \frac{\partial u_3}{\partial x_1}\frac{\partial u_3}{\partial x_2}\right) \\
&= \frac{1}{2}[2e_{12} + e_{11}(e_{12} - \omega_{21}) + e_{22}(e_{12} + \omega_{21}) \\
&\qquad + (e_{13} - \omega_{13})(e_{23} + \omega_{32})] \\
\varepsilon_{23} &= \frac{1}{2}\left(\frac{\partial u_2}{\partial x_3} + \frac{\partial u_3}{\partial x_2} + \frac{\partial u_1}{\partial x_2}\frac{\partial u_1}{\partial x_3} + \frac{\partial u_2}{\partial x_2}\frac{\partial u_2}{\partial x_3} + \frac{\partial u_3}{\partial x_2}\frac{\partial u_3}{\partial x_3}\right) \\
&= \frac{1}{2}[2e_{23} + (e_{12} - \omega_{21})(e_{13} + \omega_{13}) + e_{22}(e_{23} - \omega_{32}) \\
&\qquad + e_{33}(e_{23} + \omega_{32})]
\end{aligned} \quad (4.3.3)$$

$$\varepsilon_{31} = \frac{1}{2}\left(\frac{\partial u_3}{\partial x_1} + \frac{\partial u_1}{\partial x_3} + \frac{\partial u_1}{\partial x_3}\frac{\partial u_1}{\partial x_1} + \frac{\partial u_2}{\partial x_3}\frac{\partial u_2}{\partial x_1} + \frac{\partial u_3}{\partial x_3}\frac{\partial u_3}{\partial x_1}\right)$$

$$= \frac{1}{2}[2e_{31} + e_{11}(e_{13} + \omega_{13}) + (e_{23} - \omega_{32})(e_{12} + \omega_{21}) + e_{33}(e_{13} - \omega_{13})].$$

In Eqs. (4.3.3), if the subscripts of the ε 's are interchanged, the right-hand sides remain the same; in other words,

$$\varepsilon_{ij} = \varepsilon_{ji}. \tag{4.3.4}$$

Thus, in index notations, Eq. (4.3.2) is written:

$$(ds^*)^2 - (ds)^2 = 2\varepsilon_{ij}\,dx_i\,dx_j, \tag{4.3.5}$$

and Eqs. (4.3.3) are written:

$$\varepsilon_{ij} = \frac{1}{2}\left[\frac{\partial u_i}{\partial x_j} + \frac{\partial u_j}{\partial x_i} + \frac{\partial u_m}{\partial x_i}\frac{\partial u_n}{\partial x_j}\delta_{mn}\right]. \tag{4.3.6}$$

The elongation of the element ds per unit length is given by:

$$E_{MN} = \frac{ds^* - ds}{ds} = \frac{ds^*}{ds} - 1, \tag{4.3.7}$$

therefore,

$$ds^* = (E_{MN} + 1)ds, \tag{4.3.8}$$

and Eq. (4.3.5) becomes:

$$(ds^*)^2 - (ds)^2 = E_{MN}(E_{MN} + 2)(ds)^2 = 2\varepsilon_{ij}\,dx_i\,dx_j. \tag{4.3.9}$$

Dividing both sides of Eq. (4.3.9) by $2(ds)^2$, we obtain:

$$\begin{aligned}\frac{1}{2}\left[\left(\frac{ds^*}{ds}\right)^2 - 1\right] &= E_{MN}\left(1 + \frac{E_{MN}}{2}\right) = \varepsilon_{ij}\frac{dx_i}{ds}\frac{dx_j}{ds}\\ &= \varepsilon_{11}\ell_1^2 + \varepsilon_{22}\ell_2^2 + \varepsilon_{33}\ell_3^2 + 2\varepsilon_{12}\ell_1\ell_2\\ &\quad + 2\varepsilon_{23}\ell_2\ell_3 + 2\varepsilon_{13}\ell_1\ell_3.\end{aligned} \tag{4.3.10}$$

The quantities $dx_i/ds = \ell_i$ are the direction cosines of the element ds in the system OX_1, OX_2, OX_3. Eq. (4.3.10) shows that, in order to calculate the elongation per unit length E_{MN} of an element MN, it is sufficient to know the six quantities ε_{ij} at M. These six quantities are called the components of the state of strain at the point M. They describe the state of deformation of the body.

The right-hand side of Eq. (4.3.10) is a quadratic form in the variables ℓ_1, ℓ_2, ℓ_3, with the following matrix of coefficients (see Sec. 3.12):

$$\begin{bmatrix} \varepsilon_{11} & \varepsilon_{12} & \varepsilon_{13} \\ \varepsilon_{12} & \varepsilon_{22} & \varepsilon_{23} \\ \varepsilon_{13} & \varepsilon_{23} & \varepsilon_{33} \end{bmatrix}. \tag{4.3.11}$$

This matrix is called the matrix of the state of strain.

4.4 Geometrical Meaning of the Strain Components ε_{ij}. Strain of a Line Element

Let us take the element ds of Fig. 1.2 parallel to the OX_1 axis (Fig. 4.2). After transformation, its elongation per unit length E_{x1} can be obtained from Eq. (4.3.10). Setting $\ell_1 = 1$, $\ell_2 = \ell_3 = 0$, we obtain:

$$E_{x1}\left(1 + \frac{E_{x1}}{2}\right) = \varepsilon_{11}. \tag{4.4.1}$$

Consequently,

$$E_{x1} = \sqrt{1 + 2\varepsilon_{11}} - 1. \tag{4.4.2}$$

In a similar manner,

$$E_{x2} = \sqrt{1 + 2\varepsilon_{22}} - 1 \tag{4.4.3}$$

$$E_{x3} = \sqrt{1 + 2\varepsilon_{33}} - 1, \tag{4.4.4}$$

where E_{x2} and E_{x3} are the elongations per unit length of elements initially parallel to the OX_2 and OX_3 axes, respectively. Thus, the strain components $\varepsilon_{11}, \varepsilon_{22}, \varepsilon_{33}$ characterize (or describe) the relative elongation of those line elements through M parallel to the three axes. They are called the normal components of the state of strain.

The meaning of $\varepsilon_{12}, \varepsilon_{13}, \varepsilon_{23}$ is obtained by writing the expressions after transformation for the angles between the elements dx_1, dx_2, dx_3 originally parallel to the coordinate axes. This is a simple matter since the direction cosines of the transformed elements are known. From table (4.2.8) and Fig. 4.2, we obtain:

$$\cos(M^*N^*, M^*P^*) = \frac{2\varepsilon_{12}}{(1 + E_{x1})(1 + E_{x2})} = \cos\left(\frac{\pi}{2} - \phi_{12}\right) \qquad (4.4.5)$$
$$= \sin\phi_{12},$$

where ϕ_{12} is the change in the right angle between dx_1 and dx_2. Similarly,

$$\cos(M^*N^*, M^*Q^*) = \frac{2\varepsilon_{13}}{(1 + E_{x1})(1 + E_{x3})} = \cos\left(\frac{\pi}{2} - \phi_{13}\right) \qquad (4.4.6)$$
$$= \sin\phi_{13}$$

$$\cos(M^*P^*, M^*Q^*) = \frac{2\varepsilon_{23}}{(1 + E_{x2})(1 + E_{x3})} = \cos\left(\frac{\pi}{2} - \phi_{23}\right) \qquad (4.4.7)$$
$$= \sin\phi_{23}.$$

The angles ϕ_{12}, ϕ_{13}, and ϕ_{23} are called the shear angles. Thus ε_{12}, ε_{13}, ε_{23} characterize (or describe) the change in the right angle between elements originally parallel to the axes. They are referred to as the shearing components of the state of strain. If $\varepsilon_{12} = \varepsilon_{13} = \varepsilon_{23} = 0$, the elements dx_1, dx_2, dx_3 in Fig. 4.2 remain orthogonal after transformation. This is precisely what is required for their directions to be three principal directions of the linear transformation.

The previous analysis shows that a rigid body motion is characterized by $\varepsilon_{ij} = 0$.

A comparison of Eqs. (4.4.1) to (4.4.4) to Eq. (4.3.10) shows that the latter can be written:

$$E_{MN}\left(1 + \frac{E_{MN}}{2}\right) = \varepsilon_{MN} = \varepsilon_{11}\ell_1^2 + \varepsilon_{22}\ell_2^2 + \varepsilon_{33}\ell_3^2 + 2\varepsilon_{12}\ell_1\ell_2 + 2\varepsilon_{23}\ell_2\ell_3 + 2\varepsilon_{13}\ell_1\ell_3 \qquad (4.4.8)$$

or

$$\varepsilon_{MN} = \varepsilon_{ij}\ell_i\ell_j.$$

Thus,

$$E_{MN} = \sqrt{1 + 2\varepsilon_{MN}} - 1, \qquad (4.4.9)$$

and ε_{MN} characterizes (or describes) the change in length per unit length of an element MN. ε_{MN} is called the strain at M of the element MN.

This strain is completely defined once the elements of the matrix (4.3.11) are known at the point M.

In engineering practice, the strain of an element MN is defined as the change in length per unit length of that element: In other words, it is the quantity E_{MN} and not ε_{MN} that is called the strain of MN. It will be shown in Sec. 4.8 that for small strains the two quantities are nearly equal.

4.5 Components of the State of Strain under a Change of Coordinate System

Eqs. (4.3.3) give the definitions of the components of the state of strain ε_{ij} referred to a system of coordinates OX_1, OX_2, OX_3. Let us consider another trirectangular system OX'_1, OX'_2, OX'_3, obtained by a rotation around O (see Sec. 3.5). The position of the new system is defined by the direction cosines of its axes l_{ij} with respect to the initial ones (Fig. 4.3). From Eqs. (3.5.3), we have:

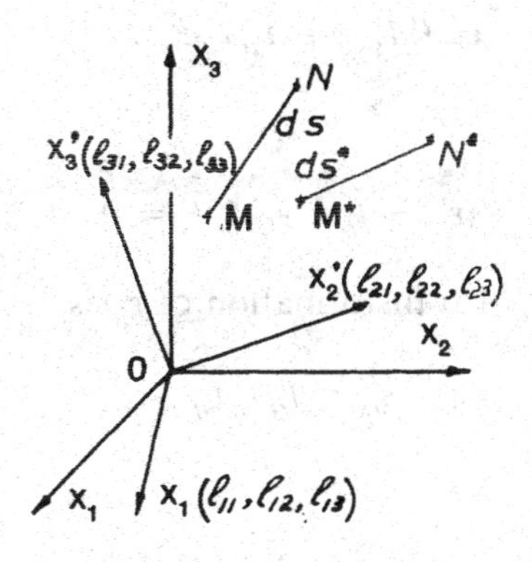

Fig. 4.3

$$dx'_i = l_{ij}\, dx_j. \tag{4.5.1}$$

Therefore,

$$dx_i = l_{ji}\, dx'_j. \tag{4.5.2}$$

Dividing both sides of Eq. (4.5.2) by ds, we obtain:

$$\frac{dx_i}{ds} = \ell_{ji}\frac{dx'_j}{ds}; \tag{4.5.3}$$

i.e.,

$$\ell_i = \ell_{ji}\ell'_j. \tag{4.5.4}$$

The left-hand sides of Eqs. (4.3.9) and (4.3.10) are obviously independent of the system of coordinate axes. Thus,

$$(ds^*)^2 - (ds)^2 = 2\varepsilon_{ij}\,dx_i\,dx_j = 2\varepsilon'_{rs}\,dx'_r\,dx'_s \tag{4.5.5}$$

and

$$\varepsilon_{MN} = \varepsilon_{ij}\frac{dx_i}{ds}\frac{dx_j}{ds} = \varepsilon_{ij}\,\ell_i\,\ell_j = \varepsilon'_{MN} = \varepsilon'_{rs}\,\ell'_r\,\ell'_s, \tag{4.5.6}$$

where ℓ'_i are the direction cosines of the element MN in the new system of coordinates. The substitution of Eq. (4.5.4) in Eq. (4.5.6) gives:

$$\varepsilon'_{rs}\,\ell'_r\,\ell'_s = \varepsilon_{ij}\,\ell_{ki}\,\ell'_k\,\ell_{mj}\,\ell'_m. \tag{4.5.7}$$

Therefore,

$$(\varepsilon'_{rs} - \ell_{ri}\,\ell_{sj}\,\varepsilon_{ij})\ell'_r\,\ell'_s = 0.$$

Since $\ell'_p \neq 0$, the rule of transformation of ε_{ij} is

$$\varepsilon'_{rs} = \ell_{ri}\,\ell_{sj}\,\varepsilon_{ij}. \tag{4.5.8}$$

For example,

$$\begin{aligned}
\varepsilon'_{11} &= \ell_{11}^2\varepsilon_{11} + \ell_{12}^2\varepsilon_{22} + \ell_{13}^2\varepsilon_{33} + 2\ell_{11}\ell_{12}\varepsilon_{12} \\
&\quad + 2\ell_{11}\ell_{13}\varepsilon_{13} + 2\ell_{12}\ell_{13}\varepsilon_{23} && (4.5.9)\\
\varepsilon'_{12} &= \ell_{11}\ell_{21}\varepsilon_{11} + \ell_{12}\ell_{22}\varepsilon_{22} + \ell_{13}\ell_{23}\varepsilon_{33} && (4.5.10)\\
&\quad + \varepsilon_{12}(\ell_{11}\ell_{22} + \ell_{12}\ell_{21}) + \varepsilon_{13}(\ell_{11}\ell_{23} + \ell_{13}\ell_{21}) \\
&\quad + \varepsilon_{23}(\ell_{12}\ell_{23} + \ell_{13}\ell_{22}),
\end{aligned}$$

and so on for the other components.

4.6 Principal Axes of Strain

The three principal axes of the linear transformation (4.1.1) are such that $\varepsilon_{12} = \varepsilon_{13} = \varepsilon_{23} = 0$ (see Sec. 4.4). Consequently, they are also called the principal axes of the strain matrix (4.3.11). When the coordinate axes are along the principal directions, the unit elongations are denoted by E_1, E_2, and E_3; the normal components of the state of strain are denoted by ε_1, ε_2, and ε_3 and are referred to as the three principal strains. Thus, in this sytem of coordinates:

$$E_1 = \sqrt{1 + 2\varepsilon_1} - 1 \tag{4.6.1}$$

$$E_2 = \sqrt{1 + 2\varepsilon_2} - 1 \tag{4.6.2}$$

$$E_3 = \sqrt{1 + 2\varepsilon_3} - 1 \tag{4.6.3}$$

$$\varepsilon_{MN} = \varepsilon_1 \ell_1^2 + \varepsilon_2 \ell_2^2 + \varepsilon_3 \ell_3^2 . \tag{4.6.4}$$

Three elements MN_1, MN_2, and MN_3 initially along the principal axes will experience unit elongations given by Eqs. (4.6.1) to (4.6.3); they will rotate in space, but will keep their orthogonality. An element MN with direction cosines ℓ_1, ℓ_2, and ℓ_3 with respect to the principal axes will experience a unit elongation,

$$E_{MN} = \sqrt{1 + 2\varepsilon_{MN}} - 1, \tag{4.6.5}$$

where ε_{MN} is given by Eq. (4.6.4). Finding the principal directions amounts to finding the system of orthogonal axes in which the quadratic form in the right-hand side of Eq. (4.3.10) is reduced to its canonical form—i.e., to the right-hand side of Eq. (4.6.4); in this system, the matrix (4.3.11), which is associated with the quadratic form, becomes a diagonal matrix. This problem has been solved in Sec. 3.12. The set of linear homogeneous equations:

$$\begin{aligned}
(\varepsilon_{11} - \varepsilon)\ell_1 + \varepsilon_{12}\ell_2 + \varepsilon_{13}\ell_3 &= 0 \\
\varepsilon_{12}\ell_1 + (\varepsilon_{22} - \varepsilon)\ell_2 + \varepsilon_{23}\ell_3 &= 0 \\
\varepsilon_{13}\ell_1 + \varepsilon_{23}\ell_2 + (\varepsilon_{33} - \varepsilon)\ell_3 &= 0,
\end{aligned} \tag{4.6.6}$$

and the condition that

$$\ell_1^2 + \ell_2^2 + \ell_3^2 = 1, \tag{4.6.7}$$

allow one to obtain the principal strains and their directions: A nontrivial solution of Eqs. (4.6.6) exists if, and only if, the determinant of the matrix formed by the coefficients is equal to zero. All the properties established in Sec. 3.6 apply here. The expansion of the determinant yields a cubic equation in ε.

$$\varepsilon^3 - I_1\varepsilon^2 + I_2\varepsilon - I_3 = 0, \tag{4.6.8}$$

whose roots are all real. I_1, I_2, and I_3 are called the three invariants of the state of strain:

$$I_1 = \varepsilon_1 + \varepsilon_2 + \varepsilon_3 = \varepsilon_{11} + \varepsilon_{22} + \varepsilon_{33} \tag{4.6.9}$$

$$\begin{aligned} I_2 &= \varepsilon_1\varepsilon_2 + \varepsilon_2\varepsilon_3 + \varepsilon_3\varepsilon_1 \\ &= \varepsilon_{11}\varepsilon_{22} + \varepsilon_{22}\varepsilon_{33} + \varepsilon_{33}\varepsilon_{11} - (\varepsilon_{12})^2 - (\varepsilon_{13})^2 - (\varepsilon_{23})^2 \end{aligned} \tag{4.6.10}$$

$$I_3 = \varepsilon_1\varepsilon_2\varepsilon_3 = \begin{vmatrix} \varepsilon_{11} & \varepsilon_{12} & \varepsilon_{13} \\ \varepsilon_{12} & \varepsilon_{22} & \varepsilon_{23} \\ \varepsilon_{13} & \varepsilon_{23} & \varepsilon_{33} \end{vmatrix}. \tag{4.6.11}$$

4.7 Volumetric Strain

Let us consider a parallelepiped with edges dx_1, dx_2, and dx_3. According to Eqs. (1.1.6), this rectangular parallelepiped is transformed to an oblique one whose three edges have projections on OX_1, OX_2, OX_3 given respectively by:

$$\begin{aligned} d\xi_1 &= \left(1 + \frac{\partial u_1}{\partial x_1}\right)dx_1 & d\xi_1 &= \frac{\partial u_1}{\partial x_2}dx_2 & d\xi_1 &= \frac{\partial u_1}{\partial x_3}dx_3 \\ d\xi_2 &= \frac{\partial u_2}{\partial x_1}dx_1 & d\xi_2 &= \left(1 + \frac{\partial u_2}{\partial x_2}\right)dx_2 & d\xi_2 &= \frac{\partial u_2}{\partial x_3}dx_3 \\ d\xi_3 &= \frac{\partial u_3}{\partial x_1}dx_1 & d\xi_3 &= \frac{\partial u_3}{\partial x_2}dx_2 & d\xi_3 &= \left(1 + \frac{\partial u_3}{\partial x_3}\right)dx_3. \end{aligned} \tag{4.7.1}$$

The volume of the transformed parallelepiped is given by the triple scalar product:

$$V^* = \begin{vmatrix} 1 + \frac{\partial u_1}{\partial x_1} & \frac{\partial u_1}{\partial x_2} & \frac{\partial u_1}{\partial x_3} \\ \frac{\partial u_2}{\partial x_1} & 1 + \frac{\partial u_2}{\partial x_2} & \frac{\partial u_2}{\partial x_3} \\ \frac{\partial u_3}{\partial x_1} & \frac{\partial u_3}{\partial x_2} & 1 + \frac{\partial u_3}{\partial x_3} \end{vmatrix} dx_1\, dx_2\, dx_3 \tag{4.7.2}$$

$$= D dx_1\, dx_2\, dx_3 .$$

Since $V = dx_1\, dx_2\, dx_3$, the change of volume per unit volume is given by:

$$\frac{V^* - V}{V} = E_v = D - 1. \tag{4.7.3}$$

E_v can be expressed in terms of the components of the state of strain by squaring the determinant of Eq. (4.7.2) and substituting the corresponding values from Eqs. (4.3.3). Thus,

$$(1 + E_v)^2 = \begin{vmatrix} 1 + 2\varepsilon_{11} & 2\varepsilon_{12} & 2\varepsilon_{13} \\ 2\varepsilon_{12} & 1 + 2\varepsilon_{22} & 2\varepsilon_{23} \\ 2\varepsilon_{13} & 2\varepsilon_{23} & 1 + 2\varepsilon_{33} \end{vmatrix} = D^2. \tag{4.7.4}$$

When expanded and the terms grouped, Eq. (4.7.4) becomes:

$$(1 + E_v)^2 = 1 + 2I_1 + 4I_2 + 8I_3, \tag{4.7.5}$$

where I_1, I_2, and I_3 are the three invariants of the state of strain. In terms of principal strains, Eq. (4.7.4) becomes:

$$(1 + E_v)^2 = \begin{vmatrix} 1 + 2\varepsilon_1 & 0 & 0 \\ 0 & 1 + 2\varepsilon_2 & 0 \\ 0 & 0 & 1 + 2\varepsilon_3 \end{vmatrix} = D^2. \tag{4.7.6}$$

Following the definition of strain used for elements of length, the volumetric strain is defined as:

$$\varepsilon_v = E_v\left(1 + \frac{E_v}{2}\right) = \frac{1}{2}(D^2 - 1) = I_1 + 2I_2 + 4I_3, \tag{4.7.7}$$

and E_v is therefore given by:

$$E_v = \sqrt{(1 + 2\varepsilon_1)(1 + 2\varepsilon_2)(1 + 2\varepsilon_3)} - 1 \tag{4.7.8}$$

or

$$E_v = \sqrt{1 + 2I_1 + 4I_2 + 8I_3} - 1. \qquad (4.7.9)$$

When the principal directions are chosen as the reference rectangular axes at a point O of the undeformed body, a unit cube is transformed to a rectangular parallelepiped with its three dimensions given by $1 + E_1$, $1 + E_2$, $1 + E_3$, respectively. The volume of the transformed unit cube is given by:

$$V^* = (1 + E_1)(1 + E_2)(1 + E_3) \qquad (4.7.10)$$

and

$$E_v = (1 + E_1)(1 + E_2)(1 + E_3) - 1. \qquad (4.7.11)$$

The expression for E_v is independent of the initial shape of the chosen element at O.

4.8 Small Strain

In engineering practice, the magnitudes of the unit elongations E_{x1}, E_{x2}, and E_{x3}, as well as the changes in right angles ϕ_{12}, ϕ_{23}, and ϕ_{13} are generally very small. For the usual engineering materials, and in most structures, the magnitudes of the unit elongations and of the strains are on the order of 10^{-3}: Such strains certainly deserve to be called small.

The assumption that the strains are small introduces substantial simplifications in Eqs. (4.4.2) to (4.4.7): If

$$E_{x1} \ll 1 < 2, \qquad (4.8.1)$$

Eq. (4.4.1) yields:

$$\varepsilon_{11} \cong E_{x1}. \qquad (4.8.2)$$

Similarly,

$$\varepsilon_{22} \cong E_{x2}, \qquad \varepsilon_{33} \cong E_{x3} \qquad (4.8.3)$$

For any element MN,

$$\varepsilon_{MN} \cong E_{MN}. \qquad (4.8.4)$$

and the definition of strain given in Sec. 4.4 does not differ appreciably from the engineering definition.

In Eq. (4.4.5), the sine of the angle can be replaced by the angle itself; also, the denominator in its right-hand side can be replaced by unity. Therefore,

$$\varepsilon_{12} \cong \frac{\phi_{12}}{2}. \tag{4.8.5}$$

Similarly,

$$\varepsilon_{13} \cong \frac{\phi_{13}}{2}, \qquad \varepsilon_{23} \cong \frac{\phi_{23}}{2}. \tag{4.8.6}$$

The expression of the change of volume per unit volume E_v as given by Eq. (4.7.9) becomes simplified because the invariants I_2 and I_3 are of the order of E_{x1}^2 and E_{x1}^3, respectively, and can be neglected with respect to I_1, which is of the order of E_{x1}. Also, if only the first-order terms in the binomial expansion of $\sqrt{1 + 2I_1}$ are retained, then

$$E_v \cong I_1 = \varepsilon_{11} + \varepsilon_{22} + \varepsilon_{33}. \tag{4.8.7}$$

Thus, under the assumption of small strains, the first invariant is the change of volume per unit volume. Because of the order of magnitude of the strains, it can be assumed that

$$E_v \ll 1 < 2, \tag{4.8.8}$$

so that Eq. (4.7.7) yields:

$$\varepsilon_v \cong E_v.$$

Thus, the definition of volumetric strain given in Sec. 4.7 coincides with the engineering definition.

4.9 Linear Strain

A further restriction that is made in the analysis of strain involves the magnitudes of the rotation and of the cylindrical dilatation produced by the antisymmetric part of Eq. (4.1.3). The angle of rotation has the same value for all the elements at M. If this angle is assumed to be small so that its square is very small compared to unity, the cylindrical dilatation can be neglected since it is of the second order with respect to the angle (Sec. 3.9). In such a case, the changes in length of the elements at M, as well as the changes in the right angles between mutually perpendicular elements, depend on the symmetric part of Eq. (4.1.3) alone.

Now let us consider Eqs. (4.3.3). If we assume that the derivatives $\partial u_i/\partial x_j$ are small enough so that their squares and products are negligible compared to the derivatives themselves, the ε_{ij}'s become equal to the e_{ij}'s. Thus, the changes in length of the elements at M as in Eq. (4.3.9), and the changes in the right angles between mutually perpendicular elements, as in Eqs. (4.4.5) to (4.4.7), are completely determined once the e_{ij}'s are known. Therefore, the use of the e_{ij}'s in place of the ε_{ij}'s does not only imply that the derivatives $\partial u_i/\partial x_j$ are small, but also that the rotation due to the antisymmetric part of the transformation is small, and that the deformation is described solely by the symmetric part:

$$\begin{bmatrix} du_1 \\ du_2 \\ du_3 \end{bmatrix} = \begin{bmatrix} e_{11} & e_{12} & e_{13} \\ e_{12} & e_{22} & e_{23} \\ e_{13} & e_{23} & e_{33} \end{bmatrix} \begin{bmatrix} \ell_1 \\ \ell_2 \\ \ell_3 \end{bmatrix}. \tag{4.9.1}$$

The e_{ij}'s are called linear strains. e_{11}, e_{22}, and e_{33} are called the linear normal strains, and e_{12}, e_{13}, and e_{23} are called the linear tangential or shearing strains. Since linear strains are exclusively used in the classical theories of elasticity and plasticity, the word linear is usually omitted and the e_{ij}'s are referred to as normal and shearing strains.

All the properties of linear symmetric transformations deduced in Chapter 3 apply to linear strain. The equations of Secs. 3.10 to 3.16 can be rewritten here with the following changes in notation:

(a) x_1, x_2, x_3 are replaced by the direction cosines ℓ_1, ℓ_2, ℓ_3.

(b) The a_{ij}'s are replaced by e_{ij}'s.

(c) $\lambda_1, \lambda_2, \lambda_3$ are replaced by e_1, e_2, e_3, and are called major, intermediate, and minor principal strains.

(d) n and t are replaced by e_n and e_t, and are called normal and tangential (or shearing) strains.

In view of the restrictions placed on the derivatives, the linear strains are necessarily small strains and the transformation expressed by Eq. (4.9.1) is a small symmetric transformation. The geometrical meaning of the e_{ij}'s is the same as that of the a_{ij}'s of Sec. 3.15: e_{11}, e_{22}, e_{33} are the unit elongations of elements initially parallel to the three axes, and $2e_{12}, 2e_{13}, 2e_{23}$ are the changes in the right angles between those elements. The following is essentially a list of equations and results obtained by making the above substitutions in Chapter 3:

1. *Characteristic equation and invariants*

$$e^3 - I_1 e^2 + I_2 e - I_3 = 0 \tag{4.9.2}$$

$$I_1 = e_1 + e_2 + e_3 = e_{11} + e_{22} + e_{33} \tag{4.9.3}$$

$$\begin{aligned} I_2 &= e_1 e_2 + e_2 e_3 + e_3 e_1 \\ &= e_{11} e_{22} + e_{22} e_{33} + e_{33} e_{11} - e_{12}^2 - e_{13}^2 - e_{23}^2 \end{aligned} \tag{4.9.4}$$

$$\begin{aligned} I_3 = e_1 e_2 e_3 = e_{11} e_{22} e_{33} + 2e_{12} e_{23} e_{13} - e_{33} e_{12}^2 - e_{22} e_{13}^2 \\ - e_{11} e_{23}^2 \end{aligned} \tag{4.9.5}$$

In a system of principal axes, Eq. (4.9.1) becomes:

$$\begin{bmatrix} du_1 \\ du_2 \\ du_3 \end{bmatrix} = \begin{bmatrix} e_1 & 0 & 0 \\ 0 & e_2 & 0 \\ 0 & 0 & e_3 \end{bmatrix} \begin{bmatrix} l_1 \\ l_2 \\ l_3 \end{bmatrix}. \tag{4.9.6}$$

2. *Normal and tangential strains of an element (Fig. 4.4)*

$$\begin{aligned} (e_{MN})_n = e_{11} l_1^2 + e_{22} l_2^2 + e_{33} l_3^2 + 2e_{12} l_1 l_2 + 2e_{13} l_1 l_3 \\ + 2e_{23} l_2 l_3 \end{aligned} \tag{4.9.7}$$

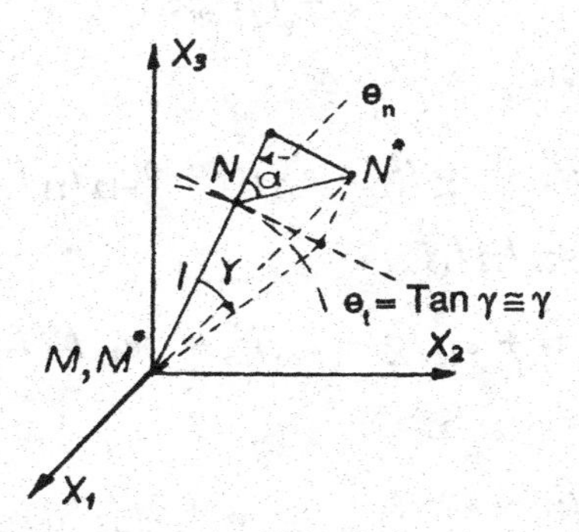

Fig. 4.4

$$(e_{MN})_t = \sqrt{(du_1)^2 + (du_2)^2 + (du_3)^2 - (e_{MN})_n^2}\ .\ (*) \tag{4.9.8}$$

3. *Mean Strain. Spherical and deviatoric components of linear strain*
The mean strain e_m is defined by:

$$e_m = \frac{e_{11} + e_{22} + e_{33}}{3}. \tag{4.9.9}$$

The right-hand side of Eq. (4.9.1) can be decomposed into two parts:

$$\begin{bmatrix} du_1 \\ du_2 \\ du_3 \end{bmatrix} = \begin{bmatrix} e_m & 0 & 0 \\ 0 & e_m & 0 \\ 0 & 0 & e_m \end{bmatrix} \begin{bmatrix} \ell_1 \\ \ell_2 \\ \ell_3 \end{bmatrix} + \begin{bmatrix} e'_{11} & e_{12} & e_{13} \\ e_{12} & e'_{22} & e_{23} \\ e_{13} & e_{23} & e'_{33} \end{bmatrix} \begin{bmatrix} \ell_1 \\ \ell_2 \\ \ell_3 \end{bmatrix}, \tag{4.9.10}$$

with

$$e'_{11} + e'_{22} + e'_{33} = 0.$$

The first matrix in the right-hand side of Eq. (4.9.10) is called the matrix of the spherical component of strain, and the second one is called the matrix of the deviatoric component of strain.

4. *Volumetric strain*

$$e_v = 3e_m = e_1 + e_2 + e_3 = e_{11} + e_{22} + e_{33} = I_1\,. \tag{4.9.11}$$

5. *Components of the linear strain in a change of coordinates*
If the direction cosines of the new system with respect to the old one are $(\ell_{11}, \ell_{12}, \ell_{13})$, $(\ell_{21}, \ell_{22}, \ell_{23})$, and $(\ell_{31}, \ell_{32}, \ell_{33})$,

$$e'_{ik} = \ell_{ij}\,\ell_{km}\,e_{jm}\,. \tag{4.9.12}$$

Thus,

$$e'_{11} = e_{11}\,\ell_{11}^2 + e_{22}\,\ell_{12}^2 + e_{33}\,\ell_{13}^2 + 2e_{12}\,\ell_{11}\,\ell_{12} + 2e_{13}\,\ell_{11}\,\ell_{13} + 2e_{23}\,\ell_{12}\,\ell_{13} \tag{4.9.13}$$

$$e'_{22} = e_{11}\,\ell_{21}^2 + e_{22}\,\ell_{22}^2 + e_{33}\,\ell_{23}^2 + 2e_{12}\,\ell_{21}\,\ell_{22} + 2e_{13}\,\ell_{21}\,\ell_{23} + 2e_{23}\,\ell_{22}\,\ell_{23} \tag{4.9.14}$$

$$e'_{33} = e_{11}\,\ell_{31}^2 + e_{22}\,\ell_{32}^2 + e_{33}\,\ell_{33}^2 + 2e_{12}\,\ell_{31}\,\ell_{32} + 2e_{13}\,\ell_{31}\,\ell_{33} + 2e_{23}\,\ell_{32}\,\ell_{33} \tag{4.9.15}$$

* See page A-24.

$$\begin{aligned} e'_{12} = &(e_{11}\ell_{11} + e_{12}\ell_{12} + e_{13}\ell_{13})\ell_{21} \\ &+ (e_{12}\ell_{11} + e_{22}\ell_{12} + e_{23}\ell_{13})\ell_{22} \\ &+ (e_{13}\ell_{11} + e_{23}\ell_{12} + e_{33}\ell_{13})\ell_{23} \end{aligned} \tag{4.9.16}$$

$$\begin{aligned} e'_{23} = &(e_{11}\ell_{21} + e_{12}\ell_{22} + e_{13}\ell_{23})\ell_{31} \\ &+ (e_{12}\ell_{21} + e_{22}\ell_{22} + e_{23}\ell_{23})\ell_{32} \\ &+ (e_{13}\ell_{21} + e_{23}\ell_{22} + e_{33}\ell_{23})\ell_{33} \end{aligned} \tag{4.9.17}$$

$$\begin{aligned} e'_{13} = &(e_{11}\ell_{11} + e_{12}\ell_{12} + e_{13}\ell_{13})\ell_{31} \\ &+ (e_{12}\ell_{11} + e_{22}\ell_{12} + e_{23}\ell_{13})\ell_{32} \\ &+ (e_{13}\ell_{11} + e_{23}\ell_{12} + e_{33}\ell_{13})\ell_{33} \end{aligned} \tag{4.9.18}$$

6. *Octahedral normal and shearing strains*

The normal strain of an element MN equally inclined to the three principal axes is called the octahedral normal strain, e_{oct}, and is obtained from Eq. (4.9.7):

$$(e_{MN})_n = e_{oct} = \frac{1}{3}(e_1 + e_2 + e_3) = \frac{I_1}{3}. \tag{4.9.19}$$

The tangential strain corresponding to the same element is obtained from Eq. (4.9.8), and is written:

$$(e_{MN})_t = \frac{\gamma_{oct}}{2} = \frac{1}{3}[(e_1 - e_2)^2 + (e_2 - e_3)^2 + (e_3 - e_1)^2]^{1/2}. \tag{4.9.20}$$

γ_{oct} is referred to as the octahedral shearing strain. When the system of reference is not a principal system,

$$e_{oct} = \frac{1}{3}(e_{11} + e_{22} + e_{33}) = \frac{I_1}{3} \tag{4.9.21}$$

and

$$\begin{aligned} \frac{\gamma_{oct}}{2} = \frac{1}{3}[&(e_{11} - e_{22})^2 + (e_{22} - e_{33})^2 + (e_{33} - e_{11})^2 \\ &+ 6e_{12}^2 + 6e_{13}^2 + 6e_{23}^2]^{1/2}. \end{aligned} \tag{4.9.22}$$

7. *Linear strain in two dimensions* (*)

* See pages A-25 to A-31 for plane strain equations.

Sec. 3.16 contains all the equations pertaining to this case of linear strain. These equations, as well as the representation on Mohr's diagram, can directly be used here after the appropriate changes in notation are made.

4.10 Compatibility Relations for Linear Strains

In Eqs. (1.2.1), we defined the components of the linear strain in terms of the components of the displacement by:

$$e_{ij} = \frac{1}{2}\left(\frac{\partial u_i}{\partial x_j} + \frac{\partial u_j}{\partial x_i}\right). \tag{4.10.1}$$

If the displacements u_1, u_2, and u_3 are prescribed continuous functions of x_1, x_2, and x_3, then the strain components e_{ij} can be uniquely determined. On the other hand, if the strain components are prescribed functions of the coordinates, it will not be possible to find unique values for the displacements because the strains represent pure deformation whereas the displacements include both deformation and rigid body motion. This difficulty is overcome by specifying the rigid body motion of a point M of the body, i.e., specifying its displacements u_i and the elements of its rotation ω_{ij}. The strain-displacement relations (4.10.1) form a system of six partial differential equations with only three unknowns, u_1, u_2, and u_3; it is obvious that some restrictions must be placed on the strains in order that Eq. (4.10.1) have a solution. These restrictions are called the compatibility relations (or conditions).

A physical interpretation of the conditions of compatibility can be obtained by examining the deformed body. Let $M(x_1, x_2, x_3)$ be a point of a continuous body at which the displacements u_i and the rotations ω_{ij} are known. The displacements u_i' of an arbitrary point $M'(x_1', x_2', x_3')$ can be obtained in terms of the known functions e_{ij} by means of a line integral along a continuous curve C joining M and M':

$$u_i' = u_i + \oint_M^{M'} du_i. \tag{4.10.2}$$

If the process of deformation does not create cracks or holes, in other words, if the body remains continuous, u_i' should be independent of the path of integration; that is, u_i' should have the same value regardless of

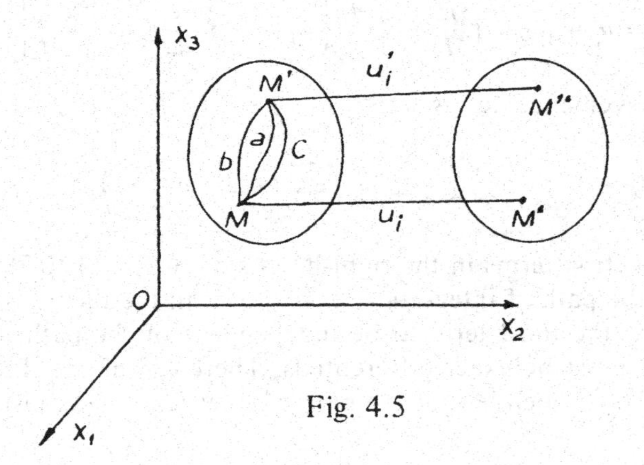

Fig. 4.5

whether the integration is along curve *a*, *b*, or any other path (Fig. 4.5). From Eqs. (1.2.1), we have:

$$du_i = \frac{\partial u_i}{\partial x_j} dx_j = (e_{ij} + \omega_{ij}) dx_j . \tag{4.10.3}$$

Therefore,

$$u_i' = u_i + \oint_M^{M'} (e_{ij} + \omega_{ij}) dx_j . \tag{4.10.4}$$

Integration by part of the second term in Eq. (4.10.4), yields:

$$u_i' = u_i + [\omega_{ij} x_j]_M^{M'} + \oint_M^{M'} \left(e_{ik} - x_j \frac{\partial \omega_{ij}}{\partial x_k} \right) dx_k , \tag{4.10.5}$$

where the dummy index j of e_{ij} has been changed to k. From Eqs. (1.2.1), it can be verified that

$$\frac{\partial \omega_{ij}}{\partial x_k} = \frac{\partial e_{ik}}{\partial x_j} - \frac{\partial e_{jk}}{\partial x_i} , \tag{4.10.6}$$

and hence Eq. (4.10.5) becomes:

$$u'_i = u_i + [\omega_{ij} x_j]_M^{M'} + \oint_M^{M'} \left[e_{ik} - x_j \left(\frac{\partial e_{ik}}{\partial x_j} - \frac{\partial e_{jk}}{\partial x_i} \right) \right] dx_k . \tag{4.10.7}$$

For convenience, let us set:

$$U_{ik} = e_{ik} - x_j \left(\frac{\partial e_{ik}}{\partial x_j} - \frac{\partial e_{jk}}{\partial x_i} \right). \tag{4.10.8}$$

The two first terms in the right-hand side of Eq. (4.10.7) are independent of the path of integration. It is shown in the theory of line integrals that, for the third term to be independent of the path, the integrands $U_{ik}\, dx_k$ must be exact differentials. Therefore, if the displacements u'_i are to be independent of the path of integration we must have:

$$\frac{\partial U_{ik}}{\partial x_\ell} = \frac{\partial U_{i\ell}}{\partial x_k}. \tag{4.10.9}$$

Now,

$$\frac{\partial U_{ik}}{\partial x_\ell} = \frac{\partial e_{ik}}{\partial x_\ell} - x_j \left(\frac{\partial^2 e_{ik}}{\partial x_j \partial x_\ell} - \frac{\partial^2 e_{jk}}{\partial x_i \partial x_\ell} \right) - \delta_{j\ell} \left(\frac{\partial e_{ik}}{\partial x_j} - \frac{\partial e_{jk}}{\partial x_i} \right) \tag{4.10.10}$$

and

$$\frac{\partial U_{i\ell}}{\partial x_k} = \frac{\partial e_{i\ell}}{\partial x_k} - x_j \left(\frac{\partial^2 e_{i\ell}}{\partial x_j \partial x_k} - \frac{\partial^2 e_{j\ell}}{\partial x_i \partial x_k} \right) - \delta_{jk} \left(\frac{\partial e_{i\ell}}{\partial x_j} - \frac{\partial e_{j\ell}}{\partial x_i} \right). \tag{4.10.11}$$

Therefore,

$$\frac{\partial U_{ik}}{\partial x_\ell} - \frac{\partial U_{i\ell}}{\partial x_k} = x_j \left(\frac{\partial^2 e_{jk}}{\partial x_i \partial x_\ell} + \frac{\partial^2 e_{i\ell}}{\partial x_j \partial x_k} - \frac{\partial^2 e_{ik}}{\partial x_j \partial x_\ell} - \frac{\partial^2 e_{j\ell}}{\partial x_i \partial x_k} \right) \tag{4.10.12}$$

$$= 0.$$

Since the x_j' s are independent, the necessary and sufficient conditions that u'_i be independent of the path of integration are:

$$\frac{\partial^2 e_{jk}}{\partial x_i \partial x_\ell} + \frac{\partial^2 e_{i\ell}}{\partial x_j \partial x_k} - \frac{\partial^2 e_{ik}}{\partial x_j \partial x_\ell} - \frac{\partial^2 e_{j\ell}}{\partial x_i \partial x_k} = 0. \tag{4.10.13}$$

These are the compatibility relations. Although Eq. (4.10.13) results in 81 equations on account of its four different subscripts, only six deserve consideration; the others are identically satisfied or are repetitions resulting from the symmetry of e_{ij}. In detailed form, we have:

$$\frac{\partial^2 e_{11}}{\partial x_2 \partial x_3} = \frac{\partial}{\partial x_1}\left(-\frac{\partial e_{23}}{\partial x_1} + \frac{\partial e_{31}}{\partial x_2} + \frac{\partial e_{12}}{\partial x_3}\right)$$

$$\frac{\partial^2 e_{22}}{\partial x_3 \partial x_1} = \frac{\partial}{\partial x_2}\left(-\frac{\partial e_{31}}{\partial x_2} + \frac{\partial e_{12}}{\partial x_3} + \frac{\partial e_{23}}{\partial x_1}\right)$$

$$\frac{\partial^2 e_{33}}{\partial x_1 \partial x_2} = \frac{\partial}{\partial x_3}\left(-\frac{\partial e_{12}}{\partial x_3} + \frac{\partial e_{23}}{\partial x_1} + \frac{\partial e_{31}}{\partial x_2}\right) \qquad (4.10.14)$$

$$2\frac{\partial^2 e_{12}}{\partial x_1 \partial x_2} = \frac{\partial^2 e_{11}}{\partial x_2^2} + \frac{\partial^2 e_{22}}{\partial x_1^2}$$

$$2\frac{\partial^2 e_{23}}{\partial x_2 \partial x_3} = \frac{\partial^2 e_{22}}{\partial x_3^2} + \frac{\partial^2 e_{33}}{\partial x_2^2}$$

$$2\frac{\partial^2 e_{31}}{\partial x_3 \partial x_1} = \frac{\partial^2 e_{33}}{\partial x_1^2} + \frac{\partial^2 e_{11}}{\partial x_3^2}. \; (*)$$

Eqs. (4.10.14) are the necessary and sufficient conditions for the strain components to give single valued displacements for a simply connected region.•

• A region of space is said to be simply connected if an arbitrary closed curve lying in the region can be shrunk to a point, by continuous deformation, without passing outside of the boundaries.

PROBLEMS

1. The displacement components at the points of a body are:

$$u_1 = c_1 x_1, \qquad u_2 = c_2 x_2, \qquad u_3 = c_3 x_3.$$

(a) Find the components ε_{ij} of the strain matrix, and the value of the three invariants of the state of strain.
(b) What is the value of the volumetric strain ε_v?
(c) If the constants c_1, c_2, and c_3 are so small that their squares and products are negligible, show that the components of the strain matrix ε_{ij} become equal to the components of the linear strain matrix e_{ij}.

2. Solve Problem 1 for displacement components given by

$$u_1 = c_1 x_2, \qquad u_2 = u_3 = 0.$$

* See pages A-32 to A-39 for more information on compatibility equations and a complete solved example.

Draw sketches showing a cubic element at a point, and with its edges parallel to the reference axes, before and after transformation.

3. Let

$$u_1 = C(2x_1 + x_2^2), \quad u_2 = C(x_1^2 - 3x_2^2), \quad u_3 = 0,$$

where $C = 10^{-2}$, be the expressions of the displacements of a certain body.
   (a) Show the distorted shape of a two-dimensional element of area whose sides dx_1 and dx_2 are initially parallel to the coordinate axes; the two elements are at a point M whose coordinates are (2, 1, 0).
   (b) Determine the coordinates of M after transformation.
   (c) Decompose the matrix of the transformation at M into its symmetric and antisymmetric components.
   (d) Find the angle of rotation and the cylindrical dilatation of the two elements dx_1 and dx_2.

4. In Problem 3, compute the strain ε_{MN} of an element MN whose direction cosines are $(1/\sqrt{3},\ 1/\sqrt{3},\ 1/\sqrt{3})$. What are the principal directions and the principal strains?

5. Given the displacement components

$$u_1 = cx_1(x_2 + x_3)^2, \quad u_2 = cx_2(x_3 + x_1)^2, \quad u_3 = cx_3(x_1 + x_2)^2$$

where c is a constant:
   (a) Find the components of the linear strain.
   (b) Find the components of the rotation.
   (c) Find the principal elongations per unit length E_1, E_2, and E_3 at a point M whose coordinates are (1, 1, 1).

6. The components of linear strain in a body are given by:

$$[e_{ij}] = \begin{bmatrix} 0 & 0 & -cx_2 \\ 0 & 0 & cx_1 \\ -cx_2 & cx_1 & 0 \end{bmatrix},$$

where c is a constant. Find the principal strains and the principal directions at the point (1, 2, 4).

7. Determine the volumetric strain ε_v for the following state of strain:

$$[\varepsilon_{ij}] = \begin{bmatrix} 0.5 & 1 & 0 \\ 1 & 2 & 0.5 \\ 0 & 0.5 & 0 \end{bmatrix}.$$

Compare the result to the unit change of volume E_v, and to the first invariant.

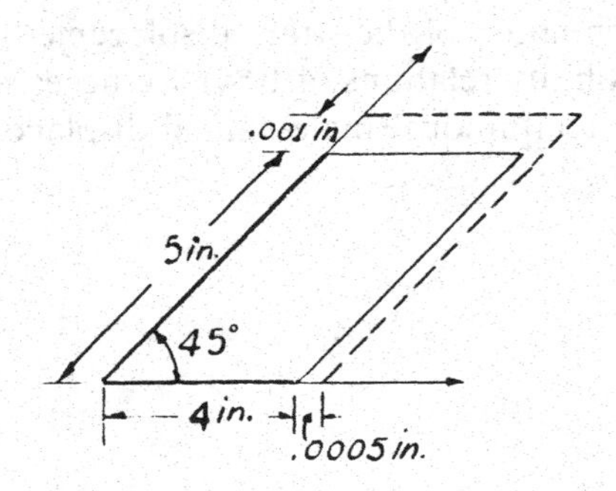

Fig. 4.6

8. A plate whose thickness is ⅛ in. is stretched as shown in Fig. 4.6. Find the principal strains, e_1, e_2, and the maximum shearing strain in the plate.

9. In a two-dimensional state of strain,

$$e_{11} = 800 \times 10^{-6},\ e_{22} = 100 \times 10^{-6},\ e_{12} = -800 \times 10^{-6}.$$

Find the magnitude and direction of the principal strains, e_1 and e_2, both analytically and through the use of Mohr's diagram. Draw a sketch showing the deformation of a unit square with edges initially along OX_1 and OX_2.

10. If

$$e_{11} = -800 \times 10^{-6},\ e_{22} = -200 \times 10^{-6},\ e_{12} = -600 \times 10^{-6},$$

show in a suitable sketch the position of the axes with which the maximum shearing strain is associated.

11. Are the following states of strain possible?

$$e_{11} = C(x_1^2 + x_2^2) \qquad e_{11} = Cx_3(x_1^2 + x_2^2)$$

$$e_{22} = Cx_2^2 \qquad e_{22} = Cx_2^2 x_3$$

$$e_{12} = 2Cx_1 x_2 \qquad e_{12} = 2Cx_1 x_2 x_3$$

$$e_{33} = e_{13} = e_{23} = 0 \qquad e_{33} = e_{13} = e_{23} = 0$$

C is a constant.

12. Show by differentiation of the strain-displacement relations (4.10.1) that the compatibility relations (4.10.4) are necessary conditions for the existence of continuous single-valued displacements. (*)

---

* See pages A-39 to A-42 for additional problems.

CHAPTER 5

CARTESIAN TENSORS

5.1 Introduction

A tensor is a quantity which describes a physical state or a physical phenomenon and which is invariant, i.e., remains unchanged when the frame of reference within which the quantity is defined is changed. In this chapter, we shall limit ourselves to cartesian frames of reference. If the value of the quantity at a point in space can be described by a single number, the quantity is a scalar or a tensor of rank zero; if three numbers are needed, the quantity is a vector or a tensor of rank one; if nine numbers are needed, the quantity is a tensor of rank two. In general, if 3^n numbers are needed to describe the value of the quantity at a point in space, the quantity is a tensor of rank n.

Before we proceed to obtain some of the important properties of tensors, the reasons for which the tensor concept is being introduced at this stage are worth giving: (1) Since tensors are quantities describing the same phenomenon regardless of the coordinate system used, they provide an important guide in the formulation of the correct form of physical laws. Equations describing physical laws must be tensorially homogeneous, which means that every term of the equation must be a tensor of the same rank. (2) The tensor concept provides a convenient means of transforming an equation from one system of coordinates to another. (3) A decisive advantage of the use of cartesian tensors is that once the properties of a tensor of a certain rank have been established, they hold for all such tensors regardless of the physical phenomena they represent. In the study of mechanics, for example, one cannot help but notice that principal directions, invariants, and Mohr's representation appear in the analyses of strain, stress, inertia properties of rigid bodies, and curvature of plates so that there must be a bond common to all: The bond is that they all are symmetric tensors of rank two. In Chapter 3, it was mentioned that linear transformation was the bond uniting the quantities mentioned above: Indeed, underlying all the operations in Chapter 3 is the concept of tensor. It is generally agreed that this concept is the most adequate analytical tool for the study of deformation; if this is the case, then why did we go through Chapters 3 and 4 without mentioning it at all? The first reason is that the use of linear transformation makes it possible to give a geometrical interpretation to linear operations in a language easily understood by engineering students; the process of deformation as expressed by sets of linear equations is more readily visualized and such a visualization is of primary importance in the study of mechanics. The second reason is that engineering students do not find any difficulty manipulating matrices; such is not the case when it comes to a mathematical being called tensor, written in a condensed notation, and defined through a rule of coordinate transformation. However, having used it implicitly and having established some of its important properties in preceding chapters, they have no reluctance accepting it as a necessary part of the study of mechanics.

5.2 Scalars and Vectors

Under a transformation of coordinate axes, a scalar, such as the density or the temperature, remains unchanged. This means that a scalar is an invariant under a coordinate transformation. Scalars are called *tensors of zero rank*.

Now consider a vector $\bar{x}$ whose components in a system of axes OX_1, OX_2, OX_3 are x_1, x_2, x_3. In a new system of rectangular axes OX'_1, OX'_2, OX'_3, the components of $\bar{x}$ are given by x'_1, x'_2, x'_3, with

$$\begin{aligned} x'_1 &= \ell_{11}x_1 + \ell_{12}x_2 + \ell_{13}x_3 \\ x'_2 &= \ell_{21}x_1 + \ell_{22}x_2 + \ell_{23}x_3 \\ x'_3 &= \ell_{31}x_1 + \ell_{32}x_2 + \ell_{33}x_3 \end{aligned} \tag{5.2.1}$$

or

$$x'_i = \ell_{ij}x_j \quad (i,j = 1,2,3).\ (*) \tag{5.2.2}$$

---

* See page A-9, Eq. (3.4.14).

ℓ_{ij} are the direction cosines of the axes of the new system with respect to the old one. By definition, quantities which transform according to the relationship (5.2.2) are vectors or *tensors of the first rank*. Tensors of the first rank need only one subscript for their representation. Clearly, the multiplication of a first-rank tensor by a zero-rank tensor (i.e., the multiplication of a vector by a scalar) yields a first-rank tensor. Thus,

$$mx'_i = m\ell_{ij}x_j = \ell_{ij}(mx_j). \tag{5.2.3}$$

Also, if ξ_i and x_i are two tensors of the first rank, then

$$\xi'_i + x'_i = \ell_{ij}\xi_j + \ell_{ij}x_j = \ell_{ij}(\xi_j + x_j). \tag{5.2.4}$$

Therefore, the sum of two tensors of the first rank is a tensor of the first rank since according to Eq. (5.2.4) it transforms as one.

5.3 Higher Rank Tensors

Consider two tensors of the first rank, u_i with components u_1, u_2, u_3, and v_j with components v_1, v_2, v_3. In a new system of coordinates OX'_1, OX'_2, OX'_3, the product

$$u'_i v'_j = (\ell_{ik} u_k)(\ell_{jm} v_m) = \ell_{ik}\ell_{jm} u_k v_m. \tag{5.3.1}$$

Eq. (5.3.1) can be written as:

$$w'_{ij} = \ell_{ik}\ell_{jm} w_{km}. \tag{5.3.2}$$

The quantities $w'_{ij} = u'_i v'_j$ and $w_{km} = u_k v_m$ represent *the general product* of the first-rank tensors u_k and v_m in the (OX'_1, OX'_2, OX'_3) system and in the (OX_1, OX_2, OX_3) system, respectively. However, products of two first-rank tensors are not the only quantities satisfying the rule of Eq. (5.3.2). In general, a set of nine quantities w_{km} referred to a set of axes and which transforms to another set according to Eq. (5.3.2) is defined as *a tensor of the second rank*. For example, Eq. (4.5.8), giving the transformation law of the components of the state of strain, is of the same form as Eq. (5.3.2). Thus, these components are the components of a tensor of the second rank. The same can be said about the components of the linear transformation matrix and the linear strain matrix since Eqs. (3.5.8a) and (4.9.12) are of the same form as Eq. (5.3.2).

In a similar fashion, *a tensor of the third rank* can be formed by multiplying together three first-rank tensors or one first-rank tensor and one second-rank tensor. In general, however, a set of 27 quantities w_{rst} referred to a set of axes, and which transforms to another set according to

$$w'_{ijk} = \ell_{ir}\ell_{js}\ell_{kt}w_{rst}, \tag{5.3.3}$$

is defined as a tensor of the third rank.

In the theory of elasticity, we shall use *tensors of the fourth rank*. These tensors can be generated by multiplying together a number of lower-rank tensors which are such that the sum of their rank is equal to four. In general, however, a set of 81 quantities S_{mnpq}, which transforms according to

$$S'_{ijk\ell} = \ell_{im}\ell_{jn}\ell_{kp}\ell_{\ell q}S_{mnpq}, \tag{5.3.4}$$

is defined as a tensor of the fourth rank.

5.4 On Tensors and Matrices

There are great similarities between the rules governing the behavior of square matrices and those governing the behavior of tensors. Yet, while a matrix is nothing but an array of elements arranged in rows and columns, the components of a tensor must satisfy specific conditions when passing from one coordinate system to another. For example, a tensor of the second rank, w_{ik}, can be symbolically represented by a square matrix of the third order:

$$\begin{bmatrix} w_{11} & w_{12} & w_{13} \\ w_{21} & w_{22} & w_{23} \\ w_{31} & w_{32} & w_{33} \end{bmatrix}; \tag{5.4.1}$$

but not every square matrix is the matrix of an even-rank tensor. In Sec. 5.2, we have seen that adding two tensors amounts to adding their corresponding components to obtain another tensor, and that the multiplication of a tensor by a scalar amounts to multiplying each of its components by the same scalar; the same is true for matrices, as we have seen in Chapter 2. In Sec. 2.4, it was shown that a square matrix can be split into a symmetric and an antisymmetric component; the same can be said about a second-rank tensor: If, in Eq. (5.3.2), we interchange i and j, we get:

$$w'_{ji} = \ell_{jk}\,\ell_{im}\,w_{km}. \tag{5.4.2}$$

Since k and m are repeated indices—in other words, dummies—they can be interchanged. Thus,

$$w'_{ji} = \ell_{jm}\,\ell_{ik}\,w_{mk}; \tag{5.4.3}$$

w_{mk} is seen to transform according to the same rules as w_{km}, and is therefore a tensor of the second rank. The tensor w_{mk} is said to be the conjugate of w_{km}. The two sets $\frac{1}{2}(w_{km} + w_{mk})$ and $\frac{1}{2}(w_{km} - w_{mk})$ are also tensors of the second rank. The set $\frac{1}{2}(w_{km} + w_{mk})$ is unaltered if k and m are interchanged, and is called a *symmetric tensor*. The set $\frac{1}{2}(w_{km} - w_{mk})$ has its components reversed in sign when k and m are interchanged, and is called an *antisymmetric tensor*. The sum of the two sets is equal to w_{km}. Thus, we can consider any tensor of the second rank as the sum of a symmetric tensor and of an antisymmetric one. Since the antisymmetric tensor has only three components, it can be associated with a vector. (This, in fact, was done in the study of antisymmetric transformations in Chapter 3.)

The general product of two first-rank tensors can also be presented in matrix form. For example, the general product $u_k v_m = w_{km}$ of the two vectors $\bar{u}(u_1, u_2, u_3)$ and $\bar{v}(v_1, v_2, v_3)$ can be written as:

$$\begin{bmatrix} u_1 \\ u_2 \\ u_3 \end{bmatrix} [v_1 \quad v_2 \quad v_3] = \begin{bmatrix} u_1 v_1 & u_1 v_2 & u_1 v_3 \\ u_2 v_1 & u_2 v_2 & u_2 v_3 \\ u_3 v_1 & u_3 v_2 & u_3 v_3 \end{bmatrix} = \begin{bmatrix} w_{11} & w_{12} & w_{13} \\ w_{21} & w_{22} & w_{23} \\ w_{31} & w_{32} & w_{33} \end{bmatrix}, \tag{5.4.4}$$

and yields a tensor of the second rank.

Finally, it must be remembered that a tensor is defined by a given formula of transformation and is attached to a specific point in a given space; its components are all related to a given system of coordinates in this space and do not straddle on two or more systems. A set of nine quantities like the ℓ_{ij}'s defining the position of one coordinate system with respect to the other, are not the components of a second-rank tensor.

5.5 The Kronecker Delta and the Alternating Symbol. Isotropic Tensors

In Chapter 2 we introduced two symbols: δ_{ij}, which we called the Kronecker Delta; and ε_{ijk}, which we called the Alternating Symbol. These two symbols were used to simplify the writing of some equations.

In the following, we shall prove that δ_{ij} is a tensor of the second rank while ε_{ijk} is of the third rank. Let us first consider δ_{ij}: We know that $\delta_{ij} = 1$ for $i = j$ and $\delta_{ij} = 0$ for $i \neq j$. If δ_{ij} was a tensor of the second rank, it would transform according to Eq. (5.3.2) in a change of cartesian coordinates. Also, in the new system, δ'_{ij} would still be such that

$$\delta'_{ij} = \begin{cases} 1 \text{ for } i = j, \\ 0 \text{ for } i \neq j. \end{cases} \tag{5.5.1}$$

From Eq. (5.3.2), we would have:

$$\delta'_{ij} = \ell_{ik}\,\ell_{jm}\,\delta_{km} = \ell_{ik}\,\ell_{jk} = \delta_{ij} = \begin{cases} 1 \text{ for } i = j, \\ 0 \text{ for } i \neq j. \end{cases} \tag{5.5.2}$$

Thus, the quantity δ_{ij} transforms into itself and is a tensor of the second rank. δ_{ij} is also called the substitution tensor.

To prove that ε_{ijk} is a tensor of the third rank, we must prove that it transforms according to the general equation:

$$w'_{ijk} = \ell_{ir}\,\ell_{js}\,\ell_{kt}\,w_{rst}\,.$$

In other words, we must prove that

$$\varepsilon'_{ijk} = \ell_{ir}\,\ell_{js}\,\ell_{kt}\,\varepsilon_{rst}\,. \tag{5.5.3}$$

Here, too, we find by writing the detailed form of Eq. (5.5.3)

$\varepsilon'_{ijk} = 0$, when two of i, j, k are equal;

$\varepsilon'_{ijk} = 1$, when i, j, k are different and in cycle order;

$\varepsilon'_{ijk} = -1$, when i, j, k are different and not in cycle order .

Thus, ε_{ijk} transforms into itself in a change of cartesian coordinates and is a third-rank tensor.

The substitution and the alternating tensors are the exceptions to the rule that tensors must describe a physical phenomenon. Their components remain unchanged during a transformation of coordinates. Any tensor whose components remain unchanged during a transformation of coordinates is called an *isotropic tensor*. Such a tensor possesses no directional properties. Therefore, a vector can never be isotropic, but tensors of any rank other than one can be.

5.6 Function of a Tensor. Invariants

A general property of tensors is the following: If $f(w_{ij})$ is a function of a tensor w_{ij}, and if $q_{ij} = \partial f/\partial w_{ij}$, then q_{ij} is also a tensor which in a change of coordinates is given by $q'_{ij} = \partial f/\partial w'_{ij}$. This property can be proved as follows:

$$\frac{\partial f(w_{ij})}{\partial w'_{ij}} = \frac{\partial f(w_{ij})}{\partial w_{km}} \frac{\partial w_{km}}{\partial w'_{ij}}. \tag{5.6.1}$$

Also, from Eq. (5.3.2), we have:

$$w'_{ij} = \ell_{ik}\,\ell_{jm}\,w_{km}.$$

Multiplying both sides of this equation by $\ell_{ir}\,\ell_{js}$, we obtain:

$$\ell_{ir}\,\ell_{js}\,w'_{ij} = \ell_{ir}\,\ell_{ik}\,\ell_{js}\,\ell_{jm}\,w_{km} = \delta_{rk}\,\delta_{sm}\,w_{km} = w_{rs}. \tag{5.6.2}$$

Thus,

$$\frac{\partial w_{km}}{\partial w'_{ij}} = \ell_{ik}\,\ell_{jm} \tag{5.6.3}$$

and

$$\frac{\partial f(w_{ij})}{\partial w'_{ij}} = q'_{ij} = \ell_{ik}\,\ell_{jm}\,q_{km}, \tag{5.6.4}$$

which proves the property.

In the study of linear transformations (Chapter 3), we established the existence of three invariants. The same can be done for second-rank tensors. We shall prove that any tensor of the second rank, T_{ij}, has three invariants which do not change when we pass from a system of cartesian coordinates OX_1, OX_2, OX_3 to another system of cartesian coordinates OX'_1, OX'_2, OX'_3. These invariants are:

$$T_{ii} = T'_{ii} \tag{5.6.5}$$

$$T_{ij}\,T_{ji} = T'_{ij}\,T'_{ji} \tag{5.6.6}$$

$$T_{ij}\,T_{jk}\,T_{ki} = T'_{ij}\,T'_{jk}\,T'_{ki}. \tag{5.6.7}$$

Eq. (5.6.5) is easily proved by writing:

$$T'_{ij} = \ell_{ik}\,\ell_{j\ell}\,T_{k\ell},$$

and setting

$$i = j$$

$$T'_{ii} = \ell_{jk}\,\ell_{j\ell}\,T_{k\ell} = \delta_{k\ell}\,T_{k\ell} = T_{ii}\,.$$

Eq. (5.6.6) is proved by writing:

$$\begin{aligned} T'_{ij}T'_{ji} &= (\ell_{ik}\,\ell_{j\ell}\,T_{k\ell})(\ell_{jm}\,\ell_{in}\,T_{mn}) = \ell_{ik}\,\ell_{in}\,\ell_{j\ell}\,\ell_{jm}\,T_{k\ell}\,T_{mn} \\ &= \delta_{kn}\,\delta_{\ell m}\,T_{k\ell}\,T_{mn} = T_{mn}\,T_{nm} = T_{ij}\,T_{ji}\,. \end{aligned}$$

Eq. (5.6.7) is proved by writing:

$$\begin{aligned} T'_{ij}\,T'_{jk}\,T'_{ki} &= (\ell_{im}\,\ell_{jn}\,T_{mn})(\ell_{jr}\,\ell_{ks}\,T_{rs})(\ell_{kt}\,\ell_{ip}\,T_{tp}) \\ &= \delta_{mp}\,\delta_{nr}\,\delta_{st}\,T_{mn}\,T_{rs}\,T_{tp} = T_{mn}\,T_{ns}\,T_{sm} \\ &= T_{ij}\,T_{jk}\,T_{ki}\,. \end{aligned}$$

In Sec. 3.7, combinations of coefficients of the linear transformations were found invariant in a change of coordinates. These combinations are equivalent to those of Eqs. (5.6.5), (5.6.6), and (5.6.7). Indeed, the coefficients a_{ij} in the linear transformation were shown to be components of a second-rank tensor. The same can be said about the invariants of the state of strain in Eqs. (4.9.3), (4.9.4), and (4.9.5). They can be expressed in a form similar to Eqs. (5.6.5), (5.6.6), and (5.6.7), as follows:

$$e_{ii} = I_1, \quad \frac{1}{2}e_{ij}\,e_{ji} = \frac{1}{2}I_1^2 - I_2,$$

$$\frac{1}{3}e_{ij}\,e_{jk}\,e_{ki} = \frac{1}{3}I_1^3 - I_1\,I_2 + I_3\,.$$

5.7 Contraction

An operation which is often done in tensor manipulations is the operation of contraction. It simply consists of setting two free indices equal to each other, thus dropping the rank of the tensor by two. The free indices become dummies. For example, a second-rank tensor, T_{ij}, becomes T_{ii} upon contraction, where

$$T_{ii} = T_{11} + T_{22} + T_{33}\,.$$

In other words, it is reduced to a scalar or a tensor of rank zero. Now consider two second-rank tensors, A and B. The general product of A by B gives a tensor of the fourth rank with 81 components, $A_{ij}B_{km}(i,j,k,m = 1,2,3)$. If this general product of two tensors is contracted, a second-rank tensor will result. This second-rank tensor may have any of the four forms:

$$A_{ij}B_{ki},\ A_{ij}B_{ik},\ A_{ij}B_{kj},\ A_{ij}B_{jk}. \tag{5.7.1}$$

The contractions $A_{ii}B_{km}$ and $A_{ij}B_{kk}$ are products of the scalars A_{ii} and B_{kk} and the tensors B_{km} and A_{ij}, respectively. The index notation makes quite clear which contraction is involved; however, the matrix notation is sometimes quite useful. The nine components of the product $A_{ij}B_{jk} = C_{ik}$ can be written in matrix notation as $[A]\,[B]$. The four forms of Eq. (5.7.1) can thus be written:

$$[B][A],\ [A]'[B],\ [A][B]',\ [A][B]. \tag{5.7.2}$$

In the study of linear transformations, we were continuously faced with the product of the transformation matrix $[a]$ by the vector $\{\overline{OM}\}$. The transformation matrix $[a]$ was shown to be the matrix of a second-rank tensor (see Sec. 5.3). The general product of the second-rank tensor, a_{ij}, and the vector, x_k, is a third-rank tensor, $a_{ij}x_k$. Upon contraction, we obtain a first-rank tensor or a vector. This contraction would either give $a_{ij}x_j$ or $a_{ij}x_i$ ($a_{ii}x_k$ is the product of a vector by a scalar). In matrix notation, the components of the product $a_{ij}x_j = \xi_i$ can be written $[a]\{\bar{x}\} = \{\bar{\xi}\}$. Thus, it appears that all the properties of linear transformations studied in Chapter 3 are also those of the contracted or inner product of a second-rank tensor by a first-rank tensor. All the subjects discussed in Chapter 3 apply to a second-rank tensor and can be generalized to include higher even-rank tensors: Existence of principal directions, characteristic equations and eigenvalues, invariants and invariant directions, antisymmetric and symmetric transformations, Mohr's diagram, etc. ... can be discussed directly within the framework of tensor analysis. Chapter 3 can practically be reread substituting the word tensor for the word matrix. Finally, it will be recalled that in Eqs. (3.14.4) and (4.9.10) the matrix of the transformation was decomposed into two parts referred to as spherical and deviatoric; the same can be done with any tensor T_{ij}:

$$T_{ij} = \frac{1}{3}\delta_{ij}T_{\alpha\alpha} + T'_{ij}. \tag{5.7.3}$$

$\delta_{ij} T_{\alpha\alpha}$ is the same in any system of coordinates and, as such, is an isotropic tensor. T'_{ij} can be symbolically represented by a matrix with zero trace, and is called a deviator.

5.8 The Quotient Rule of Tensors

Suppose we know nine quantities a_{ij}, and we wish to establish whether they are the components of a tensor of rank two or not, without going to the trouble of determining the law of transformation. In many cases, the quotient rule of tensors is a convenient method to use for this purpose.

Let x_i be an arbitrary tensor of rank one. If the product $a_{ij} x_j$ is known to yield a tensor of rank one, ξ_i, then the a_{ij}'s are the components of a tensor of rank two. The proof of this statement is obtained by making a rotation of coordinates and showing that a_{ij} transforms according to Eq. (5.3.2):

$$a'_{ij} x'_j = \xi'_i = \ell_{ik} \xi_k = \ell_{ik} a_{km} x_m . \tag{5.8.1}$$

But

$$x_m = \ell_{nm} x'_n ,$$

so that

$$a'_{ij} x'_j = \ell_{ik} \ell_{nm} a_{km} x'_n , \tag{5.8.2}$$

and

$$(a'_{in} - \ell_{ik} \ell_{nm} a_{km}) x'_n = 0. \tag{5.8.3}$$

Since x_i is arbitrary, then

$$a'_{in} = \ell_{ik} \ell_{nm} a_{km} , \tag{5.8.4}$$

which shows that a_{ij} is a tensor of rank two. (This is another proof that the matrices of the linear transformations studied in Chapter 3 are the matrices of tensors of rank two.) In the same way, we can prove that if the product $a_{ij} x_i x_j$ is known to yield a tensor of rank zero (a scalar), a_{ij} is a tensor of rank two. Indeed this has been done with ϵ_{ij} in Sec. 4.5. The quotient rule of tensors holds for tensors of any rank.

It was shown in Chapter 4 and mentioned in Section 5.3 that the e_{ij}'s and the ϵ_{ij}'s all followed Eq. (5.3.2) in a change of coordinate axes. The e_{ij}'s and ϵ_{ij}'s are the components of symmetric tensors of rank 2. Mohr's representation can be used for both since it applies to all symmetric tensors of rank 2. It must be noticed, however, that while the e_{ij}'s are the components of a symmetric matrix in a linear transformation (4.9.1), no such transformation exists for the ϵ_{ij}'s.

PROBLEMS

1. Find the components of the tensor of rank two $c_{ij} = b_{ij} + d_{ij}$, when

$$b_{ij} = \begin{bmatrix} 2 & 8 & 6 \\ -3 & 0 & -3 \\ 4 & 6 & -2 \end{bmatrix} \text{ and } d_{ij} = \begin{bmatrix} -7 & 2 & 8 \\ 2 & 0 & -2 \\ -7 & 9 & 1 \end{bmatrix}.$$

2. Find the components of the tensor w_{ij} resulting from the general product of the two tensors of rank one $v_i = (1, -2, 3)$ and $u_i = (-2, 3, 4)$.

3. Verify the $\varepsilon - \delta$ identity

$$\varepsilon_{ijk}\,\varepsilon_{imn} = \delta_{jm}\,\delta_{kn} - \delta_{jn}\,\delta_{km}.$$

4. Show that
(a) $\varepsilon_{ijk}\,\delta_{jk} = 0$
(b) $\varepsilon_{ijk}\,\varepsilon_{jk\ell} = 2\delta_{i\ell}$
(c) $\varepsilon_{ijk}\,x_j\,x_k = 0$.

5. Find the components of the tensor t_{ik} resulting from the contracted product $t_{ik} = b_{ij}\,d_{jk}$ when b_{ij} and d_{ij} are the same as in Problem 1.

6. Show that the scalar product of two vectors is nothing but the contracted product of these two vectors.

7. By writing down the expression of the cosine of the angle between two lines whose direction cosines are ℓ_i and m_i, show that δ_{ij} is a tensor of rank two.

8. Show that the vector product c_i of two vectors a_i and b_i can be written as $c_i = \varepsilon_{ijk}\,a_j\,b_k$.

9. Let the three vectors a_i, b_i, and c_i form the three edges of a parallelepiped. By writing the expression of the volume, show that ε_{ijk} is a tensor of rank three.

10. Given the tensor of rank two

$$a_{ij} = \begin{bmatrix} 2 & -2 & 3 \\ 1 & 1 & 1 \\ 1 & 3 & -1 \end{bmatrix}:$$

(a) Find its symmetric and antisymmetric components.
(b) Find the invariants and the principal directions of its symmetric component.
(c) Decompose the symmetric component into its isotropic and deviatoric components. What are the principal directions of the deviatoric component?

CHAPTER 6

ORTHOGONAL CURVILINEAR COORDINATES

6.1 Introduction

In many problems of mechanics, geometry suggests the use of nonrectilinear coordinate systems. For example, it seems natural to study the mechanical behavior of axisymmetric and spherical objects using cylindrical and spherical coordinates, respectively: The formulation of the problems and their solutions are substantially simplified. These two systems are special cases of the general curvilinear coordinate systems.

The aim of this chapter is to present in a simple way the basic operations involved in the use of orthogonal curvilinear coordinates. The expressions of such quantities as gradient, divergence, curl, and Laplacian are obtained, and the general expressions of the components of the strain tensor and the strain-displacement relations are established. Those expressions and relations are continuously referred to in future chapters. Since we have limited ourselves to orthogonal systems, the notions of covariance and contravariance, Christoffel's symbol, and Riemann's tensor, commonly used in the study of curvilinear coordinates, are not needed. To avoid unnecessary difficulties, they will neither be introduced nor defined.

6.2 Curvilinear Coordinates

Let us refer a region of space to a set of orthogonal cartesian axes OX_1, OX_2, OX_3. The coordinates of any point P in the space are

x_1, x_2, x_3 (Fig. 6.1). If we make a transformation of coordinates from this cartesian system to another system, functional relations between the two must be given. Let these relations be:

$$\begin{aligned} y_1 &= y_1(x_1, x_2, x_3) \\ y_2 &= y_2(x_1, x_2, x_3) \\ y_3 &= y_3(x_1, x_2, x_3). \end{aligned} \tag{6.2.1}$$

We shall assume that the functions $y_i(x_1, x_2, x_3)$ are single valued and continuously differentiable at all points of the region, and that Eqs. (6.2.1) can be solved to yield the inverse transformation:

$$\begin{aligned} x_1 &= x_1(y_1, y_2, y_3) \\ x_2 &= x_2(y_1, y_2, y_3) \\ x_3 &= x_3(y_1, y_2, y_3), \end{aligned} \tag{6.2.2}$$

in which the functions $x_i(y_1, y_2, y_3)$ are single valued and continuously differentiable with respect to the variables y_i. The passage from Eqs. (6.2.1) to (6.2.2) and vice versa requires that the jacobian $|\partial y_i / \partial x_j| \neq 0$. Coordinate transformations with the above properties are called admissible transformations. If we set $y_1 = c_1$ in Eqs. (6.2.1), where c_1 is a constant, the equation

$$y_1(x_1, x_2, x_3) = c_1 \tag{6.2.3}$$

represents a surface S_1. Similarly,

$$y_2(x_1, x_2, x_3) = c_2 \tag{6.2.4}$$

and

$$y_3(x_1, x_2, x_3) = c_3 \tag{6.2.5}$$

represent surfaces S_2 and S_3. These surfaces (Fig. 6.1) intersect at the point whose coordinates are obtained by solving Eqs. (6.2.3), (6.2.4), and (6.2.5). The surfaces S_i are called the coordinate surfaces and their intersection pair by pair are the coordinate lines Y_1, Y_2 and Y_3. The Y_1 coordinate line is the intersection of the two surfaces $y_2 = c_2$ and $y_3 = c_3$. Along this line, the only variable that changes is y_1. Similarly,

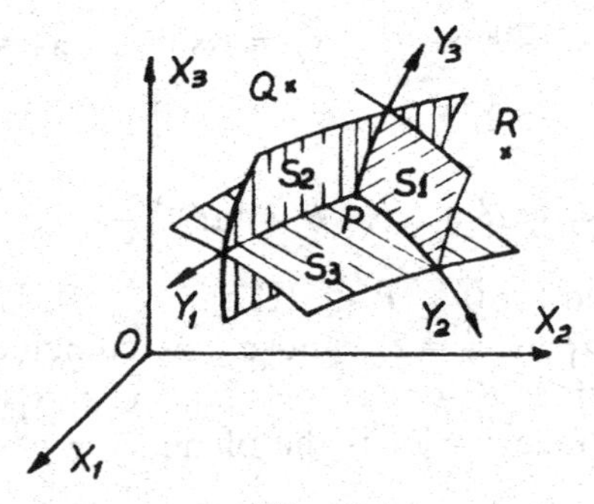

Fig. 6.1

along the Y_2 and the Y_3 coordinate lines the only variables that change are y_2 and y_3, respectively. By changing the values of the constants c_1, c_2, and c_3, other points such as Q and R can be located in the Y_i coordinate system.

As an example, *consider the transformation to a cylindrical coordinates system* with variables

$$y_1 = r, \quad y_2 = \theta, \quad y_3 = z. \tag{6.2.6}$$

$\bar{e}_r$, $\bar{e}_\theta$ and $\bar{e}_z$ are the three-unit vectors in the radial, tangential, and axial directions (Fig. 6.2).

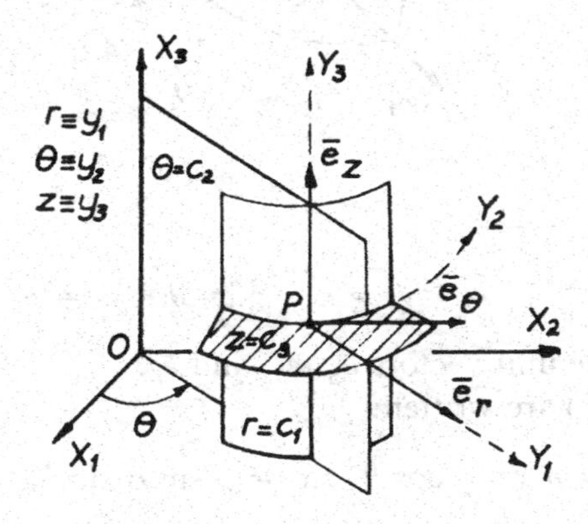

Fig. 6.2

Eqs. (6.2.2) are written:

$$x_1 = r \cos \theta, \quad x_2 = r \sin \theta, \quad x_3 = z. \tag{6.2.7}$$

The inverse of Eqs. (6.2.7) is:

$$r = \sqrt{x_1^2 + x_2^2}, \quad \theta = \tan^{-1}\frac{x_2}{x_1}, \quad z = x_3, \tag{6.2.8}$$

and is single valued for $0 \leq \theta < 2\Pi$ and $r > 0$. The surface $r = c_1$ is a circular cylinder $x_1^2 + x_2^2 = c_1^2$ whose axis coincides with the OX_3 axis (Fig. 6.2). The surface $\theta = c_2$ is the plane $x_2 = x_1 \tan c_2$ containing the OX_3 axis. The surface $z = c_3$ is the plane $x_3 = c_3$ perpendicular to the OX_3 axis.

As another example, *consider the transformation to a spherical polar coordinates system* with variables (Fig. 6.3)

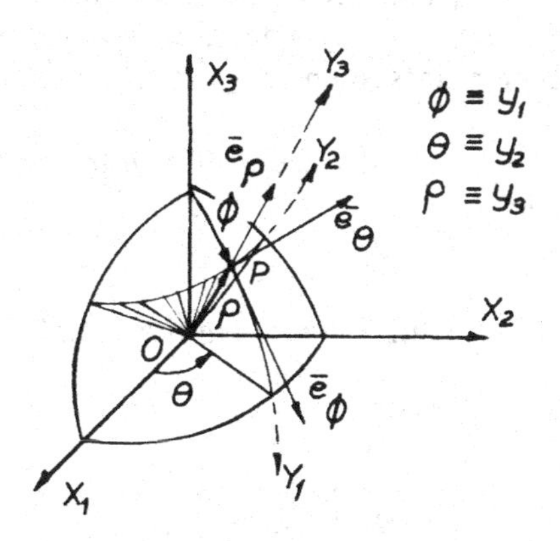

Fig. 6.3

$$y_1 = \phi, \quad y_2 = \theta, \quad y_3 = \rho, \tag{6.2.9}$$

and the three unit vectors $\bar{e}_\phi$, $\bar{e}_\theta$, and $\bar{e}_\rho$.

Eqs. (6.2.2) are written:

$$x_1 = \rho \sin \phi \cos \theta, \quad x_2 = \rho \sin \phi \sin \theta, \quad x_3 = \rho \cos \phi. \tag{6.2.10}$$

The inverse of Eqs. (6.2.10) is:

$$\rho = \sqrt{x_1^2 + x_2^2 + x_3^2}, \quad \phi = \tan^{-1}\frac{\sqrt{x_1^2 + x_2^2}}{x_3}, \quad \theta = \tan^{-1}\frac{x_2}{x_1}, \tag{6.2.11}$$

and is single valued for $\rho > 0$, $0 < \phi < \Pi$, $0 \leq \theta < 2\Pi$. The surface $\rho = c_1$ is a sphere. The surface $\phi = c_2$ is a cone. The surface $\theta = c_3$ is a plane. The coordinate lines are the meridians, the parallels, and the radial lines.

Consider a scalar function U defined in a cartesian system of coordinates OX_1, OX_2, OX_3, as well as in a curvilinear system of coordinates Y_1, Y_2, Y_3. Let the functional relation between the two systems [Eqs. (6.2.1), (6.2.2)] be known. We have:

$$dU = \frac{\partial U}{\partial y_1} dy_1 + \frac{\partial U}{\partial y_2} dy_2 + \frac{\partial U}{\partial y_3} dy_3 = \frac{\partial U}{\partial y_j} dy_j. \tag{6.2.12}$$

Since each of y_1, y_2, and y_3 is a function of x_1, x_2, x_3, therefore,

$$\frac{\partial U}{\partial x_i} = \frac{\partial U}{\partial y_j} \frac{\partial y_j}{\partial x_i}. \tag{6.2.13}$$

We thus have the following relation for the operator $\partial/\partial x_i$:

$$\frac{\partial}{\partial x_i} = \frac{\partial y_j}{\partial x_i} \frac{\partial}{\partial y_j}. \tag{6.2.14}$$

To find the second derivatives, we write:

$$\frac{\partial}{\partial x_m}\left(\frac{\partial}{\partial x_i}\right) = \frac{\partial^2}{\partial x_m \partial x_i} = \frac{\partial}{\partial x_m}\left(\frac{\partial y_j}{\partial x_i} \frac{\partial}{\partial y_j}\right). \tag{6.2.15}$$

Therefore,

$$\frac{\partial^2}{\partial x_m \partial x_i} = \frac{\partial^2 y_j}{\partial x_m \partial x_i} \frac{\partial}{\partial y_j} + \frac{\partial y_s}{\partial x_m} \frac{\partial y_j}{\partial x_i} \frac{\partial^2}{\partial y_s \partial y_j}. \tag{6.2.16}$$

For example, in the passage to cylindrical coordinates we use the functional relations expressed by Eqs. (6.2.6) to (6.2.8), and obtain:

$$\frac{\partial}{\partial x_1} = \cos\theta \frac{\partial}{\partial r} - \frac{\sin\theta}{r} \frac{\partial}{\partial \theta} \tag{6.2.17}$$

$$\frac{\partial}{\partial x_2} = \sin\theta \frac{\partial}{\partial r} + \frac{\cos\theta}{r} \frac{\partial}{\partial \theta} \tag{6.2.18}$$

$$\frac{\partial}{\partial x_3} = \frac{\partial}{\partial z} \tag{6.2.19}$$

$$\frac{\partial^2}{\partial x_1^2} = \cos^2\theta \frac{\partial^2}{\partial r^2} - \frac{2 \sin\theta \cos\theta}{r} \frac{\partial^2}{\partial r \partial\theta} + \frac{2 \sin\theta \cos\theta}{r^2} \frac{\partial}{\partial\theta} \tag{6.2.20}$$

$$+ \frac{\sin^2\theta}{r} \frac{\partial}{\partial r} + \frac{\sin^2\theta}{r^2} \frac{\partial^2}{\partial\theta^2}$$

$$\frac{\partial^2}{\partial x_2^2} = \sin^2\theta \frac{\partial^2}{\partial r^2} + \frac{2 \sin\theta \cos\theta}{r} \frac{\partial^2}{\partial r \partial\theta} - \frac{2 \sin\theta \cos\theta}{r^2} \frac{\partial}{\partial\theta} \tag{6.2.21}$$

$$+ \frac{\cos^2}{r} \frac{\partial}{\partial r} + \frac{\cos^2\theta}{r^2} \frac{\partial^2}{\partial\theta^2}$$

$$\frac{\partial^2}{\partial x_1 \partial x_2} = \frac{1}{2} \sin 2\theta \left(\frac{\partial^2}{\partial r^2} - \frac{1}{r} \frac{\partial}{\partial r} - \frac{1}{r^2} \frac{\partial^2}{\partial\theta^2} \right) \tag{6.2.22}$$

$$+ \cos 2\theta \left(\frac{1}{r} \frac{\partial^2}{\partial r \partial\theta} - \frac{1}{r^2} \frac{\partial}{\partial\theta} \right)$$

$$\frac{\partial^2}{\partial x_1 \partial x_3} = \left(\cos\theta \frac{\partial^2}{\partial r \partial z} - \frac{\sin\theta}{r} \frac{\partial^2}{\partial\theta \partial z} \right) \tag{6.2.23}$$

$$\frac{\partial^2}{\partial x_2 \partial x_3} = \left(\sin\theta \frac{\partial^2}{\partial r \partial z} + \frac{\cos\theta}{r} \frac{\partial^2}{\partial\theta \partial z} \right). \tag{6.2.24}$$

Similar relations can be written for spherical coordinates.

6.3 Metric Coefficients

In a coordinate system, the most important thing to know is how to measure lengths. This information is given by the metric coefficients. Let $P_1(x_1, x_2, x_3)$ be any point referred to a set of cartesian axes OX_i. The position vector $\bar{r}$ of P_1 is written:

$$\bar{r} = x_1 \bar{i}_1 + x_2 \bar{i}_2 + x_3 \bar{i}_3, \tag{6.3.1}$$

where $\bar{i}_1$, $\bar{i}_2$, $\bar{i}_3$ are the unit base vectors (Fig. 6.4). The square of the element of arc ds along a curve c between $P_1(x_1, x_2, x_3)$ and $P_2(x_1 + dx_1, x_2 + dx_2, x_3 + dx_3)$ is given by

$$(ds)^2 = (dx_1)^2 + (dx_2)^2 + (dx_3)^2 = dx_i \, dx_i. \tag{6.3.2}$$

The vector $d\bar{r}$ between P_1 and P_2 is given by

$$d\bar{r} = dx_1 \bar{i}_1 + dx_2 \bar{i}_2 + dx_3 \bar{i}_3. \tag{6.3.3}$$

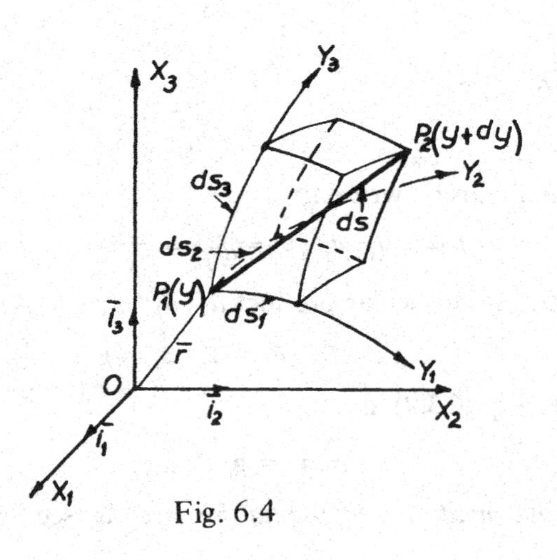

Fig. 6.4

Therefore,

$$(ds)^2 = d\bar{r} \cdot d\bar{r}. \tag{6.3.4}$$

This scalar product is, of course, an invariant independent of the coordinate system used. In a new system Y_1, Y_2, Y_3, defined by

$$y_i = y_i(x_1, x_2, x_3), \tag{6.3.5}$$

the two points P_1 and P_2 have coordinates $P_1(y_1, y_2, y_3)$ and $P_2(y_1 + dy_1, y_2 + dy_2, y_3 + dy_3)$ and

$$d\bar{r} = \frac{\partial \bar{r}}{\partial y_1} dy_1 + \frac{\partial \bar{r}}{\partial y_2} dy_2 + \frac{\partial \bar{r}}{\partial y_3} dy_3 = \frac{\partial \bar{r}}{\partial y_i} dy_i. \tag{6.3.6}$$

The symbol $\partial \bar{r}/\partial y_i$ denotes the derivative of $\bar{r}$ with respect to a particular variable $y_i (i = 1, 2, 3)$ when the remaining variables are held constant. Thus, if we fix the variables y_2 and y_3, $\bar{r}$ becomes a function of y_1 alone and the terminus of $\bar{r}$ is constrained to move along the Y_1 coordinate line in the Y_i coordinate system determined by Eq. (6.3.5). Consequently, the vector $\partial \bar{r}/\partial y_1$ is tangent to the coordinate line Y_1. Similarly, the vectors $\partial \bar{r}/\partial y_2$ and $\partial \bar{r}/\partial y_3$ are tangent to the Y_2 and Y_3 coordinate lines, respectively (Fig. 6.4). If we denote these vectors by $\bar{a}_i$, so that

$$\bar{a}_i = \frac{\partial \bar{r}}{\partial y_i}, \tag{6.3.7}$$

then from (6.3.6)

$$d\bar{r} = \bar{a}_i \, dy_i, \tag{6.3.8}$$

and Eq. (6.3.4) can be written:

$$(ds)^2 = (\bar{a}_i \, dy_i) \cdot (\bar{a}_j \, dy_j) = \bar{a}_i \cdot \bar{a}_j \, dy_i \, dy_j. \tag{6.3.9}$$

If we now define the scalar product $\bar{a}_i \cdot \bar{a}_j$ by

$$g_{ij} = \bar{a}_i \cdot \bar{a}_j = \bar{a}_j \cdot \bar{a}_i = g_{ji}, \tag{6.3.10}$$

we can write Eq. (6.3.9) as

$$(ds)^2 = g_{ij} \, dy_i \, dy_j. \tag{6.3.11}$$

Expanded, *this quadratic differential form* reads (see Sec. 3.12):

$$\begin{aligned}(ds)^2 = g_{11}(dy_1)^2 + g_{22}(dy_2)^2 + g_{33}(dy_3)^2 + 2g_{12}\, dy_1\, dy_2 \\ + 2g_{13}\, dy_1\, dy_3 + 2g_{23}\, dy_2\, dy_3.\end{aligned} \tag{6.3.12}$$

The coefficients g_{ij} are called *metric coefficients*. As can be seen from Eq. (6.3.12), they are the link between the length of an element and the differentials dy_i. These coefficients are the components of a second-rank

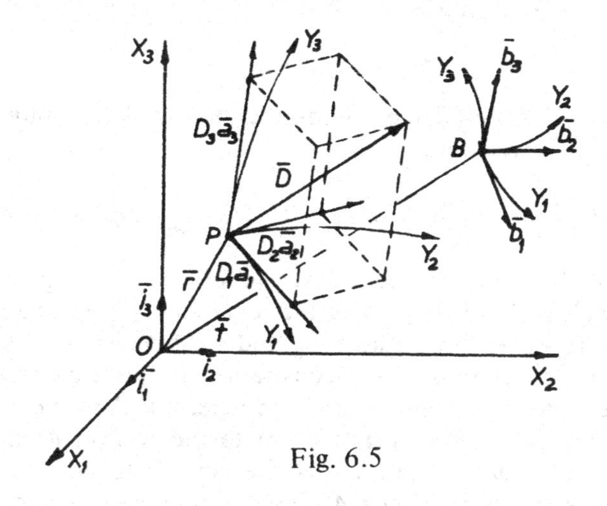

Fig. 6.5

tensor called the *metric tensor* (see Sec. 5.8). The vectors $\bar{a}_i$, which were found to be tangent to the coordinate lines Y_i at a given point P, are called base vectors in the curvilinear system Y_i. Any vector $\bar{D}$ (Fig. 6.5) with its origin at P can be resolved into three components along $\bar{a}_1$, $\bar{a}_2$, and $\bar{a}_3$:

$$\bar{D} = D_1\bar{a}_1 + D_2\bar{a}_2 + D_3\bar{a}_3. \tag{6.3.13}$$

$D_1\bar{a}_1$, $D_2\bar{a}_2$, and $D_3\bar{a}_3$ form the edges of a parellelepiped whose diagonal is $\bar{D}$. Thus, in the Y_i system the base vectors $\bar{a}_1$, $\bar{a}_2$, $\bar{a}_3$ play the same role as the vectors $\bar{i}_1$, $\bar{i}_2$, $\bar{i}_3$ play in the OX_i system. *However, while the magnitude and directions of the cartesian base vectors are fixed, the base vectors $\bar{a}_i$, in general, vary from point to point in space.* For example, point B (Fig. 6.5) is defined by the vector $\bar{i}$ and a set of constants c_1, c_2, c_3 [see Eqs. (6.2.3) to (6.2.5)] different from the ones used to locate point P. The base vectors at B could be called $\bar{b}_i$ and are given by $\partial\bar{i}/\partial y_i$; they are different from $\bar{a}_i$ in both magnitude and direction. The components of the metric tensor at B are therefore different from the components of the metric tensor at P.

From the definition of Eq. (6.3.10), we see on setting $i = j = 1$ that the length of $\bar{a}_1$ is

$$|\bar{a}_1| = \sqrt{g_{11}}\,. \tag{6.3.14}$$

Similarly, for $i = j = 2, 3$, we get:

$$|\bar{a}_2| = \sqrt{g_{22}} \tag{6.3.15}$$

$$|\bar{a}_3| = \sqrt{g_{33}}\,. \tag{6.3.16}$$

These vectors are orthogonal if, and only if,

$$\begin{aligned} g_{12} &= g_{21} = \bar{a}_1 \cdot \bar{a}_2 = 0 \\ g_{13} &= g_{31} = \bar{a}_1 \cdot \bar{a}_3 = 0 \\ g_{23} &= g_{32} = \bar{a}_2 \cdot \bar{a}_3 = 0. \end{aligned} \tag{6.3.17}$$

A curvilinear system for which these relations hold is called orthogonal. In such a system:

$$(ds)^2 = g_{11}(dy_1)^2 + g_{22}(dy_2)^2 + g_{33}(dy_3)^2. \tag{6.3.18}$$

To obtain the meaning of the coefficients g_{11}, g_{22}, and g_{33} we chose the element ds along the Y_1 coordinate line. Therefore, dy_2 and dy_3 are equal to zero since y_2 and y_3 do not change along the Y_1 line. In this case, Eq. (6.3.12) gives:

$$(ds_1)^2 = g_{11}(dy_1)^2.$$

Therefore,

$$ds_1 = \sqrt{g_{11}}\, dy_1. \tag{6.3.19}$$

Thus, the length of the element of arc ds_1 along the Y_1 coordinate line is obtained by multiplying the differential of y_1 by $\sqrt{g_{11}}$. Similarly, the differentials of arc along the Y_2 and Y_3 coordinate lines are (Fig. 6.4):

$$ds_2 = \sqrt{g_{22}}\, dy_2 \tag{6.3.20}$$

$$ds_3 = \sqrt{g_{33}}\, dy_3 \tag{6.3.21}$$

Since both the ds_i ' s and the dy_i ' s are real, we conclude that:

$$g_{11} > 0, \quad g_{22} > 0, \quad g_{33} > 0.$$

If $\bar{e}_1, \bar{e}_2, \bar{e}_3$ are unit vectors along $\bar{a}_1, \bar{a}_2, \bar{a}_3$, then Eq. (6.3.13) can be written:

$$\bar{D} = \bar{e}_1\sqrt{g_{11}}\, D_1 + \bar{e}_2\sqrt{g_{22}}\, D_2 + \bar{e}_3\sqrt{g_{33}}\, D_3. \tag{6.3.22}$$

In case of an orthogonal system of coordinates in which the normal projections of $\bar{D}$ on the vectors $\bar{a}_i$ are d_1, d_2, and d_3, Eq. (6.3.22) is written:

$$\bar{D} = \bar{e}_1 d_1 + \bar{e}_2 d_2 + \bar{e}_3 d_3 \tag{6.3.23}$$

and

$$D_i = \frac{d_i\cdot}{\sqrt{g_{ii}}} \text{ (no sum on i)}. \tag{6.3.24}$$

An element of volume dv in general curvilinear coordinates is given by the triple scalar product:

$$dv = |\bar{a}_1 \cdot \bar{a}_2 \times \bar{a}_3|\, dy_1\, dy_2\, dy_3.$$

For an orthogonal system, this equation reduces to:

$$dv = \sqrt{g_{11} g_{22} g_{33}}\, dy_1\, dy_2\, dy_3 . \tag{6.3.25}$$

When a curvilinear coordinate system Y_i is determined by a relation such as Eq. (6.3.5), the inverse is written $x_i = x_i(y_1, y_2, y_3)$ and

$$dx_i = \frac{\partial x_i}{\partial y_k}\, dy_k .$$

But, in cartesian coordinates, $(ds)^2 = dx_i\, dx_i$; therefore,

$$(ds)^2 = \frac{\partial x_i}{\partial y_k}\, dy_k\, \frac{\partial x_i}{\partial y_j}\, dy_j = \left(\frac{\partial x_i}{\partial y_k} \frac{\partial x_i}{\partial y_j} \right) dy_k\, dy_j . \tag{6.3.26}$$

Comparing Eqs. (6.3.11) and (6.3.26), we conclude that

$$g_{kj} = \frac{\partial x_i}{\partial y_k} \frac{\partial x_i}{\partial y_j} . \tag{6.3.27}$$

This formula allows one to calculate the metric coefficients.

Example 1. Case of Cylindrical Coordinates

From Eqs. (6.2.6), (6.2.7), and (6.3.27), we obtain:

$$g_{11} = \left(\frac{\partial x_1}{\partial y_1} \right)^2 + \left(\frac{\partial x_2}{\partial y_1} \right)^2 + \left(\frac{\partial x_3}{\partial y_1} \right)^2 = \cos^2\theta + \sin^2\theta + 0 = 1 \tag{6.3.28}$$

$$g_{22} = \left(\frac{\partial x_1}{\partial y_2} \right)^2 + \left(\frac{\partial x_2}{\partial y_2} \right)^2 + \left(\frac{\partial x_3}{\partial y_2} \right)^2 = r^2 \sin^2\theta + r^2 \cos^2\theta + 0 = r^2 \tag{6.3.29}$$

$$g_{33} = \left(\frac{\partial x_1}{\partial y_3} \right)^2 + \left(\frac{\partial x_2}{\partial y_3} \right)^2 + \left(\frac{\partial x_3}{\partial y_3} \right)^2 = 0 + 0 + 1 = 1 \tag{6.3.30}$$

$$g_{12} = g_{13} = g_{23} = 0.$$

The expression for $(ds)^2$ is

$$(ds)^2 = g_{ij}\, dy_i\, dy_j = (dr)^2 + r^2 (d\theta)^2 + (dz)^2 . \tag{6.3.31}$$

The element of volume is given by

$$dv = r\, dr\, d\theta\, dz \tag{6.3.32}$$

Example 2. Case of Spherical Polar Coordinates

From Eqs. (6.2.9), (6.2.10), and (6.3.27), we obtain:

$$g_{11} = \rho^2, \quad g_{22} = \rho^2 \sin^2\phi, \quad g_{33} = 1 \tag{6.3.33}$$

$$g_{12} = g_{13} = g_{23} = 0. \tag{6.3.34}$$

The expression for $(ds)^2$ is

$$(ds)^2 = (d\rho)^2 + \rho^2(d\phi)^2 + \rho^2\sin^2\phi(d\theta)^2. \tag{6.3.35}$$

The element of volume is given by

$$dv = \rho^2 \sin\phi \, d\rho \, d\phi \, d\theta. \tag{6.3.36}$$

To simplify the writing, it is sometimes convenient for orthogonal systems to write:

$$g_{ii} = h_i^2 \text{ (no sum)}, \tag{6.3.37}$$

where the h_i ' s are called the scale factors.

Remark

Spaces in which it is possible to construct a coordinate system such that the quadratic differential form (6.3.11) reduces to a sum of squares of the coordinate differentials are called Euclidean spaces. In other words, in a Euclidean space the length of a line segment can always be given by the formula of Pythagoras. If one specifies the rule for the measurement between points (i.e., if one specifies g_{ij}), the space is called metric.

Spaces in which no coordinate system can be found such that the formula of Pythagoras can be applied to the length of a line segment, are called Non-Euclidean spaces. A surface represents the only variety of Non-Euclidean space capable of actual visualization. Elements of the theory of surfaces will be presented in Chapter 18 within the scope of the theory of thin shells.

6.4 Gradient, Divergence, Curl, and Laplacian in Orthogonal Curvilinear Coordinates

Assume an orthogonal curvilinear coordinate system defined by

$$x_i = x_i(y_1, y_2, y_3),$$

where the variables x_i are cartesian. Since y_i are orthogonal, then

$$(ds)^2 = g_{11}(dy_1)^2 + g_{22}(dy_2)^2 + g_{33}(dy_3)^2. \tag{6.4.1}$$

We denote the unit base vectors along the tangents to the Y_i axes at P by $\bar{e}_1$, $\bar{e}_2$, and $\bar{e}_3$, and express a vector $\overline{M}$ at P by

$$\overline{M} = \bar{e}_1 m_1 + \bar{e}_2 m_2 + \bar{e}_3 m_3. \tag{6.4.2}$$

m_1, m_2. and m_3 are the orthogonal projections of M on the unit base vectors.

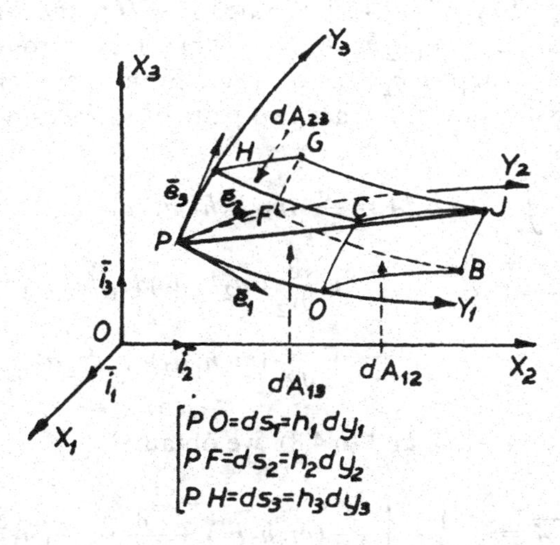

Fig.6.6

The volume element dv, formed by the coordinate surfaces $y_i =$ constant and $y_i + dy_i =$ constant (Fig. 6.6), has the shape of a curvilinear parallelepiped with edges $ds_i = \sqrt{g_{ii}}\, dy_i$ (no sum on i). The areas of its faces are given by

$$dA_{12} = \sqrt{g_{11} g_{22}}\, dy_1 dy_2 \tag{6.4.3}$$

$$dA_{13} = \sqrt{g_{11} g_{33}}\, dy_1 dy_3 \tag{6.4.4}$$

$$dA_{23} = \sqrt{g_{22} g_{33}}\, dy_2 dy_3, \tag{6.4.5}$$

and its volume is

$$dv = \sqrt{g_{11} g_{22} g_{33}}\, dy_1\, dy_2\, dy_3 . \tag{6.4.6}$$

To compute the divergence of a vector $\overline{M}$ at a point P, we use the definition of the divergence at a point:

$$div\overline{M} = \lim_{dv \to 0} \left[\frac{1}{dv} \int\int_A \overline{M} \cdot \overline{n}\, dA \right], \tag{6.4.7}$$

where $\overline{n}$ is the outward unit vector normal to dA. The contribution to the integral $\int\int_A \overline{M} \cdot \overline{n}\, dA$ through the area $PFGH$ in the direction of the outward normal is $-m_1 h_2 h_3 dy_2 dy_3$, while that through $JCOB$ is $m_1 h_2 h_3 dy_2 dy_3 + \partial/\partial y_1 (m_1 h_2 h_3) dy_2 dy_3 dy_1$. From these, and the corresponding expressions for the other two pairs of surfaces, we have:

$$\begin{aligned} \int\int_A \overline{M} \cdot \overline{n}\, dA = & \frac{\partial}{\partial y_1}(m_1 h_2 h_3) dy_1\, dy_2\, dy_3 \\ & + \frac{\partial}{\partial y_2}(m_2 h_1 h_3) dy_1\, dy_2\, dy_3 \\ & + \frac{\partial}{\partial y_3}(m_3 h_1 h_2) dy_1\, dy_2\, dy_3 . \end{aligned} \tag{6.4.8}$$

From Eqs. (6.4.6), (6.4.7), and (6.4.8), we obtain:

$$\begin{aligned} div\overline{M} = \frac{1}{h_1 h_2 h_3} \Bigg[& \frac{\partial}{\partial y_1}(m_1 h_2 h_3) + \frac{\partial}{\partial y_2}(m_2 h_1 h_3) \\ & + \frac{\partial}{\partial y_3}(m_3 h_1 h_2) \Bigg]. \end{aligned} \tag{6.4.9}$$

To compute the curl of a vector $\overline{M}$ at a point P, we use Stokes' theorem:

$$\oint \overline{M} \cdot \overline{ds} = \int\int_A curl\overline{M} \cdot \overline{n}\, dA, \tag{6.4.10}$$

where $\oint$ is the integral taken along a closed contour. The first component of the curl of $\overline{M}$ is obtained by applying Stokes' theorem to the surface $PFGH$ (Fig. 6.6):

$$\oint \overline{M} \cdot \overline{ds} = \int_P^F \overline{M} \cdot \overline{ds} + \int_F^G \overline{M} \cdot \overline{ds} + \int_G^H \overline{M} \cdot \overline{ds} + \int_H^P \overline{M} \cdot \overline{ds}$$

$$= m_2 h_2 dy_2 + \left[m_3 h_3 dy_3 + \frac{\partial}{\partial y_2}(m_3 h_3) dy_2 dy_3 \right] \tag{6.4.11}$$

$$- \left[m_2 h_2 dy_2 + \frac{\partial}{\partial y_3}(m_2 h_2) dy_2 dy_3 \right] - m_3 h_3 dy_3$$

$$= \left[\frac{\partial}{\partial y_2}(m_3 h_3) - \frac{\partial}{\partial y_3}(m_2 h_2) \right] dy_2 dy_3 .$$

By Stokes' theorem, Eq. (6.4.11) is equal to the first component of the curl along $\overline{e}_1$ multiplied by the area $PFGH$. Hence,

$$(\text{curl } \overline{M})_1 = \frac{1}{h_2 h_3} \left[\frac{\partial}{\partial y_2}(m_3 h_3) - \frac{\partial}{\partial y_3}(m_2 h_2) \right]. \tag{6.4.12}$$

The two other components along $\overline{e}_2$ and along $\overline{e}_3$ can be obtained by a similar reasoning or directly by a cyclic permutation of the indices. Thus,

$$\begin{aligned} \text{curl } \overline{M} &= \overline{e}_1 \frac{1}{h_2 h_3} \left[\frac{\partial (h_3 m_3)}{\partial y_2} - \frac{\partial (h_2 m_2)}{\partial y_3} \right] \\ &+ \overline{e}_2 \frac{1}{h_1 h_3} \left[\frac{\partial (h_1 m_1)}{\partial y_3} - \frac{\partial (h_3 m_3)}{\partial y_1} \right] \\ &+ \overline{e}_3 \frac{1}{h_1 h_2} \left[\frac{\partial (h_2 m_2)}{\partial y_1} - \frac{\partial (h_1 m_1)}{\partial y_2} \right]. \end{aligned}$$

This expression can be written as:

$$\text{curl } \overline{M} = \frac{1}{h_1 h_2 h_3} \begin{vmatrix} h_1 \overline{e}_1 & h_2 \overline{e}_2 & h_3 \overline{e}_3 \\ \frac{\partial}{\partial y_1} & \frac{\partial}{\partial y_2} & \frac{\partial}{\partial y_3} \\ h_1 m_1 & h_2 m_2 & h_3 m_3 \end{vmatrix} . \tag{6.4.13}$$

To compute the gradient of a scalar $U(y_1, y_2, y_3)$, we use the definition of the gradient. Thus, the component in the direction of $\overline{e}_1$ which is tangent to the element PO is (Fig. 6.6):

$$(\text{grad } U)_1 = \lim_{dy_1 \to 0} \frac{U(O) - U(P)}{h_1 \, dy_1} = \frac{1}{h_1} \frac{\partial U}{\partial y_1}. \tag{6.4.14}$$

The two other components along $\bar{e}_2$ and $\bar{e}_3$ can be obtained by a similar reasoning. Thus,

$$\text{grad } U = \frac{\bar{e}_1}{h_1} \frac{\partial U}{\partial y_1} + \frac{\bar{e}_2}{h_2} \frac{\partial U}{\partial y_2} + \frac{\bar{e}_3}{h_3} \frac{\partial U}{\partial y_3}. \tag{6.4.15}$$

In vector analysis a certain vector differential operator $\overline{\nabla}$ (read del or nabla), defined in orthogonal cartesian coordinates by

$$\overline{\nabla} = \bar{i}_1 \frac{\partial}{\partial x_1} + \bar{i}_2 \frac{\partial}{\partial x_2} + \bar{i}_3 \frac{\partial}{\partial x_3}, \tag{6.4.16}$$

plays a prominent role. The gradient of a scalar U is written:

$$\overline{\nabla} U = \bar{i}_1 \frac{\partial U}{\partial x_1} + \bar{i}_2 \frac{\partial U}{\partial x_2} + \bar{i}_3 \frac{\partial U}{\partial x_3}; \tag{6.4.17}$$

and the divergence of a vector M is written:

$$\overline{\nabla} \cdot \overline{M} = \frac{\partial m_1}{\partial x_1} + \frac{\partial m_2}{\partial x_2} + \frac{\partial m_3}{\partial x_3}. \tag{6.4.18}$$

The Laplace equation is written:

$$\overline{\nabla} \cdot \overline{\nabla} U = \frac{\partial^2 U}{\partial x_1^2} + \frac{\partial^2 U}{\partial x_2^2} + \frac{\partial^2 U}{\partial x_3^2} = \nabla^2 U = 0. \tag{6.4.19}$$

Comparing Eqs. (6.4.15) and (6.4.17), we see that in curvilinear orthogonal coordinates *the differential operator* $\overline{\nabla}$ *is written*:

$$\overline{\nabla} = \frac{\bar{e}_1}{h_1} \frac{\partial}{\partial y_1} + \frac{\bar{e}_2}{h_2} \frac{\partial}{\partial y_2} + \frac{\bar{e}_3}{h_3} \frac{\partial}{\partial y_3}. \tag{6.4.20}$$

Inasmuch as

$$\text{div}(\text{grad } U) = \overline{\nabla} \cdot \overline{\nabla} U = \nabla^2 U, \tag{6.4.21}$$

the expression for $\nabla^2 U$ is obtained by substituting $\overline{\nabla} U$ instead of $\overline{M}$ in Eq. (6.4.9). Thus,

$$\nabla^2 U = \frac{1}{h_1 h_2 h_3} \left[\frac{\partial}{\partial y_1} \left(\frac{h_2 h_3}{h_1} \frac{\partial U}{\partial y_1} \right) + \frac{\partial}{\partial y_2} \left(\frac{h_1 h_3}{h_2} \frac{\partial U}{\partial y_2} \right) + \frac{\partial}{\partial y_3} \left(\frac{h_1 h_2}{h_3} \frac{\partial U}{\partial y_3} \right) \right]. \tag{6.4.22}$$

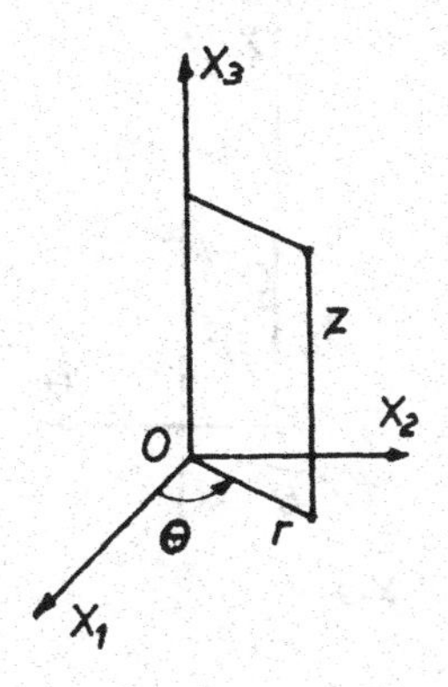

Fig. 6.7

Example 1. Cylindrical Coordinates (Fig. 6.7)

In this case, we have (see Sec. 6.3):

$$g_{11} = 1 \quad g_{22} = r^2 \quad g_{33} = 1 \tag{6.4.23}$$

$$h_1 = 1 \quad h_2 = r \quad h_3 = 1 \tag{6.4.24}$$

Therefore, the divergence of a vector $\overline{M}(m_r, m_\theta, m_z)$ is given by:

$$\operatorname{div} \overline{M} = \frac{1}{r}\frac{\partial}{\partial r}(rm_r) + \frac{1}{r}\frac{\partial m_\theta}{\partial \theta} + \frac{\partial m_z}{\partial z}. \tag{6.4.25}$$

The components of the curl of a vector $\overline{M}$ are given by:

$$\begin{gathered}(\operatorname{curl} \overline{M})_r = \frac{1}{r}\frac{\partial m_z}{\partial \theta} - \frac{\partial m_\theta}{\partial z}, \quad (\operatorname{curl} \overline{M})_\theta = \frac{\partial m_r}{\partial z} - \frac{\partial m_z}{\partial r} \\ (\operatorname{curl} \overline{M})_z = \frac{1}{r}\left[\frac{\partial}{\partial r}(rm_\theta) - \frac{\partial m_r}{\partial \theta}\right].\end{gathered} \tag{6.4.26}$$

The components of the gradient of a scalar $U(r, \theta, z)$ are given by:

$$(\operatorname{grad} U)_r = \frac{\partial U}{\partial r}, \quad (\operatorname{grad} U)_\theta = \frac{1}{r}\frac{\partial U}{\partial \theta}, \quad (\operatorname{grad} U)_z = \frac{\partial U}{\partial z}. \tag{6.4.27}$$

The Laplacian of a scalar $U(r, \theta, z)$ is given by:

$$\nabla^2 U = \frac{1}{r}\frac{\partial}{\partial r}\left(r\frac{\partial U}{\partial r}\right) + \frac{1}{r^2}\frac{\partial^2 U}{\partial \theta^2} + \frac{\partial^2 U}{\partial z^2}. \tag{6.4.28}$$

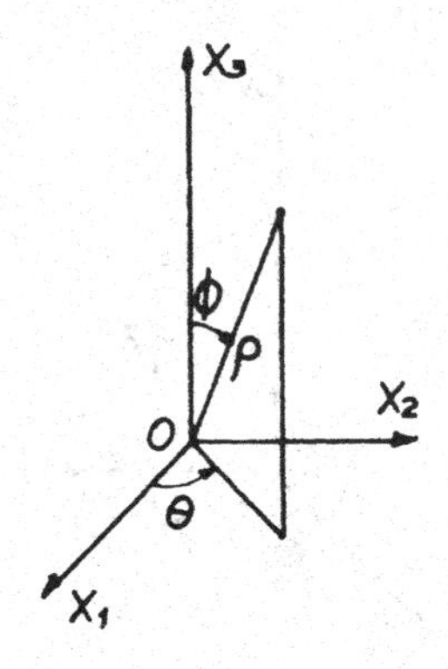

Fig. 6.8

Example 2. Spherical Polar Coordinates (Fig. 6.8)

In this case, we have (see Sec. 6.3):

$$g_{11} = \rho^2 \quad g_{22} = \rho^2 \sin^2\phi \quad g_{33} = 1 \tag{6.4.29}$$

$$h_1 = \rho \quad h_2 = \rho \sin\phi \quad h_3 = 1 \tag{6.4.30}$$

Therefore, the divergence of a vector $\overline{M}(m_\phi, m_\theta, m_\rho)$ is given by:

$$\operatorname{div} \overline{M} = \frac{1}{\rho^2}\frac{\partial}{\partial \rho}(\rho^2 m_\rho) + \frac{1}{\rho \sin\phi}\frac{\partial}{\partial \phi}(m_\phi \sin\phi) + \frac{1}{\rho \sin\phi}\frac{\partial m_\theta}{\partial \theta}. \tag{6.4.31}$$

The components of the curl of a vector $\overline{M}(m_\phi, m_\theta, m_\rho)$ are given by:

$$\begin{aligned}
(\operatorname{curl} \overline{M})_\phi &= \frac{1}{\rho}\left[\frac{1}{\sin\phi}\frac{\partial m_\rho}{\partial \theta} - \frac{\partial(\rho m_\theta)}{\partial \rho}\right] \\
(\operatorname{curl} \overline{M})_\theta &= \frac{1}{\rho}\left[\frac{\partial}{\partial \rho}(\rho m_\phi) - \frac{\partial m_\rho}{\partial \phi}\right] \\
(\operatorname{curl} \overline{M})_\rho &= \frac{1}{\rho \sin\phi}\left[\frac{\partial}{\partial \phi}(m_\theta \sin\phi) - \frac{\partial m_\phi}{\partial \theta}\right].
\end{aligned} \tag{6.4.32}$$

The components of the gradient of a scalar $U(\phi, \theta, \rho)$ are given by:

$$\begin{aligned}
&(\operatorname{grad} U)_\phi = \frac{1}{\rho}\frac{\partial U}{\partial \phi}, \quad (\operatorname{grad} U)_\theta = \frac{1}{\rho \sin\phi}\frac{\partial U}{\partial \theta}, \\
&(\operatorname{grad} U)_\rho = \frac{\partial U}{\partial \rho}.
\end{aligned} \tag{6.4.33}$$

The Laplacian of a scalar $U(\phi, \theta, \rho)$ is given by:

$$\nabla^2 U = \frac{1}{\rho^2}\frac{\partial}{\partial \rho}\left(\rho^2 \frac{\partial U}{\partial \rho}\right) + \frac{1}{\rho^2 \sin\phi}\frac{\partial}{\partial \phi}\left(\sin\phi \frac{\partial U}{\partial \phi}\right) + \frac{1}{\rho^2 \sin^2\phi}\frac{\partial^2 U}{\partial \theta^2}. \tag{6.4.34}$$

6.5 Rate of Change of the Vectors $\bar{a}_i$ and of the Unit Vectors $\bar{e}_i$ in an Orthogonal Curvilinear Coordinate System

In Fig. 6.6, the unit vectors $\bar{e}_1$, $\bar{e}_2$, and $\bar{e}_3$ (in the directions of increasing y_1, y_2, and y_3) being mutually perpendicular, satisfy the same relations among themselves as the unit vectors $\bar{i}_1$, $\bar{i}_2$, and $\bar{i}_3$ of the cartesian system, namely:

$$\bar{e}_1 \cdot \bar{e}_1 = \bar{e}_2 \cdot \bar{e}_2 = \bar{e}_3 \cdot \bar{e}_3 = 1 \tag{6.5.1}$$

$$\bar{e}_1 \cdot \bar{e}_2 = \bar{e}_2 \cdot \bar{e}_3 = \bar{e}_3 \cdot \bar{e}_1 = 0 \tag{6.5.2}$$

$$\bar{e}_1 \times \bar{e}_2 = \bar{e}_3, \quad \bar{e}_2 \times \bar{e}_3 = \bar{e}_1, \quad \bar{e}_3 \times \bar{e}_1 = \bar{e}_2 \tag{6.5.3}$$

$$\bar{e}_1 \times \bar{e}_1 = \bar{e}_2 \times \bar{e}_2 = \bar{e}_3 \times \bar{e}_3 = 0. \tag{6.5.4}$$

Referring ourselves to Fig. 6.1, $\bar{e}_1$ is normal to the surface S_1, $\bar{e}_2$ is normal to the surface S_2, and $\bar{e}_3$ is normal to the surface S_3. For the vectors $\bar{a}_1$, $\bar{a}_2$, and $\bar{a}_3$, we can also write:

$$\bar{a}_1 \cdot \bar{a}_2 = \bar{a}_2 \cdot \bar{a}_3 = \bar{a}_3 \cdot \bar{a}_1 = 0. \tag{6.5.5}$$

Differentiating Eqs. (6.5.5) with respect to y_3, y_1, and y_2, respectively, we get:

$$\frac{\partial \bar{a}_1}{\partial y_3} \cdot \bar{a}_2 + \bar{a}_1 \cdot \frac{\partial \bar{a}_2}{\partial y_3} = 0 \tag{6.5.6}$$

$$\frac{\partial \bar{a}_2}{\partial y_1} \cdot \bar{a}_3 + \bar{a}_2 \cdot \frac{\partial \bar{a}_3}{\partial y_1} = 0 \tag{6.5.7}$$

$$\frac{\partial \bar{a}_3}{\partial y_2} \cdot \bar{a}_1 + \bar{a}_3 \cdot \frac{\partial \bar{a}_1}{\partial y_2} = 0. \tag{6.5.8}$$

Subtracting Eq. (6.5.8) from (6.5.7), we get:

$$\bar{a}_3 \cdot \left(\frac{\partial \bar{a}_2}{\partial y_1} - \frac{\partial \bar{a}_1}{\partial y_2} \right) + \bar{a}_2 \cdot \frac{\partial \bar{a}_3}{\partial y_1} - \frac{\partial \bar{a}_3}{\partial y_2} \cdot \bar{a}_1 = 0. \qquad (6.5.9)$$

But, from Eq. (6.3.7):

$$\frac{\partial \bar{a}_i}{\partial y_j} = \frac{\partial}{\partial y_j}\left(\frac{\partial \bar{r}}{\partial y_i} \right) = \frac{\partial}{\partial y_i}\left(\frac{\partial \bar{r}}{\partial y_j} \right) = \frac{\partial \bar{a}_j}{\partial y_i}; \qquad (6.5.10)$$

therefore,

$$\frac{\partial \bar{a}_2}{\partial y_1} = \frac{\partial \bar{a}_1}{\partial y_2}, \qquad (6.5.11)$$

and Eq. (6.5.9) becomes

$$\bar{a}_2 \cdot \frac{\partial \bar{a}_3}{\partial y_1} - \frac{\partial \bar{a}_3}{\partial y_2} \cdot \bar{a}_1 = 0. \qquad (6.5.12)$$

Comparing Eqs. (6.5.12) and (6.5.6), and taking Eq. (6.5.10) into account, we conclude that

$$a_1 \cdot \frac{\partial \bar{a}_2}{\partial y_3} = \bar{a}_2 \cdot \frac{\partial \bar{a}_3}{\partial y_1} = \bar{a}_3 \cdot \frac{\partial \bar{a}_1}{\partial y_2} = 0. \qquad (6.5.13)$$

Now, by differentiating,

$$\bar{a}_1 \cdot \bar{a}_1 = h_1^2, \qquad (6.5.14)$$

with respect to y_1, y_2, and y_3, and using Eq. (6.5.10), we get:

$$\bar{a}_1 \cdot \frac{\partial \bar{a}_1}{\partial y_1} = h_1 \frac{\partial h_1}{\partial y_1} \qquad (6.5.15)$$

$$\bar{a}_1 \cdot \frac{\partial \bar{a}_1}{\partial y_2} = \bar{a}_1 \cdot \frac{\partial \bar{a}_2}{\partial y_1} = h_1 \frac{\partial h_1}{\partial y_2} \qquad (6.5.16)$$

$$\bar{a}_1 \cdot \frac{\partial \bar{a}_1}{\partial y_3} = \bar{a}_1 \cdot \frac{\partial \bar{a}_3}{\partial y_1} = h_1 \frac{\partial h_1}{\partial y_3}. \qquad (6.5.17)$$

Differentiating Eq. (6.5.5) with respect to y_1, and using Eqs. (6.5.16) and (6.5.17), we get:

$$\bar{a}_2 \cdot \frac{\partial \bar{a}_1}{\partial y_1} = -\bar{a}_1 \cdot \frac{\partial \bar{a}_2}{\partial y_1} = h_1 \frac{\partial h_1}{\partial y_2} \qquad (6.5.18)$$

$$\bar{a}_3 \cdot \frac{\partial \bar{a}_1}{\partial y_1} = -\bar{a}_1 \cdot \frac{\partial \bar{a}_3}{\partial y_1} = -h_1 \frac{\partial h_1}{\partial y_3} \tag{6.5.19}$$

Two sets of equations similar to Eqs. (6.5.18) and (6.5.19) can be derived by differentiating $\bar{a}_2 \cdot \bar{a}_2$ and $\bar{a}_3 \cdot \bar{a}_3$. These relations can, however, be directly written down by cyclical permutation of Eqs. (6.5.18) and (6.5.19):

$$\bar{a}_3 \cdot \frac{\partial \bar{a}_2}{\partial y_2} = -\bar{a}_2 \cdot \frac{\partial \bar{a}_3}{\partial y_2} = -h_2 \frac{\partial h_2}{\partial y_3} \tag{6.5.20}$$

$$\bar{a}_1 \cdot \frac{\partial \bar{a}_2}{\partial y_2} = -\bar{a}_2 \cdot \frac{\partial \bar{a}_1}{\partial y_2} = -h_2 \frac{\partial h_2}{\partial y_1}, \tag{6.5.21}$$

and

$$\bar{a}_1 \cdot \frac{\partial \bar{a}_3}{\partial y_3} = -\bar{a}_3 \cdot \frac{\partial \bar{a}_1}{\partial y_3} = -h_3 \frac{\partial h_3}{\partial y_1} \tag{6.5.22}$$

$$\bar{a}_2 \cdot \frac{\partial \bar{a}_3}{\partial y_3} = -\bar{a}_3 \cdot \frac{\partial \bar{a}_2}{\partial y_3} = -h_3 \frac{\partial h_3}{\partial y_2}. \tag{6.5.23}$$

Having established the previous relations, the derivatives of $\bar{a}_i$ can now be obtained.

Consider the vector $\partial \bar{a}_1 / \partial y_1$; its three components along $\bar{e}_1$, $\bar{e}_2$, and $\bar{e}_3$ are

$$\frac{\partial \bar{a}_1}{\partial y_1} \cdot \bar{e}_1, \quad \frac{\partial \bar{a}_1}{\partial y_1} \cdot \bar{e}_2, \quad \frac{\partial \bar{a}_1}{\partial y_1} \cdot \bar{e}_3 \tag{6.5.24}$$

or

$$\frac{\partial \bar{a}_1}{\partial y_1} \cdot \frac{\bar{a}_1}{h_1}, \quad \frac{\partial \bar{a}_1}{\partial y_1} \cdot \frac{\bar{a}_2}{h_2}, \quad \frac{\partial \bar{a}_1}{\partial y_1} \cdot \frac{\bar{a}_3}{h_3}, \tag{6.5.25}$$

which by virtue of Eqs. (6.5.15), (6.5.18), and (6.5.19) are equal to

$$\frac{\partial h_1}{\partial y_1}, \quad -\frac{h_1}{h_2} \frac{\partial h_1}{\partial y_2}, \quad -\frac{h_1}{h_3} \frac{\partial h_1}{\partial y_3}. \tag{6.5.26}$$

Hence, we may write:

$$\frac{\partial \bar{a}_1}{\partial y_1} = \frac{\partial h_1}{\partial y_1} \frac{\bar{a}_1}{h_1} - \frac{h_1}{h_2^2} \frac{\partial h_1}{\partial y_2} \bar{a}_2 - \frac{h_1}{h_3^2} \frac{\partial h_1}{\partial y_3} \bar{a}_3. \tag{6.5.27}$$

Repeating the previous steps, two equations similar to Eq. (6.5.27) can be derived: They can, however, be directly written down by cyclical permutation of Eq. (6.5.27):

$$\frac{\partial \bar{a}_2}{\partial y_2} = \frac{\partial h_2}{\partial y_2}\frac{\bar{a}_2}{h_2} - \frac{h_2}{h_3^2}\frac{\partial h_2}{\partial y_3}\bar{a}_3 - \frac{h_2}{h_1^2}\frac{\partial h_2}{\partial y_1}\bar{a}_1 \tag{6.5.28}$$

$$\frac{\partial \bar{a}_3}{\partial y_3} = \frac{\partial h_3}{\partial y_3}\frac{\bar{a}_3}{h_3} - \frac{h_3}{h_1^2}\frac{\partial h_3}{\partial y_1}\bar{a}_1 - \frac{h_3}{h_2^2}\frac{\partial h_3}{\partial y_2}\bar{a}_2. \tag{6.5.29}$$

In the same way, $\partial \bar{a}_2/\partial y_3$ can be resolved into the three components:

$$\frac{\partial \bar{a}_2}{\partial y_3}\cdot\bar{e}_1, \quad \frac{\partial \bar{a}_2}{\partial y_3}\cdot\bar{e}_2, \quad \frac{\partial \bar{a}_2}{\partial y_3}\cdot\bar{e}_3, \tag{6.5.30}$$

which, because of Eqs. (6.5.13), (6.5.20), and (6.5.23), become:

$$0, \quad \frac{\partial h_2}{\partial y_3}, \quad \frac{\partial h_3}{\partial y_2}. \tag{6.5.31}$$

Hence,

$$\frac{\partial \bar{a}_2}{\partial y_3} = \frac{\partial h_2}{\partial y_3}\frac{\bar{a}_2}{h_2} + \frac{\partial h_3}{\partial y_2}\frac{\bar{a}_3}{h_3}. \tag{6.5.32}$$

By cyclical permutation of Eq. (6.5.32), we obtain the two other relations:

$$\frac{\partial \bar{a}_3}{\partial y_1} = \frac{\partial h_3}{\partial y_1}\frac{\bar{a}_3}{h_3} + \frac{\partial h_1}{\partial y_3}\frac{\bar{a}_1}{h_1} \tag{6.5.33}$$

$$\frac{\partial \bar{a}_1}{\partial y_2} = \frac{\partial h_1}{\partial y_2}\frac{\bar{a}_1}{h_1} + \frac{\partial h_2}{\partial y_1}\frac{\bar{a}_2}{h_2}. \tag{6.5.34}$$

The derivatives of $\bar{e}_i$ can easily be deduced from the derivatives of $\bar{a}_i$. Thus,

$$\frac{\partial \bar{e}_1}{\partial y_1} = \frac{\partial}{\partial y_1}\left(\frac{\bar{a}_1}{h_1}\right) = \frac{1}{h_1^2}\left[h_1\frac{\partial \bar{a}_1}{\partial y_1} - \bar{a}_1\frac{\partial h_1}{\partial y_1}\right] = -\frac{\partial h_1}{\partial y_2}\frac{\bar{e}_2}{h_2} - \frac{\partial h_1}{\partial y_3}\frac{\bar{e}_3}{h_3} \tag{6.5.35}$$

$$\frac{\partial \bar{e}_1}{\partial y_2} = \frac{\partial h_2}{\partial y_1}\frac{\bar{e}_2}{h_1}, \quad \frac{\partial \bar{e}_1}{\partial y_3} = \frac{\partial h_3}{\partial y_1}\frac{\bar{e}_3}{h_1}, \tag{6.5.36}$$

$$\frac{\partial \bar{e}_2}{\partial y_1} = \frac{\partial h_1}{\partial y_2}\frac{\bar{e}_1}{h_2}, \quad \frac{\partial \bar{e}_2}{\partial y_2} = -\frac{\partial h_2}{\partial y_3}\frac{\bar{e}_3}{h_3} - \frac{\partial h_2}{\partial y_1}\frac{\bar{e}_1}{h_1}, \quad \frac{\partial \bar{e}_2}{\partial y_3} = \frac{\partial h_3}{\partial y_2}\frac{\bar{e}_3}{h_2} \tag{6.5.37}$$

$$\frac{\partial \bar{e}_3}{\partial y_1} = \frac{\partial h_1}{\partial y_3}\frac{\bar{e}_1}{h_3}, \quad \frac{\partial \bar{e}_3}{\partial y_2} = \frac{\partial h_2}{\partial y_3}\frac{\bar{e}_2}{h_3}, \quad \frac{\partial \bar{e}_3}{\partial y_3} = -\frac{\partial h_3}{\partial y_1}\frac{\bar{e}_1}{h_1} - \frac{\partial h_3}{\partial y_2}\frac{\bar{e}_2}{h_2} \qquad (6.5.38)$$

6.6 The Strain Tensor in Orthogonal Curvilinear Coordinates

The problem formulated in Sec. 1.1, and studied in Chapter 4 in cartesian coordinates, will now be analyzed using curvilinear coordinates. In Fig. 6.9, consider the point-to-point transformation of the body B to B^*. Points M and N, which are infintesimally near one

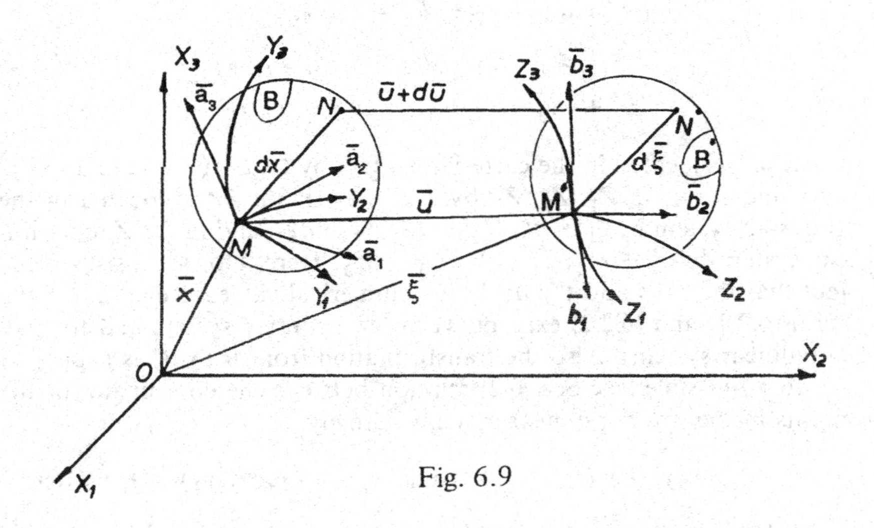

Fig. 6.9

another, are transformed to M^* and N^*, respectively. The vector $d\bar{x}$ of length ds is transformed to the vector $d\bar{\xi}$ of length ds^* . In addition to being located in the cartesian system OX_1, OX_2, OX_3 by x_1, $x_2 x_3$, point M is also located in a curvilinear system Y_1, Y_2, Y_3 by y_1, y_2, y_3. Point N is located by $x_1 + dx_1$, $x_2 + dx_2$, $x_3 + dx_3$ in the cartesian system, and by $y_1 + dy_1, y_2 + dy_2, y_3 + dy_3$ in the curvilinear system. It is important to keep in mind that, when we say that the coordinates of M are y_1, y_2, y_3, this means that three numerical values are allocated to M and that when these values are introduced in Eqs. (6.2.3) to (6.2.5) we obtain three surfaces which can be drawn in the cartesian system and whose intersection gives M. If these numerical values are c_1, c_2, c_3, the three surfaces are:

$$y_1 = c_1 = y_1(x_1, x_2, x_3), \quad y_2 = c_2 = y_2(x_1, x_3, x_3),$$
$$y_3 = c_3 = y_3(x_1, x_2, x_3). \tag{6.6.1}$$

Also, when we say that the coordinates of N are $y_1 + dy_1$, $y_2 + dy_2$, $y_3 + dy_3$, this means that three numerical values $c_1 + dc_1$, $c_2 + dc_2$, $c_3 + dc_3$, slightly larger than c_1, c_2, c_3, are allocated to N, and that when these values are introduced in Eqs. (6.2.3) to (6.2.5) we obtain three surfaces whose intersection gives N at a close distance from M: These surfaces are:

$$y_1 + dy_1 = c_1 + dc_1 = y_1(x_1, x_2 x_3),$$
$$y_2 + dy_2 = c_2 + dc_2 = y_2(x_1, x_2, x_3) \tag{6.6.2}$$
$$y_3 + dy_3 = c_3 + dc_3 = y_3(x_1, x_2, x_3).$$

Point M^* is located in the cartesian system by ξ_1, ξ_2, ξ_3, and in another curvilinear system Z_1, Z_2, Z_3 by z_1, z_2, z_3. Point N^* is located in the cartesian system by $\xi_1 + d\xi_1$, $\xi_2 + d\xi_2$, $\xi_3 + d\xi_3$, and in the Z_i curvilinear system by $z_1 + dz_1$, $z_2 + dz_2$, $z_3 + dz_3$. Here, too, we insist on the fact that the z_i' s and the dz_i' s are numerical values. Relations of the form (6.2.1) and (6.2.2) exist between the cartesian system and the two curvilinear systems. Also the transformation from B to B^* is expressed by an admissible (see Sec. 6.2) relation between the coordinates of the points in the two curvilinear systems, namely:

$$y_i = y_i(z_1, z_2, z_3) \quad \text{and} \quad z_i = z_i(y_1, y_2, y_3). \tag{6.6.3}$$

At M, in the system Y_i, the base vectors and the metric tensor are $\bar{a}_1$, $\bar{a}_2$, $\bar{a}_3$, and g_{ij}. At M^*, in the system Z_i, the base vectors and the metric tensor are $\bar{b}_1$, $\bar{b}_2$, $\bar{b}_3$, and G_{ij}. The square of the length of MN in the original configuation B is

$$(ds)^2 = g_{ij}\, dy_i\, dy_j, \tag{6.6.4}$$

where g_{ij} is evaluated at M in the system Y_i. The square of the length of M^*N^* in the transformed configuration B^* is

$$(ds^*)^2 = G_{ij}\, dz_i\, dz_j, \tag{6.6.5}$$

where G_{ij} is evaluated at M^* in the system Z_i. As stated in Sec. 1.2, the Langrangian Method will be used; in other words, we shall attempt to express our variables in terms of the conditions prior to transformation. From Eq. (6.6.3), we have:

$$dz_i = \frac{\partial z_i}{\partial y_1} dy_1 + \frac{\partial z_i}{\partial y_2} dy_2 + \frac{\partial z_i}{\partial y_3} dy_3 = \frac{\partial z_i}{\partial y_j} dy_j, \tag{6.6.6}$$

so that Eq. (6.6.5) becomes:

$$(ds^*)^2 = G_{ij} \frac{\partial z_i}{\partial y_k} dy_k \frac{\partial z_j}{\partial y_m} dy_m . \tag{6.6.7}$$

After changing some of the symbols for dummy indices, the change in length of the element MN can be expressed by

$$(ds^*)^2 - (ds)^2 = \left(G_{rt} \frac{\partial z_r}{\partial y_i} \frac{\partial z_t}{\partial y_j} - g_{ij} \right) dy_i \, dy_j . \tag{6.6.8}$$

If we set the bracketed quantity in the right-hand side of Eq. (6.6.8) equal to $2\gamma_{ij}$, then

$$(ds^*)^2 - (ds)^2 = 2\gamma_{ij} \, dy_i \, dy_j . \tag{6.6.9}$$

The quantity $[(ds^*)^2 - (ds)^2]$ is an invariant and $\gamma_{ij} = \gamma_{ji}$ is a symmetric tensor of the second rank called the *strain tensor.*

So far we have used two systems of curvilinear coordinates, one at M and one at M^* connected by Eqs. (6.6.3). Let us now distort the Z_i frame of reference in the transformed configuration of the body in a way such that the coordinates z_1, z_2, z_3 of a point there, have the same numerical values y_1, y_2, y_3 as in the original configuration. For point M^*, for example, we would have $z_1 = y_1 = c_1, z_2 = y_2 = c_2, z_3 = y_3 = c_3$. Also, the differentials dz_i and dy_i would be equal and the partial derivatives $\partial z_r / \partial y_i$ would be equal to δ_{ri}, the Kronecker Delta. When such an operation is performed, all the information regarding the change in distances between adjacent points and the change in angle between the various elements is contained in the change of the metric tensor from g_{ij} to G_{ij} as the body is being transformed. Eq. (6.6.5) now becomes:

$$(ds^*)^2 = G_{ij} \, dy_i \, dy_j \tag{6.6.10}$$

and

$$(ds^*)^2 - (ds)^2 = (G_{ij} - g_{ij})dy_i\,dy_j. \tag{6.6.11}$$

The expression for the strain tensor becomes:

$$2\gamma_{ij} = G_{ij} - g_{ij}. \tag{6.6.12}$$

In the following, we shall consider that the Y_i system in the original configuration is chosen to be orthogonal.

Since $G_{ij} = \bar{b}_i \cdot \bar{b}_j$ and $\bar{g}_{ij} = \bar{a}_i \cdot \bar{a}_j$, we have:

$$2\gamma_{ij} = (|\bar{b}_i|)(|\bar{b}_j|)\cos(\theta_{ij})_b - (|\bar{a}_i|)(|\bar{a}_j|)\cos(\theta_{ij})_a \text{ (no sum)}, \tag{6.6.13}$$

where $(\theta_{ij})_a$ and $(\theta_{ij})_b$ are the angles between the base vectors in the initial and the transformed states. The change in length per unit length of MN is given by

$$E_{MN} = \frac{ds^* - ds}{ds} = \frac{|d\bar{\xi}| - |d\bar{x}|}{|d\bar{x}|}. \tag{6.6.14}$$

Thus,

$$|d\bar{\xi}| = (1 + E_{MN})|d\bar{x}|. \tag{6.6.15}$$

If the element MN is now along a base vector $\bar{a}_1$, for example, its lengths before and after transformation are given by Eqs. (6.6.4) and (6.6.10) in which we set $i = j = 1$. Thus, $ds = \sqrt{g_{11}}\,dy_1$ and $ds^* = \sqrt{G_{11}}\,dy_1$. Consequently, the change in length per unit length of this element is

$$E_{a1} = \frac{(\sqrt{G_{11}} - \sqrt{g_{11}})dy_1}{\sqrt{g_{11}}\,dy_1}. \tag{6.6.16}$$

Therefore,

$$\sqrt{G_{11}} = (1 + E_{a1})\sqrt{g_{11}} \tag{6.6.17}$$

or

$$|\bar{b}_1| = (1 + E_{a1})|\bar{a}_1|. \tag{6.6.18}$$

Similarly,

$$|\bar{b}_2| = (1 + E_{a2})|\bar{a}_2|, \quad |\bar{b}_3| = (1 + E_{a3})|\bar{a}_3|. \tag{6.6.19}$$

Eq. (6.6.13) can now be written as follows:

$$2\gamma_{ij} = (1 + E_{ai})(1 + E_{aj})(|\bar{a}_i|)(|\bar{a}_j|)\cos(\theta_{ij})_b - (|\bar{a}_i|)(|\bar{a}_j|)\cos(\theta_{ij})_a \text{ (no sum).} \tag{6.6.20}$$

Therefore,

$$\frac{2\gamma_{ij}}{\sqrt{g_{ii}}\sqrt{g_{jj}}} = (1 + E_{ai})(1 + E_{aj})\cos(\theta_{ij})_b - \cos(\theta_{ij})_a \text{ (no sum).} \tag{6.6.21}$$

Eq. (6.6.21) will now be used to obtain a physical meaning for the quantities γ_{ij}. In this equation, setting $i = j$ and taking into account the orthogonality of the $\bar{a}_i$ ' s, we get,

$$\frac{2\gamma_{ii}}{g_{ii}} = (1 + E_{ai})^2 - 1 \text{ (no sum)} \tag{6.6.22}$$

or

$$E_{ai} = \sqrt{1 + \frac{2\gamma_{ii}}{g_{ii}}} - 1 \text{ (no sum).} \tag{6.6.23}$$

Eq. (6.6.23) shows that $\gamma_{11}, \gamma_{22}, \gamma_{33}$ characterize the change in length per unit length of those elements through M along the three base vectors $\bar{a}_1, \bar{a}_2, \bar{a}_3$. Let us now set $(\theta_{ij})_b = \pi/2 - \phi_{ij}$, so that ϕ_{ij} represents the change in the right angle between $\bar{a}_i$ and $\bar{a}_j$ after transformation. Eq. (6.6.21) gives:

$$\sin \phi_{ij} = \frac{2\gamma_{ij}}{(1 + E_{ai})\sqrt{g_{ii}}\,(1 + E_{aj})\sqrt{g_{jj}}} = \frac{2\gamma_{ij}}{\sqrt{G_{ii}}\sqrt{G_{jj}}} \text{ (no sum, } i \neq j). \tag{6.6.24}$$

Using Eq. (6.6.22), Eq. (6.6.24) can be written as follows [compare with Eqs. (4.4.5) to (4.4.7)]:

$$\sin \phi_{ij} = \frac{2\gamma_{ij}}{\sqrt{g_{ii} + 2\gamma_{ii}}\sqrt{g_{jj} + 2\gamma_{jj}}} \text{ (no sum, } i \neq j). \tag{6.6.25}$$

Eq. (6.6.25) shows that $\gamma_{12}, \gamma_{13}, \gamma_{23}$ characterize the change in the right angle between elements along the three base vectors (see Sec. 4.4). $\gamma_{11}, \gamma_{22}, \gamma_{33}$ are called the normal components of the strain tensor, and $\gamma_{12}, \gamma_{13}, \gamma_{23}$ are called the shearing components of the strain tensor.

From Secs. 4.3 and 4.4, the strain ε_{MN} in any direction was given by

$$\varepsilon_{MN} = \frac{1}{2}\frac{(ds^*)^2 - (ds)^2}{(ds)^2}. \tag{6.6.26}$$

Using the same definition for curvilinear coordinates, we have from Eq. (6.6.11)

$$\varepsilon_{MN} = \gamma_{ij}\frac{dy_i}{ds}\frac{dy_j}{ds}. \tag{6.6.27}$$

The derivatives dy_1/ds, dy_2/ds, dy_3/ds can be expressed in terms of the direction cosines ℓ_1, ℓ_2, ℓ_3 of the element MN relative to the orthogonal unit base vectors $\bar{e}_1$, $\bar{e}_2$, $\bar{e}_3$. Indeed, since

$$\ell_1 = h_1\frac{dy_1}{ds}, \quad \ell_2 = h_2\frac{dy_2}{ds}, \quad \ell_3 = h_3\frac{dy_3}{ds}, \tag{6.6.28}$$

Eq. (6.6.27) can be written as:

$$\begin{aligned}\varepsilon_{MN} = {} & \gamma_{11}\left(\frac{\ell_1}{h_1}\right)^2 + \gamma_{22}\left(\frac{\ell_2}{h_2}\right)^2 + \gamma_{33}\left(\frac{\ell_3}{h_3}\right)^2 + 2\gamma_{12}\left(\frac{\ell_1\ell_2}{h_1h_2}\right) \\ & + 2\gamma_{13}\left(\frac{\ell_1\ell_3}{h_1h_3}\right) + 2\gamma_{23}\left(\frac{\ell_2\ell_3}{h_2h_3}\right).\end{aligned} \tag{6.6.29}$$

Let us set

$$\begin{aligned} \varepsilon_{11} &= \frac{\gamma_{11}}{h_1^2}, & \varepsilon_{22} &= \frac{\gamma_{22}}{h_2^2}, & \varepsilon_{33} &= \frac{\gamma_{33}}{h_3^2}, \\ \varepsilon_{12} &= \frac{\gamma_{12}}{h_1h_2}, & \varepsilon_{13} &= \frac{\gamma_{13}}{h_1h_3}, & \varepsilon_{23} &= \frac{\gamma_{23}}{h_2h_3}. \end{aligned} \tag{6.6.30}$$

Eq. (6.6.29) can now be written in the same form as Eq. (4.4.8), namely

$$\varepsilon_{MN} = \varepsilon_{11}\ell_1^2 + \varepsilon_{22}\ell_2^2 + \varepsilon_{33}\ell_3^2 + 2\varepsilon_{12}\ell_1\ell_2 + 2\varepsilon_{13}\ell_1\ell_3 + 2\varepsilon_{23}\ell_2\ell_3. \tag{6.6.31}$$

6.7 Strain-Displacement Relations in Orthogonal Curvilinear Coordinates

In Fig. 6.9, the displacement of point M is given by

$$\bar{u} = \bar{\xi} - \bar{x}. \tag{6.7.1}$$

Recalling the distortion of the reference frame at M^* and its consequences, which were discussed in the previous section, we have:

$$\frac{\partial \bar{u}}{\partial y_i} = \frac{\partial \bar{\xi}}{\partial y_i} - \frac{\partial \bar{x}}{\partial y_i} = \bar{b}_i - \bar{a}_i. \tag{6.7.2}$$

The metric tensor G_{ij} of the transformed configuration can be obtained in terms of the displacement vector $\bar{u}$ from the scalar product of $\bar{b}_i$ by $\bar{b}_j$. Thus,

$$\begin{aligned} G_{ij} = \bar{b}_i \cdot \bar{b}_j &= \left(\bar{a}_i + \frac{\partial \bar{u}}{\partial y_i}\right) \cdot \left(\bar{a}_j + \frac{\partial \bar{u}}{\partial y_j}\right) \\ &= g_{ij} + \bar{a}_i \cdot \frac{\partial \bar{u}}{\partial y_j} + a_j \cdot \frac{\partial \bar{u}}{\partial y_i} + \frac{\partial \bar{u}}{\partial y_i} \cdot \frac{\partial \bar{u}}{\partial y_j} \end{aligned} \tag{6.7.3}$$

and

$$2\gamma_{ij} = G_{ij} - g_{ij} = \bar{a}_i \cdot \frac{\partial \bar{u}}{\partial y_j} + \bar{a}_j \cdot \frac{\partial \bar{u}}{\partial y_i} + \frac{\partial \bar{u}}{\partial y_i} \cdot \frac{\partial \bar{u}}{\partial y_j}. \tag{6.7.4}$$

The vector $\bar{u}$ can be written in the orthogonal coordinate system Y_i at M as follows:

$$\bar{u} = u_1 \bar{e}_1 + u_2 \bar{e}_2 + u_3 \bar{e}_3 = u_i \bar{e}_i, \tag{6.7.5}$$

where $\bar{e}_i$ is the unit vector along $\bar{a}_i$. Substituting in Eq. (6.7.4), we obtain:

$$2\gamma_{ij} = \bar{a}_i \cdot \frac{\partial}{\partial y_j}(u_k \bar{e}_k) + \bar{a}_j \cdot \frac{\partial}{\partial y_i}(u_s \bar{e}_s) + \frac{\partial}{\partial y_i}(u_r \bar{e}_r) \cdot \frac{\partial}{\partial y_j}(u_t \bar{e}_t) \tag{6.7.6}$$

or

$$\begin{aligned} 2\gamma_{ij} = \bar{a}_i \cdot &\frac{\partial}{\partial y_j}(u_1 \bar{e}_1 + u_2 \bar{e}_2 + u_3 \bar{e}_3) \\ &+ \bar{a}_j \cdot \frac{\partial}{\partial y_i}(u_1 \bar{e}_1 + u_2 \bar{e}_2 + u_3 \bar{e}_3) \\ &+ \frac{\partial}{\partial y_i}(u_1 \bar{e}_1 + u_2 \bar{e}_2 + u_3 \bar{e}_3) \cdot \frac{\partial}{\partial y_j}(u_1 \bar{e}_1 + u_2 \bar{e}_2 + u_3 \bar{e}_3). \end{aligned} \tag{6.7.7}$$

Using Eqs. (6.5.35) to (6.5.38), we obtain:

$$\gamma_{11} = h_1 \frac{\partial u_1}{\partial y_1} + \frac{h_1 u_2}{h_2} \frac{\partial h_1}{\partial y_2} + \frac{h_1 u_3}{h_3} \frac{\partial h_1}{\partial y_3} + \frac{1}{2}\left(\frac{\partial u_1}{\partial y_1} + \frac{u_2}{h_2} \frac{\partial h_1}{\partial y_2} + \frac{u_3}{h_3} \frac{\partial h_1}{\partial y_3} \right)^2 + \frac{1}{2}\left(\frac{\partial u_2}{\partial y_1} - \frac{u_1}{h_2} \frac{\partial h_1}{\partial y_2} \right)^2 + \frac{1}{2}\left(\frac{\partial u_3}{\partial y_1} - \frac{u_1}{h_3} \frac{\partial h_1}{\partial y_3} \right)^2 \tag{6.7.8}$$

$$\gamma_{22} = h_2 \frac{\partial u_2}{\partial y_2} + \frac{h_2 u_3}{h_3} \frac{\partial h_2}{\partial y_3} + \frac{h_2 u_1}{h_1} \frac{\partial h_2}{\partial y_1} + \frac{1}{2}\left(\frac{\partial u_2}{\partial y_2} + \frac{u_3}{h_3} \frac{\partial h_2}{\partial y_3} + \frac{u_1}{h_1} \frac{\partial h_2}{\partial y_1} \right)^2 + \frac{1}{2}\left(\frac{\partial u_3}{\partial y_2} - \frac{u_2}{h_3} \frac{\partial h_2}{\partial y_3} \right)^2 + \frac{1}{2}\left(\frac{\partial u_1}{\partial y_2} - \frac{u_2}{h_1} \frac{\partial h_2}{\partial y_1} \right)^2 \tag{6.7.9}$$

$$\gamma_{33} = h_3 \frac{\partial u_3}{\partial y_3} + \frac{h_3 u_1}{h_1} \frac{\partial h_3}{\partial y_1} + \frac{h_3 u_2}{h_2} \frac{\partial h_3}{\partial y_2} + \frac{1}{2}\left(\frac{\partial u_3}{\partial y_3} + \frac{u_1}{h_1} \frac{\partial h_3}{\partial y_1} + \frac{u_2}{h_2} \frac{\partial h_3}{\partial y_2} \right)^2 + \frac{1}{2}\left(\frac{\partial u_1}{\partial y_3} - \frac{u_3}{h_1} \frac{\partial h_3}{\partial y_1} \right)^2 + \frac{1}{2}\left(\frac{\partial u_2}{\partial y_3} - \frac{u_3}{h_2} \frac{\partial h_3}{\partial y_2} \right)^2 \tag{6.7.10}$$

$$\begin{aligned} \gamma_{12} = {} & \frac{1}{2}\left(h_1 \frac{\partial u_1}{\partial y_2} + h_2 \frac{\partial u_2}{\partial y_1} - u_2 \frac{\partial h_2}{\partial y_1} - u_1 \frac{\partial h_1}{\partial y_2} \right) \\ & + \frac{1}{2}\left(\frac{\partial u_1}{\partial y_2} - \frac{u_2}{h_1} \frac{\partial h_2}{\partial y_1} \right)\left(\frac{\partial u_1}{\partial y_1} + \frac{u_2}{h_2} \frac{\partial h_1}{\partial y_2} + \frac{u_3}{h_3} \frac{\partial h_1}{\partial y_3} \right) \\ & + \frac{1}{2}\left(\frac{\partial u_2}{\partial y_1} - \frac{u_1}{h_2} \frac{\partial h_1}{\partial y_2} \right)\left(\frac{\partial u_2}{\partial y_2} + \frac{u_1}{h_1} \frac{\partial h_2}{\partial y_1} + \frac{u_3}{h_3} \frac{\partial h_2}{\partial y_3} \right) \\ & + \frac{1}{2}\left(\frac{\partial u_3}{\partial y_1} - \frac{u_1}{h_3} \frac{\partial h_1}{\partial y_3} \right)\left(\frac{\partial u_3}{\partial y_2} - \frac{u_2}{h_3} \frac{\partial h_2}{\partial y_3} \right) \end{aligned} \tag{6.7.11}$$

$$
\begin{aligned}
\gamma_{13} = {} & \frac{1}{2}\left(h_3 \frac{\partial u_3}{\partial y_1} + h_1 \frac{\partial u_1}{\partial y_3} - u_1 \frac{\partial h_1}{\partial y_3} - u_3 \frac{\partial h_3}{\partial y_1}\right) \\
& + \frac{1}{2}\left(\frac{\partial u_1}{\partial y_3} - \frac{u_3}{h_1}\frac{\partial h_3}{\partial y_1}\right)\left(\frac{\partial u_1}{\partial y_1} + \frac{u_3}{h_3}\frac{\partial h_1}{\partial y_3} + \frac{u_2}{h_2}\frac{\partial h_1}{\partial y_2}\right) \\
& + \frac{1}{2}\left(\frac{\partial u_3}{\partial y_1} - \frac{u_1}{h_3}\frac{\partial h_1}{\partial y_3}\right)\left(\frac{\partial u_3}{\partial y_3} + \frac{u_1}{h_1}\frac{\partial h_3}{\partial y_1} + \frac{u_2}{h_2}\frac{\partial h_3}{\partial y_2}\right) \\
& + \frac{1}{2}\left(\frac{\partial u_2}{\partial y_1} - \frac{u_1}{h_2}\frac{\partial h_1}{\partial y_2}\right)\left(\frac{\partial u_2}{\partial y_3} - \frac{u_3}{h_2}\frac{\partial h_3}{\partial y_2}\right)
\end{aligned}
\qquad (6.7.12)
$$

$$
\begin{aligned}
\gamma_{23} = {} & \frac{1}{2}\left(h_3 \frac{\partial u_3}{\partial y_2} + h_2 \frac{\partial u_2}{\partial y_3} - u_2 \frac{\partial h_2}{\partial y_3} - u_3 \frac{\partial h_3}{\partial y_2}\right) \\
& + \frac{1}{2}\left(\frac{\partial u_2}{\partial y_3} - \frac{u_3}{h_2}\frac{\partial h_3}{\partial y_2}\right)\left(\frac{\partial u_2}{\partial y_2} + \frac{u_3}{h_3}\frac{\partial h_2}{\partial y_3} + \frac{u_1}{h_1}\frac{\partial h_2}{\partial y_1}\right) \\
& + \frac{1}{2}\left(\frac{\partial u_3}{\partial y_2} - \frac{u_2}{h_3}\frac{\partial h_2}{\partial y_3}\right)\left(\frac{\partial u_3}{\partial y_3} + \frac{u_2}{h_2}\frac{\partial h_3}{\partial y_2} + \frac{u_1}{h_1}\frac{\partial h_3}{\partial y_1}\right) \\
& + \frac{1}{2}\left(\frac{\partial u_1}{\partial y_2} - \frac{u_2}{h_1}\frac{\partial h_2}{\partial y_1}\right)\left(\frac{\partial u_1}{\partial y_3} - \frac{u_3}{h_1}\frac{\partial h_3}{\partial y_1}\right).
\end{aligned}
\qquad (6.7.13)
$$

Eqs. (6.7.8) to (6.7.13) can be sutstituted in Eq. (6.6.29) to give the value of ε_{MN}. The expressions for ε_{ij} are obtained by dividing the γ_{ij} ' s by $h_i h_j$ as indicated by Eqs. (6.6.30).

Example 1. Cartesian Coordinates

In this case, $h_1 = h_2 = h_3 = 1$ and $\gamma_{ij} = \varepsilon_{ij}$. Eqs. (6.7.8) to (6.7.13) reduce to Eqs. (4.3.3).

Example 2. Cylindrical Coordinates (Fig. 6.2)

In this case, $h_1 = 1$, $h_2 = r$, and $h_3 = 1$. The subscripts r, θ, z are substituted for 1, 2, 3. The displacement vector $\bar{u}$ is written $\bar{u} = u_r \bar{e}_r + u_\theta \bar{e}_\theta + u_z \bar{e}_z$. Eq. (6.6.11) becomes:

$$
\begin{aligned}
\frac{1}{2}[(ds^*)^2 - (ds)^2] = {} & \gamma_{rr}(dr)^2 + \gamma_{\theta\theta}(d\theta)^2 + \gamma_{zz}(dz)^2 + 2\gamma_{r\theta}\, dr\, d\theta \\
& + 2\gamma_{rz}\, dr\, dz + 2\gamma_{\theta z}\, d\theta\, dz.
\end{aligned}
\qquad (6.7.14)
$$

Substituting (6.6.30) into the previous equation, we obtain:

$$\frac{1}{2}[(ds^*)^2 - (ds)^2] = \varepsilon_{rr}(dr)^2 + \varepsilon_{\theta\theta}(r\,d\theta)^2 + \varepsilon_{zz}(dz)^2 + 2\varepsilon_{r\theta}(r\,dr\,d\theta) + 2\varepsilon_{rz}(dr\,dz) + 2\varepsilon_{\theta z}(r\,d\theta\,dz). \tag{6.7.15}$$

Equation (6.6.31) becomes:

$$\varepsilon_{MN} = \varepsilon_{rr}\,\ell_r^2 + \varepsilon_{\theta\theta}\,\ell_\theta^2 + \varepsilon_{zz}\,\ell_z^2 + 2\varepsilon_{r\theta}\,\ell_r\,\ell_\theta + 2\varepsilon_{rz}\,\ell_r\,\ell_z + 2\varepsilon_{\theta z}\,\ell_\theta\,\ell_z. \tag{6.7.16}$$

Using Eqs. (6.6.30), the set of relations (6.7.8) to (6.7.13) becomes:

$$\varepsilon_{rr} = \frac{\gamma_{rr}}{(1)(1)} = \frac{\partial u_r}{\partial r} + \frac{1}{2}\left[\left(\frac{\partial u_r}{\partial r}\right)^2 + \left(\frac{\partial u_\theta}{\partial r}\right)^2 + \left(\frac{\partial u_z}{\partial r}\right)^2\right] \tag{6.7.17}$$

$$\varepsilon_{\theta\theta} = \frac{\gamma_{\theta\theta}}{(r)(r)} = \frac{1}{r}\frac{\partial u_\theta}{\partial \theta} + \frac{u_r}{r} + \frac{1}{2}\left[\left(\frac{1}{r}\frac{\partial u_\theta}{\partial \theta} + \frac{u_r}{r}\right)^2 + \left(\frac{1}{r}\frac{\partial u_r}{\partial \theta} - \frac{u_\theta}{r}\right)^2 + \left(\frac{1}{r}\frac{\partial u_z}{\partial \theta}\right)^2\right] \tag{6.7.18}$$

$$\varepsilon_{zz} = \frac{\gamma_{zz}}{(1)(1)} = \frac{\partial u_z}{\partial z} + \frac{1}{2}\left[\left(\frac{\partial u_r}{\partial z}\right)^2 + \left(\frac{\partial u_\theta}{\partial z}\right)^2 + \left(\frac{\partial u_z}{\partial z}\right)^2\right] \tag{6.7.19}$$

$$\varepsilon_{r\theta} = \frac{\gamma_{r\theta}}{(1)(r)} = \frac{1}{2}\left[\frac{1}{r}\frac{\partial u_r}{\partial \theta} + \frac{\partial u_\theta}{\partial r} - \frac{u_\theta}{r} + \frac{\partial u_r}{\partial r}\left(\frac{1}{r}\frac{\partial u_r}{\partial \theta} - \frac{u_\theta}{r}\right) + \frac{\partial u_\theta}{\partial r}\left(\frac{1}{r}\frac{\partial u_\theta}{\partial \theta} + \frac{u_r}{r}\right) + \frac{\partial u_z}{\partial r}\frac{1}{r}\frac{\partial u_z}{\partial \theta}\right] \tag{6.7.20}$$

$$\varepsilon_{rz} = \frac{\gamma_{rz}}{(1)(1)} = \frac{1}{2}\left[\frac{\partial u_r}{\partial z} + \frac{\partial u_z}{\partial r} + \frac{\partial u_r}{\partial r}\frac{\partial u_r}{\partial z} + \frac{\partial u_\theta}{\partial r}\frac{\partial u_\theta}{\partial z} + \frac{\partial u_z}{\partial r}\frac{\partial u_z}{\partial z}\right] \tag{6.7.21}$$

$$\varepsilon_{\theta z} = \frac{\gamma_{\theta z}}{(r)(1)} = \frac{1}{2}\left[\frac{\partial u_\theta}{\partial z} + \frac{1}{r}\frac{\partial u_z}{\partial \theta} + \frac{\partial u_r}{\partial z}\left(\frac{1}{r}\frac{\partial u_r}{\partial \theta} - \frac{u_\theta}{r}\right) + \frac{\partial u_\theta}{\partial z}\left(\frac{1}{r}\frac{\partial u_\theta}{\partial \theta} + \frac{u_r}{r}\right) + \frac{\partial u_z}{\partial z}\frac{1}{r}\frac{\partial u_z}{\partial \theta}\right]. \tag{6.7.22}$$

If, in Eqs. (6.7.17) to (6.7.22), the second-order terms are neglected, we obtain *the expressions used in the linear theory*:

$$e_{rr} = \frac{\partial u_r}{\partial r}, \quad e_{\theta\theta} = \frac{1}{r}\frac{\partial u_\theta}{\partial \theta} + \frac{u_r}{r}, \quad e_{zz} = \frac{\partial u_z}{\partial z} \tag{6.7.23}$$

$$e_{r\theta} = \frac{1}{2}\left(\frac{1}{r}\frac{\partial u_r}{\partial \theta} + \frac{\partial u_\theta}{\partial r} - \frac{u_\theta}{r}\right), \quad e_{rz} = \frac{1}{2}\left(\frac{\partial u_r}{\partial z} + \frac{\partial u_z}{\partial r}\right),$$
$$e_{\theta z} = \frac{1}{2}\left(\frac{\partial u_\theta}{\partial z} + \frac{1}{r}\frac{\partial u_z}{\partial \theta}\right). \tag{6.7.24}$$

Example 3. Spherical Polar Coordinates (Fig. 6.3)

In this case, $h_1 = \rho$, $h_2 = \rho \sin\phi$, and $h_3 = 1$. The subscripts ϕ, θ, ρ are substituted for 1, 2, 3. The displacement vector $\bar{u}$ is written $\bar{u} = u_\phi \bar{e}_\phi + u_\theta \bar{e}_\theta + u_\rho \bar{e}_\rho$. Eq. (6.6.11) becomes:

$$\frac{1}{2}[(ds^*)^2 - (ds)^2] = \gamma_{\phi\phi}(d\phi)^2 + \gamma_{\theta\theta}(d\theta)^2 + \gamma_{\rho\rho}(d\rho)^2 + 2\gamma_{\phi\theta}\, d\phi\, d\theta + 2\gamma_{\rho\phi}\, d\phi\, d\rho + 2\gamma_{\theta\rho}\, d\theta\, d\rho. \tag{6.7.25}$$

Substituting Eq. (6.6.30) in Eq. (6.7.25), we obtain:

$$\frac{1}{2}[(ds^*)^2 - (ds)^2] = \varepsilon_{\phi\phi}(\rho\, d\phi)^2 + \varepsilon_{\theta\theta}(\rho \sin\phi\, d\theta)^2 + \varepsilon_{\rho\rho}(d\rho)^2 + 2\varepsilon_{\phi\theta}\rho^2 \sin\phi\, d\phi\, d\theta + 2\varepsilon_{\phi\rho}\rho\, d\phi\, d\rho + 2\varepsilon_{\theta\rho}\rho \sin\phi\, d\theta\, d\rho. \tag{6.7.26}$$

Eq. (6.6.31) becomes:

$$\varepsilon_{MN} = \varepsilon_{\phi\phi}\ell_\phi^2 + \varepsilon_{\theta\theta}\ell_\theta^2 + \varepsilon_{\rho\rho}\ell_\rho^2 + 2\varepsilon_{\phi\theta}\ell_\phi\ell_\theta + 2\varepsilon_{\phi\rho}\ell_\phi\ell_\rho + 2\varepsilon_{\theta\rho}\ell_\theta\ell_\rho. \tag{6.7.27}$$

Using Eqs. (6.6.30), the set of relations (6.7.8) to (6.7.13) becomes:

$$\varepsilon_{\phi\phi} = \frac{1}{\rho}\frac{\partial u_\phi}{\partial \phi} + \frac{u_\rho}{\rho} + \frac{1}{2}\left[\left(\frac{1}{\rho}\frac{\partial u_\rho}{\partial \phi} - \frac{u_\phi}{\rho}\right)^2 + \left(\frac{1}{\rho}\frac{\partial u_\phi}{\partial \phi} + \frac{u_\rho}{\rho}\right)^2 + \left(\frac{1}{\rho}\frac{\partial u_\theta}{\partial \phi}\right)^2\right] \tag{6.7.28}$$

$$\varepsilon_{\theta\theta} = \frac{1}{\rho \sin\phi}\frac{\partial u_\theta}{\partial \theta} + \frac{u_\rho}{\rho} + \frac{u_\phi}{\rho}\cot\phi + \frac{1}{\sin^2\phi}\left[\left(\frac{1}{\rho}\frac{\partial u_\rho}{\partial \theta} - \frac{u_\theta}{\rho}\sin\phi\right)^2 + \left(\frac{1}{\rho}\frac{\partial u_\phi}{\partial \theta} - \frac{u_\theta}{\rho}\cos\phi\right)^2 + \left(\frac{1}{\rho}\frac{\partial u_\theta}{\partial \theta} + \frac{u_\rho}{\rho}\sin\phi + \frac{u_\phi}{\rho}\cos\phi\right)^2\right] \tag{6.7.29}$$

$$\varepsilon_{\rho\rho} = \frac{\partial u_\rho}{\partial \rho} + \frac{1}{2}\left[\left(\frac{\partial u_\rho}{\partial \rho}\right)^2 + \left(\frac{\partial u_\phi}{\partial \rho}\right)^2 + \left(\frac{\partial u_\theta}{\partial \rho}\right)^2\right] \tag{6.7.30}$$

$$\begin{aligned}\varepsilon_{\phi\theta} = \frac{1}{2}\Bigg\{ & \frac{1}{\rho \sin\phi}\frac{\partial u_\phi}{\partial \theta} - \cot\phi \frac{u_\theta}{\rho} + \frac{1}{\rho}\frac{\partial u_\theta}{\partial \phi} \\ & + \frac{1}{\sin\phi}\left[\left(\frac{1}{\rho}\frac{\partial u_\rho}{\partial \phi} - \frac{u_\phi}{\rho}\right)\left(\frac{1}{\rho}\frac{\partial u_\rho}{\partial \theta} - \sin\phi\frac{u_\theta}{\rho}\right)\right. \\ & + \left(\frac{1}{\rho}\frac{\partial u_\phi}{\partial \phi} + \frac{u_\rho}{\rho}\right)\left(\frac{1}{\rho}\frac{\partial u_\phi}{\partial \theta} - \cot\phi\frac{u_\theta}{\rho}\right) \\ & \left. + \left(\frac{1}{\rho}\frac{\partial u_\theta}{\partial \phi}\right)\left(\frac{1}{\rho}\frac{\partial u_\theta}{\partial \theta} + \sin\phi\frac{u_\rho}{\rho} + \cos\phi\frac{u_\phi}{\rho}\right)\right]\Bigg\}\end{aligned} \tag{6.7.31}$$

$$\begin{aligned}\varepsilon_{\phi\rho} = \frac{1}{2}\Bigg\{ & \frac{1}{\rho}\frac{\partial u_\rho}{\partial \phi} - \frac{u_\phi}{\rho} + \frac{\partial u_\phi}{\partial \rho} + \left[\frac{\partial u_\rho}{\partial \rho}\left(\frac{1}{\rho}\frac{\partial u_\rho}{\partial \phi} - \frac{u_\phi}{\rho}\right)\right. \\ & \left. + \frac{\partial u_\phi}{\partial \rho}\left(\frac{1}{\rho}\frac{\partial u_\phi}{\partial \phi} + \frac{u_\rho}{\rho}\right) + \frac{\partial u_\theta}{\partial \rho}\frac{1}{\rho}\frac{\partial u_\theta}{\partial \phi}\right]\Bigg\}\end{aligned} \tag{6.7.32}$$

$$\begin{aligned}\varepsilon_{\theta\rho} = \frac{1}{2}\Bigg\{ & \frac{\partial u_\theta}{\partial \rho} + \frac{1}{\rho\sin\phi}\frac{\partial u_\rho}{\partial \theta} - \frac{u_\theta}{\rho} \\ & + \frac{1}{\sin\phi}\left[\left(\frac{\partial u_\rho}{\partial \rho}\right)\left(\frac{1}{\rho}\frac{\partial u_\rho}{\partial \theta} - \frac{u_\theta}{\rho}\sin\phi\right)\right. \\ & + \left(\frac{\partial u_\phi}{\partial \rho}\right)\left(\frac{1}{\rho}\frac{\partial u_\phi}{\partial \theta} - \frac{u_\theta}{\rho}\cos\phi\right) \\ & \left. + \left(\frac{\partial u_\theta}{\partial \rho}\right)\left(\frac{1}{\rho}\frac{\partial u_\theta}{\partial \theta} + \frac{u_\rho}{\rho}\sin\phi + \frac{u_\phi}{\rho}\cos\phi\right)\right]\Bigg\}.\end{aligned} \tag{6.7.33}$$

If, in Eqs. (6.7.28) to (6.7.33), the second-order terms are neglected, we obtain *the expressions used in the linear theory*:

$$\begin{aligned} & e_{\phi\phi} = \frac{1}{\rho}\frac{\partial u_\phi}{\partial \phi} + \frac{u_\rho}{\rho}, \quad e_{\theta\theta} = \frac{1}{\rho\sin\phi}\frac{\partial u_\theta}{\partial \theta} + \frac{u_\rho}{\rho} + \frac{u_\phi}{\rho}\cot\phi, \\ & e_{\rho\rho} = \frac{\partial u_\rho}{\partial \rho}\end{aligned} \tag{6.7.34}$$

$$e_{\phi\theta} = \frac{1}{2}\left(\frac{1}{\rho \sin \phi}\frac{\partial u_\phi}{\partial \theta} - \frac{u_\theta}{\rho}\cot \phi + \frac{1}{\rho}\frac{\partial u_\theta}{\partial \phi}\right),$$

$$e_{\phi\rho} = \frac{1}{2}\left(\frac{1}{\rho}\frac{\partial u_\rho}{\partial \phi} - \frac{u_\phi}{\rho} + \frac{\partial u_\phi}{\partial \rho}\right), \tag{6.7.35}$$

$$e_{\theta\rho} = \frac{1}{2}\left(\frac{\partial u_\theta}{\partial \rho} + \frac{1}{\rho \sin \phi}\frac{\partial u_\rho}{\partial \theta} - \frac{u_\theta}{\rho}\right).$$

6.8 Components of the Rotation in Orthogonal Curvilinear Coordinates

From the definition given in Eq. (1.2.1), we notice that the three components of the rotation ω_{32}, ω_{13}, and ω_{21} are nothing but one-half of the components of the curl of the displacement vector $\bar{u}$. If (u_1, u_2, u_3) are the components of $\bar{u}$ in any orthogonal curvilinear system of coordinates, then, from Eq. (6.4.13), we have:

$$\frac{1}{2}\,\text{Curl}\ \bar{u} = \frac{1}{2h_1 h_2 h_3}\begin{vmatrix} h_1 \bar{e}_1 & h_2 \bar{e}_2 & h_3 \bar{e}_3 \\ \dfrac{\partial}{\partial y_1} & \dfrac{\partial}{\partial y_2} & \dfrac{\partial}{\partial y_3} \\ h_1 u_1 & h_2 u_2 & h_3 u_3 \end{vmatrix}. \tag{6.8.1}$$

For a *cylindrical system of coordinates*:

$$\omega_r = \frac{1}{2}\left(\frac{1}{r}\frac{\partial u_z}{\partial \theta} - \frac{\partial u_\theta}{\partial z}\right), \quad \omega_\theta = \frac{1}{2}\left(\frac{\partial u_r}{\partial z} - \frac{\partial u_z}{\partial r}\right),$$
$$\omega_z = \frac{1}{2}\left[\frac{1}{r}\frac{\partial}{\partial r}(r u_\theta) - \frac{1}{r}\frac{\partial u_r}{\partial \theta}\right]. \tag{6.8.2}$$

For a *spherical system of coordinates*:

$$\omega_\phi = \frac{1}{2\rho}\left[\frac{1}{\sin \phi}\frac{\partial u_\rho}{\partial \theta} - \frac{\partial}{\partial \rho}(\rho u_\theta)\right], \quad \omega_\theta = \frac{1}{2\rho}\left[\frac{\partial}{\partial \rho}(\rho u_\phi) - \frac{\partial u_\rho}{\partial \phi}\right],$$
$$\omega_\rho = \frac{1}{2\rho \sin \phi}\left[\frac{\partial}{\partial \phi}(u_\theta \sin \phi) - \frac{\partial u_\phi}{\partial \theta}\right]. \tag{6.8.3}$$

6.9 Equations of Compatibility for Linear Strains in Orthogonal Curvilinear Coordinates

The classical method of obtaining the compatibility relations in curvilinear coordinates involves the use of the Reimann tensor about which nothing has been said in this chapter. The introduction of this tensor and the discussion of its properties fall outside the scope of this text. However, one can still obtain the compatibility relations in any system of orthogonal curvilinear coordinates first by writing the six expressions of the strains in terms of the displacements, then by eliminating from these expressions the three components of the displacements. For example, if we eliminate u_r, u_θ and u_z from Eqs. (6.7.24), we obtain the *compatibility relations in cylindrical coordinates.* These relations are written as follows:

$$\frac{\partial^2 e_{\theta\theta}}{\partial r^2} + \frac{1}{r^2}\frac{\partial^2 e_{rr}}{\partial \theta^2} + \frac{2}{r}\frac{\partial e_{\theta\theta}}{\partial r} - \frac{1}{r}\frac{\partial e_{rr}}{\partial r} = 2\left(\frac{1}{r}\frac{\partial^2 e_{r\theta}}{\partial r \partial \theta} + \frac{1}{r^2}\frac{\partial e_{r\theta}}{\partial \theta}\right) \tag{6.9.1}$$

$$\frac{\partial^2 e_{\theta\theta}}{\partial z^2} + \frac{1}{r^2}\frac{\partial^2 e_{zz}}{\partial \theta^2} + \frac{1}{r}\frac{\partial e_{zz}}{\partial r} = 2\left(\frac{1}{r}\frac{\partial^2 e_{\theta z}}{\partial z \partial \theta} + \frac{1}{r}\frac{\partial e_{zr}}{\partial z}\right) \tag{6.9.2}$$

$$\frac{\partial^2 e_{zz}}{\partial r^2} + \frac{\partial^2 e_{rr}}{\partial z^2} = 2\frac{\partial^2 e_{rz}}{\partial z \partial r} \tag{6.9.3}$$

$$\frac{1}{r}\frac{\partial^2 e_{zz}}{\partial r \partial \theta} - \frac{1}{r^2}\frac{\partial e_{zz}}{\partial \theta} = \frac{\partial}{\partial z}\left(\frac{1}{r}\frac{\partial e_{zr}}{\partial \theta} + \frac{\partial e_{\theta z}}{\partial r} - \frac{\partial e_{r\theta}}{\partial z}\right) - \frac{\partial}{\partial z}\left(\frac{e_{\theta z}}{r}\right) \tag{6.9.4}$$

$$\frac{1}{r}\frac{\partial^2 e_{rr}}{\partial \theta \partial z} = \frac{\partial}{\partial r}\left(\frac{1}{r}\frac{\partial e_{zr}}{\partial \theta} - \frac{\partial e_{\theta z}}{\partial r} + \frac{\partial e_{r\theta}}{\partial z}\right) - \frac{\partial}{\partial r}\left(\frac{e_{\theta z}}{r}\right) + \frac{2}{r}\frac{\partial e_{r\theta}}{\partial z} \tag{6.9.5}$$

$$\frac{\partial^2 e_{\theta\theta}}{\partial r \partial z} - \frac{1}{r}\frac{\partial e_{rr}}{\partial z} + \frac{1}{r}\frac{\partial e_{\theta\theta}}{\partial z} = \frac{1}{r}\frac{\partial}{\partial \theta}\left(-\frac{1}{r}\frac{\partial e_{zr}}{\partial \theta} + \frac{\partial e_{\theta z}}{\partial r} + \frac{\partial e_{r\theta}}{\partial z}\right) + \frac{1}{r}\frac{\partial}{\partial \theta}\left(\frac{e_{\theta z}}{r}\right). \tag{6.9.6}$$

PROBLEMS

1. Elliptical cylindrical coordinates may be defined by

$$x_1 = a\cosh y_1 \cos y_2, \quad x_2 = a \sinh y_1 \sin y_2, \quad x_3 = y_3,$$

where $y_1 \geq 0$ and $0 \leq y_2 < 2\Pi$.

(a) Show that this system of coordinates is orthogonal.

(b) Show that, in the OX_1, OX_2 plane, a curve y_1 = constant is an ellipse with semi-axes $(a \cosh y_1)$ in the OX_1 direction, and $(a \sinh y_1)$ in the OX_2 direction.

(c) Show that a curve y_2 = constant is half of one branch of an hyperbola with semi-axes $(a \cos y_2)$ and $(a \sin y_2)$.

(d) Using the metric coefficients appropriate to this system of coordinates, obtain the expressions of the gradient, the divergence, the curl, and the Laplacian.

(e) Write down the strain-displacement relations.

2. Parabolic cylindrical coordinates may be defined by

$$x_1 = \frac{1}{2}(y_1^2 - y_2^2), \quad x_2 = y_1 y_2, \quad x_3 = y_3,$$

where $-\infty < y_1 < \infty$ and $y_2 \geq 0$.

(a) Show that this system is orthogonal.

(b) Show that in the OX_1, OX_2 plane a curve y_2 = constant is a parabola symmetrical with respect to the OX_1 axis and opening to the right, while a curve y_1 = constant is one-half of a similar parabola opening to the left.

(c) Using the metric coefficients appropriate to this system of coordinates, obtain the expressions of the gradient, the divergence, the curl, and the Laplacian.

(d) Write down the strain-displacement relations.

PART II

THEORY OF STRESS

CHAPTER 7

ANALYSIS OF STRESS

7.1 Introduction

When a body is subjected to external forces, its behavior depends upon the magnitude of the forces, upon their direction, and upon the inherent strength of the material of which it is made. Structural and mechanical construction units are usually subjected to various combinations of forces, some having more detrimental effects than others. It is therefore necessary to consider how forces are transmitted through the material constituting these units.

In this chapter, the concepts of stress vector on a surface and state of stress at a point will be introduced. It will be shown that the components of the stress vector can be obtained through a linear symmetric transformation with a matrix whose elements are the components of a tensor of rank two called the stress tensor. All the properties of linear symmetric transformations will be applied to stress the same way they have been applied to linear strain. However, while the components of the linear strain tensor have to satisfy six compatibility relations of the second order, it will be shown that the components of the stress tensor must satisfy three partial differential equations of the first order, called the differential equations of equilibrium. These equations will be derived in both cartesian and orthogonal curvilinear coordinate systems.

7.2 Stress on a Plane at a Point. Notation and Sign Convention

Let us consider a body in equilibrium under a system of external forces $\overline{Q}_1 \ldots \overline{Q}_n$, and let us pass a fictitious plane P through a point O

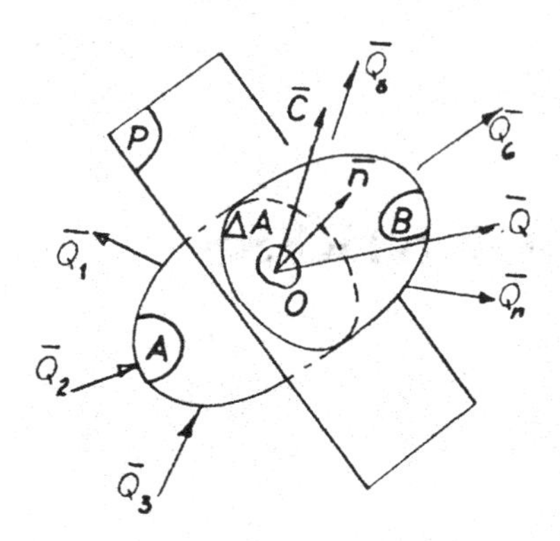

Fig. 7.1

in the interior of this body (Fig. 7.1). Part A of the body is in equilibrium under $\overline{Q}_1$, $\overline{Q}_2$, $\overline{Q}_3$, and the effect of part B. We shall assume this effect is continuously distributed over the surface of intersection. Around the point O, let us consider a small surface ΔA and an outward unit normal vector $\overline{n}$. The effect of B on this small surface can be reduced to a force $\overline{Q}$ and a couple $\overline{C}$. Now let ΔA shrink in size toward zero in a manner such that point O always remains inside and $\overline{n}$ remains the normal vector. It will be assumed that $\overline{Q}/\Delta A$ tends to a definite limit $\overline{\sigma}$ and that $\overline{C}/\Delta A$ tends to zero as ΔA tends to zero. Thus,

$$\lim_{\Delta A \to 0} \left(\frac{\overline{Q}}{\Delta A} \right) = \overline{\sigma} \tag{7.2.1}$$

$$\lim_{\Delta A \to 0} \left(\frac{\overline{C}}{\Delta A} \right) = 0. \tag{7.2.2}$$

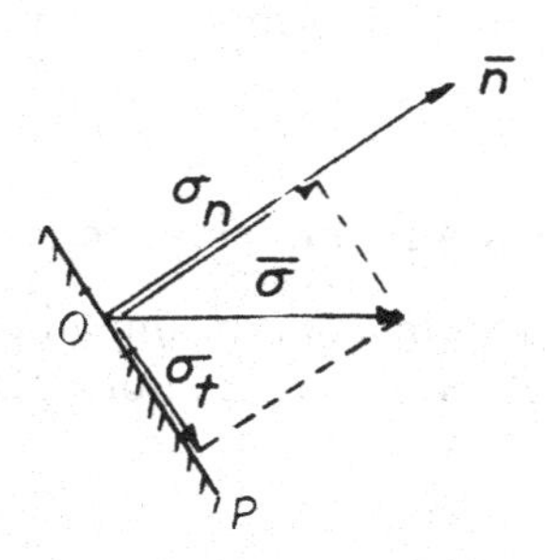

Fig. 7.2

The vector $\bar{\sigma}$ is called *the stress vector on P at O*. The projection of $\bar{\sigma}$ on the normal $\bar{n}$ is called the normal stress σ_n (Fig. 7.2). The projection of $\bar{\sigma}$ on the plane P, in the plane of $\bar{n}$ and $\bar{\sigma}$, is called the tangential or shearing stress σ_t. Therefore,

$$(|\bar{\sigma}|)^2 = \sigma_n^2 + \sigma_t^2 . \tag{7.2.3}$$

σ_t can, in turn, be projected on two orthogonal directions in the plane P. A stress in the direction of the outward normal is considered positive and is called a tensile stress. A stress in the opposite direction is considered negative and is called a compressive stress.

Forces like $\bar{Q}_1, \bar{Q}_2, \ldots \bar{Q}_n$ acting over the surface of a body are called surface forces. Loads applied to a body are never idealized point forces; they are, in reality, forces per unit area applied over some finite area. These external forces per unit area are called *tractions*. Forces distributed across the volume of a body such as gravitational forces and magnetic forces are called *body forces*.

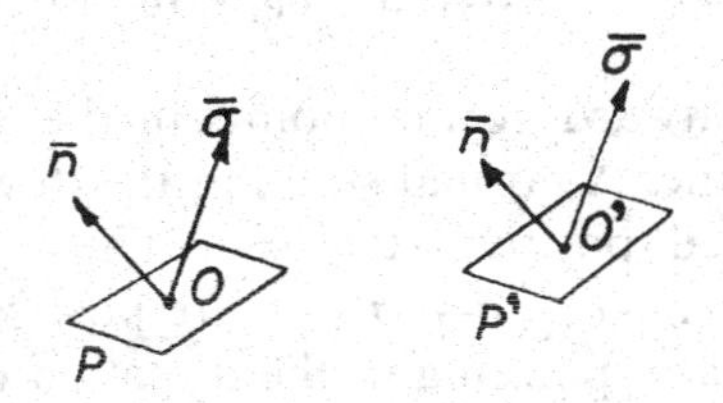

Fig. 7.3

If P and P' are any two parallel planes through any two points O and O' of a continuous body, and if the stress on P at O is equal to the stress on P' at O', the state of stress in the body is said to be a *homogeneous state of stress* (Fig. 7.3).

In a trirectangular system of coordinates (Fig. 7.4a), the stress component normal to a plane that is perpendicular to the OX_1 axis is denoted by σ_{11}. The first subscript indicates the direction of the axis perpendicular to the plane in question, and the second indicates the direction of the stress. On the same plane, σ_{12} indicates the tangential stress in the direction of the OX_2 axis, and σ_{13} the tangential stress in the direction of the OX_3 axis. The same convention applies to two planes perpendicular to the OX_2 and the OX_3 axes, respectively (Fig.

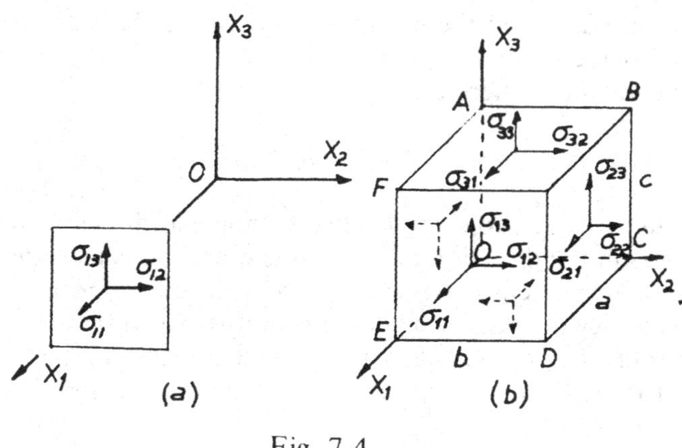

Fig. 7.4

7.4b): σ_{22} is parallel to OX_2, σ_{21} is parallel to OX_1, and σ_{23} is parallel to OX_3; also σ_{33} is parallel to OX_3, σ_{31} is parallel to OX_1, and σ_{32} is parallel to OX_2.

A plane whose outward normal points in the direction of a positive axis is a positive plane. A normal stress in the direction of this outward normal is considered positive (tension). Thus, in Fig. 7.4b, all the normal stresses shown—σ_{11}, σ_{22}, σ_{33} — are positive. A stress tangential to a positive plane, and pointing in the direction of a positive axis, is a positive tangential stress. On the other hand, a stress tangential to a negative plane, pointing in the direction of a positive axis, is a negative tangential stress. In Fig. 7.4b, all the tangential stresses shown are positive. The sign convention previously defined will be adhered to throughout this text.

7.3 State of Stress at a Point. The Stress Tensor

In Sec. 7.2, we have seen that, on a plane P (Fig. 7.1) passing through O, there acts a stress vector defined by Eq. (7.2.1). On another plane through O a different stress vector will act. We shall prove that the stress vector on any plane through O can be obtained once the stress vectors on three planes normal to the coordinate axes and passing through O are known (Fig. 7.5). According to the convention established in Sec. 7.2, the stress vector on the plane P_1 has three compo-

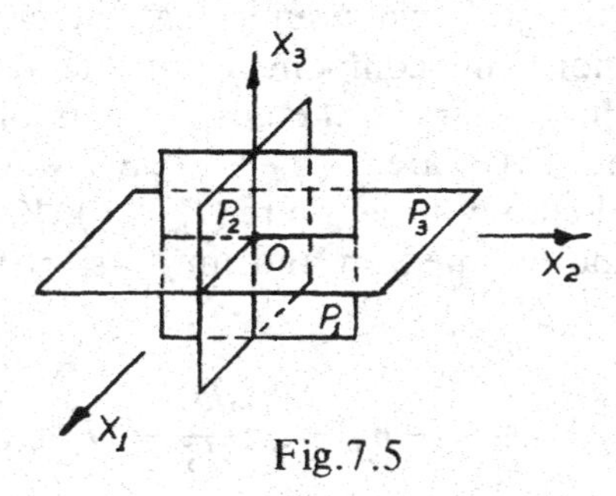

Fig.7.5

nents, σ_{11}, σ_{12} and σ_{13}; the stress vector on the plane P_2 has three components, σ_{21}, σ_{22}, and σ_{23}; and the stress vector on the plane P_3 has three components, σ_{31}, σ_{32}, and σ_{33}. The matrix

$$\begin{bmatrix} \sigma_{11} & \sigma_{21} & \sigma_{31} \\ \sigma_{12} & \sigma_{22} & \sigma_{32} \\ \sigma_{13} & \sigma_{23} & \sigma_{33} \end{bmatrix}, \tag{7.3.1}$$

whose columns are the components of the three stress vectors, is called the matrix of the state of stress at O.

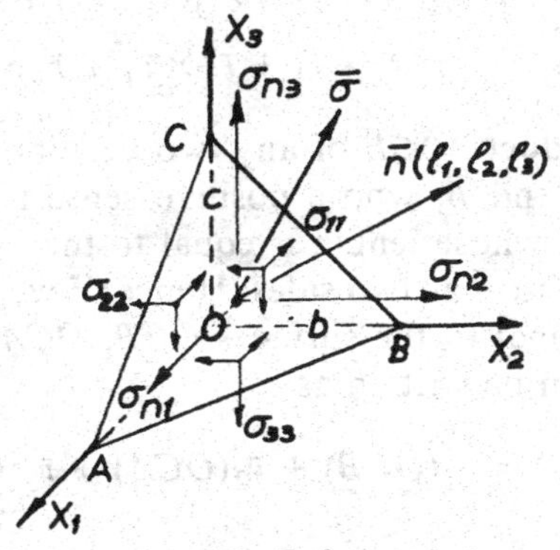

Fig.7.6

To find the stress vector $\bar{\sigma}(\sigma_{n1}, \sigma_{n2}, \sigma_{n3})$ on an oblique plane whose normal is $\bar{n}$, let us isolate from the continuous body a small tetrahedron $OABC$ (Fig. 7.6), and write that the forces acting on it are in equilibrium. The plane ABC is normal to $\bar{n}$ and at a small distance h from O. Let the areas ABC, OCB, OCA, and OAB be denoted by dS, dS_1, dS_2, and dS_3, respectively. If we assume that the stress vector varies in a continuous fashion, the components of the force on ABC are $(\sigma_{ni} + \varepsilon_i)dS$, where $\lim_{h\to o}\varepsilon_i = 0$; the components of the forces acting on OCB, OCA, and OAB are $(-\sigma_{ji} + \varepsilon_{ji})dS_j$, where $\lim_{h\to o}\varepsilon_{ij} = 0$; the components of the body force $\bar{F}$ are $(F_i + \varepsilon_i')\Delta V$, where $\Delta V = \frac{1}{3}h(dS)$ is the volume of the tetrahedron and $\lim_{h\to 0}\varepsilon_i' = 0$. For equilibrium, we must have:

$$(\sigma_{ni} + \varepsilon_i) + (-\sigma_{ji} + \varepsilon_{ji})\frac{dS_j}{dS} + (F_i + \varepsilon_i')\frac{h}{3} = 0. \tag{7.3.2}$$

The relation between the areas of the triangles ABC, OCB, OCA, and OAB can be obtained as follows (Fig. 7.6): Let us write, vectorially,

$$\overline{OA} = \bar{r}_1, \quad \overline{OB} = \bar{r}_2, \quad \overline{OC} = \bar{r}_3,$$

then

$$\overline{AB} = \bar{r}_2 - \bar{r}_1, \quad \overline{AC} = \bar{r}_3 - \bar{r}_1, \quad \overline{BC} = \bar{r}_3 - \bar{r}_2,$$

and we have:

$$\begin{aligned} \overline{AB} \times \overline{AC} &= (\bar{r}_2 - \bar{r}_1) \times (\bar{r}_3 - \bar{r}_1) \\ &= \bar{r}_2 \times \bar{r}_3 + \bar{r}_3 \times \bar{r}_1 + \bar{r}_1 \times \bar{r}_2 . \end{aligned} \tag{7.3.3}$$

Now the vector product $\bar{A} \times \bar{B}$ of any two vectors $\bar{A}$ and $\bar{B}$ is a vector perpendicular to $\bar{A}$ and $\bar{B}$, whose positive sense is determined by the right-hand rule, and whose length is equal to the area of the parallelogram formed by $\bar{A}$ and $\bar{B}$ as two sides. Hence, if we denote by $\bar{\nu}_1$, $\bar{\nu}_2$, $\bar{\nu}_3$ the unit vectors normal to the surfaces OCB, OCA, and OAB, respectively, Eq. (7.3.3) can be written as:

$$\bar{n}(ABC) = \bar{\nu}_1(OCB) + \bar{\nu}_2(OCA) + \bar{\nu}_3(OAB) \tag{7.3.4}$$

and

$$\bar{n} = \bar{\nu}_1\left(\frac{OCB}{ABC}\right) + \bar{\nu}_2\left(\frac{OCA}{ABC}\right) + \bar{\nu}_3\left(\frac{OAB}{ABC}\right). \tag{7.3.5}$$

If the direction cosines of $\bar{n}$ are ℓ_1, ℓ_2, and ℓ_3, then

$$\frac{OCB}{ABC} = \ell_1, \quad \frac{OCA}{ABC} = \ell_2, \quad \frac{OAB}{ABC} = \ell_3 \tag{7.3.6}$$

or

$$\frac{dS_j}{dS} = \ell_j. \tag{7.3.7}$$

Substituting Eq. (7.3.7) into Eq. (7.3.2), and passing to the limit as $h \to 0$, we get:

$$\sigma_{ni} = \sigma_{ji} \ell_j. \tag{7.3.8}$$

In matrix notation, Eq. (7.3.8) is written:

$$\begin{bmatrix} \sigma_{n1} \\ \sigma_{n2} \\ \sigma_{n3} \end{bmatrix} = \begin{bmatrix} \sigma_{11} & \sigma_{21} & \sigma_{31} \\ \sigma_{12} & \sigma_{22} & \sigma_{32} \\ \sigma_{13} & \sigma_{23} & \sigma_{33} \end{bmatrix} \begin{bmatrix} \ell_1 \\ \ell_2 \\ \ell_3 \end{bmatrix}. \tag{7.3.8a}$$

Thus, knowing the matrix of the state of stress at O, we can find the stress vector on any plane whose normal $\bar{n}$ has direction cosines ℓ_1, ℓ_2, and ℓ_3. Eq. (7.3.8) shows that the matrix of the state of stress is the matrix of a tensor of rank two, called the stress tensor (see Sec. 5.8). This matrix transforms the vector $\bar{n}$ to the vector $\bar{\sigma}$. In the next section, it will be shown that this linear transformation is symmetric; in other words, that the stress tensor is a symmetric tensor.

The magnitude of the stress vector is:

$$|\bar{\sigma}| = \sqrt{(\sigma_{n1})^2 + (\sigma_{n2})^2 + (\sigma_{n3})^2}\,. \tag{7.3.9}$$

The normal component of the stress vector is given by:

$$\sigma_n = \sigma_{n1} \ell_1 + \sigma_{n2} \ell_2 + \sigma_{n3} \ell_3 = \sigma_{ij} \ell_i \ell_j. \tag{7.3.10}$$

The tangential component of the stress vector is given by:

$$\sigma_t^2 = (|\bar{\sigma}|)^2 - \sigma_n^2. \tag{7.3.11}$$

Eq. (7.3.8) is known as Cauchy's formula. (*)

7.4 Equations of Equilibrium. Symmetry of the Stress Tensor. Boundary Conditions

So far we have considered the state of stress at a point. If it is desired to move from one point to another, the stress components will change in intensities and it is necessary to investigate the conditions which

* See pages A-43 and A-44 for information on the use of Cauchy's equations.

control the way in which they vary. The requirement that the laws of equilibrium must be obeyed gives us the means for determining how the stresses vary from point to point. Consider an arbitrary closed surface S within a body in equilibrium. The external forces on the volume V enclosed in S consist of surface forces and body forces (Fig. 7.7). The projection of the resultant body force vector on the OX_1 axis is:

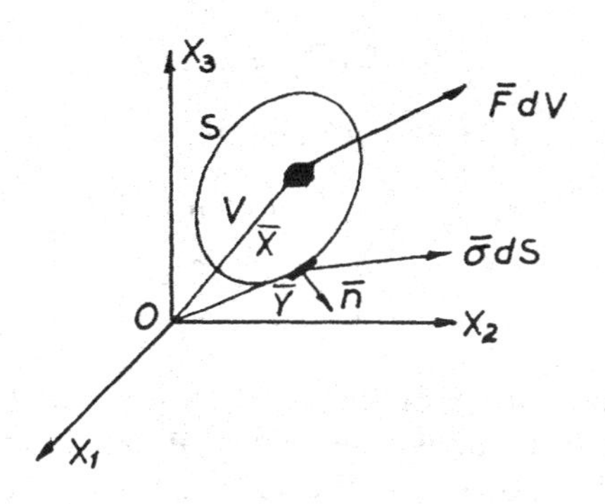

Fig. 7.7

$$\iiint_V F_1 \, dV. \tag{7.4.1}$$

The projection of the tractions on the OX_1 axis is:

$$\iint_S \sigma_{n1} \, dS = \iint_S (\ell_1 \sigma_{11} + \ell_2 \sigma_{21} + \ell_3 \sigma_{31}) \, dS. \tag{7.4.2}$$

Since the resultant force on a body in equilibrium must be equal to zero, then

$$\iiint_V F_1 \, dV + \iint_S (\ell_1 \sigma_{11} + \ell_2 \sigma_{21} + \ell_3 \sigma_{31}) \, dS = 0. \tag{7.4.3}$$

Applying the divergence theorem to the surface integral, we obtain:

$$\iiint_V \left(\frac{\partial \sigma_{11}}{\partial x_1} + \frac{\partial \sigma_{21}}{\partial x_2} + \frac{\partial \sigma_{31}}{\partial x_3} + F_1 \right) dV = 0. \tag{7.4.4}$$

Since this equation applies for any volume V of the body, the integrand must vanish identically; that is,

$$\frac{\partial \sigma_{11}}{\partial x_1} + \frac{\partial \sigma_{21}}{\partial x_2} + \frac{\partial \sigma_{31}}{\partial x_3} + F_1 = 0. \tag{7.4.5}$$

In a similar manner, summations of forces in the OX_2 and OX_3 directions yield two more equations. The three equations thus obtained are called the differential equations of equilibrium of a deformable continuous body. They are written:

$$\begin{aligned}
\frac{\partial \sigma_{11}}{\partial x_1} + \frac{\partial \sigma_{21}}{\partial x_2} + \frac{\partial \sigma_{31}}{\partial x_3} + F_1 = 0 \\
\frac{\partial \sigma_{12}}{\partial x_1} + \frac{\partial \sigma_{22}}{\partial x_2} + \frac{\partial \sigma_{32}}{\partial x_3} + F_2 = 0 \\
\frac{\partial \sigma_{13}}{\partial x_1} + \frac{\partial \sigma_{23}}{\partial x_2} + \frac{\partial \sigma_{33}}{\partial x_3} + F_3 = 0
\end{aligned} \tag{7.4.6}$$

or

$$\frac{\partial \sigma_{ji}}{\partial x_j} + F_i = 0. \tag{7.4.7}$$

In case of motion, Eq. (7.4.7) becomes

$$\frac{\partial \sigma_{ji}}{\partial x_j} + F_i = \rho A_i, \tag{7.4.8}$$

where A_1, A_2, A_3 are the components of the acceleration vector, and ρ is the mass per unit volume of the body. Eqs. (7.4.6) can also be derived by summation of the forces that act on the faces of an elementary parallelepiped and consideration of the variation of the stresses through the element. This will be done in a later section for the analysis in terms of curvilinear coordinates.

Let us now consider the equilibrium of moments. The moment of the body forces with respect to the origin is given by the integral over the volume V of the vector product $\overline{X} \times \overline{F} dV$ (see Fig. 7.7); in index notation, this vector product is written (see Problem 5.8) $\varepsilon_{ijk} x_j F_k dV$, where x_1, x_2, x_3 are the coordinates of points inside the volume V. The moment of the tractions with respect to the origin is given by the integral over the surface S of the vector product $\overline{Y} \times \overline{\sigma} dS$ (see Fig. 7.7); in index notation, this vector product is written $\varepsilon_{ijk} y_j \sigma_{nk} dS$, where σ_{n1}, σ_{n2}, and σ_{n3} are the components of $\overline{\sigma}$, and y_1, y_2, y_3 are the coordinates of points on the surface S. Since for equilibrium the resultant moment due to body and surface forces must vanish, then

$$\iint_S \varepsilon_{ijk} y_j \sigma_{nk}\, dS + \iiint_V \varepsilon_{ijk} x_j F_k\, dV = 0. \tag{7.4.9}$$

Substituting Eq. (7.3.8) in the surface integral, and using the divergence theorem, we get:

$$\begin{aligned}\iint_S \varepsilon_{ijk} y_j \sigma_{nk}\, dS &= \iint_S \varepsilon_{ijk} y_j \sigma_{rk}\, \ell_r\, dS \\ &= \iiint_V \frac{\partial}{\partial x_r}(\varepsilon_{ijk} x_j \sigma_{rk})\, dV \\ &= \iiint_V \varepsilon_{ijk}\left(x_j \frac{\partial \sigma_{rk}}{\partial x_r} + \delta_{jr}\sigma_{rk}\right) dV.\end{aligned} \tag{7.4.10}$$

But from Eq. (7.4.7),

$$\frac{\partial \sigma_{rk}}{\partial x_r} = -F_k,$$

therefore,

$$\iint_S \varepsilon_{ijk} y_j \sigma_{nk}\, dS = \iiint_V \varepsilon_{ijk}(-x_j F_k + \sigma_{jk})\, dV. \tag{7.4.11}$$

Eq. (7.4.9) now becomes:

$$\iiint_V \varepsilon_{ijk} \sigma_{jk}\, dV = 0. \tag{7.4.12}$$

Since the stress tensor varies in a continuous fashion and the volume V is arbitrary, then

$$\varepsilon_{ijk} \sigma_{jk} = 0. \tag{7.4.13}$$

In expanded form, Eq. (7.4.13) yields:

$$\sigma_{12} = \sigma_{21}, \quad \sigma_{13} = \sigma_{31}, \quad \sigma_{23} = \sigma_{32}, \tag{7.4.14}$$

or

$$\sigma_{ij} = \sigma_{ji},$$

which means that the stress tensor is symmetric. On account of this symmetry, the state of stress at every point—in other words, the stress field—is specified by six instead of nine functions of position.

In summary, the six components of the state of stress must satisfy three partial differential equations (7.4.6) within the body, and the three equations (7.3.8a) on the bounding surface. Eqs. (7.3.8a) are called the boundary conditions. It is obvious that the three equations of equilibrium do not suffice for the determination of the six functions that specify the stress field. This may be expressed by the statement that the stress field is statically indeterminate. To determine the stress field, the equations of equilibrium must be supplemented by other relations that cannot be obtained from statics considerations.

Eqs. (7.4.14) show that the linear transformation (7.3.8a) is symmetric. Therefore, all the properties of linear symmetric transformations studied in Chapter 3 can be applied to the study of the state of stress at a point, the same way they have been applied to the study of linear strain. Reciprocity, principal directions, invariants, Mohr's representation, etc. . . . are presented in the next section within the framework of stress.(*)

Remark

The components of the state of stress exist only in the deformed state of the body. Therefore, all the equations of Sec. 7.4 are referred to the deformed body. However, when stresses are studied within the framework of the classical theory of elasticity, no distinction is made between the predeformation and postdeformation values of the magnitudes and directions of the areas on which they act. This kind of approximation is quite consistent with the theory of linear strain. Indeed, in this theory it was assumed that the edges of the deformed element of volume undergo negligible rotations and that the lengths of the edges in the deformed state differ only by a very small amount from their original lengths. These factors make the deformed volume element indistinguishable from the undeformed volume element as far as the analysis of stress is concerned.

7.5 Application of the Properties of Linear Symmetric Transformations to the Analysis of Stress

Let $\bar{n}_1$ and $\bar{n}_2$ be the outward unit normals to two planes through a point O at which the state of stress is known (Fig. 7.8). If $(\bar{\sigma})_1$ and $(\bar{\sigma})_2$ are the stress vectors on these two planes, then, by the property of

* See page A-44.

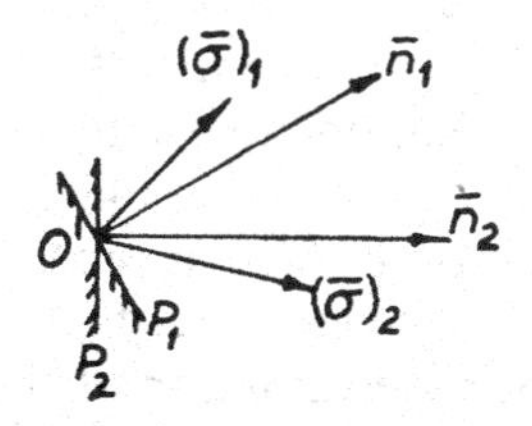

Fig.7.8

reciprocity, the projection of $(\bar{\sigma})_1$ on $\bar{n}_2$ is equal to the projection of $(\bar{\sigma})_2$ on $\bar{n}_1$:

$$(\bar{\sigma})_1 \cdot \bar{n}_2 = (\bar{\sigma})_2 \cdot \bar{n}_1 . \tag{7.5.1}$$

The principal directions of the linear transformation (*7.3.8*) *are at the same time invariant directions.* While in the study of strains we were searching for those directions which are associated solely with normal strains, we are here searching for the directions (principal directions) and the planes they define (principal planes) such that those planes are subjected only to normal stresses (the tangential stresses vanish). Let the stress on a principal plane be called σ. The direction cosines of the normal to this plane can be obtained by substitution in Eq. (7.3.8), as follows:

$$\begin{bmatrix} \sigma_{n1} \\ \sigma_{n2} \\ \sigma_{n3} \end{bmatrix} = \begin{bmatrix} \sigma \ell_1 \\ \sigma \ell_2 \\ \sigma \ell_3 \end{bmatrix} = \begin{bmatrix} \sigma_{11} & \sigma_{12} & \sigma_{13} \\ \sigma_{12} & \sigma_{22} & \sigma_{23} \\ \sigma_{13} & \sigma_{23} & \sigma_{33} \end{bmatrix} \begin{bmatrix} \ell_1 \\ \ell_2 \\ \ell_3 \end{bmatrix}. \tag{7.5.2}$$

For a nontrivial solution, the determinant of the coefficients of the homogeneous system (7.5.2) must be equal to zero:

$$\begin{vmatrix} \sigma_{11} - \sigma & \sigma_{12} & \sigma_{13} \\ \sigma_{12} & \sigma_{22} - \sigma & \sigma_{23} \\ \sigma_{13} & \sigma_{23} & \sigma_{33} - \sigma \end{vmatrix} = 0. \tag{7.5.3}$$

Eq. (7.5.3) is a cubic equation in σ:

$$\begin{aligned}&\sigma^3 - \sigma^2(\sigma_{11} + \sigma_{22} + \sigma_{33})\\ &\quad + \sigma(\sigma_{11}\sigma_{22} + \sigma_{22}\sigma_{33} + \sigma_{33}\sigma_{11} - \sigma_{12}^2 - \sigma_{23}^2 - \sigma_{13}^2)\\ &\quad - (\sigma_{11}\sigma_{22}\sigma_{33} + 2\sigma_{12}\sigma_{23}\sigma_{13} - \sigma_{11}\sigma_{23}^2 - \sigma_{22}\sigma_{13}^2 - \sigma_{33}\sigma_{12}^2)\\ &\quad = 0.\end{aligned} \tag{7.5.4}$$

The three roots of this equation are the principal stresses σ_1, σ_2, and σ_3. The roots are independent of the system of reference axes and so are the coefficients of Eq. (7.5.4). Those coefficients are called *the invariants of the state of stress*, and are written:

$$J_1 = \sigma_{11} + \sigma_{22} + \sigma_{33} = \sigma_1 + \sigma_2 + \sigma_3 \tag{7.5.5}$$

$$\begin{aligned}J_2 &= \sigma_{11}\sigma_{22} + \sigma_{22}\sigma_{33} + \sigma_{33}\sigma_{11} - \sigma_{12}^2 - \sigma_{23}^2 - \sigma_{13}^2\\ &= \sigma_1\sigma_2 + \sigma_2\sigma_3 + \sigma_3\sigma_1\end{aligned} \tag{7.5.6}$$

$$\begin{aligned}J_3 &= \sigma_{11}\sigma_{22}\sigma_{33} + 2\sigma_{12}\sigma_{23}\sigma_{13} - \sigma_{11}\sigma_{23}^2 - \sigma_{22}\sigma_{13}^2 - \sigma_{33}\sigma_{12}^2\\ &= \sigma_1\sigma_2\sigma_3.\end{aligned} \tag{7.5.7}$$

If one chooses the coordinate axes along the principal directions, the matrix of Eq. (7.3.8) becomes a diagonal matrix. In this frame of reference,

$$\begin{bmatrix}\sigma_{n1}\\ \sigma_{n2}\\ \sigma_{n3}\end{bmatrix} = \begin{bmatrix}\sigma_1 & 0 & 0\\ 0 & \sigma_2 & 0\\ 0 & 0 & \sigma_3\end{bmatrix}\begin{bmatrix}\ell_1\\ \ell_2\\ \ell_3\end{bmatrix}. \tag{7.5.8}$$

The major principal stress acts on the major principal plane, the intermediate principal stress acts on the intermediate principal plane, and the minor principal stress acts on the minor principal plane. In terms of principal stresses, Eqs. (7.3.10) and (7.3.11) are written as follows:

$$\sigma_n = \sigma_1\ell_1^2 + \sigma_2\ell_2^2 + \sigma_3\ell_3^2 \tag{7.5.9}$$

$$\sigma_t^2 = (\sigma_1 - \sigma_2)^2\ell_1^2\ell_2^2 + (\sigma_2 - \sigma_3)^2\ell_2^2\ell_3^2 + (\sigma_3 - \sigma_1)^2\ell_3^2\ell_1^2, \tag{7.5.10}$$

with

$$\ell_1^2 + \ell_2^2 + \ell_3^2 = 1. \tag{7.5.11}$$

At a point O, *Mohr's diagram* allows us to represent graphically the normal and tangential stresses on any plane whose normal has direction cosines ℓ_1, ℓ_2 and ℓ_3. The unit sphere (see Sec. 3.13) is centered at O (Fig. 7.9). The unit element $\overline{OH}$ along the normal to the plane under

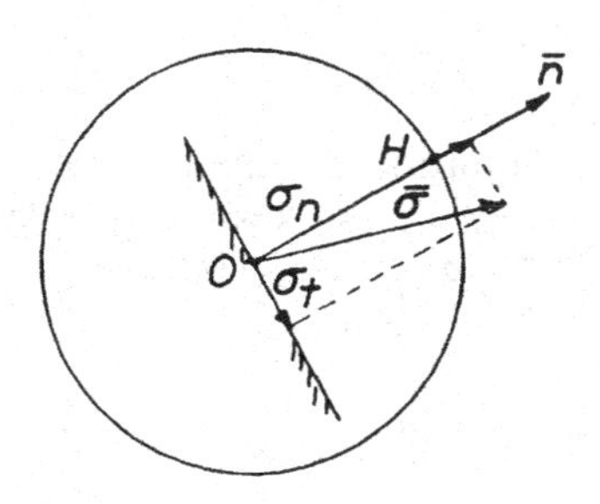

Fig. 7.9

consideration has direction cosines ℓ_1, ℓ_2, and ℓ_3. The coordinates of H are ℓ_1, ℓ_2, and ℓ_3. Under the linear transformation (7.3.8), $\overline{OH}$ is transformed to $\bar{\sigma}$, whose normal and tangential components along $\overline{OH}$ and the plane are given by Eqs. (7.5.9) and (7.5.10), respectively. The formulas of Sec. 3.13 are duplicated here:
Solving Eqs. (7.5.9), (7.5.10), and (7.5.11) simultaneously, one obtains:

$$\ell_1^2 = \frac{\sigma_t^2 + (\sigma_n - \sigma_2)(\sigma_n - \sigma_3)}{(\sigma_1 - \sigma_2)(\sigma_1 - \sigma_3)} \tag{7.5.12}$$

$$\ell_2^2 = \frac{\sigma_t^2 + (\sigma_n - \sigma_3)(\sigma_n - \sigma_1)}{(\sigma_2 - \sigma_3)(\sigma_2 - \sigma_1)} \tag{7.5.13}$$

$$\ell_3^2 = \frac{\sigma_t^2 + (\sigma_n - \sigma_1)(\sigma_n - \sigma_2)}{(\sigma_3 - \sigma_1)(\sigma_3 - \sigma_2)} . \tag{7.5.14}$$

These equations are similar to Eqs. (3.13.6). The entire discussion in Sec. 3.13 of the correspondence between points on the sphere representing different directions of planes through O and points on Mohr's diagram, can be repeated here and the same conclusions drawn (Fig. 7.10):

1. At a given point, and for a given state of stress characterized by the nine σ_{ij} ' s, the largest normal stress is the major principal stress and the smallest normal stress is the minor principal stress.

2. The planes subjected to the highest shearing stresses bisect the angles between the principal planes. There are three such planes and the

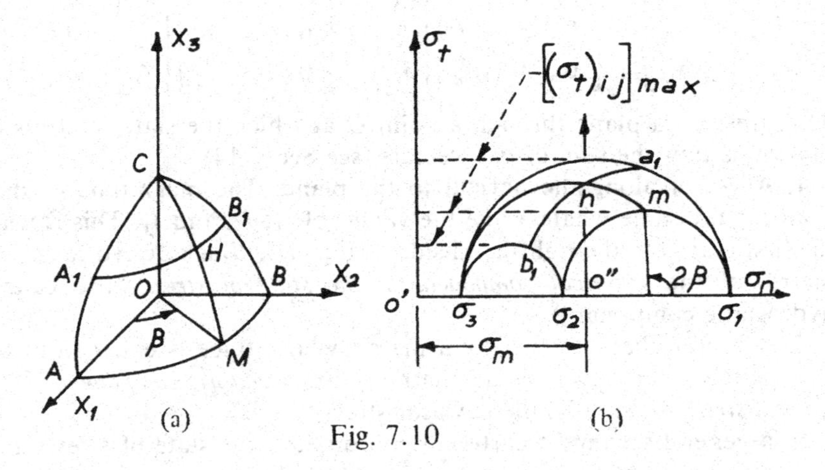

Fig. 7.10

maximum shearing stress acts on the plane bisecting the angle between the major and minor principal planes. The magnitudes of those shearing stresses, which are called the principal shearing stresses, are:

$$[(\sigma_t)_{12}]_{\max} = \frac{\sigma_1 - \sigma_2}{2}, \quad [(\sigma_t)_{23}]_{\max} = \frac{\sigma_2 - \sigma_3}{2},$$
$$[(\sigma_t)_{13}]_{\max} = \frac{\sigma_1 - \sigma_3}{2}. \tag{7.5.15}$$

3. If the same quantity σ_0 is added to the three principal stresses, the Mohr circles do not change in size; they are just shifted along the $o'\sigma_n$ axis. The σ_t ' s do not change. If we define the mean stress by:

$$\sigma_m = \frac{\sigma_1 + \sigma_2 + \sigma_3}{3} = \frac{\sigma_{11} + \sigma_{22} + \sigma_{33}}{3} = \frac{\sigma_{ii}}{3}, \tag{7.5.16}$$

then

$$\sigma_1 = \sigma_m + \sigma_1', \quad \sigma_2 = \sigma_m + \sigma_2', \quad \sigma_3 = \sigma_m + \sigma_3' \tag{7.5.17}$$

and

$$\sigma_1' + \sigma_2' + \sigma_3' = 0. \tag{7.5.18}$$

Eq. (7.5.8) can now be written:

$$\begin{bmatrix} \sigma_{n1} \\ \sigma_{n2} \\ \sigma_{n3} \end{bmatrix} = \begin{bmatrix} \sigma_m & 0 & 0 \\ 0 & \sigma_m & 0 \\ 0 & 0 & \sigma_m \end{bmatrix} \begin{bmatrix} \ell_1 \\ \ell_2 \\ \ell_3 \end{bmatrix} + \begin{bmatrix} \sigma'_1 & 0 & 0 \\ 0 & \sigma'_2 & 0 \\ 0 & 0 & \sigma'_3 \end{bmatrix} \begin{bmatrix} \ell_1 \\ \ell_2 \\ \ell_3 \end{bmatrix}. \tag{7.5.19}$$

The stress on a plane through a point O at which the state of stress is known, is thus the sum of two vectors (see Sec. 3.14):

1. A vector along the normal to the plane: The magnitude of this vector is the same whatever be the values of ℓ_1, ℓ_2, and ℓ_3. This is why the first matrix in the right-hand side of Eq. (7.5.19) is referred to as the matrix of the *spherical component of the state of stress* (also called hydrostatic component).

2. A vector characterized by a matrix whose trace is equal to zero: This matrix is referred to as the matrix of the *deviatoric component of the state of stress* or, simply, the deviator stress matrix.

In a general system of cartesian coordinates, the state of stress at a point can be expressed as:

$$\begin{bmatrix} \sigma_{11} & \sigma_{12} & \sigma_{13} \\ \sigma_{12} & \sigma_{22} & \sigma_{23} \\ \sigma_{13} & \sigma_{23} & \sigma_{33} \end{bmatrix} = \begin{bmatrix} \sigma_m & 0 & 0 \\ 0 & \sigma_m & 0 \\ 0 & 0 & \sigma_m \end{bmatrix} + \begin{bmatrix} \sigma'_{11} & \sigma_{12} & \sigma_{13} \\ \sigma_{12} & \sigma'_{22} & \sigma_{23} \\ \sigma_{13} & \sigma_{23} & \sigma'_{33} \end{bmatrix} \tag{7.5.20}$$

with

$$\sigma'_{11} + \sigma'_{22} + \sigma'_{33} = 0. \tag{7.5.21}$$

In index notation, Eq. (7.5.20) is written:

$$\sigma_{ij} = \frac{1}{3}\delta_{ij}\sigma_{mm} + \sigma'_{ij}. \tag{7.5.22}$$

The deviatoric component of the state of stress at a point is represented on Mohr's diagram by Fig. 7.10b, but with the origin shifted to o'' so that $o'o'' = \sigma_m$. Both the spherical and the deviatoric components have three invariants each. In terms of the invariants of the state of stress defined in Eqs. (7.5.5), (7.5.6), and (7.5.7), *the invariants of the spherical components* are written:

$$J_{s1} = J_1, \quad J_{s2} = \frac{1}{3}(J_1)^2, \quad J_{s3} = \frac{1}{27}(J_1)^3. \tag{7.5.23}$$

The invariants of the deviatoric components are written:

$$J_{d1} = 0, \quad J_{d2} = J_2 - \frac{1}{3}(J_1)^2, \quad J_{d3} = J_3 - \frac{1}{3}(J_1 J_2) + \frac{2}{27}(J_1)^3 \; (*) \tag{7.5.24}$$

* See pages A-44 to A-46 for a solved example.

7.6 Stress Quadric

In a trirectangular system of coordinates, OX_1, OX_2, OX_3, consider the equation,

$$\sigma_{ij} x_i x_j = \pm K^2, \tag{7.6.1}$$

where K is a constant. This equation represents a quadric surface with its center at the origin O (Fig. 7.11). This quadric is called the stress

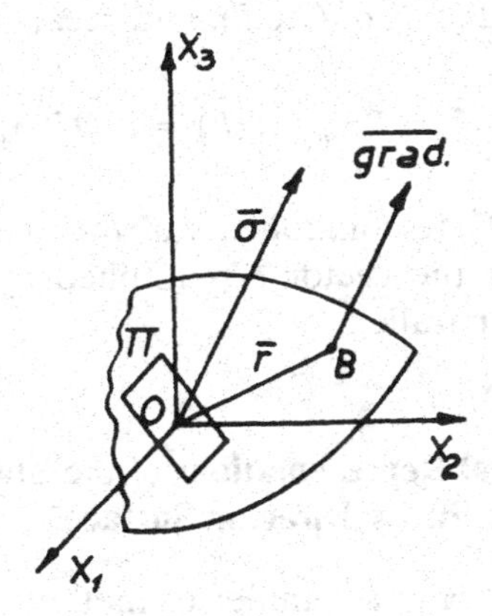

Fig. 7.11

quadric and is completely determined once the state of stress σ_{ij} at a point P is known. Let $\bar{r}$ be the radius vector to any point B (x_1, x_2, x_3) on the quadric. Then, the length OB is a measure of the normal stress on the plane Π passing through P, and whose normal is in the direction of $\bar{r}$: Indeed, since the direction cosines of $\bar{r}$ are:

$$\ell_1 = \frac{x_1}{OB}, \quad \ell_2 = \frac{x_2}{OB}, \quad \ell_3 = \frac{x_3}{OB}, \quad \ell_i = \frac{x_i}{OB}, \tag{7.6.2}$$

the corresponding normal stress is given by Eq. (7.3.10) as:

$$\sigma_n = \sigma_{ij} \frac{x_i}{OB} \frac{x_j}{OB} = \pm \frac{K^2}{(OB)^2}. \tag{7.6.3}$$

Another interesting property of the stress quadric is that the normal to this surface at the end of the vector $\bar{r}$ (Fig. 7.11) is parallel to the stress vector $\bar{\sigma}$ acting on the plane Π. To prove this property, let us write

Eq. (7.6.1) in the form:

$$A = \sigma_{ij} x_i x_j \mp K^2 = 0. \tag{7.6.4}$$

Then the direction of the normal to the quadric is given by the gradient of the scalar function A. The components of the gradient are:

$$\frac{\partial A}{\partial x_m} = \sigma_{ij} \delta_{im} x_j + \sigma_{ij} x_i \delta_{jm} = 2\sigma_{mj} x_j . \tag{7.6.5}$$

Substituting Eq. (7.6.2) in Eq. (7.6.5), and recalling Eq. (7.3.8), we get:

$$\frac{\partial A}{\partial x_m} = 2\sigma_{mj} \ell_j (OB) = 2(OB)\sigma_{nm} . \tag{7.6.6}$$

Eq. (7.6.6) shows that the components of the gradient vector are equal to the components of the vector $\bar{\sigma}$ multiplied by $2(OB)$, which means that both vectors are parallel.

7.7 Further Graphical Representations of the State of Stress at a Point. Stress Ellipsoid. Stress Director Surface

Besides Mohr's diagram, a number of methods have been devised to help visualize the state of stress at a point and to compute the stresses on oblique planes:

a. *The Stress Ellipsoid* (Lamé's Ellipsoid)

Let the axes of reference OX_1, OX_2, OX_3 be taken in the direction of the principal stresses at a point P. Also let the three components of the stress vector $\bar{\sigma}(\sigma_{n1}, \sigma_{n2}, \sigma_{n3})$ on a plane Π through this point be measured along these axes. Eq. (7.5.8) can be written as:

$$x_1 = \sigma_{n1} = \ell_1 \sigma_1, \quad x_2 = \sigma_{n2} = \ell_2 \sigma_2, \quad x_3 = \sigma_{n3} = \ell_3 \sigma_3 . \tag{7.7.1}$$

Substituting Eq. (7.7.1) into Eq. (7.5.11), we obtain:

$$\frac{x_1^2}{\sigma_1^2} + \frac{x_2^2}{\sigma_2^2} + \frac{x_3^2}{\sigma_3^2} = 1. \tag{7.7.2}$$

Eq. (7.7.2) means that for each inclined plane through O, the stress is represented by a vector from O with components σ_{n1}, σ_{n2}, and σ_{n3}, the

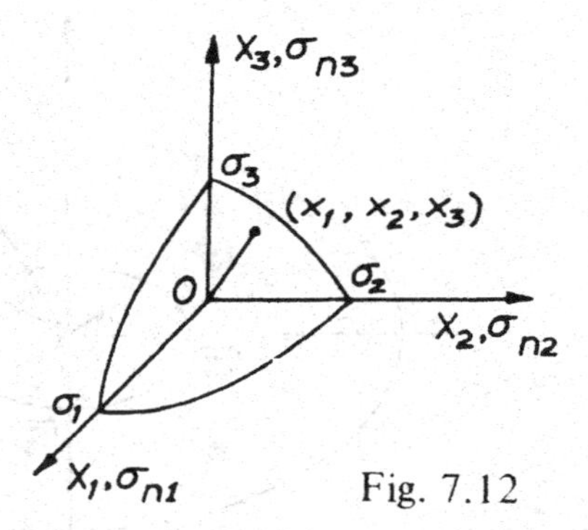

Fig. 7.12

end of which lies on the surface of an ellipsoid (Fig. 7.12). This ellipsoid is called the stress ellipsoid and its semi-axes are the principal stresses. From this, it can be concluded that the maximum stress at a point is the major principal stress. If two principal stresses are numerically equal, the ellipsoid is of revolution. If all the principal stresses are numerically equal, the ellipsoid becomes a sphere and any three perpendicular directions can be taken as principal axes. When one of the principal stresses is zero, the ellipsoid reduces to an ellipse. When two principal stresses are equal to zero, the ellipsoid reduces to a line.

b. *The Stress Director Surface*

The radii of the stress ellipsoid represent the stress on one of the planes through point *P* at which the state of stress is known. To find the plane corresponding to a given radius, we use the stress director surface defined by the equation:

$$\frac{x_1^2}{\sigma_1} + \frac{x_2^2}{\sigma_2} + \frac{x_3^2}{\sigma_3} = 1. \tag{7.7.3}$$

It can be shown that the stress represented by a radius of the stress ellipsoid acts on a plane parallel to the plane tangent to the stress director surface at the point it is intersected by this radius. This is illustrated in Fig. 7.13 for the case in which one of the principal stresses, namely σ_3, is equal to zero.

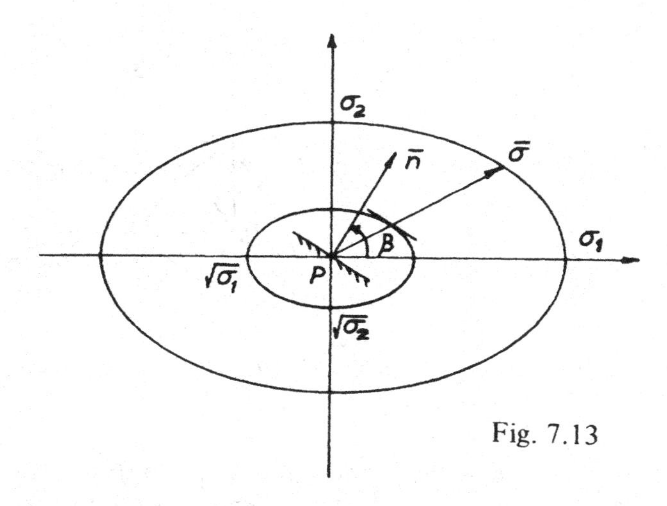

Fig. 7.13

7.8 The Octahedral Normal and Octahedral Shearing Stresses

An octahedral plane is a plane equally inclined on the directions of the three principal stresses. The directions cosines of the normal to this plane are $1/\sqrt{3}$, $1/\sqrt{3}$, $1/\sqrt{3}$ (Fig. 7.14). The expression for the octahedral normal stress is obtained by substituting these direction cosines in Eq. (7.5.9):

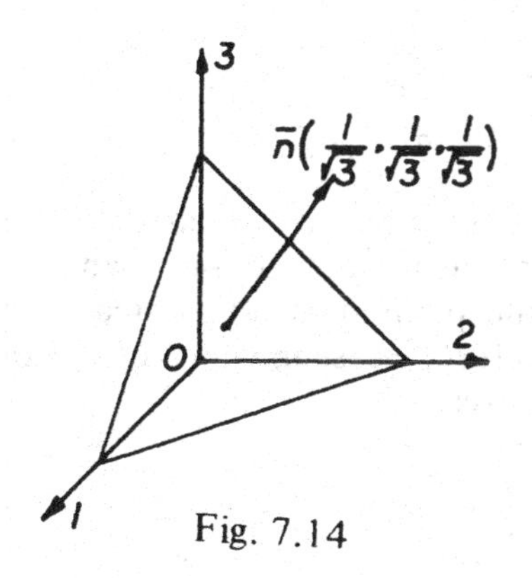

Fig. 7.14

$$\sigma_{oct} = \frac{\sigma_1 + \sigma_2 + \sigma_3}{3} = \frac{\sigma_{11} + \sigma_{22} + \sigma_{33}}{3} = \sigma_m = \frac{J_1}{3}. \qquad (7.8.1)$$

The expression for the octahedral shearing stress is obtained by substituting these direction cosines in Eq. (7.5.10):

$$\begin{aligned}\tau_{oct} &= \frac{1}{3}\sqrt{(\sigma_1 - \sigma_2)^2 + (\sigma_2 - \sigma_3)^2 + (\sigma_3 - \sigma_1)^2} \\ &= \frac{1}{3}\sqrt{2J_1^2 - 6J_2} = \sqrt{-\tfrac{2}{3}J_{d2}}\,.\end{aligned} \qquad (7.8.2)$$

In a general system of cartesian coordinates, the expression for τ_{oct} is written as follows:

$$\tau_{oct} = \frac{1}{3}\sqrt{(\sigma_{11} - \sigma_{22})^2 + (\sigma_{22} - \sigma_{33})^2 + (\sigma_{33} - \sigma_{11})^2 + 6\sigma_{12}^2 + 6\sigma_{23}^2 + 6\sigma_{13}^2}. \qquad (7.8.3)$$

7.9 The Haigh-Westergaard Stress Space

A state of stress defined by three principal stresses, σ_1, σ_2, and σ_3, is represented in a stress space by a point having cartesian coordinates σ_1, σ_2, σ_3 (Fig. 7.15). The equation of the trisectrix Δ is $\sigma_1 = \sigma_2 = \sigma_3$.

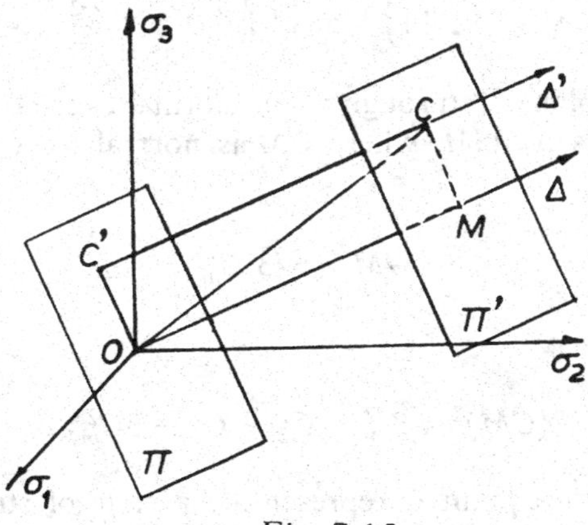

Fig. 7.15

Consider a point C whose coordinates, C_1, C_2, C_3, represent a state of stress. The mean stress C_m and the three components of the deviator are given by:

$$C_m = \frac{C_1 + C_2 + C_3}{3} \tag{7.9.1}$$

and

$$\begin{bmatrix} C_1 - C_m & 0 & 0 \\ 0 & C_2 - C_m & 0 \\ 0 & 0 & C_3 - C_m \end{bmatrix} = \begin{bmatrix} \frac{2C_1 - C_2 - C_3}{3} & 0 & 0 \\ 0 & \frac{2C_2 - C_1 - C_3}{3} & 0 \\ 0 & 0 & \frac{2C_3 - C_1 - C_2}{3} \end{bmatrix}. \tag{7.9.2}$$

In the stress space, the three components of the deviator have a resultant equal to

$$\left\{\frac{1}{9}[(2C_1 - C_2 - C_3)^2 + (2C_2 - C_1 - C_3)^2 + (2C_3 - C_1 - C_2)^2]\right\}^{1/2} = \left\{C_1^2 + C_2^2 + C_3^2 - 3C_m^2\right\}^{1/2}. \tag{7.9.3}$$

Now consider a plane Π' through C and normal to the trisectrix Δ. Such a plane intersects Δ at M, where CM is normal to Δ. From analytic geometry,

$$OM = \sqrt{3}\, C_m \tag{7.9.4}$$

and

$$(CM)^2 = C_1^2 + C_2^2 + C_3^2 - 3C_m^2. \tag{7.9.5}$$

Therefore, if from a point C representing a state of stress we draw a perpendicular CM to the trisectrix, the length of CM gives the magnitude of the deviator stress, and the length of OM divided by $\sqrt{3}$ gives the magnitude of the mean stress. All the points on a line Δ' through C have the same deviator and all the points on the plane Π' (which is an octahedral plane) have the same mean stress.

The previous considerations point to a graphical method that can be used to decompose the state of stress represented by a point $C(C_1, C_2, C_3)$ into its spherical and its deviatoric components:

(a) Draw an octahedral plane Π through O and a line Δ' through C parallel to Δ. The point of intersection of Δ' and Π gives the point C', and the vector OC' represents the deviator stress.

(b) Draw a plane Π' parallel to Π through C. The point of intersection of Π' and Δ gives the point M, and OM represents the mean stress according to Eq. (7.9.4).

In the study of the theory of plasticity, the yielding of various materials is often expressed by a relation among principal stresses. Such relations may be graphically represented by means of surfaces in the stress space. Since spherical states of stress are known to have little or no influence on the yielding of common metals, a representation of the deviator alone is sought. Such a representation is obtained by projecting the various possible states of stress on an octahedral plane through the origin. The directions of the three principal stresses, $O\sigma_1$, $O\sigma_2$, $O\sigma_3$, when projected on the octahedral plane Π make an angle of 120 ° with respect to each other (Fig. 7.16). Points on the trisectrix are projected at the origin O. Point C is projected to C', and OC' represents the deviator stress. $OC' = \rho$ is called the intensity of the deviator and ϕ is called the phase angle; ϕ can be measured from any of the three axes $O1$, $O2$, and

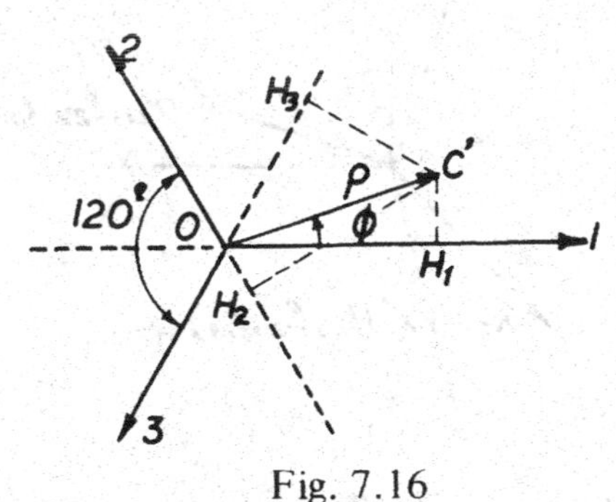

Fig. 7.16

$O3$ (Fig. 7.16). If the three components of the deviator are called S_1, S_2, and S_3,

$$OC' = \rho = (S_1^2 + S_2^2 + S_3^2)^{1/2}, \tag{7.9.6}$$

and the 3 normal projections of OC' (Fig. 7.16) are given by:

$$OH_1 = \sqrt{\frac{3}{2}}\, S_1, \quad OH_2 = \sqrt{\frac{3}{2}}\, S_2, \quad OH_3 = \sqrt{\frac{3}{2}}\, S_3. \qquad (7.9.7)$$

From the geometry of the figure, we see that:

$$OH_1 + OH_2 + OH_3 = 0 \text{ (in magnitude and sign)}, \qquad (7.9.8)$$

which is consistent with the fact that $S_1 + S_2 + S_3 = 0$. Projections on octahedral planes are also used to represent deviator strains and to express geometrically relations between stresses and strains.(*)

7.10 Components of the State of Stress at a Point in a Change of Coordinates

The tensor character of stress was established in Sec. 7.3. Therefore, in a rotation of the reference system of coordinate axes, the components of the state of stress are given by (Fig. 7.17):

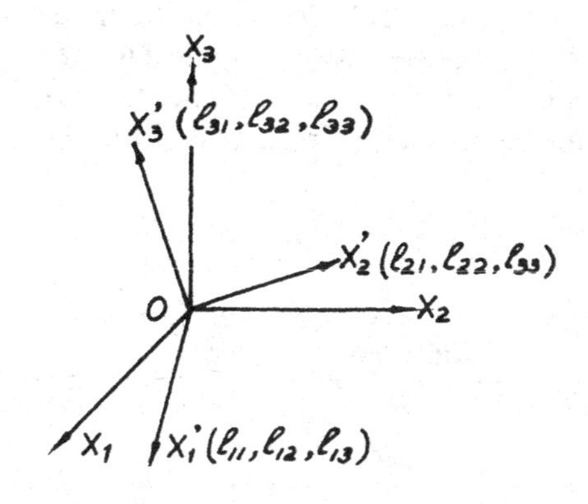

Fig. 7.17

$$\sigma'_{ik} = \ell_{ij}\,\ell_{km}\,\sigma_{jm}. \qquad (7.10.1)$$

Thus,

$$\begin{aligned}\sigma'_{11} = \sigma_{11}\,\ell_{11}^2 + \sigma_{22}\,\ell_{12}^2 + \sigma_{33}\,\ell_{13}^2 + 2\sigma_{12}\,\ell_{11}\,\ell_{12} + 2\sigma_{13}\,\ell_{11}\,\ell_{13}\\ + 2\sigma_{23}\,\ell_{12}\,\ell_{13}\end{aligned} \qquad (7.10.2)$$

* See page A-46 for an alternative method.

$$\sigma'_{22} = \sigma_{11}\ell_{21}^2 + \sigma_{22}\ell_{22}^2 + \sigma_{33}\ell_{23}^2 + 2\sigma_{12}\ell_{21}\ell_{22} + 2\sigma_{13}\ell_{21}\ell_{23} + 2\sigma_{23}\ell_{22}\ell_{23} \quad (7.10.3)$$

$$\sigma'_{33} = \sigma_{11}\ell_{31}^2 + \sigma_{22}\ell_{32}^2 + \sigma_{33}\ell_{33}^2 + 2\sigma_{12}\ell_{31}\ell_{32} + 2\sigma_{13}\ell_{31}\ell_{33} + 2\sigma_{23}\ell_{32}\ell_{33} \quad (7.10.4)$$

$$\begin{aligned} \sigma'_{12} = {} & (\sigma_{11}\ell_{11} + \sigma_{21}\ell_{12} + \sigma_{31}\ell_{13})\ell_{21} \\ & + (\sigma_{12}\ell_{11} + \sigma_{22}\ell_{12} + \sigma_{32}\ell_{13})\ell_{22} \\ & + (\sigma_{13}\ell_{11} + \sigma_{23}\ell_{12} + \sigma_{33}\ell_{13})\ell_{23} \end{aligned} \quad (7.10.5)$$

$$\begin{aligned} \sigma'_{23} = {} & (\sigma_{11}\ell_{21} + \sigma_{21}\ell_{22} + \sigma_{31}\ell_{23})\ell_{31} \\ & + (\sigma_{12}\ell_{21} + \sigma_{22}\ell_{22} + \sigma_{32}\ell_{23})\ell_{32} \\ & + (\sigma_{13}\ell_{21} + \sigma_{23}\ell_{22} + \sigma_{33}\ell_{23})\ell_{33} \end{aligned} \quad (7.10.6)$$

$$\begin{aligned} \sigma'_{13} = {} & (\sigma_{11}\ell_{11} + \sigma_{21}\ell_{12} + \sigma_{31}\ell_{13})\ell_{31} \\ & + (\sigma_{12}\ell_{11} + \sigma_{22}\ell_{12} + \sigma_{32}\ell_{13})\ell_{32} \\ & + (\sigma_{13}\ell_{11} + \sigma_{23}\ell_{12} + \sigma_{33}\ell_{13})\ell_{33}. \end{aligned} \quad (7.10.7)$$

7.11 Stress Analysis in Two Dimensions

This section parallels Sec. 3.16. Let Π be a principal plane through a point O, and let OX_1 and OX_2 be two reference axes in this plane (Fig. 7.18); OX_3 is the principal direction normal to Π. In this system of axes, σ_{13} and σ_{23} are equal to zero, and any plane passing through OX_3 is

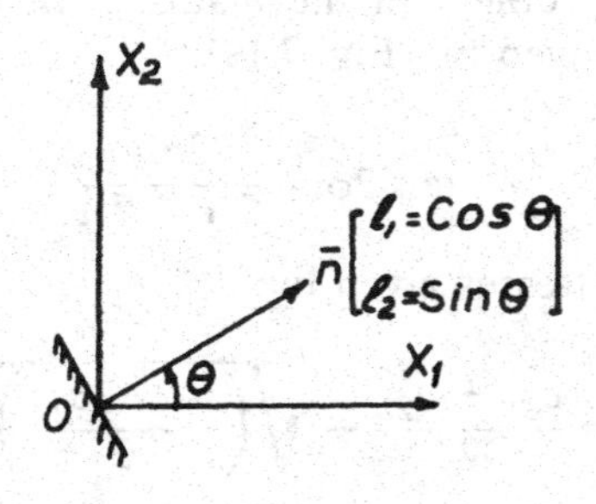

Fig. 7.18

subjected to a stress vector located in the plane Π: This is obvious since the stress vector results from the transformation of a normal $\bar{n}$ located in an invariant plane. The stress vector on any plane through OX_3, and whose normal $\bar{n}$ has direction cosines $\ell_1 = \cos\theta$, $\ell_2 = \sin\theta$, and $\ell_3 = 0$, is such that

$$\begin{bmatrix} \sigma_{n1} \\ \sigma_{n2} \end{bmatrix} = \begin{bmatrix} \sigma_{11} & \sigma_{12} \\ \sigma_{12} & \sigma_{22} \end{bmatrix} \begin{bmatrix} \cos\theta \\ \sin\theta \end{bmatrix}, \tag{7.11.1}$$

and $\sigma_{n3} = 0$. The corresponding normal and tangential stresses on this plane are given by:

$$\begin{aligned} \sigma_n &= \sigma_{11}\cos^2\theta + \sigma_{22}\sin^2\theta + 2\sigma_{12}\sin\theta\cos\theta \\ &= \frac{\sigma_{11}+\sigma_{22}}{2} + \frac{\sigma_{11}-\sigma_{22}}{2}\cos 2\theta + \sigma_{12}\sin 2\theta \end{aligned} \tag{7.11.2}$$

$$\sigma_t = -\frac{\sigma_{11}-\sigma_{22}}{2}\sin 2\theta + \sigma_{12}\cos 2\theta. \tag{7.11.3}$$

If the reference axes are rotated an angle θ around OX_3 (Fig. 7.19), the σ_{ij} ' s in Eq. (7.11.1) become:

$$\begin{aligned} \sigma'_{11} &= \sigma_{11}\cos^2\theta + \sigma_{22}\sin^2\theta + 2\sigma_{12}\sin\theta\cos\theta \\ &= \frac{\sigma_{11}+\sigma_{22}}{2} + \frac{\sigma_{11}-\sigma_{22}}{2}\cos 2\theta + \sigma_{12}\sin 2\theta \end{aligned} \tag{7.11.4}$$

$$\begin{aligned} \sigma'_{22} &= \sigma_{11}\sin^2\theta + \sigma_{22}\cos^2\theta - 2\sigma_{12}\sin\theta\cos\theta \\ &= \frac{\sigma_{11}+\sigma_{22}}{2} - \frac{\sigma_{11}-\sigma_{22}}{2}\cos 2\theta - \sigma_{12}\sin 2\theta \end{aligned} \tag{7.11.5}$$

$$\sigma'_{12} = -\frac{\sigma_{11}-\sigma_{22}}{2}\sin 2\theta + \sigma_{12}\cos 2\theta. \tag{7.11.6}$$

The eigenvalue problem in the plane Π yields two principal directions, $O1$ and $O2$, given by (Fig. 7.19):

$$\tan 2\phi = \frac{2\sigma_{12}}{\sigma_{11}-\sigma_{22}}, \tag{7.11.7}$$

and two principal stresses given by:

$$\sigma_1 = \frac{\sigma_{11}+\sigma_{22}}{2} + \sqrt{\left(\frac{\sigma_{11}-\sigma_{22}}{2}\right)^2 + \sigma_{12}^2} \tag{7.11.8}$$

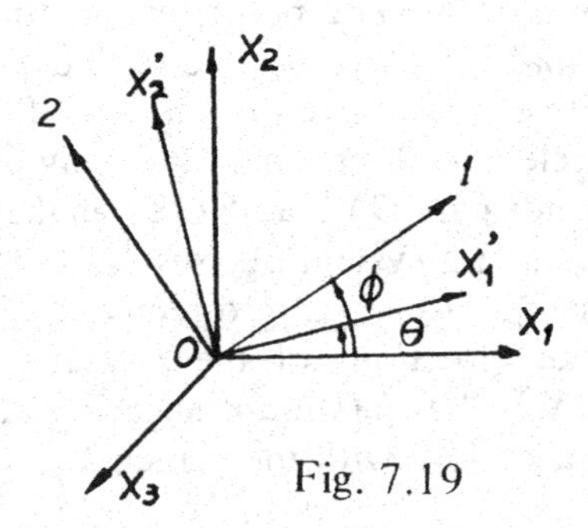

Fig. 7.19

$$\sigma_2 = \frac{\sigma_{11} + \sigma_{22}}{2} - \sqrt{\left(\frac{\sigma_{11} - \sigma_{22}}{2}\right)^2 + \sigma_{12}^2} \,. \qquad (7.11.9)$$

In the representation on Mohr's diagram, only one circle is of interest (Fig. 7.20b). Let us assume that we are given the components of the

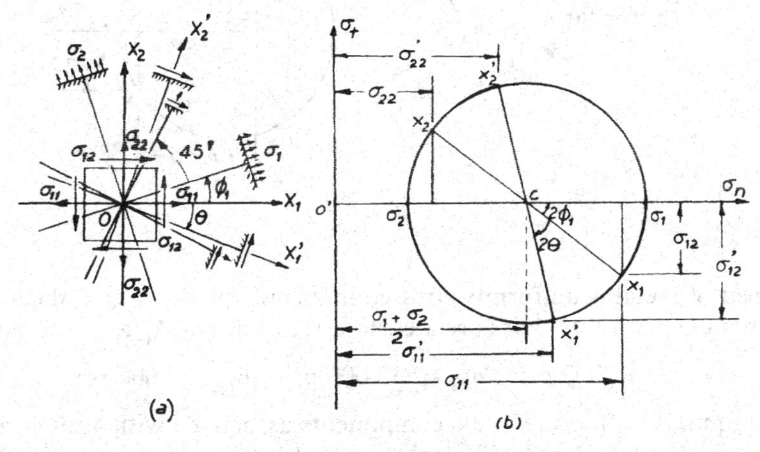

Fig. 7.20

state of stress σ_{11}, σ_{22}, and σ_{12} at a point O of a body related to a system of coordinates OX_1, OX_2. The convention for the normal stresses is that they are plotted positive to the right of the origin on the $o'\sigma_n$ axis and negative to its left. The convention for the tangential or shearing stresses is as follows: Consider the two planes normal to the two axes OX_1 and OX_2, and assume that one always goes from OX_1 to OX_2 through a counterclockwise motion. The point representing the stress σ_{12} on the plane whose normal is OX_1 is plotted at a distance σ_{12} below the $o'\sigma_n$ axis if σ_{12} is positive, and above the $o'\sigma_n$ axis if σ_{12} is negative. Fig. 7.20 shows Mohr's circle and the normal and tangential stresses on the pair of planes whose normals OX'_1 and OX'_2 make an angle θ (clockwise) with OX_1 and OX_2. In a system of axes OX'_1, OX'_2, the components of the state of stress σ'_{11}, σ'_{22}, σ'_{12} are directly read in magnitude and sign on the circle. The directions of the principal stresses make ϕ_1 and $\phi_1 + 90°$ with OX_1. The maximum shearing stresses occur on planes whose normals make 45 ° with the principal directions.

Example

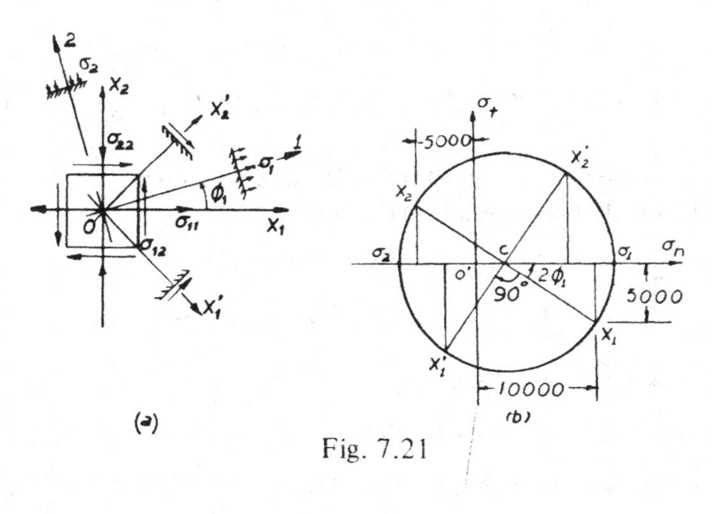

Fig. 7.21

A sheet of metal is uniformly stressed in its own plane, so that the stress components at all its points related to a set of axes OX_1 and OX_2 are:

$$\sigma_{11} = 10{,}000 \text{ psi}, \quad \sigma_{22} = -5{,}000 \text{ psi}, \quad \sigma_{12} = 5{,}000 \text{ psi} .$$

It is required to find the stress components associated with a set of axes OX'_1 and OX'_2, inclined 45° clockwise to the OX_1, OX_2 set as shown in Fig. 7.21. It is also required to find the principal stresses and the directions of the principal axes.

Fig. 7.2lb shows Mohr's circle constructed from the given data. The magnitude of the stresses on the various planes can be directly read on the circle, and their actual direction as well as the direction of the shearing and normal stresses can be plotted (Fig. 7.2la) following the previously established sign conventions. Thus,

$$\sigma'_{11} = -2{,}500 \text{ psi}, \quad \sigma'_{22} = 7{,}500 \text{ psi}, \quad \sigma'_{12} = +7{,}500 \text{ psi}$$

and

$$\sigma_1 = 11{,}520 \text{ psi}, \quad \sigma_2 = -6{,}520, \quad \tan 2\phi_1 = \frac{2}{3}.\text{(*)}$$

7.12 Equations of Equilibrium in Orthogonal Curvilinear Coordinates

The equations of equilibrium will be obtained by considering an elementary curvilinear parallelepiped and writing the equilibrium of the forces acting on it (Fig. 7.22). Let the stress vectors acting on the faces

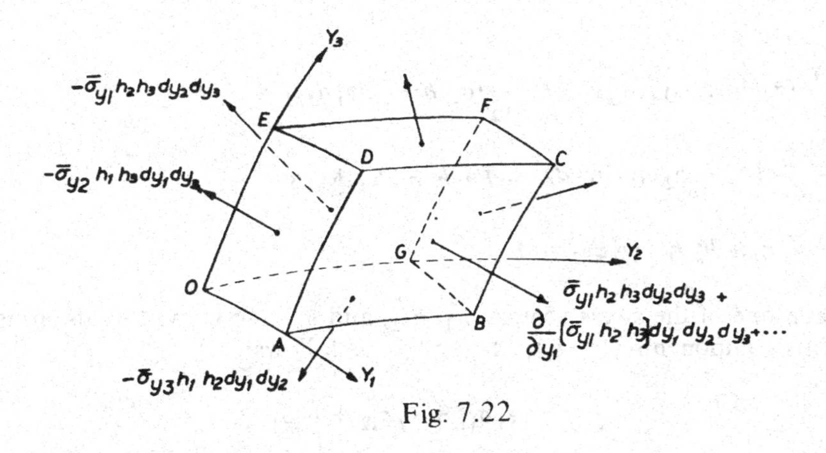

Fig. 7.22

OGFE, *OADE*, and *OABG* be denoted by $\bar{\sigma}_{y1}$, $\bar{\sigma}_{y2}$, and $\bar{\sigma}_{y3}$, respectively. The force acting on *OGFE* is equal to $-\bar{\sigma}_{y1} h_2 h_3 \, dy_2 \, dy_3$. The force acting on *ABCD* is equal to

$$\left[\bar{\sigma}_{y1} + \frac{\partial \bar{\sigma}_{y1}}{\partial y_1} dy_1\right]\left[h_2 h_3 \, dy_2 \, dy_3 + \frac{\partial}{\partial y_1}(h_2 h_3) dy_2 \, dy_3 \, dy_1\right]$$
$$= \bar{\sigma}_{y1} h_2 h_3 \, dy_2 \, dy_3 + \bar{\sigma}_{y1} \frac{\partial}{\partial y_1}(h_2 h_3) dy_1 \, dy_2 \, dy_3$$
$$+ h_2 h_3 \frac{\partial \bar{\sigma}_{y1}}{\partial y_1} dy_1 \, dy_2 \, dy_3 + \frac{\partial \bar{\sigma}_{y1}}{\partial y_1} \frac{\partial (h_2 h_3)}{\partial y_1} (dy_1)^2 dy_2 \, dy_3 .$$

* See pages A-47 to A-49 for additional information.

The last term is an infinitesimal of the fourth order and, as such, is negligible. Therefore, the force on face $ABCD$ is equal to

$$\bar{\sigma}_{y1} h_2 h_3 \, dy_2 \, dy_3 + \frac{\partial}{\partial y_1}(\bar{\sigma}_{y1} h_2 h_3) dy_1 \, dy_2 \, dy_3 .$$

The same steps can be repeated for the forces acting on the four other faces of the parallelepiped. The body force vector is $(\bar{F} h_1 h_2 h_3 \, dy_1 \, dy_2 \, dy_3)$, and the inertia force vector in case of motion is $(\bar{A} \rho h_1 h_2 h_3 \, dy_1 \, dy_2 \, dy_3)$. $\bar{A}$ is the acceleration vector and ρ the mass per unit volume of the body. Therefore, the equations of equilibrium in vector form are written:

$$\begin{aligned} &\frac{\partial}{\partial y_1}(\bar{\sigma}_{y1} h_2 h_3) dy_1 \, dy_2 \, dy_3 + \frac{\partial}{\partial y_2}(\bar{\sigma}_{y2} h_1 h_3) dy_1 \, dy_2 \, dy_3 \\ &+ \frac{\partial}{\partial y_3}(\bar{\sigma}_{y3} h_1 h_2) dy_1 \, dy_2 \, dy_3 + \bar{F} h_1 h_2 h_3 \, dy_1 \, dy_2 \, dy_3 \\ &- \bar{A} \rho h_1 h_2 h_3 \, dy_1 \, dy_2 \, dy_3 = 0. \end{aligned} \tag{7.12.1}$$

Each one of the stress vectors $\bar{\sigma}_{y1}$, $\bar{\sigma}_{y2}$, and $\bar{\sigma}_{y3}$ can be written in terms of its components along Y_1, Y_2, and Y_3 as follows:

$$\begin{aligned} \bar{\sigma}_{y1} &= \bar{e}_1 \sigma_{11} + \bar{e}_2 \sigma_{12} + \bar{e}_3 \sigma_{13} \\ \bar{\sigma}_{y2} &= \bar{e}_1 \sigma_{21} + \bar{e}_2 \sigma_{22} + \bar{e}_3 \sigma_{23} \\ \bar{\sigma}_{y3} &= \bar{e}_1 \sigma_{31} + \bar{e}_2 \sigma_{32} + \bar{e}_3 \sigma_{33} . \end{aligned} \tag{7.12.2}$$

Substituting Eqs. (7.12.2) into Eq. (7.12.1), we get:

$$\begin{aligned} &\frac{\partial}{\partial y_1}[h_2 h_3 (\bar{e}_1 \sigma_{11} + \bar{e}_2 \sigma_{12} + \bar{e}_3 \sigma_{13})] \\ &+ \frac{\partial}{\partial y_2}[h_1 h_3 (\bar{e}_1 \sigma_{21} + \bar{e}_2 \sigma_{22} + \bar{e}_3 \sigma_{23})] \\ &+ \frac{\partial}{\partial y_3}[h_1 h_2 (\bar{e}_1 \sigma_{31} + \bar{e}_2 \sigma_{32} + \bar{e}_3 \sigma_{33})] + h_1 h_2 h_3 (\bar{F} - \rho \bar{A}) = 0. \end{aligned} \tag{7.12.3}$$

The equations established in Sec. 6.5 are now used in taking the partial derivatives of $\bar{e}_1$, $\bar{e}_2$, and $\bar{e}_3$: Thus,

$$\begin{aligned}
\frac{\partial}{\partial y_1}&\left[h_2 h_3(\bar{e}_1\sigma_{11} + \bar{e}_2\sigma_{12} + \bar{e}_3\sigma_{13})\right] \\
&= \bar{e}_1\left[\frac{\partial}{\partial y_1}(\sigma_{11}h_2h_3) + h_3\sigma_{12}\frac{\partial h_1}{\partial y_2} + h_2\sigma_{13}\frac{\partial h_1}{\partial y_3}\right] \\
&+ \bar{e}_2\left[\frac{\partial}{\partial y_1}(\sigma_{12}h_2h_3) - h_3\sigma_{11}\frac{\partial h_1}{\partial y_2}\right] \\
&+ \bar{e}_3\left[\frac{\partial}{\partial y_1}(\sigma_{13}h_2h_3) - h_2\sigma_{11}\frac{\partial h_1}{\partial y_3}\right]
\end{aligned} \tag{7.12.4}$$

$$\begin{aligned}
\frac{\partial}{\partial y_2}&\left[h_1 h_3(\bar{e}_1\sigma_{21} + \bar{e}_2\sigma_{22} + \bar{e}_3\sigma_{23})\right] \\
&= \bar{e}_1\left[\frac{\partial}{\partial y_2}(\sigma_{21}h_1h_3) - h_3\sigma_{22}\frac{\partial h_2}{\partial y_1}\right] \\
&+ \bar{e}_2\left[\frac{\partial}{\partial y_2}(\sigma_{22}h_1h_3) + h_3\sigma_{21}\frac{\partial h_2}{\partial y_1} + h_1\sigma_{23}\frac{\partial h_2}{\partial y_3}\right] \\
&+ \bar{e}_3\left[\frac{\partial}{\partial y_2}(\sigma_{23}h_1h_3) - h_1\sigma_{22}\frac{\partial h_2}{\partial y_3}\right]
\end{aligned} \tag{7.12.5}$$

$$\begin{aligned}
\frac{\partial}{\partial y_3}&\left[h_1 h_2(\bar{e}_1\sigma_{31} + \bar{e}_2\sigma_{32} + \bar{e}_3\sigma_{33})\right] \\
&= \bar{e}_1\left[\frac{\partial}{\partial y_3}(h_1h_2\sigma_{31}) - h_2\sigma_{33}\frac{\partial h_3}{\partial y_1}\right] \\
&+ \bar{e}_2\left[\frac{\partial}{\partial y_3}(h_1h_2\sigma_{32}) - h_1\sigma_{33}\frac{\partial h_2}{\partial y_2}\right] \\
&+ \bar{e}_3\left[\frac{\partial}{\partial y_3}(h_1h_2\sigma_{33}) + h_2\sigma_{31}\frac{\partial h_3}{\partial y_1} + h_1\sigma_{32}\frac{\partial h_3}{\partial y_2}\right].
\end{aligned} \tag{7.12.6}$$

By substituting Eqs. (7.12.4) to (7.12.6) into Eq. (7.12.3), and factoring $\bar{e}_1$, $\bar{e}_2$, and $\bar{e}_3$, one obtains a vector equation of equilibrium in a form which expresses the projection of the forces in the $\bar{e}_1$ direction, the $\bar{e}_2$ direction, and the $\bar{e}_3$ direction. This allows us to write the three scalar equations of equilibrium along the tangents to the three curvilinear coordinates as follows:

$$\frac{\partial}{\partial y_1}(\sigma_{11} h_2 h_3) + \frac{\partial}{\partial y_2}(\sigma_{21} h_1 h_3) + \frac{\partial}{\partial y_3}(\sigma_{31} h_1 h_2) + \sigma_{12} h_3 \frac{\partial h_1}{\partial y_2}$$
$$+ \sigma_{13} h_2 \frac{\partial h_1}{\partial y_3} - \sigma_{22} h_3 \frac{\partial h_2}{\partial y_1} - \sigma_{33} h_2 \frac{\partial h_3}{\partial y_1} + h_1 h_2 h_3 (F_1 - \rho A_1) = 0 \tag{7.12.7}$$

$$\frac{\partial}{\partial y_1}(\sigma_{12} h_2 h_3) + \frac{\partial}{\partial y_2}(\sigma_{22} h_1 h_3) + \frac{\partial}{\partial y_3}(\sigma_{32} h_1 h_2) + \sigma_{23} h_1 \frac{\partial h_2}{\partial y_3}$$
$$+ \sigma_{21} h_3 \frac{\partial h_2}{\partial y_1} - \sigma_{33} h_1 \frac{\partial h_3}{\partial y_2} - \sigma_{11} h_3 \frac{\partial h_1}{\partial y_2} + h_1 h_2 h_3 (F_2 - \rho A_2) = 0 \tag{7.12.8}$$

$$\frac{\partial}{\partial y_1}(\sigma_{13} h_2 h_3) + \frac{\partial}{\partial y_2}(\sigma_{23} h_1 h_3) + \frac{\partial}{\partial y_3}(\sigma_{33} h_1 h_2) + \sigma_{31} h_2 \frac{\partial h_3}{\partial y_1}$$
$$+ \sigma_{32} h_1 \frac{\partial h_3}{\partial y_2} - \sigma_{11} h_2 \frac{\partial h_1}{\partial y_3} - \sigma_{22} h_1 \frac{\partial h_2}{\partial y_3} + h_1 h_2 h_3 (F_3 - \rho A_3) = 0. \tag{7.12.9}$$

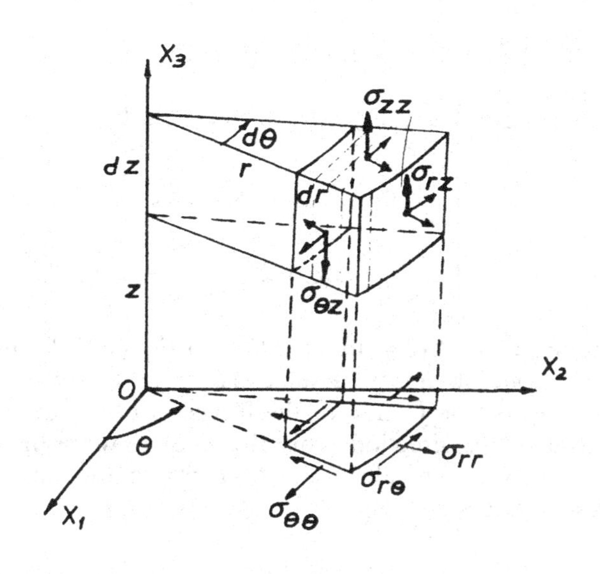

Fig. 7.23

Example 1. Cylindrical Coordinates (Fig. 7.23)
In this case, $h_1 = 1$, $h_2 = r$, and $h_3 = 1$. The subscripts r, θ, z are substituted for 1, 2, 3, and dr, $d\theta$, dz are substituted for dy_1, dy_2, dy_3, respectively, in Eqs. (7.12.7) to (7.12.9). These equations become:

$$\begin{aligned}
&\frac{\partial\sigma_{rr}}{\partial r} + \frac{1}{r}\frac{\partial\sigma_{\theta r}}{\partial\theta} + \frac{\partial\sigma_{zr}}{\partial z} + \frac{1}{r}(\sigma_{rr} - \sigma_{\theta\theta}) + F_r - \rho A_r = 0 \\
&\frac{\partial\sigma_{r\theta}}{\partial r} + \frac{1}{r}\frac{\partial\sigma_{\theta\theta}}{\partial\theta} + \frac{\partial\sigma_{z\theta}}{\partial z} + \frac{2}{r}\sigma_{r\theta} + F_\theta - \rho A_\theta = 0 \\
&\frac{\partial\sigma_{rz}}{\partial r} + \frac{1}{r}\frac{\partial\sigma_{\theta z}}{\partial\theta} + \frac{\partial\sigma_{zz}}{\partial z} + \frac{1}{r}\sigma_{rz} + F_z - \rho A_z = 0.
\end{aligned} \tag{7.12.10}$$

Example 2. Spherical Polar Coordinates (Fig. 7.24)

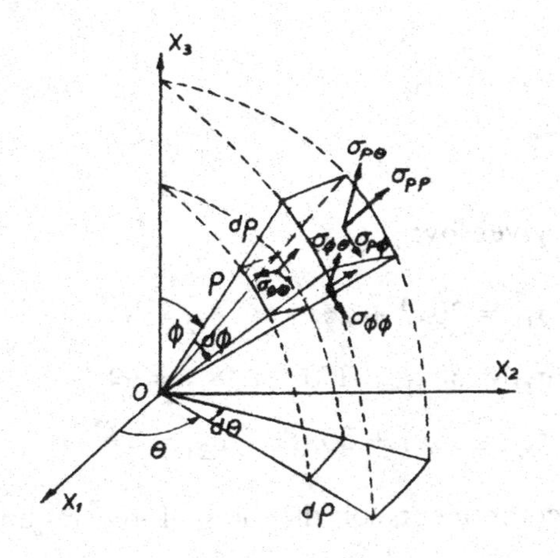

Fig. 7.24

In this case, $h_1 = \rho$, $h_2 = \rho \sin \phi$, and $h_3 = 1$. The subscripts ϕ, θ, ρ are substituted for 1, 2, 3, and $d\phi$, $d\theta$, $d\rho$ are substituted for dy_1, dy_2, dy_3, respectively, in Eqs. (7.12.7) to (7.12.9). The mass per unit volume here is called γ. Eqs. (7.12.7) to (7.12.9) become:

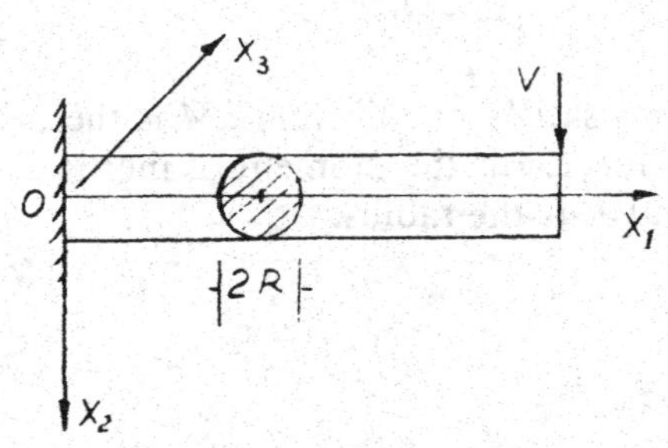

Fig . 7.25

$$\frac{\partial \sigma_{\rho\phi}}{\partial \rho} + \frac{1}{\rho}\frac{\partial \sigma_{\phi\phi}}{\partial \phi} + \frac{1}{\rho \sin \phi}\frac{\partial \sigma_{\theta\phi}}{\partial \theta} + \frac{3}{\rho}\sigma_{\rho\phi} + \frac{\cos \phi}{\rho \sin \phi}(\sigma_{\phi\phi} - \sigma_{\theta\theta}) + F_\phi - \gamma A_\phi = 0$$

$$\frac{\partial \sigma_{\rho\theta}}{\partial \rho} + \frac{1}{\rho}\frac{\partial \sigma_{\phi\theta}}{\partial \phi} + \frac{1}{\rho \sin \phi}\frac{\partial \sigma_{\theta\theta}}{\partial \theta} + \frac{3}{\rho}\sigma_{\rho\theta} + \frac{2\cos \phi}{\rho \sin \phi}\sigma_{\phi\theta} + F_\theta - \gamma A_\theta = 0 \qquad (7.12.11)$$

$$\frac{\partial \sigma_{\rho\rho}}{\partial \rho} + \frac{1}{\rho}\frac{\partial \sigma_{\phi\rho}}{\partial \phi} + \frac{1}{\rho \sin \phi}\frac{\partial \sigma_{\theta\rho}}{\partial \theta} + \frac{\cos \phi}{\rho \sin \phi}\sigma_{\rho\phi} + \frac{1}{\rho}[2\sigma_{\rho\rho} - \sigma_{\phi\phi} - \sigma_{\theta\theta}] + F_\rho - \gamma A_\rho = 0.$$

PROBLEMS

1. A stress field is given by:

$$\sigma_{11} = 20x_1^3 + x_2^2 \qquad \sigma_{12} = x_3$$
$$\sigma_{22} = 30x_1^3 + 200 \qquad \sigma_{13} = x_2^3$$
$$\sigma_{33} = 30x_2^2 + 30x_3^3 \qquad \sigma_{23} = x_1^3.$$

What are the components of the body force required to insure equilibrium?

2. The usual engineering equations for the stresses due to the bending of a circular beam are (Fig. 7.25):

$$\sigma_{11} = \frac{Mx_2}{I} \qquad \sigma_{12} = \frac{V(R^2 - x_2^2)}{3I}$$
$$\sigma_{22} = 0 \qquad \sigma_{13} = 0 \qquad I = \frac{\Pi R^4}{4}$$
$$\sigma_{33} = 0 \qquad \sigma_{23} = 0$$

Do these equations satisfy equilibrium? M is the bending moment, V is the shearing force, I is the moment of inertia about a diameter of the section, and R is the radius.

3. The stress field in a continuous body is given by:

$$[\sigma_{ij}] = 10^3 \begin{bmatrix} 1 & 0 & 2x_2 \\ 0 & 1 & 4x_1 \\ 2x_2 & 4x_1 & 1 \end{bmatrix} \text{ psi.}$$

Find the stress vector $\bar{\sigma}$ at a point M (1, 2, 3), acting on a plane $x_1 + x_2 + x_3 = 6$.

4. The state of stresses at a point is given by:

$$[\sigma_{ij}] = 10^2 \begin{bmatrix} 10 & 5 & -10 \\ 5 & 20 & -15 \\ -10 & -15 & -10 \end{bmatrix} \text{ psi.}$$

Find the magnitude and direction of the stress vector acting on a plane whose normal has direction cosines $(1/2, 1/2, 1/\sqrt{2})$; what are the normal and tangential stresses acting on this plane?

5. In a solid circular shaft subjected to pure torsion, the stress field is given by:

$$[\sigma_{ij}] = \begin{bmatrix} 0 & 0 & -Cx_2 \\ 0 & 0 & Cx_1 \\ -Cx_2 & Cx_1 & 0 \end{bmatrix},$$

where C is a constant. At the point whose coordinates are (1, 2, 4), find:

(a) the principal stresses
(b) the principal directions
(c) the maximum shearing stress and the plane on which it acts.

6. At a point M of a continuous body, the components of the stress tensor are:

$$[\sigma_{ij}] = 10^3 \begin{bmatrix} 1 & -3 & \sqrt{2} \\ -3 & 4 & -\sqrt{2} \\ \sqrt{2} & -\sqrt{2} & 4 \end{bmatrix} \text{ psi.}$$

(a) Find the principal stresses and the principal directions.
(b) Draw Mohr's circles, and obtain the normal and tangential stresses on a plane whose normal has direction cosines

$(1/\sqrt{3}, 1/\sqrt{3}, 1/\sqrt{3})$ with respect to the reference axes.

(c) Find the octahedral normal and shearing stresses.

(d) What are the invariants of the spherical and the deviatoric components of this stress tensor?

(e) What is the equation of the stress quadric?

7. Find the components of the stress tensor of Problem 4 in a system of coordinates whose axes have direction cosines (0, 0, 1), $(1/\sqrt{2}, 1/\sqrt{2}, 0)$, $(1/\sqrt{2}, -1/\sqrt{2}, 0)$.

8. A very thin plate is uniformly loaded as shown in Fig. 7.26. Among all the planes that are normal to the plane of the plate, which ones are the principal planes and what is the value of the stresses to which they are subjected?

9. For the following states of stress at a point, use Mohr's circle to obtain the magnitude and directions of the principal stresses:

$(a)\sigma_{11} = 4{,}000$ psi
$\sigma_{22} = 0$
$\sigma_{12} = 8{,}000$ psi
$\sigma_{13} = \sigma_{23} = \sigma_{33} = 0$

$(b)\sigma_{11} = 14{,}000$ psi
$\sigma_{22} = 5{,}000$ psi
$\sigma_{12} = -6{,}000$ psi
$\sigma_{13} = \sigma_{23} = \sigma_{33} = 0$

$(c)\sigma_{11} = 12{,}000$ psi
$\sigma_{22} = 5{,}000$ psi
$\sigma_{12} = 10{,}000$ psi
$\sigma_{13} = \sigma_{23} = \sigma_{33} = 0.$

10. Obtain the equations of equilibrium in the two systems of coordinates defined in Problems 1 and 2 of Chapter 6.(*)

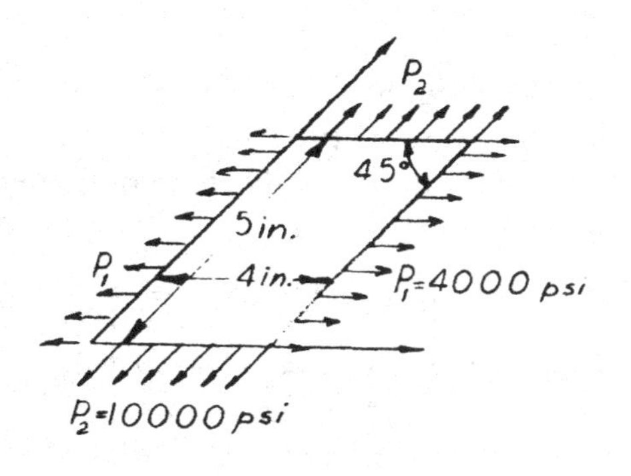

Fig. 7.26

* See pages A-50 and A-51 for additional problems.

PART III
THE THEORY OF ELASTICITY APPLICATIONS TO ENGINEERING PROBLEMS

CHAPTER 8

ELASTIC STRESS-STRAIN RELATIONS AND FORMULATION OF ELASTICITY PROBLEMS

8.1 Introduction

In the preceding chapters, the study of strain and the study of stress were pursued independently. Although certain engineering problems can be solved without relating the stresses to the strains, many require the simultaneous consideration of stress and strain. Constitutive relations connecting the stresses to the strains are therefore needed to solve this class of problems. In this chapter, the elastic relations will be developed. They will contain experimentally determined constants and for simplicity will be restricted in their applicability to linear strains. This restriction is not as drastic as it may appear since experiments have shown that, in their working range, a large number of structural materials behave in a linearly elastic way with deformations very adequately described by the components of linear strain.

8.2 Work, Energy, and the Existence of a Strain Energy Function

A body subjected to external forces (body forces and surface forces) will deform. These forces will do some work and the body will acquire an internal energy which will depend upon its shape and its temperature distribution. This internal energy will be calculated with reference to a standard state of chosen uniform temperature and zero strain. If the deformation occurs under adiabatic or isothermal conditions, the internal energy is identified with the strain energy [4].

Consider a body occupying a volume V, enclosed in a surface S, and in a deformed state of equilibrium. During the deformation, let dW_B and dW_S denote the work done on the volume V by the body forces and by the surface forces, respectively. If the process of deformation is adiabatic, the first law of thermodynamics yields:

$$dW_B + dW_S = d\iiint_V U\,dV = \iiint_V (dU)\,dV, \tag{8.2.1}$$

where U is called the strain energy density. The work done by the body forces is expressed by the formula (see Fig. 7.7):

$$dW_B = \iiint_V [F_1\,du_1 + F_2\,du_2 + F_3\,du_3]\,dV. \tag{8.2.2}$$

The work done by the surface forces is expressed by the formula:

$$dW_S = \iint_S (\sigma_{n1}\,du_1 + \sigma_{n2}\,du_2 + \sigma_{n3}\,du_3)\,dS. \tag{8.2.3}$$

Substituting Eq. (7.3.8) into Eq. (8.2.3), we obtain:

$$\begin{aligned} dW_S = \iint_S [(&\ell_1\sigma_{11} + \ell_2\sigma_{21} + \ell_3\sigma_{31})du_1 \\ &+ (\ell_1\sigma_{12} + \ell_2\sigma_{22} + \ell_3\sigma_{32})du_2 + \\ &(\ell_1\sigma_{13} + \ell_2\sigma_{23} + \ell_3\sigma_{33})du_3]\,dS. \end{aligned} \tag{8.2.4}$$

By the divergence theorem, this integral can be transformed into the following volume integral:

$$\begin{aligned} dW_S = \iiint_V \Big[&\frac{\partial}{\partial x_1}(\sigma_{11}\,du_1) + \frac{\partial}{\partial x_2}(\sigma_{21}\,du_1) + \frac{\partial}{\partial x_3}(\sigma_{31}\,du_1) \\ &+ \frac{\partial}{\partial x_1}(\sigma_{12}\,du_2) + \frac{\partial}{\partial x_2}(\sigma_{22}\,du_2) + \frac{\partial}{\partial x_3}(\sigma_{32}\,du_2) \\ &+ \frac{\partial}{\partial x_1}(\sigma_{13}\,du_3) + \frac{\partial}{\partial x_2}(\sigma_{23}\,du_3) + \frac{\partial}{\partial x_3}(\sigma_{33}\,du_3)\Big]\,dV. \end{aligned} \tag{8.2.5}$$

Substituting Eqs. (8.2.2) to (8.2.5) into Eq. (8.2.1), and using Eqs. (1.2.1) and (7.4.6), we obtain:

$$\iiint_V (dU)\,dV = \iiint_V (\sigma_{11}\,de_{11} + \sigma_{22}\,de_{22} + \sigma_{33}\,de_{33} + \sigma_{12}\,de_{12} + \sigma_{21}\,de_{21} + \sigma_{13}\,de_{13} + \sigma_{31}\,de_{31} + \sigma_{23}\,de_{23} + \sigma_{32}\,de_{32})\,dV = \iiint_V (\sigma_{ij}\,de_{ij})\,dV, \tag{8.2.6}$$

where de_{ij} represents the increments of the components of the linear strain. Therefore,

$$dU = \sigma_{11}\,de_{11} + \sigma_{22}\,de_{22} \ldots \sigma_{32}\,de_{32} = \sigma_{ij}\,de_{ij}. \tag{8.2.7}$$

Thus, the expression on the right-hand side is an exact differential and a function $U(e_{ij})$ exists such that,

$$\sigma_{11} = \frac{\partial U}{\partial e_{11}}, \quad \sigma_{22} = \frac{\partial U}{\partial e_{22}}, \quad \ldots, \quad \sigma_{32} = \frac{\partial U}{\partial e_{32}}. \tag{8.2.8}$$

Since the stress tensor is symmetric, then

$$\sigma_{ij} = \frac{1}{2}\left(\frac{\partial U}{\partial e_{ij}} + \frac{\partial U}{\partial e_{ji}}\right). \tag{8.2.9}$$

If, under the application of the external forces, the change in state is isothermal, it can be shown through the use of the second law of thermodynamics that a function U with the properties expressed by Eqs. (8.2.8) still exists. The function U is called the strain energy density function. *Therefore, the assumption that a process is of a reversible adiabatic or isothermal nature is implicit in the use of a strain energy density function.* Loads applied very slowly represent nearly isothermal conditions, and loads applied very rapidly represent adiabatic conditions [4].

The use of a strain energy density function implies perfect elasticity, in other words a one-to-one relationship between states of stress and strain, reversibility and path independence of stresses and strains, and complete satisfaction of the laws of thermodynamics. This perfect elasticity is referred to as Green's elasticity or Hyperelasticity [4]. It does not necessarily require that the relations between stress and strain be linear. Linearity is the added assumption on which Hooke's law is based. It will be shown in Sec. 8.7 that this law can be arrived at by neglecting all the terms higher than the quadratic in the expression of the strain energy density in terms of the linear strains.

8.3 The Generalized Hooke's Law

For a large number of hard solids, the measured strain is proportional to the load over a wide range of loads. This means that when the load increases, the measured strain increases in the same ratio, and when the load decreases, the measured strain decreases in the same ratio. Also, when the load is reduced to zero, the strain disappears. These experimental facts lead by inductive reasoning to the generalized Hooke's law of the proportionality of the stress and strain. The general form of the law is expressed by the statement: Each of the components of the state of stress at a point is a linear function of the components of the state of strain at the point. Mathematically, this is expressed by:

$$\sigma_{k\ell} = C_{k\ell mn} e_{mn}, \tag{8.3.1}$$

where the $C_{k\ell mn}$ are elasticity constants. There are 81 such constants corresponding to the indices k, ℓ, m, n taking values equal to 1, 2, and 3. For example,

$$\begin{aligned}\sigma_{12} = {} & C_{1211} e_{11} + C_{1222} e_{22} + C_{1233} e_{33} + C_{1212} e_{12} + C_{1221} e_{21} \\ & + C_{1213} e_{13} + C_{1231} e_{31} + C_{1223} e_{23} + C_{1232} e_{32}.\end{aligned} \tag{8.3.2}$$

Now, since the stress tensor is symmetric, i.e., since

$$\sigma_{k\ell} = \sigma_{\ell k}, \tag{8.3.3}$$

then

$$\sigma_{k\ell} = C_{k\ell mn} e_{mn} = \sigma_{\ell k} = C_{\ell kmn} e_{mn}. \tag{8.3.4}$$

Therefore,

$$C_{k\ell mn} = C_{\ell kmn}, \tag{8.3.5}$$

and the first pair of indices can be freely interchanged. It is also possible to prove that the second pair of indices can be freely interchanged. For that, let us assume that a body is in a state of strain such that the only strain component different from zero is $e_{12} = e_{21}$. For this special situation, Eq. (8.3.1) is written:

$$\sigma_{k\ell} = C_{k\ell 12} e_{12} + C_{k\ell 21} e_{21} \tag{8.3.6}$$

or

$$\sigma_{k\ell} = (C_{k\ell 12} + C_{k\ell 21})e_{12}. \tag{8.3.7}$$

Let us introduce the new constant $C'_{k\ell 12}$ defined by

$$C'_{k\ell 12} = \frac{1}{2}(C_{k\ell 12} + C_{k\ell 21}). \tag{8.3.8}$$

We see that $C'_{k\ell 12}$ is symmetric with respect to the two last indices. In terms of this new coefficient, Eq. (8.3.7) is written as follows:

$$\sigma_{k\ell} = 2C'_{k\ell 12}e_{12} = C'_{k\ell 12}e_{12} + C'_{k\ell 21}e_{21}. \tag{8.3.9}$$

From Eqs. (8.3.6) and (8.3.9), we see that one can always consider that the constants $C_{k\ell mn}$ are also symmetric with respect to the two last indices: In other words, these indices can be freely interchanged. This reduces the 81 elastic constants to 36. For example, Eq. (8.3.2) can also be written as follows:

$$\begin{aligned}\sigma_{12} = {} & C_{1211}e_{11} + C_{1222}e_{22} + C_{1233}e_{33} \\ & + 2(C_{1212}e_{12} + C_{1213}e_{13} + C_{1223}e_{23}).\end{aligned} \tag{8.3.10}$$

The existence of a strain energy density function, when the system is adiabatic or isothermal, allows us to go one step further. Indeed, if such a function exists, then according to Eqs. (8.2.8),

$$\begin{aligned}\frac{\partial U}{\partial e_{11}} &= \sigma_{11} = C_{1111}e_{11} + C_{1122}e_{22} + \ldots + C_{1132}e_{32} \\ \frac{\partial U}{\partial e_{22}} &= \sigma_{22} = C_{2211}e_{11} + C_{2222}e_{22} + \ldots + C_{2232}e_{32}.\end{aligned} \tag{8.3.11}$$

Hence,

$$\frac{\partial^2 U}{\partial e_{11}\,\partial e_{22}} = C_{1122} = C_{2211}$$

and, in general,

$$\frac{\partial^2 U}{\partial e_{k\ell}\,\partial e_{mn}} = C_{k\ell mn} = C_{mnk\ell}. \tag{8.3.12}$$

Eq. (8.3.12) shows that the elastic constants $C_{k\ell mn}$ are symmetric; in other words,

$$C_{k\ell mn} = C_{mnk\ell}. \tag{8.3.13}$$

Accordingly, the number of independent elastic coefficients for the general anisotropic linearly elastic material is reduced to 21. In addition, if certain symmetries exist in the material, this number will be further reduced. The generalized Hooke's law can now be written in matrix notation as follows:

$$\begin{bmatrix} \sigma_{11} \\ \sigma_{22} \\ \sigma_{33} \\ \sigma_{12} \\ \sigma_{13} \\ \sigma_{23} \end{bmatrix} = \begin{bmatrix} C_{1111} & C_{1122} & C_{1133} & C_{1112} & C_{1113} & C_{1123} \\ C_{2211} & C_{2222} & C_{2233} & C_{2212} & C_{2213} & C_{2223} \\ C_{3311} & C_{3322} & C_{3333} & C_{3312} & C_{3313} & C_{3323} \\ C_{1211} & C_{1222} & C_{1233} & C_{1212} & C_{1213} & C_{1223} \\ C_{1311} & C_{1322} & C_{1333} & C_{1312} & C_{1313} & C_{1323} \\ C_{2311} & C_{2322} & C_{2333} & C_{2312} & C_{2313} & C_{2323} \end{bmatrix} \begin{bmatrix} e_{11} \\ e_{22} \\ e_{33} \\ 2e_{12} \\ 2e_{13} \\ 2e_{23} \end{bmatrix}, \tag{8.3.14}$$

with $C_{k\ell mn} = C_{mnk\ell}$. The matrix of the elastic coefficients is a symmetric matrix. It is called the stiffness matrix.

Since the components of the stress and strain tensors are functions of the orientation of the system of reference axes, the elastic coefficients in Eq. (8.3.1) are also functions of this orientation. By the quotient rule, $C_{k\ell mn}$ is a tensor of rank four called the stiffness tensor. Therefore, in a new system of coordinates OX'_1, OX'_2, OX'_3 (Fig. 8.1):

$$C'_{prst} = \ell_{pk}\,\ell_{r\ell}\,\ell_{sm}\,\ell_{tn}\,C_{k\ell mn}. \tag{8.3.15}$$

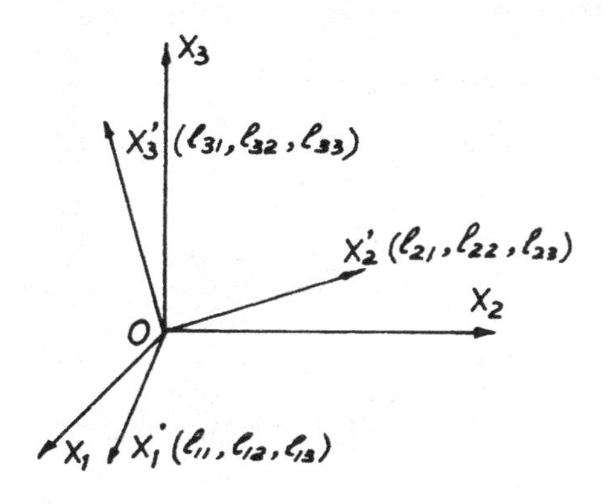

Fig. 8.1

The stress-strain relations given by Eq. (8.3.1) can be expressed in the inverted form,

$$e_{kl} = S_{klmn}\sigma_{mn}, \tag{8.3.16}$$

where S_{klmn} are constants. It is evident that S_{klmn} has the same symmetry properties as C_{klmn} and it transforms according to Eq. (5.3.4). S_{klmn} are the components of a tensor of the fourth order called the compliance tensor. In matrix notation, the generalized Hooke's law can be written using the compliance matrix as follows:

$$\begin{bmatrix} e_{11} \\ e_{22} \\ e_{33} \\ e_{12} \\ e_{13} \\ e_{23} \end{bmatrix} = \begin{bmatrix} S_{1111} & S_{1122} & S_{1133} & S_{1112} & S_{1113} & S_{1123} \\ S_{2211} & S_{2222} & S_{2233} & S_{2212} & S_{2213} & S_{2223} \\ S_{3311} & S_{3322} & S_{3333} & S_{3312} & S_{3313} & S_{3323} \\ S_{1211} & S_{1222} & S_{1233} & S_{1212} & S_{1213} & S_{1223} \\ S_{1311} & S_{1322} & S_{1333} & S_{1312} & S_{1313} & S_{1323} \\ S_{2311} & S_{2322} & S_{2333} & S_{2312} & S_{2313} & S_{2323} \end{bmatrix} \begin{bmatrix} \sigma_{11} \\ \sigma_{22} \\ \sigma_{33} \\ 2\sigma_{12} \\ 2\sigma_{13} \\ 2\sigma_{23} \end{bmatrix}, \tag{8.3.17}$$

with $S_{klmn} = S_{mnkl}$.

8.4 Elastic Symmetry

A type of symmetry is expressed by the statement that the coefficients C_{klmn} (or S_{klmn}) remain invariant under the transformation of coordinates which describes this symmetry. We shall consider the following cases: (1) symmetry with respect to a plane, (2) symmetry with respect to two mutually perpendicular planes, (3) symmetry of rotation with respect to one axis, and (4) symmetry of rotation with respect to two mutually perpendicular axes—in other words, isotropy.

(1) *Symmetry with Respect to One Plane*: A material which exhibits symmetry of its elastic properties with respect to one plane is called a monoclinic material. Let us take this plane to be the OX_1, OX_2 plane (Fig. 8.2). This symmetry is expressed by the requirement that the elastic constants do not change under a change from the system OX_1, OX_2, OX_3 to the systems OX'_1, OX'_2, OX'_3. The direction cosines of the new axes with respect to the initial ones are (1, 0, 0), (0, 1, 0), and (0, 0,-1). From Eq. (8.3.15), we must, for example, have:

$$C'_{1111} = \ell_{1k}\ell_{1l}\ell_{1m}\ell_{1n}C_{klmn} = C_{1111}, \tag{8.4.1}$$

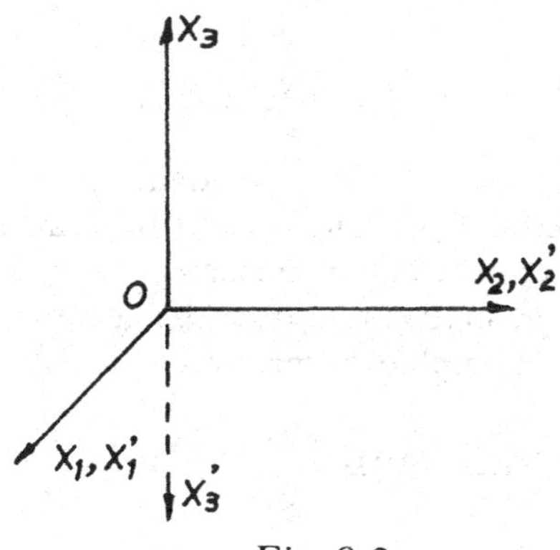

Fig. 8.2

which is true. The expansion in Eq. (8.4.1) is simplified, since we have only three non-zero direction cosines, namely:

$$\ell_{11} = 1, \quad \ell_{22} = 1, \quad \ell_{33} = -1. \tag{8.4.2}$$

In a similar way, for this type of symmetry, we must have:

$$C'_{1123} = \ell_{1k}\,\ell_{1l}\,\ell_{2m}\,\ell_{3n}\,C_{klmn} = C_{1123}, \tag{8.4.3}$$

which is impossible since

$$\ell_{1k}\,\ell_{1l}\,\ell_{2m}\,\ell_{3n}\,C_{klmn} = -C_{1123}. \tag{8.4.4}$$

Therefore, C_{1123} must be equal to zero. A similar reasoning will show that the number of elements of the stiffness matrix is reduced to thirteen. The matrix is written as follows:

$$\begin{bmatrix} C_{1111} & C_{1122} & C_{1133} & C_{1112} & 0 & 0 \\ C_{1122} & C_{2222} & C_{2233} & C_{2212} & 0 & 0 \\ C_{1133} & C_{2233} & C_{3333} & C_{3312} & 0 & 0 \\ C_{1112} & C_{2212} & C_{3312} & C_{1212} & 0 & 0 \\ 0 & 0 & 0 & 0 & C_{1313} & C_{1323} \\ 0 & 0 & 0 & 0 & C_{1323} & C_{2323} \end{bmatrix}. \tag{8.4.5}$$

It is to be noticed that any subsequent rotation of axes will bring in non-zero terms in the matrix (8.4.5). The matrix will, however, remain symmetric and the number of independent coefficients will remain 13. A similar reasoning in terms of compliance leads to a matrix of the same form as (8.4.5). In this case, the stress-strain relations are written as follows:

$$\begin{bmatrix} e_{11} \\ e_{22} \\ e_{33} \\ e_{12} \\ e_{13} \\ e_{23} \end{bmatrix} = \begin{bmatrix} S_{1111} & S_{1122} & S_{1133} & S_{1112} & 0 & 0 \\ S_{1122} & S_{2222} & S_{2233} & S_{2212} & 0 & 0 \\ S_{1133} & S_{2233} & S_{3333} & S_{3312} & 0 & 0 \\ S_{1112} & S_{2212} & S_{3312} & S_{1212} & 0 & 0 \\ 0 & 0 & 0 & 0 & S_{1313} & S_{1323} \\ 0 & 0 & 0 & 0 & S_{1323} & S_{2323} \end{bmatrix} \begin{bmatrix} \sigma_{11} \\ \sigma_{22} \\ \sigma_{33} \\ 2\sigma_{12} \\ 2\sigma_{13} \\ 2\sigma_{23} \end{bmatrix}. \tag{8.4.6}$$

Eq. (8.4.6) shows what type of strains result when each one of the components of the stress tensor is applied individually. For example, the application of a stress σ_{33} along the OX_3 axis, results in three normal strains e_{11}, e_{22}, e_{33} and one shear strain e_{12}; both e_{13} and e_{23} are equal to zero. The application of a stress σ_{13} will cause no normal strain; just shear strains e_{13} and e_{23}.

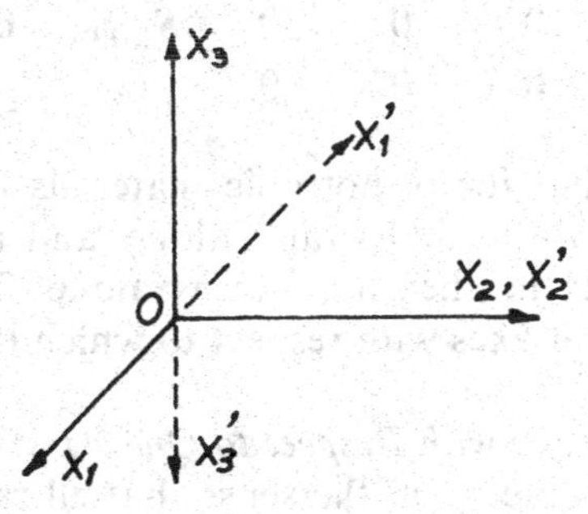

Fig. 8.3

(2) *Symmetry with Respect to Two Orthogonal Planes*: A material which exhibits symmetry of its elastic properties with respect to two orthogonal planes is called an orthotropic material. Let the two planes be the OX_1, OX_2 plane and the OX_2, OX_3 plane (Fig. 8.3). The direction cosines of the new axes with respect to the initial ones are in this case

(-1, 0, 0), (0, 1, 0), and (0, 0, -1). As was done in the case of monoclinic materials, the application of the transformation law (8.3.15) with the given direction cosines will lead to contradictions of the type shown in Eqs. (8.4.3) and (8.4.4). These contradictions are again resolved by setting the elastic constants equal to zero. The number of elastic constants is reduced to nine. The stiffness matrix is written as follows:

$$\begin{bmatrix} C_{1111} & C_{1122} & C_{1133} & 0 & 0 & 0 \\ C_{1122} & C_{2222} & C_{2233} & 0 & 0 & 0 \\ C_{1133} & C_{2233} & C_{3333} & 0 & 0 & 0 \\ 0 & 0 & 0 & C_{1212} & 0 & 0 \\ 0 & 0 & 0 & 0 & C_{1313} & 0 \\ 0 & 0 & 0 & 0 & 0 & C_{2323} \end{bmatrix}. \tag{8.4.7}$$

The compliance matrix has the same form as the stiffness matrix. Using the compliance matrix, the stress-strain relations are written as follows:

$$\begin{bmatrix} e_{11} \\ e_{22} \\ e_{33} \\ e_{12} \\ e_{13} \\ e_{23} \end{bmatrix} = \begin{bmatrix} S_{1111} & S_{1122} & S_{1133} & 0 & 0 & 0 \\ S_{1122} & S_{2222} & S_{2233} & 0 & 0 & 0 \\ S_{1133} & S_{2233} & S_{3333} & 0 & 0 & 0 \\ 0 & 0 & 0 & S_{1212} & 0 & 0 \\ 0 & 0 & 0 & 0 & S_{1313} & 0 \\ 0 & 0 & 0 & 0 & 0 & S_{2323} \end{bmatrix} \begin{bmatrix} \sigma_{11} \\ \sigma_{22} \\ \sigma_{33} \\ 2\sigma_{12} \\ 2\sigma_{13} \\ 2\sigma_{23} \end{bmatrix}.$$

This equation shows that for orthotropic materials the application of normal stresses results in normal strains alone, and the application of shearing stresses results in shearing strains alone. This is only true, however, in the system of axes with respect to which the symmetries are defined.

(3) *Symmetry of Rotation with Respect to One Axis*: A material which possesses an axis of symmetry, in the sense that all rays at right angles to this axis are equivalent, is called transversely isotropic or cross anisotropic. The symmetry is expressed by the requirement that the elastic constants are unaltered in any rotation θ around the axis of symmetry (Fig. 8.4). Taking OX_3 as the axis of symmetry, the direction cosines of the new axes with respect to the initial ones are ($\cos\theta, \sin\theta$, 0), ($-\sin\theta, \cos\theta$, 0), and (0, 0, 1).

Instead of starting from Eq. (8.3.15), we shall start from the equations

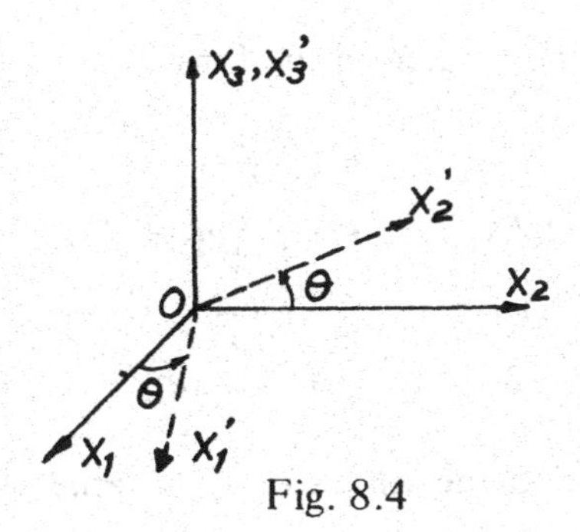

Fig. 8.4

of elasticity (8.3.1) in the initial and the transformed system and keep the elasticity coefficients unchanged. In the OX_1, OX_2, OX_3 system, the elastic stress-strain relations are written:

$$\sigma_{kl} = C_{klmn} e_{mn}, \tag{8.4.8}$$

and in the OX'_1, OX'_2, OX'_3 system, the elastic stress-strain relations are written:

$$\sigma'_{pr} = C_{prst} e'_{st}. \tag{8.4.9}$$

Notice that in Eq. (8.4.9) C_{prst} is unprimed. For a rotation of axes around the OX_3 axis:

$$\begin{aligned}
e'_{11} &= \cos^2\theta(e_{11}) + 2\cos\theta\sin\theta(e_{12}) + \sin^2\theta(e_{22}) \\
e'_{22} &= \sin^2\theta(e_{11}) - 2\cos\theta\sin\theta(e_{12}) + \cos^2\theta(e_{22}) \\
e'_{33} &= e_{33} \\
e'_{12} &= (e_{22} - e_{11})\cos\theta\sin\theta + e_{12}(\cos^2\theta - \sin^2\theta) \\
e'_{13} &= \cos\theta(e_{13}) + \sin\theta(e_{23}) \\
e'_{23} &= -\sin\theta(e_{13}) + \cos\theta(e_{23}).
\end{aligned} \tag{8.4.10}$$

The components of the stress tensor σ_{kl} will transform exactly in the same way. Let us, for instance, consider

$$\sigma'_{33} = \sigma_{33}, \tag{8.4.11}$$

a relation which may be written in the form:

$$\begin{aligned} &C_{3311} e'_{11} + C_{3322} e'_{22} + C_{3333} e'_{33} \\ &\quad + 2(C_{3312} e'_{12} + C_{3313} e'_{13} + C_{3323} e'_{23}) \\ &\quad = C_{3311} e_{11} + C_{3322} e_{22} + C_{3333} e_{33} \\ &\qquad + 2(C_{3312} e_{12} + C_{3313} e_{13} + C_{3323} e_{23}). \end{aligned} \tag{8.4.12}$$

Inserting in this equation the values of e'_{ij} obtained from Eqs. (8.4.10), we find that

$$\begin{aligned} &C_{3311}[\cos^2\theta(e_{11}) + 2 \cos \theta \sin \theta(e_{12}) + \sin^2\theta(e_{22})] \\ &\quad + C_{3322}[\sin^2\theta(e_{11}) - 2 \cos \theta \sin \theta(e_{12}) + \cos^2\theta(e_{22})] \\ &\quad + 2C_{3312}[(e_{22} - e_{11})\cos \theta \sin \theta + e_{12}(\cos^2\theta - \sin^2\theta)] \\ &\quad + 2C_{3313}[\cos \theta(e_{13})] + \sin \theta(e_{23})] \\ &\quad + 2C_{3323}[-\sin \theta(e_{13}) + \cos \theta(e_{23})] \\ &\quad = C_{3311} e_{11} + C_{3322} e_{22} + 2(C_{3312} e_{12} + C_{3313} e_{13} + C_{3323} e_{23}). \end{aligned} \tag{8.4.13}$$

Equating to zero the sum of the coefficients of e_{11} in this equation, we find that for all values of θ,

$$(C_{3311} - C_{3322})\sin^2\theta + 2 \sin \theta \cos \theta C_{3312} = 0, \tag{8.4.14}$$

from which it follows that

$$C_{3311} = C_{3322}, \quad C_{3312} = 0. \tag{8.4.15}$$

If we equate the sum of the coefficients of e_{22} and e_{12} to zero, we obtain the same results. If we equate the sum of the coefficients of e_{13} and e_{23} to zero, we find that

$$C_{3313} = C_{3323} = 0. \tag{8.4.16}$$

If we perform similar calculations for σ_{13} and σ'_{13}, we find, on equating the sum of the coefficients of e_{11} to zero, that

$$C_{1311} = C_{1322} = C_{1312} = C_{2311} = 0. \qquad (8.4.17)$$

The sum of the coefficients of e_{12}, when equated to zero, leads to

$$C_{1311} = C_{1322}, \quad C_{1312} = C_{2312} = 0; \qquad (8.4.18)$$

while the sum of the coefficients of e_{13} yields the relation:

$$C_{2313} = 0. \qquad (8.4.19)$$

Repeating these calculations for σ_{11} and σ'_{11}, we find, on equating to zero the sum of the coefficients of e_{11}, e_{22}, and e_{33}, that

$$C_{1211} = C_{1233} = 0, \quad C_{1133} = C_{2233}, \quad C_{1111} = C_{2222}. \qquad (8.4.20)$$

The sum of the coefficients of e_{12}, when equated to zero, yields:

$$C_{1212} = \tfrac{1}{2}(C_{1111} - C_{1122}), \quad C_{1222} = 0. \qquad (8.4.21)$$

Finally, if we consider the sum of the coefficients of e_{13} in the equation obtained from σ_{23} and σ'_{23}, we get

$$C_{2323} = C_{1313}. \qquad (8.4.22)$$

Substituting Eqs. (8.4.15) to (8.4.22) in the stiffness matrix, we get:

$$\begin{bmatrix} C_{1111} & C_{1122} & C_{1133} & 0 & 0 & 0 \\ C_{1122} & C_{1111} & C_{1133} & 0 & 0 & 0 \\ C_{1133} & C_{1133} & C_{3333} & 0 & 0 & 0 \\ 0 & 0 & 0 & \tfrac{1}{2}(C_{1111} - C_{1122}) & 0 & 0 \\ 0 & 0 & 0 & 0 & C_{1313} & 0 \\ 0 & 0 & 0 & 0 & 0 & C_{1313} \end{bmatrix}. \qquad (8.4.23)$$

The number of independent coefficients is reduced to five. A matrix of the same form can be written for the compliances. Using the compliance matrix, the stress-strain relations are written as follows:

$$\begin{bmatrix} e_{11} \\ e_{22} \\ e_{33} \\ e_{12} \\ e_{13} \\ e_{23} \end{bmatrix} = \begin{bmatrix} S_{1111} & S_{1122} & S_{1133} & 0 & 0 & 0 \\ S_{1122} & S_{1111} & S_{1133} & 0 & 0 & 0 \\ S_{1133} & S_{1133} & S_{3333} & 0 & 0 & 0 \\ 0 & 0 & 0 & \frac{1}{2}(S_{1111} - S_{1122}) & 0 & 0 \\ 0 & 0 & 0 & 0 & S_{1313} & 0 \\ 0 & 0 & 0 & 0 & 0 & S_{1313} \end{bmatrix} \begin{bmatrix} \sigma_{11} \\ \sigma_{22} \\ \sigma_{33} \\ 2\sigma_{12} \\ 2\sigma_{13} \\ 2\sigma_{23} \end{bmatrix}$$

(4) *Isotropy*: An isotropic material possesses elastic properties which are independent of the orientation of the axes. In other words, it is a material which possesses a rotational symmetry with respect to two perpendicular axes. By repeating the arguments of the previous subsection, we see that the elastic constants of an isotropic body are given by a matrix similar to Eq. (8.4.23) but with

$$C_{1313} = \tfrac{1}{2}(C_{1111} - C_{1122}), \quad C_{3333} = C_{1111}, \quad C_{1133} = C_{1122}, \qquad (8.4.24)$$

so that, in fact, we only have two independent constants. The stiffness matrix is written:

$$\begin{bmatrix} C_{1111} & C_{1122} & C_{1122} & 0 & 0 & 0 \\ C_{1122} & C_{1111} & C_{1122} & 0 & 0 & 0 \\ C_{1122} & C_{1122} & C_{1111} & 0 & 0 & 0 \\ 0 & 0 & 0 & \frac{1}{2}(C_{1111} - C_{1122}) & 0 & 0 \\ 0 & 0 & 0 & 0 & \frac{1}{2}(C_{1111} - C_{1122}) & 0 \\ 0 & 0 & 0 & 0 & 0 & \frac{1}{2}(C_{1111} - C_{1122}) \end{bmatrix} \qquad (8.4.25)$$

A matrix of the same form can be written for the compliances. The stress-strain relations in terms of the compliance matrix are written as follows:

$$\begin{bmatrix} e_{11} \\ e_{22} \\ e_{33} \\ e_{12} \\ e_{13} \\ e_{23} \end{bmatrix} = \begin{bmatrix} S_{1111} & S_{1122} & S_{1122} & 0 & 0 & 0 \\ S_{1122} & S_{1111} & S_{1122} & 0 & 0 & 0 \\ S_{1122} & S_{1122} & S_{1111} & 0 & 0 & 0 \\ 0 & 0 & 0 & \frac{1}{2}(S_{1111} - S_{1122}) & 0 & 0 \\ 0 & 0 & 0 & 0 & \frac{1}{2}(S_{1111} - S_{1122}) & 0 \\ 0 & 0 & 0 & 0 & 0 & \frac{1}{2}(S_{1111} - S_{1122}) \end{bmatrix} \begin{bmatrix} \sigma_{11} \\ \sigma_{22} \\ \sigma_{33} \\ 2\sigma_{12} \\ 2\sigma_{13} \\ 2\sigma_{23} \end{bmatrix} \tag{8.4.26}$$

The isotropic case is examined in detail in the following sections.(*)

8.5 Elastic Stress-Strain Relations for Isotropic Media

The elastic constants in the matrix (8.4.25) are usually written in the notation:

$$C_{1122} = \lambda, \quad C_{1212} = \tfrac{1}{2}(C_{1111} - C_{1122}) = \mu, \quad C_{1111} = \lambda + 2\mu. \tag{8.5.1}$$

The pair of constants λ and μ are called *Lamé's constants*, and μ is referred to as the *shear modulus* (also called G). The stress-strain relations for an isotropic material are now written as follows:

$$\begin{bmatrix} \sigma_{11} \\ \sigma_{22} \\ \sigma_{33} \\ \sigma_{12} \\ \sigma_{13} \\ \sigma_{23} \end{bmatrix} = \begin{bmatrix} \lambda + 2\mu & \lambda & \lambda & 0 & 0 & 0 \\ \lambda & \lambda + 2\mu & \lambda & 0 & 0 & 0 \\ \lambda & \lambda & \lambda + 2\mu & 0 & 0 & 0 \\ 0 & 0 & 0 & \mu & 0 & 0 \\ 0 & 0 & 0 & 0 & \mu & 0 \\ 0 & 0 & 0 & 0 & 0 & \mu \end{bmatrix} \begin{bmatrix} e_{11} \\ e_{22} \\ e_{33} \\ 2e_{12} \\ 2e_{13} \\ 2e_{23} \end{bmatrix}. \tag{8.5.2}$$

In index notation, these equations are written:

$$\sigma_{ij} = 2\mu\, e_{ij} + \lambda\, \delta_{ij}\, e_{nn}. \tag{8.5.3}$$

* See pages A-52 to A-54 for a discussion on anisotropy.

Eq. (8.5.3) can be solved to yield the expressions of the strains in terms of the stresses:

$$e_{ij} = \frac{-\lambda\, \delta_{ij}}{2\mu(3\lambda + 2\mu)}\sigma_{nn} + \frac{1}{2\mu}\sigma_{ij}. \tag{8.5.4}$$

Obviously, we must require that $\mu \neq 0$ and $3\lambda + 2\mu \neq 0$. In matrix form, we have:

$$\begin{bmatrix} e_{11} \\ e_{22} \\ e_{33} \\ e_{12} \\ e_{13} \\ e_{23} \end{bmatrix} = \begin{bmatrix} \frac{\lambda+\mu}{\mu(3\lambda+2\mu)} & \frac{-\lambda}{2\mu(3\lambda+2\mu)} & \frac{-\lambda}{2\mu(3\lambda+2\mu)} & 0 & 0 & 0 \\ \frac{-\lambda}{2\mu(3\lambda+2\mu)} & \frac{\lambda+\mu}{\mu(3\lambda+2\mu)} & \frac{-\lambda}{2\mu(3\lambda+2\mu)} & 0 & 0 & 0 \\ \frac{-\lambda}{2\mu(3\lambda+2\mu)} & \frac{-\lambda}{2\mu(3\lambda+2\mu)} & \frac{\lambda+\mu}{\mu(3\lambda+2\mu)} & 0 & 0 & 0 \\ 0 & 0 & 0 & \frac{1}{4\mu} & 0 & 0 \\ 0 & 0 & 0 & 0 & \frac{1}{4\mu} & 0 \\ 0 & 0 & 0 & 0 & 0 & \frac{1}{4\mu} \end{bmatrix} \begin{bmatrix} \sigma_{11} \\ \sigma_{22} \\ \sigma_{33} \\ 2\sigma_{12} \\ 2\sigma_{13} \\ 2\sigma_{23} \end{bmatrix} \tag{8.5.5}$$

Comparing Eqs. (8.5.5) and (8.4.26), we see that

$$S_{1111} = \frac{\lambda + \mu}{\mu(3\lambda + 2\mu)}, \quad S_{1122} = \frac{-\lambda}{2\mu(3\lambda + 2\mu)}, \quad \tfrac{1}{2}(S_{1111} - S_{1122}) = \frac{1}{4\mu}. \tag{8.5.6}$$

Eq. (8.5.5) shows that, for isotropic linearly elastic materials, a spherical state of stress results in a spherical state of strain. Indeed, if we set $\sigma_{11} = \sigma_{22} = \sigma_{33} = \sigma_m$, we obtain:

$$e_{11} = e_{22} = e_{33} = e_m = \frac{e_v}{3} = \frac{\sigma_m}{3\lambda + 2\mu}. \tag{8.5.7}$$

Therefore, the change of volume per unit volume due a spherical stress σ_m is given by

$$e_v = \frac{3\sigma_m}{3\lambda + 2\mu} = \frac{\sigma_m}{K}, \tag{8.5.8}$$

where

$$K = \frac{3\lambda + 2\mu}{3} \tag{8.5.9}$$

is called the *Bulk Modulus*. In case we have a hydrostatic compression of magnitude P, Eq. (8.5.8) is written:

$$e_v = -\frac{P}{K}. \tag{8.5.10}$$

Eq. (8.5.5) also shows that for isotropic linearly elastic materials, a shear stress results in nothing but a shear strain. In general, a deviator stress will cause a deviator strain. Indeed, if, in Eq. (8.5.5), we set $\sigma_{33} = -(\sigma_{11} + \sigma_{22})$, so that the sum of the three normal stresses is equal to zero (see the definition of a deviator stress), we obtain the following values for the normal strains:

$$e_{11} = \frac{(\lambda + \mu)\sigma_{11}}{\mu(3\lambda + 2\mu)} - \frac{\lambda\,\sigma_{22}}{2\mu(3\lambda + 2\mu)} + \frac{\lambda(\sigma_{11} + \sigma_{22})}{2\mu(3\lambda + 2\mu)} = \frac{\sigma_{11}}{2\mu} \tag{8.5.11}$$

$$e_{22} = \frac{-\lambda\,\sigma_{11}}{2\mu(3\lambda + 2\mu)} + \frac{(\lambda + \mu)\sigma_{22}}{\mu(3\lambda + 2\mu)} + \frac{\lambda(\sigma_{11} + \sigma_{22})}{2\mu(3\lambda + 2\mu)} = \frac{\sigma_{22}}{2\mu} \tag{8.5.12}$$

$$e_{33} = \frac{-\lambda\,\sigma_{11}}{2\mu(3\lambda + 2\mu)} - \frac{\lambda\,\sigma_{22}}{2\mu(3\lambda + 2\mu)} - \frac{(\lambda + \mu)(\sigma_{11} + \sigma_{22})}{\mu(3\lambda + 2\mu)} = -\frac{(\sigma_{11} + \sigma_{22})}{2\mu}, \tag{8.5.13}$$

whose sum is equal to zero. The elastic stress-strain relations can thus be split into two sets: First a relation between the spherical components of the stress and strain tensors, and second a relation between the deviatoric components of the stress and strain tensors. The two following equations can therefore replace Eq. (8.5.3):

$$\sigma_{nn} = 3K\, e_{nn} \tag{8.5.14}$$

$$\sigma'_{mn} = 2\mu\, e'_{mn}. \tag{8.5.15}$$

Notice that Eq. (8.5.14) is a scalar equation. In terms of μ and the bulk modulus K, Eq. (8.5.4) can also be written as follows:

$$e_{ij} = \frac{1}{2\mu}\left[\sigma_{ij} - \frac{1}{3}\delta_{ij}\sigma_{nn}\right] + \frac{1}{9K}\delta_{ij}\sigma_{nn}. \tag{8.5.16}$$

The two independent constants in engineering terminology are E, *Young's modulus* and ν, *Poisson's ratio*. Under a uniaxial state of stress σ_{11},

$$\begin{aligned} e_{11} &= \frac{\lambda + \mu}{\mu(3\lambda + 2\mu)}\sigma_{11} = \frac{\sigma_{11}}{E} \\ e_{22} &= \frac{-\lambda}{2\mu(3\lambda + 2\mu)}\sigma_{11} = -\frac{\nu}{E}\sigma_{11} \\ e_{33} &= \frac{-\lambda}{2\mu(3\lambda + 2\mu)}\sigma_{11} = -\frac{\nu}{E}\sigma_{11}. \end{aligned} \tag{8.5.17}$$

Poisson's ratio ν is thus the ratio between the lateral contraction and the axial elongation under a uniaxial stress condition. Eqs. (8.5.17) define E and ν as

$$E = \frac{\mu(3\lambda + 2\mu)}{\lambda + \mu}, \quad \nu = \frac{\lambda}{2(\lambda + \mu)}. \tag{8.5.18}$$

Eqs. (8.5.18) can be solved for λ and μ to give:

$$\lambda = \frac{\nu E}{(1 + \nu)(1 - 2\nu)}, \quad \mu = \frac{E}{2(1 + \nu)}. \tag{8.5.19}$$

Eq. (8.5.3) can be rewritten in terms of E and ν as follows:

$$\sigma_{ij} = \frac{E}{1 + \nu}e_{ij} + \frac{\nu E}{(1 + \nu)(1 - 2\nu)}\delta_{ij}e_{nn} \tag{8.5.20}$$

$$\sigma_{11} = \frac{E}{(1 + \nu)(1 - 2\nu)}[(1 - \nu)e_{11} + \nu\, e_{22} + \nu\, e_{33}] \tag{8.5.21}$$

$$\sigma_{22} = \frac{E}{(1 + \nu)(1 - 2\nu)}[\nu\, e_{11} + (1 - \nu)e_{22} + \nu\, e_{33}] \tag{8.5.22}$$

$$\sigma_{33} = \frac{E}{(1+\nu)(1-2\nu)}[\nu\, e_{11} + \nu\, e_{22} + (1-\nu)e_{33}] \tag{8.5.23}$$

$$\sigma_{12} = \frac{E}{1+\nu}e_{12}, \quad \sigma_{13} = \frac{E}{1+\nu}e_{13}, \quad \sigma_{23} = \frac{E}{1+\nu}e_{23}. \tag{8.5.24}$$

Eq. (8.5.4) can also be rewritten in terms of E and ν as follows:

$$e_{ij} = \frac{1}{E}[(1+\nu)\sigma_{ij} - \nu\, \delta_{ij}\sigma_{nn}] \tag{8.5.25}$$

or

$$e_{11} = \frac{1}{E}(\sigma_{11} - \nu\, \sigma_{22} - \nu\, \sigma_{33}) \tag{8.5.26}$$

$$e_{22} = \frac{1}{E}(-\nu\sigma_{11} + \sigma_{22} - \nu\, \sigma_{33}) \tag{8.5.27}$$

$$e_{33} = \frac{1}{E}(-\nu\, \sigma_{11} - \nu\, \sigma_{22} + \sigma_{33}) \tag{8.5.28}$$

$$e_{12} = \frac{1+\nu}{E}\sigma_{12}, \quad e_{13} = \frac{1+\nu}{E}\sigma_{13}, \quad e_{23} = \frac{1+\nu}{E}\sigma_{23}. \tag{8.5.29}$$

We next give in tabular form the relations between the various elastic constants:

Constant	Basic Pair		
	$\lambda, \mu = G$	E, ν	K, μ
λ	λ	$\frac{\nu E}{(1+\nu)(1-2\nu)}$	$\frac{3K - 2\mu}{3}$
$\mu = G$	μ, G	$\frac{E}{2(1+\nu)}$	μ
K	$\frac{3\lambda + 2\mu}{3}$	$\frac{E}{3(1-2\nu)}$	K
E	$\frac{\mu(3\lambda + 2\mu)}{\lambda + \mu}$	E	$\frac{9K\mu}{3K + \mu}$
ν	$\frac{\lambda}{2(\lambda + \mu)}$	ν	$\frac{3K - 2\mu}{6K + 2\mu}$

(8.5.30)

For the shear modulus, the letters μ and G will be used interchangeably in the coming sections. Since K is positive for all physical substances, and E is positive, it follows that $\nu < 1/2$. For a totally incompressible material $\nu = 1/2$ and $\mu = E/3$. Eq. (8.5.2) shows that μ is a positive constant, therefore $1 + \nu > 0$, and $\nu > -1$. Thus, the value of Poisson's ratio seems to be limited at one end to 0.5 for incompressible materials and at the other end to -1. Negative values of Poisson's ratio are unknown in reality.

Remarks

All the equations established in this section are also valid in a cylindrical or a spherical system of coordinates. In a cylindrical system, the subscripts r, θ, z are substituted for 1, 2, 3, respectively, and in a spherical system, the subscripts ϕ, θ, ρ are substituted for 1, 2, 3, respectively.

8.6 Thermoelastic Stress-Strain Relations for Isotropic Media

Unequal heating of the various parts of an elastic continuous solid produces stresses. These stresses would not be present if each element of the solid were allowed to expand freely; the continuity of the material prevents free expansion and this in turn results in thermal stresses. Let the temperature T of an elastic isotropic body in an arbitrary zero configuration be increased by a small amount (ΔT). Since the body is isotropic, all infinitesimal line elements in the volume undergo equal expansions. Also, all line elements maintain their initial directions. Therefore, if α is the coefficient of thermal expansion, the strain components due to the temperature change (ΔT) are

$$e_{11} = e_{22} = e_{33} = \alpha(\Delta T), \quad e_{12} = e_{13} = e_{23} = 0, \tag{8.6.1}$$

or

$$e_{ij} = \delta_{ij}\,\alpha(\Delta T). \tag{8.6.2}$$

The thermally induced strains can be superimposed to the stress induced strains of Eqs. (8.5.26) to (8.5.29) to give:

$$e_{11} = \frac{1}{E}(\sigma_{11} - \nu\,\sigma_{22} - \nu\,\sigma_{33}) + \alpha(\Delta T) \tag{8.6.3}$$

$$e_{22} = \frac{1}{E}(-\nu\,\sigma_{11} + \sigma_{22} - \nu\,\sigma_{33}) + \alpha(\Delta T) \tag{8.6.4}$$

$$e_{33} = \frac{1}{E}(-\nu\,\sigma_{11} - \nu\,\sigma_{22} + \sigma_{33}) + \alpha(\Delta T) \tag{8.6.5}$$

$$e_{12} = \frac{1+\nu}{E}\sigma_{12}, \quad e_{13} = \frac{1+\nu}{E}\sigma_{13}, \quad e_{23} = \frac{1+\nu}{E}\sigma_{23}. \tag{8.6.6}$$

In index notation, this set of equations is written:

$$e_{ij} = \frac{1}{E}[(1 + \nu)\sigma_{ij} - \nu\,\delta_{ij}\,\sigma_{nn}] + \alpha\,\delta_{ij}(\Delta T). \qquad (8.6.7)$$

The inversion of Eqs. (8.6.3) to (8.6.6) gives:

$$\sigma_{11} = \frac{E}{(1 + \nu)(1 - 2\nu)}[(1 - \nu)e_{11} + \nu\,e_{22} + \nu\,e_{33}] - \frac{E}{1 - 2\nu}\alpha(\Delta T) \qquad (8.6.8)$$

$$\sigma_{22} = \frac{E}{(1 + \nu)(1 - 2\nu)}[\nu\,e_{11} + (1 - \nu)e_{22} + \nu\,e_{33}] - \frac{E}{1 - 2\nu}\alpha(\Delta T) \qquad (8.6.9)$$

$$\sigma_{33} = \frac{E}{(1 + \nu)(1 - 2\nu)}[\nu\,e_{11} + \nu\,e_{22} + (1 - \nu)e_{33}] - \frac{E}{1 - 2\nu}\alpha(\Delta T) \qquad (8.6.10)$$

$$\sigma_{12} = \frac{E}{1 + \nu}e_{12}, \quad \sigma_{13} = \frac{E}{1 + \nu}e_{13}, \quad \sigma_{23} = \frac{E}{1 + \nu}e_{23}, \qquad (8.6.11)$$

where the term $[(E/1 - 2\nu)\alpha(\Delta T)]$ is the thermal stress induced by the temperature change. Another way of writing the stresses in terms of the strains and the temperature change is provided by Eq. (8.5.3):

$$\sigma_{ij} = 2\mu\,e_{ij} + \lambda\,\delta_{ij}\,e_{nn} - (3\lambda + 2\mu)\delta_{ij}\,\alpha(\Delta T). \qquad (8.6.12)$$

The stress-strain relation (8.6.12) is called the Duhamel-Neumann law.

Remarks

(a) In a cylindrical system of coordinates, the subscripts r, θ, z are substituted for 1, 2, 3, respectively; and in a spherical system of coordinates, the subscripts ϕ, θ, ρ are substituted for 1, 2, 3, respectively.

(b) In thermoelastic problems, the temperature distribution must first be determined from Fourier's heat conduction equation.

8.7 Strain Energy Density

Let us assume that the strain energy density $U(e_{ij})$ defined by Eqs. (8.2.8) can be expanded in a power series in terms of the e_{ij} ' s. Then

$$U = C_o + C_{mn}\, e_{mn} + \frac{1}{2} C_{pqrs}\, e_{pq}\, e_{rs} + \ldots, \tag{8.7.1}$$

in which the C 's are constants. The constant C_o can be disregarded since the energy is measured from any arbitrary level. From Eq. (8.2.9), and neglecting the terms higher than the quadratic in Eq. (8.7.1), we have:

$$\begin{aligned} \frac{\partial U}{\partial e_{ij}} = \sigma_{ij} &= C_{mn}\, \delta_{mi}\, \delta_{nj} + \frac{1}{2} C_{pqrs}[e_{rs}\, \delta_{pi}\, \delta_{qj} + e_{pq}\, \delta_{ri}\, \delta_{sj}] \\ &= C_{ij} + \frac{1}{2} C_{ijrs}\, e_{rs} + \frac{1}{2} C_{pqij}\, e_{pq} \end{aligned} \tag{8.7.2}$$

or

$$\sigma_{ij} = C_{ij} + \frac{1}{2}[C_{ijrs} + C_{rsij}]e_{rs}\,. \tag{8.7.3}$$

For the stress to vanish in the absence of strain, the constant C_{ij} must be equal tc zero. Thus the expression for the strain energy density reduces to

$$U = \frac{1}{2} C_{ijrs}\, e_{ij}\, e_{rs}, \tag{8.7.4}$$

and that of the stress to

$$\sigma_{ij} = \frac{1}{2}(C_{ijrs} + C_{rsij})e_{rs}\,. \tag{8.7.5}$$

Eq. (8.7.5) can be rewritten as follows:

$$\sigma_{ij} = C_{ijrs}\, e_{rs}, \tag{8.7.6}$$

where the coefficients C_{ijrs} are symmetrized. Therefore, starting from the assumption that a strain energy density function satisfying Eqs. (8.2.8) exists, and neglecting the terms higher than the quadratic, we have arrived at Hooke's law and proved the symmetry of the stiffness matrix. Substituting Eq. (8.7.6) into Eq. (8.7.4), we get:

$$U = \frac{1}{2}\sigma_{ij}\, e_{ij}\,. \tag{8.7.7}$$

This formula for the strain energy density is known as the *Clapeyron Formula*. When the stress-strain law is written, as in Eq. (8.3.16):

$$e_{ij} = S_{ijmn}\sigma_{mn}, \tag{8.7.8}$$

the formula of Clapeyron gives:

$$U = \frac{1}{2}S_{ijmn}\sigma_{ij}\sigma_{mn}, \tag{8.7.9}$$

so that

$$\frac{\partial U}{\partial \sigma_{ij}} = S_{ijmn}\sigma_{mn} = e_{ij}. \tag{8.7.10}$$

This formula is known as *Castigliano's Formula*. Notice that, while Eqs. (8.2.8) do not imply linearity, Eq. (8.7.10) does. The total strain energy in a body is

$$U_t = \iiint_V U\, dV. \tag{8.7.11}$$

In the following, we give several alternative forms of Eq. (8.7.7) for isotropic bodies. For example, substituting Eq. (8.5.25) into Eq. (8.7.7), we get:

$$U = \frac{1}{2E}[(1+\nu)\sigma_{ij}\sigma_{ij} - \nu(\sigma_{nn})^2]. \tag{8.7.12}$$

Explicitly, Eq. (8.7.12) has the form:

$$\begin{aligned} U = \frac{1}{2E}[(\sigma_{11}^2 + \sigma_{22}^2 + \sigma_{33}^2) - 2\nu(\sigma_{11}\sigma_{22} + \sigma_{22}\sigma_{33} + \sigma_{33}\sigma_{11}) \\ + 2(1+\nu)(\sigma_{12}^2 + \sigma_{23}^2 + \sigma_{31}^2)]. \end{aligned} \tag{8.7.13}$$

Substituting Eq. (8.5.16) into Eq. (8.7.7), we get:

$$U = \frac{1}{4\mu}[\sigma_{ij}\sigma_{ij} - \frac{1}{3}(\sigma_{nn})^2] + \frac{1}{18K}(\sigma_{nn})^2. \tag{8.7.14}$$

Substituting Eq. (8.5.3) into Eq. (8.7.7), we obtain the expression of the strain energy density in terms of the strains:

$$U = \mu\, e_{ij}e_{ij} + \frac{1}{2}\lambda(e_{nn})^2 \tag{8.7.15}$$

Explicitly, Eq. (8.7.15) has the form:

$$U = \frac{1}{2}\lambda(e_{11} + e_{22} + e_{33})^2 + \mu(e_{11}^2 + e_{22}^2 + e_{33}^2) + 2\mu(e_{12}^2 + e_{23}^2 + e_{13}^2). \tag{8.7.16}$$

In case of temperature changes, Equation (8.7.16) must be modified to read

$$U = \mu\, e_{ij}e_{ij} + \frac{1}{2}\lambda(e_{nn})^2 - (3\lambda + 2\mu)\alpha(\Delta T)(e_{nn}) + \frac{3}{2}(\alpha\Delta T)^2(3\lambda + 2\mu)$$

We sometimes need to know the strain energy associated with the deviatoric components and that associated with the spherical components of the stress and strain tensors. It has been shown in Sec. 8.5 that, for isotropic materials, spherical stresses result in spherical strains and deviator stresses result in deviator strains. For this reason, the energy associated with the spherical components is called *energy of dilatation* U_s, and the energy associated with the deviatoric components is called *energy of distortion* U_d. In terms of spherical and deviatoric components, the expressions for the stress and strain tensors have been shown to be:

$$\sigma_{ij} = \sigma'_{ij} + \frac{1}{3}\delta_{ij}\sigma_{nn} \tag{8.7.17}$$

$$e_{ij} = e'_{ij} + \frac{1}{3}\delta_{ij}e_{pp}. \tag{8.7.18}$$

Substituting Eqs. (8.7.17) and (8.7.18) into Eq. (8.7.7), we get:

$$U = \frac{1}{6}\sigma_{nn}e_{pp} + \frac{1}{2}\sigma'_{ij}e'_{ij}. \tag{8.7.19}$$

We notice that the first term of Eq. (8.7.19) is the energy of dilatation U_s, and the second term is the energy of distortion U_d. Thus,

$$U = U_s + U_d, \tag{8.7.20}$$

and the total energy is the sum of the energy due to the spherical components alone and that due to the deviators alone. This superposition, however, does not hold for any general partitioning of the states of stress or of strain. The expressions of U_s and U_d can be written in terms either of the stresses or of the strains. Using Eqs. (8.5.14) and (8.5.15), we get the following expressions in terms of the stresses:

$$U_s = \frac{1}{6}\sigma_{nn} e_{pp} = \frac{(\sigma_{nn})^2}{18K} = \frac{(\sigma_{11} + \sigma_{22} + \sigma_{33})^2}{18K} \tag{8.7.21}$$

$$U_d = \frac{1}{2}\sigma'_{ij} e'_{ij} = \frac{1}{4\mu}(\sigma'_{ij}\sigma'_{ij}) = \frac{1}{4\mu}\left[\sigma_{ij}\sigma_{ij} - \frac{1}{3}(\sigma_{nn})^2\right]. \tag{8.7.22}$$

Explicitly, Eq. (8.7.22) is written:

$$U_d = \frac{1}{12\mu}[(\sigma_{11} - \sigma_{22})^2 + (\sigma_{22} - \sigma_{33})^2 + (\sigma_{33} - \sigma_{11})^2 + 6\sigma_{12}^2 + 6\sigma_{23}^2 + 6\sigma_{13}^2] \tag{8.7.23}$$

or

$$U_d = -\frac{J_{d2}}{2G} = \frac{3\tau_{oct}^2}{4G}. \tag{8.7.24}$$

In terms of the strains,

$$U_s = \frac{1}{6}\sigma_{nn} e_{pp} = \frac{K(e_{nn})^2}{2} = \frac{K}{2}(e_{11} + e_{22} + e_{33})^2 \tag{8.7.25}$$

$$U_d = \frac{1}{2}\sigma'_{ij} e'_{ij} = \mu(e'_{ij} e'_{ij}) = \mu\left[e_{ij} e_{ij} - \frac{1}{3}(e_{nn})^2\right]. \tag{8.7.26}$$

Explicitly, Eq. (8.7.26) is written:

$$U_d = \frac{\mu}{3}[(e_{11} - e_{22})^2 + (e_{22} - e_{33})^2 + (e_{33} - e_{11})^2 + 6e_{12}^2 + 6e_{23}^2 + 6e_{13}^2] \tag{8.7.27}$$

or

$$U_d = -2GI_{d2} = \frac{3}{4}G\gamma_{oct}^2, \tag{8.7.28}$$

where I_{d2} is the second invariant of the deviator strain tensor. Finally, upon examination of Eqs. (8.7.4) and (8.7.16), one notices that the strain energy density is a quadratic form in the strains (see Sec. 3.12). For every set of values of the e_{ij}'s, it takes only positive values. The same can be said of U_s and U_d. Thus, U is a positive definite quadratic form in the strains. This will be used in establishing the uniqueness of the solution of elasticity problems.

Remark

When writing energy expressions in a cylindrical system of coordinates, the subscripts r, θ, z are substituted for 1, 2, 3, respectively. In a spherical system of coordinates, the subscripts ϕ, θ, ρ are substituted for 1, 2, 3, respectively.

8.8 Formulation of Elasticity Problems. Boundary-Value Problems of Elasticity

In general, a problem of elasticity consists of finding the stresses and the displacements in an elastic body subjected to surface forces, surface displacements, and body forces. There are six components of the state of stress at each point, and the three equations of equilibrium are not sufficient to obtain the solution of the problem. The six stress-strain relations are therefore introduced together with the six strain-displacement relations (1.2.1). In all, we have 15 equations and 15 unknowns $(\sigma_{ij}, e_{ij}, u_i)$. To insure a unique value of the displacement components at each point, the strains must satisfy the compatibility relations.

The stresses and displacements are functions of the coordinates. When the coordinates of the points on the surface of the body are substituted into the expressions of the stresses and displacements, the resulting values must coincide with the externally imposed ones. We are thus faced with two types of boundary value problems:

Problem 1 Determine the expressions of the stresses and displacements at all the points in the interior of an elastic body in equilibrium when the body forces are known and the surface forces are prescribed.

Problem 2 Determine the expressions of the stresses and displacements at all the points in the interior of an elastic body in equilibrium when the body forces are known and the surface displacements are prescribed.

Sometimes the forces are prescribed on one portion of the boundary and the displacements on the remaining one. This case is referred to as the mixed boundary value problem. If stresses alone are imposed on the boundary of the elastic body, it becomes desirable to express all the equations in terms of stresses. If displacements alone are imposed to the boundary, then a formulation of the equations in terms of displacements is generally more useful. Both approaches will be examined in the following sections.

8.9 Elasticity Equations in Terms of Displacements

The original 15 equations which are to be solved in the analysis of an elasticity problem can be reduced to three equations in terms of the components of the displacements. To obtain these equations, the set of Eqs. (8.5.2) is substituted into the equations of equilibrium, then the strains are written in terms of the displacements. The operations are most conveniently made in index notation. We have:
Stress-Strain:

$$\sigma_{ij} = 2\mu\, e_{ij} + \lambda\, \delta_{ij}\, e_{nn} = 2\mu\, e_{ij} + \lambda\, \delta_{ij}\, e_v. \tag{8.9.1}$$

Equilibrium:

$$\frac{\partial \sigma_{ji}}{\partial x_j} + F_i = 0. \tag{8.9.2}$$

Strain-displacement:

$$e_{ij} = \frac{1}{2}\left(\frac{\partial u_i}{\partial x_j} + \frac{\partial u_j}{\partial x_i}\right). \tag{8.9.3}$$

Substituting Eqs. (8.9.1) and (8.9.3) into Eq. (8.9.2), we get:

$$\mu\, \nabla^2 u_i + (\lambda + \mu)\frac{\partial e_v}{\partial x_i} + F_i = 0, \tag{8.9.4}$$

where $\nabla^2 = \partial^2/\partial x_1^2 + \partial^2/\partial x_2^2 + \partial^2/\partial x_3^2$ is Laplace's operator. In explicit form, Eq. (8.9.4) is written:

$$(\lambda + \mu)\frac{\partial e_v}{\partial x_1} + \mu\, \nabla^2 u_1 + F_1 = 0 \tag{8.9.5}$$

$$(\lambda + \mu)\frac{\partial e_v}{\partial x_2} + \mu\, \nabla^2 u_2 + F_2 = 0 \tag{8.9.6}$$

$$(\lambda + \mu)\frac{\partial e_v}{\partial x_3} + \mu\, \nabla^2 u_3 + F_3 = 0. \tag{8.9.7}$$

Eqs. (8.9.5), (8.9.6), and (8.9.7) are called Navier's equations of elasticity. The boundary conditions (7.3.8) can be written in terms of the displacements and strains as follows:

$$\sigma_{ni} = \mu\left(\frac{\partial u_i}{\partial x_j} + \frac{\partial u_j}{\partial x_i}\right)\ell_j + \lambda\, \ell_i e_v. \tag{8.9.8}$$

Once one has found a solution satisfying Eqs. (8.9.5) to (8.9.7), the strains and the stresses can easily be obtained. The boundary conditions must be satisfied. There is no need to check compatibility since the strains are obtained from the displacements and not vice versa. In case of temperature changes, Eqs. (8.9.4) and (8.9.8) become:

$$\mu\nabla^2 u_i + (\lambda + \mu)\frac{\partial e_v}{\partial x_i} - (3\lambda + 2\mu)\alpha\frac{\partial}{\partial x_i}(\Delta T) + F_i = 0 \tag{8.9.9}$$

$$\sigma_{ni} + (3\lambda + 2\mu)\alpha(\Delta T)\ell_i = \mu\left(\frac{\partial u_i}{\partial x_j} + \frac{\partial u_j}{\partial x_i}\right)\ell_j + \lambda\ell_i e_v. \tag{8.9.10}$$

Eqs. (8.9.5) to (8.9.7) can be written under the form of one vector equation. Since e_v is nothing but the divergence of the displacement $\bar{u}$, and $\nabla^2 u_1$, $\nabla^2 u_2$, and $\nabla^2 u_3$ are the components of a vector $\nabla^2 \bar{u}$, then

$$(\lambda + \mu)\text{ grad }(\text{div }\bar{u}) + \mu\, \nabla^2\bar{u} + \bar{F} - (3\lambda + 2\mu)\alpha\text{ grad }(\Delta T) = 0. \tag{8.9.11}$$

This vector equation can easily be translated into components in any system of orthogonal curvilinear coordinates by using the relations of Sec. 6.4.

8.10 Elasticity Equations in Terms of Stresses

Not every solution of the equations of equilibrium corresponds to a possible state of strain because the components of the strain must satisfy the equation of compatibility (4.10.14) to insure the existence of single valued displacements. Let us consider, for example, the compatibility equation:

$$2\frac{\partial^2 e_{23}}{\partial x_2\,\partial x_3} = \frac{\partial^2 e_{22}}{\partial x_3^2} + \frac{\partial^2 e_{33}}{\partial x_2^2}. \tag{8.10.1}$$

Here we shall use the notation:

$$\theta = J_1 = \sigma_{11} + \sigma_{22} + \sigma_{33}. \tag{8.10.2}$$

Substituting Eqs. (8.5.26) to (8.5.29) into Eq. (8.10.1), we get:

$$(1+\nu)\left(\frac{\partial^2 \sigma_{22}}{\partial x_3^2}+\frac{\partial^2 \sigma_{33}}{\partial x_2^2}\right)-\nu\left(\frac{\partial^2 \theta}{\partial x_3^2}+\frac{\partial^2 \theta}{\partial x_2^2}\right)$$
$$=2(1+\nu)\frac{\partial^2 \sigma_{23}}{\partial x_2\,\partial x_3}. \tag{8.10.3}$$

From the second and third equations of equilibrium, we have:

$$\frac{\partial \sigma_{32}}{\partial x_3}=\frac{\partial \sigma_{23}}{\partial x_3}=-\frac{\partial \sigma_{22}}{\partial x_2}-\frac{\partial \sigma_{12}}{\partial x_1}-F_2 \tag{8.10.4}$$

$$\frac{\partial \sigma_{23}}{\partial x_2}=\frac{\partial \sigma_{32}}{\partial x_2}=-\frac{\partial \sigma_{33}}{\partial x_3}-\frac{\partial \sigma_{13}}{\partial x_1}-F_3. \tag{8.10.5}$$

Differentiating Eq. (8.10.4) with respect to x_2 and Eq. (8.10.5) with respect to x_3, adding them together, and introducing the first equation of equilibrium, we get:

$$2\frac{\partial^2 \sigma_{23}}{\partial x_2\,\partial x_3}=\frac{\partial^2 \sigma_{11}}{\partial x_1^2}-\frac{\partial^2 \sigma_{22}}{\partial x_2^2}-\frac{\partial^2 \sigma_{33}}{\partial x_3^2}+\frac{\partial F_1}{\partial x_1}-\frac{\partial F_2}{\partial x_2}-\frac{\partial F_3}{\partial x_3}. \tag{8.10.6}$$

Substituting Eq. (8.10.6) into Eq. (8.10.3), we get:

$$(1+\nu)\left(\nabla^2\theta-\nabla^2\sigma_{11}-\frac{\partial^2 \theta}{\partial x_1^2}\right)-\nu\left(\nabla^2\theta-\frac{\partial^2 \theta}{\partial x_1^2}\right)$$
$$=(1+\nu)\left(\frac{\partial F_1}{\partial x_1}-\frac{\partial F_2}{\partial x_2}-\frac{\partial F_3}{\partial x_3}\right). \tag{8.10.7}$$

Using the two other relations of the type Eq. (8.10.1) in Eq. (4.10.14), two equations similar to Eq. (8.10.7) can be obtained. They can directly be deduced from Eq. (8.10.7) by cyclical permutation. Adding these three equations together, we find that

$$\nabla^2\theta=-\frac{1+\nu}{1-\nu}\left(\frac{\partial F_1}{\partial x_1}+\frac{\partial F_2}{\partial x_2}+\frac{\partial F_3}{\partial x_3}\right). \tag{8.10.8}$$

Substituting this expression for $\nabla^2\theta$ into Eq. (8.10.7), we finally obtain:

$$\nabla^2\sigma_{11}+\frac{1}{1+\nu}\frac{\partial^2 \theta}{\partial x_1^2}=-\frac{\nu}{1-\nu}\left(\frac{\partial F_1}{\partial x_1}+\frac{\partial F_2}{\partial x_2}+\frac{\partial F_3}{\partial x_3}\right)-2\frac{\partial F_1}{\partial x_1}. \tag{8.10.9}$$

Two similar equations are obtained by circular permutation.

In a similar manner, the remaining three compatibility equations can be transformed into equations of the form

$$\nabla^2 \sigma_{23} + \frac{1}{1+\nu}\frac{\partial^2 \theta}{\partial x_2\, \partial x_3} = -\left(\frac{\partial F_3}{\partial x_2} + \frac{\partial F_2}{\partial x_3}\right). \tag{8.10.10}$$

Two similar equations are again obtained by circular permutation. Gathering the results, the stresses are obtained by solving the system of six equations with 6 unknowns:

$$\nabla^2 \sigma_{11} + \frac{1}{1+\nu}\frac{\partial^2 \theta}{\partial x_1^2} = -\frac{\nu}{1-\nu}\operatorname{div}\bar{F} - 2\frac{\partial F_1}{\partial x_1} \tag{8.10.11}$$

$$\nabla^2 \sigma_{22} + \frac{1}{1+\nu}\frac{\partial^2 \theta}{\partial x_2^2} = -\frac{\nu}{1-\nu}\operatorname{div}\bar{F} - 2\frac{\partial F_2}{\partial x_2} \tag{8.10.12}$$

$$\nabla^2 \sigma_{33} + \frac{1}{1+\nu}\frac{\partial^2 \theta}{\partial x_3^2} = -\frac{\nu}{1-\nu}\operatorname{div}\bar{F} - 2\frac{\partial F_3}{\partial x_3} \tag{8.10.13}$$

$$\nabla^2 \sigma_{12} + \frac{1}{1+\nu}\frac{\partial^2 \theta}{\partial x_1\, \partial x_2} = -\left(\frac{\partial F_1}{\partial x_2} + \frac{\partial F_2}{\partial x_1}\right) \tag{8.10.14}$$

$$\nabla^2 \sigma_{23} + \frac{1}{1+\nu}\frac{\partial^2 \theta}{\partial x_2\, \partial x_3} = -\left(\frac{\partial F_2}{\partial x_3} + \frac{\partial F_3}{\partial x_2}\right) \tag{8.10.15}$$

$$\nabla^2 \sigma_{31} + \frac{1}{1+\nu}\frac{\partial^2 \theta}{\partial x_3\, \partial x_1} = -\left(\frac{\partial F_3}{\partial x_1} + \frac{\partial F_1}{\partial x_3}\right). \tag{8.10.16}$$

In index notation, these equations are written:

$$\nabla^2 \sigma_{ij} + \frac{1}{1+\nu}\frac{\partial^2 \theta}{\partial x_i\, \partial x_j} = -\frac{\nu}{1-\nu}\delta_{ij}\operatorname{div}\bar{F} - \left(\frac{\partial F_i}{\partial x_j} + \frac{\partial F_j}{\partial x_i}\right). \tag{8.10.17}$$

In case of temperature changes, Eq. (8.10.17) becomes:

$$\begin{aligned}\nabla^2 \sigma_{ij} &+ \frac{1}{1+\nu}\frac{\partial^2[\theta + E\alpha(\Delta T)]}{\partial x_i\, \partial x_j} \\ &= -\frac{\nu}{1-\nu}\delta_{ij}\operatorname{div}\bar{F} - \left(\frac{\partial F_i}{\partial x_j} + \frac{\partial F_j}{\partial x_i}\right) - \delta_{ij}\frac{\nabla^2[E\alpha(\Delta T)]}{1-\nu}.\end{aligned} \tag{8.10.18}$$

Eqs. (8.10.11) to (8.10.16) are known as the *Beltrami-Michell compatibility equations*. Thus, the state of stress in the interior of an elastic body must satisfy the three equations of equilibrium, the Beltrami-Michell compatibility equations, and the boundary conditions (7.3.8).

8.11 The Principle of Superposition

Let the stresses in an elastic body subjected to surface forces Q_i and to body forces F_i be σ_{ji}. Let the stresses in the same elastic body when it is subjected to the surface forces Q'_i and to the body forces F'_i be σ'_{ji}. Then the stresses $\sigma_{ji} + \sigma'_{ji}$ will represent the stresses due to the surface forces $Q_i + Q'_i$ and the body forces $F_i + F'_i$. This holds because all the differential equations and boundary conditions are linear. Thus, adding up the two sets of equations of equilibrium for the first and second state of stress,

$$\frac{\partial \sigma_{ji}}{\partial x_j} + F_i = 0 \tag{8.11.1}$$

and

$$\frac{\partial \sigma'_{ji}}{\partial x_j} + F'_i = 0, \tag{8.11.2}$$

we get:

$$\frac{\partial}{\partial x_j}(\sigma_{ji} + \sigma'_{ji}) + F_i + F'_i = 0. \tag{8.11.3}$$

Also adding up the two sets of boundary conditions for the first and second state of stress,

$$\sigma_{ni} = \sigma_{ji}\,\ell_j \tag{8.11.4}$$

and

$$\sigma'_{ni} = \sigma'_{ji}\,\ell_j, \tag{8.11.5}$$

we get:

$$\sigma_{ni} + \sigma'_{ni} = (\sigma_{ji} + \sigma'_{ji})\ell_j. \tag{8.11.6}$$

The compatibility equations can also be combined in the same manner. The complete set of equations shows that $(\sigma_{ji} + \sigma'_{ji})$ satisfy all the requirements and conditions determining the stresses due to the surface forces $Q_i + Q'_i$ and to the body forces $F_i + F'_i$. This is the principle of superposition.

In our study, no distinction was made between the undeformed and the deformed shapes of the elements in equilibrium. Consequently, the principle of superposition will only be valid for the cases of small displacements. Again, all this is in line with the use of the linear theory of strain.

8.12 Existence and Uniqueness of the Solution of an Elasticity Problem

The rigorous proof of the existence of solution is too lengthy and will not be examined here; also it is to be remembered that we are dealing with linear strains and small displacements. In order to establish the uniqueness of the solution of an elasticity problem, let us assume that it is possible to obtain two solutions,

$$\sigma'_{11}, \sigma'_{22}, \ldots, u'_1, u'_2, u'_3 \tag{8.12.1}$$

$$\sigma''_{11}, \sigma''_{22}, \ldots, u''_1, u''_2, u''_3, \tag{8.12.2}$$

which satisfy the 15 elasticity equations and the same set of boundary conditions. Therefore, for the first set of stresses, the equations of equilibrium,

$$\frac{\partial \sigma'_{ji}}{\partial x_j} + F_i = 0, \tag{8.12.3}$$

are satisfied as well as the following boundary conditions,

$$\sigma_{ni} = \sigma'_{ji}\, \ell_j, \tag{8.12.4}$$

if the surface forces are prescribed, or

$$u_i = u'_i \tag{8.12.5}$$

if the boundary displacements are prescribed. For the second state of stress, we have:

$$\frac{\partial \sigma''_{ji}}{\partial x_j} + F_i = 0 \tag{8.12.6}$$

$$\sigma_{ni} = \sigma''_{ji}\,\ell_j \tag{8.12.7}$$

$$u_i = u''_i . \tag{8.12.8}$$

If we subtract Eqs. (8.12.6), (8.12.7), and (8.12.8) from Eqs. (8.12.3), (8.12.4), and (8.12.5), respectively, we find that the stress distribution defined by $\sigma'_{ij} - \sigma''_{ij}$ satisfied the equations:

$$\frac{\partial}{\partial x_j}(\sigma'_{ji} - \sigma''_{ji}) = 0 \tag{8.12.9}$$

$$(\sigma'_{ji} - \sigma''_{ji})\ell_j = 0 \tag{8.12.10}$$

$$u'_i - u''_i = 0. \tag{8.12.11}$$

This is a new "difference" stress distribution in which all the external forces, the body forces, and the boundary displacements vanish. If there are no external forces or boundary displacements, there is no work done and the strain energy stored in the body must be equal to zero. It has been established in Sec. 8.7 that the strain energy density is a positive definite quadratic form in the strain components. It cannot vanish unless all the strains vanish. Therefore, if the strain energy stored in the body is zero, the strain components and consequently the stress components must be zero everywhere. Consequently, the difference state of stress $\sigma'_{ji} - \sigma''_{ji}$ must be zero and the two solutions must be identical.

8.13 Saint-Venant's Principle

In 1885, B. de Saint-Venant in his memoir on torsion proposed a principle which can be stated as follows: If a system of forces acting on a small portion of the surface of an elastic body is replaced by another statically equivalent system of forces acting on the same portion of the surface, the redistribution of loading produces substantial changes in the stresses only in the immediate neighborhood of the loading, and the stresses are essentially the same in the part of the body which are at large distances in comparison with the linear dimension of the surface on which the forces are changed. By statically equivalent, we mean that the two distributions of forces have the same resultant force and moment. The principle of Saint-Venant allows us to simplify the solution of many problems by altering the boundary conditions while keeping the systems of applied forces statically equivalent. A satisfactory approximate solution can thus be obtained.

8.14 One Dimensional Elasticity

Let us assume that the body forces are negligible. There are two states to consider: a) a one dimensional state of stress and b) a one dimensional state of deformation.

a) *One dimensional state of stress.* A one dimensional state of stress exists in a body, if, at all its points, the stress matrix is of the form:

$$\begin{bmatrix} 0 & 0 & 0 \\ 0 & 0 & 0 \\ 0 & 0 & \sigma_{33} \end{bmatrix}, \tag{8.14.1}$$

in which σ_{33} is a function of x_3 alone. The principal directions are the OX_3 direction and all directions in the OX_1, OX_2 plane. The two first equations of equilibrium are identically satisfied and the third equation yields:

$$\sigma_{33} = \text{constant} = \sigma_o. \tag{8.14.2}$$

The stress-strain relations (8.5.25) become:

$$e_{11} = e_{22} = -\frac{\nu}{E}\sigma_o, \quad e_{33} = \frac{\sigma_o}{E}, \tag{8.14.3}$$

$$e_{12} = e_{13} = e_{23} = 0. \tag{8.14.4}$$

The displacements are obtained by integration of Eqs. (1.2.1):

$$\frac{\partial u_1}{\partial x_1} = e_{11} = -\frac{\nu}{E}\sigma_o \tag{8.14.5}$$

$$\frac{\partial u_2}{\partial x_2} = e_{22} = -\frac{\nu}{E}\sigma_o \tag{8.14.6}$$

$$\frac{\partial u_3}{\partial x_3} = e_{33} = \frac{\sigma_o}{E} \tag{8.14.7}$$

$$\frac{\partial u_1}{\partial x_2} + \frac{\partial u_2}{\partial x_1} = \frac{\partial u_1}{\partial x_3} + \frac{\partial u_3}{\partial x_1} = \frac{\partial u_2}{\partial x_3} + \frac{\partial u_3}{\partial x_2} = 0. \tag{8.14.8}$$

For no rigid body displacement u_1, u_2, and u_3 are given by:

$$u_1 = -\frac{\nu}{E}\sigma_o x_1, \quad u_2 = -\frac{\nu}{E}\sigma_o x_2, \quad u_3 = \frac{\sigma_o}{E}x_3. \tag{8.14.9}$$

b) *One dimensional state of deformation.* A one dimensional state of deformation exists in a body, if, at all its points,

$$u_1 = 0, \quad u_2 = 0, \tag{8.14.10}$$

$$u_3 = u_3(x_3) \text{ is a function of } x_3 \text{ alone .} \tag{8.14.11}$$

From Eqs. (1.2.1), we get:

$$e_{11} = e_{22} = e_{12} = e_{13} = e_{23} = 0 \tag{8.14.12}$$

$$e_{33} = e_{33}(x_3) \text{ is a function of } x_3 \text{ alone .} \tag{8.14.13}$$

The stress-strain relations (8.5.21) to (8.5.24) give:

$$\sigma_{11} = \sigma_{22} = \frac{E\,\nu}{(1+\nu)(1-2\nu)}e_{33} \tag{8.14.14}$$

$$\sigma_{33} = \frac{E(1-\nu)}{(1+\nu)(1-2\nu)}e_{33} \text{ is a function of } x_3 \text{ alone} \tag{8.14.15}$$

$$\sigma_{12} = \sigma_{13} = \sigma_{23} = 0. \tag{8.14.16}$$

The two first equations of equilibrium are identically satisfied and the third one gives:

$$\sigma_{33} = \text{constant} = \sigma_o. \tag{8.14.17}$$

Therefore,

$$e_{33} = \text{constant} = e_o. \tag{8.14.18}$$

For no rigid body displacement,

$$u_1 = u_2 = 0 \tag{8.14.19}$$

$$u_3 = e_o x_3. \tag{8.14.20}$$

8.15 Plane Elasticity

In a body that is being elastically deformed, let us consider, for example, an axis OX_3: If all the planes initially normal to OX_3 remain normal to it after deformation, and if all the straight lines initially parallel to OX_3 remain parallel to it after deformation, *a state of plane deformation* is said to exist in the body. Analytically, these conditions are expressed by:

$$u_1 = u_1(x_1, x_2), \quad u_2 = u_2(x_1, x_2), \quad u_3 = u_3(x_3).$$

If $u_3 = u_3(x_3) = 0$, the body is said to be in a *state of plane strain* in the OX_1, OX_2 plane. Thus, the state of plane deformation results from the superposition of a state of plane strain and a state of one dimensional deformation.

If the state of stress in a body is such that $\sigma_{13} = \sigma_{23} = \sigma_{33} = 0$, and

$$\sigma_{11} = \sigma_{11}(x_1, x_2), \quad \sigma_{22} = \sigma_{22}(x_1, x_2), \quad \sigma_{12} = \sigma_{12}(x_1, x_2),$$

the body is said to be in *a state of plane stress.*

If a constant strain e_o in the OX_3 direction is superimposed to a state of plane strain without changing the stresses in the OX_1, OX_2 planes, the body is said to be in a *state of generalized plane strain.*

In the case of a thin plate, we sometimes seek the average values across the thickness of the components of the displacement vector and of the stress tensor. The equations in terms of these averages are called the equations of *generalized plane stress.*

8.16 State of Plane Strain (Fig. 8.5)

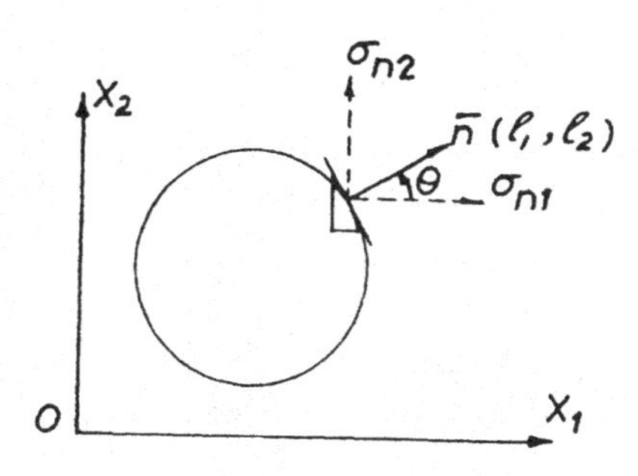

Fig. 8.5

As indicated in Sec. 8.15, a state of plane strain is characterized by $u_1 = u_1(x_1, x_2)$, $u_2 = u_2(x_1, x_2)$, $u_3 = 0$. Thus,

$$e_{11} = e_{11}(x_1, x_2) = \frac{\partial u_1}{\partial x_1}, \quad e_{22} = e_{22}(x_1, x_2) = \frac{\partial u_2}{\partial x_2} \tag{8.16.1}$$

$$e_{12} = e_{12}(x_1, x_2) = \frac{1}{2}\left(\frac{\partial u_1}{\partial x_2} + \frac{\partial u_2}{\partial x_1}\right) \tag{8.16.2}$$

$$e_{33} = e_{13} = e_{23} = 0. \tag{8.16.3}$$

The stress-strain relations (8.5.3) become:

$$\sigma_{11} = 2\mu\, e_{11} + \lambda(e_{11} + e_{22}) \tag{8.16.4}$$

$$\sigma_{22} = 2\mu\, e_{22} + \lambda(e_{11} + e_{22}) \tag{8.16.5}$$

$$\sigma_{33} = \lambda(e_{11} + e_{22}) \tag{8.16.6}$$

$$\sigma_{12} = 2\mu\, e_{12}, \quad \sigma_{13} = \sigma_{23} = 0. \tag{8.16.7}$$

The stress-strain relations (8.5.21) to (8.5.24) become:

$$\sigma_{11} = \frac{E}{(1+\nu)(1-2\nu)}[(1-\nu)e_{11} + \nu\, e_{22}] \tag{8.16.8}$$

$$\sigma_{22} = \frac{E}{(1+\nu)(1-2\nu)}[\nu\, e_{11} + (1-\nu)e_{22}] \tag{8.16.9}$$

$$\sigma_{33} = \frac{\nu\, E}{(1+\nu)(1-2\nu)}(e_{11} + e_{22}) \tag{8.16.10}$$

$$\sigma_{12} = \frac{E}{1+\nu} e_{12}, \quad \sigma_{13} = \sigma_{23} = 0. \tag{8.16.11}$$

The stress-strain relations (8.5.26) to (8.5.29) become:

$$e_{11} = \frac{1+\nu}{E}[(1-\nu)\sigma_{11} - \nu\,\sigma_{22}] \tag{8.16.12}$$

$$e_{22} = \frac{1+\nu}{E}[(1-\nu)\sigma_{22} - \nu\,\sigma_{11}] \tag{8.16.13}$$

$$e_{12} = \frac{1+\nu}{E}\sigma_{12}, \quad e_{33} = e_{13} = e_{23} = 0. \tag{8.16.14}$$

Eq. (8.5.28) shows that for the case of plane strain, σ_{33} is proportional to $(\sigma_{11} + \sigma_{22})$ since

$$\sigma_{33} = \nu(\sigma_{11} + \sigma_{22}). \tag{8.16.15}$$

Eq. (8.16.6) shows that σ_{33} is a function only of x_1 and x_2.

The equations of equilibrium become:

$$\frac{\partial \sigma_{11}}{\partial x_1} + \frac{\partial \sigma_{21}}{\partial x_2} + F_1 = 0 \tag{8.16.16}$$

$$\frac{\partial \sigma_{12}}{\partial x_1} + \frac{\partial \sigma_{22}}{\partial x_2} + F_2 = 0 \tag{8.16.17}$$

$$F_3 = 0. \tag{8.16.18}$$

Consequently, in plane strain, the body forces are such that $F_3 = 0$ and F_1 and F_2 are independent of x_3.

Five out of the six compatibility relations are identically satisfied and only one, namely,

$$2\frac{\partial^2 e_{12}}{\partial x_1 \partial x_2} = \frac{\partial^2 e_{11}}{\partial x_2^2} + \frac{\partial^2 e_{22}}{\partial x_1^2}, \tag{8.16.19}$$

is to be considered.

The equations of elasticity (8.9.5) to (8.9.7) become:

$$(\lambda + G)\frac{\partial(e_{11} + e_{22})}{\partial x_1} + G\nabla^2 u_1 + F_1 = 0 \tag{8.16.20}$$

$$(\lambda + G)\frac{\partial(e_{11} + e_{22})}{\partial x_2} + G\nabla^2 u_2 + F_2 = 0 \tag{8.16.21}$$

$$F_3 = 0, \tag{8.16.22}$$

where

$$\nabla^2 = \left(\frac{\partial^2}{\partial x_1^2} + \frac{\partial^2}{\partial x_2^2}\right). \tag{8.16.23}$$

Using the stress-strain relations and the equilibrium equations, Eq. (8.16.19) can be written in terms of stresses as follows:

$$\left(\frac{\partial^2}{\partial x_1^2} + \frac{\partial^2}{\partial x_2^2}\right)(\sigma_{11} + \sigma_{22}) = -\frac{1}{1-\nu}\left(\frac{\partial F_1}{\partial x_1} + \frac{\partial F_2}{\partial x_2}\right). \tag{8.16.24}$$

The boundary conditions (7.3.8) become (Fig. 8.5):

$$\begin{aligned}\sigma_{n1} &= \ell_1\sigma_{11} + \ell_2\sigma_{21}\\ \sigma_{n2} &= \ell_1\sigma_{12} + \ell_2\sigma_{22}\\ \sigma_{n3} &= \ell_3\sigma_{33}\,.\end{aligned} \tag{8.16.25}$$

For constant body forces,

$$\nabla^2(\sigma_{11} + \sigma_{22}) = 0, \tag{8.16.26}$$

which indicates that $(\sigma_{11} + \sigma_{22})$ is harmonic.

In a system of cylindrical coordinates the strain-displacement relations are given by Eqs. (6.7.23) and (6.7.24):

$$e_{rr} = \frac{\partial u_r}{\partial r}, \quad e_{\theta\theta} = \frac{1}{r}\frac{\partial u_\theta}{\partial \theta} + \frac{u_r}{r} \tag{8.16.27}$$

$$e_{r\theta} = \frac{1}{2}\left(\frac{1}{r}\frac{\partial u_r}{\partial \theta} + \frac{\partial u_\theta}{\partial r} - \frac{u_\theta}{r}\right) \tag{8.16.28}$$

$$e_{zz} = e_{rz} = e_{\theta z} = 0. \tag{8.16.29}$$

All the relations in Eqs. (8.16.4) to (8.16.15) remain the same except that r, θ, and z are substituted for 1, 2, and 3, respectively. The equations of equilibrium are given by Eqs. (7.12.10):

$$\frac{\partial \sigma_{rr}}{\partial r} + \frac{1}{r}\frac{\partial \sigma_{r\theta}}{\partial \theta} + \frac{1}{r}(\sigma_{rr} - \sigma_{\theta\theta}) + F_r = 0 \tag{8.16.30}$$

$$\frac{\partial \sigma_{r\theta}}{\partial r} + \frac{1}{r}\frac{\partial \sigma_{\theta\theta}}{\partial \theta} + \frac{2}{r}\sigma_{r\theta} + F_\theta = 0 \tag{8.16.31}$$

$$F_z = 0. \tag{8.16.32}$$

The equation of compatibility (8.16.19) becomes:

$$\frac{\partial^2 e_{\theta\theta}}{\partial r^2} + \frac{1}{r^2}\frac{\partial^2 e_{rr}}{\partial \theta^2} + \frac{2}{r}\frac{\partial e_{\theta\theta}}{\partial r} - \frac{1}{r}\frac{\partial e_{rr}}{\partial r} = 2\left(\frac{1}{r}\frac{\partial^2 e_{r\theta}}{\partial r\partial \theta} + \frac{1}{r^2}\frac{\partial e_{r\theta}}{\partial \theta}\right). \tag{8.16.33}$$

The equations of elasticity in terms of displacements become

$$\begin{aligned}&(\lambda + G)\frac{\partial}{\partial r}\left[\frac{1}{r}\frac{\partial}{\partial r}(ru_r) + \frac{1}{r}\frac{\partial u_\theta}{\partial \theta}\right]\\ &\quad + G\left[\frac{1}{r}\frac{\partial}{\partial r}\left(r\frac{\partial u_r}{\partial r}\right) + \frac{1}{r^2}\left(\frac{\partial^2 u_r}{\partial \theta^2}\right)\right] + F_r = 0\end{aligned} \tag{8.16.34}$$

$$\frac{(\lambda + G)}{r}\frac{\partial}{\partial\theta}\left[\frac{1}{r}\frac{\partial}{\partial r}(ru_r) + \frac{1}{r}\frac{\partial u_\theta}{\partial\theta}\right] + G\left[\frac{1}{r}\frac{\partial}{\partial r}\left(r\frac{\partial u_\theta}{\partial r}\right) + \frac{1}{r^2}\left(\frac{\partial^2 u_\theta}{\partial\theta^2}\right)\right] + F_\theta = 0 \tag{8.16.35}$$

$$F_z = 0. \tag{8.16.36}$$

Using the relations established in Sec. 6.2, the compatibility relation (8.16.24) becomes:

$$\left(\frac{\partial^2}{\partial r^2} + \frac{1}{r}\frac{\partial}{\partial r} + \frac{1}{r^2}\frac{\partial^2}{\partial\theta^2}\right)(\sigma_{rr} + \sigma_{\theta\theta}) = -\frac{1}{1-\nu}\left(\frac{\partial F_r}{\partial r} + \frac{1}{r}\frac{\partial F_\theta}{\partial\theta} + \frac{F_r}{r}\right). \tag{8.16.37}$$

For constant body forces, $(\sigma_{rr} + \sigma_{\theta\theta})$ is harmonic.

8.17 State of Plane Stress

As mentioned in Sec. 8.15, this state is defined by

$$\sigma_{11} = \sigma_{11}(x_1, x_2), \quad \sigma_{22} = \sigma_{22}(x_1, x_2), \quad \sigma_{12} = \sigma_{12}(x_1, x_2) \tag{8.17.1}$$

$$\sigma_{33} = \sigma_{13} = \sigma_{23} = 0. \tag{8.17.2}$$

The stress-strain relations (8.5.3) become:

$$\sigma_{11} = 2\mu\, e_{11} + \lambda(e_{11} + e_{22} + e_{33}) = 2\mu\, e_{11} + \frac{2\mu\lambda}{2\mu + \lambda}(e_{11} + e_{22}) \tag{8.17.3}$$

$$\sigma_{22} = 2\mu\, e_{22} + \lambda(e_{11} + e_{22} + e_{33}) = 2\mu\, e_{22} + \frac{2\mu\lambda}{2\mu + \lambda}(e_{11} + e_{22}) \tag{8.17.4}$$

$$e_{33} = -\frac{\lambda}{2\mu + \lambda}(e_{11} + e_{22}) \tag{8.17.5}$$

$$\sigma_{12} = 2\mu\, e_{12}, \quad \sigma_{13} = \sigma_{23} = 0. \tag{8.17.6}$$

Formally, Eqs. (8.17.3) and (8.17.4) become identical to Eqs. (8.16.4) and (8.16.5) if one replaces the constant $2\lambda\mu/(\lambda + 2\mu) \equiv \bar{\lambda}$ by λ. The stress-strain relations (8.5.21) to (8.5.24) become:

$$\sigma_{11} = \frac{E}{1 - \nu^2}(e_{11} + \nu\, e_{22}) \tag{8.17.7}$$

$$\sigma_{22} = \frac{E}{1 - \nu^2}(e_{22} + \nu\, e_{11}) \tag{8.17.8}$$

$$e_{33} = -\frac{\nu}{1 - \nu}(e_{11} + e_{22}) \tag{8.17.9}$$

$$\sigma_{12} = \frac{E}{1 + \nu} e_{12}, \quad \sigma_{13} = \sigma_{23} = 0. \tag{8.17.10}$$

The stress-strain relations (8.5.26) to (8.5.29) become:

$$e_{11} = \frac{1}{E}(\sigma_{11} - \nu\, \sigma_{22}) \tag{8.17.11}$$

$$e_{22} = \frac{1}{E}(-\nu\, \sigma_{11} + \sigma_{22}) \tag{8.17.12}$$

$$e_{33} = -\frac{\nu}{E}(\sigma_{11} + \sigma_{22}) \tag{8.17.13}$$

$$e_{12} = \frac{1 + \nu}{E}\sigma_{12}, \quad e_{13} = e_{23} = 0. \tag{8.17.14}$$

Equation (8.17.13) shows that e_{33} is only a function of x_1 and x_2. The equations of equilibrium become:

$$\frac{\partial \sigma_{11}}{\partial x_1} + \frac{\partial \sigma_{21}}{\partial x_2} + F_1 = 0 \tag{8.17.15}$$

$$\frac{\partial \sigma_{12}}{\partial x_1} + \frac{\partial \sigma_{22}}{\partial x_2} + F_2 = 0 \tag{8.17.16}$$

$$F_3 = 0. \tag{8.17.17}$$

Thus, in plane stress, the body forces are such that $F_3 = 0$, and F_1 and F_2 are independent of x_3.

Of the six compatibility equations (4.10.14), two are identically satisfied and the four remaining ones become:

$$2\frac{\partial^2 e_{12}}{\partial x_1\, \partial x_2} = \frac{\partial^2 e_{11}}{\partial x_2^2} + \frac{\partial^2 e_{22}}{\partial x_1^2} \tag{8.17.18}$$

$$\frac{\partial^2 e_{33}}{\partial x_1^2} = \frac{\partial^2 e_{33}}{\partial x_2^2} = \frac{\partial^2 e_{33}}{\partial x_1\, \partial x_2} = 0. \tag{8.17.19}$$

Eqs. (8.17.19) demand that e_{33} be given by an equation of the form

$$e_{33} = C_1 + C_2 x_1 + C_3 x_2, \tag{8.17.20}$$

where C_1, C_2 and C_3 are constants. In general, however, only Eq. (8.17.18) is taken into account in common problems, and the requirement of (8.17.20) is neglected. This results in a solution which, although approximate, is very satisfactory when the dimension of the body in the OX_3 direction is very small (thin plates).

Using the stress-strain relations (8.17.11) to (8.17.14) and the equilibrium equations (8.17.15) to (8.17.17), Eq. (8.17.18) can be written in terms of the stresses as follows:

$$\left(\frac{\partial^2}{\partial x_1^2} + \frac{\partial^2}{\partial x_2^2}\right)(\sigma_{11} + \sigma_{22}) = -(1+\nu)\left(\frac{\partial F_1}{\partial x_1} + \frac{\partial F_2}{\partial x_2}\right). \tag{8.17.21}$$

The boundary conditions (7.3.8) become:

$$\begin{aligned} \sigma_{n1} &= \ell_1 \sigma_{11} + \ell_2 \sigma_{21} \\ \sigma_{n2} &= \ell_1 \sigma_{12} + \ell_2 \sigma_{22} \\ \sigma_{n3} &= 0. \end{aligned} \tag{8.17.22}$$

From Eqs. (8.16.24) and (8.17.21) we see that, when the body forces are constant, the distribution of stresses in the (OX_1, OX_2) plane is the same for plane strain and for plane stress problems, and is governed by the equilibrium equations and Eq. (8.16.26). One must not forget however the approximate nature of the solution when Eq. (8.17.20) is neglected.

In a system of cylindrical coordinates, the strain displacements relations are given by Eqs. (6.7.23) and (6.7.24):

$$e_{rr} = \frac{\partial u_r}{\partial r}, \quad e_{\theta\theta} = \frac{1}{r}\frac{\partial u_\theta}{\partial \theta} + \frac{u_r}{r}, \quad e_{zz} = \frac{\partial u_z}{\partial z} = e_{zz}(r, \theta) \tag{8.17.23}$$

$$e_{r\theta} = \frac{1}{2}\left(\frac{1}{r}\frac{\partial u_r}{\partial \theta} + \frac{\partial u_\theta}{\partial r} - \frac{u_\theta}{r}\right), \quad e_{rz} = e_{\theta z} = 0. \tag{8.17.24}$$

All the stress-strain relations in Eqs. (8.17.3) to (8.17.14) remain the same except that r, θ, and z are substituted for 1, 2, and 3, respectively. The equations of equilibrium are given by Eqs. (8.16.30) to (8.16.32). The compatibility relation (8.17.21) becomes:

$$\left(\frac{\partial^2}{\partial r^2} + \frac{1}{r}\frac{\partial}{\partial r} + \frac{1}{r^2}\frac{\partial^2}{\partial \theta^2}\right)(\sigma_{rr} + \sigma_{\theta\theta}) = -(1+\nu)\left(\frac{\partial F_r}{\partial r} + \frac{1}{r}\frac{\partial F_\theta}{\partial \theta} + \frac{F_r}{r}\right). \tag{8.17.25}$$

8.18 State of Generalized Plane Stress

Let us consider a thin plane—in other words, a body with parallel faces and a thickness $2h$ which is very small compared to the linear dimensions of the faces. The middle surface of the plate is located halfway between the faces and is taken as the OX_1, OX_2 plane (Fig. 8.6). The generators forming the edges of the plate are normal to the two parallel faces. All the loading is applied to the edges, in planes parallel

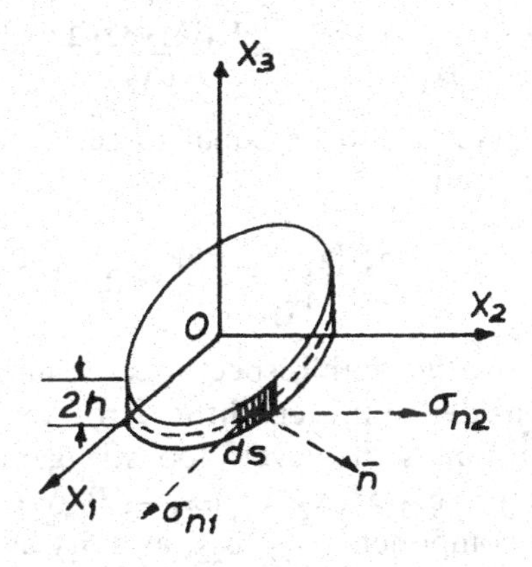

Fig. 8.6

to the OX_1, OX_2 plane. This loading is symmetrically distributed with respect to the middle surface, and so are the body forces F_1 and F_2; F_3 is equal to zero. In view of the previous assumptions, points on the middle surface do not suffer any displacement u_3 in the OX_3 direction, and for all other points u_3 is very small; also the variations of the components u_1 and u_2 of the displacement through the thickness of the plate will be small. This suggests working with average values: In reality these averages are often the only ones susceptible to experimental measurements. Thus,

$$\bar{u}_1(x_1, x_2) = \frac{1}{2h}\int_{-h}^{+h} u_1(x_1, x_2, x_3)\,dx_3 \tag{8.18.1}$$

$$\bar{u}_2(x_1, x_2) = \frac{1}{2h}\int_{-h}^{+h} u_2(x_1, x_2, x_3)\,dx_3 \tag{8.18.2}$$

$$\bar{u}_3(x_1, x_2) = \frac{1}{2h} \int_{-h}^{+h} u_3(x_1, x_2, x_3)\, dx_3, \tag{8.18.3}$$

where bars over letters denote mean values. Because of symmetry, $\bar{u}_3 = 0$. Since the faces of the plate are assumed free of external loads,

$$\sigma_{13}(x_1, x_2, \pm h) = \sigma_{23}(x_1, x_2, \pm h) = \sigma_{33}(x_1, x_2, \pm h) = 0. \tag{8.18.4}$$

Therefore,

$$\frac{\partial \sigma_{13}(x_1, x_2, \pm h)}{\partial x_1} = \frac{\partial \sigma_{23}(x_1, x_2, \pm h)}{\partial x_2} = 0, \tag{8.18.5}$$

and, since F_3 was assumed to be equal to zero, the third equation of equilibrium requires that

$$\frac{\partial \sigma_{33}(x_1, x_2, \pm h)}{\partial x_3} = 0. \tag{8.18.6}$$

Since σ_{33} and its derivative with respect to x_3 vanish on the faces of the plate, we could consider with very little error that σ_{33} is zero everywhere. The assumption is, however, too stringent and it suffices to consider that $1/2h \int_{-h}^{+h} \sigma_{33}(x_1, x_2, x_3)\, dx_3 \equiv 0$. We define the average values of the stress components σ_{11}, σ_{22}, and σ_{12} as follows:

$$\bar{\sigma}_{11} = \frac{1}{2h} \int_{-h}^{+h} \sigma_{11}\, dx_3, \quad \bar{\sigma}_{22} = \frac{1}{2h} \int_{-h}^{+h} \sigma_{22}\, dx_3,$$
$$\bar{\sigma}_{12} = \frac{1}{2h} \int_{-h}^{+h} \sigma_{12}\, dx_3. \tag{8.18.7}$$

The mean values of the body forces are defined as

$$\bar{F}_1 = \frac{1}{2h} \int_{-h}^{+h} F_1\, dx_3, \quad \bar{F}_2 = \frac{1}{2h} \int_{-h}^{+h} F_2\, dx_3,$$
$$\bar{F}_3 = \frac{1}{2h} \int_{-h}^{+h} F_3\, dx_3 = 0. \tag{8.18.8}$$

Integrating the equations of equilibrium, taking into account the definitions of the mean values, we get:

$$\frac{\partial \bar{\sigma}_{11}}{\partial x_1} + \frac{\partial \bar{\sigma}_{21}}{\partial x_2} + \bar{F}_1 = 0 \tag{8.18.9}$$

$$\frac{\partial \bar{\sigma}_{12}}{\partial x_1} + \frac{\partial \bar{\sigma}_{22}}{\partial x_2} + \bar{F}_2 = 0. \tag{8.18.10}$$

In terms of the mean values across the thickness, the stress-strain relations (8.5.3) become:

$$\bar{\sigma}_{11} = 2\mu\bar{e}_{11} + \bar{\lambda}(\bar{e}_{11} + \bar{e}_{22}) \tag{8.18.11}$$

$$\bar{\sigma}_{22} = 2\mu\bar{e}_{22} + \bar{\lambda}(\bar{e}_{11} + \bar{e}_{22}) \tag{8.18.12}$$

$$\bar{\sigma}_{12} = 2\mu\bar{e}_{12}, \tag{8.18.13}$$

where $\bar{\lambda} = 2\lambda\mu/(\lambda + 2\mu)$. These three equations, together with the two equations of equilibrium, serve to determine the five unknown $\bar{u}_1$, $\bar{u}_2$, $\bar{\sigma}_{11}$, $\bar{\sigma}_{22}$, and $\bar{\sigma}_{12}$, all a function of x_1 and x_2 only.

When Eqs. (8.18.11) to (8.18.13) are substituted into Eqs. (8.18.9) and (8.18.10), we obtain two equations of elasticity in terms of average strains and displacements:

$$(\bar{\lambda} + \mu)\frac{\partial}{\partial x_1}(\bar{e}_{11} + \bar{e}_{22}) + \mu\,\nabla^2\bar{u}_1 + \bar{F}_1 = 0 \tag{8.18.14}$$

$$(\bar{\lambda} + \mu)\frac{\partial}{\partial x_2}(\bar{e}_{11} + \bar{e}_{22}) + \mu\nabla^2\bar{u}_2 + \bar{F}_2 = 0. \tag{8.18.15}$$

These equations serve to determine the average components of the displacement $\bar{u}_1$ and $\bar{u}_2$ when the average displacements are specified on the contour.

The mean strains satisfy the compatibility equation (8.17.18). In terms of stresses, this equation becomes:

$$\nabla^2(\bar{\sigma}_{11} + \bar{\sigma}_{22}) = -\frac{2(\bar{\lambda} + \mu)}{\bar{\lambda} + 2\mu}\left(\frac{\partial \bar{F}_1}{\partial x_1} + \frac{\partial \bar{F}_2}{\partial x_2}\right). \tag{8.18.16}$$

8.19 State of Generalized Plane Strain

As stated in Sec. 8.15, this state results from the superposition of a constant strain e_o along OX_3 to a state of plane strain in the OX_1, OX_2 plane:

$$e'_{11} = \frac{1 + \nu}{E}[(1 - \nu)\sigma_{11} - \nu\,\sigma_{22}] - \nu e_o \tag{8.19.1}$$

$$e'_{22} = \frac{1+\nu}{E}[(1-\nu)\sigma_{22} - \nu\,\sigma_{11}] - \nu e_o \tag{8.19.2}$$

$$e'_{33} = e_o \tag{8.19.3}$$

$$e'_{12} = e_{12} = \frac{1+\nu}{E}\sigma_{12} \tag{8.19.4}$$

$$e_{33} = e'_{13} = e'_{23} = 0, \tag{8.19.5}$$

where the e'_{ij} ' s are the resulting strains and the σ_{ij} ' s and e_{ij} ' s are the stresses and strains of the state of plane strain. Solving for the stresses, we get:

$$\sigma_{11} = \frac{E}{(1+\nu)(1-2\nu)}[(1-\nu)e'_{11} + \nu(e'_{22} + e_o)] \tag{8.19.6}$$

$$\sigma_{22} = \frac{E}{(1+\nu)(1-2\nu)}[(1-\nu)e'_{22} + \nu(e'_{11} + e_o)] \tag{8.19.7}$$

$$\sigma_{33} = \frac{\nu E}{(1+\nu)(1-2\nu)}(e_{11} + e_{22}) + Ee_o = \nu(\sigma_{11} + \sigma_{22}) + Ee_o \tag{8.19.8}$$

$$\sigma_{12} = \frac{E}{1+\nu}e_{12} \tag{8.19.9}$$

$$\sigma_{13} = \sigma_{23} = 0. \tag{8.19.10}$$

The addition of constant terms to the equations of plane strain does not introduce any change in the differential equations of Sec. 8.16. These equations still hold in this case. The state of generalized plane strain will be encountered in the study of rotating long cylinders.

8.20 Solution of Elasticity Problems

In Sec. 8.8, it was stated that the solution of a problem of elasticity requires the solution of a system of 15 equations with 15 unknowns. A systematic elimination of some of the unknowns has allowed us to reduce the system to equations, either in terms of the displacements in Sec. 8.9, or in terms of the stresses in Sec. 8.10. This suggests various methods that can be used in solving elasticity problems. Some of these methods depend primarily on intuition, while others are based on a systematic application of the techniques of applied mathematics:

1. The Inverse Method. In this method, one guesses a solution satisfying all the requirements of the theory of elasticity and tries to find to what boundary conditions the solution corresponds [1].

2. The Semi-Inverse Method. There, one guesses a part of the solution assuming expressions for the stresses, the strains, or the displacements; enough freedom is left in the assumptions so that the differential equations and the boundary conditions can be satisfied [1].

3. The Method of Potentials. To simplify the solution of the various systems of equations describing elasticity problems, a number of potentials have been introduced. Potentials related to displacements provide solutions to Navier's equations while potentials related to stresses generate systems of equilibrating stresses. Some of the most powerful potentials will be studied in the next chapter [2,4,6].

4. Betti's Method [2,3].

5. The Integral Transform Methods [2].

6. The Complex Variables Method. Devised primarily for the solution of plane problems, this method is discussed in Chapter 19 [4,5].

7. The Variational Methods. These methods are based on the fact that the governing equations of elasticity can be obtained from the minimization of an energy expression. Thus, we may seek a solution which will minimize the energy expression and avoid the difficulties involved in the solving of the differential equations. The use of these methods will be discussed in Chapter 15.

8. The Numerical Methods [1,6].

In the following chapters the inverse method, the semi-inverse method, the method of potentials, the variational method and the complex variables method will be presented and used to solve a variety of problems of engineering importance. Betti's method, the integral transform methods and the numerical methods are not examined in this text.

PROBLEMS

1. Show that the stress-strain relations for a panel (Fig. 8.7) made of orthotropic material under a condition of plane stress can be written in the following form which involves only four independent constants:

$$\begin{bmatrix} \sigma_{11} \\ \sigma_{22} \\ \sigma_{12} \end{bmatrix} = \begin{bmatrix} H_{11} & H_{12} & 0 \\ H_{12} & H_{22} & 0 \\ 0 & 0 & 2G_{12} \end{bmatrix} \begin{bmatrix} e_{11} \\ e_{22} \\ e_{12} \end{bmatrix}.$$

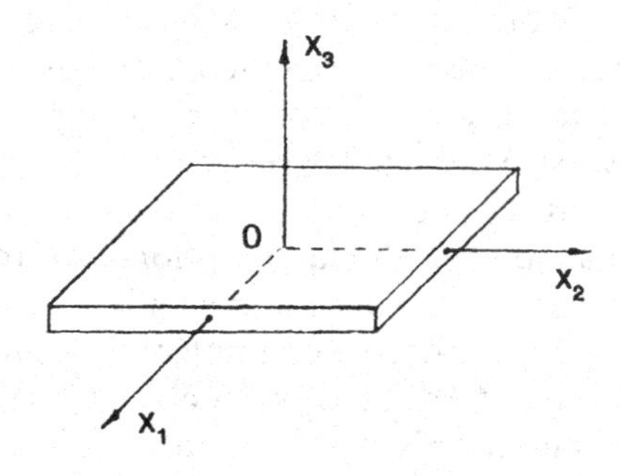

Fig. 8.7

Derive the expression of H_{11}, H_{22}, H_{12}, and G_{12} in terms of the tensor C_{ijkl}. In Fig. 8.7, the reference axes are parallel to the axes of symmetry.

2. Sometimes the components of the compliance matrix S_{ijkl} are written in terms of the constants E, ν, and G in the following manner:

$$S_{1111} = \frac{1}{E_1}, \quad S_{1212} = \frac{1}{4G_{12}}, \quad S_{2211} = -\frac{\nu_{12}}{E_1}, \ldots,$$

where E_1 is Young's modulus in the OX_1 direction, G_{12} is the shear modulus associated with the OX_1, OX_2 directions, and ν_{12} is Poisson's ratio for the strain in the OX_2 direction caused by the stress in the OX_1 direction. The inverse of the stress-strain relations in Problem 1 is written as:

$$\begin{bmatrix} e_{11} \\ e_{22} \\ e_{12} \end{bmatrix} = \begin{bmatrix} \dfrac{1}{E_1} & -\dfrac{\nu_{12}}{E_1} & 0 \\ -\dfrac{\nu_{12}}{E_1} & \dfrac{1}{E_2} & 0 \\ 0 & 0 & \dfrac{1}{2G_{12}} \end{bmatrix} \begin{bmatrix} \sigma_{11} \\ \sigma_{22} \\ \sigma_{12} \end{bmatrix}.$$

Determine H_{11}, H_{22}, and H_{12} in terms of E_1, E_2, and ν_{12}.

3. Find the coefficients of the matrix of the elastic coefficients in Problem 2, if the system of axes is rotated 30 degrees counterclockwise around the OX_3 axis.

4. A cubic material is a material in which the properties are the same along three orthogonal directions. Show that the matrix of the coefficients of elasticity contains three independent constants only: Choosing the coordinate axes along these directions, the compliance matrix can be written as follows:

$$\begin{bmatrix} S_{1111} & S_{1122} & S_{1122} & 0 & 0 & 0 \\ S_{1122} & S_{1111} & S_{1122} & 0 & 0 & 0 \\ S_{1122} & S_{1122} & S_{1111} & 0 & 0 & 0 \\ 0 & 0 & 0 & S_{1212} & 0 & 0 \\ 0 & 0 & 0 & 0 & S_{1212} & 0 \\ 0 & 0 & 0 & 0 & 0 & S_{1212} \end{bmatrix}.$$

5. Prove that in an isotropic, homogeneous, linearily elastic solid, the principal axes of the stress tensor coincide with the principal axes of the linear strain tensor.

6. Could the following stress fields be possible stress fields in an elastic solid, and, if so, under what conditions?

$$\begin{aligned} &\sigma_{11} = ax_1 + bx_2 && \sigma_{11} = ax_1^2 x_2^2 + bx_1 && \sigma_{11} = a[x_2^2 + b(x_1^2 - x_2^2)] \\ &\sigma_{22} = cx_1 + dx_2 && \sigma_{22} = cx_2^2 && \sigma_{22} = a[x_1^2 + b(x_2^2 - x_1^2)] \\ &\sigma_{12} = fx_1 + gx_2 && \sigma_{12} = dx_1 x_2 && \sigma_{33} = ab(x_1^2 + x_2^2) \\ &\sigma_{13} = \sigma_{23} = 0 && \sigma_{13} = \sigma_{23} = 0 && \sigma_{12} = 2abx_1 x_2 \\ &\sigma_{33} = 0 && \sigma_{33} = 0 && \sigma_{13} = \sigma_{23} = 0. \end{aligned}$$

a, b, c, d, f, and g are constants.

7. A cube of iron whose edges are 10 in. long is subjected to a uniform pressure of 10 tons / in^2 on two opposite faces; the other faces are prevented from moving more than 0.002 in. by lateral pressure. Determine the pressures on these faces and the maximum shearing stress in the cube. $E = 30 \times 10^6$ psi and $\nu = 0.3$.

8. A cube of Duralumin, whose edges are 5 in. long, is subjected to a uniform pressure of 15,000 psi on the four faces normal to the OX_1 and OX_2 axes. The two faces normal to the OX_3 axis are restricted to a total deformation of 0.0006 in. Determine the stress σ_{33} and the change in length of the diagonal of the cube. $E = 10^7$ psi and $\nu = 0.3$.

9. In Problem 8 of Chapter 7, find the change in length of the diagonals. $E = 30 \times 10^6$ and $\nu = 0.3$.

10. A steel pulley is to be fitted tightly around a shaft. The internal diameter of the hole in the pulley is 0.998 in., while the outside diameter of the shaft is 1.000 in. The pulley will be assembled on the shaft by heating the pulley, then allowing the assembly to reach a uniform temperature. What is the temperature change required to produce a clearance of 0.001 in. for easy assembly? For steel, $\alpha = 6.0 \times 10^{-6}/°F$, $E = 30 \times 10^6$ psi, $\nu = 0.3$.

11. A weight of 20,000 lbs. is supported on two short lengths of concentric copper and steel tubes (Fig. 8.8). The thickness of these tubes is such that both tubes have a cross-sectional area of 2 in^2. Determine the amount of load carried by each tube at room temperature and when the temperature is raised 100 ° F above room temperature. For steel, $E = 30 \times 10^6$ psi, $\nu = 0.3$, $\alpha = 6 \times 10^{-6}/°F$, and for copper, $E = 17 \times 10^6$ psi, $\nu = 0.35$, $\alpha = 9.2 \times 10^{-6}/°F$. The tubes have the same length at room temperature when unloaded.

12. A prismatic bar of length ℓ hangs under its own weight and is supported at its top by the uniform stress $\rho g \ell$ where ρg is the weight per unit volume (Fig. 8.9). Show that the solution,

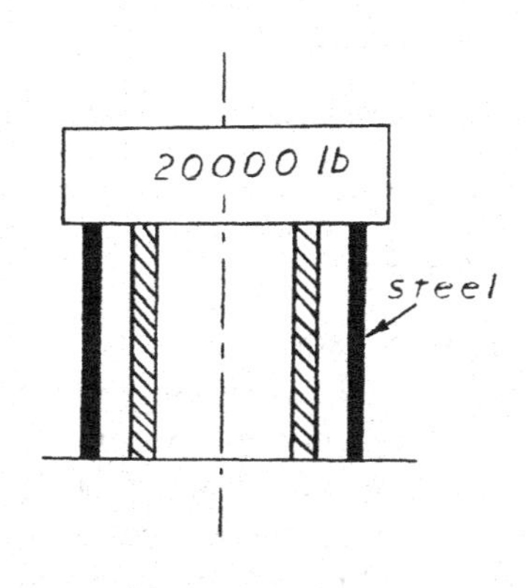

Fig. 8.8

$$\sigma_{33} = \rho g x, \quad \sigma_{11} = \sigma_{22} = \sigma_{12} = \sigma_{13} = \sigma_{23} = 0,$$

satisfies equilibrium, compatibility, and the prescribed boundary conditions. If an element at A along the OX_3 axis is fixed, find the

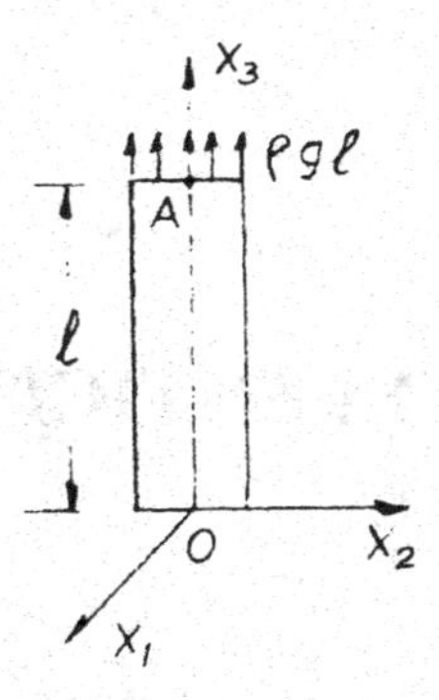

Fig. 8.9

expressions of the displacements u_1, u_2, and u_3.

13. The stress distribution in a thin disk of radius b rotating at an angular velocity ω rad./sec. is given by:

$$\sigma_{rr} = \frac{3+\nu}{8}\rho\omega^2 b^2\left(1 - \frac{r^2}{b^2}\right)$$

$$\sigma_{\theta\theta} = \frac{3+\nu}{8}\rho\omega^2 b^2\left(1 - \frac{1+3\nu}{3+\nu}\frac{r^2}{b^2}\right)$$

$$\sigma_{r\theta} = \sigma_{rz} = \sigma_{\theta z} = \sigma_{zz} = 0.$$

Neglecting gravity forces, show that this solution satisfies equilibrium, compatibility, and the prescribed boundary conditions.

14. The solution of the problem of the circular shaft fixed at one end and subjected to a twisting moment at the other is given by (see Fig. 10.5):

$$u_1 = -\alpha x_2 x_3, \quad u_2 = \alpha x_1 x_3, \quad u_3 = 0.$$

What are the conditions that this solution imposes on the applied twisting moments? Is the shaft in a state of plane strain or of plane stress? (*)

REFERENCES

[1] S. Timoshenko and J. N. Goodier, *Theory of Elasticity*, McGraw-Hill, New York, N. Y., 1970.

[2] I. N. Sneddon and D. S. Berry, "The Classical Theory of Elasticity," *Encyclopedia of Physics*, Vol. 6, Springer-Verlag, 1958.

[3] A. E. H. Love, *A Treatise on the Mathematical Theory of Elasticity*, 4th ed., Dover, New York, N. Y., 1927.

[4] Y. C. Fung, *Foundation of Solid Mechanics*, Prentice-Hall, Englewood Cliffs, N. J., 1965.

[5] N. J. Muskhelishvili, *Some Basic Problems of the Mathematical Theory of Elasticity*, Noordhoff, Groningen, 1953.

[6] I. S. Sokolnikoff, *Mathematical Theory of Elasticity*, McGraw-Hill, New York, N. Y., 1956.

---

* See pages A-54 and A-55 for additional problems.

CHAPTER 9

SOLUTION OF ELASTICITY PROBLEMS BY POTENTIALS

9.1 Introduction

In this chapter we shall describe two types of potentials:

(a) potentials related to displacements, namely, the scalar and vector potentials, the Galerkin Vectors, and the Neuber-Papkovich functions;

(b) potentials that generate systems of equilibrating stresses, namely, the Maxwell stress function, the Morera stress function, and the Airy stress function.

On the subject of potentials, references [1] and [3] of Chapter 8 give extensive details as well as a large bibliography.

We shall first summarize some results of field theory. For added clarity, a vector notation is used in most of the operations involving the displacements.

9.2 Some Results of Field Theory

Suppose that, associated with each point in a region R, there is a scalar point function

$$T = T(x_1, x_2, x_3) = T(x_i). \tag{9.2.1}$$

The resulting field is said to be a *scalar field*. An example of a scalar field is the temperature distribution in a room. In the study of the displacements of the points of a body subjected to linear transformation, there is, associated with every point, a vector

$$\bar{u} = \bar{u}(x_1, x_2, x_3). \tag{9.2.2}$$

The resulting field is said to be a *vector field.* The scalar field associated with a given point is designated by one number while the vector field is designated by three numbers.

Let ϕ be a scalar point function, $\bar{\psi}$ be a vector point function, and $\bar{X}$ be the position vector. The following identities are established in vector analysis:

$$\text{div curl } \bar{\psi} = \bar{\nabla} \cdot (\bar{\nabla} \times \bar{\psi}) = 0 \tag{9.2.3}$$

$$\text{curl grad } \phi = \bar{\nabla} \times (\bar{\nabla}\phi) = 0 \tag{9.2.4}$$

$$\text{curl curl } \bar{\psi} = \text{curl}^2 \bar{\psi} = \bar{\nabla} \times (\bar{\nabla} \times \bar{\psi}) = \bar{\nabla}(\bar{\nabla} \cdot \bar{\psi}) - \nabla^2\bar{\psi} \tag{9.2.5}$$

$$\text{div grad } \phi = \bar{\nabla} \cdot (\bar{\nabla}\phi) = \nabla^2\phi \tag{9.2.6}$$

$$\text{div Lapl } \bar{\psi} = \bar{\nabla} \cdot (\nabla^2\bar{\psi}) = \nabla^2(\bar{\nabla} \cdot \bar{\psi}) = \text{Lapl div } \bar{\psi} \tag{9.2.7}$$

$$\text{Lapl grad } \phi = \nabla^2(\bar{\nabla}\phi) = \bar{\nabla}(\nabla^2\phi) = \text{grad Lapl } \phi \tag{9.2.8}$$

$$\text{Lapl } (\bar{\psi} \cdot \bar{X}) = \nabla^2(\bar{\psi} \cdot \bar{X}) = 2\bar{\nabla} \cdot \bar{\psi} + \bar{X} \cdot \nabla^2\bar{\psi} \tag{9.2.9}$$

$$\text{div } (\phi\bar{\psi}) = \phi\bar{\nabla} \cdot \bar{\psi} + (\bar{\nabla}\phi) \cdot \bar{\psi} \tag{9.2.10}$$

$$\text{Lapl } (\phi\bar{X}) = 2\bar{\nabla}\phi + \bar{X}\nabla^2\phi. \tag{9.2.11}$$

If, throughout a region R, a vector point function $\bar{\psi}$ satisfies the conditions

$$\text{div } \bar{\psi} = \bar{\nabla} \cdot \bar{\psi} = 0, \tag{9.2.12}$$

the field is said to be solenoidal in that region. If, at every point in R,

$$\text{curl } \bar{\psi} = \bar{\nabla} \times \bar{\psi} = 0, \tag{9.2.13}$$

the field is said to be irrotational or lamellar. For a lamellar field, there exists a scalar point function ϕ say, such that

$$\bar{\psi} = \text{grad } \phi = \bar{\nabla}\phi. \tag{9.2.14}$$

ϕ is determined within an additive arbitrary constant since the gradient of the latter is zero. Setting the constant component of ϕ equal to some chosen value (for example, zero) uniquely determines ϕ. ϕ is called a *scalar potential.* For a solenoidal field, it is always possible to introduce a *vector potential* $\bar{A}$, say, such that

$$\bar{\psi} = \text{curl}\ \bar{A} = \bar{\nabla} \times \bar{A}. \tag{9.2.15}$$

$\bar{A}$ is determined within an additive arbitrary vector function representing a potential field, since the curl of the latter is zero [Eq. (9.2.4)]. If the vector function $\bar{A}$ is introduced by Eq. (9.2.15) merely for convenience, that is, as an auxiliary function in the course of an analysis, this difficulty may be overcome by the further demand that

$$\text{div}\ \bar{A} = 0. \tag{9.2.16}$$

This is analogous to setting the constant component of ϕ equal to zero in Eq. (9.2.14). Any other disposition of the value of div $\bar{A}$ appropriate to the circumstances pertaining to a specific problem may likewise be made. Taking the curl of both sides of Eq. (9.2.15), we obtain:

$$\text{curl}\ \bar{\psi} = \text{curl curl}\ \bar{A} = \text{grad div}\ \bar{A} - \nabla^2 \bar{A}.$$

If div $\bar{A}$ is chosen equal to zero, the previous equation becomes:

$$\text{curl}\ \bar{\psi} = -\nabla^2 \bar{A}. \tag{9.2.17}$$

This is Poisson's equation, which can be solved to give $\bar{A}$.

Finally, we shall give the proof of a theorem which is of importance in connection with the applications of scalar and vector potential functions. This theorem is known as *Helmholtz's theorem*, and can be stated as follows:

A vector field $\bar{E}$ with known divergence and curl, none of which equal to zero, and which is finite, uniform, and vanishes at infinity, may be expressed as the sum of a lamellar vector $\bar{U}$ and a solenoidal vector $\bar{V}$,

$$\bar{E} = \bar{U} + \bar{V} \tag{9.2.18}$$

with

$$\text{curl}\ \bar{U} = 0, \quad \text{div}\ \bar{V} = 0. \tag{9.2.19}$$

To prove this statement, we shall show that both $\overline{U}$ and $\overline{V}$ can be found when $\overline{E}$ is given everywhere. Since curl $\overline{U} = 0$, a scalar potential ϕ exists such that

$$\overline{U} = \text{grad}\ \phi. \tag{9.2.20}$$

Also, since div $\overline{V} = 0$, a vector potential $\overline{\psi}$ can be introduced such that

$$\overline{V} = \text{curl}\ \overline{\psi}, \quad \text{div}\ \overline{\psi} = 0. \tag{9.2.21}$$

Returning to Eq. (9.2.18), we have:

$$\text{div}\ \overline{E} = \text{div}\ \overline{U} + \text{div}\ \overline{V} = \text{div grad}\ \phi + 0 = \nabla^2 \phi. \tag{9.2.22}$$

Also,

$$\text{curl}\ \overline{E} = \text{curl}\ \overline{U} + \text{curl}\ \overline{V} = 0 + \text{curl curl}\ \overline{\psi} = -\nabla^2 \overline{\psi}. \tag{9.2.23}$$

Eqs. (9.2.22) and (9.2.23) are two Poisson equations whose solutions give ϕ and $\overline{\psi}$. Knowing ϕ and $\overline{\psi}$, $\overline{U}$ and $\overline{V}$ can be obtained. This proves the theorem.

9.3 The Homogenous Equations of Elasticity and the Search for Particular Solutions

When the body forces are equal to zero, Eq. (8.9.4) is reduced to the homogeneous form:

$$\mu \nabla^2 u_i + (\lambda + \mu) \frac{\partial e_v}{\partial x_i} = 0. \tag{9.3.1}$$

Whenever a particular solution of Eq. (8.9.4) is found, the solution of an elasticity problem can be reduced to the solution of a set of three homogeneous equations: By particular solution, we mean a solution satisfying Eq. (8.9.4) but not the boundary conditions of the given problem. To show that this is the case, let

$$u_1' = u_1'(x_1, x_2, x_3), \quad u_2' = u_2'(x_1, x_2, x_3), \quad u_3' = u_3'(x_1, x_2, x_3) \tag{9.3.2}$$

be a particular solution of Eq. (8.9.4) in a problem where the displacements are prescribed on the boundary; the values that u_1', u_2', and u_3' take on the boundary differ from the prescribed ones and can readily be obtained by introducing the coordinates of the points on the boundary, in Eqs. (9.3.2). Let us set

$$u''_1 = u_1 - u'_1, \quad u''_2 = u_2 - u'_2, \quad u''_3 = u_3 - u'_3, \quad e''_v = e_v - e'_v, \tag{9.3.3}$$

and consider u''_1, u''_2, and u''_3 as our new unknowns. If we substitute Eqs. (9.3.2) in Eqs. (8.9.4), we get:

$$\mu\nabla^2 u'_i + (\lambda + \mu)\frac{\partial e'_v}{\partial x_i} + F_i = 0. \tag{9.3.4}$$

Subtracting Eq. (9.3.4) from Eq. (8.9.4), we obtain the equation:

$$\mu\nabla^2 u''_i + (\lambda + \mu)\frac{\partial e''_v}{\partial x_i} = 0. \tag{9.3.5}$$

The problem is now reduced to finding the solution of Eq. (9.3.5) with new boundary conditions in terms of u''_i; these boundary conditions are obtained from Eqs. (9.3.3). Once the solution of this homogeneous system is obtained, the particular solution is added to it to give the general solution.

It is often simple to find a particular solution to Eq. (8.9.4) and the main difficulty resides in finding the solution to the homogeneous system of equations. Let us, for example, consider Eq. (8.9.4) with the body forces deriving from a scalar potential ϕ, and let us find a particular solution of the form:

$$\bar{u} = \text{grad } M.$$

In this case, Eq. (8.9.4), when written in vector form, becomes:

$$\text{grad }[(\lambda + 2\mu)\nabla^2 M + \phi] = 0. \tag{9.3.6}$$

A particular solution of this equation is

$$\nabla^2 M = -\frac{\phi}{\lambda + 2\mu}, \tag{9.3.7}$$

which is a Poisson equation, for which a particular solution can be found. Let us assume the body forces to be gravity forces. Then

$$F_1 = F_2 = 0 \tag{9.3.8}$$

$$\begin{aligned} F_3 &= -\rho g \\ \phi &= -\rho g\, x_3 . \end{aligned} \tag{9.3.9}$$

Eq. (9.3.7) in this case becomes:

$$\nabla^2 M = \frac{\rho g\, x_3}{\lambda + 2\mu}. \tag{9.3.10}$$

A particular solution of Eq. (9.3.10) is:

$$M = \frac{\rho g\, x_3^3}{6(\lambda + 2\mu)}. \tag{9.3.11}$$

Thus, the particular solution of Eq. (8.9.4) is:

$$u_1 = 0, \quad u_2 = 0, \quad u_3 = \frac{\rho g\, x_3^2}{2(\lambda + 2\mu)}. \tag{9.3.12}$$

When the body forces derive from a scalar potential ϕ, Eq. (8.9.4) can be written in vector form as:

$$(\lambda + \mu) \text{ curl curl } \bar{u} + (\lambda + 2\mu)\nabla^2 \bar{u} + \text{grad } \phi = 0. \tag{9.3.13}$$

Taking the divergence of Eq. (9.3.13), we get:

$$\nabla^2[(\lambda + 2\mu) \text{ div } \bar{u} + \phi] = 0. \tag{9.3.14}$$

Therefore, $[(\lambda + 2\mu) \text{ div } \bar{u} + \phi]$ is a harmonic function. In most practical cases, ϕ is harmonic, therefore div $\bar{u} = e_v$ is harmonic. Because of Eq. (8.5.14), σ_{nn} is also a harmonic function. Taking the Laplacian of Eq. (9.3.13), we get

$$\nabla^2(\nabla^2 \bar{u}) = \nabla^4 \bar{u} = 0. \tag{9.3.15}$$

Eq. (9.3.15) shows that the components of the displacement vector $\bar{u}$ are biharmonic. The components of the states of stress and strain, being linear combinations of the first derivatives of u_1, u_2, and u_3, are also biharmonic. In summary, when the body forces F_i derive from a potential, and as a particular case when they are equal to zero, we have:

$$\nabla^2 e_v = 0, \quad \nabla^2 \sigma_{nn} = 0 \tag{9.3.16}$$

$$\nabla^4 u_i = 0, \quad \nabla^4 \sigma_{ij} = 0, \quad \nabla^4 e_{ij} = 0; \tag{9.3.17}$$

in other words, all the basic functions of the theory of elasticity are biharmonic.

In the following sections, we shall concentrate on finding a solution to the homogenous system. This solution must, of course, satisfy the boundary conditions of the problem. Once such a solution is found, the uniqueness theorem will ensure that it is the only solution to the problem at hand. There are rare cases in which Navier's equations can be directly integrated to give the displacements: One such case is encountered in the study of disks.

9.4 Scalar and Vector Potentials. Lamé's Strain Potential

In vector form, the homogeneous equation (9.3.1) is written as follows:

$$(\lambda + G)\ \text{grad}\ (\text{div}\ \bar{u}) + G\ \text{Lapl}\ \bar{u} = (\lambda + G)\bar{\nabla}(\bar{\nabla} \cdot \bar{u}) + G\nabla^2\bar{u} = 0. \tag{9.4.1}$$

According to Helmholtz's theorem, the vector field $\bar{u}$ can be written in terms of its scalar and vector potentials $\phi(x_1, x_2, x_3)$ and $\bar{\psi}(x_1, x_2, x_3)$ as follows:

$$\bar{u} = \text{grad}\ \phi + \text{curl}\ \bar{\psi} = \bar{\nabla}\phi + \bar{\nabla} \times \bar{\psi}, \tag{9.4.2}$$

with

$$\text{div}\ \bar{\psi} = 0. \tag{9.4.3}$$

If we take the divergence of both sides of Eq. (9.4.2), we get:

$$\bar{\nabla} \cdot \bar{u} = \nabla^2\phi = e_v. \tag{9.4.4}$$

If we take the curl of both sides of Eq. (9.4.2), we get:

$$\bar{\nabla} \times \bar{u} = \bar{\nabla} \times (\bar{\nabla} \times \bar{\psi}). \tag{9.4.5}$$

But the curl of $\bar{u}$ is nothing but twice the rotation vector whose components are ω_{32}, ω_{13}, and ω_{21}. Thus,

$$2\bar{\omega} = \bar{\nabla} \times (\bar{\nabla} \times \bar{\psi}) = -\nabla^2\psi. \tag{9.4.6}$$

Substituting Eq. (9.4.2) into Eq. (9.4.1), we get:

$$(\lambda + 2G)\bar{\nabla}(\nabla^2\phi) + G\bar{\nabla} \times \nabla^2\bar{\psi} = 0. \tag{9.4.7}$$

Any set of functions ϕ and $\bar{\psi}$ which satisfies Eq. (9.4.7) will produce, when substituted into Eq. (9.4.2), a displacement field $\bar{u}$ which satisfies Navier's equations. Conversely, for every $\bar{u}$ that satisfies Navier's equation at least one set of ϕ and $\bar{\psi}$ exists which satisfies Eqs. (9.4.2), (9.4.3), and (9.4.7). Obviously, since $\bar{u}$ is related to ϕ and $\bar{\psi}$ by means of first derivatives, for a given u, ϕ and $\bar{\psi}$ are not unique.

Some particular solutions of Eq. (9.4.7) are functions which satisfy the two equations:

$$\nabla^2\phi = \text{constant}, \quad \nabla^2\bar{\psi} = \text{constant}\,. \tag{9.4.8}$$

If we chose

$$\nabla^2\phi = \text{constant}, \quad \bar{\psi} = 0, \tag{9.4.9}$$

then the function ϕ is called the *Lamé strain potential*. Eq. (9.4.9) shows that any harmonic function may be used as ϕ and the resulting displacement field,

$$\bar{u} = \bar{\nabla}\phi, \tag{9.4.10}$$

will satisfy Navier's equation. For convenience, $\bar{u}$ is often written as

$$\bar{u} = \frac{1}{2G}\bar{\nabla}\phi.\ (*) \tag{9.4.11}$$

This form is the one we shall use in subsequent sections. Many solutions satisfying practical boundary conditions in cylindrical and spherical coordinates can be generated from this potential. Plane strain axisymmetric problems in particular have been studied by Lamé using Eqs. (9.4.9). Some of these problems will be examined in Chapter 11.

The following harmonic functions are helpful in the solution of practical problems:

$$\phi = A(x_1^2 - x_2^2) + 2Bx_1x_2 \tag{9.4.12}$$

$$\phi = Cr^n\cos(n\theta), \quad r^2 = x_1^2 + x_2^2 \tag{9.4.13}$$

$$\phi = C\ell n\frac{r}{K}, \quad r^2 = x_1^2 + x_2^2 \tag{9.4.14}$$

$$\phi = C\theta, \quad \theta = \tan^{-1}\frac{x_2}{x_1} \tag{9.4.15}$$

$$\phi = \frac{C}{\rho}, \quad \rho^2 = x_1^2 + x_2^2 + x_3^2 \tag{9.4.16}$$

$$\phi = C\ell n(\rho + x_3), \quad \rho^2 = x_1^2 + x_2^2 + x_3^2. \tag{9.4.17}$$

* See page A-56 for details.

One may also mention the two functions of the Poisson type:

$$\phi = Cr^2 \tag{9.4.18}$$

$$\phi = C\rho^2. \tag{9.4.19}$$

If we take the divergence of Eq. (9.4.7), we obtain:

$$\nabla^4\phi = 0 \tag{9.4.20}$$

Eqs. (9.4.4) and (9.4.20) show that e_v is a harmonic function (already proven). By taking the curl of Eq. (9.4.7), we obtain:

$$\nabla^4\bar{\psi} = 0. \tag{9.4.21}$$

Eqs. (9.4.6) and (9.4.21) show that $\bar{\omega}$ is a harmonic function. A solution of Eq. (9.4.7) must also be a solution of Eqs. (9.4.20) and (9.4.21), but the reverse is not necessarily true.

9.5 The Galerkin Vector. Love's Strain Function. Kelvin's and Cerruti's Problems

In the previous section, the displacement vector $\bar{u}$ was represented by the sum of first derivatives of two functions—namely, a scalar function ϕ and a vector function $\bar{\psi}$. The differential operator of order one, $\bar{\nabla}$, was used for that purpose in Eq. (9.4.2). In search for solutions of general applicability, it is reasonable to try differential operators of order two which would express the displacement vector $\bar{u}$ in terms of second derivatives. Two such operators are the Laplace operator, ∇^2, and the operator grad (div) = $\bar{\nabla}(\bar{\nabla}\ \cdot)$; both can be expressed in any system of coordinates and can be applied to a vector function. Thus, let us consider a vector function $\bar{V}$ which is related to the displacement vector by

$$2G\bar{u} = 2(1-\nu)\nabla^2\bar{V} - \bar{\nabla}(\bar{\nabla}\cdot\bar{V}). \tag{9.5.1}$$

The vector $\bar{V}$ is called a Galerkin vector. This vector supplies a general solution of Navier's equation. This can be proved by showing that, for any vector function $\bar{u}$, it is possible to find a vector $\bar{V}$ satisfying Eq. (9.5.1) [1]. Substituting Eq. (9.5.1) into Navier's equation (9.4.1), we obtain:

$$\nabla^2(\nabla^2\bar{V}) = 0. \tag{9.5.2}$$

Therefore, any biharmonic vector function may be used as the Galerkin vector and the resulting displacement vector given by Eq. (9.5.1) will always satisfy Navier's equations. Thus Eqs. (9.5.1) and (9.5.2) are equivalent to Navier's equation. Comparing Eqs. (9.5.1) and (9.4.2), we notice that the Galerkin vector is related to the scalar potential ϕ and the vector potential $\bar{\psi}$ by

$$\phi = -\frac{1}{2G}(\bar{\nabla} \cdot \bar{V}) \tag{9.5.3}$$

$$\bar{\nabla} \times \bar{\psi} = \frac{2(1-\nu)}{2G} \nabla^2 \bar{V}. \tag{9.5.4}$$

If $\bar{V}$ is chosen to be not only biharmonic but also harmonic, then

$$\bar{\nabla} \times \bar{\psi} = 0, \tag{9.5.5}$$

and ϕ satisfies the equation

$$\nabla^2 \phi = 0. \tag{9.5.6}$$

This ϕ is Lamé's strain potential of the previous section.

Let us consider the particular case in which $\bar{V}$ has only one component V_3. In this case, we have what is called *Love's strain function*:

$$\bar{V} = \bar{i}_3 V_3, \quad V_1 = V_2 = 0. \tag{9.5.7}$$

The governing equation (9.5.2) is now reduced to:

$$\nabla^2(\nabla^2 V_3) = 0, \tag{9.5.8}$$

and Eq. (9.5.1) becomes:

$$2G\bar{u} = 2(1-\nu)\bar{i}_3 \nabla^2 V_3 - \bar{\nabla}\left(\frac{\partial V_3}{\partial x_3}\right). \tag{9.5.9}$$

In a system of cartesian coordinates Eq. (9.5.9) is written in expanded form as follows:

$$\begin{aligned} 2Gu_1 &= -\frac{\partial^2 V_3}{\partial x_1 \partial x_3}, \quad 2Gu_2 = -\frac{\partial^2 V_3}{\partial x_2 \partial x_3}, \\ 2Gu_3 &= 2(1-\nu)\nabla^2 V_3 - \frac{\partial^2 V_3}{\partial x_3^2}. \end{aligned} \tag{9.5.10}$$

In cylindrical coordinates, Eq. (9.5.9) is written in expanded form as follows:

$$2Gu_r = -\frac{\partial^2 V_z}{\partial r \partial z}, \quad 2Gu_\theta = -\frac{1}{r}\frac{\partial^2 V_z}{\partial \theta \partial z},$$
$$2Gu_z = 2(1-\nu)\nabla^2 V_z - \frac{\partial^2 V_z}{\partial z^2}. \tag{9.5.11}$$

Using the strain-displacement relations and the stress-strain relations in cylindrical coordinates, we obtain the expressions for the stresses corresponding to Eqs. (9.5.11). They are:

$$\sigma_{rr} = \frac{\partial}{\partial z}\left(\nu\nabla^2 V_z - \frac{\partial^2 V_z}{\partial r^2}\right) \tag{9.5.12}$$

$$\sigma_{\theta\theta} = \frac{\partial}{\partial z}\left(\nu\nabla^2 V_z - \frac{1}{r}\frac{\partial V_z}{\partial r} - \frac{1}{r^2}\frac{\partial^2 V_z}{\partial \theta^2}\right) \tag{9.5.13}$$

$$\sigma_{zz} = \frac{\partial}{\partial z}\left[(2-\nu)\nabla^2 V_z - \frac{\partial^2 V_z}{\partial z^2}\right] \tag{9.5.14}$$

$$\sigma_{r\theta} = -\frac{\partial^3}{\partial r \partial \theta \partial z}\left(\frac{V_z}{r}\right) \tag{9.5.15}$$

$$\sigma_{\theta z} = \frac{1}{r}\frac{\partial}{\partial \theta}\left[(1-\nu)\nabla^2 V_z - \frac{\partial^2 V_z}{\partial z^2}\right] \tag{9.5.16}$$

$$\sigma_{zr} = \frac{\partial}{\partial r}\left[(1-\nu)\nabla^2 V_z - \frac{\partial^2 V_z}{\partial z^2}\right]. \tag{9.5.17}$$

Love introduced the strain function $V_z(r, z)$ in studying solids of revolution under axisymmetric loading. One application of Love's strain function is the problem of the single concentrated force acting in the interior of an infinite body. This problem is known as *Kelvin's problem* (Fig. 9.1):

Let $2P$ be applied at the origin in the direction of OX_3. The boundary conditions are: (a) All the stresses vanish at infinity, and (b) the stress singularity is equivalent to a concentrated force of magnitude $2P$. The concentrated force may be regarded as the limit of a system of loads applied on the surface of a small cavity at the origin. Using cylindrical coordinates, Love's strain function will be of the form:

$$V_3 = V_z = V_z(r, z), \tag{9.5.18}$$

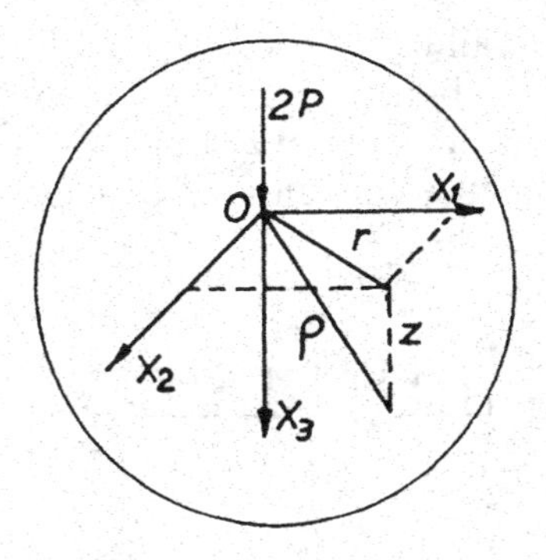

Fig. 9.1

and must be biharmonic. Its third partial derivative should define stresses that vanish at infinity and have a singularity at the origin. A function satisfying these conditions is:

$$V_z = B\rho = B(z^2 + r^2)^{1/2}. \tag{9.5.19}$$

Applying Eqs. (9.5.11) to (9.5.17), we get:

$$2Gu_r = \frac{Brz}{\rho^3}, \quad 2Gu_\theta = 0, \quad 2Gu_z = B\left[\frac{2(1-2\nu)}{\rho} + \frac{1}{\rho} + \frac{z^2}{\rho^3}\right] \tag{9.5.20}$$

$$\sigma_{rr} = B\left[\frac{(1-2\nu)z}{\rho^3} - \frac{3r^2 z}{\rho^5}\right] \tag{9.5.21}$$

$$\sigma_{\theta\theta} = \frac{(1-2\nu)Bz}{\rho^3} \tag{9.5.22}$$

$$\sigma_{zz} = -B\left[\frac{(1-2\nu)z}{\rho^3} + \frac{3z^3}{\rho^5}\right] \tag{9.5.23}$$

$$\sigma_{rz} = -B\left[\frac{(1-2\nu)r}{\rho^3} + \frac{3rz^2}{\rho^5}\right] \tag{9.5.24}$$

$$\sigma_{r\theta} = \sigma_{\theta z} = 0. \tag{9.5.25}$$

The stresses are singular at the origin and vanish at infinity. To determine the constant B, we compute the total vertical stress on two planes z = constant falling on both sides of the origin where $2P$ is acting. Let us isolate a band $z = \pm a$ and write the equilibrium of the forces in the OX_3 direction. Fig. 9.2 shows the stresses in their positive direction according to the conventions of Chapter 7. The equilibrium equation is:

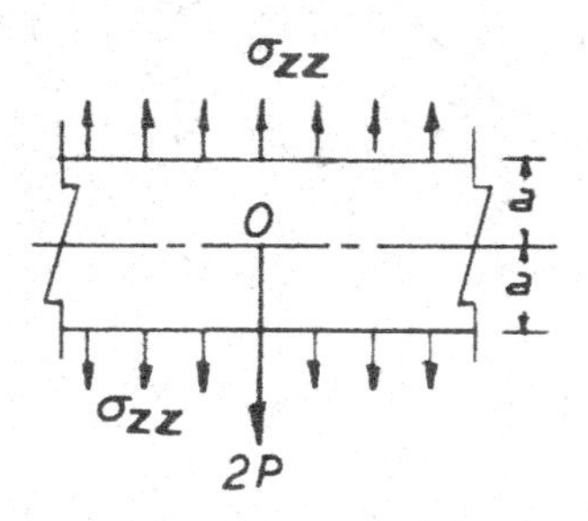

Fig. 9.2

$$2P = \int_0^{\infty} 2\Pi r\, dr(\sigma_{zz})_{z=-a} - \int_0^{\infty} 2\Pi r\, dr(\sigma_{zz})_{z=+a}. \tag{9.5.26}$$

Substituting Eq. (9.5.23) into Eq. (9.5.26), and noting that $r\,dr = \rho\, d\rho$ for a given value of z, we get:

$$2P = 4\Pi B\left[(1-2\nu)a \int_a^{\infty} \frac{\rho\, d\rho}{\rho^3} + 3a^3 \int_a^{\infty} \frac{\rho\, d\rho}{\rho^5}\right] \tag{9.5.27}$$

$$= 8\Pi B(1-\nu). \tag{9.5.28}$$

Therefore,

$$B = \frac{P}{4\Pi(1-\nu)}. \tag{9.5.29}$$

This value of B is now substituted in Eqs. (9.5.19) to (9.5.25) to yield the expressions of the strain function, the displacements, and the stresses.

Problems can also be solved by combining several Galerkin vectors or by combining Lamé's strain potentials and Galerkin vectors. For example, *Cerruti's problem* of a tangential force acting on the boundary of a semi-infinite solid (Fig. 9.3) can be solved by combining the Galerkin vector $\overline{V}$ whose components are

$$V_1 = A\rho, \quad V_2 = 0, \quad V_3 = Bx_1\, ln(\rho + x_3), \tag{9.5.30}$$

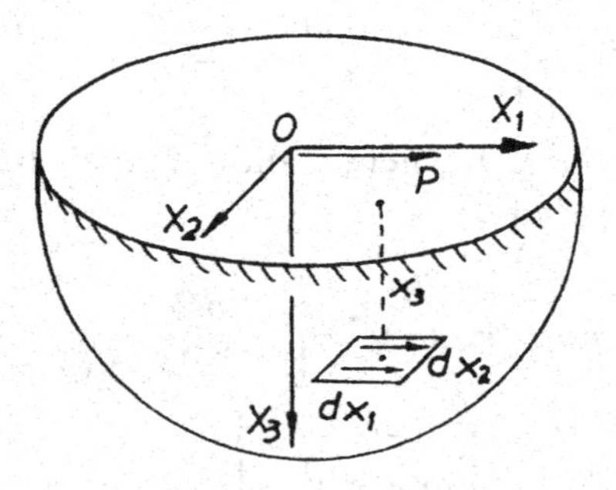

Fig. 9.3

and the Lamé strain potential:

$$\phi = \frac{C\,x_1}{\rho + x_3}. \tag{9.5.31}$$

A, B, and C are constants to be determined from the boundary conditions. The displacement vector is obtained from the superposition of Eqs. (9.4.11) and (9.5.1). Hence,

$$2G\bar{u} = \overline{\nabla}\phi + 2(1 - \nu)\nabla^2\overline{V} - \overline{\nabla}(\overline{\nabla} \cdot \overline{V}). \tag{9.5.32}$$

The strains are obtained from the strain-displacement relations (1.2.1) and the stresses from the stress-strain relations (8.5.21) to (8.5.24). The constants A, B, and C can be obtained from the following conditions:

1. At $x_3 = 0$

$$\sigma_{33} = \sigma_{23} = 0. \tag{9.5.33}$$

2. On any horizontal plane, at a depth x_3 from the surface, the sum of all the forces along OX_1 must balance P; i.e.,

$$\int_{-\infty}^{+\infty}\int_{-\infty}^{+\infty} \sigma_{13}\,dx_1\,dx_2 + P = 0. \tag{9.5.34}$$

The integration in Eq. (9.5.34) is most easily accomplished in cylindrical coordinates. The computations are lengthy but do not present any difficulty. They yield:

$$A = \frac{P}{4\Pi(1-\nu)}, \quad B = \frac{P(1-2\nu)}{4\Pi(1-\nu)}, \quad C = \frac{P(1-2\nu)}{2\Pi}. \qquad (9.5.35)$$

The values of the displacements and of the stresses can now be obtained. They are given by:

$$u_1 = \frac{P}{4\Pi G\rho}\left\{1 + \frac{x_1^2}{\rho^2} + (1-2\nu)\left[\frac{\rho}{\rho + x_3} - \frac{x_1^2}{(\rho + x_3)^2}\right]\right\} \qquad (9.5.36)$$

$$u_2 = \frac{P}{4\Pi G\rho}\left[\frac{x_1 x_2}{\rho^2} - \frac{(1-2\nu)x_1 x_2}{(\rho + x_3)^2}\right] \qquad (9.5.37)$$

$$u_3 = \frac{P}{4\Pi G\rho}\left[\frac{x_1 x_3}{\rho^2} + \frac{(1-2\nu)x_1}{\rho + x_3}\right] \qquad (9.5.38)$$

$$\sigma_{11} = \frac{Px_1}{2\Pi\rho^3}\left[-\frac{3x_1^2}{\rho^2} + \frac{1-2\nu}{(\rho + x_3)^2}\left(\rho^2 - x_2^2 - \frac{2\rho x_2^2}{\rho + x_3}\right)\right] \qquad (9.5.39)$$

$$\sigma_{22} = \frac{Px_1}{2\Pi\rho^3}\left[-\frac{3x_2^2}{\rho^2} + \frac{1-2\nu}{(\rho + x_3)^2}\left(3\rho^2 - x_1^2 - \frac{2\rho x_1^2}{\rho + x_3}\right)\right] \qquad (9.5.40)$$

$$\sigma_{33} = -\frac{3Px_1 x_3^2}{2\Pi\rho^5} \qquad (9.5.41)$$

$$\sigma_m = \frac{\sigma_{11} + \sigma_{22} + \sigma_{33}}{3} = -\frac{1+\nu}{\Pi\rho^3}Px_1 \qquad (9.5.42)$$

$$\sigma_{12} = \frac{Px_2}{2\Pi\rho^3}\left[-\frac{3x_1^2}{\rho^2} + \frac{1-2\nu}{(\rho + x_3)^2}\left(-\rho^2 + x_1^2 + \frac{2\rho x_1^2}{\rho + x_3}\right)\right] \qquad (9.5.43)$$

$$\sigma_{23} = \frac{3Px_1 x_2 x_3}{2\Pi\rho^5}, \quad \sigma_{13} = -\frac{3Px_1^2 x_3}{2\Pi\rho^5}. \qquad (9.5.44)$$

9.6 The Neuber-Papkovich Representation. Boussinesq's Problem

This representation uses a combination of harmonic functions to represent the displacement vector $\bar{u}$. We introduce the expression:

$$2G\bar{u} = \bar{A} - \bar{\nabla}\left[B + \frac{\bar{A}\cdot\bar{X}}{4(1-\nu)}\right], \qquad (9.6.1)$$

where $\bar{A}$ is a vector field, B a scalar field, and $\bar{X}$ is the position vector. Substituting Eq. (9.6.1) into Eq. (9.4.1), we get:

$$G\nabla^2\bar{A} - (\lambda + 2G)\bar{\nabla}(\nabla^2 B) - \left(\frac{\lambda + G}{2}\right)\bar{\nabla}(\bar{X} \cdot \nabla^2\bar{A}) = 0. \quad (9.6.2)$$

This equation is satisfied if:

$$\nabla^2\bar{A} = 0, \quad \nabla^2 B = 0. \quad (9.6.3)$$

Therefore, any four harmonic functions A_1, A_2, A_3, and B can be substituted in Eq. (9.6.1) and the resulting $\bar{u}$ satisfies Navier's equation. These four functions, however, are not completely independent. It can be proved ([2], [3]) that for an arbitrary three dimensional convex domain, the number of independent functions is reduced to three. The functions $\bar{A}$ and B can be deduced from Galerkin's vector $\bar{V}$ if we set

$$\bar{A} = 2(1 - \nu)\nabla^2\bar{V}, \quad B = \bar{\nabla} \cdot \bar{V} - \frac{\bar{A} \cdot \bar{X}}{4(1 - \nu)}. \quad (9.6.4)$$

A special form for $\bar{A}$ and B for problems with axial symmetry is:

$$A_r = A_\theta = 0, \quad A_z = A_z(r, z) \quad (9.6.5)$$

$$B = B(r, z). \quad (9.6.6)$$

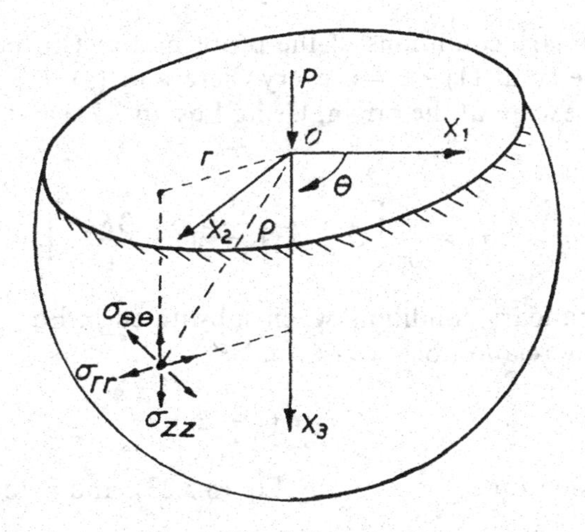

Fig. 9.4

For example, *Boussinesq's problem* (Fig. 9.4) of a force P acting in the OX_3 direction, at the origin of coordinates, on a semi-infinite elastic solid, has the solution:

$$A_r = A_\theta = 0, \quad A_z = 4(1-\nu)\frac{K}{\rho} \tag{9.6.7}$$

$$B = C\, ln(\rho + z). \tag{9.6.8}$$

To show this, let us substitute Eqs. (9.6.7) and (9.6.8) in Eq. (9.6.1):

$$\bar{u} = \frac{4(1-\nu)}{2G}\frac{K}{\rho}\bar{e}_z - \frac{\nabla}{2G}\left[C\, ln(\rho + z) + \frac{Kz}{\rho}\right], \tag{9.6.9}$$

where $\bar{e}_z$ is the unit vector in the OX_3 direction. In expanded form, in cylindrical coordinates, this expression is written as follows:

$$u_r = -\frac{Cr}{2G\rho(\rho + z)} + \frac{Kzr}{2G\rho^3} \tag{9.6.10}$$

$$u_\theta = 0 \tag{9.6.11}$$

$$u_z = \frac{(3-4\nu)K - C}{2G\rho} + \frac{Kz^2}{2G\rho^3}. \tag{9.6.12}$$

The boundary conditions of the problem are: On the surface of the semi-infinite solid (1) $\sigma_{rz} = 0$ everywhere and (2) σ_{zz} is equal to zero everywhere except at the origin. Using Eqs. (6.7.24) and Eq. (8.5.24), we get:

$$\sigma_{rz} = \frac{r}{\rho^3}\left[C - K(1-2\nu) - \frac{3Kz^2}{\rho^2}\right]. \tag{9.6.13}$$

The first boundary condition, when substituted in Eq. (9.6.13), leads to the following relation between C and K:

$$C = K(1-2\nu). \tag{9.6.14}$$

The expression for σ_{zz} is given by Eq. (8.5.23), and is found to be:

$$\sigma_{zz} = -\frac{3Kz^3}{\rho^5}. \tag{9.6.15}$$

This quantity is indeterminate at the origin O and Boussinesq did not attempt to describe the problem there. To determine K, let us consider a horizontal plane at a distance z from the origin and write that the resultant of all the vertical forces on this plane is equal to P (Fig. 9.5).

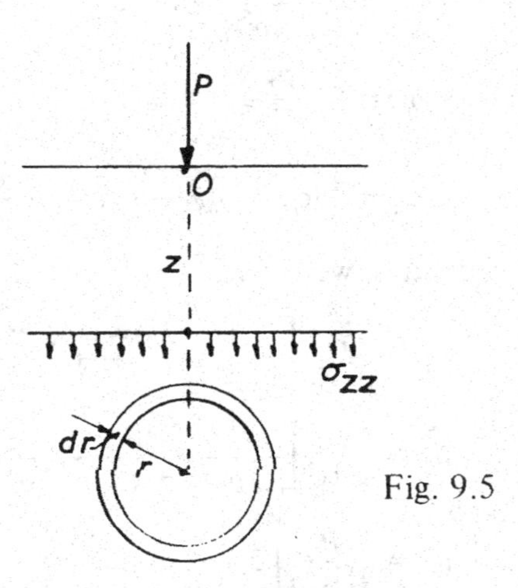

Fig. 9.5

$$P = \int_{r=0}^{r=\infty} \frac{3Kz^3}{\rho^5} 2\Pi r\, dr. \tag{9.6.16}$$

This integration is easily performed by substitution: It yields:

$$K = \frac{P}{2\Pi}. \tag{9.6.17}$$

Equation (9.6.14) yields:

$$C = \frac{P(1-2\nu)}{2\Pi}. \tag{9.6.18}$$

Summing up the results of this important problem we have:

$$u_r = \frac{P}{4\Pi G\rho}\left[\frac{rz}{\rho^2} - \frac{(1-2\nu)r}{\rho+z}\right] \tag{9.6.19}$$

$$u_\theta = 0 \tag{9.6.20}$$

$$u_z = \frac{P}{4\Pi G\rho}\left[2(1-\nu) + \frac{z^2}{\rho^2}\right] \tag{9.6.21}$$

$$\sigma_{rr} = \frac{P}{2\Pi\rho^2}\left[-\frac{3r^2 z}{\rho^3} + \frac{(1-2\nu)\rho}{\rho + z}\right] \tag{9.6.22}$$

$$\sigma_{\theta\theta} = \frac{(1-2\nu)P}{2\Pi\rho^2}\left[\frac{z}{\rho} - \frac{\rho}{\rho + z}\right] \tag{9.6.23}$$

$$\sigma_{zz} = -\frac{3Pz^3}{2\Pi\rho^5} \tag{9.6.24}$$

$$\sigma_{rz} = -\frac{3Prz^2}{2\Pi\rho^5} \tag{9.6.25}$$

$$\sigma_m = \frac{1}{3}(\sigma_{rr} + \sigma_{\theta\theta} + \sigma_{zz}) = -\frac{(1+\nu)}{3\Pi}\frac{Pz}{\rho^3}. \tag{9.6.26}$$

In cartesian coordinates, we have:

$$u_1 = \frac{P}{4\Pi G}\left[-\frac{(1-2\nu)x_1}{\rho(x_3 + \rho)} + \frac{x_3 x_1}{\rho^3}\right] \tag{9.6.27}$$

$$u_2 = \frac{P}{4\Pi G}\left[-\frac{(1-2\nu)x_2}{\rho(x_3 + \rho)} + \frac{x_3 x_2}{\rho^3}\right] \tag{9.6.28}$$

$$u_3 = \frac{P}{4\Pi G\rho}\left[2(1-\nu) + \frac{x_3^2}{\rho^2}\right] \tag{9.6.29}$$

$$\sigma_{11} = -\frac{P}{2\Pi\rho^2}\left\{\frac{3x_3 x_1^2}{\rho^3} - (1-2\nu)\left[\frac{x_3}{\rho} - \frac{\rho}{\rho + x_3} + \frac{x_1^2(2\rho + x_3)}{\rho(\rho + x_3)^2}\right]\right\} \tag{9.6.30}$$

$$\sigma_{22} = -\frac{P}{2\Pi\rho^2}\left\{\frac{3x_3 x_2^2}{\rho^3} - (1-2\nu)\left[\frac{x_3}{\rho} - \frac{\rho}{\rho + x_3} + \frac{x_2^2(2\rho + x_3)}{\rho(\rho + x_3)^2}\right]\right\} \tag{9.6.31}$$

$$\sigma_{33} = -\frac{3Px_3^3}{2\Pi\rho^5} \tag{9.6.32}$$

$$\sigma_{12} = -\frac{P}{2\Pi\rho^2}\left[\frac{3x_1 x_2 x_3}{\rho^3} - \frac{(1-2\nu)(2\rho + x_3)x_1 x_2}{\rho(\rho + x_3)^2}\right] \tag{9.6.33}$$

$$\sigma_{13} = -\frac{3Px_1 x_3^2}{2\Pi\rho^5}, \quad \sigma_{23} = -\frac{3Px_2 x_3^2}{2\Pi\rho^5}. \tag{9.6.34}$$

The extension of this problem to include loads distributed over finite areas will be given in Chapter 14.

9.7 Summary of Displacement Functions

The following chart summarizes the displacement functions presented in the previous sections, and the problems they helped solve:

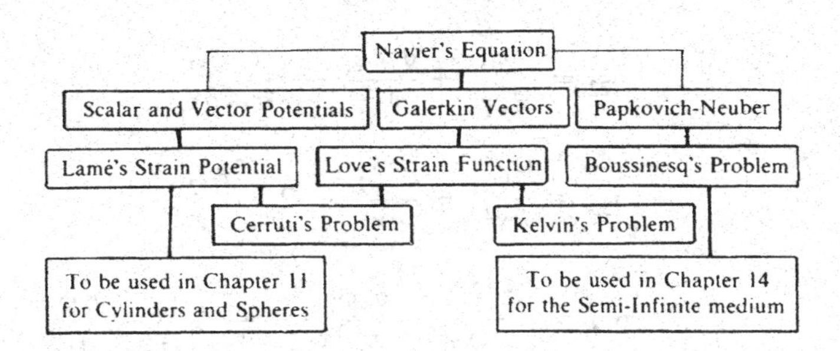

Reference [4] gives a more complete chart of displacement functions and their interrelation.

9.8 Stress Functions

In Sec. 8.10, we have seen that the stresses at different points of an elastic body are governed by the equilibrium equations, the Beltrami-Michell compatibility relations, and the boundary conditions. In a manner similar to the study of displacement functions, functions generating systems of equilibrating stresses have been examined. However, while in the case of displacement functions the quantity sought was a tensor of rank 1, namely, $\bar{u}$, in this case the quantity is a tensor of rank two, σ_{ij}. This tensor is symmetric and the stress functions must reflect this property. In the following, we shall neglect the body forces, and the equations of equilibrium to be satisfied are:

$$\frac{\partial \sigma_{11}}{\partial x_1} + \frac{\partial \sigma_{21}}{\partial x_2} + \frac{\partial \sigma_{31}}{\partial x_3} = 0 \tag{9.8.1}$$

$$\frac{\partial \sigma_{12}}{\partial x_1} + \frac{\partial \sigma_{22}}{\partial x_2} + \frac{\partial \sigma_{32}}{\partial x_3} = 0 \tag{9.8.2}$$

$$\frac{\partial \sigma_{13}}{\partial x_1} + \frac{\partial \sigma_{23}}{\partial x_2} + \frac{\partial \sigma_{33}}{\partial x_3} = 0. \tag{9.8.3}$$

Let us choose a set of arbitrary functions $\phi_{ij}(x_1, x_2, x_3)$, with $\phi_{ij} = \phi_{ji}$, and assume that:

$$\sigma_{11} = \frac{\partial^2 \phi_{22}}{\partial x_3^2} + \frac{\partial^2 \phi_{33}}{\partial x_2^2} - 2\frac{\partial^2 \phi_{23}}{\partial x_2 \, \partial x_3} \tag{9.8.4}$$

$$\sigma_{22} = \frac{\partial^2 \phi_{33}}{\partial x_1^2} + \frac{\partial^2 \phi_{11}}{\partial x_3^2} - 2\frac{\partial^2 \phi_{31}}{\partial x_3 \, \partial x_1} \tag{9.8.5}$$

$$\sigma_{33} = \frac{\partial^2 \phi_{11}}{\partial x_2^2} + \frac{\partial^2 \phi_{22}}{\partial x_1^2} - 2\frac{\partial^2 \phi_{12}}{\partial x_1 \, \partial x_2} \tag{9.8.6}$$

$$\sigma_{23} = \frac{\partial^2 \phi_{31}}{\partial x_1 \, \partial x_2} + \frac{\partial^2 \phi_{12}}{\partial x_1 \, \partial x_3} - \frac{\partial^2 \phi_{11}}{\partial x_2 \, \partial x_3} - \frac{\partial^2 \phi_{23}}{\partial x_1^2} \tag{9.8.7}$$

$$\sigma_{31} = \frac{\partial^2 \phi_{12}}{\partial x_2 \, \partial x_3} + \frac{\partial^2 \phi_{23}}{\partial x_2 \, \partial x_1} - \frac{\partial^2 \phi_{22}}{\partial x_3 \, \partial x_1} - \frac{\partial^2 \phi_{31}}{\partial x_2^2} \tag{9.8.8}$$

$$\sigma_{12} = \frac{\partial^2 \phi_{23}}{\partial x_3 \, \partial x_1} + \frac{\partial^2 \phi_{31}}{\partial x_3 \, \partial x_2} - \frac{\partial^2 \phi_{33}}{\partial x_1 \, \partial x_2} - \frac{\partial^2 \phi_{12}}{\partial x_3^2}. \tag{9.8.9}$$

We notice that the second and third relations can be obtained from the first by cyclical permutation, and that the fifth and sixth relations can be obtained from the fourth by cyclical permutation. It is easy to verify that the equilibrium Eqs. (9.8.1), (9.8.2), and (9.8.3) are formally satisfied by the previous assumptions on the values of the stresses. The six scalar functions ϕ_{ij} are not independent however. Two methods of generating complete solutions from ϕ_{ij} are the methods of Maxwell's and Morera's stress functions. On setting $\phi_{12} = \phi_{23} = \phi_{31} = 0$, we obtain the solution proposed by Maxwell; and on taking $\phi_{11} = \phi_{22} = \phi_{33} = 0$ we get the solution proposed by Morera. Each of these two solutions is complete; i.e., for every stress distribution that satisfies the equilibrium equations, it is possible to construct a set of Maxwell functions and a set of Morera functions. To prove this statement for the case of Maxwell's function, for example, consider Eqs. (9.8.4) to (9.8.9) in which ϕ_{12}, ϕ_{23}, and ϕ_{31} are set equal to zero:

$$\sigma_{11} = \frac{\partial^2 \phi_{22}}{\partial x_3^2} + \frac{\partial^2 \phi_{33}}{\partial x_2^2} \tag{9.8.10}$$

$$\sigma_{22} = \frac{\partial^2 \phi_{33}}{\partial x_1^2} + \frac{\partial^2 \phi_{11}}{\partial x_3^2} \tag{9.8.11}$$

$$\sigma_{33} = \frac{\partial^2 \phi_{11}}{\partial x_2^2} + \frac{\partial^2 \phi_{22}}{\partial x_1^2} \tag{9.8.12}$$

$$\sigma_{23} = -\frac{\partial^2 \phi_{11}}{\partial x_2 \, \partial x_3}, \quad \sigma_{31} = -\frac{\partial^2 \phi_{22}}{\partial x_3 \, \partial x_1}, \quad \sigma_{12} = -\frac{\partial^2 \phi_{33}}{\partial x_1 \, \partial x_2}. \tag{9.8.13}$$

If we integrate Eqs. (9.8.13) and substitute the values of ϕ_{11}, ϕ_{22} and ϕ_{33} thus obtained in Eqs. (9.8.10) to (9.8.12), the result should be an identity. Eqs. (9.8.13) when integrated give:

$$\begin{aligned} &\phi_{11} = -\iint \sigma_{23}\, dx_2\, dx_3, \quad \phi_{22} = -\iint \sigma_{31}\, dx_3\, dx_1, \\ &\phi_{33} = -\iint \sigma_{12}\, dx_1\, dx_2, \end{aligned} \tag{9.8.14}$$

in which the constants of integration have been omitted since ϕ_{ij} is arbitrary. Substituting Eq. (9.8.14) into Eq. (9.8.10), we get:

$$\sigma_{11} = -\int \left(\frac{\partial \sigma_{12}}{\partial x_2} + \frac{\partial \sigma_{13}}{\partial x_3} \right) dx_1 . \tag{9.8.15}$$

Differentiating Eq. (9.8.15) with respect to x_1, we get:

$$\frac{\partial \sigma_{11}}{\partial x_1} = -\frac{\partial \sigma_{12}}{\partial x_2} - \frac{\partial \sigma_{13}}{\partial x_3}, \tag{9.8.16}$$

which is an identity. In the same way, we can show that the two other Eqs. (9.8.11) and (9.8.12) become identities. The same reasoning can be repeated to prove that Morera's stress functions provide a complete solution for stress distributions satisfying equilibrium.

If we substitute Eqs. (9.8.10) to (9.8.13) into the Beltrami-Michell compatibility relations without body forces, we get:

$$(1+\nu)\nabla^2\left(\frac{\partial^2\phi_{33}}{\partial x_2^2}+\frac{\partial^2\phi_{22}}{\partial x_3^2}\right)+\frac{\partial^2}{\partial x_1^2}(\nabla^2 Q-R)=0 \quad (9.8.17)$$

$$(1+\nu)\nabla^2\left(\frac{\partial^2\phi_{11}}{\partial x_3^2}+\frac{\partial^2\phi_{33}}{\partial x_1^2}\right)+\frac{\partial^2}{\partial x_2^2}(\nabla^2 Q-R)=0 \quad (9.8.18)$$

$$(1+\nu)\nabla^2\left(\frac{\partial^2\phi_{22}}{\partial x_1^2}+\frac{\partial^2\phi_{11}}{\partial x_2^2}\right)+\frac{\partial^2}{\partial x_3^2}(\nabla^2 Q-R)=0 \quad (9.8.19)$$

$$\frac{\partial^2}{\partial x_1\,\partial x_2}[(1+\nu)\nabla^2\phi_{33}-\nabla^2 Q+R]=0 \quad (9.8.20)$$

$$\frac{\partial^2}{\partial x_2\,\partial x_3}[(1+\nu)\nabla^2\phi_{11}-\nabla^2 Q+R]=0 \quad (9.8.21)$$

$$\frac{\partial^2}{\partial x_3\,\partial x_1}[(1+\nu)\nabla^2\phi_{22}-\nabla^2 Q+R]=0, \quad (9.8.22)$$

where

$$Q=\phi_{11}+\phi_{22}+\phi_{33} \quad (9.8.23)$$

and

$$R=\frac{\partial^2\phi_{11}}{\partial x_1^2}+\frac{\partial^2\phi_{22}}{\partial x_2^2}+\frac{\partial^2\phi_{33}}{\partial x_3^2}. \quad (9.8.24)$$

Solution of Eqs. (9.8.17) to (9.8.22) satisfy both equilibrium and compatibility and are therefore possible stress states in an elastic body.

Both Maxwell's and Morera's stress functions can be particularized to generate Prandtl's stress function for torsion problems and Airy's stress function for plane problems [4]. Airy's stress function is examined in the next two sections of this chapter and Prandtl's stress function will be examined in Chapter 10.

9.9 Airy's Stress Function for Plane Strain Problems

It has been shown in Sec. 8.16 that, for plane strain problems, the equations of equilibrium are reduced to two equations, namely Eq. (8.16.16) and Eq. (8.16.17). In practice, body forces can usually be expressed by a potential function Ω such that

$$F_1 = -\frac{\partial \Omega}{\partial x_1}, \quad F_2 = -\frac{\partial \Omega}{\partial x_2}. \tag{9.9.1}$$

The equilibrium equations then become:

$$\frac{\partial}{\partial x_1}(\sigma_{11} - \Omega) + \frac{\partial \sigma_{21}}{\partial x_2} = 0 \tag{9.9.2}$$

$$\frac{\partial \sigma_{12}}{\partial x_1} + \frac{\partial}{\partial x_2}(\sigma_{22} - \Omega) = 0. \tag{9.9.3}$$

If we now choose a stress function $\phi(x_1, x_2)$ such that

$$\sigma_{11} = \Omega + \frac{\partial^2 \phi}{\partial x_2^2}, \quad \sigma_{22} = \Omega + \frac{\partial^2 \phi}{\partial x_1^2}, \quad \sigma_{12} = -\frac{\partial^2 \phi}{\partial x_1 \partial x_2}, \tag{9.9.4}$$

we see that the equilibrium equations are identically satisfied. Substituting Eqs. (9.9.4) in the compatibility Eq. (8.16.24), we get:

$$\frac{\partial^4 \phi}{\partial x_1^4} + 2\frac{\partial^4 \phi}{\partial x_1^2 \partial x_2^2} + \frac{\partial^4 \phi}{\partial x_2^4} + \frac{1 - 2\nu}{1 - \nu}\left(\frac{\partial^2 \Omega}{\partial x_1^2} + \frac{\partial^2 \Omega}{\partial x_2^2}\right) = 0, \tag{9.9.5}$$

or

$$\nabla^4 \phi + \frac{1 - 2\nu}{1 - \nu}\nabla^2 \Omega = 0. \tag{9.9.6}$$

When there are no body forces,

$$\nabla^4 \phi = 0, \tag{9.9.7}$$

which shows that ϕ is a biharmonic function. By the above analysis, the problem of elasticity in plane strain has been reduced to seeking a solution to Eq. (9.9.5) such that the stress components satisfy the boundary conditions.

In cylindrical coordinates, we shall assume that the body forces are radial and derive from a potential Ω which depends only on r. Thus,

$$F_r = -\frac{\partial \Omega}{\partial r}. \tag{9.9.8}$$

The equations of equilibrium (8.16.30) and (8.16.31) become:

$$\frac{\partial \sigma_{rr}}{\partial r} + \frac{1}{r}\frac{\partial \sigma_{r\theta}}{\partial \theta} + \frac{1}{r}(\sigma_{rr} - \sigma_{\theta\theta}) - \frac{\partial \Omega}{\partial r} = 0 \tag{9.9.9}$$

$$\frac{\partial \sigma_{r\theta}}{\partial r} + \frac{1}{r}\frac{\partial \sigma_{\theta\theta}}{\partial \theta} + \frac{2}{r}\sigma_{r\theta} = 0. \tag{9.9.10}$$

These equations are identically satisfied by a stress function $\phi(r, \theta)$ defined as:

$$\sigma_{rr} = \frac{1}{r}\frac{\partial \phi}{\partial r} + \frac{1}{r^2}\frac{\partial^2 \phi}{\partial \theta^2} + \Omega, \ \sigma_{\theta\theta} = \frac{\partial^2 \phi}{\partial r^2} + \Omega, \ \sigma_{r\theta} = -\frac{\partial}{\partial r}\left(\frac{1}{r}\frac{\partial \phi}{\partial \theta}\right). \tag{9.9.11}$$

In cylindrical coordinates Eq. (9.9.6) becomes

$$\left(\frac{\partial^2}{\partial r^2} + \frac{1}{r}\frac{\partial}{\partial r} + \frac{1}{r^2}\frac{\partial^2}{\partial \theta^2}\right)\left(\frac{\partial^2 \phi}{\partial r^2} + \frac{1}{r}\frac{\partial \phi}{\partial r} + \frac{1}{r^2}\frac{\partial^2 \phi}{\partial \theta^2}\right) + \frac{1-2\nu}{1-\nu}\left(\frac{\partial^2 \Omega}{\partial r^2} + \frac{1}{r}\frac{\partial \Omega}{\partial r}\right) = 0 \tag{9.9.12}$$

or

$$\nabla^4 \phi + \frac{1-2\nu}{1-\nu}\nabla^2 \Omega = 0. \tag{9.9.13}$$

When there are no body forces,

$$\nabla^4 \phi = 0. \tag{9.9.14}$$

9.10 Airy's Stress Function for Plane Stress Problems

If the body forces are derived from a potential function Ω such that

$$F_1 = -\frac{\partial \Omega}{\partial x_1}, \quad F_2 = -\frac{\partial \Omega}{\partial x_2}, \tag{9.10.1}$$

the equations of equilibrium (8.17.15) and (8.17.16) become:

$$\frac{\partial}{\partial x_1}(\sigma_{11} - \Omega) + \frac{\partial \sigma_{12}}{\partial x_2} = 0 \tag{9.10.2}$$

$$\frac{\partial \sigma_{12}}{\partial x_1} + \frac{\partial}{\partial x_2}(\sigma_{22} - \Omega) = 0. \tag{9.10.3}$$

As was done for the case of plane strain, we choose a function $\phi(x_1, x_2)$ such that

$$\sigma_{11} = \Omega + \frac{\partial^2 \phi}{\partial x_2^2}, \quad \sigma_{22} = \Omega + \frac{\partial^2 \phi}{\partial x_1^2}, \quad \sigma_{12} = -\frac{\partial^2 \phi}{\partial x_1 \partial x_2}. \tag{9.10.4}$$

With this choice, the equations of equilibrium are identically satisfied.

It was shown in Sec. 8.17, that two out of the six compatibility conditions are identically satisfied in plane stress. This leaves Eqs. (8.17.18) and (8.17.19) for consideration. Eq. (8.17.18) gives:

$$\frac{\partial^4 \phi}{\partial x_1^4} + 2\frac{\partial^4 \phi}{\partial x_1^2 \partial x_2^2} + \frac{\partial^4 \phi}{\partial x_2^4} + (1 - \nu)\left(\frac{\partial^2 \Omega}{\partial x_1^2} + \frac{\partial^2 \Omega}{\partial x_2^2}\right) = 0 \quad (9.10.5)$$

or

$$\nabla^4 \phi + (1 - \nu)\nabla^2 \Omega = 0. \quad (9.10.6)$$

Using Eqs. (8.17.13) and (9.10.4), we see that Eqs. (8.17.19) give:

$$2\frac{\partial^2 \Omega}{\partial x_1^2} + \frac{\partial^2}{\partial x_1^2}(\nabla^2 \phi) = 0 \quad (9.10.7)$$

$$2\frac{\partial^2 \Omega}{\partial x_2^2} + \frac{\partial^2}{\partial x_2^2}(\nabla^2 \phi) = 0 \quad (9.10.8)$$

$$2\frac{\partial^2 \Omega}{\partial x_1 \partial x_2} + \frac{\partial^2}{\partial x_1 \partial x_2}(\nabla^2 \phi) = 0, \quad (9.10.9)$$

to be satisfied by $\phi(x_1, x_2)$. For zero or constant body forces, we see that $\nabla^2 \phi$ must satisfy the condition:

$$\nabla^2 \phi = A_1 x_1 + A_2 x_2 + A_3, \quad (9.10.10)$$

where A_1, A_2, and A_3 are arbitrary constants. If one neglects the conditions that Eqs. (9.10.7) to (9.10.9) impose on ϕ, the solution is only an approximate one. When there are no body forces Eq. (9.10.6) becomes

$$\nabla^4 \phi = 0, \quad (9.10.11)$$

which is identical to Eq. (9.9.7). Both plane strain and plane stress problems have the same solution in this case.

In Cylindrical Coordinates, if the body forces are radial and derive from a potential $\Omega = \Omega(r)$, then

$$F_r = -\frac{\partial \Omega}{\partial r} \quad (9.10.12)$$

and

$$\frac{\partial \sigma_{rr}}{\partial r} + \frac{1}{r}\frac{\partial \sigma_{r\theta}}{\partial \theta} + \frac{1}{r}(\sigma_{rr} - \sigma_{\theta\theta}) - \frac{\partial \Omega}{\partial r} = 0 \tag{9.10.13}$$

$$\frac{\partial \sigma_{r\theta}}{\partial r} + \frac{1}{r}\frac{\partial \sigma_{\theta\theta}}{\partial \theta} + \frac{2}{r}\sigma_{r\theta} = 0. \tag{9.10.14}$$

These equations are identically satisfied by a stress function $\phi(r,\theta)$ defined as:

$$\sigma_{rr} = \frac{1}{r}\frac{\partial \phi}{\partial r} + \frac{1}{r^2}\frac{\partial^2 \phi}{\partial \theta^2} + \Omega,\ \sigma_{\theta\theta} = \frac{\partial^2 \phi}{\partial r^2} + \Omega,\ \sigma_{r\theta} = -\frac{\partial}{\partial r}\left(\frac{1}{r}\frac{\partial \phi}{\partial \theta}\right). \tag{9.10.15}$$

In cylindrical coordinates Eq. (9.10.6) becomes

$$\left(\frac{\partial^2}{\partial r^2} + \frac{1}{r}\frac{\partial}{\partial r} + \frac{1}{r^2}\frac{\partial^2}{\partial \theta^2}\right)\left(\frac{\partial^2 \phi}{\partial r^2} + \frac{1}{r}\frac{\partial \phi}{\partial r} + \frac{1}{r^2}\frac{\partial^2 \phi}{\partial \theta^2}\right) + (1-\nu)\left(\frac{\partial^2 \Omega}{\partial r^2} + \frac{1}{r}\frac{\partial \Omega}{\partial r}\right) = 0 \tag{9.10.16}$$

or

$$\nabla^4 \phi + (1-\nu)\nabla^2 \Omega = 0. \tag{9.10.17}$$

When there are no body forces,

$$\nabla^4 \phi = 0. \tag{9.10.18}$$

9.11 Forms of Airy's Stress Function

In looking for a suitable stress function providing the solution of plane elastic problems, one can often make a good guess for various types of boundary conditions. A common method is to use a polynomial and find the combination of terms which fit a particular set of boundary conditions. For example, in the polynomial

$$\phi = a\,x_1^2 + b\,x_1 x_2 + c\,x_2^2 + d\,x_1^3 + e\,x_1^2 x_2 + f\,x_1 x_2^2 + g\,x_2^3 + h\,x_1^4 + j\,x_1^3 x_2 + k\,x_1^2 x_2^2 + l\,x_1 x_2^3 + m\,x_2^4 + n\,x_1^5 \ldots \tag{9.11.1}$$

any term containing x_1 or x_2 up to the third power will satisfy the biharmonic equation $\nabla^4\phi = 0$. Terms containing x_1^4 or x_2^4 and higher powers must have, among their coefficients, relations satisfying the biharmonic equation. For discontinuous loads on the boundary, however, the polynomial approach has severe theoretical limitations since discontinuous boundary conditions are not representable by polynomials.

In cylindrical coordinates, the stress function is, in general, of the form:

$$\phi = f(r)\cos n\theta, \quad \phi = f(r)\sin n\theta, \tag{9.11.2}$$

where $f(r)$ is a function of r alone and n is an integer. Three other forms deserve mentioning, namely:

$$\phi = Cr\theta \cos \theta, \quad \phi = Cr\theta \sin \theta, \quad \phi = Cr^2\theta, \tag{9.11.3}$$

where C is a constant. Combinations of the previous forms can of course be used.

For axially symmetric stress distribution (about the Z axis), ϕ does not depend on θ, and the biharmonic equation $\nabla^4\phi = 0$ becomes:

$$\frac{d^4\phi}{dr^4} + \frac{2}{r}\frac{d^3\phi}{dr^3} - \frac{1}{r^2}\frac{d^2\phi}{dr^2} + \frac{1}{r^3}\frac{d\phi}{dr} = 0. \tag{9.11.4}$$

This is Euler's differential equation. The solution is:

$$\phi = C_1 r^2 \ell nr + C_2 r^2 + C_3 \ell nr + C_4, \tag{9.11.5}$$

where C_1, C_2, C_3, and C_4 are constants of integration. C_4 plays no part in the solution since the stresses are expressed in terms of the derivatives of ϕ; C_1, C_2, and C_3 are to be determined from the boundary conditions. For a state of plane stress, for example, the stresses are given by Eqs. (9.10.15). Thus,

$$\sigma_{rr} = \frac{C_3}{r^2} + C_1(1 + 2\ell nr) + 2C_2 \tag{9.11.6}$$

$$\sigma_{\theta\theta} = C_1(3 + 2\ell nr) + 2C_2 - \frac{C_3}{r^2} \tag{9.11.7}$$

$$\sigma_{r\theta} = 0. \tag{9.11.8}$$

The strains and the displacements can be obtained by making use of Eqs. (8.17.11), (8.17.12), (8.17.14) and Eqs. (8.17.23), (8.17.24). Eqs. (8.17.11) and (8.17.23) yield:

$$e_{rr} = \frac{\partial u_r}{\partial r} = \frac{1}{E}\left[\frac{1+\nu}{r^2}C_3 + 2(1-\nu)C_1 \ell nr + (1-3\nu)C_1 + 2(1-\nu)C_2\right]. \tag{9.11.9}$$

Integrating Eq. (9.11.9), we get:

$$u_r = \frac{1}{E}\left[-\frac{1+\nu}{r}C_3 + 2(1-\nu)C_1 r\ell nr - (1+\nu)C_1 r + 2(1-\nu)C_2 r\right] + f_1(\theta), \tag{9.11.10}$$

where $f_1(\theta)$ is a function of θ only. Eqs. (8.17.12) and (8.17.23) yield:

$$\frac{\partial u_\theta}{\partial \theta} = re_{\theta\theta} - u_r = \frac{4C_1 r}{E} - f_1(\theta). \tag{9.11.11}$$

Integrating Eq. (9.11.11), we get:

$$u_\theta = \frac{4C_1 r\theta}{E} - \int f_1(\theta)\,d\theta + f_2(r), \tag{9.11.12}$$

where $f_2(r)$ is a function of r only. Eqs. (8.17.14) and (8.17.24) yield:

$$e_{r\theta} = \frac{\sigma_{r\theta}}{2G} = \frac{\partial u_\theta}{\partial r} + \frac{1}{r}\frac{\partial u_r}{\partial \theta} - \frac{u_\theta}{r} = 0, \tag{9.11.13}$$

since $\sigma_{r\theta} = 0$. Substitution of Eqs. (9.11.10) and (9.11.12) into Eq. (9.11.13) gives:

$$\frac{\partial f_2(r)}{\partial r} + \frac{1}{r}\frac{\partial f_1(\theta)}{\partial \theta} + \frac{1}{r}\int f_1(\theta)\,d\theta - \frac{1}{r}f_2(r) = 0. \tag{9.11.14}$$

This equation can be separated into two equations in the variables r and θ. Thus,

$$r\frac{df_2(r)}{dr} - f_2(r) = m, \quad \text{and} \quad \frac{df_1(\theta)}{d\theta} + \int f_1(\theta)\,d\theta = -m. \tag{9.11.15}$$

The equation in θ is only satisfied if the constant $m = 0$ so that $f_2(r) = Ar$ and $f_1(\theta) = B \sin\theta + F\cos\theta$, where A, B, and F are constants. Therefore,

$$u_r = \frac{1}{E}\left[-\frac{1+\nu}{r}C_3 + 2(1-\nu)C_1 r\ell nr - (1+\nu)C_1 r + 2(1-\nu)C_2 r\right] + B\sin\theta + F\cos\theta \tag{9.11.16}$$

and

$$u_\theta = \frac{4C_1 r\theta}{E} + Ar + B\cos\theta - F\sin\theta. \tag{9.11.17}$$

The constants in Eqs. (9.11.16) and (9.11.17) are to be determined for each particular case.

PROBLEMS

1. Given the scalar and vector potentials $\phi = x_1^2 + 2x_2^2$ and $\bar{\psi} = \rho^2 \bar{i}_3$, does the displacement field generated by ϕ and $\bar{\psi}$ satisfy Navier's equations, and, if so, what is it?
2. Find the displacements and the stresses defined by the following Lamé strain potentials:

$$\phi = A(x_1^2 - x_2^2) + 2Bx_1 x_2$$

$$\phi = Cr^n \cos(n\theta).$$

3. Determine the displacements and the stresses defined by the Galerkin vectors:
   (a) $\bar{V} = C\rho^2 \bar{i}_3$
   (b) $\bar{V} = -C\rho^2 x_2 \bar{i}_1 + C\rho^2 x_1 \bar{i}_2$.
4. Find the stresses corresponding to the Lamé strain potential $\phi = C\ell n(\rho + x_3)$. What is the problem to which this potential furnishes a solution [1]?
5. Show that the solution of Boussinesq's problem can be obtained through a combination of a Galerkin vector $\bar{V} = B\rho\bar{i}_3$ and a Lamé strain potential $\phi = C\ell n(\rho + x_3)$. Show that $C = -(1 - 2\nu)B$ and $B = P/2\Pi$.

6. What are the stresses corresponding to the following Airy stress functions:

$$\phi = \frac{a}{2}x_1^2 + bx_1x_2 + \frac{c}{2}x_2^2$$

$$\phi = \frac{a}{6}x_1^3 + \frac{b}{2}x_1^2x_2 + \frac{c}{2}x_1x_2^2 + \frac{d}{6}x_2^3.$$

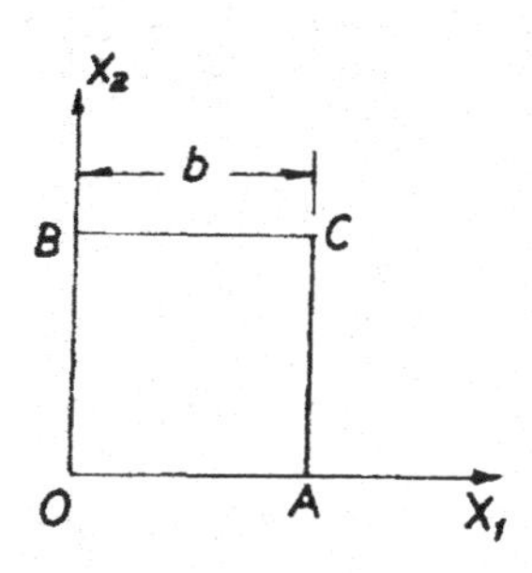

Fig. 9.6

7. A thin square plate whose sides are parallel to the OX_1 and OX_2 axes (Fig. 9.6) has in it stresses described by $\sigma_{11} = cx_2$, $\sigma_{22} = cx_1$, and possibly some shearing stresses σ_{12}. c is a constant.
   (a) Find the stress function by integration, and the most general shearing stresses which can be associated with the given σ_{11} and σ_{22}.
   (b) Obtain the strains and, by integration, deduce the expressions of the displacements u_1 and u_2.
   (c) Find the extension of the diagonal OC.

8. Show that the stress function

$$\phi = C\left[(x_1^2 + x_2^2)\tan^{-1}\frac{x_2}{x_1} - x_1x_2\right]$$

provides the solution to the problem of the semi-infinite elastic medium acted upon by a uniform pressure q on one side of the origin (Fig. 9.7).

9. Investigate what problem of plane strain is solved by the stress function $\phi = Cr\theta \sin\theta$.

10. Investigate the expression $\phi = \cos^3\theta/r$ as a possible stress function. (*)

* See pages A-56 to A-59 for additional problems.

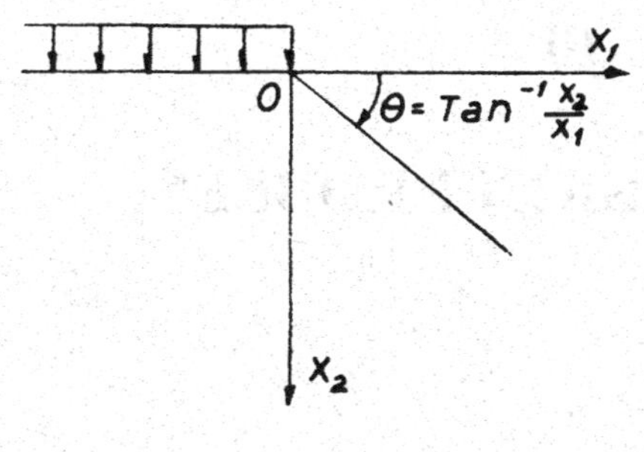

Fig. 9.7

REFERENCES

[1] H. M. Westergaard, *Theory of Elasticity and Plasticity*, Dover, New York, N. Y., 1964.

[2] E. Sternberg, "On Some Recent Developments in the Linear Theory of Elasticity," Structural Mechanics, Proceedings of the First Symposium on Naval Structural Mechanics, Pergamon Press, New York, N. Y., 1960.

[3.] R. A. Eubanks and E. Sternberg, "On the Completeness of Boussinesq-Papkovich Stress Functions," *J. Rat. Mech. Analy.* Vol. 5, p. 735, 1956.

[4] P. C. Chou and N. J. Pagano, *Elasticity*, Van Nostrand, Princeton, N. J., 1967.

CHAPTER 10

THE TORSION PROBLEM

10.1 Introduction

In this chapter, the semi-inverse method proposed by Saint-Venant is used to solve the problem of non-circular prismatic bars subjected to torque. A stress function—namely, Prandtl's stress function—is introduced, and it suggests an analogy which is utilized to obtain solutions for complicated shapes. The case of a circular prismatic bar is first treated since it represents the first step of the intuitive solution suggested by Saint-Venant.

10.2 Torsion of Circular Prismatic Bars

Let us consider a circular prismatic bar of length L and of radius a, with one end fixed and the other end acted upon by a couple whose moment, $M_{33} = M_z$, is along the OX_3 axis. The bar deforms and its

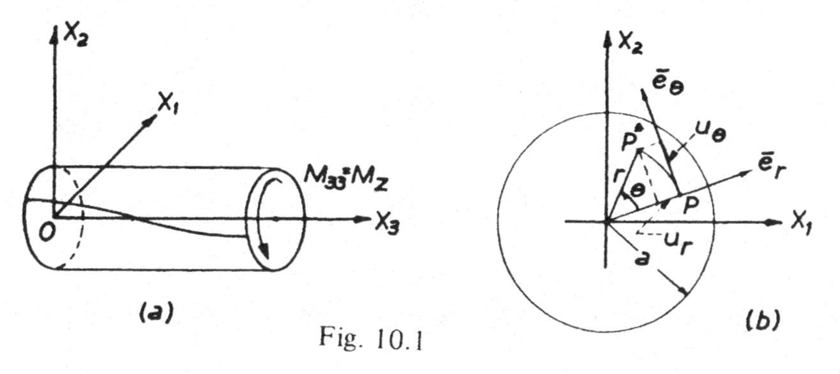

Fig. 10.1

generators are transformed from straight lines to helical curves (Fig. 10.1a). On account of symmetry [1], it is reasonable to assume that cross sections of the bar normal to the OX_3 axis remain plane after deformation, and that the couple rotates every section by an angle θ proportional to its distance from the fixed end $x_3 = 0$ (proved by experiment). Thus,

$$\theta = \alpha x_3 = \alpha z, \tag{10.2.1}$$

where α is the twist per unit length, i.e., the relative angular displacement of two cross sections a unit distance apart. From the above assumptions, it can be concluded that the displacement vector $\overline{PP^*}$ of any point P, in a cross section at a distance $x_3 = z$ from 0 (Fig. 10.1b), has the following components:

$$u_r = -r(1 - \cos\theta), \quad u_\theta = r\sin\theta, \quad u_z = 0. \tag{10.2.2}$$

For small values of θ, $\sin\theta \approx \theta$ and $\cos\theta \approx 1$ so that Eq. (10.2.2) becomes:

$$u_r = 0, \quad u_\theta = r\theta = r\alpha z, \quad u_z = 0. \tag{10.2.3}$$

Knowing the components of the displacement, the strains can be obtained from Eqs. (6.7.23) and (6.7.24):

$$e_{rr} = e_{\theta\theta} = e_{zz} = e_{r\theta} = e_{rz} = 0 \tag{10.2.4}$$

$$e_{\theta z} = \frac{\alpha r}{2}. \tag{10.2.5}$$

Substituting Eqs. (10.2.4) and (10.2.5) in the stress strain relations, we get:

$$\sigma_{rr} = \sigma_{\theta\theta} = \sigma_{zz} = \sigma_{r\theta} = \sigma_{rz} = 0 \tag{10.2.6}$$

$$\sigma_{\theta z} = G\alpha r. \tag{10.2.7}$$

This state of stress satisfies the general equations of equilibrium. It is illustrated in Fig. 10.2, where $\sigma_{\theta z}$ is the only stress component acting on the element referred to cylindrical coordinates. Every cross section including the end one is subjected to the shearing stress distribution

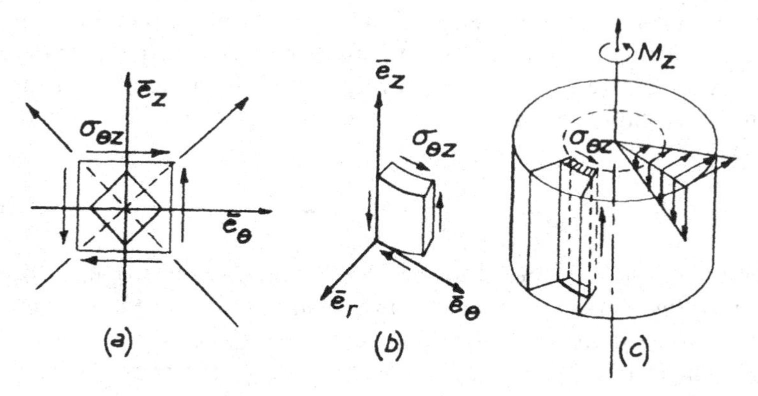

Fig. 10.2

shown in Fig. 10.2c. The radial direction is a principal direction with the principal stress $\sigma_2 = \sigma_{rr} = 0$. This state of stress is therefore a state of plane stress. The two other principal stresses, σ_1 and σ_3, lie in the tangential plane $(\bar{e}_\theta, \bar{e}_z)$. The representation on Mohr's diagram is shown in Fig. 10.3. The same state of stress exists at all points of

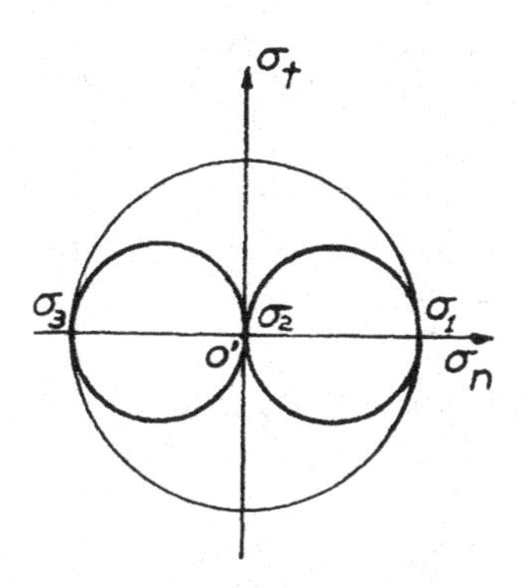

Fig. 10.3

concentric cylindrical surfaces which do not interact with each other.

It remains now to check if the proposed solution satisfies the boundary conditions at the lateral surface and at both ends of the bar.

The direction cosines of the normal to the lateral surface with respect to a system of cylindrical coordinates are (1,0,0). This surface is free of stress. The stress distribution of Eqs. (10.2.6) and (10.2.7) satisfies the boundary conditions (7.3.8). The direction cosines of the normal to the plane end of the bar are (0,0,1). This end surface is free from normal forces and consequently from normal stresses. Here, too, the stress distribution of Eqs. (10.2.6) and (10.2.7) satisfies this boundary condition. There is, however, a twisting moment M_z acting on the end of the beam, which must be in equilibrium with the resultant of the stress distribution. Thus (Fig. 10.2c),

$$M_z = \int_0^a 2\Pi r^2 \sigma_{\theta z}\, dr = \alpha G \int_0^a 2\Pi r^3\, dr = \alpha G\left(\frac{\Pi a^4}{2}\right) \qquad (10.2.8)$$

or

$$M_z = \alpha G I_z, \qquad (10.2.9)$$

where I_z is the polar moment of inertia of the cross sectional area about the axis of the bar. Therefore, for a bar of length L, which is loaded at the end by a twisting moment M_z, Eqs. (10.2.1), (10.2.7), and (10.2.9) can be written:

$$\frac{\theta G}{L} = \frac{M_z}{I_z} = \frac{\sigma_{\theta z}}{r}. \qquad (10.2.10)$$

The factor by which we divide the torque to obtain α is called the torsional rigidity.

Summarizing, we have assumed a mode of deformation and deduced the strains and then the stresses which were found to satisfy equilibrium and boundary conditions. Because of uniqueness, this assumed solution is the only solution to the problem. The only shortcoming of this solution occurs at the end of the bar where the externally applied twisting moment must be distributed according to the pattern shown in Fig. 10.2c. In practice, the stress distribution, although statically equivalent to that of Fig. 10.2c, is quite different from it. The influence of this difference decreases quite rapidly as we move away from the end and the solution presents an excellent approximation starting from a distance of one or two diameters from the plane of application of M_z (Saint-Venant's principle).

In the case of *hollow circular bars*, the value of I_z is given by

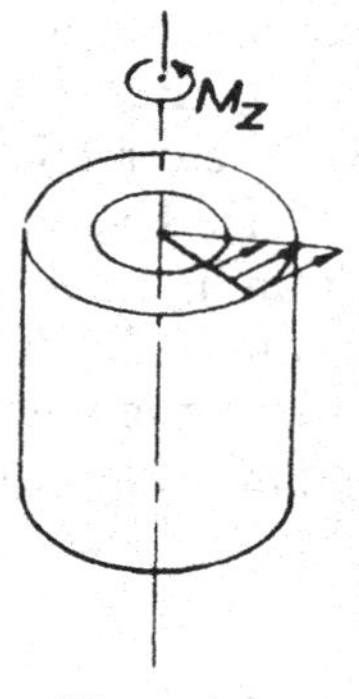

Fig. 10.4

$$I_z = \frac{\Pi a_o^4}{2}\left(1 - \frac{a_i^4}{a_o^4}\right), \tag{10.2.11}$$

where a_o and a_i are the outer and inner radii of the bar. All the equations previously given apply to this case with I_z given by Eq. (10.2.11) (Fig. 10.4).

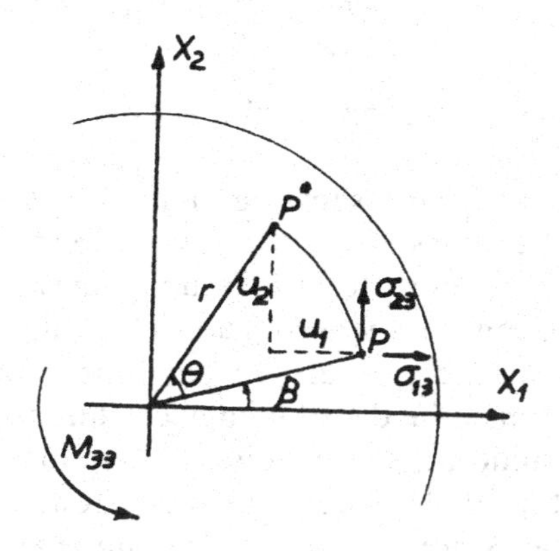

Fig. 10.5

It is of interest to write some of the previous equations in a cartesian system of coordinates. These equations will be needed in the next section; they are in fact the starting point in the search for a solution to the problem of non-circular prismatic bars. The components of the displacement vector are (Fig. 10.5):

$$u_1 = r\cos(\theta + \beta) - r\cos\beta \approx -x_2\theta = -\alpha x_2 x_3 \qquad (10.2.12)$$

$$u_2 = r\sin(\theta + \beta) - r\sin\beta \approx x_1\theta = \alpha x_1 x_3 \qquad (10.2.13)$$

$$u_3 = 0. \qquad (10.2.14)$$

The components of the state of strain are:

$$e_{11} = e_{22} = e_{33} = e_{12} = 0 \qquad (10.2.15)$$

$$e_{13} = -\frac{1}{2}\alpha x_2, \quad e_{23} = \frac{1}{2}\alpha x_1. \qquad (10.2.16)$$

The components of the state of stress are:

$$\sigma_{11} = \sigma_{22} = \sigma_{33} = \sigma_{12} = 0 \qquad (10.2.17)$$

$$\sigma_{13} = -G\alpha x_2, \quad \sigma_{23} = G\alpha x_1. \qquad (10.2.18)$$

The twisting moment M_{33} must be in equilibrium with the stresses at the end of the bar. Therefore,

$$M_{33} = \iint (G\alpha x_1^2 + G\alpha x_2^2)\,dx_1\,dx_2 = G\alpha I_3. \qquad (10.2.19)$$

The magnitude of the stress vector is given by

$$\sigma_t = G\alpha\sqrt{x_1^2 + x_2^2} = G\alpha r. (*) \qquad (10.2.20)$$

10.3 Torsion of Non-Circular Prismatic Bars

For non-circular prismatic bars, Navier tried to use the assumption made for the case of circular bars—namely, that the plane cross sections remain plane; this, led him to erroneous conclusions. In fact, it can be proved that the solution

$$\sigma_{13} = -G\alpha x_2, \quad \sigma_{23} = G\alpha x_1, \qquad (10.3.1)$$

* Note regarding the use of the angle θ. See page A-60.

obtained in the previous section, is only valid for circular prismatic bars. Indeed, the lateral surface being free from stresses, the third boundary condition of Eqs. (7.3.8) gives (Fig. 10.6):

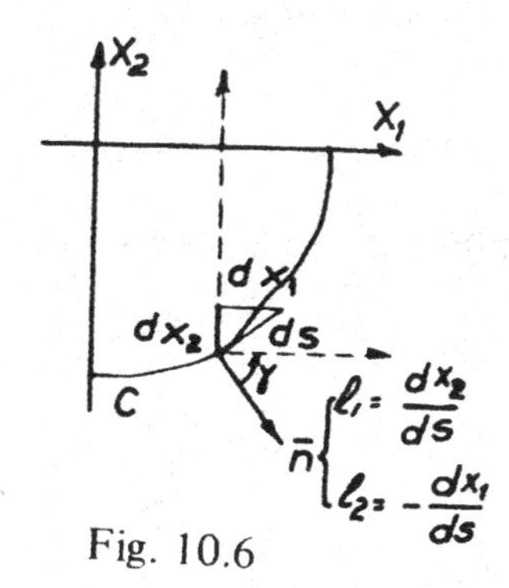

Fig. 10.6

$$0 = -G\alpha x_2 \frac{dx_2}{ds} - G\alpha x_1 \frac{dx_1}{ds} \tag{10.3.2}$$

or

$$x_2\,dx_2 + x_1\,dx_1 = 0. \tag{10.3.3}$$

This is the equation of a set of concentric circles

$$x_1^2 + x_2^2 = \text{constant} .$$

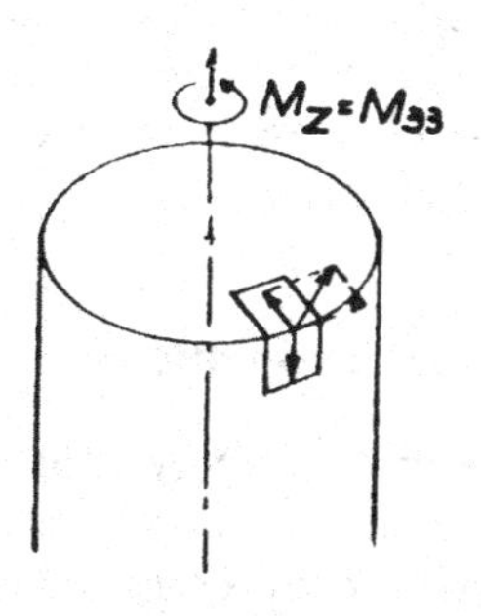

Fig. 10.7

If the values of Eqs. (10.3.1) were valid for non-circular bars, the stress vectors would have to be tangent to concentric circles, and at the boundary they would have a component tangent to the contour and one normal to it (Fig. 10.7). This last one would be associated with another shear stress on the free outside surface of the bar, which does not exist.

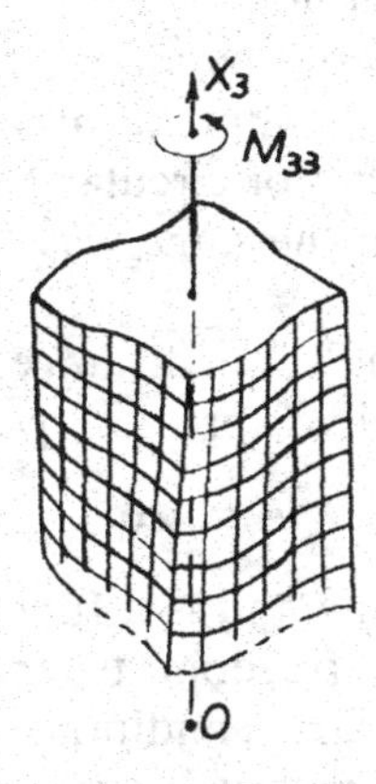

Fig. 10.8

For example, the shear stress at the corner of a rectangular bar must be zero because none of its two perpendicular components can exist. We are thus led to the assumption that for bars with non-circular sections, plane cross sections do not remain plane but are warped, and that all cross sections are warped the same way (Fig. 10.8).

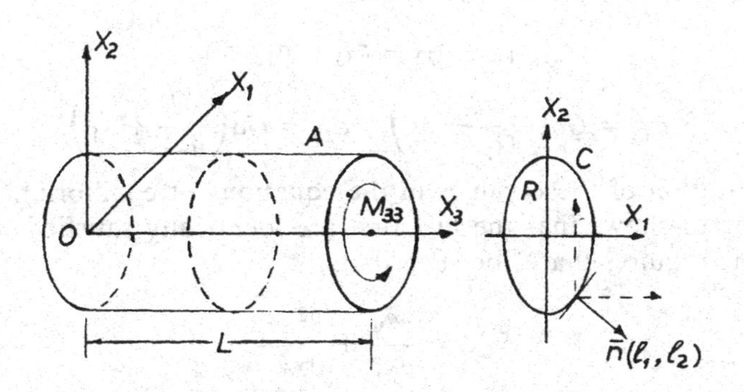

Fig. 10.9

Let us now consider a prismatic bar fixed at one end in the (OX_1, OX_2) plane, while the other end is subjected to a couple whose moment M_{33} is along the OX_3 axis (Fig. 10.9). OX_3 passes through the center of twist of each section, i.e., through the point about which each cross section will rotate. Using the semi-inverse method, Saint-Venant made the following assumptions:

1. Each cross section rotates by an angle θ proportional to its distance from the fixed end with no inplane distortion, i.e., with $e_{12} = 0$; this assumption is similar to that of circular bars.

2. All sections will warp the same way, i.e., the warping is independent of x_3.

For any cross section at a distance x_3 from the origin, these assumptions are analytically expressed by:

$$u_1 = -\alpha x_2 x_3, \quad u_2 = \alpha x_1 x_3, \quad u_3 = \alpha\psi(x_1, x_2), \tag{10.3.4}$$

where ψ defines the warping and is a function of x_1 and x_2 alone. Eqs. (10.3.4) leave us enough freedom to try to satisfy the equilibrium equations and the boundary conditions. In doing so, a number of restrictions to be imposed on ψ will result. With the above values of the displacements, the strains are given by:

$$e_{11} = e_{22} = e_{33} = e_{12} = 0 \tag{10.3.5}$$

$$e_{13} = \frac{1}{2}\left(\frac{\partial u_1}{\partial x_3} + \frac{\partial u_3}{\partial x_1}\right) = \frac{\alpha}{2}\left(\frac{\partial \psi}{\partial x_1} - x_2\right) \tag{10.3.6}$$

$$e_{23} = \frac{1}{2}\left(\frac{\partial u_2}{\partial x_3} + \frac{\partial u_3}{\partial x_2}\right) = \frac{\alpha}{2}\left(\frac{\partial \psi}{\partial x_2} + x_1\right). \tag{10.3.7}$$

The stress-strain relations (8.5.2) give:

$$\sigma_{11} = \sigma_{22} = \sigma_{33} = \sigma_{12} = 0 \tag{10.3.8}$$

$$\sigma_{13} = G\alpha\left(\frac{\partial \psi}{\partial x_1} - x_2\right), \quad \sigma_{23} = G\alpha\left(\frac{\partial \psi}{\partial x_2} + x_1\right). \tag{10.3.9}$$

A substitution of these values in the equations of equilibrium with no body forces shows that the two first are identically satisfied while the third one requires that ψ be such that

$$\nabla^2\psi = \frac{\partial^2\psi}{\partial x_1^2} + \frac{\partial^2\psi}{\partial x_2^2} = 0 \tag{10.3.10}$$

throughout each section. This is Laplace's equation.

Let us now investigate the boundary conditions, first on the lateral sides, then at the ends of the bar. The direction cosines of the normal to the contour C (Fig. 10.9) are $(dx_2/ds, -dx_1/ds, 0)$. The first two Eqs. (7.3.8) are identically satisfied and the third one gives:

$$\left(\frac{\partial\psi}{\partial x_1} - x_2\right)\frac{dx_2}{ds} - \left(\frac{\partial\psi}{\partial x_2} + x_1\right)\frac{dx_1}{ds} = 0. \tag{10.3.11}$$

On the other two boundary surfaces, i.e., the ends of the bar defined by $x_3 = 0$ and $x_3 = L$, the distribution of stresses given by Eqs. (10.3.8) and (10.3.9) must have no resultant force and must be equivalent to a torsional couple. Let us first prove that the resultant force is equal to zero. The resultant force in the OX_1 direction is given by:

$$\iint_R \sigma_{13}\, dx_1\, dx_2 = G\alpha \iint_R \left(\frac{\partial\psi}{\partial x_1} - x_2\right) dx_1\, dx_2. \tag{10.3.12}$$

Eq. (10.3.12) can be written as:

$$\iint_R \sigma_{13}\, dx_1\, dx_2 = G\alpha \iint_R \left\{\frac{\partial}{\partial x_1}\left[x_1\left(\frac{\partial\psi}{\partial x_1} - x_2\right)\right] + \frac{\partial}{\partial x_2}\left[x_1\left(\frac{\partial\psi}{\partial x_2} + x_1\right)\right]\right\} dx_1\, dx_2. \tag{10.3.13}$$

Now, using the Green-Reimann formula (see appendix to this chapter), Eq. (10.3.13) can be written:

$$\iint_R \sigma_{13}\, dx_1\, dx_3 = G\alpha \oint \left[x_1\left(\frac{\partial\psi}{\partial x_1} - x_2\right) dx_2 - x_1\left(\frac{\partial\psi}{\partial x_2} + x_1\right) dx_1\right] = 0, \tag{10.3.14}$$

where Eq. (10.3.11) has been used. In a similar way, we can prove that

$$\iint_R \sigma_{23}\, dx_1\, dx_2 = 0, \tag{10.3.15}$$

so that the resultant force acting on the ends of the bar vanishes.

The resultant torsional moment on the end of the bar due to the assumed stress distribution must be equal to M_{33}. Thus,

$$
\begin{aligned}
M_{33} &= \iint_R (x_1 \sigma_{23} - x_2 \sigma_{13})\, dx_1\, dx_2 \\
&= G\alpha \iint_R \left(x_1^2 + x_2^2 + x_1 \frac{\partial \psi}{\partial x_2} - x_2 \frac{\partial \psi}{\partial x_1} \right) dx_1\, dx_2 .
\end{aligned}
\tag{10.3.16}
$$

The integral in Eq. (10.3.16) depends on ψ, and hence on the shape of the cross section. Setting

$$
\iint_R \left(x_1^2 + x_2^2 + x_1 \frac{\partial \psi}{\partial x_2} - x_2 \frac{\partial \psi}{\partial x_1} \right) dx_1\, dx_2 = J, \tag{10.3.17}
$$

we have

$$
M_{33} = GJ\alpha. \tag{10.3.18}
$$

Summarizing, we have assumed a mode of deformation and deduced the strains and then the stresses. Those stresses were found to satisfy equilibrium and boundary conditions provided the warping function ψ satisfies Eqs. (10.3.10) and (10.3.11).

It is worthwhile noticing that Eq. (10.3.11) may be written in a form such that the torsion problem can be classified as a special case of the second boundary value problem of potential theory. This is done as follows: By definition, the gradient (also called normal derivative) of the function $\psi(x_1, x_2)$ at a point P is a vector directed along the normal $\bar{n}$ to the level curve of the function through P, provided this curve possesses a tangent at P (Fig. 10.10); the magnitude of the gradient vector is

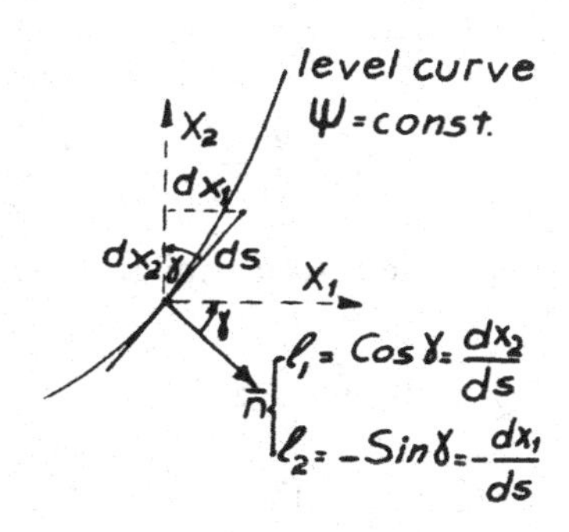

Fig. 10.10

$d\psi/dn$ and its components along the OX_1 and OX_2 axes are $\partial\psi/\partial x_1$ and $\partial\psi/\partial x_2$, respectively. The scalar product of the unit vector along $\bar{n}$ and of the gradient vector gives:

$$\frac{d\psi}{dn} = \frac{\partial\psi}{\partial x_1}\cos\gamma - \frac{\partial\psi}{\partial x_2}\sin\gamma, \tag{10.3.19}$$

which shows that

$$\frac{dx_1}{dn} = \cos\gamma = \frac{dx_2}{ds}, \quad \frac{dx_2}{dn} = -\sin\gamma = -\frac{dx_1}{ds}. \tag{10.3.20}$$

Substituting Eqs. (10.3.19) and (10.3.20) into Eq. (10.3.11), we get on the contour C (Fig. 10.11):

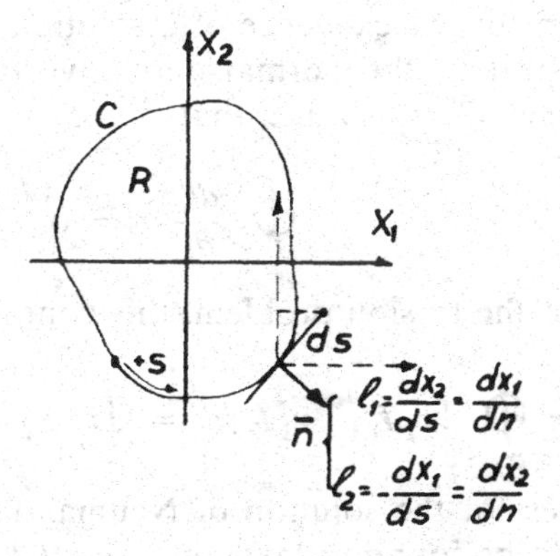

Fig. 10.11

$$\frac{d\psi}{dn} = x_2\frac{dx_2}{ds} + x_1\frac{dx_1}{ds} = \frac{1}{2}\frac{d}{ds}(x_1^2 + x_2^2). \tag{10.3.21}$$

For circular cross sections, $x_1^2 + x_2^2$ = constant and $d\psi/dn = 0$; i.e., ψ = constant. In general, however,

$$\frac{d\psi}{dn} = x_2\ell_1 - x_1\ell_2 \tag{10.3.22}$$

on the boundary. The right-hand of Eq. (10.3.22) is a function $f(s)$ of the bounding curve C. We can now state that the solution of the problem of torsion of prismatic bars amounts to finding a function $\psi(x_1, x_2)$ which satisfies the equations

$$\nabla^2 \psi = 0 \qquad \text{in } R \tag{10.3.23}$$

$$\frac{d\psi}{dn} = x_2 \ell_1 - x_1 \ell_2 = f(s) \qquad \text{on } C. \tag{10.3.24}$$

Eqs. (10.3.23) and (10.3.24) define the Neumann boundary value problem which has been extensively studied in potential theory. The statement of Neumann's problem is as follows: To determine a function $\psi(x_1, x_2)$ which is harmonic and regular in a region R and on its boundary C and such that its normal derivative $d\psi/dn$ takes on preassigned values $f(s)$ on C.*

* A function $\psi(x_1, x_2)$ is said to be regular in a region R and on its boundary C if in this region $\psi(x_1, x_2)$ is uniform and possesses second derivatives which are continuous in R and finite on C. It is obvious that under these conditions both $\psi(x_1, x_2)$ and its first derivatives are continuous in R.

The condition for the existence of a solution to Neumann's problem is that the integral of the normal derivative $d\psi/dn$ calculated over the entire boundary C vanish, i.e., that

$$\oint_C \frac{d\psi}{dn} ds = 0. \tag{10.3.25}$$

In the case of the torsion problem, this condition is satisfied since

$$\oint_C \frac{d\psi}{dn} ds = \oint_C [x_2 \ell_1 - x_1 \ell_2] ds = \oint_C x_2 \, dx_2 + x_1 \, dx_1 = 0. \tag{10.3.26}$$

The uniqueness of the solution of Neumann's problem is easily established and can be found in texts on potential theory.

10.4 Torsion of an Elliptic Bar

Knowing that the warping function $\psi(x_1, x_2)$ for any prismatic bar must satisfy Eqs. (10.3.10) and (10.3.11), we can use the inverse method by trying various expressions for ψ and finding the boundaries to which they correspond. For example, the simplest solution of Laplace's Eq. (10.3.10) is:

$$\psi = C = \text{constant}, \tag{10.4.1}$$

which was found in the previous section to solve the problem of the prism with circular cross section.

Now consider the function:

$$\psi = Kx_1 x_2, \tag{10.4.2}$$

where K is a constant. This function, when substituted in Eq. (10.3.11), gives on the boundary:

$$(Kx_2 - x_2)\frac{dx_2}{ds} - (Kx_1 + x_1)\frac{dx_1}{ds} \text{ or } \frac{d}{ds}\left(x_1^2 + \frac{1-K}{1+K}x_2^2\right) = 0. \tag{10.4.3}$$

Upon integration, Eq. (10.4.3) gives:

$$x_1^2 + \left(\frac{1-K}{1+K}\right)x_2^2 = \text{constant}, \tag{10.4.4}$$

in which x_1 and x_2 are the coordinates of any point on the boundary. The equation of an ellipse with center at the origin and semi-axes a and b is

$$x_1^2 + \frac{a^2}{b^2}x_2^2 = a^2. \tag{10.4.5}$$

Eqs. (10.4.4) and (10.4.5) become identical if:

$$\frac{a^2}{b^2} = \frac{1-K}{1+K}. \tag{10.4.6}$$

By solving for K we get:

$$K = \frac{b^2 - a^2}{b^2 + a^2}. \tag{10.4.7}$$

Therefore, the warping function for an elliptical cylinder under torsion is

$$\psi = \left(\frac{b^2 - a^2}{b^2 + a^2}\right)x_1 x_2. \tag{10.4.8}$$

The constant J of Eq. (10.3.17) is given by:

$$\begin{aligned} J &= \iint_R (x_1^2 + x_2^2 + Kx_1^2 - Kx_2^2)\,dx_1\,dx_2 \\ &= (K+1)\iint_R x_1^2\,dx_1\,dx_2 + (1-K)\iint_R x_2^2\,dx_1\,dx_2 \end{aligned} \tag{10.4.9}$$

or

$$J = (K+1)I_2 + (1-K)I_1 = \frac{\Pi a^3 b^3}{a^2 + b^2}, \tag{10.4.10}$$

where I_1 and I_2 are the moments of inertia with respect to the OX_1 and OX_2 axes, respectively. The torsional moment at the end of the bar is given by:

$$M_{33} = G\frac{\Pi a^3 b^3}{a^2 + b^2}\alpha. \tag{10.4.11}$$

Knowing the warping function ψ, the displacement vector $\bar{u}$ can be computed from Eq. (10.3.4). The three components of the displacement are:

$$u_1 = -\alpha x_2 x_3, \quad u_2 = \alpha x_3 x_1 \tag{10.4.12}$$

$$u_3 = \alpha x_1 x_2\left(\frac{b^2 - a^2}{b^2 + a^2}\right) \tag{10.4.13}$$

Eq. (10.4.13) shows that the contour lines defined by u_3 = constant are hyperbolas (Fig. 10.12). If the bar is twisted by the torque M_{33} in a

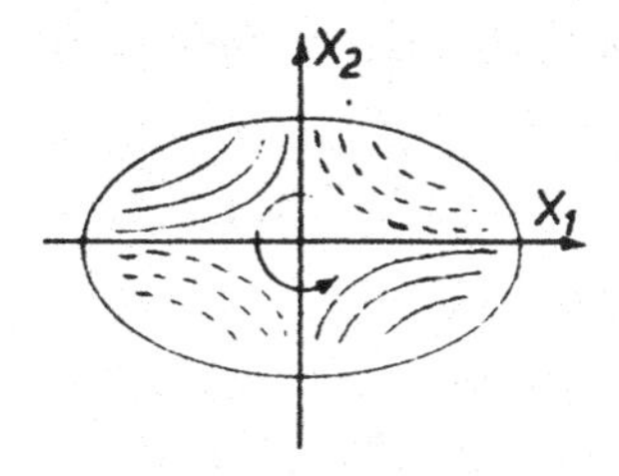

Fig. 10.12

counterclockwise direction, the parts where u_3 is positive (upward) are indicated by solid lines and the parts where u_3 is negative (downward) are indicated by dotted lines. If one end of the bar is restrained and prevented from warping, normal stresses will be induced, positive in the two quadrants which otherwise would have become concave, and negative in the two others.

The components of the state of strain are given by Eqs. (10.3.5) to (10.3.7):

$$e_{11} = e_{22} = e_{33} = e_{12} = 0 \tag{10.4.14}$$

$$e_{13} = \frac{\alpha x_2}{2}(K - 1) = -\frac{\alpha a^2 x_2}{b^2 + a^2} \tag{10.4.15}$$

$$e_{23} = \frac{\alpha x_1}{2}(K + 1) = +\frac{\alpha b^2 x_1}{b^2 + a^2}. \tag{10.4.16}$$

The components of the state of stress are given by Eqs. (10.3.8) and (10.3.9):

$$\sigma_{11} = \sigma_{22} = \sigma_{33} = \sigma_{12} = 0 \tag{10.4.17}$$

$$\sigma_{13} = -\frac{2G\alpha a^2 x_2}{b^2 + a^2} = -\frac{2M_{33} x_2}{\Pi a b^3} \tag{10.4.18}$$

$$\sigma_{23} = \frac{2G\alpha b^2 x_1}{b^2 + a^2} = \frac{2M_{33} x_1}{\Pi a^3 b}. \tag{10.4.19}$$

The resultant shearing stress at any point $P(x_1, x_2)$ is given by:

$$\sigma_t = \sqrt{(\sigma_{13})^2 + (\sigma_{23})^2} = \frac{2M_{33}}{\Pi a^2 b^2}\sqrt{\frac{b^2 x_1^2}{a^2} + \frac{a^2 x_2^2}{b^2}}, \tag{10.4.20}$$

and its direction is given by:

$$\tan\phi = -\frac{x_1 b^2}{x_2 a^2}. \tag{10.4.21}$$

Now, along every diameter of the ellipse, the ratio x_1/x_2 is constant and the direction of the tangent at any point of the ellipse's contour is given by (Fig. 10.13):

$$\left(\frac{dx_2}{dx_1}\right) = -\frac{x_1 b^2}{x_2 a^2}. \tag{10.4.22}$$

Therefore, the resultant shearing stress at points on a given diameter of the ellipse increases linearly as we move away from its origin; it is parallel to the tangent at the point of intersection of the diameter and the ellipse (Fig. 10.13).

The maximum value of σ_t occurs at the extremity B of the minor axis. One way to prove this statement is to express the stress at any point C on the boundary in terms of the coordinates of D on the conjugate diameter. From analytic geometry, it is known that the coordinates of D are:

$$x_1'' = -\frac{a x_2'}{b}, \quad x_2'' = \frac{b x_1'}{a}. \tag{10.4.23}$$

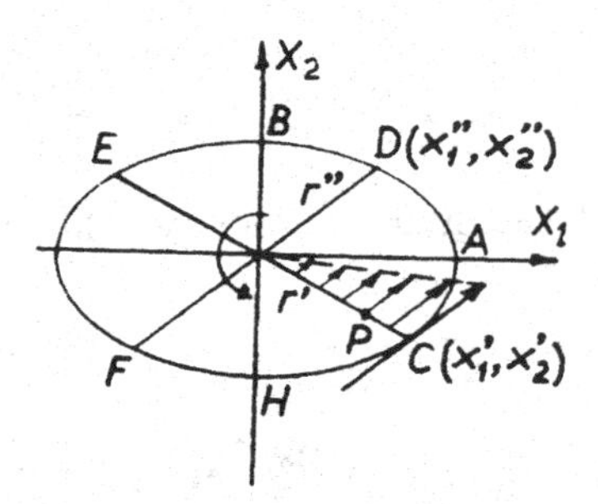

Fig. 10.13

At point C, the resultant stress σ_t can be rewritten as follows:

$$(\sigma_t)_C = \frac{2M_{33}}{\Pi a^2 b^2}\sqrt{(x''_1)^2 + (x''_2)^2} = \frac{2M_{33}}{\Pi a^2 b^2} r''. \qquad (10.4.24)$$

The expression for $(\sigma_t)_C$ is maximum when r'' is maximum and this occurs when C coincides with B or H. Therefore, the maximum shearing stress occurs at B and H.

10.5 Prandtl's Stress Function

In Sec. 10.3, we have seen that the solution of the torsion problem amounted to finding a function $\psi(x_1, x_2)$ which satisfied Laplace's equation

$$\nabla^2\psi = 0 \qquad (10.5.1)$$

in a region R, and the condition

$$\left(\frac{\partial\psi}{\partial x_1} - x_2\right)\frac{dx_2}{ds} - \left(\frac{\partial\psi}{\partial x_2} + x_1\right)\frac{dx_1}{ds} = 0 \qquad (10.5.2)$$

on the contour C (Fig. 10.11). An alternative procedure which leads to simpler boundary conditions involves the introduction of a stress function $\phi(x_1, x_2)$ called Prandtl's stress function. This function is defined in terms of ψ by the two following equations:

$$\frac{\partial \phi}{\partial x_2} = G\alpha\left(\frac{\partial \psi}{\partial x_1} - x_2\right) = \sigma_{13} \tag{10.5.3}$$

$$\frac{\partial \phi}{\partial x_1} = -G\alpha\left(\frac{\partial \psi}{\partial x_2} + x_1\right) = -\sigma_{23}. \tag{10.5.4}$$

With this definition, Eq. (10.5.1) is identically satisfied and Eq. (10.5.2) becomes:

$$\frac{\partial \phi}{\partial x_2}\frac{dx_2}{ds} + \frac{\partial \phi}{\partial x_1}\frac{dx_1}{ds} = \frac{d\phi}{ds} = 0. \tag{10.5.5}$$

Thus the function ϕ must be constant on the contour C (Fig. 10.11). This constant is arbitrary for solid bars and we shall take it equal to zero. Eliminating ψ from Eqs. (10.5.3) and (10.5.4), we get:

$$\frac{\partial^2 \phi}{\partial x_1^2} + \frac{\partial^2 \phi}{\partial x_2^2} = -2G\alpha. \tag{10.5.6}$$

This is a Poisson equation, for which a solution can always be found. The solution of the torsion problem is thus reduced to finding a function $\phi(x_1, x_2)$ such that (Fig. 10.11):

$$\nabla^2 \phi = -2G\alpha \qquad \text{in } R \tag{10.5.7}$$

$$\phi = 0 \qquad \text{on } C. \tag{10.5.8}$$

The expression of the twisting moment in terms of ϕ is obtained by substituting Eqs. (10.5.3) and (10.5.4) in Eq. (10.3.16):

$$\begin{aligned} M_{33} &= -\iint_R \left[x_1 \frac{\partial \phi}{\partial x_1} + x_2 \frac{\partial \phi}{\partial x_2}\right] dx_1\, dx_2 \\ &= -\iint_R \left[\frac{\partial}{\partial x_1}(x_1 \phi) + \frac{\partial}{\partial x_2}(x_2 \phi)\right] dx_1\, dx_2 \\ &\quad + 2\iint_R \phi\, dx_1\, dx_2. \end{aligned} \tag{10.5.9}$$

Now, by the Green-Riemann formula, the integral (Fig. 10.11)

$$-\iint_R \left[\frac{\partial}{\partial x_1}(x_1\phi) + \frac{\partial}{\partial x_2}(x_2\phi)\right] dx_1\, dx_2$$
$$= -\oint_C \phi[x_1 \cos\gamma + x_2 \sin\gamma]\, ds = 0, \tag{10.5.10}$$

since $\phi = 0$ on C. Therefore,

$$M_{33} = 2\iint_R \phi\, dx_1\, dx_2, \tag{10.5.11}$$

and one-half of the torque is due to σ_{13} while the other half is due to σ_{23}.

In some cases, it is quite advantageous to express the torsion problem in terms of Prandtl's stress functions. For example, when the equation of the boundary of the cross section is a simple function of x_1 and x_2, the stress function ϕ is chosen such that it contains the equation of the boundary and consequently its value is always equal to zero on it; arbitrary constant factors are included in ϕ so that Eq. (10.5.7) can also be satisfied. This approach to the problem, although not generally applicable, has allowed us to find solutions for a number of simple cases two of which are examined in the next section. Another interesting fact noticed by Prandtl is that Eq. (10.5.7) is the same as the differential equation for the shape of a stretched membrane originally flat which is then blown up by air pressure from the bottom. This allowed him to draw analogies between the geometrical parameters of the membrane and the stress-strain conditions of the cross section of a bar subjected to twist. The membrane analogy will be examined in detail in subsequent sections.

10.6 Two Simple Solutions Using Prandtl's Stress Function

The problem of a bar with an elliptical cross section, solved in Sec. 10.4, can be studied by starting with a stress function of the form:

$$\phi = m\left(\frac{x_1^2}{a^2} + \frac{x_2^2}{b^2} - 1\right). \tag{10.6.1}$$

This equation satisfies Eq. (10.5.8), and when substituted in Eq. (10.5.7) gives:

$$m = -\frac{G\alpha a^2 b^2}{a^2 + b^2}. \tag{10.6.2}$$

Therefore, the stress function providing the solution is:

$$\phi = -\frac{G\alpha a^2 b^2}{a^2 + b^2}\left(\frac{x_1^2}{a^2} + \frac{x_2^2}{b^2} - 1\right). \tag{10.6.3}$$

The problem of a bar with a cross section in the form of an equilateral triangle is solved by starting with a stress function of the form (Fig. 10.14):

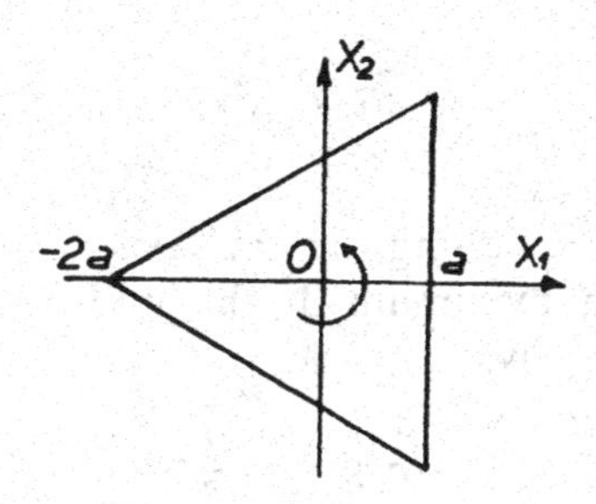

Fig. 10.14

$$\phi = m(x_1 - \sqrt{3}\, x_2 + 2a)(x_1 + \sqrt{3}\, x_2 + 2a)(x_1 - a). \tag{10.6.4}$$

This equation satisfies Eq. (10.5.8), and when substituted in Eq. (10.5.7) gives:

$$m = -\frac{G\alpha}{6a}. \tag{10.6.5}$$

Therefore, the stress function providing the solution is:

$$\phi = -\frac{G\alpha}{6a}(x_1 - \sqrt{3}\, x_2 + 2a)(x_1 + \sqrt{3}\, x_2 + 2a)(x_1 - a). \tag{10.6.6}$$

Making use of Eqs. (10.5.3) and (10.5.4), we get:

$$\sigma_{13} = \frac{G\alpha}{a}(x_1 - a)x_2 \tag{10.6.7}$$

$$\sigma_{23} = \frac{G\alpha}{2a}(x_1^2 + 2ax_1 - x_2^2). \tag{10.6.8}$$

From these equations, we see that the shearing stress component σ_{13} vanishes along the OX_1 axis, while σ_{23} becomes:

$$\sigma_{23} = \frac{G\alpha}{2a} x_1(x_1 + 2a). \tag{10.6.9}$$

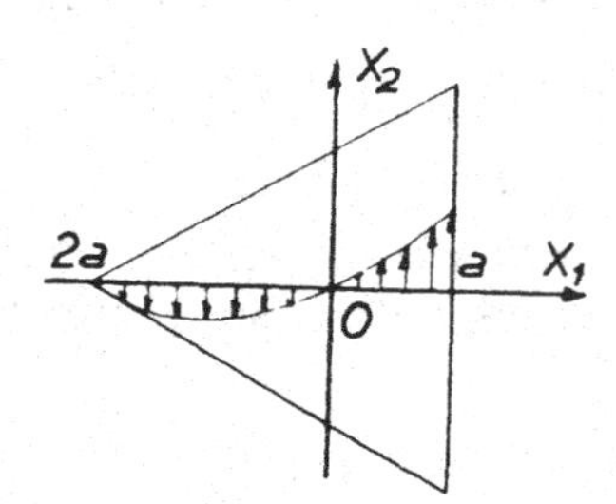

Fig. 10.15

The distribution of σ_{23} along the OX_1 axis is shown in Fig. 10.15. The shearing stress is a maximum at the midpoints of the sides of the triangle, and is given by:

$$(\sigma_t)_{\max} = \frac{3}{2} G\alpha a. \tag{10.6.10}$$

The shearing stresses vanish at the corners and at the origin O. The twisting moment is given by Eq. (10.5.11) as:

$$M_{33} = 2 \int\int -\frac{G\alpha}{6a}(x_1 - \sqrt{3}\, x_2 + 2a)(x_1 + \sqrt{3}\, x_2 + 2a) (x_1 - a)\, dx_1\, dx_2 = \frac{27}{5\sqrt{3}} G\alpha a^4 \tag{10.6.11}$$

or

$$M_{33} = \frac{3}{5} G\alpha I_3, \tag{10.6.12}$$

where $I_3 = 3\sqrt{3}\, a^4$ is the polar moment of inertia of the triangle. The warping function is given by:

$$\psi(x_1, x_2) = \frac{3x_1^2 x_2 - x_2^3}{6a} + C, \tag{10.6.13}$$

where C is a constant equal to zero for no rigid body displacement. The three components of the displacement are:

$$u_1 = -\frac{5M_{33}}{3GI_3}x_2x_3, \quad u_2 = +\frac{5M_{33}}{3GI_3}x_1x_3,$$
$$u_3 = \frac{5M_{33}}{18GaI_3}[3x_1^2x_2 - x_2^3]. \tag{10.6.14}$$

Lines of equal vertical displacement are shown in Fig. 10.16.

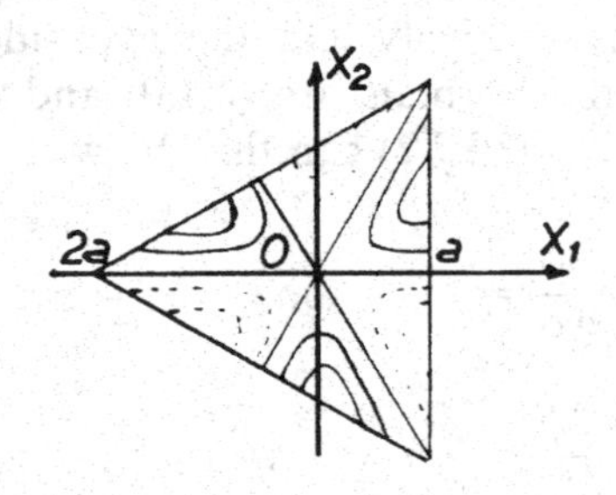

Fig. 10.16

10.7 Torsion of Rectangular Bars

Consider a bar with a rectangular cross section, with its center at the origin, and with sides $2a$ and $2b$ (Fig. 10.17). As stated in the previous sections, the solution to the problem amounts to finding a function $\psi(x_1, x_2)$ such that

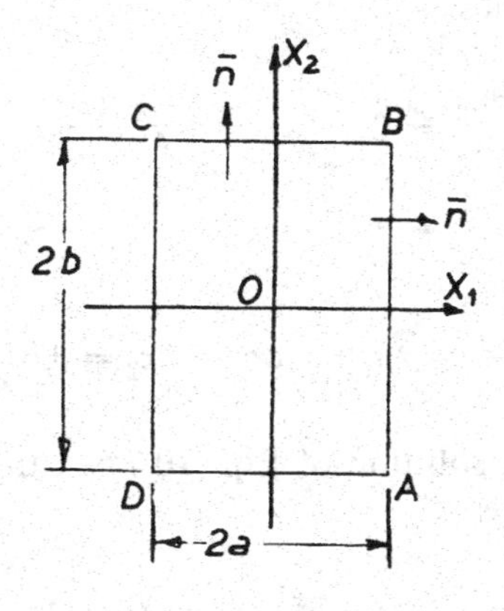

Fig. 10.17

$$\nabla^2\psi(x_1, x_2) = 0 \quad \text{in the rectangle} \tag{10.7.1}$$

and

$$\left(\frac{\partial\psi}{\partial x_1} - x_2\right)\frac{dx_2}{ds} - \left(\frac{\partial\psi}{\partial x_2} + x_1\right)\frac{dx_1}{ds} = 0 \quad \text{on the boundary .} \tag{10.7.2}$$

On the two sides AB and CD, the direction cosines of the normals are (1,0,0) and (-1,0,0), respectively. On the two sides BC and DA, the direction cosines of the normals are (0,1,0) and (0,-1,0), respectively. The boundary condition (10.7.2) can thus be written as:

$$\frac{\partial\psi}{\partial x_1} = x_2 \qquad \text{on} \qquad x_1 = \pm a \tag{10.7.3}$$

$$\frac{\partial\psi}{\partial x_2} = -x_1 \qquad \text{on} \qquad x_2 = \pm b. \tag{10.7.4}$$

If we now introduce the function $\psi_1(x_1, x_2)$ such that [2]

$$\psi_1 = x_1 x_2 - \psi, \tag{10.7.5}$$

Eq. (10.7.1) becomes

$$\nabla^2\psi_1(x_1, x_2) = 0 \quad \text{over the rectangle,} \tag{10.7.6}$$

and Eq. (10.7.2) becomes

$$\frac{\partial\psi_1}{\partial x_1} = 0 \qquad \text{on} \qquad x_1 = \pm a \tag{10.7.7}$$

and

$$\frac{\partial\psi_1}{\partial x_2} = 2x_1 \qquad \text{on} \qquad x_2 = \pm b. \tag{10.7.8}$$

Let us assume that the solution of Eq. (10.7.6) can be expressed in the form of an infinite series,

$$\psi_1(x_1, x_2) = \sum_{n=0}^{\infty} E_n(x_1) G_n(x_2), \tag{10.7.9}$$

where each term of the series satisfies the differential Eq. (10.7.6), $E_n(x_1)$ are functions of x_1 alone, and $G_n(x_2)$ are functions of x_2 alone. Substituting Eq. (10.7.9) into Eq. (10.7.6), we get:

$$E''_n(x_1)G_n(x_2) + E_n(x_1)G''_n(x_2) = 0, \tag{10.7.10}$$

for each value of n. Eq. (10.7.10) can also be written as follows:

$$\frac{E''_n(x_1)}{E_n(x_1)} = -\frac{G''_n(x_2)}{G_n(x_2)}. \tag{10.7.11}$$

This equality cannot be fulfilled unless both sides of Eq. (10.7.11) are equal to a constant. This constant will be taken equal to $-k_n^2$. This leads us to a pair of ordinary differential equations:

$$\frac{d^2 E_n}{dx_1^2} + k_n^2 E_n = 0 \tag{10.7.12}$$

$$\frac{d^2 G_n}{dx_2^2} - k_n^2 G_n = 0. \tag{10.7.13}$$

The solution of these equations is:

$$E_n = c_1 \sin(k_n x_1) + c_2 \cos(k_n x_1) \tag{10.7.14}$$

$$G_n = c_3 \sinh(k_n x_2) + c_4 \cosh(k_n x_2). \tag{10.7.15}$$

The constants c_1, c_2, c_3, and c_4 will be determined from the boundary conditions (10.7.7) and (10.7.8). Let us first consider the boundary condition (10.7.8). We see that

$$\frac{\partial \psi_1}{\partial x_2} = \sum_{n=0}^{\infty} E_n(x_1)G'_n(x_2) = 2x_1 \tag{10.7.16}$$

must have the same value for $x_2 = +b$ and $x_2 = -b$. Therefore, $G'_n(x_2)$ must be a symmetric function in x_2. Also for $x_2 = \pm b$:

$$\sum_{n=0}^{\infty} E_n(x_1)G'_n(b) = 2x_1 . \tag{10.7.17}$$

Therefore, $E_n(x_1)$ must be an antisymmetric function in x_1. From these considerations, we find that $c_2 = c_4 = 0$ in Eqs. (10.7.14) and (10.7.15).

The condition (10.7.7) is satisfied if:

$$E_n'(\pm a) = 0 \tag{10.7.18}$$

or

$$c_1 k_n \cos(k_n a) = 0. \tag{10.7.19}$$

Thus,

$$k_n = \frac{(2n + 1)\Pi}{2a}. \tag{10.7.20}$$

c_1 and c_3 being arbitrary, Eq. (10.7.9) can now be written as follows:

$$\psi_1 = \sum_{n=0}^{\infty} A_n \sin(k_n x_1)\sinh(k_n x_2), \tag{10.7.21}$$

where k_n is given by Eq. (10.7.20) and A_n is still to be determined from boundary condition (10.7.8).

$$\begin{aligned} \left(\frac{\partial \psi_1}{\partial x_2}\right)_{x_2=\pm b} &= \sum_{n=0}^{\infty} A_n k_n \cosh(k_n b)\sin(k_n x_1) = 2x_1 \\ &= \sum_{n=0}^{\infty} B_n \sin(k_n x_1), \end{aligned} \tag{10.7.22}$$

where we have set:

$$B_n = A_n k_n \cosh(k_n b). \tag{10.7.23}$$

To determine A_n, let us multiply both sides of Eq. (10.7.22) by $\sin(k_m x_1)$ and integrate with respect to x_1. We get:

$$\int_{-a}^{+a} 2x_1 \sin(k_m x_1)\,dx_1 = \sum_{n=0}^{\infty} \int_{-a}^{+a} B_n \sin(k_m x_1)\sin(k_n x_1)\,dx_1. \tag{10.7.24}$$

However, since

$$\int_{-a}^{+a} \sin(k_m x_1)\sin(k_n x_1)\,dx_1 = \begin{cases} 0 \text{ if } m \neq n \\ a \text{ if } m = n \end{cases}, \tag{10.7.25}$$

then

$$\int_{-a}^{+a} 2x_1 \sin(k_m x_1)\,dx_1 = \int_{-a}^{+a} B_m \sin^2(k_m x_1)\,dx_1 = B_m a. \tag{10.7.26}$$

Integrating Eq. (10.7.26), we get:

$$B_m = \frac{(-1)^m 16a}{\Pi^2(2m+1)^2}. \tag{10.7.27}$$

Therefore,

$$A_n = \frac{(-1)^n 32a^2}{\Pi^3(2n+1)^3 \cosh k_n b}, \tag{10.7.28}$$

so that the warping function ψ is:

$$\psi = x_1 x_2 - \frac{32a^2}{\Pi^3} \sum_{n=0}^{\infty} \frac{(-1)^n \sin(k_n x_1)\sinh(k_n x_2)}{(2n+1)^3 \cosh(k_n b)}. \tag{10.7.29}$$

The constant J is given by Eq. (10.3.17) as:

$$J = \int_{x_2=-b}^{x_2=+b} \int_{x_1=-a}^{x_1=+a} \left(x_1^2 + x_2^2 + x_1 \frac{\partial \psi}{\partial x_2} - x_2 \frac{\partial \psi}{\partial x_2} \right) dx_1\, dx_2 \tag{10.7.30}$$

or

$$J = \frac{8a^3 b}{3}\left[1 + \frac{96}{\Pi^4} \sum_{n=0}^{\infty} \frac{1}{(2n+1)^4} - \frac{384a}{\Pi^5 b} \sum_{n=0}^{\infty} \frac{\tanh(k_n b)}{(2n+1)^5}\right]. \tag{10.7.31}$$

Since

$$\sum_{n=0}^{\infty} \frac{1}{(2n+1)^4} = \frac{\Pi^4}{96}, \tag{10.7.32}$$

then

$$J = 16a^3 b\left[\frac{1}{3} - \frac{64a}{\Pi^5 b} \sum_{n=0}^{\infty} \frac{\tanh(k_n b)}{(2n+1)^5}\right] = Ka^3 b. \tag{10.7.33}$$

The series

$$\sum_{n=0}^{\infty} \frac{\tanh(k_n b)}{(2n+1)^5}$$

can be written as:

$$\sum_{n=0}^{\infty} \frac{\tanh(k_n b)}{(2n+1)^5} = \tanh\left(\frac{\Pi b}{2a}\right) + \sum_{n=1}^{\infty} \frac{\tanh(k_n b)}{(2n+1)^5}. \tag{10.7.34}$$

The first term in the right-hand side of the previous equation gives the value of the series to within 1/2 percent, and for all practical purposes:

$$J = 16a^3b\left[\frac{1}{3} - \frac{64a}{\pi^5 b}\tanh\left(\frac{\Pi b}{2a}\right)\right] = Ka^3b. \tag{10.7.35}$$

The shearing stresses are given by:

$$\begin{aligned}\sigma_{13} &= \frac{M_{33}}{J}\left(\frac{\partial\psi}{\partial x_1} - x_2\right)\\ &= -\frac{16M_{33}a}{J\Pi^2}\sum_{n=0}^{\infty}\frac{(-1)^n}{(2n+1)^2}\frac{\sinh(k_n x_2)}{\cosh(k_n b)}\cos(k_n x_1)\end{aligned} \tag{10.7.36}$$

$$\begin{aligned}\sigma_{23} &= \frac{M_{33}}{J}\left(\frac{\partial\psi}{\partial x_2} + x_1\right)\\ &= \frac{M_{33}}{J}\left[2x_1 - \frac{16a}{\Pi^2}\sum_{n=0}^{\infty}\frac{(-1)^n\cosh(k_n x_2)}{(2n+1)^2\cosh(k_n b)}\sin(k_n x_1)\right].\end{aligned} \tag{10.7.37}$$

The maximum shearing stress occurs at the mid-points of the long sides $x_1 = \pm a$ of the rectangle. Its value is given by:

$$\begin{aligned}(\sigma_t)_{max} = \sigma_{23} &= \frac{2M_{33}a}{J}\left[1 - \frac{8}{\Pi^2}\sum_{n=0}^{\infty}\frac{1}{(2n+1)^2\cosh(k_n b)}\right]\\ &= K_1\frac{M_{33}a}{J}.\end{aligned} \tag{10.7.38}$$

Substituting the value of J from Eq. (10.7.33) in Eq. (10.7.38), we get:

$$(\sigma_t)_{max} = K_2\frac{M_{33}}{a^2b}.$$

The following table gives the values of K, K_1, and K_2 for different ratios b/a:

$\frac{b}{a}$	K	K_1	K_2	$\frac{b}{a}$	K	K_1	K_2
1.0	2.250	1.350	0.600	3.0	4.208	1.970	0.468
1.2	2.656	1.518	0.571	4.0	4.496	1.994	0.443
1.5	3.136	1.696	0.541	5.0	4.656	1.998	0.430
2.0	3.664	1.860	0.508	10.0	4.992	2.000	0.401
2.5	3.984	1.936	0.484	∞	5.328	2.000	0.375

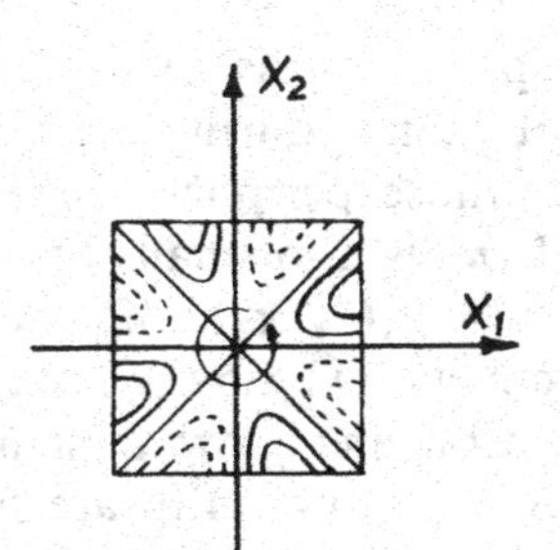

Fig. 10.18

The expression for u_3 is given by $u_3 = \alpha\psi(x_1, x_2)$, and the contour lines of the surface u_3 = constant can easily be drawn. For a square bar, these contour lines are shown in Fig. 10.18.

10.8 Prandtl's Membrane Analogy

In 1903, Prandtl observed that the differential equation (10.5.6) of the stress function is the same as the differential equation of the shape of a stretched membrane initially flat which is then blown up by air pressure from the bottom. This observation will provide us with a way of visualizing the shape of the stress function ϕ and the stress distribution. Consider a thin weightless membrane, initially with a large tension T, having the same value in all directions. It is blown up from a flat shape into a curved surface, being held at the edges by a frame having the same outline as the cross section of the bar under torsion. We shall

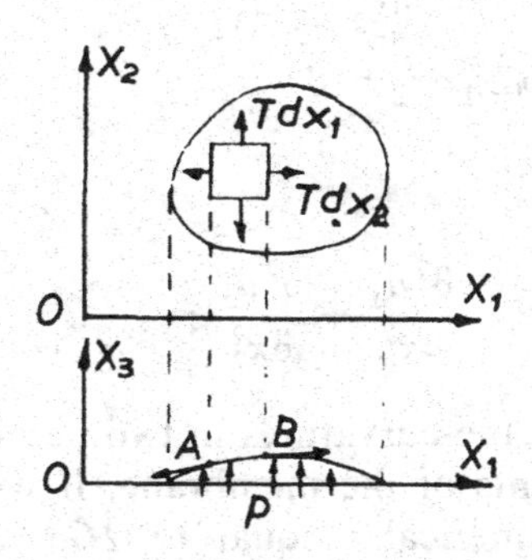

Fig. 10.19

assume that the air pressure p is so small and the initial tension T so large that T does not change during the blowing up process. Fig. 10.19 shows a membrane whose periphery is held down in the plane OX_1, OX_2. Since p is small, u_3 is also small. The equation of the membrane's surface is given by $u_3 = u_3(x_1, x_2)$, and the slopes in the OX_1 and OX_2 directions are respectively given by $\partial u_3/\partial x_1$ and $\partial u_3/\partial x_2$. These slopes will also be small. Consider now a small element dx_1, dx_2 of the membrane (Fig. 10.19). Acting on it, are the forces Tdx_1 and Tdx_2 as well as $pdx_1\,dx_2$ in the OX_3 direction. If we resolve the forces in the OX_1, OX_2, and OX_3 directions, we find, because of the smallness of the slope, that the equilibrium equations in the OX_1 and OX_2 directions are automatically satisfied. The equilibrium in the OX_3 direction gives (Fig. 10.20):

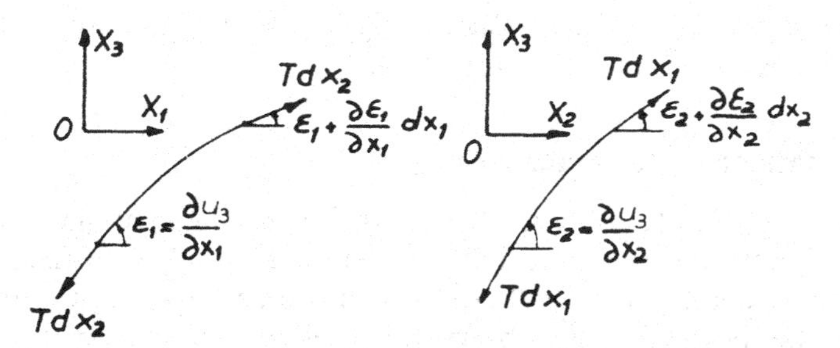

Fig. 10.20

$$
\begin{aligned}
&Tdx_2\left(\frac{\partial u_3}{\partial x_1} + \frac{\partial^2 u_3}{\partial x_1^2}dx_1\right) - Tdx_2\left(\frac{\partial u_3}{\partial x_1}\right) \\
&\quad + Tdx_1\left(\frac{\partial u_3}{\partial x_2} + \frac{\partial^2 u_3}{\partial x_2^2}dx_2\right) - Tdx_1\left(\frac{\partial u_3}{\partial x_2}\right) \\
&\quad + pdx_1\,dx_2 = 0
\end{aligned}
\tag{10.8.1}
$$

or

$$
\frac{\partial^2 u_3}{\partial x_1^2} + \frac{\partial^2 u_3}{\partial x_2^2} = -\frac{p}{T}. \tag{10.8.2}
$$

Therefore, the sum of the curvatures in two perpendicular directions is a constant for all points of the membrane. If we now adjust p and T such that p/T is numerically equal to $2G\alpha$, Eq. (10.8.2) becomes identical to Eq. (10.5.7). If we arrange the membrane so that its height u_3 remains zero at the boundary contour of the section, then the heights u_3 of the membrane are numerically equal to the stress function ϕ. If we have the deflection surface of the membrane represented by contour lines, several important conclusions regarding stress distribution in torsion can be obtained:

1. The contour lines u_3 = constant are lines following the shearing stress, i.e., tangent to the shearing stress vector (Fig. 10.21): Along a

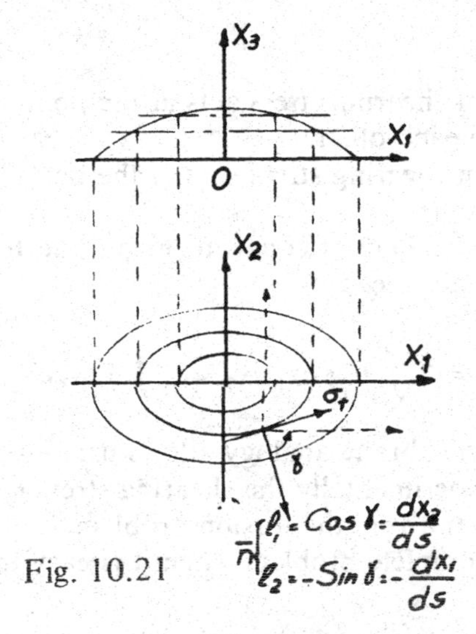

Fig. 10.21

contour line, $du_3/ds = 0$, since u_3 = constant. Therefore $d\phi/ds = 0$ and

$$\frac{d\phi}{ds} = \frac{\partial \phi}{\partial x_1}\frac{dx_1}{ds} + \frac{\partial \phi}{\partial x_2}\frac{dx_2}{ds} = 0. \tag{10.8.3}$$

Therefore,

$$\sigma_{13}\cos\gamma - \sigma_{23}\sin\gamma = 0. \tag{10.8.4}$$

In other words, the projection of the stress vector on the normal $\bar{n}$ is equal to zero. Therefore, the stress vector is tangent to the contour line through a given point of the twisted bar.

2. The shearing stress vector is given in magnitude by the slope of the membrane normal to the contour line (i.e., the maximum slope): Since the shearing stress vector is directed along the contour lines, its magnitude is given by Fig. (10.21).

$$\sigma_t = \sigma_{13}\sin\gamma + \sigma_{23}\cos\gamma = -\left(\frac{\partial\phi}{\partial x_2}\frac{dx_2}{dn} + \frac{\partial\phi}{\partial x_1}\frac{dx_1}{dn}\right) = -\frac{d\phi}{dn}$$
$$= -\frac{du_3}{dn}. \tag{10.8.5}$$

3. The maximum shearing stress acts at the points where the contour lines are closest to each other.

4. The maximum shearing stress acts at the boundary since the slope is maximum there.

5. The twisting moment is equal in magnitude to twice the volume under the membrane since

$$M_{33} = 2\int\!\!\int \phi\, dx_1\, dx_2 = 2\int\!\!\int u_3\, dx_1\, dx_2. \tag{10.8.6}$$

In summary, the membrane analogy allows us:

a. to measure experimentally the shearing stress σ_t

b. to visualize intuitively the torsion problem

c. to solve the complete problem when it is easy to find u_3.

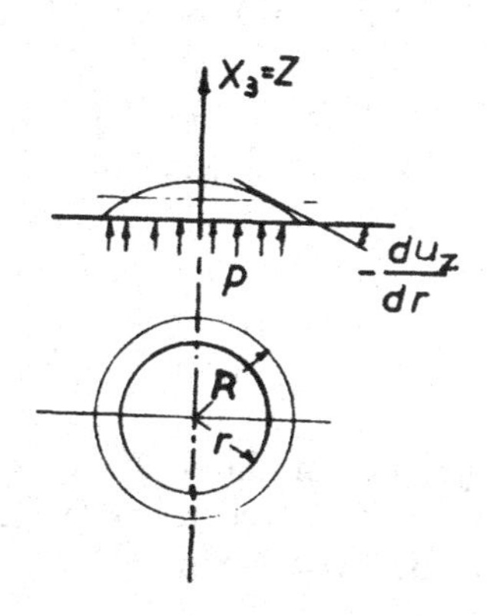

Fig. 10.22

10.9 Application of the Membrane Analogy to Solid Sections

In this section, the membrane analogy is applied to bars with circular and thin rectangular cross sections, then extended to thin open sections.

1. *Circular Cross Section*

The computations are most easily made in cylindrical coordinates. On account of symmetry, the height u_z of the membrane depends only on r. Cutting a concentric circle out of the membrane (Fig. 10.22) and writing the equilibrium in the vertical direction, we get:

$$-2\Pi r T \frac{du_z}{dr} = \Pi r^2 p \tag{10.9.1}$$

$$-\frac{du_z}{dr} = \frac{p}{2T} r. \tag{10.9.2}$$

This means that the slope of the membrane is proportional to the distance r and hence the shearing stresses follow this law. Substituting $2G\alpha$ to p/T we get:

$$\sigma_t = \sigma_{\theta z} = G\alpha r, \tag{10.9.3}$$

which was obtained in Sec. 10.2. The shape of the membrane is obtained by integrating Eq. (10.9.2).

$$u_z = -\int \frac{pr}{2T}\, dr = -\frac{pr^2}{4T} + \text{constant}\,. \tag{10.9.4}$$

On the periphery, $u_z = 0$, so that

$$u_z = \frac{p}{4T}(R^2 - r^2). \tag{10.9.5}$$

The volume under the membrane is given by $\int_0^R 2\Pi r u_z\, dr$, and the twisting moment by twice this value:

$$M_z = 2\int_0^R 2\Pi r u_z\, dr = \frac{\Pi}{4}\frac{p}{T}R^4 = G\alpha I_z\,. \tag{10.9.6}$$

Therefore,

$$\frac{M_z}{\alpha} = GI_z, \tag{10.9.7}$$

which is the result obtained in Sec. 10.2.

2. *Thin Rectangular Section*

Consider a narrow rectangular cross section with sides b and t, in which b is much larger than t (Fig. 10.23). Since b is much larger than t, we can deduce that the shape of the membrane will be the same between CC and DD. The membrane flattens up at both ends. The contour lines in the central portion are straight and parallel to the OX_2 axis. Cutting out a central piece of membrane of dimensions $2x_1$ and ℓ, the third equation of equilibrium gives:

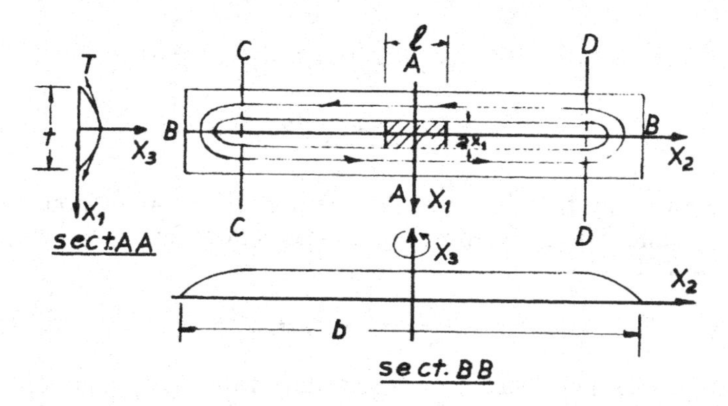

Fig. 10.23

$$-2T\ell \frac{du_3}{dx_1} = 2x_1 \ell p \tag{10.9.8}$$

$$-\frac{du_3}{dx_1} = \frac{p}{T} x_1 . \tag{10.9.9}$$

This means that the slope of the membrane is proportional to x_1, and hence the shear stress follows this law. The maximum shear stress occurs at $x_1 = \pm t/2$, and is equal to

$$(\sigma_t)_{\max} = \frac{p}{T}\frac{t}{2} = G\alpha t . \tag{10.9.10}$$

Integrating Eq. (10.9.9), we get:

$$u_3 = -\frac{p x_1^2}{2T} + \text{constant} . \tag{10.9.11}$$

On the periphery, $x_3 = 0$, so that:

$$u_3 = \frac{p}{2T}\left(\frac{t^2}{4} - x_1^2\right), \tag{10.9.12}$$

which is a parabola whose area is $pt^3/12T$. In calculating the volume under the membrane, the flattening near the edges $x_2 = \pm\frac{b}{2}$ is neglected so that the volume is $pt^3b/12T$. The twisting moment is thus given by:

$$M_{33} = 2\left(\frac{pt^3b}{12T}\right) = \frac{Gbt^3}{3}\alpha. \tag{10.9.13}$$

Therefore,

$$\frac{M_{33}}{\alpha} = \frac{Gbt^3}{3} = GJ, \tag{10.9.14}$$

where

$$J = \frac{bt^3}{3}. \tag{10.9.15}$$

The expression of the maximum shearing stress can be written in terms of the twisting moment as follows:

$$(\sigma_t)_{\max} = \frac{3M_{33}}{bt^2}. \tag{10.9.16}$$

3. *Thin Open Sections*

Suppose the narrow rectangle of Fig. 10.23 is bent by 90 degrees so that the section becomes a thin-walled angle. The shape of the membrane will not change except for local effects at the corners and the volume under the membrane will remain essentially the same. Eq. (10.9.13)

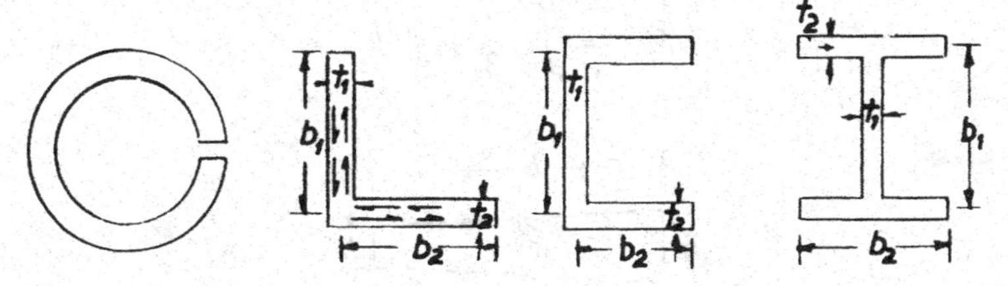

Fig. 10.24

remains valid for an angular section if b is taken as the length of both legs combined (Fig. 10.24). The same is true for T shapes, I shapes, or any section built up with rectangles; box sections will be examined in the next section. If the web and the flanges do not have the same section, Eq. (10.9.14) is applied to each part alone and the results added up. The angle α is the same for all parts of the same section and each part is subjected to a moment proportional to its torsional rigidity:

For angle sections we have, to a good approximation:

$$J \approx \frac{b_1 t_1^3 + b_2 t_2^3}{3} \tag{10.9.17}$$

$$\frac{M_{33}}{\alpha} \approx G\frac{b_1 t_1^3 + b_2 t_2^3}{3} \tag{10.9.18}$$

$$(\sigma_t)_{\max} \approx G\alpha t_i, \tag{10.9.19}$$

where t_i is the larger of t_1 and t_2.

For a channel or an I section, we have:

$$J \approx \frac{b_1 t_1^3 + 2b_2 t_2^3}{3} \tag{10.9.20}$$

$$\frac{M_{33}}{\alpha} \approx G\frac{b_1 t_1^3 + 2b_2 t_2^3}{3} \tag{10.9.21}$$

$$(\sigma_t)_{\max} \approx G\alpha t_i, \tag{10.9.22}$$

where t_i is the larger t.

For a trapezoidal section, the value of α is obtained by assuming that the surface of the deflected membrane is conical. From Eq. (10.9.13), we have (Fig. 10.25):

$$M_{33} = \int_0^b \frac{1}{3} G\alpha t^3 \, dx_2, \tag{10.9.23}$$

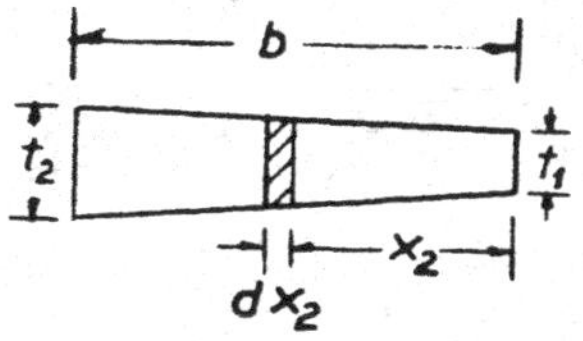

Fig. 10.25

with

$$t = t_1 + (t_2 - t_1)\frac{x_2}{b}. \tag{10.9.24}$$

Therefore,

$$\frac{M_{33}}{\alpha} = G\frac{b(t_1 + t_2)(t_1^2 + t_2^2)}{12}, \tag{10.9.25}$$

where

$$J = \frac{b(t_1 + t_2)(t_1^2 + t_2^2)}{12}. \tag{10.9.26}$$

The previous equations are directly applicable to an *I section with sloping flanges* (Fig. 10.26):

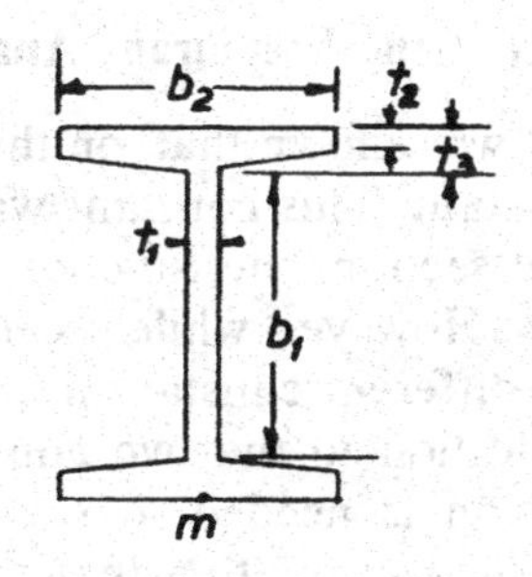

Fig. 10.26

$$J = \frac{b_1 t_1^3}{3} + 4\frac{b_2}{2}\left[\frac{(t_2 + t_3)(t_2^2 + t_3^2)}{12}\right] \tag{10.9.27}$$

$$\frac{M_{33}}{\alpha} = JG \tag{10.9.28}$$

$$(\sigma_t)_{\max} = G\alpha t_3, \tag{10.9.29}$$

occurring at point *m*.

It should be noted that in all the previous cases a considerable stress concentration takes place at the re-entrant corners, the magnitude of which depends on the radius of the fillets. For small radii of fillets, Trefftz found the following approximate solution (Fig. 10.27):

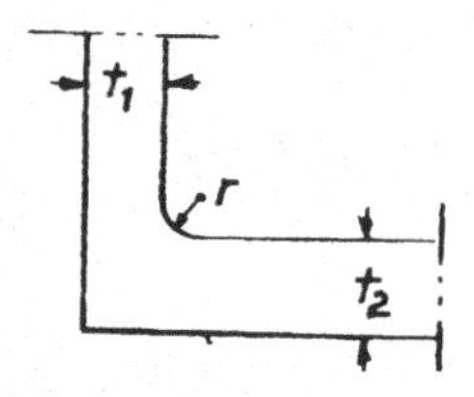

Fig. 10.27

$$(\sigma_t)\ \text{fillet} = 1.74(\sigma_t)_{max} \sqrt[3]{\frac{t}{r}},\qquad (10.9.30)$$

in which t is the larger of t_1 and t_2.(*)

10.10 Application of the Membrane Analogy to Thin Tubular Members

In Sec. 10.5, it was shown that, on the boundary of a bar subjected to torsion, ϕ is constant. This constant was chosen equal to zero. For the case of a hollow section, the function ϕ is constant on the outer and inner boundaries. However, while it can still be chosen equal to zero on the outer one, a different constant must be assigned to its value on the inner one. In addition to the two equations (10.5.7) and (10.5.8), one additional equation is needed to solve the problem. This additional equation is obtained by writing that the displacements must be single valued. Fig. 10.28 shows what is meant by this last statement: Starting

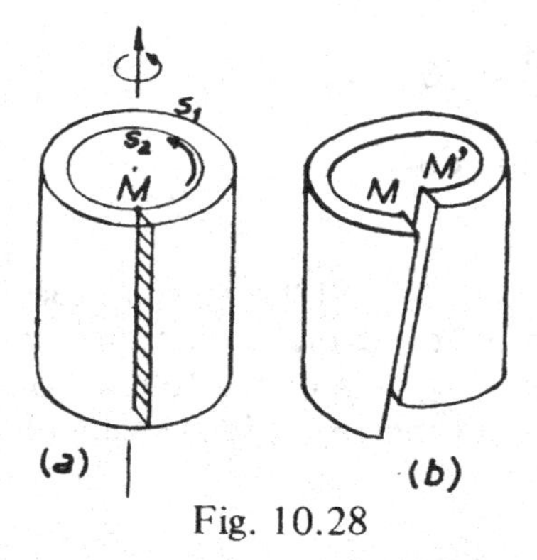

Fig. 10.28

* See pages A-60 to A-62 for a numerical example.

from point M on the contour s_2 and proceeding around the hole, we must end up with the same value of u_3 we started from. We see that this is the case in Fig. 10.28a, while the slit in the tubular member of Fig. 10.28b results in a discontinuity MM'. Mathematically the continuity or compatibility requirement is written as:

$$\oint_{s_2} du_3 = \oint_{s_2} \left(\frac{\partial u_3}{\partial x_1} dx_1 + \frac{\partial u_3}{\partial x_2} dx_2 \right) = 0. \tag{10.10.1}$$

Let us now compute the integral $\oint \sigma_t \, ds$ around the inner boundary s_2. Using Eqs. (10.3.4), (10.5.3), (10.5.4), and (10.8.5), we get:

$$\oint_{s_2} \sigma_t \, ds = \oint_{s_2} \left(\sigma_{13} \frac{dx_1}{ds} + \sigma_{23} \frac{dx_2}{ds} \right) ds \tag{10.10.2}$$

$$= G \oint_{s_2} \left(\frac{\partial u_3}{\partial x_1} dx_1 + \frac{\partial u_3}{\partial x_2} dx_2 \right) + G\alpha \oint_{s_2} (x_1 \, dx_2 - x_2 \, dx_1) \tag{10.10.3}$$

$$= G\alpha \oint_{s_2} (x_1 \, dx_2 - x_2 \, dx_1). \tag{10.10.4}$$

The integral in the right-hand side of Eq. (10.10.4) is equal to twice the area A_2 inside the contour s_2: Indeed, the first integral $\oint_{s_2} x_1 \, dx_2$ is equal to (Fig. 10.29):

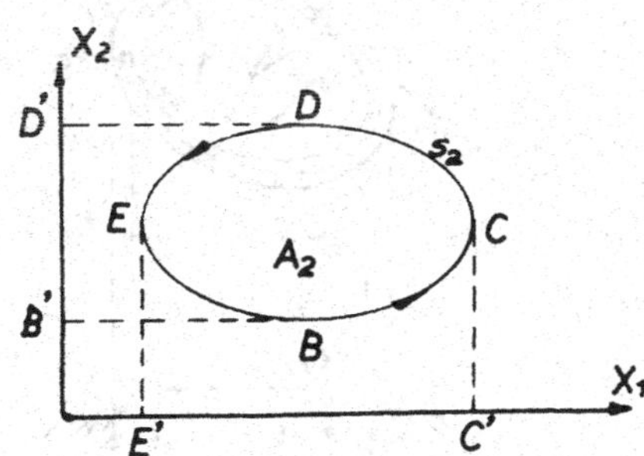

Fig. 10.29

$$\oint_{s_2} x_1 \, dx_2 = - \text{ area } D'DEBB' + \text{ area } B'BCDD' = A_2, \tag{10.10.5}$$

and the second integral $\oint_{s_2} -x_2 \, dx_1$ is equal to:

$$\oint_{s_2} -x_2 \, dx_1 = -(\text{area } EE'C'CB - \text{ area } EE'C'CD) = A_2, \tag{10.10.6}$$

so that

$$\oint_{s_2} \sigma_t \, ds = 2G\alpha A_2 . \tag{10.10.7}$$

If we translate Eq. (10.10.7) in terms of membrane by means of Eq. (10.8.5), we get:

$$\oint_{s_2} -\frac{d\phi}{dn} ds = -\oint_{s_2} \frac{du_3}{dn} ds = \frac{pA_2}{T} \tag{10.10.8}$$

or

$$-T \oint_{s_2} \frac{du_3}{dn} ds = pA_2 . \tag{10.10.9}$$

Eq. (10.10.9) represents the equation of equilibrium of a weightless flat plate covering the area A_2 and subjected to an upward pressure p and a downward pull around its contour by a membrane with tension T and slope $-du_3/dn$ (Fig. 10.30). Therefore, in the case of hollow sections,

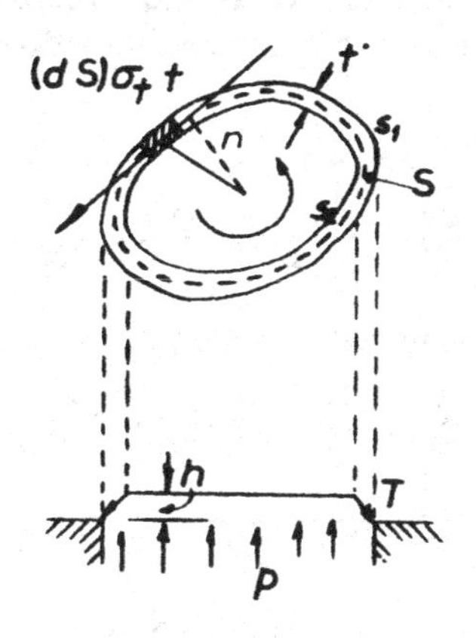

Fig. 10.30

the membrane may be considered as stretched between the outer contour s_1 and a weightless flat plate of the same shape as the inner contour s_2. If the hollow tube has thin walls, the membrane between s_1 and s_2 will be approximately straight. This means that the shearing stress is uniformly distributed across the thickness of the wall. If the height of the plate is h (Fig. 10.30), the shearing stress at any point where the thickness is t is given by:

$$\sigma_t = \frac{h}{t}. \tag{10.10.10}$$

The applied twisting moment M_{33} must be equal to the moments of the shearing stresses around the center of rotation of each section of the hollow bar. Thus (Fig. 10.30),

$$M_{33} = \oint_S \sigma_t tn\, dS = \oint_S hn\, dS = 2hA, \tag{10.10.11}$$

where S is the mean line between s_1 and s_2 and A is the area inside S. Therefore, the twisting moment is equal to twice the volume under the plate and the membrane. The shearing stress at any point where the thickness of the wall is t is given by:

$$\sigma_t = \frac{M_{33}}{2At}. \tag{10.10.12}$$

Eq. (10.10.7) gives:

$$\oint_{s_2} \frac{M_{33}}{2At}\, ds = 2G\alpha A_2. \tag{10.10.13}$$

Therefore, the torsional rigidity is given by:

$$\frac{M_{33}}{\alpha} = G\frac{4AA_2}{\oint_{s_2} \frac{ds}{t}} \approx G\frac{4A^2}{\oint_S \frac{dS}{t}} = GJ. \tag{10.10.14}$$

If t is constant, Eq. (10.10.14) becomes:

$$\frac{M_{33}}{\alpha} = G\frac{4A^2 t}{L} = GJ, \tag{10.10.15}$$

where L is the length of the mean line of the boundaries and

$$J = \frac{4A^2 t}{L}. \tag{10.10.16}$$

Combining Eqs. (10.10.15) and (10.10.12), we get the value of the angle of twist per unit length in terms of σ_t:

$$\alpha = \frac{\sigma_t L}{2AG}. \tag{10.10.17}$$

Two important conclusions can be drawn from the previous equations:
a) If a hollow tube is flattened (i.e., $A \approx 0$), the torsional rigidity tends to zero and the shearing stresses become very high.
b) For a given length of contour L, a circular shape would give the maximum possible rigidity since it corresponds to the highest A.

10.11 Application of the Membrane Analogy to Multicellular Thin Sections

If the cross section of a tubular member has more than two boundaries, the membrane analogy involves several stiff weightless plates which will have to be blown up with the air pressure p. The heights of the plates are unknown $(h_1, h_2, \ldots, h_n)$. Assuming that the thickness of the walls is small, the slope of the outside walls will be h/t while the slope of the inside walls will be $\Delta h/t$, where Δh is the difference in height of the two adjoining plates. We can write one equation of equilibrium for each plate,

$$pA_n = T \oint \frac{\Delta h}{t}\, dS, \tag{10.11.1}$$

where A_n is the area of the nth plate, the integral extends around that plate, and Δh is the height of the plate in question less the height of the neighboring plate. There are n such equations and they can be solved for the heights $h_1, h_2, \ldots, h_n$. Once the heights are known, we translate the membrane problem to the torsion problem by setting $\Delta h/t = \sigma_t$ and $p/T = 2G\alpha$. The twisting moment is given by twice the volume under the plates and membrane:

$$M_{33} = 2 \sum A_n h_n. \tag{10.11.2}$$

The following example illustrates the use of the membrane analogy for a bicellular section.

Example
In the section shown in Fig. 10.31, let

$$BCD = S_1, \quad DEB = S_2, \quad BD = S_3. \tag{10.11.3}$$

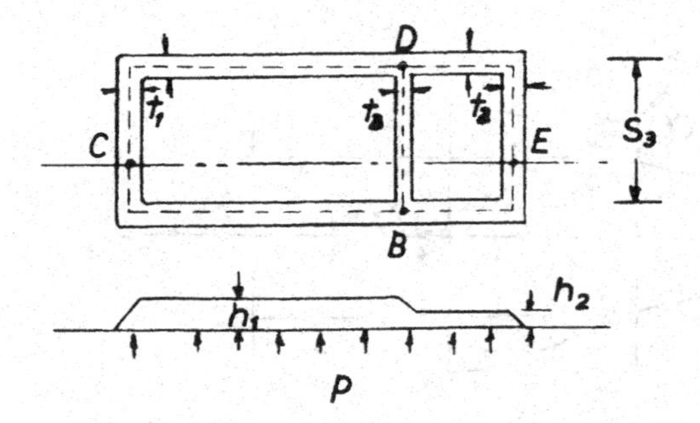

Fig. 10.31

The slopes of the membrane are given by:

$$\sigma_{t1} = \frac{h_1}{t_1}, \quad \sigma_{t2} = \frac{h_2}{t_2} \tag{10.11.4}$$

$$\sigma_{t3} = \frac{h_1 - h_2}{t_3} = \frac{t_1 \sigma_{t1} - t_2 \sigma_{t2}}{t_3}. \tag{10.11.5}$$

The twisting moment is given by twice the volume under the plates and membrane:

$$M_{33} = 2(A_1 t_1 \sigma_{t1} + A_2 t_2 \sigma_{t2}). \tag{10.11.6}$$

The vertical equilibrium of the two plates gives:

$$\frac{p}{T} A_1 = \sigma_{t1} S_1 + \sigma_{t3} S_3 = 2G\alpha A_1 \tag{10.11.7}$$

$$\frac{p}{T} A_2 = \sigma_{t2} S_2 - \sigma_{t3} S_3 = 2G\alpha A_2. \tag{10.11.8}$$

The solution of the four simultaneous Eqs. (10.11.5) to (10.11.8) gives σ_{t1}, σ_{t2}, σ_{t3}, and α. In the special case where the central wall is a plane of symmetry of the cross section, $h_1 = h_2$ and $\sigma_{t3} = 0$.(*)

10.12 Torsion of Circular Shafts of Varying Cross Section

In the study of bodies of revolution, the computations are most conveniently performed in cylindrical coordinates. Fig. 10.32 shows such a body subjected to terminal couples. The axis of the shaft will be taken as the OX_3 or the Z axis. The general equations of equilibrium are given by Eqs. (7.4.6) in which we set the body forces equal to zero. The

* See pages A-62 to A-64 for a numerical example.

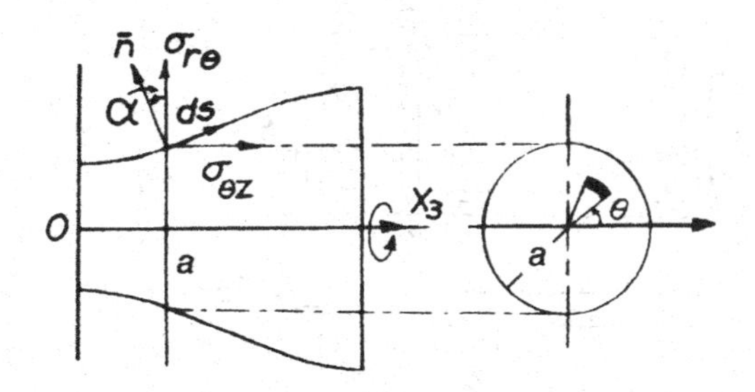

Fig. 10.32

strain-displacement relations are given by Eqs. (6.7.23) and (6.7.24). Let us assume that

$$u_r = 0, \quad u_z = 0, \tag{10.12.1}$$

and then prove that the solution based on this assumption satisfies equilibrium and boundary conditions. On account of symmetry, u_θ cannot be a function of θ and will only depend on r and z. Eqs. (6.7.23) and (6.7.24) give:

$$e_{rr} = e_{\theta\theta} = e_{zz} = e_{rz} = 0 \tag{10.12.2}$$

$$e_{r\theta} = \frac{1}{2}\left(\frac{\partial u_\theta}{\partial r} - \frac{u_\theta}{r}\right), \; e_{\theta z} = \frac{1}{2}\frac{\partial u_\theta}{\partial z}. \tag{10.12.3}$$

From Hooke's law, we have:

$$\sigma_{rr} = \sigma_{\theta\theta} = \sigma_{zz} = \sigma_{rz} = 0 \tag{10.12.4}$$

$$\sigma_{r\theta} = G\left(\frac{\partial u_\theta}{\partial r} - \frac{u_\theta}{r}\right), \; \sigma_{\theta z} = G\frac{\partial u_\theta}{\partial z}. \tag{10.12.5}$$

With these values of the stresses, two equations of equilibrium are identically satisfied while the third one gives:

$$\frac{\partial \sigma_{r\theta}}{\partial r} + \frac{\partial \sigma_{\theta z}}{\partial z} + \frac{2\sigma_{r\theta}}{r} = 0 \tag{10.12.6}$$

or

$$\frac{\partial}{\partial r}(r^2 \sigma_{r\theta}) + \frac{\partial}{\partial z}(r^2 \sigma_{\theta z}) = 0. \tag{10.12.7}$$

Eq. (10.12.7) is identically satisfied by introducing a stress function H (r, z), such that

$$\frac{\partial H}{\partial r} = r^2 \sigma_{\theta z}, \quad \frac{\partial H}{\partial z} = -r^2 \sigma_{r\theta}. \tag{10.12.8}$$

Since $e_{r\theta}$ and $e_{\theta z}$ are the only non-vanishing strains, and since both are expressed in terms of u_θ, compatibility is obtained by eliminating u_θ from the two equations of (10.12.3). This gives the following compatibility relation:

$$\frac{\partial e_{r\theta}}{\partial z} = \frac{\partial e_{\theta z}}{\partial r} - \frac{e_{\theta z}}{r}. \tag{10.12.9}$$

Using Eqs. (10.12.5) and (10.12.8), Eq. (10.12.9) is written in terms of the stress function H as follows:

$$\frac{\partial^2 H}{\partial r^2} - \frac{3}{r}\frac{\partial H}{\partial r} + \frac{\partial^2 H}{\partial z^2} = 0. \tag{10.12.10}$$

Since the lateral surface of the shaft is free from external loads, it follows that the shearing stress there must be directed along the tangent to the boundary of the axial section and its projection on the normal to the boundary must be zero. Therefore,

$$\sigma_{r\theta} \cos \alpha - \sigma_{\theta z} \sin \alpha = 0 \tag{10.12.11}$$

or

$$-\frac{1}{r^2}\frac{\partial H}{\partial z}\frac{dz}{ds} - \frac{1}{r^2}\frac{\partial H}{\partial r}\frac{dr}{ds} = \frac{1}{r^2}\frac{dH}{ds} = 0, \tag{10.12.12}$$

which gives

$$H = \text{constant on the boundary} . \tag{10.12.13}$$

The solution of the torsion problem of a circular shaft with variable diameter thus reduces to finding a function H satisfying Eqs. (10.12.10) and (10.12.13).

The magnitude of the twisting moment is given by:

$$\begin{aligned} M_z &= \int_0^a \sigma_{\theta z}\, r(2\Pi r)\, dr = 2\Pi \int_0^a \frac{\partial H}{\partial r}\, dr \\ &= 2\Pi[H(a, z) - H(0, z)]. \end{aligned} \tag{10.12.14}$$

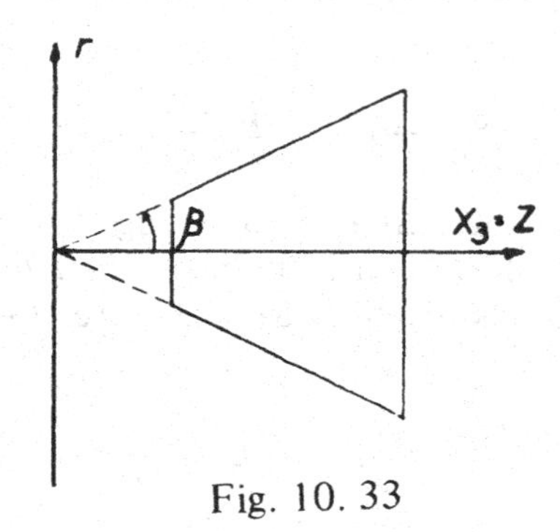

Fig. 10. 33

For the case of a *conical shaft* (Fig. 10.33), we have on the boundary

$$\frac{z}{(r^2 + z^2)^{1/2}} = \cos\beta = \text{constant}\,. \tag{10.12.15}$$

Thus, any function of this ratio will satisfy the boundary condition. One can verify that the function

$$H = C\left\{\frac{z}{(r^2 + z^2)^{1/2}} - \frac{1}{3}\left[\frac{z}{(r^2 + z^2)^{1/2}}\right]^3\right\}, \tag{10.12.16}$$

where C is a constant, satisfies Eq. (10.12.10). Substituting Eq. (10.12.16) into Eq. (10.12.14), the constant C is found to be given by:

$$C = -\frac{3M_z}{2\Pi(2 - 3\cos\beta + \cos^3\beta)}\,. \tag{10.12.17}$$

The shearing stresses $\sigma_{\theta z}$ and $\sigma_{r\theta}$ are given by:

$$\sigma_{\theta z} = -\frac{Crz}{(r^2 + z^2)^{5/2}} \tag{10.12.18}$$

$$\sigma_{r\theta} = -\frac{Cr^2}{(r^2 + z^2)^{5/2}}\,, \tag{10.12.19}$$

where C is given by Eq. (10.12.17).

10.13 Torsion of Thin-Walled Members of Open Section in which some Cross Section is Prevented from Warping

In our study of thin open sections in Sec. 10.9, it was assumed that each section was free to warp. A substantial amount of torsional rigidity can be added to such bars if one or more cross sections are prevented from warping. For bars of solid cross sections such as ellipses or rectangles, the prevention of warping has a negligible effect on the angle of twist per unit length α [3]. On the other hand, in thin open sections like I beams and channels, the prevention of warping during twist is accompanied by a bending of the flanges which may have substantial effect on the angle of twist. This problem has been analyzed by Timoshenko [3, 4].

Let us consider, for example, an I beam in which a section is prevented from warping. This can be accomplished by subjecting the beam to a torque M_{33} at each end and a torque $2M_{33}$ at the center as shown in Fig. 10.34. The center of the beam remains plane because of

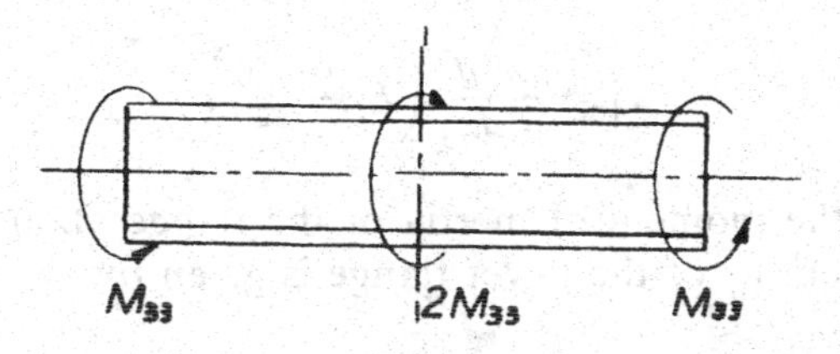

Fig. 10.34

symmetry and it may be considered as built in with the torque M_{33} applied at the other end (Fig. 10.35). Let θ be the angle of rotation of any section of the beam at a distance x_3 from the origin. The twisting moment M_{33} is transmitted from one section to the other of the beam in two ways:

(a) By the usual torsional stresses distributed over the I section.
(b) By means of shearing forces in the flanges due to bending as shown in Fig. 10.35.

Therefore,

$$M_{33} = JG\alpha + V_1 h = JG\frac{d\theta}{dx_3} + V_1 h, \tag{10.13.1}$$

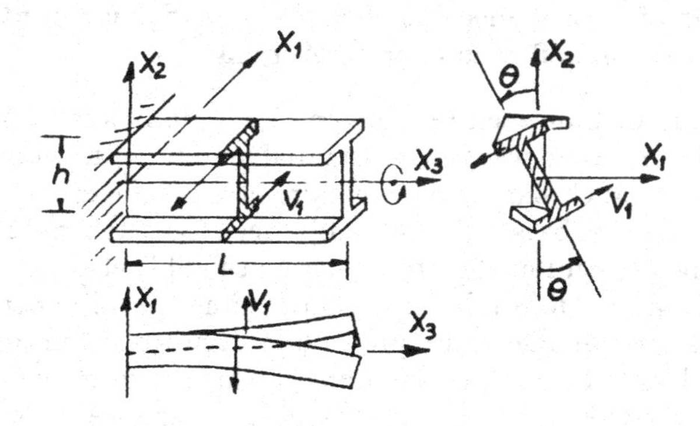

Fig. 10.35

where JG is the torsional rigidity of the I section and V_1 is the shearing force in the lower flange in the system of axes of Fig. 10.35. Considering each flange as a beam with one end built in and the other free, we have (from the elementary theory of beams):

$$V_1 = -\frac{d}{dx_3}\left(EI_f\frac{d^2u_1}{dx_3^2}\right), \tag{10.13.2}$$

in which I_f is the moment of inertia of the flange about the OX_2 axis. The displacement u_1 of the lower flange is given by:

$$u_1 = \theta\frac{h}{2}. \tag{10.13.3}$$

Substituting Eqs. (10.13.2) and (10.13.3) into Eq. (10.13.1), we get:

$$M_{33} = JG\frac{d\theta}{dx_3} - EI_f\frac{h^2}{2}\frac{d^3\theta}{dx_3^3} = JG\alpha - EI_f\frac{h^2}{2}\frac{d^2\alpha}{dx_3^2} \tag{10.13.4}$$

or

$$\frac{d^2\alpha}{dx_3^2} - \left(\frac{2GJ}{h^2EI_f}\right)\alpha = -\frac{2M_{33}}{EI_fh^2}. \tag{10.13.5}$$

This is a linear differential equation in α, whose solution is:

$$\alpha = C_1e^{bx_3} + C_2e^{-bx_3} + \frac{M_{33}}{GJ}, \tag{10.13.6}$$

where

$$b = \sqrt{\frac{2GJ}{h^2 EI_f}}. \tag{10.13.7}$$

The constants of integration C_1 and C_2 are obtained from the boundary conditions. The first boundary condition is that at $x_3 = \infty$, $d\theta/dx_3 = \alpha$ must remain finite so that $C_1 = 0$. Eq. (10.13.6) becomes:

$$\alpha = C_2 e^{-bx_3} + \frac{M_{33}}{GJ}. \tag{10.13.8}$$

The second boundary condition is that, at $x_3 = 0$, the slope $du_1/dx_3 = 0$, and hence $d\theta/dx_3 = \alpha = 0$, so that

$$C_2 = -\frac{M_{33}}{GJ}. \tag{10.13.9}$$

The value of α is now given by:

$$\alpha = \frac{d\theta}{dx_3} = \frac{M_{33}}{GJ}(1 - e^{-bx_3}). \tag{10.13.10}$$

Eq. (10.13.10) can be integrated to give θ as a function of x_3:

$$\theta = \frac{M_{33}}{GJ}\left(x_3 + \frac{1}{b}e^{-bx_3}\right) + C_3 \tag{10.13.11}$$

The constant C_3 can be obtained by setting $\theta = 0$ for $x_3 = 0$. Hence,

$$\theta = \frac{M_{33}}{GJ}\left[x_3 + \frac{1}{b}(e^{-bx_3} - 1)\right]. \tag{10.13.12}$$

Due to the presence of a second term in Eq. (10.13.10), the angle of twist per unit length varies along the length of the beam although M_{33} remains constant: In other words, the torsion is not uniform. Once α is determined, both parts in the right-hand side of Eq. (10.13.4) can be obtained and that portion due to the bending of the flanges determined. At the point $x_3 = 0$, $\alpha = 0$ so that the entire torque is balanced by the moment of the shearing forces in the flanges and $V_1 = M_{33}/h$. For very large values of x_3, Eq. (10.13.12) gives:

$$\theta = \frac{M_{33}}{GJ}\left(x_3 - \frac{1}{b}\right). \tag{10.13.13}$$

Eq. (10.13.13) states that the torsional stiffness of a long cantilever I beam is equal to the torsional stiffness of a free beam of a length shorter than the cantilever beam by an amount $1/b$.

Example

Let us consider a steel I beam with the dimensions shown in Fig. 10.36. For such an I beam, we have:

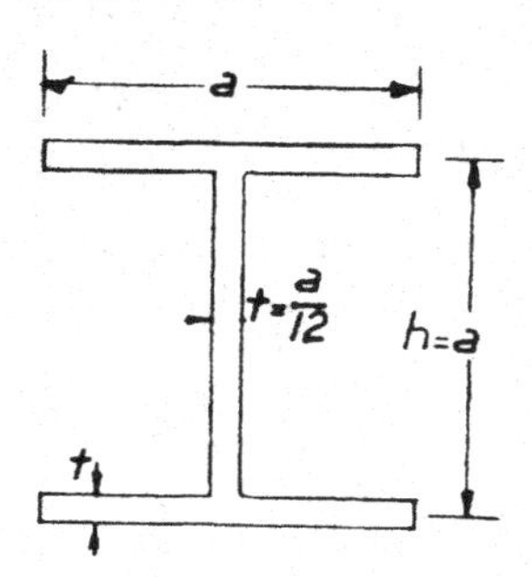

Fig. 10.36

$$EI_f = \frac{Eta^3}{12} = Et^2a^2 \qquad (10.13.14)$$

$$GJ = G\frac{3at^3}{3} = Gat^3 \qquad (10.13.15)$$

$$b = \frac{t}{a^2}\sqrt{\frac{24G}{E}} = \frac{t}{a^2}\sqrt{\frac{24 \times 12}{30}} = \frac{1}{3.9a}. \qquad (10.13.16)$$

The expression of the bending moment in the lower flange is given by:

$$M_f = -EI_f\frac{h}{2}\frac{d^2\theta}{dx_3^2} = -EI_f\frac{h}{2}\frac{d\alpha}{dx_3} = -\frac{M_{33}}{hb}e^{-bx_3}, \qquad (10.13.17)$$

with the maximum moment occurring at $x_3 = 0$. Hence,

$$(M_f)_{\max} = \frac{M_{33}}{hb} = 3.9M_{33}. \qquad (10.13.18)$$

The shearing force in the lower flange V_1 is obtained from Eq. (10.13.2). Its maximum value is given by:

$$V_1 = \frac{M_{33}}{a}. \qquad (10.13.19)$$

The maximum normal stress due to bending occurs at $x_3 = 0$, and is equal to

$$(\sigma_{33})_{\max} = \pm\frac{3.9M_{33}}{ta^2/6} = \pm\frac{280.8M_{33}}{a^3}. \qquad (10.13.20)$$

The maximum shearing stress due to bending occurs at $x_3 = 0$, and is equal to:

$$(\sigma_{13})_{\max.B} = \frac{18M_{33}}{a^3}. \qquad (10.13.21)$$

The maximum shearing stress due to twist occurs at $x_3 = L$, and is given by:

$$(\sigma_{13})_{\max T} = G\alpha t = \frac{M_{33}t}{J}(1 - e^{-bL}) = \frac{144M_{33}}{a^3}\left(1 - e^{-\frac{L}{3.9a}}\right). \qquad (10.13.22)$$

From the previous calculations, it is clear that the shearing stresses due to bending are negligible compared to those due to twist for long beams. Also, the maximum normal bending stresses do not occur at the same section where the maximum shearing stresses occur: For design purposes, the shearing stresses are potentially more dangerous. Therefore, in this example, the prevention of warping will affect the design inasmuch as it causes a shortening of the beam by $3.9a$ for purposes of torsional stiffness computations.

Remark

Additional examples of torsion of thin open sections can be found in [3].

APPENDIX TO CHAPTER 10

A-10.1 The Green-Riemann Formula

The Green-Riemann formula transforms a line integral taken around a closed contour C (Fig. 10.37) into a surface integral over the area A enclosed by C. It is written as follows:

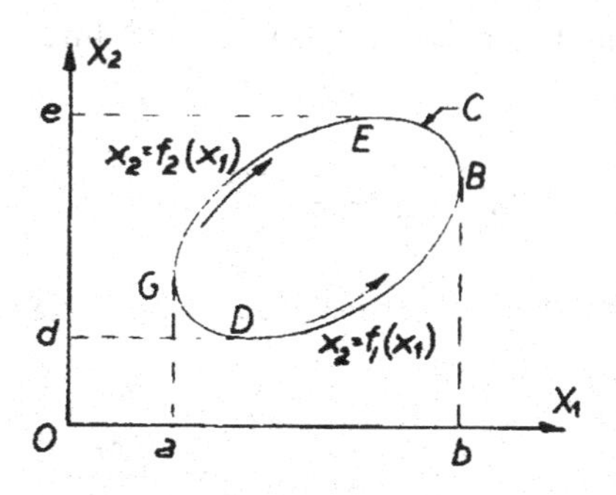

Fig. 10.37

$$\oint_C P(x_1, x_2)\, dx_1 + Q(x_1, x_2)\, dx_2 = \iint_A \left(-\frac{\partial P}{\partial x_2} + \frac{\partial Q}{\partial x_1}\right) dx_1\, dx_2 . \tag{A-10.1.1}$$

To prove this formula, let us compute the first double integral of the right-hand side of (A-10.1.1):

$$\iint_A -\frac{\partial P}{\partial x_2}\, dx_1\, dx_2 = -\int_a^b dx_1 \int_{x_2=f_1(x_1)}^{x_2=f_2(x_1)} \frac{\partial P}{\partial x_2}\, dx_2$$

or

$$\iint_A -\frac{\partial P}{\partial x^2}\, dx_1\, dx_2 = \int_a^b P[x_1, f_1(x_1)] - P[x_1, f_2(x_1)]\, dx_1$$

$$= \int_{GDB} P(x_1, x_2)\, dx_1 - \int_{GEB} P(x_1, x_2)\, dx_1 = \oint_C P(x_1, x_2)\, dx_1 .$$

In a similar way, one can prove that:

$$\iint_A \frac{\partial Q}{\partial x_1}\, dx_1\, dx_2 = \oint_C Q(x_1, x_2)\, dx_2 .$$

PROBLEMS

1. A circular shaft is made of an inner circular solid cylinder whose material has a shear modulus G_1 and an outer circular annulus whose material has a shear modulus G_2 (Fig. 10.38). The materials are perfectly bonded at the interface r_i and the shaft is subjected to a twisting moment M_z:

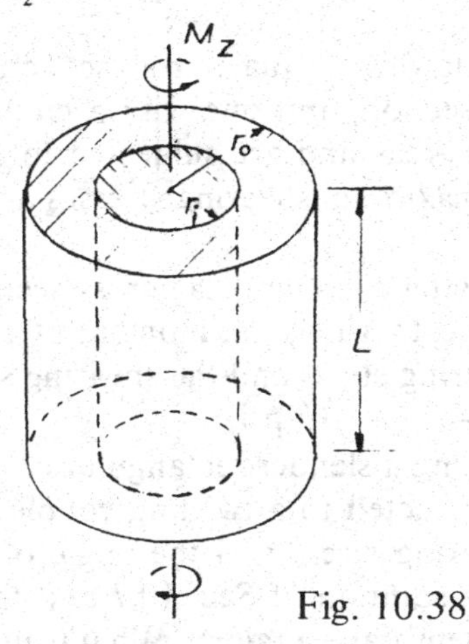

Fig. 10.38

(a) Find the expression of the angle of rotation per unit length α.
(b) Find the distribution of the shearing stresses $\sigma_{\theta z}$ in the cylinder and the annulus.
(c) How much of the total twisting moment M_z does the annulus carry?

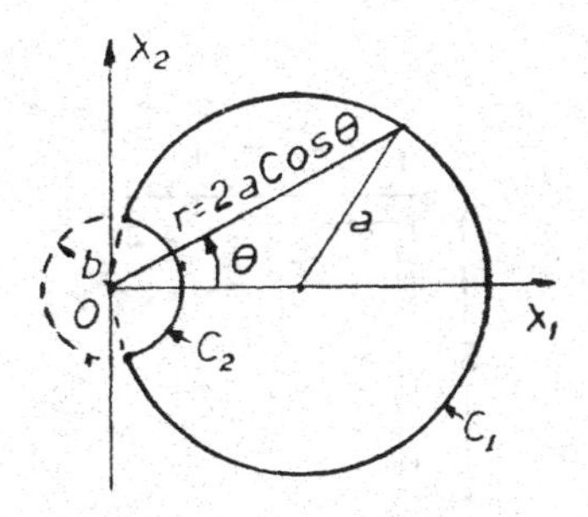

Fig. 10.39

2. Show that the Prandtl stress function

$$\phi = m(r^2 - b^2)\left(\frac{2a \cos \theta}{r} - 1\right)$$

furnishes the solution to the problem of the circular shaft with a circular groove (Fig. 10.39). Find the value of the constant m and the expressions of the stresses σ_{13} and σ_{23} on the boundaries C_1 and C_2.
3. Three bars—one with a square cross section, one with an equilateral triangle cross section, and one with a circular section—have equal cross sectional areas and are subjected to equal twisting moments. Compare the maximum shearing stresses and the torsional rigidities of the bars.
4. A steel bar having a rectangular cross section 1 in. wide and 2 in. long is subjected to a twisting moment of 1,000 lb-in. Calculate the maximum shearing stress and the shearing stress at the center of the short side. ($G = 12 \times 10^6$ psi)
5. A steel bar having a slender rectangular cross section $\frac{1}{4}$ in. wide and 6 in. long is subjected to a twisting couple of 1,500 lb-in. Find the maximum shearing stress and the angle of twist per unit length α using the exact solution of Sec. 10.7 and the approximate solution based on the membrane analogy. What is the magnitude of the error involved? ($G = 12 \times 10^6$ psi .)
6. A brass bar having the cross section shown in Fig. 10.40 is subjected to a twisting couple of 600 lb-in. This bar is 6 ft. long. What is the angle of twist per unit length and the maximum shearing stress? ($G = 4 \times 10^6$ psi .)

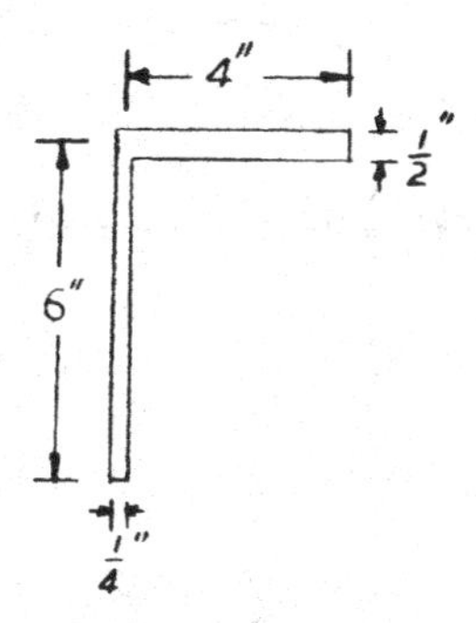

Fig. 10.40

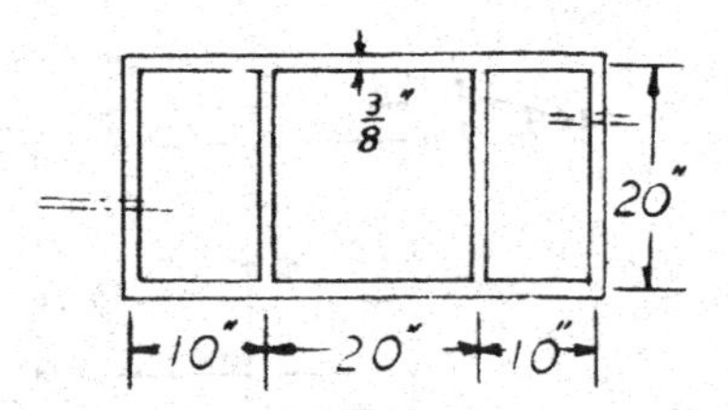

Fig. 10.41

7. A hollow cylinder having outside and inside diameters of 10 in. and 8 in., respectively, is subjected to a twisting moment M_z. Compare the angle of twist per unit length to the one the cylinder would have if it were split longitudinally along a generator.
8. A three-compartment thin-walled box section has a constant wall thickness (Fig. 10.41). Find the values of the maximum shearing stresses in the various elements if the box is subjected to a twisting moment of 1,000,000 lb-in. How much larger would the stresses in the walls of the central compartment get if two cuts were made in the walls of the side compartments? ($G = 12 \times 10^6$ psi .)
9. The section shown in Fig. 10.42 is subjected to a torque of 1,000,000 lb-in. Calculate the stresses in the different parts of the section. ($G = 12 \times 10^6$ psi .)
10. The wide-flange I beam shown in Fig. 10.43 is 20 ft. long. It is fixed at one end and free at the other. A twisting moment of 30,000 lb-in. is applied at the free end. If $E = 30 \times 10^6$ psi and $G = 12 \times 10^6$ psi,

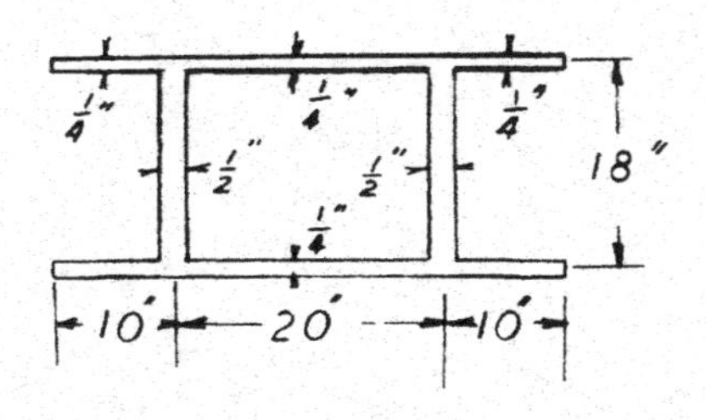

Fig. 10.42

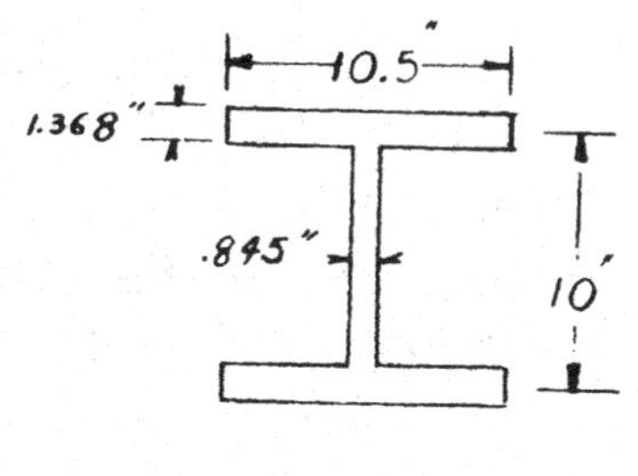

Fig. 10.43

compute the maximum normal and shearing stresses due to bending and the maximum shearing stress due to twist. What is the angle of twist of the beam at the free end? (*)

REFERENCES

[1] S. H. Crandall and N. C. Dahl, *An Introduction to the Mechanics of Solids*, McGraw-Hill, New York, N. Y., 1959.
[2] C. T. Wang, *Applied Elasticity*, McGraw-Hill, New York, N. Y., 1953.
[3] S. Timoshenko, *Strength of Materials*, Vol. 2, Van Nostrand, Princeton, N. J., 1955.
[4] S. Timoshenko and J. N. Goodier, *Theory of Elasticity*, McGraw-Hill, New York, N. Y., 1970.

* See pages A-64 to A-67 for additional problems.

CHAPTER 11

THICK CYLINDERS, DISKS, AND SPHERES

11.1 Introduction

In the study of displacement functions (Sec. 9.4), it was mentioned that particular solutions of Navier's equations are obtained by means of Lamé's strain potential ϕ; ϕ could be either a harmonic function or a function satisfying Poisson's equation. Solutions corresponding to practical boundary conditions have been generated from this potential, and it will be used in this chapter to investigate problems of cylinders, disks, and spheres. Whether a chosen $\phi(x_1, x_2, x_3)$ provides the solution to a given elasticity problem or not depends on the boundary conditions. Another method of solution we shall apply is that of Airy's stress function. The forms suggested in Sec. 9.11 are well-suited to the problems which are examined in the following sections.

11.2 Hollow Cylinder with Internal and External Pressures and Free Ends

Consider the strain function of Eq. (9.4.14).

$$\phi = C \ell n \frac{r}{K},$$

where C and K are constants. This is a harmonic function and, as such,

$$\bar{u} = \frac{1}{2G} \bar{\nabla} \phi \qquad (11.2.1)$$

is a solution of Navier's equation. The components of the deformation are given by Eqs. (6.4.27):

$$u_r = \frac{1}{2G}\frac{C}{r}, \quad u_\theta = 0, \quad u_z = 0. \tag{11.2.2}$$

The components of the state of stress are given by:

$$\sigma_{rr} = -\frac{C}{r^2} \quad \sigma_{\theta\theta} = \frac{C}{r^2} \quad \sigma_{zz} = 0 \tag{11.2.3}$$

$$\sigma_{r\theta} = 0 \quad \sigma_{\theta z} = 0 \quad \sigma_{rz} = 0. \tag{11.2.4}$$

Let a state of stress, in which $\sigma_{rr} = \sigma_{\theta\theta} = D$ and all other components are equal to zero, be superimposed to the state of stress of Eqs. (11.2.3) and (11.2.4), so that:

$$\sigma_{rr} = -\frac{C}{r^2} + D, \quad \sigma_{\theta\theta} = \frac{C}{r^2} + D, \quad \sigma_{zz} = 0 \tag{11.2.5}$$

$$\sigma_{r\theta} = 0 \quad \sigma_{\theta z} = 0 \quad \sigma_{rz} = 0. \tag{11.2.6}$$

This superposition of stress will add a strain

$$e_{rr} = e_{\theta\theta} = (1 - \nu)D/2\ (1 + \nu)G$$

and $e_{zz} = -\nu D/G(1 + \nu)$ to the original one. Consequently, the quantities $(1 - \nu)Dr/(1 + \nu)2G$ and $-\nu Dz/G(1 + \nu)$ will be added to u_r and u_z, respectively. The equations for the displacements become:

$$u_r = \frac{r}{2G}\left[\frac{C}{r^2} + \frac{(1 - \nu)}{(1 + \nu)}D\right], \quad u_\theta = 0, \quad u_z = -\frac{\nu Dz}{G(1 + \nu)}. \tag{11.2.7}$$

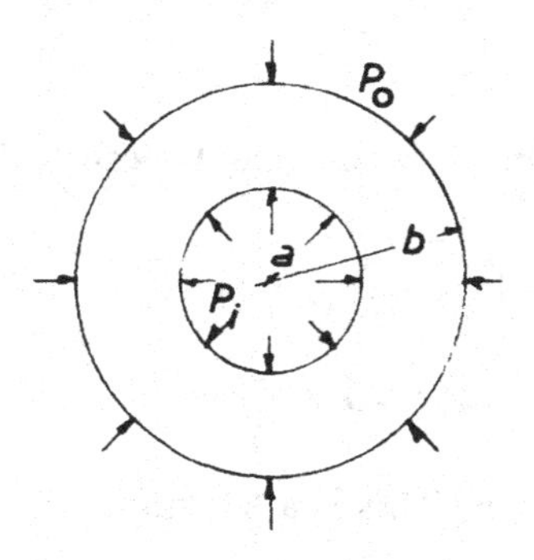

Fig. 11.1

Now consider a hollow cylinder (Fig. 11.1) with inner radius $r = a$, outer radius $r = b$, and free ends. Let a pressure P_i be applied on the inner surface, and a pressure P_o be applied on the outer surface. Eqs. (11.2.5) to (11.2.7) apply to this problem since they are such that σ_{rr} is constant for $r = a$ and $r = b$, and σ_{zz} is equal to zero; they also define values of $\sigma_{r\theta}$, $\sigma_{\theta z}$, σ_{rz}, u_θ equal to zero, and a value of e_{zz} equal to a constant as required by symmetry in the cylinder problem. The constants C and D can be obtained by setting $\sigma_{rr} = -P_i$ for $r = a$, and $\sigma_{rr} = -P_o$ for $r = b$ in Eq. (11.2.5):

$$-P_i = -\frac{C}{a^2} + D, \quad -P_o = -\frac{C}{b^2} + D. \tag{11.2.8}$$

Therefore,

$$D = \frac{P_i a^2 - P_o b^2}{b^2 - a^2}, \quad C = \frac{a^2 b^2 (P_o - P_i)}{a^2 - b^2}. \tag{11.2.9}$$

These values are now substituted into Eqs. (11.2.5) to (11.2.7). The only components of the state of stress σ_{rr} and $\sigma_{\theta\theta}$ become:

$$\begin{aligned}\sigma_{rr} &= \frac{P_i a^2 - P_o b^2}{b^2 - a^2} + \frac{1}{r^2}\frac{a^2 b^2 (P_o - P_i)}{b^2 - a^2} \\ &= -P_i \frac{\left(\frac{b}{r}\right)^2 - 1}{\left(\frac{b}{a}\right)^2 - 1} - P_o \frac{1 - \left(\frac{a}{r}\right)^2}{1 - \left(\frac{a}{b}\right)^2}\end{aligned} \tag{11.2.10}$$

$$\begin{aligned}\sigma_{\theta\theta} &= \frac{P_i a^2 - P_o b^2}{b^2 - a^2} - \frac{1}{r^2}\frac{a^2 b^2 (P_o - P_i)}{b^2 - a^2} \\ &= P_i \frac{\left(\frac{b}{r}\right)^2 + 1}{\left(\frac{b}{a}\right)^2 - 1} - P_o \frac{1 + \left(\frac{a}{r}\right)^2}{1 - \left(\frac{a}{b}\right)^2}.\end{aligned} \tag{11.2.11}$$

These are Lamé's formulas for the stresses. $\sigma_{rr} + \sigma_{\theta\theta}$ is a constant throughout the cylinder. The components of the displacement become:

$$u_r = \frac{1 - \nu}{E}\frac{a^2 P_i - b^2 P_o}{b^2 - a^2} r + \frac{1 + \nu}{E}\frac{a^2 b^2 (P_i - P_o)}{(b^2 - a^2)}\frac{1}{r} \tag{11.2.12}$$

$$= \frac{r}{2G}\left[P_i \frac{\left(\frac{b}{r}\right)^2 + \frac{1-\nu}{1+\nu}}{\left(\frac{b}{a}\right)^2 - 1} - P_o \frac{\frac{1-\nu}{1+\nu} + \left(\frac{a}{r}\right)^2}{1 - \left(\frac{a}{b}\right)^2}\right] \tag{11.2.13}$$

$$u_\theta = 0, \quad u_z = \frac{2\nu}{E}\frac{P_i a^2 - P_o b^2}{a^2 - b^2} z. \tag{11.2.14}$$

Eqs. (11.2.3) to (11.2.6) show that this is a case of plane stress, even though the cylinder is not thin.

If the outer radius b is much larger than the inner radius a, then

$$\begin{aligned} \sigma_{rr} &= -P_i\left(\frac{a}{r}\right)^2 - P_o\left[1 - \left(\frac{a}{r}\right)^2\right], \\ \sigma_{\theta\theta} &= P_i\left(\frac{a}{r}\right)^2 - P_o\left[1 + \left(\frac{a}{r}\right)^2\right] \end{aligned} \tag{11.2.15}$$

$$u_r = \frac{r}{2G}\left\{P_i\left(\frac{a}{r}\right)^2 - P_o\left[\frac{1-\nu}{1+\nu} + \left(\frac{a}{r}\right)^2\right]\right\}. \tag{11.2.16}$$

At the inner surface $r = a$, and Eqs. (11.2.15) and (11.2.16) give:

$$\sigma_{rr} = -P_i, \quad \sigma_{\theta\theta} = P_i - 2P_o, \quad u_r = \frac{a}{2G}\left(P_i - \frac{2}{1+\nu}P_o\right). \tag{11.2.17}$$

If r is very large, Eqs. (11.2.15) and (11.2.16) give:

$$\sigma_{rr} = \sigma_{\theta\theta} = -P_o, \quad u_r = -\frac{(1-\nu)}{2(1+\nu)}\frac{P_o r}{G}. \tag{11.2.18}$$

Eqs. (11.2.18) show that if values of r are kept large at the outer surface compared to the inner radius a, the outer surface may be replaced by any prismatic surface parallel to the OX_3 or Z axis and subjected to a constant pressure P_o. For example, the cross section of the outer surface may be a rectangle with edges parallel to the OX_1 and OX_2 axes with pressures $\sigma_{11} = \sigma_{22} = -P_o$. If the inside pressure $P_i = 0$, Eq. (11.2.17) shows that $\sigma_{\theta\theta} = -2P_o$, twice the pressure that would exist without the hole.

If $P_o = 0$, Eqs. (11.2.10) and (11.2.11) give:

$$\sigma_{rr} = \frac{a^2 P_i}{b^2 - a^2}\left(1 - \frac{b^2}{r^2}\right) \quad \text{(always compressive)} \tag{11.2.19}$$

$$\sigma_{\theta\theta} = \frac{a^2 P_i}{b^2 - a^2}\left(1 + \frac{b^2}{r^2}\right) \quad \text{(always tensile).} \tag{11.2.20}$$

Therefore,

$$(\sigma_{\theta\theta})_{\max} = (\sigma_{\theta\theta})_{r=a} = \frac{P_i(a^2 + b^2)}{(b^2 - a^2)} \quad \text{(always larger than } P_i\text{)} \tag{11.2.21}$$

and

$$(\sigma_{\theta\theta})_{\min} = (\sigma_{\theta\theta})_{r=b} = \frac{2a^2 P_i}{b^2 - a^2}. \tag{11.2.22}$$

If $b - a$ is very small, a good approximation (Fig. 11.2) is given by:

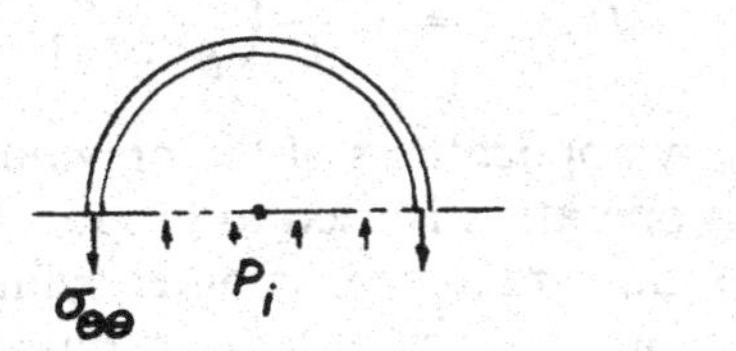

Fig. 11.2

$$\sigma_{\theta\theta} = \frac{aP_i}{b - a}.$$

The maximum shearing stress occurs at the inner surface and is given by:

$$(\sigma_t)_{\max} = \left(\frac{\sigma_{\theta\theta} - \sigma_{rr}}{2}\right)_{r=a} = \frac{P_i b^2}{b^2 - a^2}. \tag{11.2.23}$$

The radial displacement at $r = a$ is obtained from Eq. (11.2.12)

$$(u_r)_{r=a} = \frac{aP_i}{E}\left(\frac{a^2 + b^2}{b^2 - a^2} + \nu\right) \tag{11.2.24}$$

If $P_i = 0$, Eqs. (11.2.10) and (11.2.11) give:

$$\sigma_{rr} = -\frac{P_o b^2}{b^2 - a^2}\left(1 - \frac{a^2}{r^2}\right) \quad \text{(always compressive)} \tag{11.2.25}$$

$$\sigma_{\theta\theta} = -\frac{P_o b^2}{b^2 - a^2}\left(1 + \frac{a^2}{r^2}\right) \tag{11.2.26}$$

(always compressive and larger in magnitude than σ_{rr}).

Therefore,

$$(\sigma_{\theta\theta})_{\max} = (\sigma_{\theta\theta})_{r=a} = -\frac{2P_o b^2}{b^2 - a^2}, \tag{11.2.27}$$

and when b/a increases, the maximum compressive stress tends to twice the external pressure. The radial displacement at $r = b$ is obtained from Eq. (11.2.12):

$$(u_r)_{r=b} = -\frac{bP_o}{E}\left(\frac{a^2 + b^2}{b^2 - a^2} - \nu\right). \tag{11.2.28}$$

One of the many applications of the previous equations is in the area of *shrink fit*. This operation is used to produce a contact between a hub and a shaft: The inner radius of an outer cylinder is made smaller than the outer radius of an inner cylinder by a quantity δ. The outer cylinder is then heated and, upon cooling, a contact pressure is applied between the two parts. The increase in the inner radius of the outer cylinder added to the decrease in the outer radius of the inner cylinder must be equal to δ. Eqs. (11.2.24) and (11.2.28) give (Fig. 11.3):

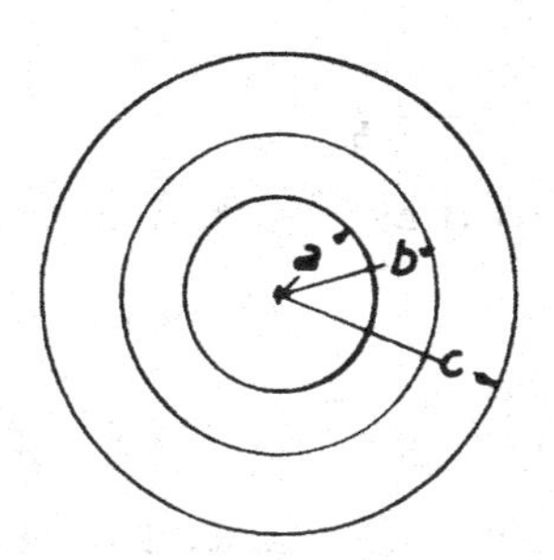

Fig. 11.3

$$\frac{bp}{E}\left(\frac{b^2 + c^2}{c^2 - b^2} + \nu\right) + \frac{bp}{E}\left(\frac{a^2 + b^2}{b^2 - a^2} - \nu\right) = \delta, \tag{11.2.29}$$

where p is the pressure between the two parts. Therefore,

$$p = \frac{E\delta}{b}\frac{(b^2 - a^2)(c^2 - b^2)}{2b^2(c^2 - a^2)}. \tag{11.2.30}$$

Knowing p, one can get the stresses at each point of the outer and inner ring. The highest stress occurs at the inner surface of the outer cylinder:

$$(\sigma_{\theta\theta})_{r=b} = \frac{p(b^2 + c^2)}{c^2 - b^2}, \quad \sigma_{rr} = -p. \tag{11.2.31}$$

The maximum shearing stress at this surface is:

$$(\sigma_t)_{\max} = \frac{pc^2}{c^2 - b^2} = \frac{E\delta c^2(b^2 - a^2)}{2b^3(c^2 - a^2)}. \tag{11.2.32}$$

If the two cylinders are different in length, stress concentrations will occur at the edges of the shorter cylinder (Fig. 11.4).

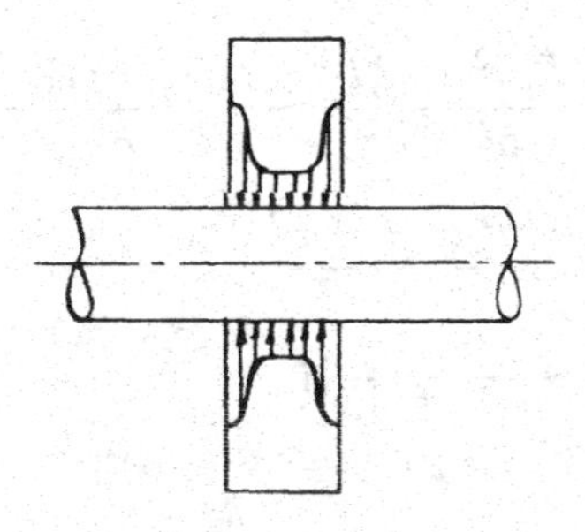

Fig. 11.4

Remark

The solution of the hollow cylinder problem in terms of Airy's stress function can be found in references [1] and [2].

11.3 Hollow Cylinder with Internal and External Pressures and Fixed Ends

The problem in this case is a problem of plane strain (Fig. 11.5). The two components of the state of stress σ_{rr} and $\sigma_{\theta\theta}$ keep the same value as in the previous case, while σ_{zz} becomes equal to $2\nu D$. As a result, e_{rr} is decreased by $2\nu^2 D/E$ and the u_r by $2\nu^2 Dr/E$. u_z and e_{zz} are equal to zero. Therefore,

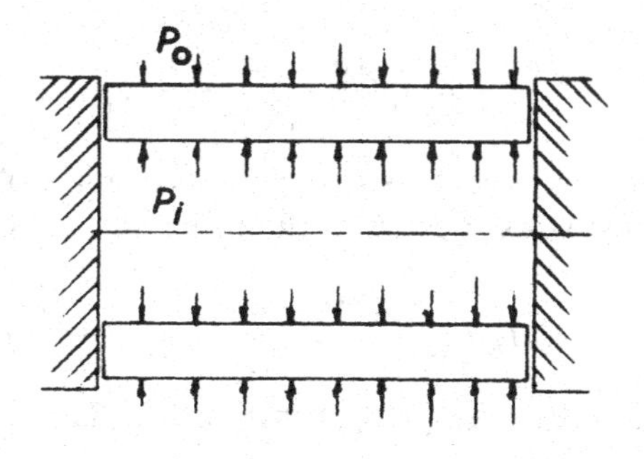

Fig. 11.5

$$\sigma_{rr} = \frac{P_i a^2 - P_o b^2}{b^2 - a^2} + \frac{1}{r^2}\frac{a^2 b^2 (P_o - P_i)}{b^2 - a^2} \tag{11.3.1}$$

$$\sigma_{\theta\theta} = \frac{P_i a^2 - P_o b^2}{b^2 - a^2} - \frac{1}{r^2}\frac{a^2 b^2 (P_o - P_i)}{b^2 - a^2} \tag{11.3.2}$$

$$\sigma_{zz} = 2\nu \frac{P_i a^2 - P_o b^2}{b^2 - a^2} \tag{11.3.3}$$

$$\sigma_{r\theta} = \sigma_{rz} = \sigma_{\theta z} = 0 \tag{11.3.4}$$

and

$$u_r = \frac{r}{2G}\left[\frac{a^2 b^2 (P_o - P_i)}{a^2 - b^2}\frac{1}{r^2} + \frac{P_i a^2 - P_o b^2}{b^2 - a^2}(1 - 2\nu)\right] \tag{11.3.5}$$

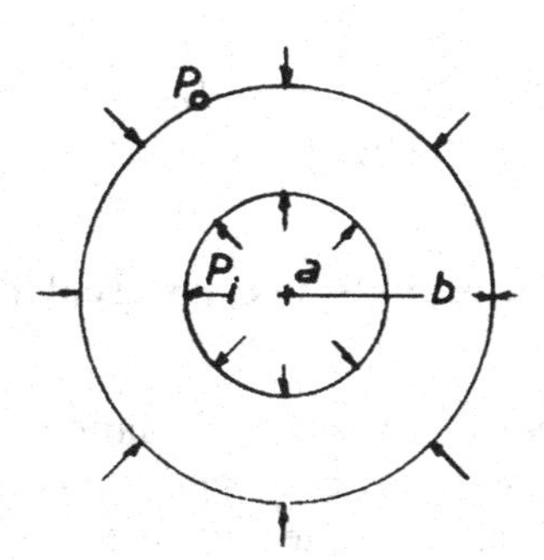

Fig. 11.6

$$u_\theta = 0, \quad u_z = 0. \tag{11.3.6}$$

Eqs. (11.3.1) to (11.3.6) could have been obtained by starting directly from the strain potential

$$\phi = C \ln \frac{r}{K} + Hr^2, \tag{11.3.7}$$

where C, K, and H are constants. Eq. (11.3.7) is a combination of Eqs. (9.4.14) and (9.4.18).

11.4 Hollow Sphere Subjected to Internal and External Pressures

Let a hollow sphere of inner radius a and outer radius b be subjected to an internal pressure P_i and an external pressure P_o (Fig. 11.6). Because of the spherical symmetry, it is advantageous to use spherical polar coordinates (ϕ, θ, ρ). In such a system of coordinates, all the shear stresses and shear strains vanish. Of the three components of the displacement vector u_ϕ, u_θ, u_ρ, only u_ρ is different from zero.

The solution to this problem is furnished by the combination of the two functions given in Eqs. (9.4.16) and (9.4.19), namely by:

$$\phi = \frac{C}{\rho} + D\rho^2. \tag{11.4.1}$$

Indeed, ϕ satisfies Poisson's equation and gives stresses and deformations fulfilling all the symmetry conditions. The displacements are given by:

$$2Gu_\rho = -\frac{C}{\rho^2} + 2D\rho \tag{11.4.2}$$

$$u_\theta = u_\phi = 0. \tag{11.4.3}$$

Eqs. (6.7.34) and (6.7.35) give:

$$e_{\theta\theta} = e_{\phi\phi} = \frac{1}{2G}\left(-\frac{C}{\rho^3} + 2D\right) \tag{11.4.4}$$

$$e_{\rho\rho} = \frac{1}{2G}\left(\frac{2C}{\rho^3} + 2D\right) \tag{11.4.5}$$

$$e_{\theta\phi} = e_{\theta\rho} = e_{\phi\rho} = 0. \tag{11.4.6}$$

The stresses are given by:

$$\sigma_{\theta\theta} = \sigma_{\phi\phi} = -\frac{C}{\rho^3} + 2\frac{1+\nu}{1-2\nu}D \tag{11.4.7}$$

$$\sigma_{\rho\rho} = \frac{2C}{\rho^3} + 2\frac{1+\nu}{1-2\nu}D. \tag{11.4.8}$$

The constants C and D are obtained by setting $\sigma_{\rho\rho} = -P_i$ for $\rho = a$ and $\sigma_{\rho\rho} = -P_o$ for $\rho = b$ in Eq. (11.4.8):

$$-P_i = \frac{2C}{a^3} + 2\frac{1+\nu}{1-2\nu}D \tag{11.4.9}$$

$$-P_o = \frac{2C}{b^3} + 2\frac{1+\nu}{1-2\nu}D. \tag{11.4.10}$$

Therefore,

$$2C = \frac{a^3 b^3 (P_o - P_i)}{b^3 - a^3} \tag{11.4.11}$$

$$D = \frac{1-2\nu}{2(1+\nu)}\frac{a^3 P_i - b^3 P_o}{b^3 - a^3}. \tag{11.4.12}$$

These values are now substituted into Eqs. (11.4.7) and (11.4.8):

$$\sigma_{\theta\theta} = \sigma_{\phi\phi} = -\frac{a^3 b^3 (P_o - P_i)}{2(b^3 - a^3)}\frac{1}{\rho^3} + \frac{a^3 P_i - b^3 P_o}{b^3 - a^3}$$

$$= \frac{P_i\left[\left(\frac{b}{\rho}\right)^3 + 2\right]}{2\left[\left(\frac{b}{a}\right)^3 - 1\right]} - \frac{P_o\left[\left(\frac{a}{\rho}\right)^3 + 2\right]}{2\left[1 - \left(\frac{a}{b}\right)^3\right]} \tag{11.4.13}$$

$$\sigma_{\rho\rho} = \frac{a^3 b^3 (P_o - P_i)}{b^3 - a^3}\frac{1}{\rho^3} + \frac{a^3 P_i - b^3 P_o}{b^3 - a^3}$$

$$= -\frac{P_i\left[\left(\frac{b}{\rho}\right)^3 - 1\right]}{\left(\frac{b}{a}\right)^3 - 1} - \frac{P_o\left[1 - \left(\frac{a}{\rho}\right)^3\right]}{1 - \left(\frac{a}{b}\right)^3}. \tag{11.4.14}$$

These are Lamé's formulas for the stresses. The components of the displacement become:

$$u_\rho = \frac{\rho}{2G}\left[-\frac{a^3b^3(P_o - P_i)}{2(b^3 - a^3)}\frac{1}{\rho^3} + \frac{(1-2\nu)}{(1+\nu)}\frac{a^3P_i - b^3P_o}{b^3 - a^3}\right] \tag{11.4.15}$$

$$= \frac{\rho}{2G}\left[P_i\frac{\frac{1}{2}\left(\frac{b}{\rho}\right)^3 + \frac{1-2\nu}{1+\nu}}{\left(\frac{b}{a}\right)^3 - 1} - P_o\frac{\frac{1-2\nu}{1+\nu} + \frac{1}{2}\left(\frac{a}{\rho}\right)^3}{1 - \left(\frac{a}{b}\right)^3}\right]. \tag{11.4.16}$$

$u_\phi = u_\theta = 0$

If b is large compared to a, we get:

$$\sigma_{\phi\phi} = \sigma_{\theta\theta} = \frac{P_i}{2}\left(\frac{a}{\rho}\right)^3 - \frac{P_o}{2}\left[\left(\frac{a}{\rho}\right)^3 + 2\right] \tag{11.4.17}$$

$$\sigma_{\rho\rho} = -P_i\left(\frac{a}{\rho}\right)^3 - P_o\left[1 - \left(\frac{a}{\rho}\right)^2\right] \tag{11.4.18}$$

$$u_\rho = \frac{\rho}{2G}\left\{\frac{P_i}{2}\left(\frac{a}{\rho}\right)^3 - P_o\left[\frac{1-2\nu}{1+\nu} + \frac{1}{2}\left(\frac{a}{\rho}\right)^3\right]\right\}. \tag{11.4.19}$$

At the inner surface $\rho = a$, and Eqs. (11.4.17) and (11.4.19) give:

$$(\sigma_{\theta\theta})_{\rho=a} = (\sigma_{\phi\phi})_{\rho=a} = \frac{P_i}{2} - \frac{3P_o}{2}, \quad (\sigma_{\rho\rho})_{\rho=a} = -P_i,$$
$$(u_\rho)_{\rho=a} = \frac{\rho}{2G}\left[\frac{P_i}{2} - P_o\frac{3(1-\nu)}{2(1+\nu)}\right]. \tag{11.4.20}$$

If ρ is very large, Eqs. (11.4.17) to (11.4.19) give:

$$\sigma_{\phi\phi} = \sigma_{\theta\theta} = \sigma_{\rho\rho} = -P_o, \quad u_\rho = -\frac{1-2\nu}{1+\nu}\frac{\rho P_o}{2G}. \tag{11.4.21}$$

Eqs. (11.4.21) show that if values of ρ are kept large at the outer surface compared to the inner radius a, the outer surface may be replaced by any surface provided the uniform external pressure P_o is maintained. For example, the outer surface may be a cube with edges parallel to the OX_1, OX_2, and OX_3 axes and with pressures $\sigma_{11} = \sigma_{22} = \sigma_{33} = -P_o$. If the inside pressure $P_i = 0$, Eqs. (11.4.20) show that $\sigma_{\theta\theta} = \sigma_{\phi\phi} = -3P_o/2$, $\frac{3}{2}$ times the pressure that would exist if there was no spherical cavity.

If $P_o = 0$, Eqs. (11.4.13) and (11.4.14) give:

$$\sigma_{\theta\theta} = \sigma_{\phi\phi} = \frac{P_ia^3}{b^3 - a^3}\left(1 + \frac{b^3}{2\rho^3}\right) > 0 \tag{11.4.22}$$

$$\sigma_{\rho\rho} = \frac{P_i a^3}{b^3 - a^3}\left(1 - \frac{b^3}{\rho^3}\right) \leqslant 0. \tag{11.4.23}$$

Therefore,

$$(\sigma_{\theta\theta})_{\max} = (\sigma_{\theta\theta})_{\rho=a} = \frac{P_i}{2}\frac{2a^3 + b^3}{b^3 - a^3}. \tag{11.4.24}$$

If $b - a$ is small, a good approximation for $\sigma_{\theta\theta}$ is given by:

$$\sigma_{\theta\theta} = \frac{P_i a}{2(b - a)}. \tag{11.4.25}$$

Similar equations can be written for $P_i = 0$.

11.5 Rotating Disks of Uniform Thickness

The problem of the stresses and deformations in disks rotating at high speed is of primary importance in the design of a wide variety of machines. Let us consider a thin disk of uniform thickness rotating with a constant angular velocity ω rad/sec. The body force is the centrifugal force (Fig. 11.7):

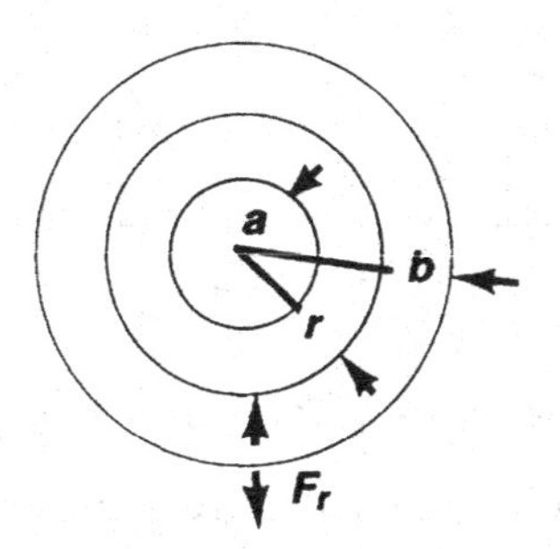

Fig. 11.7

$$F_r = \rho\omega^2 r, \tag{11.5.1}$$

where ρ is the mass per unit volume of the material of the disk. Although the method of Lamé's strain potential can be slightly modified and applied to this case [3], we shall use a stress function in seeking a solution to the problem. For a thin disk, the equations related to a state of plane stress in Sec. 9.10 apply. The force F_r derives from a potential,

$$\Omega = -\frac{\rho\omega^2 r^2}{2}. \tag{11.5.2}$$

Because of circular symmetry, the equilibrium equation (9.10.13) becomes:

$$\frac{d\sigma_{rr}}{dr} + \frac{1}{r}(\sigma_{rr} - \sigma_{\theta\theta}) + \rho\omega^2 r = 0 \tag{11.5.3}$$

or

$$\frac{d(r\sigma_{rr})}{dr} - \sigma_{\theta\theta} + \rho\omega^2 r^2 = 0, \tag{11.5.4}$$

while Eq. (9.10.14) is identically satisfied. Any stress function ϕ will be a function of r alone. It is easy to verify that the stress function defined by

$$r\sigma_{rr} = \phi, \quad \sigma_{\theta\theta} = \frac{d\phi}{dr} + \rho\omega^2 r^2, \tag{11.5.5}$$

satisfies the equation of equilibrium (11.5.4). Eqs. (6.7.23) give:

$$e_{rr} = \frac{du_r}{dr}, \quad e_{\theta\theta} = \frac{u_r}{r}, \tag{11.5.6}$$

and a simplified compatibility relation is obtained by eliminating u_r from Eqs. (11.5.6):

$$\frac{d}{dr}(re_{\theta\theta}) - e_{rr} = 0. \tag{11.5.7}$$

Using the stress-strain relations (8.17.11) and (8.17.12), and the stress function defined by Eqs. (11.5.5), Eq. (11.5.7) becomes:

$$\frac{d}{dr}\left[\frac{1}{r}\frac{d}{dr}(r\phi)\right] = -(3 + \nu)\rho\omega^2 r. \tag{11.5.8}$$

By direct integration, we get:

$$\phi = -\frac{3 + \nu}{8}\rho\omega^2 r^3 + C_1\frac{r}{2} + C_2\frac{1}{r}, \tag{11.5.9}$$

where C_1 and C_2 are constants of integration. The corresponding stress components are:

$$\sigma_{rr} = \frac{\phi}{r} = -\frac{3 + \nu}{8}\rho\omega^2 r^2 + \frac{C_1}{2} + \frac{C_2}{r^2} \tag{11.5.10}$$

$$\sigma_{\theta\theta} = \frac{d\phi}{dr} + \rho\omega^2 r^2 = -\frac{1 + 3\nu}{8}\rho\omega^2 r^2 + \frac{C_1}{2} - \frac{C_2}{r^2}. \tag{11.5.11}$$

The constants C_1 and C_2 are obtained from the boundary conditions:

a. *Solid disks*

The stress cannot be infinite at $r = 0$ so that $C_2 = 0$. At $r = b$, $\sigma_{rr} = 0$, so that $C_1 = [(3 + \nu)/4]\rho\omega^2 b^2$. Therefore (Fig. 11.8),

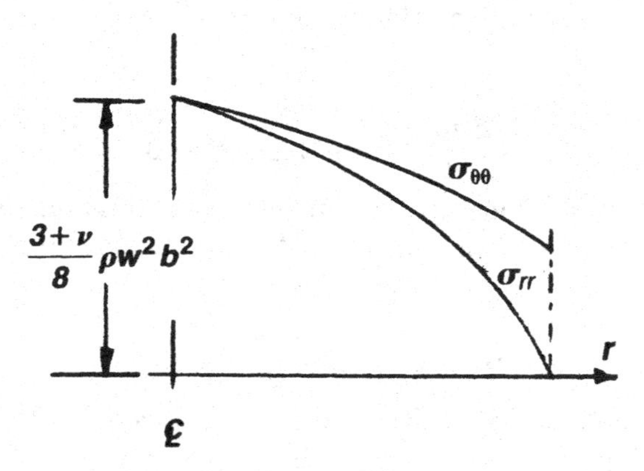

Fig. 11.8

$$\sigma_{rr} = \frac{3 + \nu}{8}\rho\omega^2 b^2\left(1 - \frac{r^2}{b^2}\right) \tag{11.5.12}$$

$$\sigma_{\theta\theta} = \frac{3 + \nu}{8}\rho\omega^2 b^2\left(1 - \frac{1 + 3\nu}{3 + \nu}\frac{r^2}{b^2}\right). \tag{11.5.13}$$

The maximum stress is at the center, where

$$\sigma_{rr} = \sigma_{\theta\theta} = \frac{3 + \nu}{8}\rho\omega^2 b^2. \tag{11.5.14}$$

The radial displacement u_r is given by:

$$\begin{aligned} u_r = r e_{\theta\theta} &= \frac{r}{E}(\sigma_{\theta\theta} - \nu\sigma_{rr}) \\ &= \frac{\rho\omega^2 r(\nu - 1)}{8E}[r^2(\nu + 1) - b^2(\nu + 3)]. \end{aligned} \tag{11.5.15}$$

b. *Disk with a circular hole of radius a*

At $r = a$, and at $r = b$, $\sigma_{rr} = 0$. Therefore,

$$C_1 = \frac{3 + \nu}{4}\rho\omega^2(b^2 + a^2), \quad C_2 = -\frac{3 + \nu}{8}\rho\omega^2 a^2 b^2 \tag{11.5.16}$$

and

$$\sigma_{rr} = \frac{3+\nu}{8}\rho\omega^2\left(b^2 + a^2 - \frac{a^2b^2}{r^2} - r^2\right) \tag{11.5.17}$$

$$\sigma_{\theta\theta} = \frac{3+\nu}{8}\rho\omega^2\left(b^2 + a^2 + \frac{a^2b^2}{r^2} - \frac{1+3\nu}{3+\nu}r^2\right). \tag{11.5.18}$$

The maximum stress occurs at the inner boundary, where

$$\sigma_{\theta\theta} = \frac{3+\nu}{4}\rho\omega^2b^2\left(1 + \frac{1-\nu}{3+\nu}\frac{a^2}{b^2}\right). \tag{11.5.19}$$

If the circular hole is very small, Eq. (11.5.19) becomes:

$$\sigma_{\theta\theta} = \frac{3+\nu}{4}\rho\omega^2b^2, \tag{11.5.20}$$

which is twice the value of the stress at the center of a solid disk [Eq. (11.5.14)]. Therefore, by making a small circular hole at the center of a rotating disk, we shall double the maximum stress. The radial displacement u_r is given by:

$$\begin{aligned} u_r = re_{\theta\theta} &= \frac{r}{E}[\sigma_{\theta\theta} - \nu\sigma_{rr}] \\ &= \frac{\rho\omega^2 r(3+\nu)(1-\nu)}{8E}\left(b^2 + a^2 + \frac{1+\nu}{1-\nu}\frac{b^2a^2}{r^2} - \frac{1+\nu}{3+\nu}r^2\right). \end{aligned} \tag{11.5.21}$$

Remarks

a) The solution of the problem could have been as easily obtained by solving Navier's equations and getting u_r.

b) The approximate nature of the plane stress solution must be kept in mind. Indeed, of the six compatibility relations (6.9.1) to (6.9.6) three are identically satisfied—namely, Eqs. (6.9.4) to (6.9.6); Eqs. (6.9.1) to (6.9.3) for the special case of this problem become:

$$\frac{d}{dr}\left[\frac{d}{dr}(re_{\theta\theta}) - e_{rr}\right] = 0 \tag{11.5.22}$$

$$\frac{1}{r}\frac{\partial e_{zz}}{\partial r} = 0, \quad \frac{\partial^2 e_{zz}}{\partial r^2} = 0. \tag{11.5.23}$$

While Eq. (11.5.22) has been satisfied, Eqs. (11.5.23) have not been considered in the solution.

11.6 Rotating Long Circular Cylinder

We shall examine two cases:

Case a. The rotating cylinder is not free to deform longitudinally
The problem is a plane strain problem. The stress strain relations (8.16.12) to (8.16.14) are the ones to be applied. Using the stress function defined in Eqs. (11.5.5), the compatibility relation (11.5.7) becomes:

$$\frac{d}{dr}\left[\frac{1}{r}\frac{d}{dr}(r\phi)\right] = -\frac{3-2\nu}{1-\nu}\rho\omega^2 r. \qquad (11.6.1)$$

This equation is seen to be different from its counterpart of Sec. 11.5. Direct integration gives:

$$\phi = -\frac{(3-2\nu)}{8(1-\nu)}\rho\omega^2 r^3 + \frac{C_1}{2}r + \frac{C_2}{r}. \qquad (11.6.2)$$

The stress components are:

$$\sigma_{rr} = \frac{\phi}{r} = -\frac{3-2\nu}{8(1-\nu)}\rho\omega^2 r^2 + \frac{C_1}{2} + \frac{C_2}{r^2} \qquad (11.6.3)$$

$$\sigma_{\theta\theta} = \frac{d\phi}{dr} + \rho\omega^2 r^2 = -\frac{1+2\nu}{8(1-\nu)}\rho\omega^2 r^2 + \frac{C_1}{2} - \frac{C_2}{r^2}. \qquad (11.6.4)$$

The constants of integration are now obtained in the same manner as in the case of the disk. Thus:

For a solid cylinder of radius b, we have:

$$\sigma_{rr} = \frac{3-2\nu}{8(1-\nu)}\rho\omega^2 b^2\left(1 - \frac{r^2}{b^2}\right) \qquad (11.6.5)$$

$$\sigma_{\theta\theta} = \frac{3-2\nu}{8(1-\nu)}\rho\omega^2 b^2\left(1 - \frac{1+2\nu}{3-2\nu}\frac{r^2}{b^2}\right) \qquad (11.6.6)$$

$$\sigma_{zz} = \nu(\sigma_{rr} + \sigma_{\theta\theta}) = \frac{\nu\rho\omega^2}{4(1-\nu)}[(3-2\nu)b^2 - 2r^2], \qquad (11.6.7)$$

and the maximum stress occurs at the center, and is

$$\sigma_{rr} = \sigma_{\theta\theta} = \frac{3-2\nu}{8(1-\nu)}\rho\omega^2 b^2. \qquad (11.6.8)$$

For a hollow cylinder of inner radius a and outer radius b, we have:

$$\sigma_{rr} = \frac{3 - 2\nu}{8(1 - \nu)} \rho\omega^2 \left(b^2 + a^2 - \frac{a^2 b^2}{r^2} - r^2 \right) \tag{11.6.9}$$

$$\sigma_{\theta\theta} = \frac{3 - 2\nu}{8(1 - \nu)} \rho\omega^2 \left(b^2 + a^2 + \frac{a^2 b^2}{r^2} - \frac{1 + 2\nu}{3 - 2\nu} r^2 \right) \tag{11.6.10}$$

$$\sigma_{zz} = \frac{3 - 2\nu}{4(1 - \nu)} \nu\rho\omega^2 \left(b^2 + a^2 - \frac{2r^2}{3 - 2\nu} \right), \tag{11.6.11}$$

and the maximum stress occurs at the inner surface, and is

$$\sigma_{\theta\theta} = \frac{3 - 2\nu}{4(1 - \nu)} \rho\omega^2 b^2 \left(1 + \frac{1 - 2\nu}{3 - 2\nu} \frac{a^2}{b^2} \right). \tag{11.6.12}$$

Here, too, we see that the maximum stress is doubled when a solid cylinder has a small hole drilled through its center. It is to be noticed that, for the previous solution to be valid, σ_{zz} must act at the ends of the cylinder according to Eqs. (11.6.7) and (11.6.11).

Case b. The rotating cylinder is free to deform longitudinally

In the above discussion, the values of the axial stress σ_{zz} were so adjusted that there were no longitudinal strains e_{zz} and, consequently, no longitudinal deformations u_z. If the cylinder is now allowed to deform freely longitudinally, then because of symmetry the strain e_{zz} must be such that every cross section remains plane. The cylinder is in a state of generalized plane strain, i.e., the radial and transverse stresses will not change while the cylinder is changing length uniformly:

For a solid cylinder:

$$\sigma_{rr} = \frac{3 - 2\nu}{8(1 - \nu)} \rho\omega^2 b^2 \left(1 - \frac{r^2}{b^2} \right) \tag{11.6.13}$$

$$\sigma_{\theta\theta} = \frac{3 - 2\nu}{8(1 - \nu)} \rho\omega^2 b^2 \left(1 - \frac{1 + 2\nu}{3 - 2\nu} \frac{r^2}{b^2} \right) \tag{11.6.14}$$

$$\sigma_{zz} = \frac{\nu\rho\omega^2}{4(1 - \nu)} [(3 - 2\nu)b^2 - 2r^2] + Ee_{zz}. \tag{11.6.15}$$

The value of e_{zz} can be obtained from the condition that there is no resultant longitudinal force on the ends. Hence,

$$\int_0^b 2\Pi r \sigma_{zz} \, dr = 0. \tag{11.6.16}$$

Substituting Eq. (11.6.15) into Eq. (11.6.16) and integrating, we get:

$$e_{zz} = e_o = -\frac{\nu}{2E} \rho\omega^2 b^2 \tag{11.6.17}$$

and

$$\sigma_{zz} = \frac{\nu\rho\omega^2}{4(1-\nu)}(b^2 - 2r^2). \qquad (11.6.18)$$

For a hollow cylinder:

$$\sigma_{rr} = \frac{3-2\nu}{8(1-\nu)}\rho\omega^2\left(b^2 + a^2 - \frac{a^2b^2}{r^2} - r^2\right) \qquad (11.6.19)$$

$$\sigma_{\theta\theta} = \frac{3-2\nu}{8(1-\nu)}\rho\omega^2\left(b^2 + a^2 + \frac{a^2b^2}{r^2} - \frac{(1+2\nu)}{3-2\nu}r^2\right) \qquad (11.6.20)$$

$$\sigma_{zz} = \frac{\nu\rho\omega^2}{4(1-\nu)}(b^2 + a^2 - 2r^2), \qquad (11.6.21)$$

and

$$e_{zz} = e_o = -\frac{\nu}{2E}\rho\omega^2(b^2 + a^2). \qquad (11.6.22)$$

11.7 Disks of Variable Thickness

Turbine disks are usually made thicker near their hub and taper down to a smaller thickness towards the periphery. The reason for this is the high stress concentration near the center of flat disks. The method used in the analysis of flat disks will be applied here with the difference that the thickness t will have to be included in the calculations. The thickness is a function of r alone (Fig. 11.9). σ_{rr} and $\sigma_{\theta\theta}$ are the mean

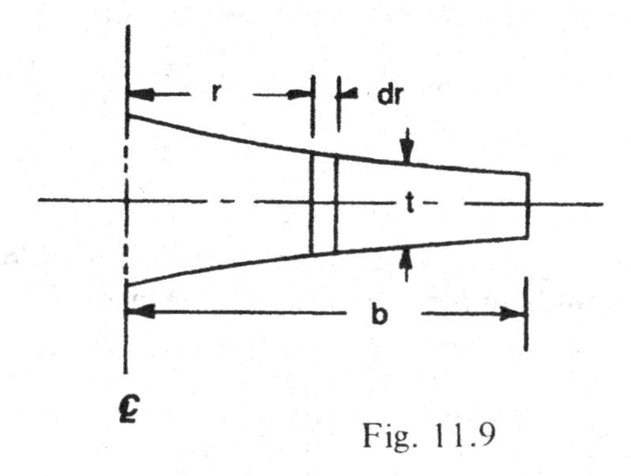

Fig. 11.9

radial and transverse stresses at any distance r from the center line. The equation of equilibrium (11.5.4) is valid for a flat disk of unit thickness. To apply to this case, all its members must be multiplied by t, so that we now have:

$$\frac{d}{dr}(tr\sigma_{rr}) - t\sigma_{\theta\theta} + \rho t\omega^2 r^2 = 0 \tag{11.7.1}$$

as our equation of equilibrium. The stress function ϕ is defined by

$$tr\sigma_{rr} = \phi, \quad t\sigma_{\theta\theta} = \frac{d\phi}{dr} + \rho\omega^2 tr^2, \tag{11.7.2}$$

and the governing equation (11.5.8) becomes:

$$r^2\frac{d^2\phi}{dr^2} + \left(1 - \frac{r}{t}\frac{dt}{dr}\right)r\frac{d\phi}{dr} + \left(\frac{\nu r}{t}\frac{dt}{dr} - 1\right)\phi = -(3+\nu)\rho\omega^2 tr^3, \tag{11.7.3}$$

from which ϕ can be found, provided $t = t(r)$ is given. For a hyperbolic shape, Eq. (11.7.3) can easily be integrated. If the thickness varies according to the law

$$t = Cr^{-p}, \tag{11.7.4}$$

where C is a constant and p is any positive number, Eq. (11.7.3) reduces to:

$$r^2\frac{d^2\phi}{dr^2} + (1+p)r\frac{d\phi}{dr} - (1+\nu p)\phi = -(3+\nu)\rho\omega^2 Cr^{3-p}. \tag{11.7.5}$$

The solution of this equation is easily obtained by substitution. This solution is:

$$\phi = C_1 r^{q_1} + C_2 r^{q_2} - \frac{3+\nu}{8-(3+\nu)p}C\rho\omega^2 r^{3-p}, \tag{11.7.6}$$

where q_1 and q_2 are the roots of the equation

$$q^2 + pq - (1+\nu p) = 0, \tag{11.7.7}$$

i.e.,

$$\left.\begin{matrix} q_1 \\ q_2 \end{matrix}\right\} = -\frac{p}{2} \pm \sqrt{\left(\frac{p}{2}\right)^2 + (1+\nu p)}\,. \tag{11.7.8}$$

The components of the stress are therefore,

$$\sigma_{rr} = \frac{\phi}{tr} = \frac{C_1}{C}r^{q_1+p-1} + \frac{C_2}{C}r^{q_2+p-1} - \frac{3+\nu}{8-(3+\nu)p}\rho\omega^2 r^2 \tag{11.7.9}$$

$$\sigma_{\theta\theta} = \frac{1}{t}\frac{d\phi}{dr} + \rho\omega^2 r^2 = \frac{C_1}{C} q_1 r^{q_1+p-1} + \frac{C_2}{C} q_2 r^{q_2+p-1} - \frac{1+3\nu}{8-(3+\nu)p}\rho\omega^2 r^2. \tag{11.7.10}$$

Eq. (11.7.8) shows that $q_2 + p$ will always be a negative quantity so that, for a solid disk, C_2 must be equal to zero since σ_{rr} and $\sigma_{\theta\theta}$ cannot be infinite for $r = 0$. For no forces acting on the boundary, $\sigma_{rr} = 0$ for $r = b$. Therefore,

$$\frac{C_1}{C} = \frac{3+\nu}{8-(3+\nu)p}\rho\omega^2 b^{3-q_1-p}. \tag{11.7.11}$$

Therefore, for a solid disk,

$$\sigma_{rr} = \frac{3+\nu}{8-(3+\nu)p}\rho\omega^2 b^2\left[\left(\frac{r}{b}\right)^{q_1+p-1} - \left(\frac{r}{b}\right)^2\right] \tag{11.7.12}$$

$$\sigma_{\theta\theta} = \frac{3+\nu}{8-(3+\nu)p}\rho\omega^2 b^2\left[q_1\left(\frac{r}{b}\right)^{q_1+p-1} - \frac{1+3\nu}{3+\nu}\left(\frac{r}{b}\right)^2\right]. \tag{11.7.13}$$

For a disk of uniform thickness $p = 0$, and Eqs. (11.7.12) and (11.7.13) reduce to Eqs. (11.5.12) and (11.5.13). Fig. 11.10 shows hyperbolic profiles for various values of p. Similar equations can be established for hollow disks. C_1/C and C_2/C are obtained by writing that $\sigma_{rr} = 0$ for $r = a$ and $r = b$, respectively.

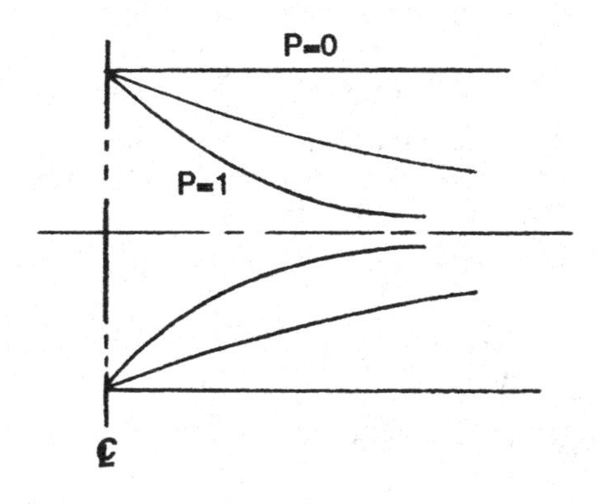

Fig. 11.10

11.8 Thermal Stresses in Thin Disks

Let us consider a thin circular disk with an uneven temperature distribution $\Delta T = \Delta T(r)$. The stress-strain relations in this case are those of a plane stress problem. Because of symmetry, the only relations of interest are provided by Eqs. (8.6.3) and (8.6.4), which become:

$$e_{rr} = \frac{1}{E}(\sigma_{rr} - \nu\sigma_{\theta\theta}) + \alpha(\Delta T) \tag{11.8.1}$$

$$e_{\theta\theta} = \frac{1}{E}(\sigma_{\theta\theta} - \nu\sigma_{rr}) + \alpha(\Delta T). \tag{11.8.2}$$

The equilibrium equation,

$$\frac{d\sigma_{rr}}{dr} + \frac{\sigma_{rr} - \sigma_{\theta\theta}}{r} = 0, \tag{11.8.3}$$

is identically satisfied if we introduce the stress function ϕ, such that

$$\sigma_{rr} = \frac{\phi}{r}, \quad \sigma_{\theta\theta} = \frac{d\phi}{dr}. \tag{11.8.4}$$

The compatibility Eq. (11.5.7) becomes:

$$\frac{d}{dr}\left[\frac{1}{r}\frac{d}{dr}(r\phi)\right] = -\alpha E\frac{d(\Delta T)}{dr}. \tag{11.8.5}$$

This equation is easily integrated to give the value of ϕ at any point of a circle of radius r. Hence,

$$\phi = -\frac{\alpha E}{r}\int_a^r (\Delta T) r\,dr + \frac{C_1 r}{2} + \frac{C_2}{r}, \tag{11.8.6}$$

where C_1 and C_2 are constants of integration and the lower limit a is equal to zero for a solid disk, and is equal to the inner radius for a hollow disk. The stresses can now be obtained by substituting Eq. (11.8.6) into Eq. (11.8.4). This gives:

$$\sigma_{rr} = -\frac{\alpha E}{r^2}\int_a^r (\Delta T) r\,dr + \frac{C_1}{2} + \frac{C_2}{r^2} \tag{11.8.7}$$

$$\sigma_{\theta\theta} = \alpha E\left[-(\Delta T) + \frac{1}{r^2}\int_a^r (\Delta T) r\,dr\right] + \frac{C_1}{2} - \frac{C_2}{r^2}. \tag{11.8.8}$$

For a *solid disk*, the stresses must be finite at the origin so that $C_2 = 0$. If there are no external forces applied on the boundary, $\sigma_{rr} = 0$ for $r = b$. Then,

$$C_1 = \frac{2\alpha E}{b^2} \int_0^b (\Delta T) r\, dr \tag{11.8.9}$$

and

$$\sigma_{rr} = \alpha E \left[\frac{1}{b^2} \int_0^b (\Delta T) r\, dr - \frac{1}{r^2} \int_0^r (\Delta T) r\, dr \right] \tag{11.8.10}$$

$$\sigma_{\theta\theta} = \alpha E \left[-(\Delta T) + \frac{1}{b^2} \int_0^b (\Delta T) r\, dr + \frac{1}{r^2} \int_0^r (\Delta T) r\, dr \right]. \tag{11.8.11}$$

For a *hollow disk* with inner radius a, $\sigma_{rr} = 0$ at $r = a$ and $r = b$. Therefore,

$$C_1 = \frac{2\alpha E}{b^2 - a^2} \int_a^b (\Delta T) r\, dr, \quad C_2 = -\frac{a^2 \alpha E}{b^2 - a^2} \int_a^b (\Delta T) r\, dr \tag{11.8.12}$$

and

$$\sigma_{rr} = \alpha E \left[-\frac{1}{r^2} \int_a^r (\Delta T) r\, dr + \frac{1}{b^2 - a^2} \int_a^b (\Delta T) r\, dr - \frac{a^2}{r^2 (b^2 - a^2)} \int_a^b (\Delta T) r\, dr \right] \tag{11.8.13}$$

$$\sigma_{\theta\theta} = \alpha E \left[-(\Delta T) + \frac{1}{r^2} \int_a^r (\Delta T) r\, dr + \frac{1}{b^2 - a^2} \int_a^b (\Delta T) r\, dr + \frac{a^2}{r^2 (b^2 - a^2)} \int_a^b (\Delta T) r\, dr \right]. \tag{11.8.14}$$

Once $\Delta T = \Delta T(r)$ is known, the value of the stresses can be computed.

11.9 Thermal Stresses in Long Circular Cylinders

Case a. The cylinder is not free to deform longitudinally

The problem is a plane strain problem. The stress-strain relations are Eqs. (8.6.3) to (8.6.5) which become:

$$e_{rr} = \frac{1+\nu}{E}[(1-\nu)\sigma_{rr} - \nu\sigma_{\theta\theta} + \alpha E(\Delta T)] \tag{11.9.1}$$

$$e_{\theta\theta} = \frac{1+\nu}{E}[(1-\nu)\sigma_{\theta\theta} - \nu\sigma_{rr} + \alpha E(\Delta T)] \tag{11.9.2}$$

$$\sigma_{zz} = \nu(\sigma_{rr} + \sigma_{\theta\theta}) - \alpha E(\Delta T). \tag{11.9.3}$$

Using the stress function defined in Eq. (11.8.4), the compatibility Eq. (11.5.7) becomes:

$$\frac{d}{dr}\left[\frac{1}{r}\frac{d}{dr}(r\phi)\right] = -\frac{\alpha E}{1-\nu}\frac{d(\Delta t)}{dr}. \tag{11.9.4}$$

Except for the coefficient of $d(\Delta T)/dr$, this equation is the same as Eq. (11.8.5). Therefore, its solution is:

$$\phi = -\frac{\alpha E}{1-\nu}\frac{1}{r}\int_a^r (\Delta T) r\,dr + \frac{C_1 r}{2} + \frac{C_2}{r}. \tag{11.9.5}$$

For a solid cylinder, $C_2 = 0$ so that the stresses will be finite at the origin, and if there are no stresses on the outer surface $r = b$,

$$C_1 = \frac{2\alpha E}{1-\nu}\frac{1}{b^2}\int_0^b (\Delta T) r\,dr. \tag{11.9.6}$$

Therefore, the stresses are:

$$\sigma_{rr} = \frac{\alpha E}{1-\nu}\left[\frac{1}{b^2}\int_0^b (\Delta T) r\,dr - \frac{1}{r^2}\int_0^r (\Delta T) r\,dr\right] \tag{11.9.7}$$

$$\sigma_{\theta\theta} = \frac{\alpha E}{1-\nu}\left[-(\Delta T) + \frac{1}{b^2}\int_0^b (\Delta T) r\,dr + \frac{1}{r^2}\int_0^r (\Delta T) r\,dr\right] \tag{11.9.8}$$

$$\sigma_{zz} = \frac{\alpha E}{1-\nu}\left[\frac{2\nu}{b^2}\int_0^b (\Delta T) r\,dr - (\Delta T)\right]. \tag{11.9.9}$$

σ_{zz} is the normal stress distribution which must be applied to keep $e_{zz} = 0$ throughout.

For a hollow cyilinder with inner radius a, $\sigma_{rr} = 0$ at $r = a$ and at $r = b$. Therefore,

$$C_1 = \frac{2\alpha E}{1-\nu}\frac{1}{b^2-a^2}\int_a^b (\Delta T) r\,dr \tag{11.9.10}$$

$$C_2 = -\frac{\alpha E}{1-\nu}\frac{a^2}{b^2-a^2}\int_a^b (\Delta T) r\,dr \tag{11.9.11}$$

and

$$\sigma_{rr} = \frac{\alpha E}{1 - \nu}\frac{1}{r^2}\left[\frac{r^2 - a^2}{b^2 - a^2}\int_a^b (\Delta T)r\,dr - \int_a^r (\Delta T)r\,dr\right] \tag{11.9.12}$$

$$\sigma_{\theta\theta} = \frac{\alpha E}{1 - \nu}\frac{1}{r^2}\left[\frac{r^2 + a^2}{b^2 - a^2}\int_a^b (\Delta T)r\,dr + \int_a^r (\Delta T)r\,dr - (\Delta T)r^2\right] \tag{11.9.13}$$

$$\sigma_{zz} = \frac{\alpha E}{1 - \nu}\left[\frac{2\nu}{b^2 - a^2}\int_a^b (\Delta T)r\,dr - (\Delta T)\right]. \tag{11.9.14}$$

σ_{zz} is the normal stress distribution which must be applied to keep $e_{zz} = 0$ throughout.

Case b. The cylinder is free to deform longitudinally

The reasoning made in Sec. 11.6 can be repeated here. The cylinder is in a condition of generalized plane strain:

For a solid cylinder:

$$\int_0^b 2\Pi r\sigma_{zz}\,dr = \int_0^b 2\Pi r\left\{\frac{\alpha E}{1 - \nu}\left[\frac{2\nu}{b^2}\int_0^b (\Delta T)r\,dr - (\Delta T)\right] + Ee_o\right\} dr = 0. \tag{11.9.15}$$

Therefore,

$$Ee_o = \frac{2\alpha E}{b^2}\int_0^b (\Delta T)r\,dr$$

and

$$\sigma_{zz} = \frac{\alpha E}{1 - \nu}\left[\frac{2}{b^2}\int_0^b (\Delta T)r\,dr - \Delta T\right]. \tag{11.9.16}$$

σ_{rr} and $\sigma_{\theta\theta}$ are given by Eqs. (11.9.7) and (11.9.8), respectively.

For a hollow cylinder:

$$Ee_o = \frac{2\alpha E}{b^2 - a^2}\int_a^b (\Delta T)r\,dr$$

and

$$\sigma_{zz} = \frac{\alpha E}{(1-\nu)}\left[\frac{2}{b^2 - a^2}\int_a^b (\Delta T) r\,dr - (\Delta T)\right].$$

σ_{rr} and $\sigma_{\theta\theta}$ are given by Eqs. (11.9.12) and (11.9.13), respectively.

11.10 Thermal Stresses in Spheres

Consider a sphere in which the temperature distribution is symmetrical with respect to the center and is therefore a function of ρ, the radial distance, alone. In a spherical system of coordinates, the nonzero stresses are $\sigma_{\rho\rho}$ and $\sigma_{\theta\theta} = \sigma_{\phi\phi}$, and the nonzero displacement is u_ρ. Only the third equation of equilibrium is of significance, the two others being identically satisfied. This equation becomes:

$$\frac{\partial \sigma_{\rho\rho}}{\partial \rho} + \frac{2}{\rho}(\sigma_{\rho\rho} - \sigma_{\theta\theta}) = 0. \tag{11.10.1}$$

The strain-displacement relations are given by Eq. (6.7.34). In our case they are:

$$e_{\rho\rho} = \frac{\partial u_\rho}{\partial \rho}, \quad e_{\phi\phi} = e_{\theta\theta} = \frac{u_\rho}{\rho}. \tag{11.10.2}$$

The equilibrium Eq. (11.10.1) can be written in terms of the displacement u_ρ. Using the stress-strain relations (8.6.9) and (8.6.10), as well as Eq. (11.10.2), we get:

$$\frac{d}{d\rho}\left[\frac{1}{\rho^2}\frac{d}{d\rho}(\rho^2 u_\rho)\right] = \frac{1+\nu}{1-\nu}\alpha\frac{d}{d\rho}(\Delta T). \tag{11.10.3}$$

The solution of Eq. (11.10.3) is:

$$u_\rho = \frac{1+\nu}{1-\nu}\frac{\alpha}{\rho^2}\int_a^\rho (\Delta T)\rho^2\,d\rho + C_1\rho + \frac{C_2}{\rho^2}, \tag{11.10.4}$$

where C_1 and C_2 are constants of integration and a is a constant taken equal to zero for a solid sphere and equal to the inner radius for a hollow sphere. Eqs. (11.10.2) give the strains. The stresses are directly obtained from the stress-strain relation. They are:

$$\sigma_{\rho\rho} = -\frac{2\alpha E}{(1-\nu)\rho^3}\int_a^\rho (\Delta T)\rho^2\,d\rho + \frac{EC_1}{1-2\nu} - \frac{2EC_2}{(1+\nu)\rho^3} \tag{11.10.5}$$

$$\sigma_{\theta\theta} = \frac{\alpha E}{(1-\nu)\rho^3}\int_a^\rho (\Delta T)\rho^2\,d\rho + \frac{EC_1}{1-2\nu} + \frac{EC_2}{(1+\nu)\rho^3} - \frac{\alpha E(\Delta T)}{1-\nu}. \tag{11.10.6}$$

For a solid sphere the stresses must be finite at the origin so that $C_2 = 0$. If there are no external forces on the boundary of the sphere, $\sigma_{\rho\rho} = 0$ for $\rho = b$. Then,

$$C_1 = \frac{2\alpha(1-2\nu)}{(1-\nu)}\frac{1}{b^3}\int_o^b (\Delta T)\rho^2\,d\rho \tag{11.10.7}$$

and

$$\sigma_{\rho\rho} = \frac{2\alpha E}{1-\nu}\left[\frac{1}{b_3}\int_0^b (\Delta T)\rho^2\,d\rho - \frac{1}{\rho^3}\int_0^\rho (\Delta T)\rho^2\,d\rho\right] \tag{11.10.8}$$

$$\sigma_{\theta\theta} = \frac{\alpha E}{1-\nu}\left[\frac{2}{b^3}\int_0^b (\Delta T)\rho^2\,d\rho + \frac{1}{\rho^3}\int_0^\rho (\Delta T)\rho^2\,d\rho - (\Delta T)\right]. \tag{11.10.9}$$

For a hollow sphere with inner radius a, $\sigma_{\rho\rho} = 0$ at $\rho = a$ and at $\rho = b$. Therefore,

$$\frac{EC_1}{1-2\nu} - \frac{2EC_2}{1+\nu}\frac{1}{a^3} = 0 \tag{11.10.10}$$

and

$$-\frac{2\alpha E}{1-\nu}\frac{1}{b^3}\int_a^b (\Delta T)\rho^2\,d\rho + \frac{EC_1}{1-2\nu} - \frac{2EC_2}{1+\nu}\frac{1}{b^3} = 0. \tag{11.10.11}$$

Solving for C_1 and C_2, and substituting the results in Eqs. (11.10.5) and (11.10.6), we get:

$$\sigma_{\rho\rho} = \frac{2\alpha E}{1-\nu}\left[\frac{\rho^3 - a^3}{(b^3 - a^3)\rho^3}\int_a^b (\Delta T)\rho^2\,d\rho - \frac{1}{\rho^3}\int_a^\rho (\Delta T)\rho^2\,d\rho\right] \tag{11.10.12}$$

$$\sigma_{\theta\theta} = \frac{2\alpha E}{1-\nu}\left[\frac{2\rho^3 + a^3}{2(b^3 - a^3)\rho^3}\int_a^b (\Delta T)\rho^2\, d\rho + \frac{1}{2\rho^3}\int_a^\rho (\Delta T)\rho^2\, d\rho - \frac{(\Delta T)}{2}\right]. \tag{11.10.13}$$

PROBLEMS

1. Find the ratio of thickness to internal diameter for a tube subjected to internal pressure when the pressure is equal in magnitude to $\frac{3}{4}$ of the maximum circumferential stress. If the internal diameter of the tube is 4 in., determine the increase in the external diameter when the internal pressure is 12,000 psi and the tube is prevented from changing length. ($E = 30 \times 10^6$ psi, $\nu = 0.3$.)
2. A solid bar of uniform circular section is subjected to uniform radial pressure. Show that the stress at any point in a plane section parallel to the axis of the bar is compressive and equal in magnitude to the radial stress.
3. A steel bar of 2 in. diameter is pressed into a steel sleeve so that, when assembled, the magnitude of the radial stress between the two is 2,000 psi, and that of the circumferential stress at the inside of the sleeve is 3,200 psi. Assuming a close fit and neglecting friction, determine the change of radial stress when the bar is subjected to an axial compressive load of 15,000 lb. ($\nu = 0.3$.)
4. A short steel rod of 2 in. diameter is subjected to an axial compressive load of 60,000 lb. It is surrounded by a sleeve $\frac{1}{2}$ in. thick, slightly shorter than the rod so that the load is carried only by the rod. Assuming a close fit before the load is applied and neglecting friction, find the pressure between the sleeve and the rod, and the maximum tensile stress in the sleeve. ($\nu = 0.3$.)
5. The external diameter of a steel hub is 10 in. and the internal diameter increases 0.005 in. when shrunk on to a solid steel shaft of 5 in. diameter. Find the reduction in diameter of the shaft, the radial pressure between the hub and the shaft, and the circumferential stress at the inner surface of the hub. (E 30×10^6 psi, $\nu = 0.3$.)
6. A steel cylinder of 8 in. external diameter and 6 in. internal diameter has another steel cylinder of 10 in. external diameter shrunk onto it. If the maximum tensile stress induced in the outer cylinder is 10,000 psi, find the radial compressive stress between the

cylinders. Determine the circumferential stresses at inner and outer diameter of both cylinders, and show on a diagram how these stresses vary with the radius. Calculate the necessary shrinkage allowance at the common radius. ($E = 30 \times 10^6$ psi, $\nu = 0.3$.)

7. A steel hollow sphere, whose inside diameter is 5 in., is subjected to an internal pressure of 5,000 psi. Determine the thickness of the material if the magnitude of the maximum stress is not to exceed 10,000 psi. Compare this thickness to that obtained from Eq. (11.4.25).
8. Find the expressions of the stresses and displacements for a hollow sphere subjected to an external pressure P_o and filled with an incompressible fluid such that its inner diameter does not change.
9. Determine the greatest value of the radial and circumferential stresses for a thin disk rotating at an angular velocity of 150 radians per sec.; the inner and outer radii of the disk are 6 in. and 12 in. respectively, and the weight per unit volume γ of the material is 0.28 lb/in.^{3}. ($E = 30 \times 10^6$ psi, $\nu = 0.3$.)
10. A solid steel shaft of 8 in. diameter has a steel cylinder of 16 in. diameter shrunk onto it. The inside diameter of the cylinder prior to the shrink fit operation was 7.992 in. ($E = 30 \times 10^6$ psi, $\nu = 0.3$.)
    (a) Determine the external pressure P_o on the outside of the cylinder which is required to reduce to zero the circumferential stress at the inner surface of the cylinder.
    (b) Determine the radial pressure on the surface of contact due to shrink fit.
    (c) Find the speed of rotation to loosen the fit. ($\gamma = 0.28$ lb/in.^{3}).
11. A solid steel shaft 36 in. in diameter is rotating at 200 rpm. If the shaft cannot deform longitudinally, calculate the total longitudinal thrust over a cross section due to rotational stresses. ($\gamma = 0.28$ lb/in.^{3}, $E = 30 \times 10^6$ psi, $\nu = 0.3$.)
12. Show that the radial displacement in a rotating solid cylinder, whose ends are free to deform, is given by:

$$u_r = \frac{\rho\omega^2(1+\nu)(1-2\nu)}{8E(1-\nu)}\left[\frac{(3-5\nu)b^2 r}{(1+\nu)(1-2\nu)} - r^3\right].$$

13. A brass rod is fitted firmly inside a steel tube whose inner and outer diameters are 1 in. and 2 in., respectively, when the materials are at a temperature of 60 ° F. If the rod and the tube are both heated to a temperature of 300° F, determine the maximum stress in the brass

and in the steel. The coefficients of expansion for steel and brass are 6×10^{-6} and 10×10^{-6} per degree Fahrenheit, respectively. Young's modulus is 30×10^{6} psi for steel and 12.5×10^{6} psi for brass. Poisson's ratio is 0.3 for steel and 0.34 for brass.

14. Find the expression of σ_{rr}, $\sigma_{\theta\theta}$, and σ_{zz} in a long hollow cylinder with fixed ends, which conducts heat in steady state according to

$$\Delta T = \frac{(\Delta T_a - \Delta T_b)\ell n \frac{b}{r}}{\ell n \frac{b}{a}}.$$

ΔT_a is a constant increase in the temperature of the inner surface of the cylinder and ΔT_b, smaller than ΔT_a, is a constant increase in the temperature of the outer surface of the cylinder.

15. Find the expression of the axial stress in the cylinder of Problem 14 when the ends are free.(*)

REFERENCES

[1] S. Timoshenko and J. N. Goodier, *Theory of Elasticity*, McGraw-Hill, New York, N. Y., 1970.

[2] C. T. Wang, *Applied Elasticity*, McGraw-Hill, New York, N. Y., 1953.

[3] H. M. Westergaard, *Theory of Elasticity and Plasticity*, Dover, New York, N. Y., 1964.

* See pages A-68 and A-69 for additional problems.

CHAPTER 12

STRAIGHT SIMPLE BEAMS

12.1 Introduction

In this chapter, the assumptions on which the elementary theory of beams is based are enumerated and the important equations listed. The inverse method, Airy's stress function, and the semi-inverse method are used to study the pure bending of a prismatic bar, two cases of narrow beams with rectangular cross section, and Saint-Venant's problem of the cantilever subjected to an end load: The results are compared with those of the elementary theory.

Recalling that a positive face (i.e., cross section) is one whose outward normal is in the positive direction, the following conventions for axial forces, shearing forces, and bending moments will be adhered to:

1) *An axial force* N is taken positive when it causes normal stresses in a positive direction on a positive face, or normal stresses in a negative direction on a negative face; N is taken negative when it causes normal stresses in a negative direction on a positive face, or normal stresses in a positive direction on a negative face (Fig. 12.1a).

2) *A shearing force* V is taken positive when it causes shearing stresses in a positive direction on a positive face, or shearing stresses in a negative direction on a negative face; V is taken negative when it causes shearing stresses in a negative direction on a positive face, or shearing stresses in a positive direction on a negative face (Fig. 12.1a).

3) *A bending moment* M, acting on a positive face, is taken positive when it causes normal stresses in the positive direction on the positive

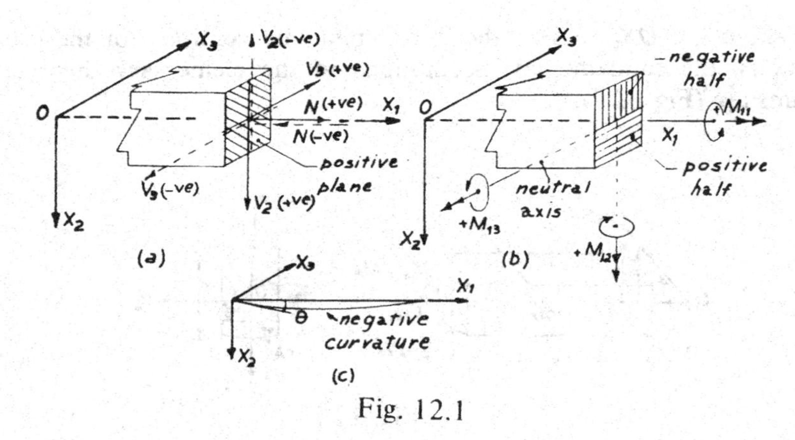

Fig. 12.1

or lower half of the face (Fig. 12.1b) and normal stresses in the negative direction on the negative or upper half of the face. The assumption here is that those halves fall above or below the neutral surface. On a *negative face, M is taken positive when it causes normal stresses in the* negative direction on the positive or lower half of the face, and normal stresses in the positive direction on the upper half of the face (Fig. 12.1b).

The previous conventions on forces and moments are based on the signs of the stresses they generate. Those stresses, in turn, follow the rules established in Sec. 7.2, which are universally used in the theory of elasticity. It is immaterial in which direction the axes of the right-hand system point, these sign conventions always hold. The displacements are positive when they are in the positive directions of the axes. In the theory of beams, the deflection which is the vertical displacement of the center line follows the same rule.

When relations are established among forces, moments, and deflections, the curvature of the beam is introduced. The center of curvature is always on the concave side of the beam. For small deflections, the curvature is given to a good approximation by the rate of change of the slope. Therefore, an upward concavity (in the system of axes of Fig. 12.1c), which is associated with positive bending moments, corresponds to a negative curvature and vice versa; hence, one can attach a sign to the radius of curvature depending on the side on which the concavity is. In summary, and with the previous sign conventions, a positive bending moment gives a negative curvature.

12.2 The Elementary Theory of Beams

Taking the OX_1 axis as the line joining the centroids of the cross sections, the fundamental assumptions in the elementary theory of beams are (Fig. 12.2):

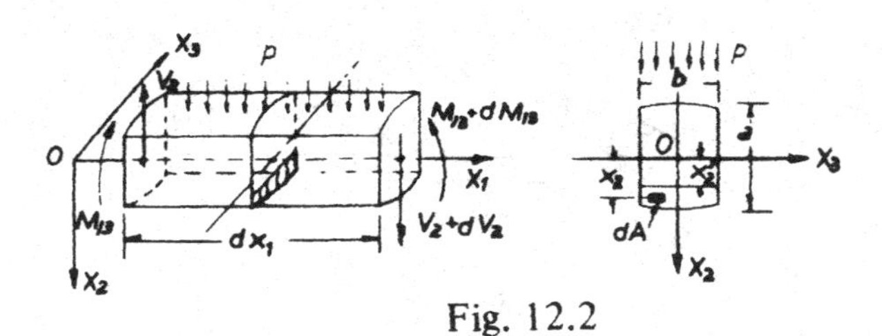

Fig. 12.2

a) The beam is prismatical and straight but no restrictions are placed on the shape of the cross section.

b) The loading and, consequently, the bending moment M_{13} are applied in a plane containing one of the principal moments of inertia of the cross section. If the applied bending moment does not lie in such a plane, it can always be split into two components each of which satisfies this assumption. Here it is applied about the OX_3 axis in the OX_1, OX_2 plane.

c) Plane cross sections in the unstressed beam remain plane during bending deformations.

d) The deflection of each beam element is assumed to be in the form of a circular arc of radius R and deflections and slopes are small enough so that the curvature C is given by

$$C = \frac{1}{R} = \frac{\dfrac{d^2u_2}{dx_1^2}}{\left[1 + \left(\dfrac{du_2}{dx_1}\right)^2\right]^{3/2}} \approx \frac{d^2u_2}{dx_1^2}. \tag{12.2.1}$$

e) The shearing stresses are assumed to be uniformly distributed across the width of the beam.

f) Lines parallel to the OX_3 axis before deformation remain parallel to this axis after deformation.

From the above assumptions and with the sign conventions and symbols shown in Figs. 12.1 and 12.2, the following well-known equations have been deduced [1]:

$$\frac{d^2 u_2}{dx_1^2} = -\frac{M_{13}}{EI_{33}}, \quad \frac{dM_{13}}{dx_1} = +V_2, \quad \frac{dV_2}{dx_1} = -p \tag{12.2.2}$$

$$\text{Slope} = \theta(x_1) = \frac{du_2}{dx_1} = -\int \frac{M_{13}}{EI_{33}}\, dx_1 + C_1 \tag{12.2.3}$$

$$\sigma_{11} = \sigma_{11}(x_1, x_2) = +\frac{M_{13} x_2}{I_{33}} \tag{12.2.4}$$

$$(\sigma_{12})_{x_2 = x_2'} = \sigma_{12}(x_1, x_2) = \frac{V_2}{I_{33} b} \int_{x_2'}^{\frac{a}{2}} x_2 \, dA = \frac{V_2 Q_3}{I_{33} b} \tag{12.2.5}$$

$$\sigma_{22} = \sigma_{33} = \sigma_{13} = \sigma_{23} = 0, \tag{12.2.6}$$

where M_{13} is the bending moment about the OX_3 axis on the face normal to OX_1, V_2 is the shearing force parallel to OX_2, I_{33} is the moment of inertia of the section about the OX_3 axis ($I_{33} = I_3$ is a principal moment of inertia), Q_3 is the static moment of the area A about the OX_3 axis, C_1 and C_2 are constants of integration to be determined from the boundary conditions, a is the depth of the beam, and b is the width of the beam at a distance x_2' from OX_3 (Fig. 12.2).

The above equations are extremely useful in design because of their simplicity. In many cases, their accuracy is quite sufficient. Under certain circumstances, however, it is necessary to use the more accurate equations provided by the theory of elasticity, since, in the derivation of these equations, compatibility is ignored and often certain boundary conditions are violated.

12.3 Pure Bending of Prismatical Bars

Let us consider a prismatic bar bent about the OX_3 axis by two opposite couples equal to M (Fig. 12.3). Let us assume that OX_2 and OX_3 are principal axes of inertia at the centroid of the cross section, and that the couples act in the principal plane OX_1, OX_2. We shall assume that

$$\sigma_{11} = \sigma_{11}(x_1, x_2, x_3) \tag{12.3.1}$$

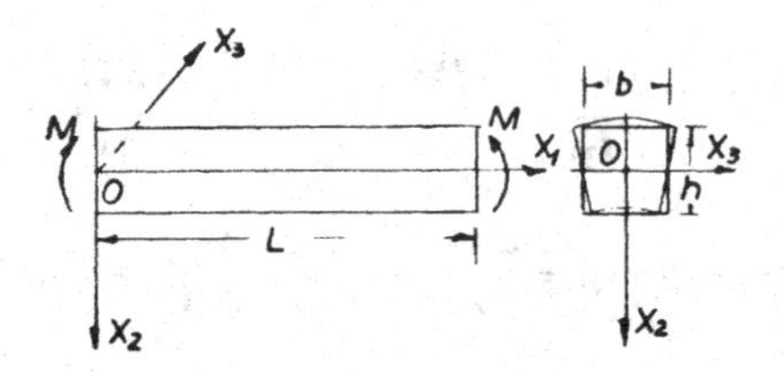

Fig. 12.3

$$\sigma_{22} = \sigma_{33} = \sigma_{13} = \sigma_{23} = \sigma_{12} = 0, \qquad (12.3.2)$$

and show that this stress field satisfies the requirements of elasticity. From Hooke's law, we have:

$$e_{11} = \frac{\sigma_{11}}{E}, \quad e_{22} = e_{33} = -\frac{\nu\sigma_{11}}{E} \qquad (12.3.3)$$

$$e_{12} = e_{23} = e_{13} = 0. \qquad (12.3.4)$$

With these relations, the general equations of equilibrium are reduced to the equation:

$$\frac{\partial \sigma_{11}}{\partial x_1} = 0, \qquad (12.3.5)$$

which means that σ_{11} is independent of x_1 and only a function of x_2 and x_3. The compatibility relations (4.10.14) reduce to:

$$\frac{\partial^2 \sigma_{11}}{\partial x_3^2} = 0 \qquad (12.3.6)$$

$$\frac{\partial^2 \sigma_{11}}{\partial x_2^2} = 0 \qquad (12.3.7)$$

$$\frac{\partial^2 \sigma_{11}}{\partial x_2 \, \partial x_3} = 0. \qquad (12.3.8)$$

Integrating Eq. (12.3.6), we get:

$$\sigma_{11} = x_3 f_1(x_2) + f_2(x_2). \tag{12.3.9}$$

Substituting Eq. (12.3.9) into Eq. (12.3.8), we get:

$$\frac{df_1}{dx_2} = 0 \quad \text{or} \quad f_1(x_2) = C_1,$$

where C_1 is a constant. Substituting Eq. (12.3.9) into Eq. (12.3.7), we get:

$$\frac{d^2 f_2}{dx_2^2} = 0 \quad \text{or} \quad f_2(x_2) = C_2 x_2 + C_3, \tag{12.3.10}$$

where C_2 and C_3 are constants. Therefore,

$$\sigma_{11} = C_1 x_3 + C_2 x_2 + C_3. \tag{12.3.11}$$

The constants C_1, C_2, and C_3 can be determined from the conditions that on any plane parallel to the OX_2, OX_3 plane the resultant force in the OX_1 direction must be equal to zero, that the resultant moment about the OX_3 axis must be equal to M, and that the resultant moment about the OX_2 axis must be equal to zero:

$$\iint \sigma_{11}\, dA = 0, \quad \iint \sigma_{11} x_2\, dA = M, \quad \iint \sigma_{11} x_3\, dA = 0. \tag{12.3.12}$$

Substituting Eq. (12.3.11) into Eq. (12.3.12), the first condition requires that

$$C_3 = 0, \tag{12.3.13}$$

and the last condition requires that

$$I_{22} C_1 + I_{23} C_2 = 0, \tag{12.3.14}$$

where I_{22} is the moment of inertia of the cross section of the bar about the OX_2 axis and I_{23} is the product of inertia. Since OX_2 and OX_3 are principal axes of inertia, $I_{23} = 0$ and, consequently, $C_1 = 0$. From the second condition we get:

$$C_2 I_{33} = M \quad \text{or} \quad C_2 = \frac{M}{I_{33}}, \tag{12.3.15}$$

where I_{33} is the moment of inertia about the OX_3 axis. Hence,

$$\sigma_{11} = \frac{Mx_2}{I_{33}}. \tag{12.3.16}$$

This result agrees with that of the simple theory of beams.

Let us now consider the displacements due to simple bending. The strain displacement relations (1.2.1), Eqs. (12.3.3), (12.3.4), and (12.3.16) give:

$$\frac{\partial u_1}{\partial x_1} = \frac{M}{EI_{33}} x_2 \tag{12.3.17}$$

$$\frac{\partial u_2}{\partial x_2} = -\frac{\nu M}{EI_{33}} x_2 \tag{12.3.18}$$

$$\frac{\partial u_3}{\partial x_3} = -\frac{\nu M}{EI_{33}} x_2 \tag{12.3.19}$$

$$\frac{\partial u_1}{\partial x_2} + \frac{\partial u_2}{\partial x_1} = \frac{\partial u_2}{\partial x_3} + \frac{\partial u_3}{\partial x_2} = \frac{\partial u_1}{\partial x_3} + \frac{\partial u_3}{\partial x_1} = 0. \tag{12.3.20}$$

The integration of Eqs. (12.3.17) to (12.3.20) to give u_1, u_2, and u_3 does not present any difficulty. It yields:

$$u_1 = \frac{Mx_1 x_2}{EI_{33}} + d_2 x_3 + d_3 x_2 + d_4 \tag{12.3.21}$$

$$u_2 = -\frac{M}{2EI_{33}}(x_1^2 - \nu x_3^2 + \nu x_2^2) - d_3 x_1 + d_1 x_3 + d_6 \tag{12.3.22}$$

$$u_3 = -\frac{\nu M}{EI_{33}} x_2 x_3 - d_2 x_1 - d_1 x_2 + d_5, \tag{12.3.23}$$

where d_1, d_2, d_3, d_4, d_5, and d_6 are constants of integration. To determine these constants, we shall assume that the centroid of the bar together with an element of the OX_1 axis and an element of the OX_1, OX_2 plane are fixed at the origin. Therefore at $x_1 = x_2 = x_3 = 0$:

$$u_1 = u_2 = u_3 = 0, \quad \frac{\partial u_3}{\partial x_1} = \frac{\partial u_3}{\partial x_2} = \frac{\partial u_2}{\partial x_1} = 0, \tag{12.3.24}$$

from which we find:

$$d_1 = d_2 = d_3 = d_4 = d_5 = d_6 = 0. \tag{12.3.25}$$

The displacements are, therefore,

$$u_1 = \frac{Mx_1 x_2}{EI_{33}} \qquad (12.3.26)$$

$$u_2 = -\frac{M}{2EI_{33}}(x_1^2 - \nu x_3^2 + \nu x_2^2) \qquad (12.3.27)$$

$$u_3 = -\frac{\nu M}{EI_{33}} x_2 x_3. \qquad (12.3.28)$$

In the plane $x_2 = 0$, both u_1 and u_3 are equal to zero. The deflection of the axis of the bar is obtained from Eq. (12.3.27), in which x_2 and x_3 are set equal to zero. The equation of the deflection curve is

$$u_2 = -\frac{M}{2EI_{33}} x_1^2, \qquad (12.3.29)$$

which is the equation of a parabola. The curvature C is:

$$C = \frac{\dfrac{d^2 u_2}{dx_1^2}}{\left[1 + \left(\dfrac{du_2}{dx_1}\right)^2\right]^{3/2}} \approx -\frac{M}{EI_{33}}, \qquad (12.3.30)$$

if ML/EI_{33} is a small quantity. This is the result that is obtained from the elementary theory.

Now consider a plane $x_1 = K$. After deformation, points on this cross section will have the following coordinates:

$$x_1^* = K + u_1 = K\left(1 + \frac{M}{EI_{33}} x_2\right) \qquad (12.3.31)$$

$$x_2^* = x_2 + u_2 = x_2 - \frac{M}{2EI_{33}}(K^2 - \nu x_3^2 + \nu x_2^2). \qquad (12.3.32)$$

Combining these two relations, and neglecting the terms containing degrees of M/EI_{33} higher than the first, we get:

$$x_1^* \approx K\left(1 + \frac{M}{EI_{33}} x_2^*\right). \qquad (12.3.33)$$

Therefore, in pure bending, a plane cross section remains plane as assumed in the elementary theory. Let us now assume that the cross section is rectangular, and consider the sides $x_3 = \pm b/2$ (Fig. 12.3). After bending, we have:

$$x_3^* = \pm\frac{b}{2} + u_3 = \pm\frac{b}{2}\left(1 - \frac{\nu M}{EI_{33}}x_2\right) \approx \pm\frac{b}{2}\left(1 - \frac{\nu M}{EI_{33}}x_2^*\right). \quad (12.3.34)$$

Therefore, in pure bending the two sides remain straight but become inclined to their original position as shown in Fig. 12.3. The two sides $x_2 = \pm h/2$ become:

$$\begin{aligned} x_2^* &= \pm\frac{h}{2} + u_2 = \pm\frac{h}{2} - \frac{M}{2EI_{33}}\left[K^2 + \nu\left(\frac{h^2}{4} - \nu x_3^2\right)\right] \\ &\approx \pm\frac{h}{2} - \frac{M}{2EI_{33}}\left\{K^2 + \nu\left[\frac{h^2}{4} - \nu(x_3^*)^2\right]\right\}. \end{aligned} \quad (12.3.35)$$

These sides are, therefore, bent into parabolic curves as shown in Fig. 12.3. This effect is called anticlastic curvature.

12.4 Bending of a Narrow Rectangular Cantilever by an End Load

Consider a cantilever having a narrow rectangular cross section and bent by an end load P (Fig. 12.4). The problem can be treated as one of plane stress with

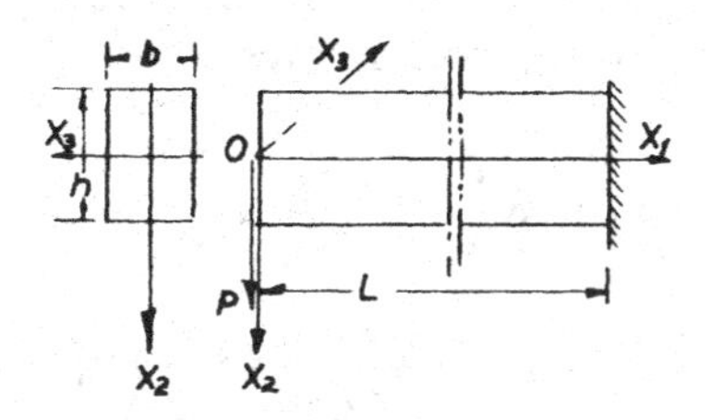

Fig. 12.4

$$\sigma_{33} = \sigma_{13} = \sigma_{23} = 0. \tag{12.4.1}$$

The origin O is taken at the centroid of the cross section at the end where the load is applied. This load will be looked upon as a shearing force which is distributed in the OX_2 direction in the same way as the shear stress σ_{12} is distributed in the beam. We shall invoke Saint-Venant's principle and consider that any disturbance of the stress pattern which otherwise satisfies the problem, dies out very rapidly away from the loadings point (at a distance of h to $2h$). Neglecting the body forces, the solution to this problem is given by the Airy stress function:

$$\phi = A\left(x_1 x_2^3 - \frac{3}{4} x_1 x_2 h^2\right), \tag{12.4.2}$$

. where A is a contant and h is the depth of the beam. From Eq. (9.10.4) we get:

$$\sigma_{11} = \frac{\partial^2 \phi}{\partial x_2^2} = 6Ax_1 x_2, \quad \sigma_{22} = 0, \quad \sigma_{12} = -A\left(3x_2^2 - \frac{3}{4}h^2\right). \tag{12.4.3}$$

The expression for σ_{12} leaves the longitudinal sides free from stress as required by the problem. At the end $x_1 = 0$, we must have:

$$\int_{-\frac{h}{2}}^{+\frac{h}{2}} b\sigma_{12}\, dx_2 = -\int_{-\frac{h}{2}}^{+\frac{h}{2}} bA\left(3x_2^2 - \frac{3}{4}h^2\right) dx_2 = -P, \tag{12.4.4}$$

from which

$$A = -\frac{2P}{bh^3} = -\frac{P}{6I_{33}}. \tag{12.4.5}$$

I_{33} is the moment of inertia of the cross section about OX_3. Eqs. (12.4.2) and (12.4.3) now become:

$$\phi = -\frac{P}{6I_{33}}\left(x_1 x_2^3 - \frac{3}{4} x_1 x_2 h^2\right) \tag{12.4.6}$$

$$\sigma_{11} = -\frac{Px_1 x_2}{I_{33}}, \quad \sigma_{22} = 0 \tag{12.4.7}$$

$$\sigma_{12} = -\frac{P}{2I_{33}}\left(\frac{h^2}{4} - x_2^2\right) = \frac{V_2}{2I_{33}}\left(\frac{h^2}{4} - x_2^2\right). \tag{12.4.8}$$

These equations coincide with the equations of the elementary theory. It should be remembered, however, that P must be distributed according to Eq. (12.4.8) at the loaded end.

Let us now obtain the displacement components corresponding to the state of stress deduced above. The stress-strain relations (8.17.11), (8.17.12), and (8.17.14) give:

$$e_{11} = -\frac{Px_1 x_2}{EI_{33}}, \quad e_{22} = \frac{\nu P x_1 x_2}{EI_{33}}, \quad e_{12} = -\frac{P}{4GI_{33}}\left(\frac{h^2}{4} - x_2^2\right). \qquad (12.4.9)$$

The components of the displacement u_1 and u_2 are obtained by integrating the previous equations. In doing so, we shall assume that u_1 and u_2 are only functions of x_1 and x_2. Thus,

$$u_1 = -\frac{Px_2 x_1^2}{2EI_{33}} + f_1(x_2), \quad u_2 = \frac{\nu P x_1 x_2^2}{2EI_{33}} + f_2(x_1), \qquad (12.4.10)$$

where the functions f_1 and f_2 are unknown. Substituting Eqs. (12.4.10) into the third of Eqs. (12.4.9), we get:

$$-\frac{Px_1^2}{2EI_{33}} + f_1'(x_2) + \frac{\nu P x_2^2}{2EI_{33}} + f_2'(x_1) = -\frac{P}{2GI_{33}}\left(\frac{h^2}{4} - x_2^2\right). \qquad (12.4.11)$$

In this equation, some terms are functions of x_1 alone, some are functions of x_2 alone, and one is a constant. Denoting these terms by $F(x_1)$, $G(x_2)$, and K, we have:

$$F(x_1) = -\frac{Px_1^2}{2EI_{33}} + f_2'(x_1) \qquad (12.4.12)$$

$$G(x_2) = f_1'(x_2) + \frac{\nu P x_2^2}{2EI_{33}} - \frac{Px_2^2}{2GI_{33}} \qquad (12.4.13)$$

$$K = -\frac{Ph^2}{8GI_{33}}, \qquad (12.4.14)$$

and Eq. (12.4.11) is written:

$$F(x_1) + G(x_2) = K. \qquad (12.4.15)$$

Such an equation means that $F(x_1)$ must be some constant d and $G(x_2)$ some constant e. Thus,

$$d + e = K = -\frac{Ph^2}{8GI_{33}} \qquad (12.4.16)$$

and

$$f_2' = \frac{Px_1^2}{2EI_{33}} + d, \quad f_1' = -\frac{\nu Px_2^2}{2EI_{33}} + \frac{Px_2^2}{2GI_{33}} + e. \tag{12.4.17}$$

The functions $f_2(x_1)$ and $f_1(x_2)$ are then:

$$f_2(x_1) = \frac{Px_1^3}{6EI_{33}} + dx_1 + h \tag{12.4.18}$$

$$f_1(x_2) = -\frac{\nu Px_2^3}{6EI_{33}} + \frac{Px_2^3}{6GI_{33}} + ex_2 + g. \tag{12.4.19}$$

Eqs. (12.4.10) now become:

$$u_1 = -\frac{Px_2 x_1^2}{2EI_{33}} - \frac{\nu Px_2^3}{6EI_{33}} + \frac{Px_2^3}{6GI_{33}} + ex_2 + g \tag{12.4.20}$$

$$u_2 = \frac{\nu Px_1 x_2^2}{2EI_{33}} + \frac{Px_1^3}{6EI_{33}} + dx_1 + h. \tag{12.4.21}$$

The constants e, g, d, and h can be determined using Eq. (12.4.16) and the three conditions which are necessary to prevent the beam from moving in the OX_1, OX_2 plane. If we assume that point A at the end section is fixed, then

$$u_1 = u_2 = 0 \text{ at } x_1 = L \text{ and } x_2 = 0,$$

and Eqs. (12.4.20) and (12.4.21) give:

$$g = 0, \quad h = -\frac{PL^3}{6EI_{33}} - dL. \tag{12.4.22}$$

The constant d can be determined from the third constraint at A. Here, we have two possibilities:

a) An element of the axis of the beam is fixed at the end A. In this case (Fig. 12.5):

$$\left(\frac{\partial u_2}{\partial x_1}\right)_{\substack{x_1=L\\x_2=0}} = 0 \tag{12.4.23}$$

and Eq. (12.4.21) gives:

$$d = -\frac{PL^2}{2EI_{33}}. \tag{12.4.24}$$

From Eq. (12.4.16),

$$e = \frac{PL^2}{2EI_{33}} - \frac{Ph^2}{8GI_{33}}. \tag{12.4.25}$$

Substituting the constants in the expressions of u_1 and u_2, we get:

$$u_1 = -\frac{Px_2x_1^2}{2EI_{33}} - \frac{\nu Px_2^3}{6EI_{33}} + \frac{Px_2^3}{6GI_{33}} + x_2\left(\frac{PL^2}{2EI_{33}} - \frac{Ph^2}{8GI_{33}}\right) \tag{12.4.26}$$

$$u_2 = \frac{\nu Px_1x_2^2}{2EI_{33}} + \frac{Px_1^3}{6EI_{33}} - \frac{PL^2x_1}{2EI_{33}} + \frac{PL^3}{3EI_{33}}. \tag{12.4.27}$$

The equation of the deflection curve is:

$$(u_2)_{x_2=0} = \frac{Px_1^3}{6EI_{33}} - \frac{PL^2x_1}{2EI_{33}} + \frac{PL^3}{3EI_{33}}. \tag{12.4.28}$$

At the loaded end we have:

$$(u_2)_{\substack{x_2=0\\x_1=0}} = \frac{PL^3}{3EI_{33}}. \tag{12.4.29}$$

This coincides with the result based on the elementary theory of beams. To illustrate the distortion of the cross section produced by the shearing stresses, consider the displacement and the slope at the fixed end $x_1 = L$. Eq. (12.4.26) gives:

$$(u_1)_{x_1=L} = -\frac{\nu Px_2^3}{6EI_{33}} + \frac{Px_2^3}{6I_{33}G} - \frac{Px_2h^2}{8GI_{33}} \tag{12.4.30}$$

$$(e_{12})_{x_1=L} = \frac{1}{2}\left(\frac{\partial u_1}{\partial x_2}\right)_{x_1=L} = \frac{1}{2}\left[-\frac{\nu Px_2^2}{2EI_{33}} + \frac{Px_2^2}{2I_{33}G} - \frac{Ph^2}{8GI_{33}}\right] \tag{12.4.31}$$

$$(e_{12})_{\substack{x_1=L\\x_2=0}} = -\frac{Ph^2}{16GI_{33}}. \tag{12.4.32}$$

The shape of the cross section after distortion is as shown in Fig. 12.5.

b) A vertical element of the cross section is fixed at the end A. In this case (Fig. 12.6):

$$\left(\frac{\partial u_1}{\partial x_2}\right)_{\substack{x_1=L\\x_2=0}} = 0 \tag{12.4.33}$$

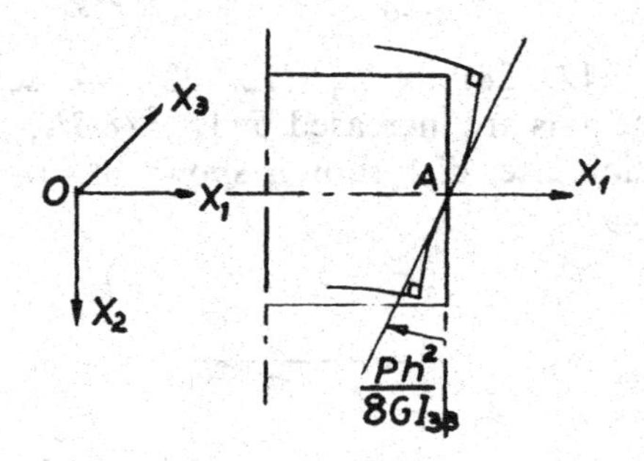

Fig. 12.5

leading to

$$e = \frac{PL^2}{2EI_{33}} \tag{12.4.34}$$

and

$$d = -\frac{PL^2}{2EI_{33}} - \frac{Ph^2}{8GI_{33}}. \tag{12.4.35}$$

Substituting the constants in the expressions of u_1 and u_2, we get:

$$u_1 = -\frac{Px_2x_1^2}{2EI_{33}} - \frac{\nu Px_2^3}{6EI_{33}} + \frac{Px_2^3}{6GI_{33}} + \frac{PL^2x_2}{2EI_{33}} \tag{12.4.36}$$

$$u_2 = \frac{\nu Px_1x_2^2}{2EI_{33}} + \frac{Px_1^3}{6EI_{33}} - x_1\left[\frac{PL^2}{2EI_{33}} + \frac{Ph^2}{8GI_{33}}\right] + \frac{PL^3}{3EI_{33}} + \frac{Ph^2L}{8GI_{33}}. \tag{12.4.37}$$

Therefore,

$$(u_2)_{x_2=0} = \frac{Px_1^3}{6EI_{33}} - x_1\left[\frac{PL^2}{2EI_{33}} + \frac{Ph^2}{8GI_{33}}\right] + \frac{PL^3}{3EI_{33}} + \frac{Ph^2L}{8GI_{33}} \tag{12.4.38}$$

and

$$(u_2)_{\substack{x_1=0\\x_2=0}} = \frac{PL^3}{3EI_{33}} + \frac{Ph^2L}{8GI_{33}}. \tag{12.4.39}$$

Comparing Eq. (12.4.28) to Eq. (12.4.38), we see that the vertical deflections of the axis are increased by $[Ph^2/8GI_{33}][L - x_1]$. The shape of the cross section after distortion is shown in Fig. 12.6.

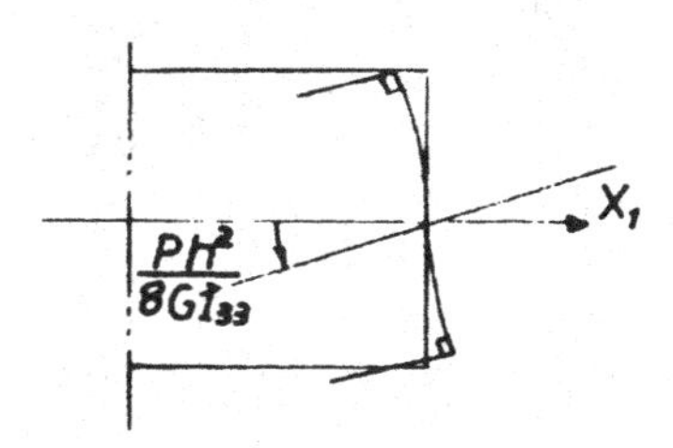

Fig. 12.6

In practice, the conditions are different from those shown in Figs. 12.5 and 12.6. The cross section is usually not free to warp. However, by invoking Saint-Venant's principle, the distribution represented by case (a) or (b) holds for all sections at a small distance from the support (about h).

Although the previous solutions represent a substantial improvement over the elementary theory, they suffer from the weakness that some of the compatibility relations have not been considered. In addition, e_{33} and u_3 have not been examined and both u_1 and u_2 were assumed to be independent of u_3, thus neglecting the anticlastic curvature.

12.5 Bending of a Narrow Rectangular Beam by a Uniform Load

Consider a narrow beam of height h and width b, which is subjected to a uniform loading of intensity q per unit length (Fig. 12.7). The problem can be treated as one of plane stress with

$$\sigma_{33} = \sigma_{13} = \sigma_{23} = 0. \tag{12.5.1}$$

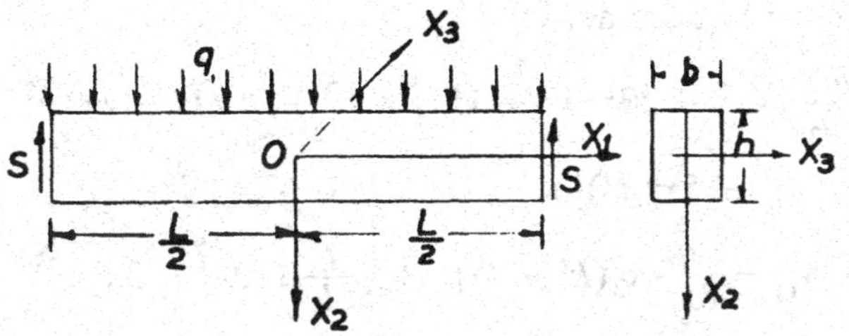

Fig. 12.7

The following conditions must apply:

$$\text{at } x_2 = \frac{h}{2}, \quad \sigma_{22} = \sigma_{12} = 0 \tag{12.5.2}$$

$$\text{at } x_2 = -\frac{h}{2}, \quad \sigma_{22} = -\frac{q}{b}, \quad \sigma_{12} = 0 \tag{12.5.3}$$

$$\text{at } x_1 = \pm\frac{L}{2}, \quad \int_{-\frac{h}{2}}^{+\frac{h}{2}} \sigma_{12}\, b\, dx_2 = \mp S = \mp\frac{qL}{2},$$

$$\int_{-\frac{h}{2}}^{+\frac{h}{2}} \sigma_{11}\, b\, dx_2 = 0, \quad \int_{-\frac{h}{2}}^{+\frac{h}{2}} \sigma_{11}\, b x_2\, dx_2 = 0. \tag{12.5.4}$$

The fifth power stress function,

$$\phi = A(-4x_2^5 + 20x_1^2 x_2^3 - 15x_1^2 x_2 h^2 - 5x_2^3 L^2 + 2x_2^3 h^2 + 5x_1^2 h^3), \tag{12.5.5}$$

provides the solution to this problem. It satisfies the biharmonic Eq. (9.10.11), and when substituted into Eq. (9.10.4) it gives:

$$\sigma_{11} = A\left[-80x_2^3 + 120x_2\left(x_1^2 - \frac{L^2}{4}\right) + 12x_2 h^2\right] \tag{12.5.6}$$

$$\sigma_{22} = A[40x_2^3 - 30x_2 h^2 + 10h^3] \tag{12.5.7}$$

$$\sigma_{12} = A[-120x_1 x_2^2 + 30x_1 h^2]. \tag{12.5.8}$$

Condition (12.5.3) gives:

$$A = \frac{-q}{20h^3 b}, \tag{12.5.9}$$

and the second and third conditions of Eqs. (12.5.4) are satisfied. If the moment of inertia of the cross section about the OX_3 axis is I_{33}, Eqs. (12.5.5) to (12.5.8) become:

$$\phi = \frac{q}{240I_{33}}(4x_2^5 - 20x_1^2x_2^3 + 15x_1^2x_2h^2 + 5x_2^3L^2 - 2x_2^3h^2 - 5x_1^2h^3) \tag{12.5.10}$$

$$\sigma_{11} = \frac{q}{8I_{33}}x_2(L^2 - 4x_1^2) + \frac{q}{60I_{33}}(20x_2^3 - 3x_2h^2) \tag{12.5.11}$$

$$\sigma_{22} = -\frac{q}{24I_{33}}(h^3 - 3h^2x_2 + 4x_2^3) \tag{12.5.12}$$

$$\sigma_{12} = \frac{q}{8I_{33}}(4x_2^2 - h^2)x_1 \tag{12.5.13}$$

The first term in the expression of σ_{11} [i.e., $(q/8I_{33})x_2(L^2 - 4x_1^2)$] is the usual elementary theory term. The second one is a correction term which comes from the consideration of σ_{22} acting on the surface $x_2 = -h/2$, and from the fact that the compatibility equations have been partly satisfied (Sec. 9.10). But this term is not important for long beams in which the span is large in comparison with the depth, it does not contain x_1 and thus is constant along the beam. It is plotted in Fig. 12.8. σ_{22} does not depend on x_1, and decreases from its maximum value at the top edge to zero at the bottom edge as shown in Fig. 12.8.

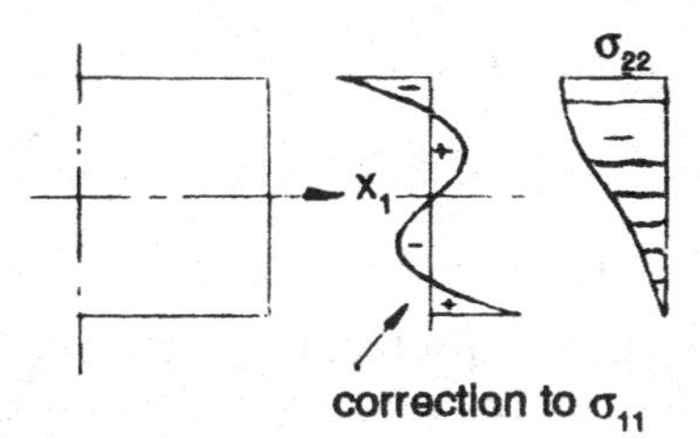

Fig. 12.8

Strictly speaking, the ends $x_1 = \pm L/2$ should be without stress and they are not. However, the resultant force and the resultant moment over the entire section are equal to zero. By Saint-Venant's principle (Sec. 8.13), the effect of the stress dies down at a short distance from the end so that the stress distribution of Eqs. (12.5.11) to (12.5.13) does represent the actual stresses in the large center portion of the beam.

The expressions for the strains and the displacements are obtained by following steps similar to those of Sec. 12.3. The boundary conditions may be given as:

$$\text{at } x_1 = x_2 = 0, \quad u_1 = 0, \quad u_2 = +\delta, \quad \frac{\partial u_2}{\partial x_1} = 0 \qquad (12.5.14)$$

$$\text{at } x_1 = \pm\frac{L}{2} \text{ and } x_2 = 0, \quad u_2 = 0. \qquad (12.5.15)$$

The components of the displacement are then given by:

$$u_1 = \frac{q}{2EI_{33}}\left[\left(\frac{L^2x_1}{4} - \frac{x_1^3}{3}\right)x_2 + x_1\left(\frac{2}{3}x_2^3 - \frac{h^2x_2}{10}\right) + \nu x_1\left(\frac{1}{3}x_2^3 - \frac{h^2}{4}x_2 + \frac{h^3}{12}\right)\right] \qquad (12.5.16)$$

$$u_2 = -\frac{q}{2EI_{33}}\left\{\frac{x_2^4}{12} - \frac{h^2x_2^2}{8} + \frac{h^3x_2}{12} + \nu\left[\left(\frac{L^2}{4} - x_1^2\right)\frac{x_2^2}{2} + \frac{x_2^4}{6} - \frac{h^2x_2^2}{20}\right]\right\} - \frac{q}{2EI_{33}}\left[\frac{L^2x_1^2}{8} - \frac{x_1^4}{12} - \frac{h^2x_1^2}{20} + \left(1 + \frac{1}{2}\nu\right)\frac{h^2x_1^2}{4}\right] + \delta. \qquad (12.5.17)$$

Eq. (12.5.16) shows that the center line has a tensile strain equal to $\nu q/2Eb$. This is due to the compressive stress $(\sigma_{22})_{x_2=0} = -q/2b$. The value of δ is obtained by substituting Eqs. (12.5.15) into Eq. (12.5.17). This gives:

$$\delta = \frac{5}{384}\frac{qL^4}{EI_{33}}\left[1 + \frac{12}{5}\frac{h^2}{L^2}\left(\frac{4}{5} + \frac{\nu}{2}\right)\right]. \qquad (12.5.18)$$

The first term of Eq. (12.5.18) corresponds to the maximum deflection as calculated from the elementary theory. The second term arises from our taking into account the distribution of σ_{22} in the OX_2 direction. This term has an appreciable value only for beams that are very short and deep where the value of h approaches the length of the beam.

The previous solution suffers from the same drawbacks mentioned at the end of Sec. 12.4.

12.6 Cantilever Prismatic Bar of Irregular Cross Section Subjected to a Transverse End Force

The bending of a narrow rectangular cantilever by an end load was studied in a previous section, and it was shown that, in addition to normal stresses proportional to the bending moment, shearing stresses proportional to the shearing force act on each section. Let us now consider the general case of the bending of a cantilever by a force P applied at the end and parallel to one of the principal axes of the cross section. The beam is prismatic in form and no restrictions are applied as to its shape. The OX_1 axis is taken such that it connects the centroids of the cross sections at all points (centroidal axis), and OX_2 and OX_3 are principal axes of inertia. The solution of this problem is due to Saint-Venant: Together with the torsion problem it illustrates the use of the semi-inverse method in the theory of elasticity.

The bending moment on each cross section is (Fig. 12.9):

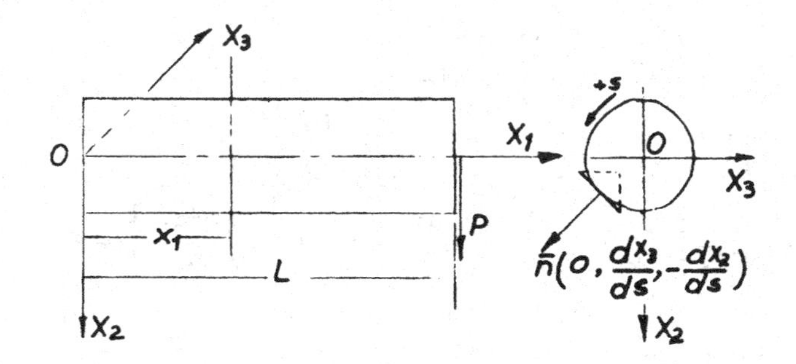

Fig. 12.9

$$M_{13} = -P(L - x_1) \tag{12.6.1}$$

Following the elementary theory, let us assume that the normal stresses over a cross section at a distance x_1 from the fixed end are given by:

$$\sigma_{11} = -\frac{P(L - x_1)x_2}{I_{33}}, \quad \sigma_{22} = \sigma_{33} = 0, \tag{12.6.2}$$

and that

$$\sigma_{23} = 0. \tag{12.6.3}$$

With these assumptions, it will be shown that a solution can be reached which satisfies all the equations of elasticity. Neglecting the body forces, the equilibrium equations become:

$$\frac{\partial \sigma_{11}}{\partial x_1} + \frac{\partial \sigma_{12}}{\partial x_2} + \frac{\partial \sigma_{13}}{\partial x_3} = \frac{Px_2}{I_{33}} + \frac{\partial \sigma_{12}}{\partial x_2} + \frac{\partial \sigma_{13}}{\partial x_3} = 0 \qquad (12.6.4)$$

$$\frac{\partial \sigma_{12}}{\partial x_1} + \frac{\partial \sigma_{22}}{\partial x_2} + \frac{\partial \sigma_{23}}{\partial x_3} = \frac{\partial \sigma_{12}}{\partial x_1} = 0 \qquad (12.6.5)$$

$$\frac{\partial \sigma_{13}}{\partial x_1} + \frac{\partial \sigma_{23}}{\partial x_2} + \frac{\partial \sigma_{33}}{\partial x_3} = \frac{\partial \sigma_{13}}{\partial x_1} = 0. \qquad (12.6.6)$$

From Eqs. (12.6.5) and (12.6.6), we conclude that the shearing stresses σ_{12} and σ_{13} are independent of x_1, which means that the total shearing force on the cross section of the beam is constant as required from the loading conditions. Let us now consider the boundary conditions (7.3.8) and apply them to the lateral surface of the bar which is free from external forces (Fig. 12.9). The three conditions reduce to one, namely:

$$\sigma_{n1} = \sigma_{12}\frac{dx_3}{ds} - \sigma_{13}\frac{dx_2}{ds} = 0. \qquad (12.6.7)$$

Turning now to the Beltrami-Michell compatibility equations (8.10.11) to (8.10.16), we see that the three first, as well as the fifth, of these equations are identically satisfied, while the fourth and sixth give:

$$\nabla^2 \sigma_{12} = -\frac{1}{1+\nu}\frac{P}{I_{33}}$$

$$\nabla^2 \sigma_{13} = 0, \qquad (12.6.8)$$

Therefore, the solution of the problem of bending of a prismatic cantilever bar of any section is reduced to finding σ_{12} and σ_{13} functions of x_2 and x_3, satisfying the equilibrium Eq. (12.6.4), the boundary condition (12.6.7), and the compatibility relations (12.6.8). The solution can be obtained using a stress function $\phi(x_2, x_3)$ defined by the two equations:

$$\sigma_{12} = \frac{\partial \phi}{\partial x_3} - \frac{Px_2^2}{2I_{33}} + f(x_3) \qquad (12.6.9)$$

$$\sigma_{13} = -\frac{\partial \phi}{\partial x_2}, \qquad (12.6.10)$$

where $f(x_3)$ is only a function of x_3 to be determined from the boundary conditions. The equation of equilibrium (12.6.4) is identically satisfied, and the two compatibility relations become:

$$\nabla^2\left(\frac{\partial\phi}{\partial x_3}\right) = \frac{\partial}{\partial x_3}(\nabla^2\phi) = \frac{\nu}{1+\nu}\frac{P}{I_{33}} - \frac{d^2 f(x_3)}{dx_3^2} \qquad (12.6.11)$$

$$\nabla^2\left(\frac{\partial\phi}{\partial x_2}\right) = \frac{\partial}{\partial x_2}(\nabla^2\phi) = 0. \qquad (12.6.12)$$

Eq. (12.6.12) shows that $\nabla^2\phi$ must be independent of x_2. Therefore, integrating Eq. (12.6.11) with respect to x_3, we get:

$$\nabla^2\phi = \frac{\nu}{1+\nu}\frac{Px_3}{I_{33}} - \frac{d}{dx_3}f(x_3) + C, \qquad (12.6.13)$$

where C is a constant of integration. The meaning of this constant is obtained as follows: The rotation of an element of area in the plane of the cross section (from OX_2 to OX_3) is given by:

$$\omega_{32} = \frac{1}{2}\left(\frac{\partial u_3}{\partial x_2} - \frac{\partial u_2}{\partial x_3}\right). \qquad (12.6.14)$$

The rate of change of this rotation in the direction of the OX_1 axis is

$$\begin{aligned}\frac{\partial\omega_{32}}{\partial x_1} &= \frac{1}{2}\left[\frac{\partial}{\partial x_1}\left(\frac{\partial u_3}{\partial x_2} - \frac{\partial u_2}{\partial x_3}\right)\right] \\ &= \frac{1}{2}\left[\frac{\partial}{\partial x_2}\left(\frac{\partial u_3}{\partial x_1} + \frac{\partial u_1}{\partial x_3}\right) - \frac{\partial}{\partial x_3}\left(\frac{\partial u_2}{\partial x_1} + \frac{\partial u_1}{\partial x_2}\right)\right].\end{aligned} \qquad (12.6.15)$$

Therefore,

$$\begin{aligned}\frac{\partial\omega_{32}}{\partial x_1} &= \frac{\partial e_{13}}{\partial x_2} - \frac{\partial e_{12}}{\partial x_3} = \frac{1}{2G}\left[\frac{\partial\sigma_{13}}{\partial x_2} - \frac{\partial\sigma_{12}}{\partial x_3}\right] \\ &= \frac{1}{2G}\left[-\frac{\partial^2\phi}{\partial x_2^2} - \frac{\partial^2\phi}{\partial x_3^2} - \frac{df(x_3)}{dx_3}\right],\end{aligned} \qquad (12.6.16)$$

and, from Eq. (12.6.13), we get:

$$2G\frac{\partial\omega_{32}}{\partial x_1} = -\frac{\nu}{1+\nu}\frac{Px_3}{I_{33}} - C. \qquad (12.6.17)$$

Thus the rate of rotation about the OX_1 axis consists of two parts: a) A constant rate represented by C, which corresponds to the uniform twist of a cylindrical rod under pure torsion, and b) a rate which is a function of x_3 and which corresponds to a distortion of the cross section in the OX_2, OX_3 plane. This second rate is similar to the anticlastic curvature of Sec. 12.3. In what follows, we shall position the force P in the OX_2, OX_3 plane, parallel to the principal axis OX_2 and such that the couple it exerts on the end cross section cancels out the twist represented by C. Therefore,

$$C = 0,$$

and Eq. (12.6.13) becomes:

$$\nabla^2 \phi = \frac{\nu}{1+\nu} \frac{Px_3}{I_{33}} - \frac{d}{dx_3} f(x_3). \tag{12.6.18}$$

Substituting Eqs. (12.6.9) and (12.6.10) into the boundary condition (12.6.7), we get:

$$\frac{\partial \phi}{\partial x_3} \frac{dx_3}{ds} + \frac{\partial \phi}{\partial x_2} \frac{dx_2}{dx} = \frac{d\phi}{ds} = \left[\frac{Px_2^2}{2I_{33}} - f(x_3) \right] \frac{dx_3}{ds}. \tag{12.6.19}$$

The form of ϕ on the boundary can be computed from Eq. (12.6.19) if $f(x_3)$ is suitably chosen. We shall choose the function $f(x_3)$ such that *the right-hand side of Eq. (12.6.19) is equal to zero for each particular* case. ϕ is then constant on the boundary. If this constant is chosen equal to zero, the solution of the bending problem reduces to solving Eq. (12.6.18) with $\phi = 0$ on the boundary. The problem is similar to that met in the study of torsion.

In the following, we shall examine the cases of the circular and the elliptic cross sections. Other cross sections can be found in reference [2].

1) *Circular Cross Section*

The boundary of the cross section (Fig. 12.10) is given by the equation:

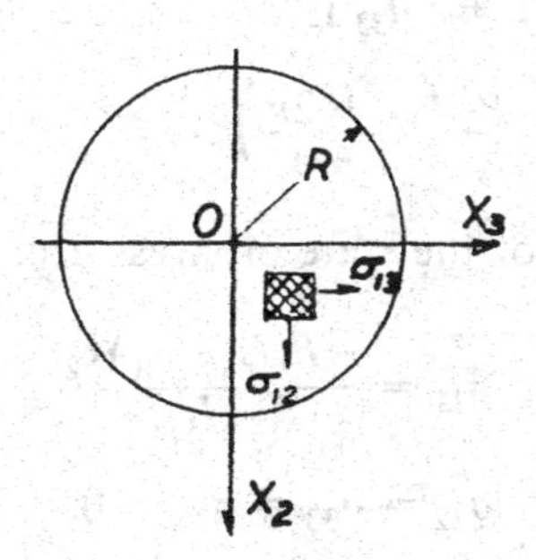

Fig. 12.10

$$x_2^2 + x_3^2 = R^2. \tag{12.6.20}$$

The right-hand side of Eq. (12.6.19) vanishes if we take

$$f(x_3) = \frac{P}{2I_{33}}(R^2 - x_3^2). \tag{12.6.21}$$

Substituting Eq. (12.6.21) into Eq. (12.6.18), the stress function is determined by:

$$\nabla^2\phi = \frac{1 + 2\nu}{1 + \nu}\frac{Px_3}{I_{33}}, \tag{12.6.22}$$

and the condition that $\phi = 0$ on the boundary. Let us take

$$\phi = mx_3(x_2^2 + x_3^2 - R^2), \tag{12.6.23}$$

where m is a constant. ϕ is then equal to zero on the boundary and satisfies Eq. (12.6.22), provided

$$m = \frac{1 + 2\nu}{1 + \nu}\frac{P}{8I_{33}}. \tag{12.6.24}$$

Eq. (12.6.23) now becomes:

$$\phi = \frac{1 + 2\nu}{1 + \nu}\frac{P}{8I_{33}}x_3(x_2^2 + x_3^2 - R^2). \tag{12.6.25}$$

From Eqs. (12.6.9) and (12.6.10) we have:

$$\sigma_{12} = \frac{(3 + 2\nu)P}{8(1 + \nu)I_{33}}\left[R^2 - x_2^2 - \frac{1 - 2\nu}{3 + 2\nu}x_3^2\right] \tag{12.6.26}$$

$$\sigma_{13} = -\frac{P}{4I_{33}}\left(\frac{1 + 2\nu}{1 + \nu}\right)x_2 x_3. \tag{12.6.27}$$

The other components of the state of stress are:

$$\sigma_{11} = -\frac{P(L - x_1)x_2}{I_{33}} \tag{12.6.28}$$

$$\sigma_{22} = \sigma_{33} = \sigma_{23} = 0. \tag{12.6.29}$$

The distribution of stresses σ_{12} and σ_{13} on any cross section gives a resultant along the vertical diameter OX_2:

$$\int \sigma_{12}\, dA = P \tag{12.6.30}$$

$$\int \sigma_{13}\, dA = 0. \tag{12.6.31}$$

Along the horizontal diameter of the cross section ($x_2 = 0$), we find:

$$\sigma_{12} = \frac{(3 + 2\nu)P}{8(1 + \nu)I_{33}}\left[R^2 - \frac{1 - 2\nu}{3 + 2\nu}x_3^2\right] \tag{12.6.32}$$

$$\sigma_{13} = 0. \tag{12.6.33}$$

The maximum value of the shearing stress occurs along the OX_1 axis and is equal to

$$(\sigma_{12})_{\max} = \frac{(3 + 2\nu)P}{8(1 + \nu)I_{33}}R^2. \tag{12.6.34}$$

The shearing stress at the end of a horizontal diameter ($x_3 = \pm R$) is:

$$(\sigma_{12})_{x_3=\pm R} = \frac{(1 + 2\nu)}{4(1 + \nu)}\frac{PR^2}{I_{33}}. \tag{12.6.35}$$

One notices that the magnitude of the shearing stress depends on Poisson's ratio. For $\nu = 0.3$ Eqs. (12.6.34) and (12.6.35) become:

$$(\sigma_{12})_{\max} = \frac{1.38P}{A}, \quad (\sigma_{12})_{x_3=\pm R} = \frac{1.23P}{A}, \tag{12.6.36}$$

where A is the cross sectional area of the beam. The elementary theory based on the assumption that the shearing stress σ_{12} is uniformly distributed along a horizontal diameter of the cross section gives:

$$\sigma_{12} = \frac{4}{3}\frac{P}{A}. \tag{12.6.37}$$

Thus, the elementary beam theory, in spite of the fact that it violates both compatibility and boundary conditions, is in error only by 3 to 4 percent.

2) *Elliptic Cross Section*

The boundary of the cross section (Fig. 12.11) is given by the equation:

$$\frac{x_2^2}{a^2} + \frac{x_3^2}{b^2} = 1. \tag{12.6.38}$$

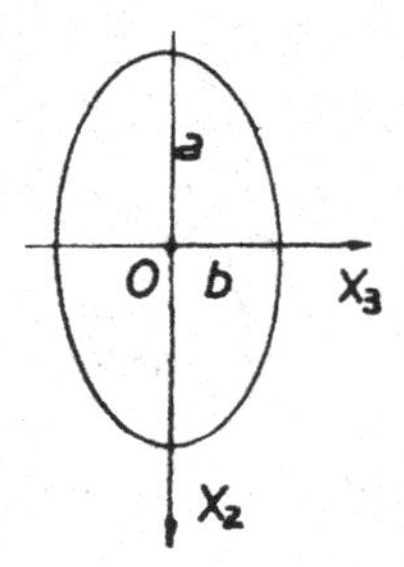

Fig. 12.11

The right-hand side of Eq. (12.6.19) vanishes if we take

$$f(x_3) = -\frac{P}{2I_{33}}\left(\frac{a^2}{b^2}x_3^2 - a^2\right). \tag{12.6.39}$$

Eq. (12.6.13) now becomes:

$$\nabla^2\phi = \frac{Px_3}{I_{33}}\left(\frac{a^2}{b^2} + \frac{\nu}{1+\nu}\right). \tag{12.6.40}$$

Eq. (12.6.40) and the condition that $\phi = 0$ on the boundary determine the stress function ϕ. Let us take

$$\phi = mx_3\left(x_2^2 + \frac{a^2}{b^2}x_3^2 - a^2\right), \tag{12.6.41}$$

where m is a constant. ϕ is then equal to zero on the boundary and satisfies Eq. (12.6.40), provided

$$m = \frac{P}{I_{33}}\frac{(1+\nu)a^2 + \nu b^2}{2(1+\nu)(3a^2+b^2)}. \tag{12.6.42}$$

Eq. (12.6.41) now becomes:

$$\phi = \frac{(1+\nu)a^2 + \nu b^2}{2(1+\nu)(3a^2+b^2)}\frac{Px_3}{I_{33}}\left(x_2^2 + \frac{a^2}{b^2}x_3^2 - a^2\right). \tag{12.6.43}$$

The stress components are obtained from Eqs. (12.6.9) and (12.6.10):

$$\sigma_{12} = \frac{2(1+\nu)a^2 + b^2}{(1+\nu)(3a^2+b^2)}\frac{P}{2I_{33}}\left[a^2 - x_2^2 - \frac{(1-2\nu)a^2}{2(1+\nu)a^2+b^2}x_3^2\right] \tag{12.6.44}$$

$$\sigma_{13} = -\frac{(1+\nu)a^2 + \nu b^2}{(1+\nu)(3a^2+b^2)} \frac{Px_2x_3}{I_{33}}. \tag{12.6.45}$$

The other components of the state of stress are:

$$\sigma_{11} = -\frac{P(L - x_1)x_2}{I_{33}} \tag{12.6.46}$$

$$\sigma_{22} = \sigma_{33} = \sigma_{23} = 0. \tag{12.6.47}$$

Along the horizontal diameter of the cross section ($x_2 = 0$), we find:

$$\sigma_{12} = \frac{2(1+\nu)a^2 + b^2}{(1+\nu)(3a^2+b^2)} \frac{P}{2I_{33}} \left[a^2 - \frac{(1-2\nu)a^2}{2(1+\nu)a^2 + b^2} x_3^2 \right] \tag{12.6.48}$$

$$\sigma_{13} = 0. \tag{12.6.49}$$

The maximum value of the shearing stress occurs along the OX_1 axis, and is equal to:

$$(\sigma_{12})_{max} = \frac{Pa^2}{2I_{33}} \left[1 - \frac{a^2 + \dfrac{\nu b^2}{(1+\nu)}}{3a^2 + b^2} \right]. \tag{12.6.50}$$

If b is small compared to a, the terms containing b^2/a^2 can be neglected, and

$$(\sigma_{12})_{max} = \frac{Pa^2}{3I_{33}} = \frac{4}{3}\frac{P}{A}, \tag{12.6.51}$$

which coincides with the elementary theory. If b is very large in comparison with a, then

$$(\sigma_{12})_{max} = \frac{2P}{(1+\nu)A}. \tag{12.6.52}$$

The stress at the ends of the horizontal diameter, when b is very large compared to a, is

$$(\sigma_{12})_{\substack{x_2=0\\x_3=\pm b}} = \frac{4\nu P}{(1+\nu)A}. \tag{12.6.53}$$

The distribution along a horizontal diameter is considerably far from being uniform in this case, and for a Poisson ratio $\nu = 0.3$ we get

$$(\sigma_{12})_{max} = 1.54\frac{P}{A} \text{ and } (\sigma_{12})_{\substack{x_2=0\\x_3=\pm b}} = 0.92\frac{P}{A}. \tag{12.6.54}$$

12.7 Shear Center

If a beam's cross section has two axes of symmetry, and if the load is applied in such a manner that the load line passes through the centroid of the cross section, the beam will deflect without twisting (Fig. 12.12a).

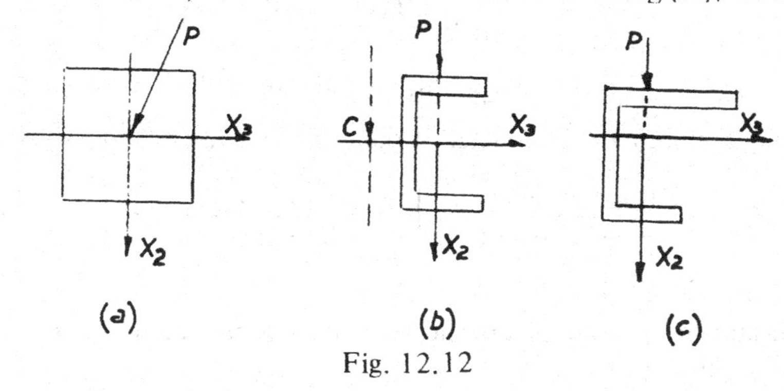

Fig. 12.12

If the beam's cross section has only a single axis of symmetry and the plane of the loading is such that it does not contain the axis of symmetry even though the load line passes through the centroid of the cross section, the beam will be subjected to twisting (Fig. 12.12.b), but it is possible to locate a point C on the axis of symmetry, through which the load must pass in order to eliminate the tendency for twist. This point is called the shear center. If the beam's cross section has no axis of symmetry (Fig. 12.12c) it is also possible to locate the shear center.

The shear center may be generally defined as the point on the cross sectional plane of a beam through which the resultant of the transverse load (shear) must be applied in order that the stresses in the beam may be determined only from the theories of pure bending and transverse shear. In discussing the cantilever problem of Sec. 12.6, it was assumed that the force P was parallel to the OX_2 axis and at such a distance from the centroid that twisting of the bar did not occur. This distance determines the location of the center of shear and can be obtained once the shear stresses σ_{12} and σ_{13} (computed without allowing for any torsional twist) are known. For this purpose, we evaluate the moment about the centroid produced by σ_{12} and σ_{13} and this moment must be equal to that of P. Therefore, (Fig. 12.13),

$$M_{11} = \int\int (-\sigma_{12}x_3 + \sigma_{13}x_2)dx_2\,dx_3 \tag{12.7.1}$$

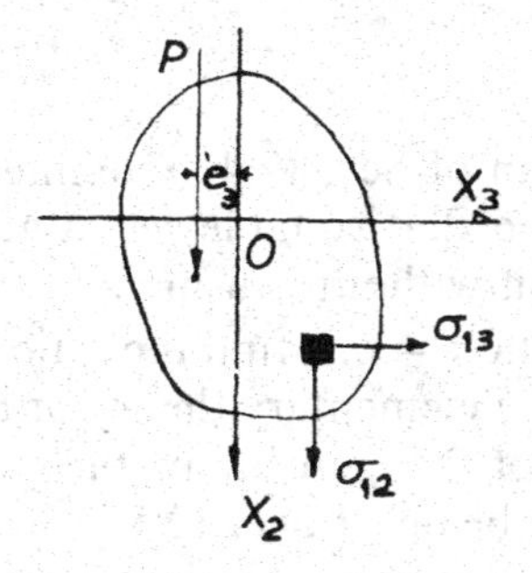

Fig. 12.13

and

$$e_3 = \frac{M_{11}}{P}. \tag{12.7.2}$$

For M_{11} positive, e_3 must be taken in the negative direction of OX_3.

For sections that are loaded in planes not parallel to a principal plane of inertia (Fig. 12.14), we first locate the principal axes OX'_2 and OX'_3, then decompose the shearing force P along these two axes. The problem is then solved independently for each component leading to the values of e'_2 and e'_3 which locate the shear center.

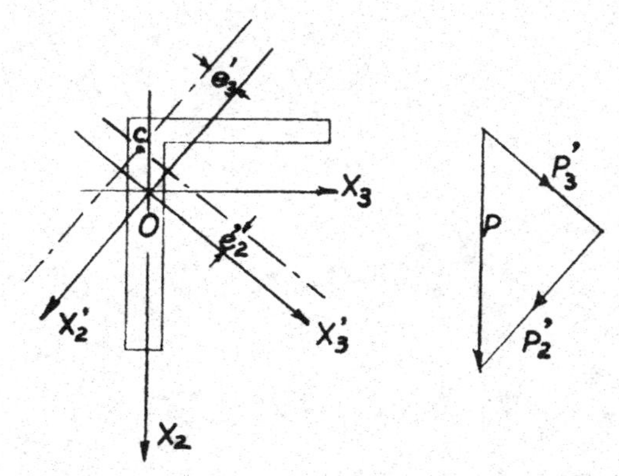

Fig. 12.14

PROBLEMS

1. Solve the problem of Sec. 12.4 assuming that the rectangular beam is not narrow and that suitable restraints prevent displacements in the OX_3 direction without causing any friction on the lateral faces.
2. For a Poisson ratio $\nu = 0.3$, find the ratio of length to height in order that the use of the elementary theory of beams does not produce an error in excess of 2.5 percent in the prediction of the maximum deflection of the beam of Sec. 12.5.

REFERENCES

[1] S. Timoshenko, *Strength of Materials*, Vol. 1, Van Nostrand, Princeton, N. J., 1955.
[2] S. Timoshenko and J. N. Goddier, *Theory of Elasticity*, McGraw Hill, New York, N. Y., 1970.

CHAPTER 13

CURVED BEAMS

13.1 Introduction

In the previous chapter, only straight beams were considered, which means that all the axial fibers were of the same length before bending. In this chapter, we shall study cases in which the line joining the centroids of the cross sections, i.e., the center line, has an initial curvature. The center line is a plane curve and the cross sections have an axis of symmetry in this plane. The beams have a constant cross section. We shall first summarize the results of the simplified theory due to Winkler, then examine more accurate solutions satisfying the conditions of the theory of elasticity. The notation and sign conventions defined in Sec. 12.1 will be followed in this chapter.

13.2 The Simplified Theory of Curved Beams

Taking the OX_1 axis tangent to the line joining the centroids of the cross sections, the following assumptions are made (Fig. 13.1):

a - The cross section has an axis of symmetry and the transverse loadings, (consequently, the bending moments) are applied in a plane containing the axis of the symmetry.

b - Transverse cross sections which were originally plane and normal to the center line remain plane after bending and axial deformation.

c - There is no lateral pressure between the longitudinal fibers.

d - The distribution of shearing stresses over the cross section is the same as for straight beams.

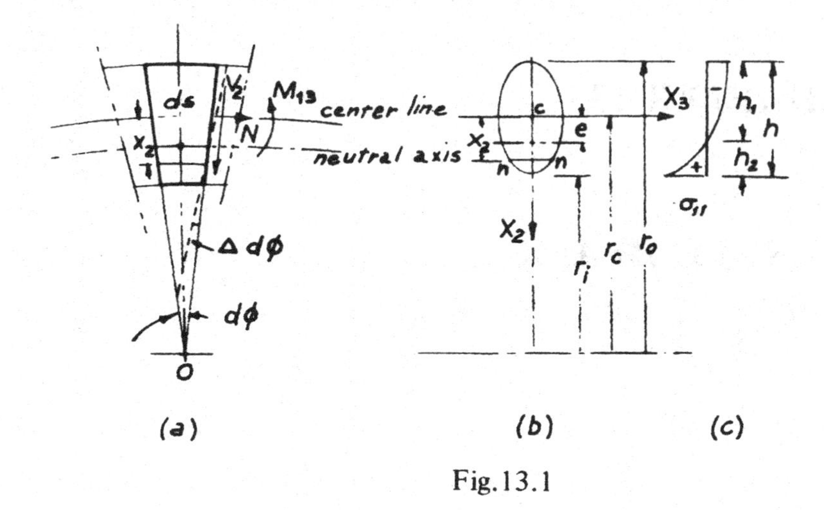

Fig. 13.1

With the above assumptions and the same sign conventions defined in Fig. 12.1 (and 13.1), the following well-known equations have been deduced [1]:

$$\sigma_{11} = \frac{M_{13}(x_2 - e)}{Ae(r_c - x_2)} + \frac{N}{A} \tag{13.2.1}$$

$$\Delta d\phi = \frac{M_{13}\, ds}{er_c AE} - \frac{N ds}{AEr_c} \tag{13.2.2}$$

$$e = r_c \frac{m}{m+1} \quad \text{or} \quad m = \frac{e}{r_c - e} \tag{13.2.3}$$

$$mA = \int \frac{x_2\, dA}{r_c - x_2} \tag{13.2.4}$$

$$\sigma_{12} = \frac{V_2 Q_3}{I_{33} b}, \tag{13.2.5}$$

where

V_2 = the shearing force in the direction of OX_2
Q_3 = the static moment about the OX_3 axis (see Sec. 12.2)
N = the longitudinal force on the cross section
M_{13} = the bending moment about the OX_3 axis
e = the distance between the center line and the neutral axis
A = the cross sectional area
r_c = the radius of curvature of the center line

mA = the modified area
E = Young's modulus
$\Delta d\phi$ = the rotation of the cross section due to bending and to N (see Fig. 13.2)

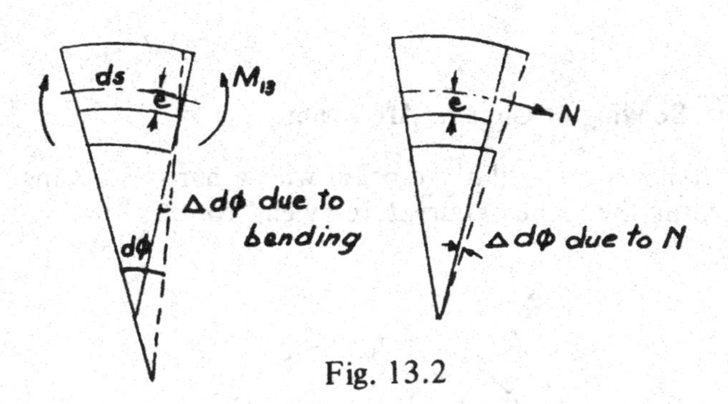

Fig. 13.2

The distribution of σ_{11} due to the effect of bending is hyperbolic as shown in Fig. 13.lc, and the neutral axis is displaced towards the center of curvature by the amount e. The quantity m can easily be computed for any cross section [1]. For rectangular and trapezoidal cross sections, we respectively have (Fig. 13.3a,b):

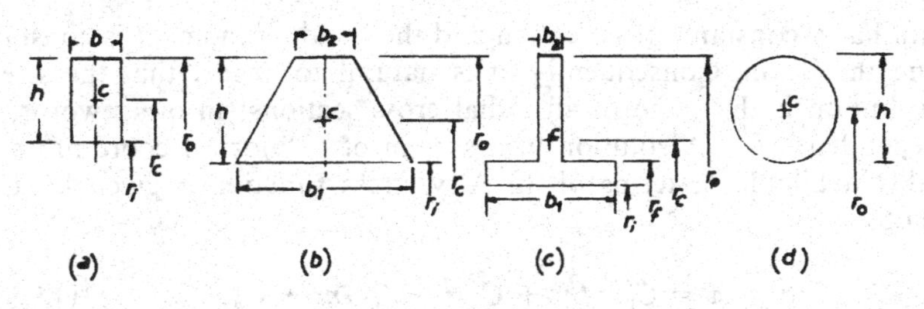

Fig. 13.3

$$m = \frac{r_c}{h} \ln \frac{r_o}{r_i} - 1,$$
$$m = \frac{r_c}{A} \left\{ \left[b_2 + \frac{r_o(b_1 - b_2)}{h} \right] \ln \frac{r_o}{r_i} - (b_1 - b_2) \right\} - 1. \qquad (13.2.6)$$

For tee sections and circular sections, we have (Fig. 13.3c,d):

$$m = \frac{r_c}{A}\left(b_1 \ln\frac{r_f}{r_i} + b_2 \ln\frac{r_o}{r_f}\right) - 1,$$
$$Am = 2\Pi r_c\left(r_c - \sqrt{r_c^2 - \frac{h^2}{4}}\right) - A. \tag{13.2.7}$$

13.3 Pure Bending of Circular Arc Beams

Let us consider a circular arc beam with a narrow rectangular cross section bent in the plane of curvature by end couples M (Fig. 13.4). The

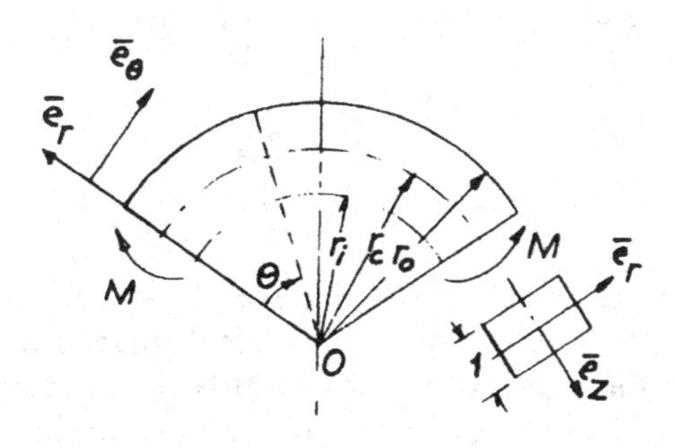

Fig. 13.4

beam has a constant cross section and the bending moment is constant along the beam. Consequently, it is natural to expect that the stress distribution is the same in all radial cross sections; in other words, is independent of θ. A solution in a system of cylindrical coordinates is readily available in terms of an Airy stress function ϕ given by Eq. (9.11.5):

$$\phi = C_1 r^2 \ln r + C_2 r^2 + C_3 \ln r + C_4, \tag{13.3.1}$$

where C_1, C_2, C_3, and C_4 are constants of integration to be determined from the boundary conditions. The body forces are neglected and since the assumption has been made that the cross section is narrow, plane stress conditions will apply. For simplicity, the width of the cross section will be assumed equal to unity. The boundary conditions are:

1) $\sigma_{rr} = 0$ for $r = r_i$ and $r = r_o$
2) $\int_{r_i}^{r_o} \sigma_{\theta\theta}\, dr = 0$ over any cross section

3) $\int_{r_i}^{r_o} r\sigma_{\theta\theta}\,dr = -M$ over any cross section
4) $\sigma_{r\theta} = 0$ at the boundary.

The stresses are given by Eq. (9.11.6) to (9.11.8). The first boundary condition gives:

$$\frac{C_3}{r_i^2} + C_1(1 + 2\ell n r_i) + 2C_2 = 0 \qquad (13.3.2)$$

$$\frac{C_3}{r_o^2} + C_1(1 + 2\ell n r_o) + 2C_2 = 0. \qquad (13.3.3)$$

The second boundary condition gives:

$$r_o\left[\frac{C_3}{r_o^2} + C_1(1 + 2\ell n r_o) + 2C_2\right] - r_i\left[\frac{C_3}{r_i^2} + C_1(1 + 2\ell n r_i) + 2C_2\right] = 0, \qquad (13.3.4)$$

which is a combination of the two previous equations. This shows that the second boundary condition will be satisfied if the first one is. The third boundary condition gives:

$$\int_{r_i}^{r_o} r\sigma_{\theta\theta}\,dr = \int_{r_i}^{r_o} \frac{d^2\phi}{dr^2}\,r\,dr = \int_{r_i}^{r_o} r\,d\left(\frac{d\phi}{dr}\right) = \left[r\frac{d\phi}{dr}\right]_{r_i}^{r_o} - \left[\phi\right]_{r_i}^{r_o} = -M. \qquad (13.3.5)$$

But, because of Eqs. (13.3.2) and (13.3.3):

$$\left[r\frac{d\phi}{dr}\right]_{r_i}^{r_o} = 0.$$

Therefore,

$$\left[\phi\right]_{r_i}^{r_o} = C_3\,\ell n\frac{r_o}{r_i} + C_1(r_o^2\,\ell n r_o - r_i^2\,\ell n r_i) + C_2(r_o^2 - r_i^2) = M. \qquad (13.3.6)$$

The fourth boundary condition is identically satisfied. Eqs. (13.3.2), (13.3.3), and (13.3.6) completely determine the three constants C_1, C_2, and C_3:

$$C_1 = \frac{2(r_o^2 - r_i^2)}{4r_i^2 r_o^2\left(\ln\frac{r_o}{r_i}\right)^2 - (r_o^2 - r_i^2)^2} M \tag{13.3.7}$$

$$C_2 = \frac{2(r_i^2 \ln r_i - r_o^2 \ln r_o) - (r_o^2 - r_i^2)}{4r_i^2 r_o^2\left(\ln\frac{r_o}{r_i}\right)^2 - (r_o^2 - r_i^2)^2} M \tag{13.3.8}$$

$$C_3 = \frac{4r_i^2 r_o^2 \ln\frac{r_o}{r_i}}{4r_i^2 r_o^2\left(\ln\frac{r_o}{r_i}\right)^2 - (r_o^2 - r_i^2)^2} M. \tag{13.3.9}$$

Substituting these values into Eq. (13.3.1), yields the expression of ϕ. The stresses are now obtained from Eqs. (9.11.6) to (9.11.8). Setting

$$D = 4r_i^2 r_o^2\left(\ln\frac{r_o}{r_i}\right)^2 - (r_o^2 - r_i^2)^2, \tag{13.3.10}$$

we get:

$$\sigma_{rr} = \frac{4M}{D}\left(\frac{r_i^2 r_o^2}{r^2}\ln\frac{r_o}{r_i} - r_i^2 \ln\frac{r}{r_i} - r_o^2 \ln\frac{r_o}{r}\right) \tag{13.3.11}$$

$$\sigma_{\theta\theta} = \frac{4M}{D}\left(-\frac{r_i^2 r_o^2}{r^2}\ln\frac{r_o}{r_i} - r_i^2 \ln\frac{r}{r_i} - r_o^2 \ln\frac{r_o}{r} + r_o^2 - r_i^2\right) \tag{13.3.12}$$

$$\sigma_{r\theta} = 0. \tag{13.3.13}$$

Since this stress field satisfies equilibrium, compatibility, and boundary conditions, it gives the solution of the problem provided the moment M is applied to the end of the beam by external forces corresponding to the $\sigma_{\theta\theta}$ distribution of Eq. (13.3.12). This distribution is shown in Fig. 13.5 together with that of σ_{rr}. If, however, the moment is applied so as to give a distribution different from that shown in Fig. 13.5, we know from Saint-Venant's principle that at a distance from the ends of the beam corresponding to the beam's depth $r_o - r_i$, Eqs. (13.3.11) to (13.3.13) will hold with a high degree of accuracy.

The strains are obtained from the stress-strain relations (8.17.11), (8.17.12), and (8.17.14). The displacements u_r and u_θ are obtained from

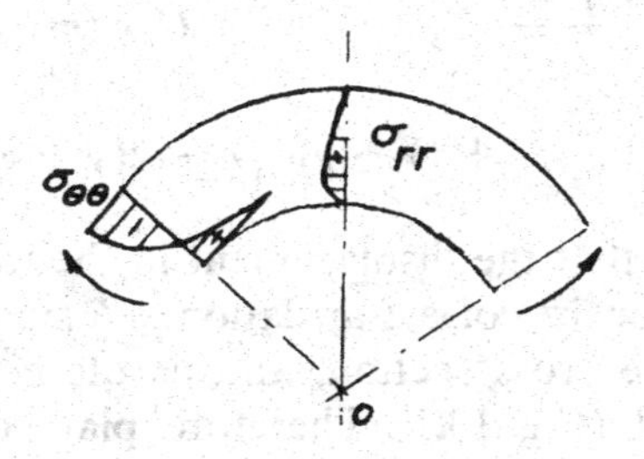

Fig. 13.5

Eqs. (9.11.16) and (9.11.17), in which C_1, C_2, and C_3 are given by Eqs. (13.3.7) to (13.3.9). To determine the constants A, B, and F, let us consider the centroid of the cross section from which θ is measured, and choose the element of radius at this point to be fixed. Thus,

$$\text{at } r = \frac{r_o + r_i}{2} \text{ and } \theta = 0,\ u_r = 0 \text{ and } \frac{\partial u_\theta}{\partial r} = 0.$$

From these boundary conditions, the following equations result:

$$A = B = 0 \tag{13.3.14}$$

$$F = \frac{1}{E}\left[\frac{1+\nu}{r_c}C_3 - 2(1-\nu)C_1 r_c \ell n r_c + (1+\nu)C_1 r_c - 2(1-\nu)C_2 r_c\right]. \tag{13.3.15}$$

Thus,

$$u_r = \frac{1}{E}\left[-\frac{(1+\nu)}{r}C_3 + 2(1-\nu)C_1 r \ell n r - (1+\nu)C_1 r + 2(1-\nu)C_2 r\right]$$
$$+ \frac{\cos\theta}{E}\left[\frac{1+\nu}{r_c}C_3 - 2(1-\nu)C_1 r_c \ell n r_c + (1+\nu)C_1 r_c - 2(1-\nu)C_2 r_c\right] \tag{13.3.16}$$

$$u_\theta = \frac{4C_1 r\theta}{E} - \frac{\sin\theta}{E}\left[\frac{1+\nu}{r_c}C_3 - 2(1-\nu)C_1 r_c \ln r_c \right. \\ \left. + (1+\nu)C_1 r_c - 2(1-\nu)C_2 r_c\right]. \tag{13.3.17}$$

Eq. (13.3.17) shows that the displacement in the transverse direction of any cross section consists of a translation $-F \sin\theta$, which is the same for all points of the cross section, and a rotation $4C_1\theta/E$ about the center of curvature 0 (Fig. 13.1). Therefore, plane cross sections remain plane as is assumed in the simplified theory.

Finally, it must be remembered that some compatibility equations have been ignored and all the quantities assumed independent of x_3.

13.4 Circular Arc Cantilever Beam Bent by a Force at the End

Let the beam have a rectangular narrow cross section, which for simplicity will be assumed equal to unity. The bending moment at any cross section *mn* (Fig. 13.6) is proportional to $\sin\theta$ and according to the

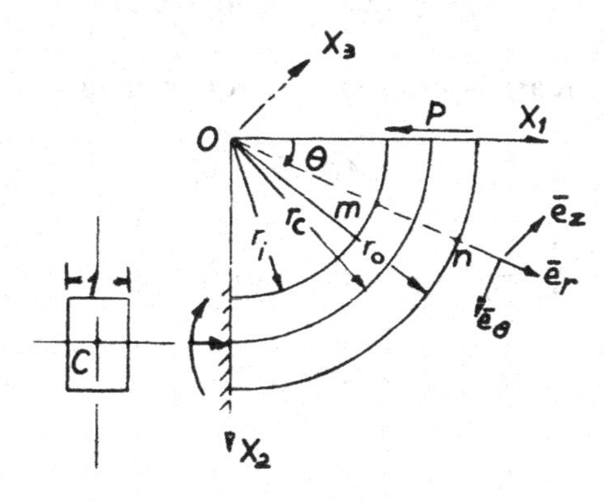

Fig. 13.6

simplified theory the normal stress $\sigma_{\theta\theta}$ is proportional to the bending moment. An Airy stress function of the form

$$\phi = f(r)\sin\theta \tag{13.4.1}$$

will therefore be tried. It will be shown that this stress function does indeed provide us with the solution to the problem. Eq. (13.4.1), when substituted into the biharmonic $\nabla^4\phi = 0$ and the $\sin\theta$ divided out, yields the ordinary differential equation:

$$\left(\frac{d^2}{dr^2} + \frac{1}{r}\frac{d}{dr} - \frac{1}{r^2}\right)\left(\frac{d^2 f}{dr^2} + \frac{1}{r}\frac{df}{dr} - \frac{f}{r^2}\right) = 0. \tag{13.4.2}$$

The general solution is:

$$f(r) = C_1 r^3 + \frac{C_2}{r} + C_3 r + C_4 r \ln r, \tag{13.4.3}$$

in which C_1, C_2, C_3, and C_4 are constants to be determined from the boundary conditions. The stress function ϕ is now given by:

$$\phi = \left(C_1 r^3 + \frac{C_2}{r} + C_3 r + C_4 r \ln r\right)\sin\theta. \tag{13.4.4}$$

Using Eqs. (9.10.15), we find the following expressions for the stress components:

$$\sigma_{rr} = \left(2C_1 r - \frac{2C_2}{r^3} + \frac{C_4}{r}\right)\sin\theta \tag{13.4.5}$$

$$\sigma_{\theta\theta} = \left(6C_1 r + \frac{2C_2}{r^3} + \frac{C_4}{r}\right)\sin\theta \tag{13.4.6}$$

$$\sigma_{r\theta} = -\left(2C_1 r - \frac{2C_2}{r^3} + \frac{C_4}{r}\right)\cos\theta. \tag{13.4.7}$$

The boundary conditions are:

$$1)\ \sigma_{rr} = \sigma_{r\theta} = 0 \qquad \text{for } r = r_i \text{ and } r = r_o \tag{13.4.8}$$

$$2)\ \int_{r_i}^{r_o} \sigma_{r\theta}\, dr = P. \qquad \text{for } \theta = 0. \tag{13.4.9}$$

The first conditions give:

$$2C_1 r_i - \frac{2C_2}{r_i^3} + \frac{C_4}{r_i} = 0, \quad 2C_1 r_o - \frac{2C_2}{r_o^3} + \frac{C_4}{r_o} = 0. \tag{13.4.10}$$

The second condition gives:

$$\left[-C_1 r^2 - \frac{C_2}{r^2} - C_4 \ell n r\right]_{r_i}^{r_o} = P = -C_1(r_o^2 - r_i^2) + C_2 \frac{(r_o^2 - r_i^2)}{r_i^2 r_o^2} - C_4 \ell n \frac{r_o}{r_i}. \quad (13.4.11)$$

From Eqs. (13.4.10) and (13.4.11), we get:

$$C_1 = \frac{P}{2N}, \quad C_2 = -\frac{P r_i^2 r_o^2}{2N}, \quad C_4 = -\frac{P}{N}(r_i^2 + r_o^2), \quad (13.4.12)$$

where

$$N = r_i^2 - r_o^2 + (r_i^2 + r_o^2)\ell n \frac{r_o}{r_i}. \quad (13.4.13)$$

The expressions for the stresses now become:

$$\sigma_{rr} = \frac{P}{N}\left(r + \frac{r_i^2 r_o^2}{r^3} - \frac{r_i^2 + r_o^2}{r}\right)\sin\theta \quad (13.4.14)$$

$$\sigma_{\theta\theta} = \frac{P}{N}\left(3r - \frac{r_i^2 r_o^2}{r^3} - \frac{r_i^2 + r_o^2}{r}\right)\sin\theta \quad (13.4.15)$$

$$\sigma_{r\theta} = -\frac{P}{N}\left(r + \frac{r_i^2 r_o^2}{r^3} - \frac{r_i^2 + r_o^2}{r}\right)\cos\theta. \quad (13.4.16)$$

For the upper end of the bar $\theta = 0$, then,

$$\sigma_{rr} = \sigma_{\theta\theta} = 0$$

$$\sigma_{r\theta} = -\frac{P}{N}\left(r + \frac{r_i^2 r_o^2}{r^3} - \frac{r_i^2 + r_o^2}{r}\right). \quad (13.4.17)$$

For the lower end of the bar $\theta = \Pi/2$, then,

$$\sigma_{rr} = \frac{P}{N}\left(r + \frac{r_i^2 r_o^2}{r^3} - \frac{r_i^2 + r_o^2}{r}\right)$$

$$\sigma_{\theta\theta} = \frac{P}{N}\left(3r - \frac{r_i^2 r_o^2}{r^3} - \frac{r_i^2 + r_o^2}{r}\right)$$

$$\sigma_{r\theta} = 0.$$

Eqs. (13.4.14) to (13.4.16) constitute an exact solution of the problem only when the forces at the end of the curved bar are distributed according to Eqs. (13.4.17). In any other distribution of forces, the solution will be valid at some distance from the ends according to Saint-Venant's principle. Here, too, numerical computations show that the results of the simplified theory are very close to those given by the exact theory.

Let us now consider the displacements produced by the force P. The elastic stress-strain relations for plane stress and the strain displacements relations give:

$$\frac{\partial u_r}{\partial r} = \frac{\sin\theta}{E}\left[2C_1 r(1-3\nu) - \frac{2C_2}{r^3}(1+\nu) + \frac{C_4}{r}(1-\nu)\right] \quad (13.4.18)$$

$$\frac{\partial u_\theta}{\partial \theta} = r e_{\theta\theta} - u_r \quad (13.4.19)$$

$$2e_{r\theta} = \frac{\partial u_r}{r\partial\theta} + \frac{\partial u_\theta}{\partial r} - \frac{u_\theta}{r}. \quad (13.4.20)$$

Integration of Eq. (13.4.18) yields:

$$u_r = \frac{\sin\theta}{E}\left[C_1 r^2(1-3\nu) + \frac{C_2}{r^2}(1+\nu) + C_4(1-\nu)\ell nr\right] + f_1(\theta), \quad (13.4.21)$$

where $f_1(\theta)$ is a function of θ only. Substituting Eqs. (13.4.21) and (6.7.23) into Eq. (13.4.19) and integrating, we get:

$$u_\theta = -\frac{\cos\theta}{E}\left[C_1 r^2(5+\nu) + \frac{C_2}{r^2}(1+\nu) - C_4(1-\nu)\ell nr + C_4(1-\nu)\right] - \int f_1(\theta)\,d\theta + f_2(r), \quad (13.4.22)$$

in which $f_2(r)$ is a function of r only. Substituting Eqs. (13.4.21) and (13.4.22) into Eq. (13.4.20), we obtain the equation:

$$\int f_1(\theta)\,d\theta + f_1'(\theta) + rf_2'(r) - f_2(r) = -\frac{4C_4\cos\theta}{E}. \quad (13.4.23)$$

Separating the variables, we get the two equations:

$$\int f_1(\theta)\,d\theta + f_1'(\theta) = -\frac{4C_4 \cos\theta}{E} \tag{13.4.24}$$

$$r f_2'(r) - f_2(r) = 0, \tag{13.4.25}$$

which are satisfied by the following functions:

$$f_1(\theta) = -\frac{2C_4\theta\cos\theta}{E} + K\sin\theta + L\cos\theta \tag{13.4.26}$$

$$f_2(r) = Hr. \tag{13.4.27}$$

K, L, and H are constants to be determined from the boundary conditions on the displacements. Eqs. (13.4.21) and (13.4.22) can now be written as follows:

$$u_r = \frac{\sin\theta}{E}\left[C_1 r^2(1-3\nu) + \frac{C_2}{r^2}(1+\nu) + C_4(1-\nu)\ell nr\right] - \frac{2C_4\theta\cos\theta}{E} + K\sin\theta + L\cos\theta \tag{13.4.28}$$

$$u_\theta = \frac{2C_4}{E}\theta\sin\theta - \frac{\cos\theta}{E}\left[C_1(5+\nu)r^2 + \frac{C_2(1+\nu)}{r^2} - C_4(1-\nu)\ell nr\right] + \frac{C_4(1+\nu)}{E}\cos\theta + K\cos\theta - L\sin\theta + Hr. \tag{13.4.29}$$

Taking the centroid of the cross section for $\theta = \Pi/2$, and also an element of the radius at this point as rigidly fixed, the boundary conditions on the displacements are:

$$u_r = u_\theta = 0, \quad \frac{\partial u_\theta}{\partial r} = 0 \qquad \text{for } \theta = \frac{\Pi}{2} \text{ and } r = r_c = \frac{r_i + r_o}{2}.$$

Applying these to expressions (13.4.28) and (13.4.29), we get:

$$H = 0, \quad L = \frac{C_4\Pi}{E}, \tag{13.4.30}$$

$$K = -\frac{1}{E}\left[C_1 r_c^2(1-3\nu) + \frac{C_2}{r_c^2}(1+\nu) + C_4(1-\nu)\ell nr_c\right]. \tag{13.4.31}$$

The deflection at the upper end is

$$(u_r)_{\theta=0} = \frac{C_4 \Pi}{E} = -\frac{P\Pi(r_i^2 + r_o^2)}{E\left[(r_i^2 - r_o^2) + (r_i^2 + r_o^2)\ln\frac{r_o}{r_i}\right]}. \tag{13.4.32}$$

If $(r_o - r_i)$ is small compared to r_i, Eq. (13.4.32) becomes:

$$(u_r)_{\theta=0} = -\frac{3\Pi r_i^3 P}{E(r_o - r_i)^3}, \tag{13.4.33}$$

which coincides with the expression obtained from the elementary theory [1].

The previous solution gives the stresses and displacements for the practical case of the crane hook lifting a load P (Fig. 13.7). In all the

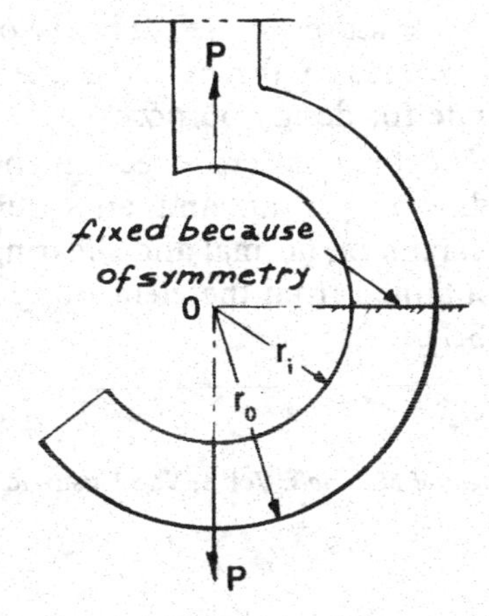

Fig. 13.7

equations, however, the sign of P is to be reversed. Additional solutions can be obtained using superposition.

PROBLEMS

1. To evaluate the approximation involved in the use of the simplified theory of curved beams, consider the case studied in Sec. 13.3. Compute and tabulate the maximum and minimum values of $\sigma_{\theta\theta}$ and σ_{11} from Eqs. (13.3.12) and (13.2.1), respectively, for the

following ratios of r_o/r_i: 1.2, 1.5, 2.0, 2.5. How adequate is the simplified theory for the design purposes, and how much is there to be gained by using the results of Sec. 13.3?

2. Knowing the results of Secs. 13.3 and 13.4, show the superposition scheme to be used when the cantilever beam of Fig. 13.6 is subjected at its upper end to a couple and a vertical force. Prove that the solution of this problem can be obtained using the Airy stress function, $\phi = f(r)\cos\theta$.
3. A curved bar of square cross section 3 in. by 3 in., and of mean radius of curvature r_c 4.5 in., is initially unstressed. If a bending moment of 60,000 lb-in. is applied to the bar in order to straighten it, find the maximum and minimum circumferential stresses using the equations of Sec. 13.3 and those of the elementary theory. Assuming that the angle subtended at the origin is 60°, and knowing that the solution of Sec. 13.3 is only valid at a certain distance from the ends of the beam, is the use of the elementary theory appropriate for design purposes?
4. A crane hook of rectangular cross section and of unit thickness has an internal radius $r_i = 3$ in. and an external radius $r_o = 4$ in. Determine the maximum normal and shearing stresses and compare them to those obtained from the elementary theory for a load P of 5000 lb. (Fig. 13.7).

REFERENCES

[1] S. Timoshenko, *Strength of Materials*, Vol. 1, Van Nostrand, Princeton, N. J., 1955.

CHAPTER 14

THE SEMI-INFINITE ELASTIC MEDIUM AND RELATED PROBLEMS

14.1 Introduction

Many problems in stress and strain analysis, which are of practical importance, are concerned with the effect in semi-infinite media of stresses acting on their straight boundaries. In theory, the solution of such problems can be obtained by integration from the results of Boussinesq and Cerruti which were presented in Secs. 9.5 and 9.6. In two dimensional cases, some solutions can readily be obtained through the use of an Airy stress function. Since the integration of the results of Boussinesq and Cerruti can sometimes become extremely tedious, solutions involving stress functions become quite valuable, at least from a practical point of view. They fail, however, to show the close relationship between the space problem and the two dimensional problem.

The problems examined in the following sections have results that are extensively used by engineers under the forms of tables and charts [1]. Many have been photoelastically confirmed and shown to be applicable to finite bodies as long as one remains far enough from the boundaries. The aim of this chapter is to discuss and show how some of these results have been obtained.

14.2 Uniform Pressure Distributed over a Circular Area on the Surface of a Semi-Infinite Solid

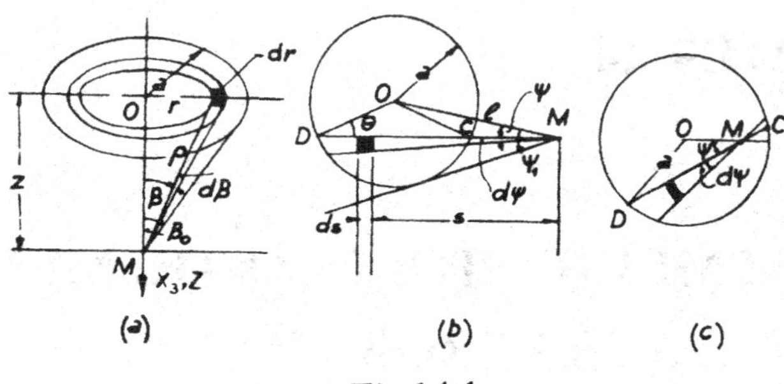

Fig.14.1

We shall first derive the expressions of the displacements at points under the center of the loaded area (Fig. 14.1a) and at the surface of the semi-infinite solid (Fig. 14.1bc). q is the uniformly distributed pressure, and Q is the total load equal to $\Pi a^2 q$.

At a point M under the center of the loaded area (Fig. 14.1a), we have, because of symmetry:

$$u_r = u_\theta = 0. \tag{14.2.1}$$

Using Eq. (9.6.21), the vertical displacement of M is given by:

$$\begin{aligned} u_z &= \int_0^{\beta_o} \frac{2\Pi r\,(dr) q \sin\beta}{4\Pi G r}[2(1-\nu)+\cos^2\beta] \\ &= \int_0^{\beta_o} \frac{qz}{2G}\left[2(1-\nu)\frac{\sin\beta}{\cos^2\beta}+\sin\beta\right] d\beta. \end{aligned} \tag{14.2.2}$$

Integration of Eq. (14.2.2), yields

$$u_z = \frac{qa}{2G}\left[\frac{\sqrt{a^2+z^2}}{a}-\frac{z}{a}\right]\left[2(1-\nu)+\frac{z}{\sqrt{a^2+z^2}}\right]. \tag{14.2.3}$$

Therefore,

$$(u_z)_{z=0} = \frac{qa(1-\nu)}{G} = \frac{2qa}{E}(1-\nu^2). \tag{14.2.4}$$

At the surface of the semi-infinite solid, two cases must be considered: a) The point M is outside the loaded area (Fig. 14.b). Setting $z = 0$ and $s = \rho$ in Eq. (9.6.21), we get:

$$(u_z)_{z=0} = \frac{P(1 - \nu^2)}{\Pi E s}. \tag{14.2.5}$$

Now consider the deflection in the OX_3 direction at a point M on the surface of the solid, and at a distance ℓ from the center of the loaded area. Take a small element of loaded area bounded by the two radii enclosing the angle $d\psi$ and two arcs of circle with radii s and $s + ds$ centered at M. The load on this element is $qs(ds)d\psi$ and, using Eq. (14.2.5), the deflection of the point M is:

$$(u_z)_M = \frac{(1 - \nu^2)q}{\Pi E} \int\int d\psi\, ds. \tag{14.2.6}$$

The distance s varies between MD and MC, and the length of the chord CD is $2\sqrt{a^2 - \ell^2 \sin^2\psi}$. Therefore, Eq. (14.2.6) becomes:

$$(u_z)_M = \frac{4(1 - \nu^2)q}{\Pi E} \int_0^{\psi_1} \sqrt{a^2 - \ell^2 \sin^2\psi}\, d\psi, \tag{14.2.7}$$

in which ψ_1 is the maximum value of ψ. The calculation of the integral in Eq. (14.2.7) is simplified by introducing the variable θ, where (Fig. 14.1b)

$$a \sin\theta = \ell \sin\psi. \tag{14.2.8}$$

θ varies from 0 to $\Pi/2$ when ψ varies from 0 to ψ_1. With this substitution, Eq. (14.2.7) becomes:

$$(u_z)_M = \frac{4(1 - \nu^2)}{\Pi E} q\ell \left[\int_0^{\frac{\Pi}{2}} \sqrt{1 - \left(\frac{a}{\ell}\right)^2 \sin^2\theta}\, d\theta - \left(1 - \frac{a^2}{\ell^2}\right) \int_0^{\frac{\Pi}{2}} \frac{d\theta}{\sqrt{1 - \left(\frac{a}{\ell}\right)^2 \sin^2\theta}} \right]. \tag{14.2.9}$$

These are elliptic integrals, and their values for various a/ℓ can be found in tables [2]. At the boundary $\ell = a$, and

$$(u_z)_{\ell=a} = \frac{4(1-\nu^2)}{\Pi E} qa. \tag{14.2.10}$$

b) The point M is within the loaded area (Fig. 14.1c). We consider the deflection in the OX_3 direction of point M due to the load $qs(ds)(d\psi)$ acting on the shaded area. This deflection is given by:

$$(u_z)_M = \frac{(1-\nu^2)}{\Pi E} q \int\int d\psi\, ds. \tag{14.2.11}$$

The distance s varies between C and D, and the length CD is equal to $2a \cos \theta$. The angle ψ varies between 0 and Π, and

$$a \sin \theta = \ell \sin \psi.$$

Therefore,

$$(u_z)_M = \frac{2(1-\nu^2)qa}{\Pi E} \int_0^{\Pi} \sqrt{1 - \left(\frac{\ell}{a}\right)^2 \sin^2\psi}\; d\psi. \tag{14.2.12}$$

The symmetry in Fig. 14.1c makes it possible to change the limits of the previous integral to 0 and $\Pi/2$, so that

$$(u_z)_M = \frac{4(1-\nu^2)qa}{\Pi E} \int_0^{\frac{\Pi}{2}} \sqrt{1 - \left(\frac{\ell}{a}\right)^2 \sin^2\psi}\; d\psi. \tag{14.2.13}$$

The deflection can be computed for any ratio ℓ/a by using tables of elliptic integrals. For $\ell = 0$, Eq. (14.2.4) is obtained. The average deflection of the loaded area is given by:

$$(u_z)_{\text{aver.}} = \frac{\int_0^a u_z 2\Pi r\, dr}{\Pi a^2} = \frac{1.7qa(1-\nu^2)}{E} = \frac{0.54Q(1-\nu^2)}{aE}. \tag{14.2.14}$$

Let us now compute the stresses under the center of the loaded area (Fig. 14.2).

From Eq. (9.6.24), we have:

$$\sigma_{zz} = -\int_{r=0}^{r=a} 2\Pi r\, dr \frac{3q}{2\Pi} \frac{z^3}{\rho^5}. \tag{14.2.15}$$

The integration is easily accomplished by setting $r = z \tan \beta$. Hence,

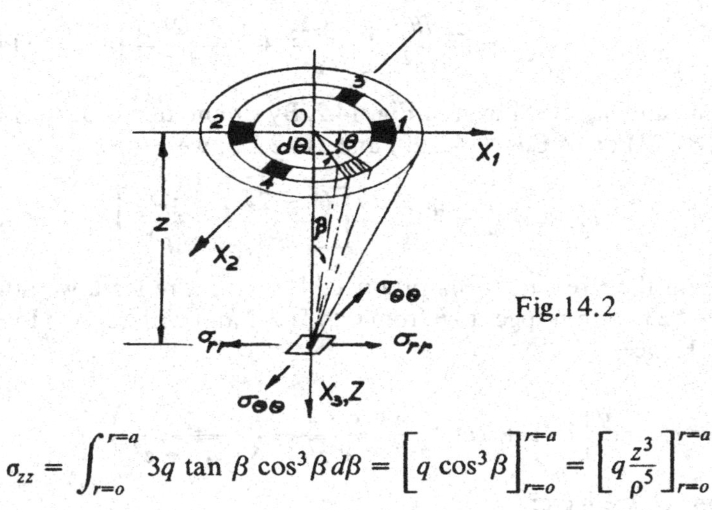

Fig. 14.2

$$\sigma_{zz} = \int_{r=o}^{r=a} 3q \tan \beta \cos^3 \beta \, d\beta = \left[q \cos^3 \beta \right]_{r=o}^{r=a} = \left[q \frac{z^3}{\rho^5} \right]_{r=o}^{r=a}$$

or

$$\sigma_{zz} = q\left[-1 + \frac{z^3}{(a^2 + z^2)^{3/2}} \right]. \tag{14.2.16}$$

Because of symmetry,

$$\sigma_{rr} = \sigma_{\theta\theta} \tag{14.2.17}$$

and

$$\sigma_{r\theta} = \sigma_{\theta z} = \sigma_{rz} = 0. \tag{14.2.18}$$

To compute σ_{rr}, we proceed as follows [8]: According to Eqs. (9.6.22) and (9.6.23), the two elements 1 and 2 in Fig. 14.2 give at M:

$$(d\sigma_{rr})_{1,2} = \frac{qr\, d\theta\, dr}{\Pi \rho^2} \left[-\frac{3r^2 z}{\rho^3} + \frac{(1 - 2\nu)\rho}{\rho + z} \right] \tag{14.2.19}$$

$$(d\sigma_{\theta\theta})_{1,2} = \frac{(1 - 2\nu) qr\, d\theta\, dr}{\Pi \rho^2} \left[\frac{z}{\rho} - \frac{\rho}{\rho + z} \right]. \tag{14.2.20}$$

Also the two elements 3 and 4 give at M:

$$(d\sigma_{rr})_{3,4} = \frac{(1 - 2\nu) qr\, d\theta\, dr}{\Pi \rho^2} \left[\frac{z}{\rho} - \frac{\rho}{\rho + z} \right] \tag{14.2.21}$$

$$(d\sigma_{\theta\theta})_{3,4} = \frac{qr\,d\theta\,dr}{\Pi\rho^2}\left[-\frac{3r^2z}{\rho^2} + \frac{(1-2\nu)\rho}{\rho+z}\right], \tag{14.2.22}$$

where σ_{rr} and $\sigma_{\theta\theta}$ are shown in Fig. 14.2. By summation of Eqs. (14.2.19) and (14.2.21) or of Eqs. (14.2.20) and (14.2.22), we get:

$$d\sigma_{rr} = d\sigma_{\theta\theta} = \frac{qr\,dr\,d\theta}{\Pi}\left[\frac{(1-2\nu)z}{\rho^3} - \frac{3r^2z}{\rho^5}\right]. \tag{14.2.23}$$

To obtain the stress $\sigma_{rr} = \sigma_{\theta\theta}$ produced by the entire load, we integrate Eq. (14.2.23) with respect to θ from 0 to $\Pi / 2$ and with respect to r from 0 to a. Hence,

$$\sigma_{rr} = \sigma_{\theta\theta} = \frac{q}{2}\left[-(1+2\nu) + \frac{2(1+\nu)z}{\sqrt{a^2+z^2}} - \left(\frac{z}{\sqrt{a^2+z^2}}\right)^3\right]. \tag{14.2.24}$$

At point O, we have:

$$\sigma_{zz} = -q, \quad \sigma_{rr} = \sigma_{\theta\theta} = -\frac{q(1+2\nu)}{2}. \tag{14.2.25}$$

The maximum shearing stress at any point on the OX_3 axis is given by:

$$\frac{1}{2}(\sigma_{\theta\theta} - \sigma_{zz}) = \frac{q}{2}\left[\frac{1-2\nu}{2} + (1+\nu)\frac{z}{\sqrt{a^2+z^2}} - \frac{3}{2}\left(\frac{z}{\sqrt{a^2+z^2}}\right)^3\right]. \tag{14.2.26}$$

This expression becomes a maximum for

$$z = a\sqrt{\frac{2(1+\nu)}{7-2\nu}} \tag{14.2.27}$$

and

$$\left[\frac{1}{2}(\sigma_{\theta\theta} - \sigma_{zz})\right]_{\max} = \frac{q}{2}\left[\frac{1-2\nu}{2} + \frac{2}{9}(1+\nu)\sqrt{2(1+\nu)}\right]. \tag{14.2.28}$$

If we set $\nu = 0.3$ in Eqs. (14.2.27) and (14.2.28), we get:

$$z = 0.638a, \quad \left[\frac{1}{2}(\sigma_{\theta\theta} - \sigma_{zz})\right]_{\max} = 0.33q. \tag{14.2.29}$$

Thus, the maximum shearing stress occurs at a depth approximately equal to two-thirds of the radius of the loaded circle a, and the magnitude of this maximum is about one-third of the applied uniform pressure q.

Remark

The equations given in this section have been used by Newmark [3, 4] to develop charts that can be used to compute stresses and displacements at any point of a semi-infinite elastic medium due to uniformly distributed stresses acting on areas of any shape.

14.3 Uniform Pressure Distributed over a Rectangular Area

The general expressions for the stresses and displacements produced by loads distributed over rectangular areas can be found in [5]. In this section, we shall only give the expression of the vertical stress under the corner of a rectangular area whose dimensions are a and b (Fig. 14.3), and that of the average vertical displacement of the surface. *The vertical*

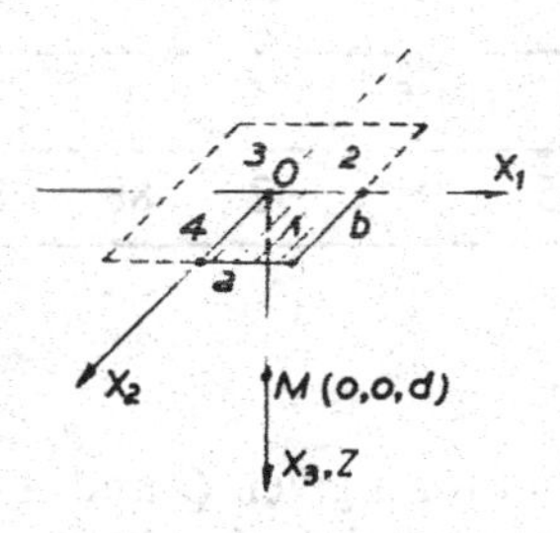

Fig. 14.3

stress at M (0,0,d) due to a uniformly distributed pressure q is given by Eq. (14.3.1):

$$(\sigma_{zz})_M = -\frac{q}{4\Pi}\left[\frac{2BV}{V^2 + B^2}\cdot\frac{V^2 + 1}{V^2} + \tan^{-1}\frac{2BV}{V^2 - B^2}\right], \quad (14.3.1)$$

where

$$V^2 = \frac{a^2 + b^2 + d^2}{d^2}, \quad B = \frac{ab}{d^2}. \quad (14.3.2)$$

One may find the stress σ_{zz} at any point under a rectangular area by dividing the area into smaller rectangles, such that the point beneath which we are seeking the stresses is a corner common to all smaller rectangles (Fig. 14.3). Eq. (14.3.1) can then be applied to each of these smaller rectangles and the sum of the individual results yields the total stress.

The average vertical displacement of the surface of the uniformly loaded medium is given by:

$$(u_3)_{\text{aver.}} = m\frac{Q(1-\nu^2)}{E\sqrt{A}}, \tag{14.3.3}$$

where Q is the total load, A is the magnitude of the loaded area, and m is a numerical factor depending on the ratio a / b. The following table gives the value of m for various a / b. For comparison purposes, m for a circular area is also included.

	Circle	Rectangles with various a/b						
		1	1.5	2	3	5	10	100
m =	0.96	.95	.94	.92	.88	.82	.71	0.37

The above table shows that for a given Q and A the deflection increases when the ratio of the perimeter of the loaded area to the area decreases.

14.4 Rigid Die in the Form of a Circular Cylinder

In this case, the displacements are given and it is necessary to find the corresponding distribution of pressures on the boundary plane. The vertical displacement $(u_z)_{z=0}$ at the surface is a constant under the die, but the distribution of pressure (Fig. 14.4) is not; its intensity is given by [6]:

$$q = \frac{Q}{2\Pi a\sqrt{a^2 - r^2}}. \tag{14.4.1}$$

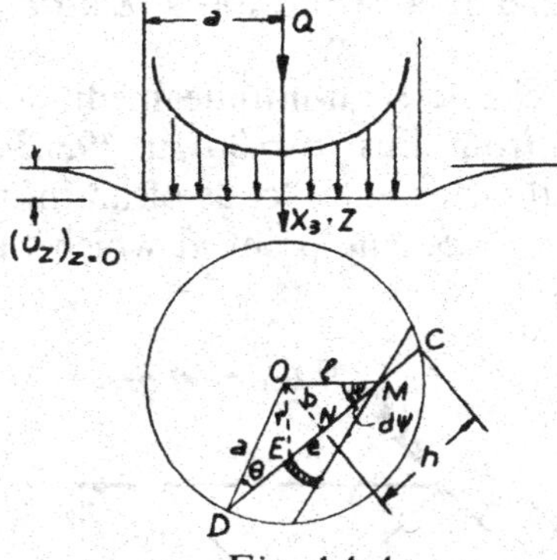

Fig. 14.4

The pressure has its smallest value at the center, and is infinite on the edges. In actual cases, we shall have a yielding of the material along the boundary. The displacements of the die $(u_z)_{z=0}$ corresponding to the distribution of Eq. (14.4.1) is given by:

$$(u_z)_{z=0} = \frac{(1 - \nu^2)}{\Pi E} \int\int q \, d\psi \, ds, \tag{14.4.2}$$

and is the same for any point M (Fig. 14.4): In Eq. (14.4.2), ψ varies from 0 to Π, and s varies between C and D. Now

$$s = \ell \cos \psi + e \tag{14.4.3}$$

and

$$r^2 = b^2 + e^2, \tag{14.4.4}$$

so that Eq. (14.4.2) can be written as follows:

$$(u_z)_{z=0} = \frac{Q(1 - \nu^2)}{2\Pi^2 aE} \int_0^{\Pi} \left[2 \int_0^h \frac{de}{\sqrt{h^2 - e^2}} \right] d\psi. \tag{14.4.5}$$

Eqs. (14.4.5) is easily integrated to give:

$$(u_z)_{z=0} = \frac{Q(1 - \nu^2)}{2aE}.$$

The value of $(u_z)_{z=0}$ is not very different from that of Eq. (14.2.14).

14.5 Vertical Line Load on a Semi-Infinite Elastic Medium

The stresses in an elastic semi-infinite medium subjected to a line load can be derived from Eqs. (9.6.30) to (9.6.34) by summing the stresses produced by the elemental loads of an infinite system of point loads as shown in Fig. 14.5. The point at which the stresses are to be

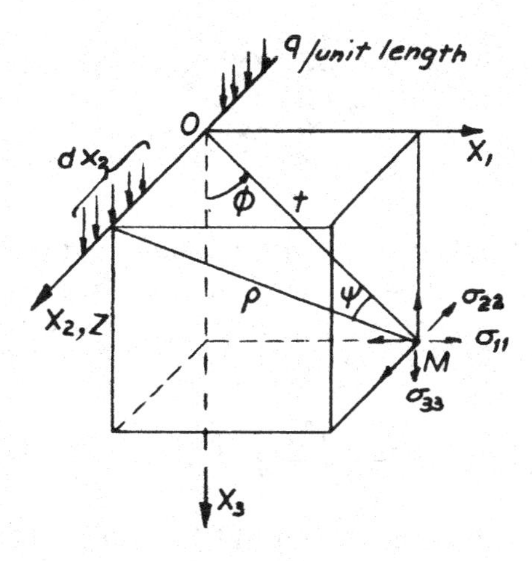

Fig. 14.5

computed is placed in the plane OX_1, OX_3 for convenience. Since the line load extends to infinity on both sides of the origin, this choice imposes no restrictions on the solution. The only components of the state of stress at a point M are σ_{11}, σ_{22}, σ_{33}, and σ_{13}; σ_{12} and σ_{23} are equal to zero because of symmetry. The problem is a plane strain problem with no displacements along the OX_2 direction. The body forces are neglected. From Eqs. (9.6.30) to (9.6.34), we get:

$$\sigma_{33} = -\frac{3qx_3^3}{2\Pi}\int_{-\infty}^{+\infty}\frac{dx_2}{\rho^5} = -\frac{3qx_3^3}{\Pi}\int_0^{\frac{\Pi}{2}}\frac{t\sec^2\psi\,d\psi}{t^5\sec^5\psi} = -\frac{2qx_3^3}{\Pi t^4} \tag{14.5.1}$$

$$= -\frac{2q\cos^4\phi}{\Pi x_3}$$

$$\sigma_{13} = -\frac{3qx_1x_3^2}{2\Pi}\int_{-\infty}^{+\infty}\frac{dx_2}{\rho^5} = -\frac{2qx_1x_3^2}{\Pi t^4} = -\frac{2q}{\Pi x_3}\sin\phi\cos^3\phi \tag{14.5.2}$$

$$\sigma_{11} = \int_{-\infty}^{+\infty} [\sigma_{11}]_M \, dx_2 = -\frac{2qx_1^2 x_3}{\Pi t^4} = -\frac{2q}{\Pi x_3} \sin^2\phi \cos^2\phi \tag{14.5.3}$$

$$\sigma_{22} = \int_{-\infty}^{+\infty} [\sigma_{22}]_M \, dx_2 = \nu(\sigma_{11} + \sigma_{33}) = -\frac{2\nu q x_3}{\Pi t^2}$$

$$= -\frac{2\nu q}{\Pi x_3} \cos^2\phi \tag{14.5.4}$$

$$\sigma_{12} = \sigma_{23} = 0. \tag{14.5.5}$$

In these equations, $t^2 = x_3^2 + x_1^2$; in other words, t is the radial distance from O to the point M in the OX_1, OX_3 plane. The stresses σ_{12} and σ_{13} vanish because of symmetry. The notation $[\sigma_{11}]_M$ in Eq. (14.5.3) denotes the stress from Eq. (9.6.30). It is not evident from inspection why σ_{11} in the plane strain case is independent of Poisson's ratio, however, the actual integration shows that the terms which are multiplied by $(1 - 2\nu)$ in Eq. (9.6.30) do not make any contribution to the result in the plane strain case. Similarly, the integration for σ_{22} yields $\sigma_{22} = \nu(\sigma_{11} + \sigma_{33})$.

In cylindrical coordinates, with r replacing t and θ replacing ϕ (Fig. 14.6), the components of the state of stress are written as follows:

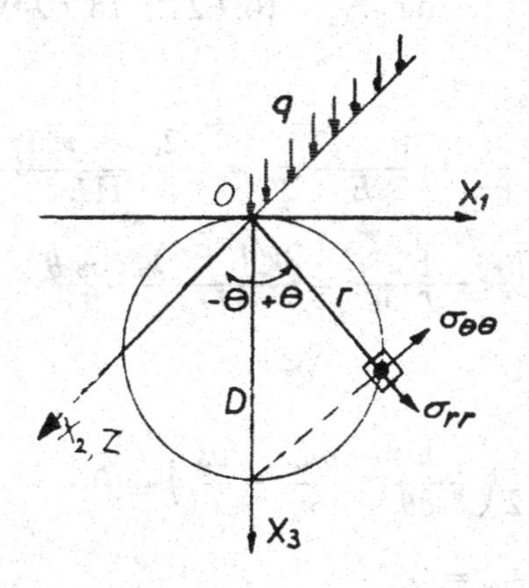

Fig. 14.6

$$\sigma_{rr} = -\frac{2q}{\Pi} \frac{\cos\theta}{r} \tag{14.5.6}$$

$$\sigma_{\theta\theta} = 0 \tag{14.5.7}$$

$$\sigma_{zz} = -\frac{2q\nu}{\Pi}\frac{\cos\theta}{r} \tag{14.5.8}$$

$$\sigma_{r\theta} = \sigma_{rz} = \sigma_{\theta z} = 0. \tag{14.5.9}$$

The previous equations show that the radial direction is a principal direction, and that all points on any circle of diameter D centered on the vertical axis and passing through the point of loading have the same principal stress $\sigma_1 = -2q/\Pi D$ and the same maximum shearing stress $\frac{1}{2}(\sigma_1 - \sigma_2) = -q/\Pi D$. Eqs. (14.5.6) to (14.5.9) can also be obtained through the use of the Airy stress function:

$$\phi = cr\theta \sin\theta \tag{14.5.10}$$

with

$$c = -\frac{q}{\Pi}. \tag{14.5.11}$$

Having the stresses, the strains and displacements can be computed at any point. It is more convenient to use cylindrical coordinates for such computations. Using Eqs. (6.7.23), (6.7.24), and (8.16.12) to (8.16.14), we get:

$$e_{rr} = \frac{\partial u_r}{\partial r} = \frac{1-\nu^2}{E}\sigma_{rr} = -\frac{2(1-\nu^2)q}{\Pi E}\frac{\cos\theta}{r} \tag{14.5.12}$$

$$e_{\theta\theta} = \frac{u_r}{r} + \frac{1}{r}\frac{\partial u_\theta}{\partial\theta} = \frac{2\nu(1+\nu)q}{\Pi E}\frac{\cos\theta}{r} \tag{14.5.13}$$

$$e_{zz} = 0 \tag{14.5.14}$$

$$e_{r\theta} = \frac{1}{2}\left(\frac{1}{r}\frac{\partial u_r}{\partial\theta} + \frac{\partial u_\theta}{\partial r} - \frac{u_\theta}{r}\right) = 0 \tag{14.5.15}$$

$$e_{rz} = e_{\theta z} = 0. \tag{14.5.16}$$

Integration of Eq. (14.5.12) yields:

$$u_r = -\frac{2(1-\nu^2)}{\Pi E}\cos\theta\, \ell nr + f_1(\theta). \tag{14.5.17}$$

Substituting Eq. (14.5.17) into Eq. (14.5.13) and integrating, we get:

$$u_\theta = \frac{2\nu(1+\nu)q}{\Pi E}\sin\theta + \frac{2(1-\nu^2)q}{\Pi E}\ell nr\sin\theta - \int f_1(\theta)\,d\theta + f_2(r), \tag{14.5.18}$$

where $f_1(\theta)$ and $f_2(r)$ are functions of θ and r, respectively. Eqs. (14.5.17) and (14.5.18) are now substituted in Eq. (14.5.15), to give:

$$\frac{2(1-\nu^2)q}{\Pi E}\sin\theta\frac{\ell nr}{r} + \frac{1}{r}\frac{d}{d\theta}f_1(\theta) + \frac{2(1-\nu^2)q}{\Pi E}\frac{\sin\theta}{r} + \frac{d}{dr}f_2(r) - \frac{2\nu(1+\nu)q}{\Pi E}\frac{\sin\theta}{r} - \frac{2(1-\nu^2)q}{\Pi E}\ell nr\frac{\sin\theta}{r} + \frac{1}{r}\int f_1(\theta)\,d\theta - \frac{1}{r}f_2(r) = 0. \tag{14.5.19}$$

Separating the variables and solving, Eq. (14.5.19) yields:

$$f_1(\theta) = A\sin\theta + B\cos\theta - \frac{q}{\Pi E}(1-2\nu)(1+\nu)\theta\sin\theta \tag{14.5.20}$$

$$f_2(r) = Cr, \tag{14.5.21}$$

where A, B, and C are constants of integration which are to be determined from the boundary conditions. Substituting the values of $f_1(\theta)$ and $f_2(r)$ in Eqs. (14.5.17) and (14.5.18), we get the equations of the displacements:

$$u_r = -\frac{2(1-\nu^2)q}{\Pi E}\cos\theta\,\ell n\,r - \frac{q}{\Pi E}(1-2\nu)(1+\nu)\theta\sin\theta + A\sin\theta + B\cos\theta \tag{14.5.22}$$

$$u_\theta = \frac{2\nu(1+\nu)q}{\Pi E}\sin\theta + \frac{2(1-\nu^2)q}{\Pi E}\ell n\,r\sin\theta + A\cos\theta - B\sin\theta + \frac{q}{\Pi E}(1-2\nu)(1+\nu)[\sin\theta - \theta\cos\theta] + Cr \tag{14.5.23}$$

$$u_z = 0. \tag{14.5.24}$$

Let us consider two sets of boundary conditions:

a) All points on the OX_3 axis do not have any lateral displacement, and one point at a distance d on this axis does not move vertically.

Therefore, $u_\theta = 0$ for $\theta = 0$, and $u_r = 0$ for $\theta = 0$ and $r = d$. The first condition leads to

$$A = C = 0, \tag{14.5.25}$$

and the second condition leads to

$$B = \frac{2(1 - \nu^2)}{\Pi E} q \ell n\, d. \tag{14.5.26}$$

The equations of the displacements in this case become:

$$u_r = \frac{2(1 - \nu^2)}{\Pi E} q \cos\theta\, \ell n \frac{d}{r} - \frac{q}{\Pi E}(1 - 2\nu)(1 + \nu)\theta \sin\theta \tag{14.5.27}$$

$$u_\theta = \frac{q(1 + \nu)}{\Pi E} \sin\theta - \frac{2(1 - \nu^2)q}{\Pi E} \ell n \frac{d}{r} \sin\theta - \frac{q}{\Pi E}(1 - 2\nu)(1 + \nu)\theta \cos\theta \tag{14.5.28}$$

$$u_z = 0. \tag{14.5.29}$$

On the surface of the straight boundary, we have:

$$(u_r)_{\theta = \pm\frac{\Pi}{2}} = -\frac{q}{2E}(1 - 2\nu)(1 + \nu) \tag{14.5.30}$$

$$(u_\theta)_{\theta = \frac{\Pi}{2}} = -\frac{2(1 - \nu^2)}{\Pi E} q \ell n \frac{d}{r} + \frac{q}{\Pi E}(1 + \nu) \tag{14.5.31}$$

b) All points on the OX_3 axis do not have any lateral displacement, and a point at a distance d along the OX_1 axis does not move vertically. Therefore, $u_\theta = 0$ for $\theta = 0$, and $u_\theta = 0$ for $\theta = \Pi / 2$ and $r = d$. The first condition leads to

$$A = C = 0, \tag{14.5.32}$$

and the second condition leads to

$$B = \frac{q(1 + \nu)}{\Pi E} + \frac{2(1 - \nu^2)}{\Pi E} q \ell n\, d. \tag{14.5.33}$$

The equations of the displacements therefore become:

$$u_r = \frac{2(1-\nu^2)}{\Pi E} q \cos\theta \, \ell n \frac{d}{r} - \frac{q}{\Pi E}(1-2\nu)(1+\nu)\theta \sin\theta + \frac{q(1+\nu)}{\Pi E} \cos\theta \qquad (14.5.34)$$

$$u_\theta = \frac{2\nu(1+\nu)}{\Pi E} q \sin\theta - \frac{2(1-\nu^2)}{\Pi E} q \ell n \frac{d}{r} \sin\theta - \frac{q(1+\nu)}{\Pi E} \sin\theta + \frac{q}{\Pi E}(1-2\nu)(1+\nu)(\sin\theta - \theta\cos\theta) \qquad (14.5.35)$$

$$u_z = 0. \qquad (14.5.36)$$

On the surface of the straight boundary, we have:

$$(u_r)_{\theta=\pm\frac{\Pi}{2}} = -\frac{q}{2E}(1-2\nu)(1+\nu) \qquad (14.5.37)$$

$$(u_\theta)_{\theta=\frac{\Pi}{2}} = \frac{-2(1-\nu^2)q}{\Pi E} \ell n \frac{d}{r} \qquad (14.5.38)$$

For both sets of boundary conditions, the quantity d is indeterminate and there is nothing in the analysis by which it can be found. It is usually taken to be very large. Both Eqs. (14.5.30) and (14.5.37) indicate a displacement of the material on the surface towards the origin. We may regard such a displacement as a physical possibility if we remove a cylindrical surface of small radius around the line of application of q (Fig. 14.7a) and substitute to q an equivalent system of stresses. Actually, in this portion, the material is plastically deformed and permits the displacements of Eqs. (14.5.30) and (14.5.37). The solutions presented in this section are subject to the same restriction imposed on the Boussinesq solution of the point load—namely, that their validity starts at a small distance from the point of application of the load.

Finally, it is of interest to examine the shape of the lines (they are actually surfaces extending in the OX_2 direction from $-\infty$ to $+\infty$) in the medium, which at each of their points are tangent to the principal stresses. Such lines are called principal stress trajectories and from Eqs. (14.5.6) to (14.5.9) are seen to be straight lines converging at the point of application of the load and concentric circles with centers at this point (Fig. 14.7a). A second set of lines, which is of interest in the study of the theory of plasticity, is formed by lines which, at each point, are tangent to the directions of the maximum shearing stresses. Such lines are called maximum shearing stress trajectories and are inclined 45° to

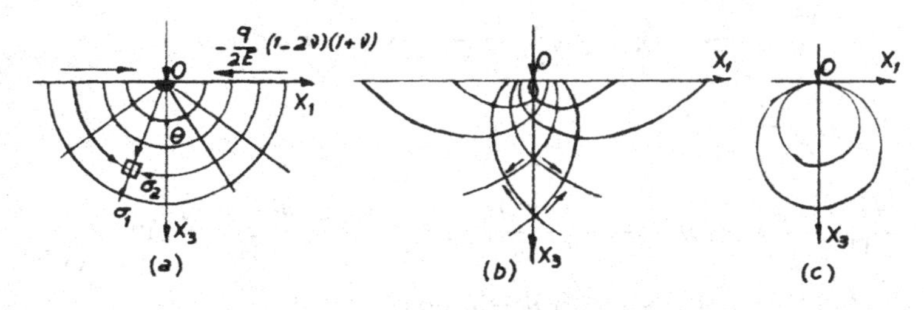

Fig. 14.7

the principal stress trajectories. They are therefore logarithmic spirals (Fig. 14.7b) whose equation is:

$$r = Ce^{\theta}.$$

A third set of lines called isochromatics consists of the loci of equal maximum shearing stresses. Those loci have already been shown to be circles passing through 0 and centered on OX_3 (Fig. 14.7c).

14.6 Vertical Line Load on a Semi-Infinite Elastic Plate

The plate is assumed to be of unit width so that the load is equal to q (Fig. 14.8). The problem is a plane stress problem. Since the body forces are neglected, the stresses are the same as those obtained in the previous case except that $\sigma_{22} = 0$. Thus,

$$\sigma_{33} = -\frac{2q\cos^4\theta}{\Pi x_3}, \quad \sigma_{11} = -\frac{2q}{\Pi x_3}\sin^2\theta\cos^2\theta, \quad \sigma_{22} = 0 \tag{14.6.1}$$

$$\sigma_{13} = -\frac{2q}{\Pi x_3}\sin\theta\cos^3\theta, \quad \sigma_{23} = \sigma_{12} = 0. \tag{14.6.2}$$

And in cylindrical coordinates:

$$\sigma_{rr} = -\frac{2q}{\Pi}\frac{\cos\theta}{r} \tag{14.6.3}$$

$$\sigma_{\theta\theta} = \sigma_{zz} = \sigma_{r\theta} = \sigma_{rz} = \sigma_{\theta z} = 0. \tag{14.6.4}$$

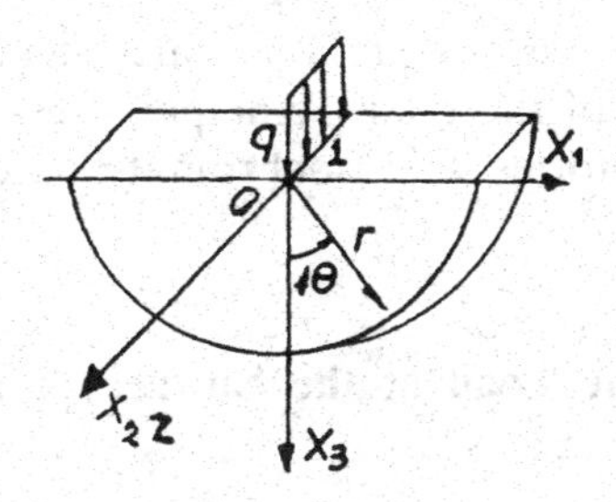

Fig. 14.8

The strains are given by:

$$e_{rr} = \frac{\partial u_r}{\partial r} = -\frac{2q}{\Pi E}\frac{\cos\theta}{r} \tag{14.6.5}$$

$$e_{\theta\theta} = \frac{u_r}{r} + \frac{1}{r}\frac{\partial u_\theta}{\partial\theta} = \frac{2q\nu}{\Pi E}\frac{\cos\theta}{r} \tag{14.6.6}$$

$$e_{r\theta} = \frac{1}{2}\left(\frac{1}{r}\frac{\partial u_r}{\partial\theta} + \frac{\partial u_\theta}{\partial r} - \frac{u_\theta}{r}\right) = 0 \tag{14.6.7}$$

$$e_{rz} = e_{\theta z} = 0. \tag{14.6.8}$$

The integration of Eqs. (14.6.5) and (14.6.8) proceeds along the same lines followed in the previous section. If we assume that a point along the OX_3 axis and at a depth d is fixed, the equations of the displacements are:

$$u_r = \frac{2q}{\Pi E}\cos\theta\,\ell n\frac{d}{r} - \frac{(1-\nu)q}{\Pi E}\theta\sin\theta \tag{14.6.9}$$

$$u_\theta = \frac{q(1+\nu)}{\Pi E}\sin\theta - \frac{2q}{\Pi E}\sin\theta\,\ell n\frac{d}{r} - \frac{(1-\nu)q}{\Pi E}\theta\cos\theta. \tag{14.6.10}$$

On the straight boundary of the plate, we have:

$$(u_r)_{\theta=\pm\frac{\Pi}{2}} = -\frac{(1-\nu)q}{2E} \tag{14.6.11}$$

$$(u_\theta)_{\theta=\frac{\Pi}{2}} = \frac{q(1+\nu)}{\Pi E} - \frac{2q}{\Pi E}\ell n\frac{d}{r}. \tag{14.6.12}$$

The validity of the previous equations starts at a small distance from the point of application of the load. Principal stress trajectories, maximum shearing stress trajectories, and isochromatics are as shown in Fig. 14.7.

14.7 Tangential Line Load at the Surface of a Semi-Infinite Elastic Medium

The stress resulting from a tangential line load at the surface of a semi-infinite medium can be derived from Eqs. (9.5.39) to (9.5.44) by a summation of the stresses for the elemental horizontal point loads (Fig. 14.9). The steps in such a summation are similar to those of Sec. 14.5, except that the angle ϕ is measured from the OX_1 axis.

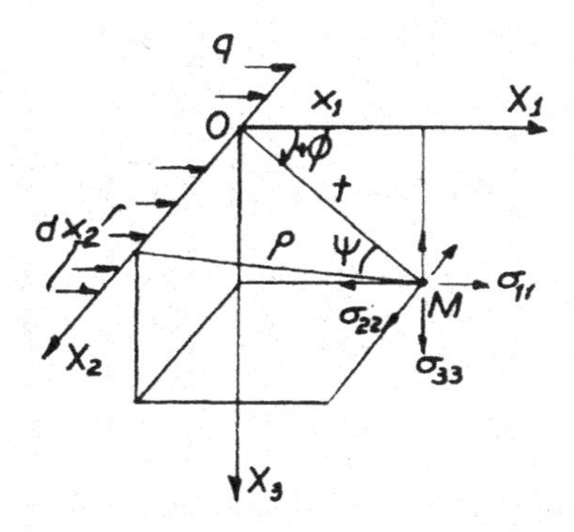

Fig. 14.9

$$\sigma_{33} = -\frac{3qx_1x_3^2}{\Pi}\int_0^{\infty}\frac{dx_2}{\rho^5} = -\frac{2qx_1x_3^2}{\Pi t^4} = \frac{-2q}{\Pi x_3}\sin^3\phi\cos\phi \qquad (14.7.1)$$

$$\sigma_{13} = \frac{-qx_1^2x_3}{\Pi t^4} = \frac{-q}{\Pi x_3}\sin^2\phi\cos^2\phi \qquad (14.7.2)$$

$$\sigma_{11} = \int_{-\infty}^{+\infty}[\sigma_{11}]_M\,dx_2 = \frac{-2qx_1^3}{\Pi t^4} = \frac{-2q}{\Pi x_3}\cos^3\phi\sin\phi \qquad (14.7.3)$$

$$\sigma_{22} = \nu(\sigma_{11}+\sigma_{33}) = \frac{-2\nu qx_1}{\Pi t^2} = \frac{-2\nu q}{\Pi x_3}\cos\phi\sin\phi \qquad (14.7.4)$$

$$\sigma_{12} = \sigma_{23} = 0. \qquad (14.7.5)$$

In these equations, $t^2 = x_3^2 + x_1^2$.

In cylindrical coordinates, with r replacing t and θ replacing ϕ, the components of the state of stress are written as follows (Fig. 14.10):

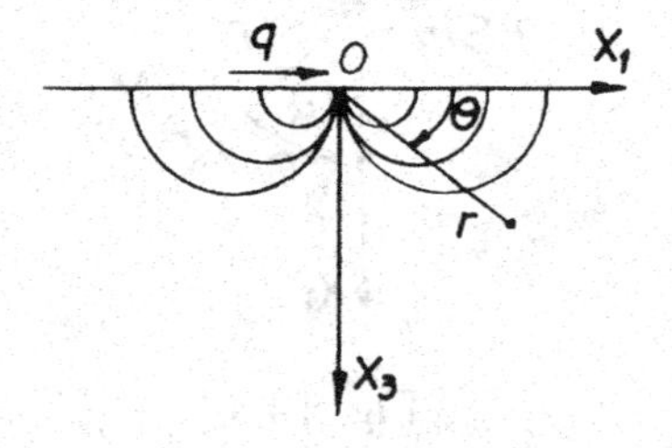

Fig. 14.10

$$\sigma_{rr} = -\frac{2q}{\Pi}\frac{\cos\theta}{r}, \quad \sigma_{\theta\theta} = 0, \quad \sigma_{zz} = -\frac{2\nu q}{\Pi}\frac{\cos\theta}{r} \tag{14.7.6}$$

$$\sigma_{r\theta} = \sigma_{rz} = \sigma_{\theta z} = 0. \tag{14.7.7}$$

These equations are the same as Eqs. (14.5.6) to (14.5.9) with the difference that θ is measured from the horizontal, i.e., from the direction of the load. Once the boundary conditions are chosen, the displacements can be computed in a manner similar to that of Sec. 14.5.

The principal stress trajectories are radial lines converging at the point of application of the load and concentric circular arcs with centers at the point of application of the load. The maximum shearing stress trajectories are logarithmic spirals similar to those shown in Fig. 14.7b. The isochromatics are semi-circles, the centers of which lie on OX_1 (Fig. 14.10).

Remark

The stresses and displacements in a semi-infinite elastic medium subjected to *inclined loads* can be obtained by superposition of the vertical and horizontal cases. If the components of the line load are $q \cos \alpha$ and $q \sin \alpha$ (Fig. 14.11), the stresses at a point M in cylindrical coordinates are given by:

$$\sigma_{rr} = -\frac{2q}{\Pi r}\cos(\theta - \alpha), \quad \sigma_{\theta\theta} = 0, \quad \sigma_{zz} = -\frac{2\nu q}{\Pi r}\cos(\theta - \alpha) \tag{14.7.8}$$

$$\sigma_{r\theta} = \sigma_{rz} = \sigma_{\theta z} = 0. \tag{14.7.9}$$

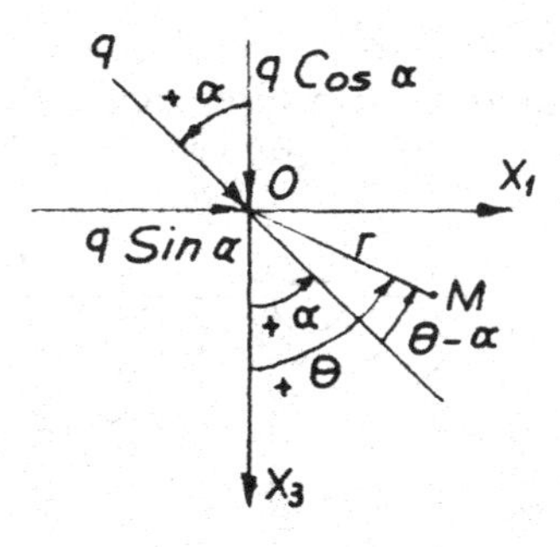

Fig. 14.11

The difference between the three groups of Eqs. (14.5.6) to (14.5.9), (14.7.6) to (14.7.7), and (14.7.8) to (14.7.9) is in the datum line from which the angle is measured. This datum line in each case is given by the direction of the applied line load.

14.8 Tangential Line Load on a Semi-Infinite Elastic Plate

The plate is assumed to be of unit width so that the load is equal to q (Fig. 14.12). The problem is a plane stress problem. The stresses are

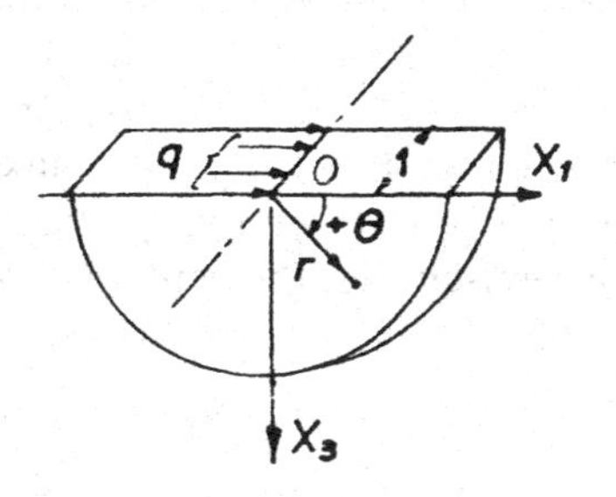

Fig. 14.12

given by Eqs. (14.7.1) to (14.7.5) in cartesian coordinates, and by Eqs. (14.7.6) and (14.7.7) in cylindrical coordinates but with $\sigma_{22} = \sigma_{zz} = 0$. Here, too, once the boundary conditions are chosen, the displacements can be computed in a manner similar to that of Sec. 14.5. The principal stress trajectories, the maximum shearing stress trajectories, and the isochromatics are the same as those of Sec. 14.7.

The effect of inclined loads can be obtained by superposition of the horizontal and vertical cases. The expression of the stresses in cylindrical coordinate is the same in all cases provided the angle is measured from the direction of the load.

14.9 Uniformly Distributed Vertical Pressure on Part of the Boundary of a Semi-Infinite Elastic Medium

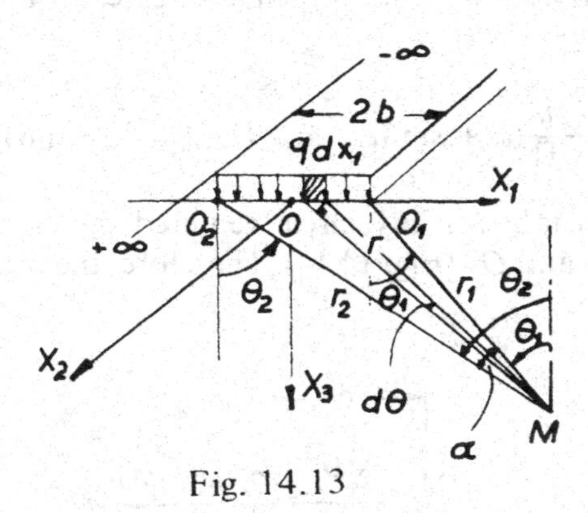

Fig. 14.13

In Fig. 14.13, the loaded strip extends to infinity on both sides of the origin along the OX_2 axis. The problem is a plane strain problem. The stresses at any point M defined by θ_1 and θ_2 are obtained by integration of Eqs. (14.5.6) to (14.5.9). From Fig. 14.13:

$$dx_1 = \frac{r\,d\theta}{\cos\theta}. \tag{14.9.1}$$

Using the equations of transformation (7.11.4) to (7.11.6), together with Eqs. (14.5.6) to (14.5.9), we get:

$$\begin{aligned}\sigma_{33} &= -\frac{2q}{\Pi}\int_{\theta_1}^{\theta_2}\cos^2\theta\,d\theta \\ &= -\frac{q}{2\Pi}[2(\theta_2-\theta_1)+(\sin 2\theta_2-\sin 2\theta_1)]\end{aligned} \tag{14.9.2}$$

$$\begin{aligned}\sigma_{11} &= -\frac{2q}{\Pi}\int_{\theta_1}^{\theta_2}\sin^2\theta\,d\theta \\ &= -\frac{q}{2\Pi}[2(\theta_2-\theta_1)-(\sin 2\theta_2-\sin 2\theta_1)]\end{aligned} \tag{14.9.3}$$

$$\sigma_{13} = -\frac{q}{\Pi}\int_{\theta_1}^{\theta_2} \sin 2\theta \, d\theta = -\frac{q}{2\Pi}[\cos 2\theta_1 - \cos 2\theta_2] \quad (14.9.4)$$

$$\sigma_{22} = \nu(\sigma_{11} + \sigma_{33}) = -\frac{2\nu q}{\Pi}(\theta_2 - \theta_1) = -\frac{2\nu q\alpha}{\Pi} \quad (14.9.5)$$

$$\sigma_{12} = \sigma_{23} = 0, \quad (14.9.6)$$

where α is equal to $\theta_2 - \theta_1$. The principal stresses at the point M are given by:

$$\sigma_1 = -\frac{q}{\Pi}(\alpha + \sin\alpha), \quad \sigma_3 = -\frac{q}{\Pi}(\alpha - \sin\alpha). \quad (14.9.7)$$

The angle α is constant for any circle centered on the OX_3 axis and passing through O_1 and O_2 (Fig. 14.14). Therefore, the principal stresses

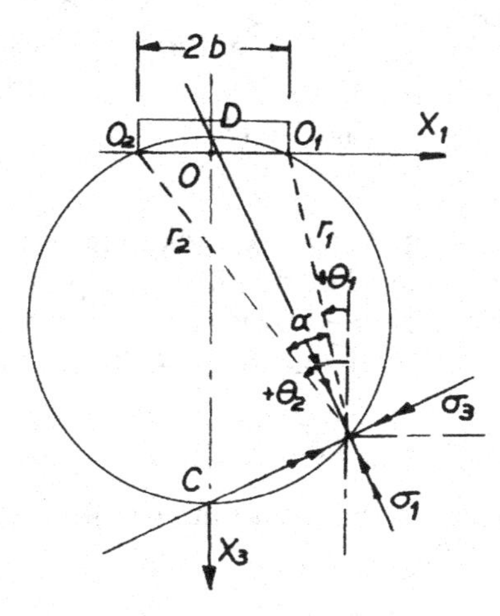

Fig. 14.14

and their difference are the same for all points falling on this circle. The angle made by the principal stresses and the OX_1 axis is given by Eq. (7.11.7) as

$$\tan 2\phi = \frac{2\sigma_{13}}{(\sigma_{11} - \sigma_{33})} = \frac{\cos 2\theta_1 - \cos 2\theta_2}{\sin 2\theta_1 - \sin 2\theta_2} = -\tan(\theta_1 + \theta_2). \quad (14.9.8)$$

Therefore,

$$\phi_1 = \frac{\Pi}{2} - \frac{1}{2}(\theta_1 + \theta_2) \quad \text{and} \quad \phi_2 = \Pi - \frac{1}{2}(\theta_1 + \theta_2). \qquad (14.9.9)$$

Eqs. (14.9.9) show that at all points on the circle, the direction of principal stresses is given by the two lines passing through C and D; in other words, the directions of the principal stresses bisect the angle between the two radii r_1 and r_2. Therefore, the principal stress trajectories are families of confocal hyperbolas and confocal ellipses (Fig. 14.15a) with focii at O_1 and O_2. The isochromatics are circles centered

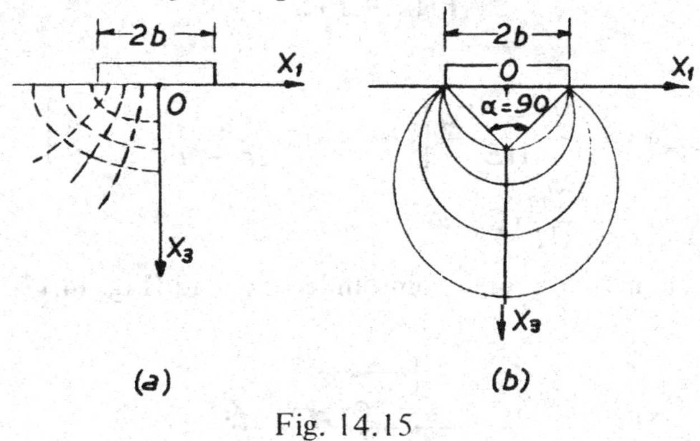

Fig. 14.15

on OX_3 and passing through O_1 and O_2 (Fig. 14.15b). On each of these circles, the maximum shearing stress is given by:

$$\frac{1}{2}(\sigma_1 - \sigma_3) = -\frac{q}{\Pi} \sin \alpha. \qquad (14.9.10)$$

The maximum value in Eq. (14.9.10) is for $\alpha = \Pi / 2$ corresponding to a circle centered at O.

The deflection of points at the surface of the semi-infinite solid can be obtained by integration of Eq. (14.5.31) or Eq. (14.5.38), depending on the boundary conditions chosen. Let us consider Eq. (14.5.31), for example. If the point whose vertical deflection is sought is outside the loaded area, we have (Fig. 14.16):

$$(u_\theta)_{\theta=\frac{\Pi}{2}} = -\frac{2(1-\nu^2)}{\Pi E} q \int_a^{a+2b} \ln\frac{d}{r}\, dr + \frac{1+\nu}{\Pi E} q \int_a^{a+2b} dr \qquad (14.9.11)$$

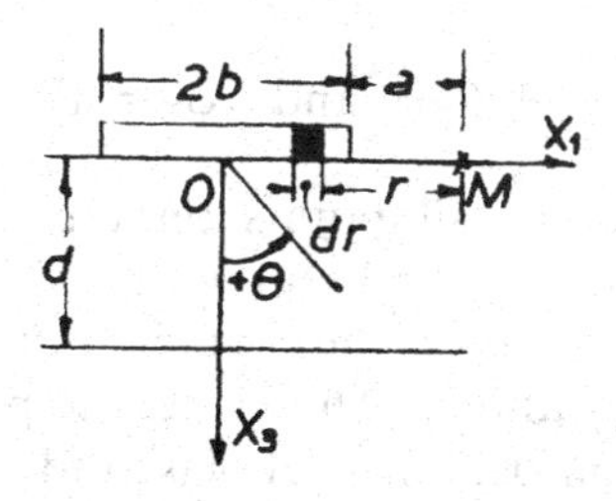

Fig. 14.16

or

$$(u_\theta)_{\theta=\frac{\Pi}{2}} = -\frac{2(1-\nu^2)q}{\Pi E}\left[(2b+a)\ln\frac{d}{2b+a} - a\ln\frac{d}{a}\right] - \frac{2bq}{\Pi E}(1+\nu)(1-2\nu). \tag{14.9.12}$$

In the same manner, for a point under the load (Fig. 14.17):

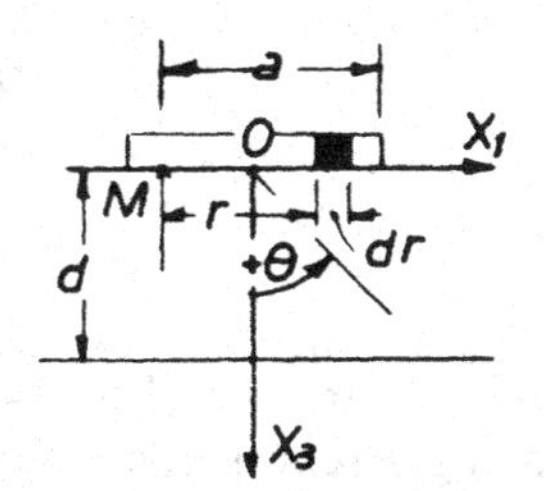

Fig. 14.17

$$(u_\theta)_{\theta=\frac{\Pi}{2}} = -\frac{2(1-\nu^2)}{\Pi E}q\left[(2b-a)\ln\frac{d}{2b-a} - a\ln\frac{d}{a}\right] - \frac{2bq}{\Pi E}(1+\nu)(1-2\nu), \tag{14.9.13}$$

d is the distance from O of a plane parallel to the surface and whose vertical displacement is equal to zero.

A number of problems related to the semi-infinite elastic medium in plane strain have been solved by Holl [7] who considered both horizontal and inclined surface planes. Timoshenko [8] has given the stress functions providing the solution for various loading patterns on the horizontal surface of a semi-infinite medium or plate.

14.10 Uniformly Distributed Vertical Pressure on Part of the Boundary of a Semi-Infinite Elastic Plate

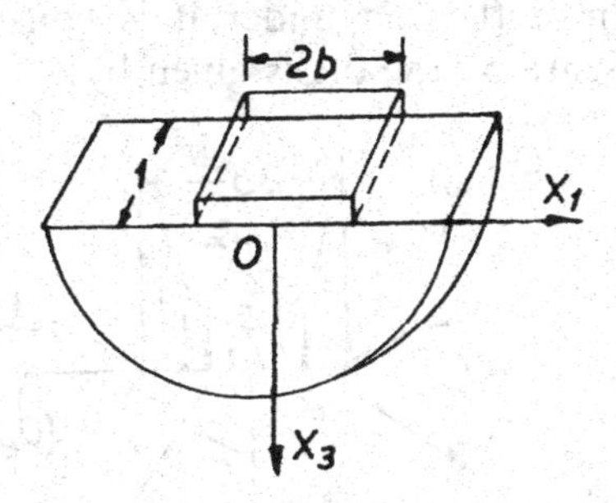

Fig. 14.18

The plate is assumed to be of unit width (Fig. 14.18). The problem is a plane stress problem. The stresses are the same as those given in Sec. 14.9, except for σ_{22} which is equal to zero. The displacements of points on the surface of the plate can be obtained by integration of Eq. (14.6.12). If the point whose vertical deflection is sought is outside the loaded area (Fig. 14.16), we have:

$$(u_\theta)_{\theta=\frac{\Pi}{2}} = \frac{1+\nu}{\Pi E} q \int_a^{a+2b} dr - \frac{2q}{\Pi E} \int_a^{a+2b} \ell n \frac{d}{r}\, dr \quad (14.10.1)$$

or

$$(u_\theta)_{\theta=\frac{\Pi}{2}} = -\frac{2q}{\Pi E}\left[(2b+a)\ell n \frac{d}{2b+a} - a\,\ell n \frac{d}{a}\right] - \frac{2(1-\nu)}{\Pi E} qb. \quad (14.10.2)$$

In the same manner, for a point under the load (Fig. 14.17):

$$(u_\theta)_{\theta=\frac{\Pi}{2}} = -\frac{2q}{\Pi E}\left[(2b-a)\ell n \frac{d}{2b-a} + a\,\ell n \frac{d}{a}\right] - \frac{2(1-\nu)}{\Pi E} qb; \quad (14.10.3)$$

d is the distance from O to a point on the OX_3 axis whose displacement is equal to zero.

Frocht [9] has provided a photoelastic confirmation for many of the solutions related to semi-infinite plates.

14.11 Rigid Strip at the Surface of a Semi-Infinite Elastic Medium

It can be shown that, when a load Q is applied through a rigid flat die [10] so that the deflection under it is constant (Fig. 14.19), the distribution of pressure on the die is given by:

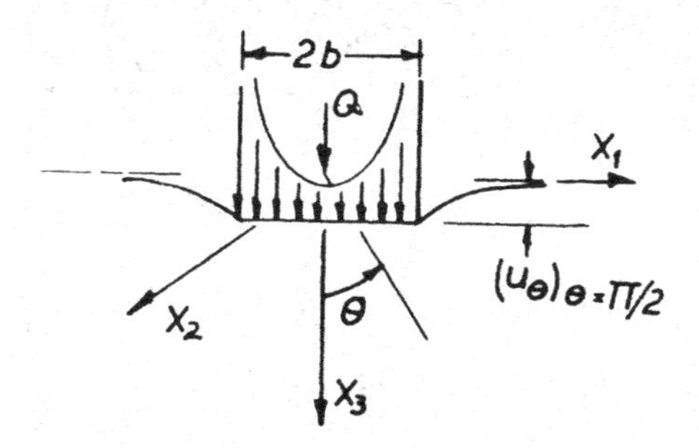

Fig. 14.19

$$q = \frac{Q}{\Pi\sqrt{b^2 - x_1^2}}. \tag{14.11.1}$$

This expression shows that $q = Q/\Pi b$ when $x_1 = 0$, and becomes infinite when $x_1 = b$. In actual cases, we shall have a yielding of the material along the boundary. The displacement $(u_\theta)_{\theta=\Pi/2}$ of the die corresponding to the distribution (14.11.1) is obtained by integrating Eq. (14.5.31). For example, under the center of the die we have:

$$(u_\theta)_{\theta=\frac{\Pi}{2}} = \frac{-4(1-\nu^2)}{\Pi E}\int_0^b q\,\ell n\frac{d}{r}\,dr + \frac{2(1+\nu)}{\Pi E}\int_0^b q\,dr, \tag{14.11.2}$$

where q is given by Eq. (14.11.1). Thus,

$$(u_\theta)_{\theta=\frac{\Pi}{2}} = -\frac{Q}{\Pi E}\left[2(1-\nu^2)\ell n\frac{2d}{b} - (1+\nu)\right]. \tag{14.11.3}$$

The same result is obtained for any point under the die.

14.12 Rigid Die at the Surface of a Semi-Infinite Elastic Plate

The distribution of pressure on the die is given by Eq. (14.11.1), namely (Fig. 14.20):

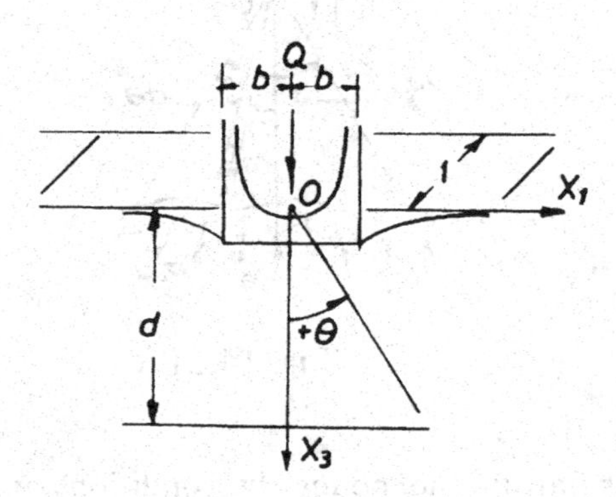

Fig. 14.20

$$q = \frac{Q}{\Pi\sqrt{b^2 - x_1^2}}. \qquad (14.12.1)$$

The displacement of the surface corresponding to the distribution (14.12.1) is obtained by integrating Eq. (14.6.12). Thus,

$$(u_\theta)_{\theta=\frac{\Pi}{2}} = -\frac{Q}{\Pi E}\left[2\ell n\frac{2d}{b} - (1+\nu)\right]. \qquad (14.12.2)$$

14.13 Radial Stresses in Wedges

The stresses in a wedge of infinite length subjected to a vertical load at the apex (Fig. 14.21) can be obtained from the Airy stress function:

$$\phi = Cr\theta \sin\theta \qquad (14.13.1)$$

The components of the state of stress are:

$$\sigma_{rr} = \frac{2C\cos\theta}{r}, \quad \sigma_{\theta\theta} = 0, \quad \sigma_{zz} = \frac{2C\nu\cos\theta}{r} \qquad (14.13.2)$$

$$\sigma_{r\theta} = \sigma_{\theta z} = \sigma_{rz} = 0. \qquad (14.13.3)$$

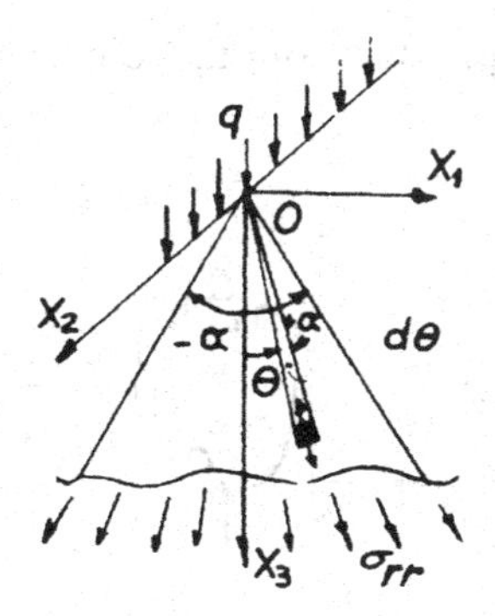

Fig. 14.21

These stresses satisfy the boundary conditions in the wedge: They vanish at infinity and leave the straight edges free of normal and shear stresses. In order to determine the constant C in terms of the load q, we set out a sector of the wedge as a free body and write the equilibrium of the forces in the OX_3 direction. Since the stress distribution for both plane strain and plane stress is the same as far as σ_{rr}, $\sigma_{\theta\theta}$, and $\sigma_{r\theta}$ are concerned, we shall consider a wedge of unit length in the OX_2 direction. Thus,

$$q + \int_{-\alpha}^{+\alpha} \sigma_{rr} \cos\theta (r\, d\theta) = q + 2C \int_{-\alpha}^{+\alpha} \cos^2\theta\, d\theta = 0 \tag{14.13.4}$$

and

$$C = -\frac{q}{2\alpha + \sin 2\alpha}. \tag{14.13.5}$$

Therefore, the components of the state of stress in the wedge are:

$$\sigma_{rr} = -\frac{2q\cos\theta}{r(2\alpha + \sin 2\alpha)}, \quad \sigma_{\theta\theta} = 0, \quad \sigma_{zz} = -\frac{2q\nu\cos\theta}{r(2\alpha + \sin 2\alpha)} \tag{14.13.6}$$

$$\sigma_{r\theta} = \sigma_{\theta z} = \sigma_{rz} = 0. \tag{14.13.7}$$

For α equal to $\Pi / 2$, Eqs. (14.5.6) to (14.5.9) are obtained.

If the load is normal to the axis of the wedge (Fig. 14.22), the same stress function can be used provided θ is measured from the direction

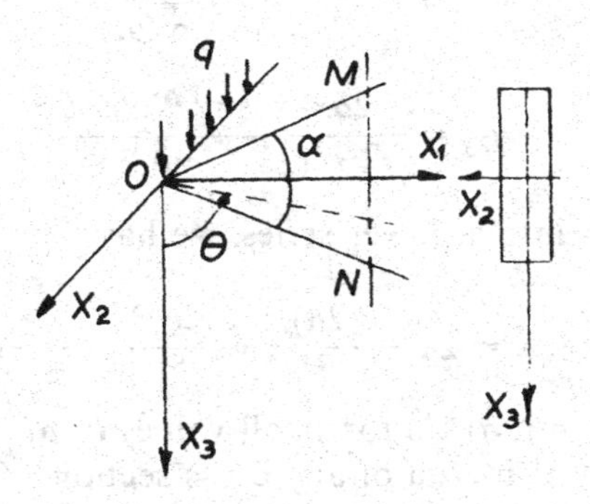

Fig. 14.22

of the force. The constant C is found from the equation of equilibrium:

$$q + \int_{\frac{\pi}{2}-\alpha}^{\frac{\pi}{2}+\alpha} \sigma_{rr} \cos\theta (r\,d\theta) = 0. \tag{14.13.8}$$

Thus,

$$C = -\frac{q}{2\alpha - \sin 2\alpha} \tag{14.13.9}$$

and

$$\sigma_{rr} = -\frac{2q\cos\theta}{r(2\alpha - \sin 2\alpha)}, \quad \sigma_{\theta\theta} = 0, \quad \sigma_{zz} = -\frac{2q\nu\cos\theta}{r(2\alpha - \sin 2\alpha)} \tag{14.13.10}$$

$$\sigma_{r\theta} = \sigma_{\theta z} = \sigma_{rz} = 0. \tag{14.13.11}$$

The case of inclined loads at the apex of a wedge can easily be studied by the superposition of the two previous cases.

If a wedge is cut from a thin plate, the sets of Eqs. (14.13.6), (14.13.7), (14.13.10), and (14.13.11) are valid except that $\sigma_{zz} = 0$. It is of interest to compare the equations in this case to those of the elementary theory of beams: In cartesian coordinates, we have [see Eqs. (7.11.4) to (7.11.6)]

$$\sigma_{11} = -\frac{2qx_1 x_3 \sin^4\theta}{x_1^3(2\alpha - \sin 2\alpha)} \tag{14.13.12}$$

$$\sigma_{13} = -\frac{2qx_3^2\sin^4\theta}{x_1^3(2\alpha - \sin 2\alpha)} \tag{14.13.13}$$

$$\sigma_{33} = -\frac{2qx_1 x_3\sin^2\theta \cos^2\theta}{x_1^3(2\alpha - \sin 2\alpha)}. \tag{14.13.14}$$

Expanding sin 2α into a power series, we have:

$$\sin 2\alpha = 2\alpha - \frac{(2\alpha)^3}{3!} + \frac{(2\alpha)^5}{5!} - - -,$$

so that $2\alpha - \sin 2\alpha = (2\alpha)^3/6$ for small values of α.
If I_{22} is the moment of inertia of any cross section MN, Eqs. (14.13.12) and (14.13.13) can be written as:

$$\sigma_{11} = -\frac{qx_1 x_3}{I_{22}}\left(\frac{\tan\alpha}{\alpha}\right)^3 \sin^4\theta \tag{14.13.15}$$

$$\sigma_{13} = -\frac{qx_3^2}{I_{22}}\left(\frac{\tan\alpha}{\alpha}\right)^3 \sin^4\theta. \tag{14.13.16}$$

For small values of α, θ is nearly equal to $\Pi\ /\ 2$ and the factor $(\tan \alpha/\alpha)^3 \sin^4\theta$ is nearly equal to unity. The expression for σ_{11} becomes equal to that of the elementary theory. The maximum shearing stress occurs at M and N and is twice as large as the shearing stress the elementary theory gives for the center of a triangular beam with a rectangular cross section. Frocht [9] gives a series of graphs illustrating this situation.

14.14 M. Levy's Problems of the Triangular and Rectangular Retaining Walls

Let us first consider a triangular retaining wall or dam subjected to a pressure linearly increasing with depth (Fig. 14.23). The boundary conditions are:

1) On O B,

$$x_1 = 0 \quad \sigma_{13} = 0. \quad \sigma_{11} = -\gamma x_3 \tag{14.14.1}$$

2) On O A,

$$x_1 = x_3 \tan\beta, \quad \sigma_n = 0, \quad \sigma_t = 0, \tag{14.14.2}$$

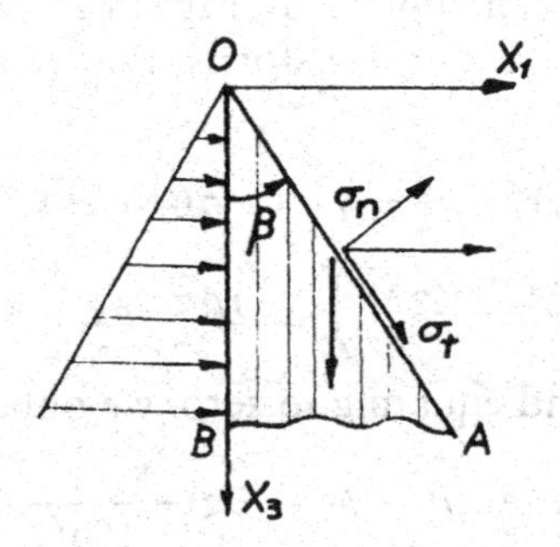

Fig. 14.23

where γ can be looked upon as the unit weight of the liquid. Since the number of boundary conditions is four, a polynomial of the third degree would provide a suitable stress function (see Sec. 9.11). Consider the Airy stress function:

$$\phi = \frac{dx_1^3}{6} + \frac{ex_1^2 x_3}{2} + \frac{fx_1 x_3^2}{2} + \frac{kx_3^3}{6}. \tag{14.14.3}$$

From Eqs. (9.9.4), we have:

$$\sigma_{11} = fx_1 + kx_3 - \rho g x_3 \tag{14.14.4}$$

$$\sigma_{33} = dx_1 + ex_3 - \rho g x_3 \tag{14.14.5}$$

$$\sigma_{13} = -ex_1 - fx_3, \tag{14.14.6}$$

where ρg is the weight per unit volume of the material of the wall. The first set of boundary conditions (14.14.1) gives:

$$f = 0 \tag{14.14.7}$$

$$k = -\gamma + \rho g. \tag{14.14.8}$$

Thus,

$$\sigma_{11} = -\gamma x_3 \tag{14.14.9}$$

$$\sigma_{33} = dx_1 + ex_3 - \rho g x_3 \tag{14.14.10}$$

$$\sigma_{13} = -ex_1. \tag{14.14.11}$$

The direction cosines of the normal $\bar{n}$ (Fig. 14.23) with OX_1 and OX_3 are $(\cos\beta, -\sin\beta)$. Applying the transformation of axes of Sec. 7.11, we get:

$$\sigma_n = (dx_1 + ex_3 - \rho g x_3)\sin^2\beta - \gamma x_3 \cos^2\beta + 2ex_1 \sin\beta\cos\beta \quad (14.14.12)$$

$$\sigma_t = \left(\frac{-\gamma x_3 - dx_1 - ex_3 + \rho g x_3}{2}\right)\sin 2\beta - ex_1 \cos 2\beta. \quad (14.14.13)$$

Setting $x_1 = x_3 \tan\beta$ and equating to zero, we obtain the two equations:

$$d\tan\beta + 3e = \rho g + \frac{\gamma}{\tan^2\beta} \quad (14.14.14)$$

$$d\tan\beta + e\left(\frac{2\tan\beta}{\tan 2\beta} + 1\right) = +\rho g - \gamma. \quad (14.14.15)$$

Solving, we get:

$$e = \frac{\gamma}{\tan^2\beta}, \quad d = \frac{\rho g}{\tan\beta} - \frac{2\gamma}{\tan^3\beta}. \quad (14.14.16)$$

The stresses now assume the final form:

$$\sigma_{11} = -\gamma x_3 \quad (14.14.17)$$

$$\sigma_{33} = \left(\frac{\rho g}{\tan\beta} - \frac{2\gamma}{\tan^3\beta}\right)x_1 + \left(\frac{\gamma}{\tan^2\beta} - \rho g\right)x_3 \quad (14.14.18)$$

$$\sigma_{13} = -\frac{\gamma x_1}{\tan^2\beta} \quad (14.14.19)$$

$$\sigma_{22} = \nu(\sigma_{11} + \sigma_{33}) \quad (14.14.20)$$

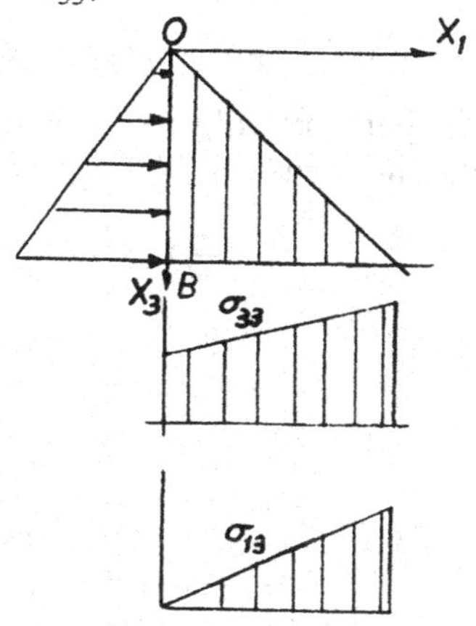

Fig. 14.24

$$\sigma_{12} = \sigma_{23} = 0. \tag{14.14.21}$$

The diagrams of stresses σ_{33} and σ_{13} over any horizontal section are shown in Fig. 14.24. The elementary theory gives the same answer for σ_{33}, but an essentially different one for σ_{13}.

For a rectangular retaining wall (Fig. 14.25), the boundary conditions are:

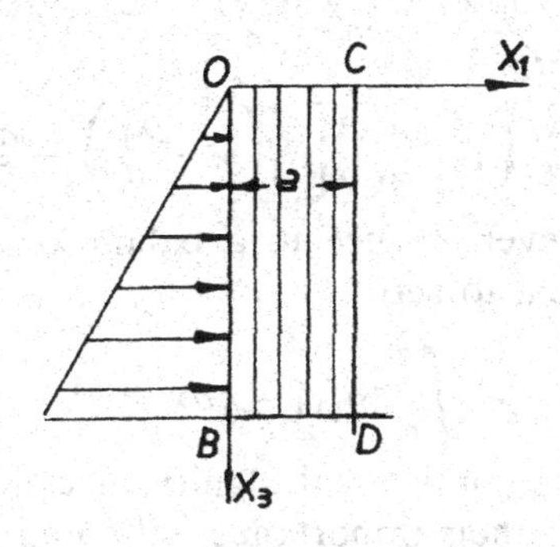

Fig. 14.25

1) on O B,

$$x_1 = 0 \quad \sigma_{13} = 0, \quad \sigma_{11} = -\gamma x_3 \tag{14.14.22}$$

2) on C D,

$$x_1 = a \quad \sigma_{13} = 0, \quad \sigma_{11} = 0 \tag{14.14.23}$$

3) on O C,

$$x_3 = 0 \quad \sigma_{33} = 0, \quad \sigma_{13} = 0. \tag{14.14.24}$$

By taking a stress function in the form of a polynomial of the sixth degree, M. Levy [11] deduced the following expressions for the stresses:

$$\sigma_{11} = -\gamma x_3 \left(1 - \frac{x_1}{a}\right)^2 \left(1 + \frac{2x_1}{a}\right) \tag{14.14.25}$$

$$\sigma_{33} = -\rho g x_3 - \frac{\gamma x_3^3}{a^2}\left(\frac{2x_1}{a} - 1\right) + \gamma x_3 \left(\frac{4x_1^3}{a^3} - \frac{6x_1^2}{a^2} + \frac{12x_1}{5a} - \frac{1}{5}\right) \tag{14.14.26}$$

$$\sigma_{13} = -\frac{\gamma x_1}{a}\left(1 - \frac{x_1}{a}\right)\left[\frac{3x_3^2}{a} - \frac{a}{5} + x_1\left(1 - \frac{x_1}{a}\right)\right] \tag{14.14.27}$$

$$\sigma_{22} = \nu(\sigma_{11} + \sigma_{33}) \tag{14.14.28}$$

$$\sigma_{23} = \sigma_{12} = 0. \tag{14.14.29}$$

Eq. (14.14.27) does not satisfy the last boundary condition, which specifies that at the top of the wall $\sigma_{13} = 0$. Indeed, if we set $x_3 = 0$ in Eq. (14.14.27), we get:

$$(\sigma_{13})_{x_3=0} = -\frac{\gamma x_1}{a}\left(1 - \frac{x_1}{a}\right)\left[x_1\left(1 - \frac{x_1}{a}\right) - \frac{a}{5}\right] \neq 0. \tag{14.14.30}$$

These stresses, however, reduce to a balanced system of forces since their resultant is equal to zero:

$$\int_0^a (\sigma_{13})_{x_3=0}\, dx_1 = 0. \tag{14.14.31}$$

They have, therefore, only local significance according to Saint-Venant's principle, and their importance is limited in practice since they act at the top of the wall where a stress analysis is not usually required. The expression of σ_{33} obtained from the elementary theory of beams is:

$$\sigma_{33} = -\rho g x_3 - \frac{\gamma x_3^3}{a^2}\left(\frac{2x_1}{a} - 1\right), \tag{14.14.32}$$

which contains only the two first terms of Eq. (14.14.26).

REFERENCES

[1] R. F. Scott, *Principles of Soil Mechanics*, Addison-Wesley, Reading, Mass., 1963.

[2] E. Jahnke and F. Emde, *Tables of Functions*, Dover, New York, N. Y., 1945.

[3] N. M. Newmark, "Influence Charts for Computation of Stresses in Elastic Foundations," *Engineering Experiment Station Bulletin*, Series No. 338, University of Illinois, Urbana, Illinois, 1942.

[4] N. M. Newmark, Influence Charts for Computation of Vertical Displacements in Elastic Foundations, *Engineering Experiment Station Bulletin*, Series No. 367, University of Illinois, Urbana, Illinois, 1947.

[5] A. E. H. Love, "The Stress Produced in a Semi-Infinite Solid by Pressure on Part of the Boundary," *Transactions of the Royal Society*, London, Series A, Vol. 228, 1929.

[6] J. Boussinesq, *Application des Potentiels*, Gauthier-Villars, Paris, 1885.

[7] D. L. Holl, "Plane Distribution of Stress in Elastic Media," *Engineering Experiment Station Bulletin*, 148, Iowa State College, Ames, Iowa, 1941.

[8] S. Timoshenko and J. N. Goodier, *Theory of Elasticity*, McGraw-Hill, New York, N. Y., 1970.

[9] M. M. Frocht, *Photoelasticity*, Vol. 2, John Wiley & Son, New York, N. Y., 1948.

[10] M. Sadowsky, *Zietschriftangew. Math. Mech.*, Vol. 8, p. 107, 1928.

[11] M. Levy, *Comptes Rendus*, Vol. 126, p. 1235, 1898.

CHAPTER 15

ENERGY PRINCIPLES AND INTRODUCTION TO VARIATIONAL METHODS

15.1 Introduction

In Sec. 8.20, it was stated that the solution of an elasticity problem amounted to solving a system of 15 equations with 15 unknowns. Various methods of solution were listed, and among them the variational methods which are based on the fact that the governing operations of elasticity can be obtained as a direct consequence of the minimization of an energy expression. Energy is an invariant, i.e., a quantity independent of the coordinate system of reference. It is a scalar and as such is easy to manipulate. The use of methods based on energy avoids the task of having to solve, in a direct way, the fifteen partial differential equations of elasticity. The basis of the variational formulation is the principle of virtual work enunciated by John Bernoulli in 1717. This principle states that if a particle is in equilibrium under n forces Q_1, Q_2, $\ldots$, Q_n, the total virtual work done during any arbitrary virtual displacement of the particle is zero. A solid body at rest may be considered as consisting of a system of particles in equilibrium under the action of surface and body forces. The difference between a particle and a solid body is that during the virtual displacements the continuity of the material as well as the boundary constraints must be observed. By expressing the virtual displacements in terms of continuous functions, the condition of continuity of the material is satisfied. The boundary constraints generally depend on the type of structure. The ends of a simple beam on two supports, for example, cannot move in the transverse direction so that the virtual displacement there must be taken as zero.

In this chapter, various energy theorems for a solid continuous body will be derived, and a number of simple examples will be presented to illustrate their use. A brief introduction to the calculus of variations is also included.

15.2 Work, Strain and Complementary Energies. Clapeyron's Law.

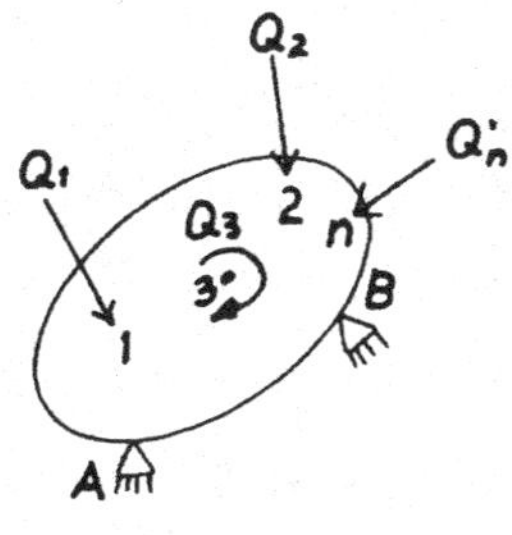

Fig. 15.1

Let us consider a deformable body fixed at points A and B (Fig. 15.1). The body is acted upon by a system of generalized forces Q_1, Q_2, ..., Q_n. As a result of the generalized forces, generalized displacements q_1, q_2, ..., q_n are produced. The term "generalized displacements" is used to mean both linear displacements along the forces and angular rotations about a line perpendicular to the plane of the couples. The product of a generalized force by a generalized displacement represents work: When both generalized force and generalized displacement have the same direction, the work is positive; when they have opposite directions, the work is negative.

Let us now consider a body acted upon by one force Q (Fig. 15.2) The dashed line represents the deformed shape of the body after Q has reached its final magnitude. The relationship between Q and q is shown

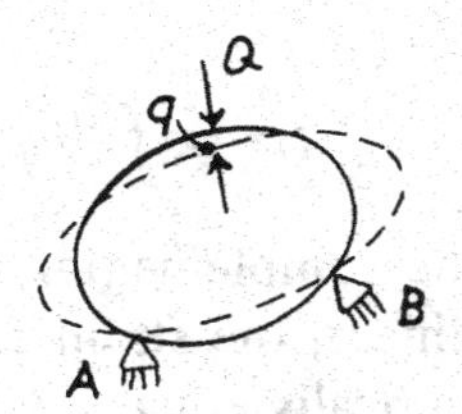

Fig. 15.2

in Fig. 15.3. For an infinitesimal increment dq, the increment in work done by Q is

$$dW = Qdq. \tag{15.2.1}$$

Assuming negligible temperature changes, this work is stored under the form of internal strain energy, so that

$$dU_t = dW = Qdq, \tag{15.2.2}$$

where U_t is the strain energy. The total strain energy,

$$U_t = \int_0^{q_1} Q\,dq, \tag{15.2.3}$$

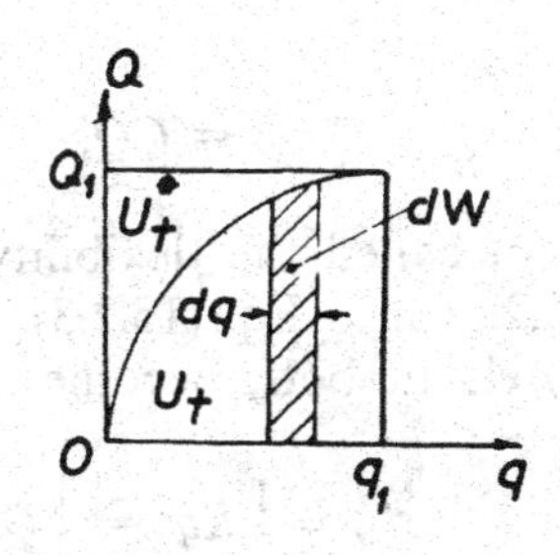

Fig. 15.3

is given by the area under the curve in Fig. 15.3. The area over the curve is given by

$$U_t^* = \int_0^{Q_1} q\,dQ, \tag{15.2.4}$$

where U_t^* is called the complementary strain energy. If the relation between Q and q is linear, the strain energy and the complementary strain energy are numerically equal (Fig. 15.4). In such a case, we can write:

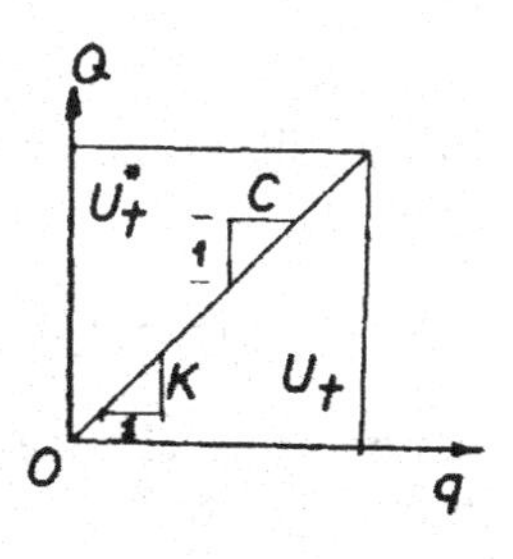

Fig. 15.4

$$Q = Kq, \tag{15.2.5}$$

where K is a constant called the stiffness coefficient. Alternatively, we can write:

$$q = CQ, \tag{15.2.6}$$

where C is a constant called the flexibility or compliance coefficient. Substituting Eq. (15.2.5) into Eq. (15.2.3), the expression of the strain energy for a linearly elastic body becomes:

$$U_t = \frac{1}{2}Kq^2 = \frac{1}{2}Qq. \tag{15.2.7}$$

Eq. (15.2.7) is known as Clapeyron's Law. [Compare to Eq. (8.7.7)]. If we substitute Eq. (15.2.6) into Eq. (15.2.4), the expression of the complementary strain energy for a linearly elastic body becomes:

$$U_t^* = \frac{1}{2}CQ^2. \tag{15.2.8}$$

Eqs. (15.2.6), (15.2.7), and (15.2.8) show that

$$K = C^{-1}. \tag{15.2.9}$$

The previous equations can be extended to a linearly elastic body subjected to a system of N generalized forces. The increment in strain energy dU_t is given by:

$$dU_t = \sum_{m=1}^{N} Q_m \, dq_m. \tag{15.2.10}$$

Each one of the generalized forces causes a generalized displacement at its point of application as well as at all other points of the body. Therefore, we can write:

$$q_n = q_n(Q_1, Q_2 - - - Q_n) \tag{15.2.11}$$

and

$$dq_n = \frac{\partial q_n}{\partial Q_1} dQ_1 + \frac{\partial q_n}{\partial Q_2} dQ_2 + - - - \frac{\partial q_n}{\partial Q_N} dQ_N. \tag{15.2.12}$$

Since the body is linearly elastic, the partial derivatives are constants. We shall call them the flexibility or compliance influence coefficients:

$$C_{nm} = \frac{\partial q_n}{\partial Q_m}. \tag{15.2.13}$$

We see that C_{nm} is the displacement at the point n caused by a unit force at the point m. Eq. (15.2.12) can be integrated to give:

$$q_n = C_{n1} Q_1 + C_{n2} Q_2 - - - C_{nN} Q_N = \sum_{m=1}^{N} C_{nm} Q_m. \tag{15.2.14}$$

Alternatively, we can write:

$$Q_n = Q_n(q_1, q_2 - - - q_N) \tag{15.2.15}$$

and

$$Q_n = \sum_{m=1}^{N} K_{nm} q_m, \tag{15.2.16}$$

where

$$K_{nm} = \frac{\partial Q_n}{\partial q_m}. \tag{15.2.17}$$

The K_{nm} ' s are called the stiffness influence coefficients. K_{nm} represents the force acting at n which causes a unit displacement at m. We shall prove in a later section that $C_{nm} = C_{mn}$ and that $K_{nm} = K_{mn}$. Eqs. (15.2.14) and (15.2.16) can, respectively, be written in matrix form as follows:

$$\begin{bmatrix} q_1 \\ q_2 \\ - \\ q_N \end{bmatrix} = \begin{bmatrix} C_{11} & C_{12} & --- & C_{1N} \\ C_{21} & C_{22} & --- & C_{2N} \\ - & - & --- & - \\ C_{N1} & C_{N2} & --- & C_{NN} \end{bmatrix} \begin{bmatrix} Q_1 \\ Q_2 \\ - \\ Q_N \end{bmatrix} \tag{15.2.18}$$

or

$$\{q\} = [C]\{Q\}; \tag{15.2.19}$$

$$\begin{bmatrix} Q_1 \\ Q_2 \\ - \\ Q_N \end{bmatrix} = \begin{bmatrix} K_{11} & K_{12} & --- & K_{1N} \\ K_{21} & K_{22} & --- & K_{2N} \\ - & - & --- & - \\ K_{N1} & K_{N2} & --- & K_{NN} \end{bmatrix} \begin{bmatrix} q_1 \\ q_2 \\ - \\ q_N \end{bmatrix} \tag{15.2.20}$$

or

$$\{Q\} = [K]\{q\}. \tag{15.2.21}$$

From Eqs. (15.2.19) and (15.2.21), we deduce that

$$[C]^{-1} = [K]. \tag{15.2.22}$$

$[C]$ and $[K]$ are called the flexibility matrix and the stiffness matrix, respectively.

The total strain energy can be evaluated by integrating Eq. (15.2.10), keeping in mind that the body under load behaves in a linear and conservative way. This strain energy depends only on the final state of force and displacement and not on the way this state has been reached: In other words, it is independent of the path of integration. The integration is simplified by setting:

$$q_n = B_n q, \tag{15.2.23}$$

where B_n is a factor of proportionality and q is a function varying from 0 to 1. The linearity of the behavior justifies the use of Eq. (15.2.23), since the generalized displacements q_n maintain the same proportion at all times. Substituting Eqs. (15.2.16) and (15.2.23) into Eq. (15.2.10) and integrating, we get:

$$U_t = \frac{1}{2} \sum_{m=1}^{N} \sum_{n=1}^{N} K_{mn} B_m B_n q^2 = \frac{1}{2} \sum_{m=1}^{N} \sum_{n=1}^{N} K_{mn} q_m q_n . \quad (15.2.24)$$

Therefore,

$$U_t = \frac{1}{2} \sum_{n=1}^{N} Q_n q_n . \quad (15.2.25)$$

This formula is Clapeyron's Law when the linearly elastic body is subjected to a system of generalized forces and displacements.

Following a similar reasoning, the complementary strain energy is given by:

$$U^*_t = \frac{1}{2} \sum_{m=1}^{N} \sum_{n=1}^{N} C_{mn} Q_m Q_n . \quad (15.2.26)$$

Clapeyron's Law can be stated as follows: The work done by a system of forces acting on a linearly elastic body is independent of the rate at which the forces increase and the order in which they are applied. The work done, which is stored as strain energy, is equal to one-half the product of the final magnitudes of the generalized forces and their corresponding generalized displacement. The body must be initially stress free and not subjected to temperature change.

15.3 Principle of Virtual Work

Consider a body which is in a state of static equilibrium under the action of surface forces and body forces (Fig. 15.5). These forces result in stresses and deformations which satisfy the differential equations of equilibrium, the strain-displacement relations, and the stress-strain relations which are not necessarily elastic. Now suppose that the body is subjected to a small change in shape caused by some source other than the applied forces. Such a change in shape is called a virtual distortion, and the word virtual is used here to indicate that the distortion is independent of the actual system of forces acting on the body. Owing to this change in shape, an element of volume inside the body is deformed, translated, and rotated. The stresses acting on this

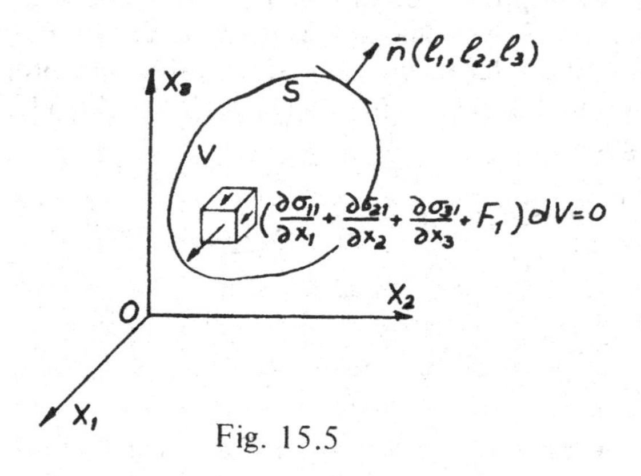

Fig. 15.5

element move and, therefore, do virtual work. The displacement of each element must be consistent with the geometrical constraints of the body. Let u_1, u_2, and u_3 be the components of the actual displacement of an element due to the surface and body forces, and let δu_1, δu_2, and δu_3 be the components of its virtual displacement. These components of the virtual displacement are assumed to be infinitesimally small quantities satisfying the conditions of continuity of the deformation, i.e., they are continuous functions of x_1, x_2, and x_3. Additionally, the virtual displacement cannot affect the equilibrium of the external forces and their internally induced stresses. Since the element is in equilibrium, the resultant of the forces acting on it is equal to zero. In a virtual displacement $\delta \bar{u}$, these forces do work, but the net result is zero. Therefore,

$$\begin{aligned} \iiint_V \Bigg[&\left(\frac{\partial \sigma_{11}}{\partial x_1} + \frac{\partial \sigma_{21}}{\partial x_2} + \frac{\partial \sigma_{31}}{\partial x_3} + F_1 \right) \delta u_1 \\ &+ \left(\frac{\partial \sigma_{12}}{\partial x_1} + \frac{\partial \sigma_{22}}{\partial x_2} + \frac{\partial \sigma_{32}}{\partial x_3} + F_2 \right) \delta u_2 \\ &+ \left(\frac{\partial \sigma_{13}}{\partial x_1} + \frac{\partial \sigma_{23}}{\partial x_2} + \frac{\partial \sigma_{33}}{\partial x_3} + F_3 \right) \delta u_3 \Bigg] \, dV = 0. \end{aligned} \tag{15.3.1}$$

Eq. (15.3.1) can be rewritten as follows:

$$\iiint_V \left[\frac{\partial}{\partial x_1}(\sigma_{11}\,\delta u_1 + \sigma_{12}\,\delta u_2 + \sigma_{13}\,\delta u_3) + \frac{\partial}{\partial x_2}(\sigma_{21}\,\delta u_1 + \sigma_{22}\,\delta u_2 + \sigma_{23}\,\delta u_3) + \frac{\partial}{\partial x_3}(\sigma_{31}\,\delta u_1 + \sigma_{32}\,\delta u_2 + \sigma_{33}\,\delta u_3)\right] dV + \iiint_V (F_1\,\delta u_1 + F_2\,\delta u_2 + F_3\,\delta u_3)\,dV \tag{15.3.2}$$

$$= \iiint_V \left[\left(\sigma_{11}\frac{\partial}{\partial x_1} + \sigma_{21}\frac{\partial}{\partial x_2} + \sigma_{31}\frac{\partial}{\partial x_3}\right)\delta u_1 + \left(\sigma_{12}\frac{\partial}{\partial x_1} + \sigma_{22}\frac{\partial}{\partial x_2} + \sigma_{32}\frac{\partial}{\partial x_3}\right)\delta u_2 + \left(\sigma_{13}\frac{\partial}{\partial x_1} + \sigma_{23}\frac{\partial}{\partial x_2} + \sigma_{33}\frac{\partial}{\partial x_3}\right)\delta u_3\right] dV.$$

Using the divergence theorem and recalling Eqs. (7.3.8), the left-hand side of Eq. (15.3.2) can be written:

$$\iint_S (\sigma_{n1}\,\delta u_1 + \sigma_{n2}\,\delta u_2 + \sigma_{n3}\,\delta u_3)\,dS + \iiint_V (F_1\,\delta u_1 + F_2\,\delta u_2 + F_3\,\delta u_3)\,dV.$$

Recalling the definition of the linear strains given by Eqs. (1.2.1), we can write:

$$\delta e_{ij} = \frac{1}{2}\left(\frac{\partial(\delta u_i)}{\partial x_j} + \frac{\partial(\delta u_j)}{\partial x_i}\right), \tag{15.3.3}$$

and the right-hand side of Eq. (15.3.2) becomes:

$$\iiint_V [\sigma_{11}\,\delta e_{11} + \sigma_{21}\,\delta e_{21} + \sigma_{31}\,\delta e_{31} + \sigma_{12}\,\delta e_{12} + \sigma_{22}\,\delta e_{22} + \sigma_{32}\,\delta e_{32} + \sigma_{13}\,\delta e_{13} + \sigma_{23}\,\delta e_{23} + \sigma_{33}\,\delta e_{33}]\,dV.$$

Eq. (15.3.2) now assumes the form (in index notation):

$$\iint_S (\sigma_{ni}\,\delta u_i)\,dS + \iiint_V (F_i\,\delta u_i)\,dV = \iiint_V (\sigma_{ij}\,\delta e_{ij})\,dV. \tag{15.3.4}$$

The left-hand side of Eq. (15.3.4) is physically interpreted as the virtual work done by the external forces, and the right-hand side as the virtual work done by the internal stresses. Although the body forces act within the volume of the body, they are considered as external forces in the sense that they cause stresses and strains. Eq. (15.3.4) is the mathematical statement of the principle of virtual work. In words, this principle may be stated as follows: If a body is in equilibrium and remains in equilibrium while it is subjected to a virtual distortion compatible with the geometrical constraints, the virtual work done by the external forces is equal to the virtual work done by the internal stresses. It is to be noticed that this principle holds independently of the stress-strain relations of the material of the body.

15.4 Variational Problems and Euler's Equations

Let us consider the problem of determining a function of $y(x)$ which makes the integral

$$I(y) = \int_a^b F(x, y, y')\,dx \tag{15.4.1}$$

an extremum, i.e., stationary, and which satisfies the prescribed end conditions,

$$y(x)_{x=a} = y(a) = y_a \tag{15.4.2}$$

$$y(x)_{x=b} = y(b) = y_b\,. \tag{15.4.3}$$

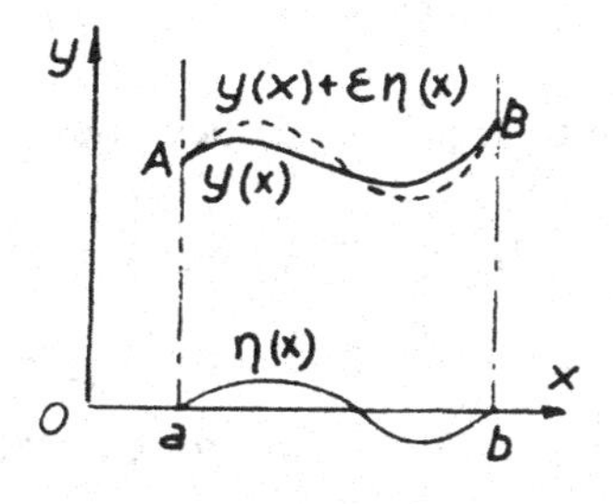

Fig. 15.6

The value of the integral in Eq. (15.4.1) depends on the choice of $y = y(x)$, hence the notation $I(y)$. The integrand $F(x,y,y')$ is a function of x,y and the first derivative $dy/dx = y'$. Let us assume that $y = y(x)$ is the actual minimizing function (Fig. 15.6), and find out what happens to the integral I when neighboring functions are used in F. These functions are constructed by adding to $y(x)$ a function $\varepsilon\eta(x)$, where ε is a constant which can take different quantitative values and $\eta(x)$ is an arbitrary function of x which vanishes at $x = a$ and at $x = b$; that is:

$$\eta(a) = 0 \quad \text{and} \quad \eta(b) = 0. \tag{15.4.4}$$

Therefore, $y(x) + \varepsilon\eta(x)$ will satisfy the end conditions. If we now replace $y(x)$ by $y(x) + \varepsilon\eta(x)$, the integral I becomes a function of ε once $\eta(x)$ is chosen, since y is the actual minimizing function. Therefore, we can write:

$$I(\varepsilon) = \int_a^b F(x, y + \varepsilon\eta, y' + \varepsilon\eta')\, dx \tag{15.4.5}$$

This integral takes its minimum value when ε is zero, but this is possible only if

$$\frac{dI(\varepsilon)}{d\varepsilon} = 0, \qquad \text{when } \varepsilon = 0. \tag{15.4.6}$$

To simplify the computations, we set:

$$F_e = F(x, y + \varepsilon\eta, y' + \varepsilon\eta') \tag{15.4.7}$$

and

$$Y(x) = y(x) + \varepsilon\eta(x). \tag{15.4.8}$$

Differentiating Y with respect to x, and F_e with respect to ε, we get:

$$Y' = y'(x) + \varepsilon\eta'(x) \tag{15.4.9}$$

and

$$\frac{dF_e}{d\varepsilon} = \frac{\partial F_e}{\partial Y}\eta(x) + \frac{\partial F_e}{\partial Y'}\eta'(x). \tag{15.4.10}$$

Therefore, from Eqs. (15.4.5) and (15.4.6), we have:

$$\frac{dI(\varepsilon)}{d\varepsilon} = \int_a^b \left[\frac{\partial F_e}{\partial Y} \eta(x) + \frac{\partial F_e}{\partial Y'} \eta'(x) \right] dx = 0. \qquad (15.4.11)$$

The second term of Eq. (15.4.11) can be transformed, through integration by parts, as follows:

$$\int_a^b \frac{\partial F_e}{\partial Y'} \eta'(x)\, dx = \left[\frac{\partial F_e}{\partial Y'} \eta(x) \right]_a^b - \int_a^b \eta(x) \frac{d}{dx} \left(\frac{\partial F_e}{\partial Y'} \right) dx. \qquad (15.4.12)$$

Since $\eta(a) = \eta(b) = 0$,

$$\left[\frac{\partial F_e}{\partial Y'} \eta(x) \right]_a^b = 0, \qquad (15.4.13)$$

and Eq. (15.4.11) becomes:

$$\frac{dI(\varepsilon)}{d\varepsilon} = \int_a^b \left[\frac{\partial F_e}{\partial Y} \eta(x) - \frac{d}{dx} \left(\frac{\partial F_e}{\partial Y'} \right) \eta(x) \right] dx = 0. \qquad (15.4.14)$$

According to Eq. (15.4.6), this derivative must vanish when ε approaches 0. But as ε approaches 0, F_e approaches F, Y approaches y, and Y' approaches y'. Therefore, Eq. (15.4.14) becomes:

$$\int_a^b \left[\frac{\partial F}{\partial y} - \frac{d}{dx} \frac{\partial F}{\partial y'} \right] \eta(x)\, dx = 0. \qquad (15.4.15)$$

It is possible to prove rigorously that, if Eq. (15.4.15) is true for any function $\eta(x)$ which is twice differentiable in the interval (a,b) and zero at the ends of that interval, the coefficient of $\eta(x)$ in the integrand must be zero everywhere in (a,b) [1]. Thus,

$$\frac{\partial F}{\partial y} - \frac{d}{dx} \left(\frac{\partial F}{\partial y'} \right) = 0. \qquad (15.4.16)$$

Eq. (15.4.16) is called Euler's equation and is a necessary condition for $y(x)$ to make I a minimum or a maximum. In computing d/dx it must be remembered that y and y' are functions of x. Thus,

$$\frac{d}{dx} \left(\frac{\partial F}{\partial y'} \right) = \frac{\partial^2 F}{\partial x \partial y'} + \frac{\partial^2 F}{\partial y \partial y'} \frac{dy}{dx} + \frac{\partial^2 F}{\partial y'^2} \frac{d^2 y}{dx^2}.$$

Euler's equation becomes:

$$\frac{\partial F}{\partial y} - \frac{\partial^2 F}{\partial x \partial y'} - \frac{\partial^2 F}{\partial y \partial y'}\frac{dy}{dx} - \frac{\partial^2 F}{\partial y'^2}\frac{d^2 y}{dx^2} = 0. \tag{15.4.17}$$

This is a second-order differential equation. Its solution contains two arbitrary constants. They are to be determined by the requirement that the curve passes through the end-points A and B.

If the integrand in Eq. (15.4.1) contains the second derivative y'', Euler's equation becomes [1]:

$$\frac{\partial F}{\partial y} - \frac{d}{dx}\frac{\partial F}{\partial y'} + \frac{d^2}{dx^2}\frac{\partial F}{\partial y''} = 0. \tag{15.4.18}$$

Eq. (15.4.18) can be generalized to include higher derivatives of y.

Let us now introduce the notation of variation and establish the analogy between the differential calculus and the calculus of variations. While in the first we deal with the differential of a function along a particular curve, in the second we deal with a variation of a functional from curve to curve [1]. To define a functional, let us consider a set S of functions satisfying certain conditions. Any quantity which takes on a specific numerical value corresponding to each function in S is said to be a functional on the set S. Thus,

$$I = \int_a^b F(x, y, y')\, dx \tag{15.4.19}$$

is a functional since, corresponding to any function $y\,(x)$, I takes a definite numerical value. Within the same context, it is justifiable to call such quantities as

$$f[y(x)], \quad g[x, y(x), y'(x), \ldots, y^{(n)}(x)]$$

functionals in those cases when the variable x is considered as fixed in a given discussion and the function $y\,(x)$ is varied. Thus, in Eq. (15.4.1), we have considered an integrand of the form:

$$F = F(x, y, y'), \tag{15.4.20}$$

which for a fixed value of x depends on the function $y\,(x)$ and its derivative. It is, therefore, a functional.

In Eq. (15.4.8), the change $\varepsilon\eta(x)$ in y (x) is called the variation of y, and is conventionally denoted by δy,

$$\delta y = \varepsilon\eta(x). \tag{15.4.21}$$

Corresponding to the change δy in the function y, F changes by an amount ΔF, where

$$\Delta F = F(x, y + \varepsilon\eta, y' + \varepsilon\eta') - F(x, y, y'). \tag{15.4.22}$$

If the right-hand member is expanded in powers of ε, there follows:

$$\Delta F = \frac{\partial F}{\partial y}\varepsilon\eta + \frac{\partial F}{\partial y'}\varepsilon\eta' + (\text{terms with higher powers of } \varepsilon). \tag{15.4.23}$$

By analogy with the definition of the differential, the first two terms in the right-hand member of Eq. (15.4.23) are defined to be the variation of F,

$$\delta F = \frac{\partial F}{\partial y}\varepsilon\eta + \frac{\partial F}{\partial y'}\varepsilon\eta'. \tag{15.4.24}$$

When $F = y$, this definition is consistent with Eq. (15.4.21) and when $F = y'$, it yields:

$$\delta y' = \varepsilon\eta', \tag{15.4.25}$$

so that Eq. (15.4.24) can be written in the form:

$$\delta F = \frac{\partial F}{\partial y}\delta y + \frac{\partial F}{\partial y'}\delta y'. \tag{15.4.26}$$

For a complete analogy with the definition of the differential, one would have anticipated the definition:

$$\delta F = \frac{\partial F}{\partial x}\delta x + \frac{\partial F}{\partial y}\delta y + \frac{\partial F}{\partial y'}\delta y'. \tag{15.4.27}$$

But, in the manipulations of F, x is not varied, so that:

$$\delta x = 0 \tag{15.4.28}$$

and the analogy between differential and variation is complete. From the definition, it follows that the laws of variation of sums, products, ratios, and so forth, are completely analogous to those of differentiation. Thus, for example,

$$\delta(F_1 F_2) = F_1\,\delta F_2 + F_2\,\delta F_1 \tag{15.4.29}$$

$$\delta\frac{F_1}{F_2} = \frac{F_2\,\delta F_1 - F_1\,\delta F_2}{F_2^2}, \tag{15.4.30}$$

where F_1 and F_2 are two different functions of x, y, and y'. From Eqs. (15.4.21) and (15.4.25), we deduce that

$$\frac{d}{dx}(\delta y) = \delta\left(\frac{dy}{dx}\right). \tag{15.4.31}$$

That is, the operators δ and d/dx are commutative. Let us now turn to Eq. (15.4.1), and show that the necessary condition for I to be stationary is that its first variation vanishes; that is,

$$\delta I = \int_a^b \delta F(x, y, y')\,dx = 0. \tag{15.4.32}$$

Indeed,

$$\int_a^b \delta F\,dx = \int_a^b \left[\frac{\partial F}{\partial y}\delta y + \frac{\partial F}{\partial y'}\delta y'\right] dx. \tag{15.4.33}$$

Integrating the second term by parts, we have:

$$\begin{aligned} \int_a^b \frac{\partial F}{\partial y'}\delta\left(\frac{dy}{dx}\right) dx &= \int_a^b \frac{\partial F}{\partial y'}\frac{d}{dx}(\delta y)\,dx \\ &= \left[\frac{\partial F}{\partial y'}\delta y\right]_a^b - \int_a^b \frac{d}{dx}\left(\frac{\partial F}{\partial y'}\right)\delta y\,dx. \end{aligned} \tag{15.4.34}$$

Thus,

$$\int_a^b \delta F\,dx = \int_a^b \left[\frac{\partial F}{\partial y} - \frac{d}{dx}\frac{\partial F}{\partial y'}\right]\delta y\,dx + \left[\frac{\partial F}{\partial y'}\delta y\right]_a^b = 0, \tag{15.4.35}$$

and this is precisely what was obtained in Eq. (15.4.15) [see Eq. (15.4.13)]. Thus, a stationary function for an integral function is one for which the variation of that integral is zero, just as a stationary point of a function is one at which the differential of the function is zero.

In the more general case, when the function to be minimized or maximized is of the form

$$I = \iint_R F(x, y, u, v, u_x, u_y, v_x, v_y)\, dx\, dy, \tag{15.4.36}$$

where u_x, u_y, v_x, v_y indicate differentations of u and v with respect to x and y, the condition for an extremum is again

$$\delta I = 0. \tag{15.4.37}$$

Here x and y are independent variables; the region R could be a body in a state of plane stress so that u and v could be the displacements. The condition (15.4.37) then becomes:

$$\begin{aligned}\delta I = \iint_R \Bigg[\left(\frac{\partial F}{\partial u}\delta u + \frac{\partial F}{\partial u_x}\delta u_x + \frac{\partial F}{\partial u_y}\delta u_y\right) \\ + \left(\frac{\partial F}{\partial v}\delta v + \frac{\partial F}{\partial v_x}\delta v_x + \frac{\partial F}{\partial v_y}\delta v_y\right)\Bigg]\, dx\, dy = 0\end{aligned} \tag{15.4.38}$$

Here the variations δu and δv are to be continuously differentiable over the region R and are to vanish over its boundary when u and v are prescribed on S, but otherwise are completely arbitrary. To transform the terms involving the variations of the derivatives, we make use of Green-Riemann's theorem (see Appendix A-10.1). The general procedure may be illustrated by considering the treatment of a typical term. If ℓ_1 and ℓ_2 are the direction cosines of the normal $\bar{n}$ to the boundary S (Fig. 15.7):

$$\begin{aligned}\iint_R \frac{\partial F}{\partial u_x}\delta\left(\frac{\partial u}{\partial x}\right) dx\, dy &= \iint_R \frac{\partial F}{\partial u_x}\frac{\partial}{\partial x}(\delta u)\, dx\, dy \\ &= \iint_R \frac{\partial}{\partial x}\left(\frac{\partial F}{\partial u_x}\delta u\right) dx\, dy - \iint_R \frac{\partial}{\partial x}\left(\frac{\partial F}{\partial u_x}\right)\delta u\, dx\, dy \\ &= \oint_S \frac{\partial F}{\partial u_x}\ell_1\, \delta u\, dS - \iint_R \frac{\partial}{\partial x}\left(\frac{\partial F}{\partial u_x}\right)\delta u\, dx\, dy.\end{aligned} \tag{15.4.39}$$

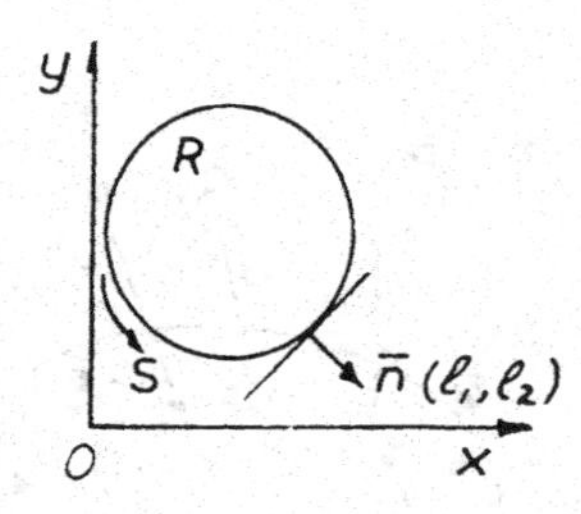

Fig. 15.7

Following the same procedure for the other terms, Eq. (15.4.38) becomes:

$$\delta I = \oint_S \left[\left(\frac{\partial F}{\partial u_x} \ell_1 + \frac{\partial F}{\partial u_y} \ell_2 \right) \delta u + \left(\frac{\partial F}{\partial v_x} \ell_1 + \frac{\partial F}{\partial v_y} \ell_2 \right) \delta v \right] dS$$
$$+ \iint_R \left\{ \left[\frac{\partial F}{\partial u} - \frac{\partial}{\partial x} \left(\frac{\partial F}{\partial u_x} \right) - \frac{\partial}{\partial y} \left(\frac{\partial F}{\partial u_y} \right) \right] \delta u \right. \tag{15.4.40}$$
$$\left. + \left[\frac{\partial F}{\partial v} - \frac{\partial}{\partial x} \left(\frac{\partial F}{\partial v_x} \right) - \frac{\partial}{\partial y} \left(\frac{\partial F}{\partial v_y} \right) \right] \delta v \right\} dx\, dy = 0.$$

If the variations δu and δv are independent of each other—that is, if u and v can be varied independently—the coefficients of δu and δv must each vanish identically in R, giving the two Euler equations:

$$\frac{\partial}{\partial x} \left(\frac{\partial F}{\partial u_x} \right) + \frac{\partial}{\partial y} \left(\frac{\partial F}{\partial u_y} \right) - \frac{\partial F}{\partial u} = 0 \tag{15.4.41}$$

$$\frac{\partial}{\partial x} \left(\frac{\partial F}{\partial v_x} \right) + \frac{\partial}{\partial y} \left(\frac{\partial F}{\partial v_y} \right) - \frac{\partial F}{\partial v} = 0. \tag{15.4.42}$$

When u is not prescribed on S,

$$\frac{\partial F}{\partial u_x} \ell_1 + \frac{\partial F}{\partial u_y} \ell_2 = 0 \text{ on } S, \tag{15.4.43}$$

and when v is not prescribed on S,

$$\frac{\partial F}{\partial v_x} \ell_1 + \frac{\partial F}{\partial v_y} \ell_2 = 0 \text{ on } S. \tag{15.4.44}$$

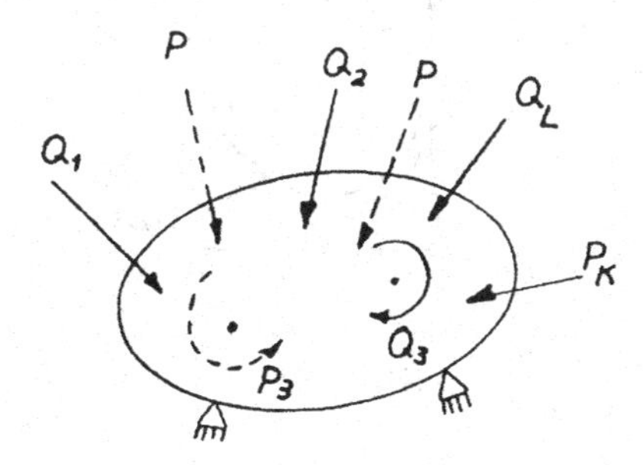

Fig. 15.8

15.5 The Reciprocal Laws of Betti and Maxwell

Consider a linearly elastic body which will be loaded in turn by two different systems of generalized forces: the Q system and the P system (Fig. 15.8). The number of the Q forces is L and that of the P forces is K. The displacements of the points of application of the Q forces caused by the P forces is denoted by $\Delta_l^{(P)}$. Similarly, the deflections of the points of application of the P forces caused by the Q forces is denoted by $\Delta_k^{(Q)}$. Let us apply the principle of virtual work to one system of forces at a time, while assuming that the virtual surface displacements and the virtual internal strains are caused by the other system. Thus, the virtual work done by the Q system is:

$$\sum_{l=1}^{l=L} Q_l \Delta_l^{(P)} = \iiint \sigma_{mn}^{(Q)} \delta e_{mn}^{(P)} \, dV, \tag{15.5.1}$$

where $\delta e_{mn}^{(P)}$ are the internal virtual strains caused by the P forces and $\sigma_{mn}^{(Q)}$ are the internal stresses caused by the Q forces. Similarly,

$$\sum_{k=1}^{K} P_k \Delta_k^{(Q)} = \iiint \sigma_{mn}^{(P)} \delta e_{mn}^{(Q)} \, dV, \tag{15.5.2}$$

where $\delta e_{mn}^{(Q)}$ are the internal virtual strains caused by the Q forces and $\sigma_{mn}^{(P)}$ are the internal stresses caused by the P forces. From Eq. (8.3.16), we have:

$$\delta e_{mn}^{(P)} = S_{mnrs} \sigma_{rs}^{(P)} \tag{15.5.3}$$

and

$$\delta e_{mn}^{(Q)} = S_{mnrs}\,\sigma_{rs}^{(Q)}. \tag{15.5.4}$$

The two Eqs. (15.5.1) and (15.5.2) become:

$$\sum_{l=1}^{L} Q_l\,\Delta_l^{(P)} = \iiint \sigma_{mn}^{(Q)}\,S_{mnrs}\,\sigma_{rs}^{(P)}\,dV \tag{15.5.5}$$

$$\sum_{k=1}^{K} P_k\,\Delta_k^{(Q)} = \iiint \sigma_{mn}^{(P)}\,S_{mnrs}\,\sigma_{rs}^{(Q)}\,dV. \tag{15.5.6}$$

The two integrals in the two last equations are the same because of the symmetry property:

$$S_{mnrs} = S_{rsmn}. \tag{15.5.7}$$

Therefore,

$$\sum_{l=1}^{L} Q_l\,\Delta_l^{(P)} = \sum_{k=1}^{K} P_k\,\Delta_k^{(Q)}. \tag{15.5.8}$$

Eq. (15.5.8) is the mathematical statement of Betti's law: In any body which is linearly elastic, the work done by a system of Q forces under a distortion caused by a system of P forces is equal to the work done by the system of P forces under a distortion caused by the system of Q forces. If we have one Q force and one P force which are numerically equal, we get (Fig. 15.9):

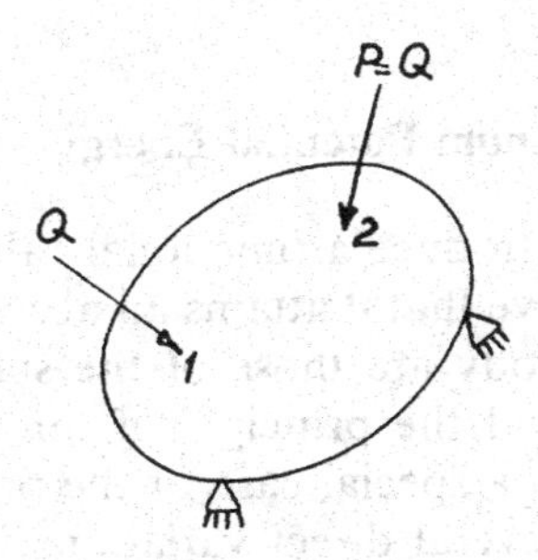

Fig. 15.9

$$Q\Delta_1^{(P)} = Q\Delta_2^{(Q)}. \tag{15.5.9}$$

Thus, if we denote the deflection at a point m caused by a load at point n by Δ_{mn}, and the deflection at point n caused by the same magnitude of load at point m by Δ_{nm}, then Eq. (15.5.9) becomes, in general:

$$\Delta_{mn} = \Delta_{nm}. \tag{15.5.10}$$

This is the general formula for Maxwell's law of reciprocal deflections: In any body which is linearly elastic, the generalized deflection at point m caused by a generalized load Q at point n is numerically equal to the generalized deflection at point n caused by the same magnitude of generalized force Q at point m.

If we now turn to Eq. (15.2.18) we see, for example, that C_{12} is the displacement at point 1 due to a unit load at point 2, and that C_{21} is the displacement at point 2 due to a unit load at point 1. According to Maxwell's law, these two displacements must be equal; so that,

$$C_{12} = C_{21}. \tag{15.5.11}$$

In the same way,

$$C_{13} = C_{31}, \quad C_{14} = C_{41}, \quad \text{etc.}; \tag{15.5.12}$$

in other words, the matrix $[C]$ is symmetric; consequently, the matrix $[K]$ is also symmetric.

15.6 Principle of Minimum Potential Energy

In this section, we introduce a functional called the potential energy of deformation and prove that it attains an absolute minimum when the displacements of the body are those of the state of equilibrium. This principle, which is called the principle of minimum potential energy, may be looked upon as a special case of the principle of virtual work: It lies at the basis of several direct variational methods of solution of elastostatic problems. We shall use the symbol δ to mean the variation of a function according to the calculus of variations (Sec. 15.4). Let a body be in equilibrium under the action of specified body and surface forces. At a given point (x_1, x_2, x_3), σ_{ij} and e_{ij} are the states of stress and strain respectively. δe_{ij} represents any small change of the function e_{ij} (x_1, x_2, x_3) satisfying the compatibility and boundary conditions. If a strain energy density function $U(e_{ij})$ exists, so that (see Sec. 8.2),

$$\sigma_{ij} = \frac{\partial U}{\partial e_{ij}}, \tag{15.6.1}$$

then the right-hand term of Eq. (15.3.4):

$$\begin{aligned}\iiint_V (\sigma_{ij}\,\delta e_{ij})\,dV &= \iiint_V \left(\frac{\partial U}{\partial e_{ij}}\delta e_{ij}\right) dV \\ &= \delta\left(\iiint_V U\,dV\right) = \delta U_t.\end{aligned} \tag{15.6.2}$$

In Eq. (15.3.4) σ_{ni} and F_i are constants, so that the principle of virtual work can now be written as:

$$\begin{aligned}\delta\Pi_p = \delta\left[\iiint_V U\,dV - \iiint_V F_i u_i\,dV - \iint_S \sigma_{ni} u_i\,dS\right] \\ = 0.\end{aligned} \tag{15.6.3}$$

It is customary to set:

$$W = \iiint_V F_i u_i\,dV + \iint_S \sigma_{ni} u_i\,dS, \tag{15.6.4}$$

so that Eq. (15.6.3) becomes:

$$\delta\Pi_p = \delta(U_t - W) = 0. \tag{15.6.5}$$

The scalar Π_p defined by Eq. (15.6.3) is called the potential energy of the body:

$$\begin{aligned}\Pi_p = U_t - W = \iiint_V U\,dV - \iiint_V F_i u_i\,dV \\ - \iint_S \sigma_{ni} u_i\,dS.\end{aligned} \tag{15.6.6}$$

Eq. (15.6.5) states that: Among all the geometrically possible states of displacement satisfying the given boundary conditions, those which satisfy the equations of equilibrium result in a stationary value of the potential energy. To show that Π_p must be a minimum for a state of stable equilibrium, consider a neighboring state Π'_p characterized by strains $e_{ij} + \delta e_{ij}$ and the corresponding displacements $u_i + \delta u_i$. Hence,

$$\Pi'_p - \Pi_p = \iiint_V [U(e_{ij} + \delta e_{ij}) - U(e_{ij})]\,dV - \iiint_V F_i\,\delta u_i\,dV - \iint_S \sigma_{ni}\,\delta u_i\,dS. \tag{15.6.7}$$

Expanding $U(e_{ij} + \delta e_{ij})$ into a power series, and neglecting terms with an order higher than the second, we get:

$$U(e_{ij} + \delta e_{ij}) = U(e_{ij}) + \frac{\partial U}{\partial e_{ij}}\delta e_{ij} + \frac{1}{2}\frac{\partial^2 U}{\partial e_{ij}\,\partial e_{kl}}\delta e_{ij}\,\delta e_{kl}. \tag{15.6.8}$$

A substitution into Eq. (15.6.7) yields:

$$\Pi'_p - \Pi_p = \iiint_V \frac{\partial U}{\partial e_{ij}}\delta e_{ij}\,dV - \iiint_V F_i\,\delta u_i\,dV - \iint_S \sigma_{ni}\,\delta u_i\,dS + \iiint \frac{1}{2}\frac{\partial^2 U}{\partial e_{ij}\,\partial e_{kl}}\delta e_{ij}\,\delta e_{kl}. \tag{15.6.9}$$

The sum of the three first terms in the right-hand side of Eq. (15.6.9) vanishes on account of Eq. (15.3.4). The last term is positive for sufficiently small values of δe_{ij}. This can be seen as follows: Let us set $e_{ij} = 0$ in Eq. (15.6.8). The constant term can be disregarded since we are only interested in derivatives of U. $\partial U/\partial e_{ij} = \sigma_{ij}$ must vanish when $e_{ij} = 0$. Therefore, up to the second order:

$$U(\delta e_{ij}) = \frac{1}{2}\frac{\partial^2 U}{\partial e_{ij}\,\partial e_{kl}}\delta e_{ij}\,\delta e_{kl}, \tag{15.6.10}$$

and if $U(\delta e_{ij})$ is positive definite, then $1/2\ (\partial^2 U/\partial e_{ij}\,\partial e_k)\ \delta e_{ij}\,\delta e_{kl}$ is positive definite. Therefore $\Pi'_p - \Pi_p$ is positive and Π_p is a minimum.

The principle of minimum potential energy can now be stated as follows: Among all the geometrically possible states of displacement satisfying the given boundary conditions, those which satisfy the equations of equilibrium result in a minimum value of the potential energy.

It is of interest to show that the variational principle (15.6.3) gives the equations of elasticity. Indeed,

$$\delta\Pi_p = \iiint_V \left(\frac{\partial U}{\partial e_{ij}} \delta e_{ij}\right) dV - \iiint_V (F_i \delta u_i)\, dV - \iint_S (\sigma_{ni}\, \delta u_i)\, dS = 0. \tag{15.6.11}$$

But

$$\iiint_V \left(\frac{\partial U}{\partial e_{ij}} \delta e_{ij}\right) dV = \frac{1}{2} \iiint_V \sigma_{ij} \left(\delta \frac{\partial u_i}{\partial x_j} + \delta \frac{\partial u_j}{\partial x_i}\right) dV = \iiint_V \frac{\partial}{\partial x_j} (\sigma_{ji}\, \delta u_i)\, dV - \iiint_V \left(\frac{\partial \sigma_{ji}}{\partial x_j} \delta u_i\right) dV. \tag{15.6.12}$$

Using the divergence theorem, we get:

$$\iiint_V \left(\frac{\partial U}{\partial e_{ij}} \delta e_{ij}\right) dV = \iint_S (\sigma_{ji}\, \ell_j\, \delta u_i)\, dS - \iiint_V \left(\frac{\partial \sigma_{ji}}{\partial x_j} \delta u_i\right) dV. \tag{15.6.13}$$

Hence,

$$\delta\Pi_p = \iiint_V \left(\frac{\partial \sigma_{ji}}{\partial x_j} + F_i\right) \delta u_i\, dV + \iint_S (\sigma_{ni} - \sigma_{ji}\, \ell_j) \delta u_i\, dS = 0, \tag{15.6.14}$$

which can be satisfied for arbitrary δu_i if

$$\frac{\partial \sigma_{ji}}{\partial x_j} + F_i = 0 \quad \text{in } V, \tag{15.6.15}$$

and either

$$\delta u_i = 0 \quad \text{on } S \text{ (where the surface displacements are prescribed)} \tag{15.6.16}$$

or

$$\sigma_{ni} = \sigma_{ji} \ell_j \quad \text{on } S \text{ (where the surface stresses are prescribed)} . \tag{15.6.17}$$

The boundary conditions which are prescribed in the problem at hand are called the forced boundary conditions: For example at the fixed end of a cantilever beam, the forced boundary conditions are mathematically expressed by writing $\delta u_2 = 0$ and $\delta(du_2/dx_1) = 0$. On the other hand, some boundary conditions may be deduced as necessary conditions for minimum potential energy. Such conditions are called natural boundary conditions (See Problem 8).

Thus we have demonstrated the one-to-one correspondence between the differential equations of equilibrium and the variational equation: We have derived Eq. (15.6.3) from the equations of equilibrium [starting from Eq. (15.3.1)] and then have shown that, conversely, Eqs. (15.6.15) to (15.6.17) necessarily follow Eq. (15.6.3). It is to be noticed that at no time was linearity of the stress-strain relations of the body invoked: All that was required was the existence of the strain energy density function. Thus, the principle is valid for any elastic body, linear or nonlinear.

If the body is linearily elastic, then:

$$\sigma_{ij} = 2\mu e_{ij} + \lambda \delta_{ij} e_{nn} \tag{15.6.18}$$

and

$$U_t = \iiint_V \left[\mu e_{ij} e_{ij} + \frac{\lambda}{2} (e_{nn})^2 \right] dV. \tag{15.6.19}$$

To satisfy the condition of continuity of displacements over the entire body, the function e_{ij} must be expressed in terms of the displacement functions u_i by

$$e_{ij} = \frac{1}{2}\left(\frac{\partial u_i}{\partial x_j} + \frac{\partial u_j}{\partial x_i} \right), \tag{15.6.20}$$

and the variation of the strain energy U_t must be carried out with respect to the displacement functions.

15.7 Castigliano's First Theorem

Consider a body in equilibrium under a set of generalized external forces Q_m. If the strain energy U_t is expressed as a function of the corresponding displacements q_m, the principle of virtual work can be used to show that

$$\frac{\partial U_t}{\partial q_m} = Q_m \qquad m = 1, 2, \ldots, N. \tag{15.7.1}$$

The proof consists of allowing a virtual displacement δq to take place in the body in such a manner that δq is continuous everywhere, but vanishes at all points of loading except under Q_m. Due to δq, the strain energy changes by an amount δU_t while the work done by the external forces is the product of Q_m times δq_m, i.e., $Q_m \delta q_m$. According to the principle of virtual work,

$$\delta U_t = Q_m \delta q_m. \tag{15.7.2}$$

When rewritten in differential form, Eq. (15.7.2) becomes Eq. (15.7.1). Therefore, Castigliano's first theorem states that: When the strain energy U_t can be expressed as a function of a system of generalized displacements q_m, then the generalized force Q_m is given by $\partial U_t / \partial q_m$. It is important to notice that this theorem has been proved using the assumption of the existence of $U_t(q_m)$ and using the principle of virtual work as applied to a state of equilibrium. Linearity has not been invoked, and the theorem is applicable to elastic bodies that follow a nonlinear load-displacement relationship.

15.8 Principle of Virtual Complementary Work

In contrast to Sec. 15.3 where we had assumed a variation of the deformations (virtual deformations), let us assume here a variation of the stresses (virtual change in the stresses) in a body held in equilibrium under the body forces per unit volume F_i and the surface forces per unit area σ_{ni}. Let σ_{ij} be the actual stress field which satisfies the equations of equilibrium and the boundary conditions.

$$\frac{\partial \sigma_{ji}}{\partial x_j} + F_i = 0 \qquad \text{in } V \tag{15.8.1}$$

$$\sigma_{ni} = \sigma_{ji} \ell_j \qquad \text{on the surface } S. \tag{15.8.2}$$

Let us consider a system of stress variations which also satisfy the equations of equilibrium and the stress boundary conditions,

$$\frac{\partial(\delta\sigma_{ji})}{\partial x_j} + \delta F_i = 0 \tag{15.8.3}$$

$$(\delta\sigma_{ji})\ell_j = \delta\sigma_{ni}. \tag{15.8.4}$$

The complementary virtual work is given by:

$$\begin{aligned}\iiint_V (u_i\,\delta F_i)\,dV + \iint_S u_i\,\delta\sigma_{ni}\,dS \\ = -\iiint_V u_i \frac{\partial}{\partial x_j}(\delta\sigma_{ji})\,dV + \iint_S u_i(\delta\sigma_{ji})\ell_j\,dS,\end{aligned} \tag{15.8.5}$$

where u_i are the components of the real displacements. Thus,

$$\begin{aligned}\iiint_V (u_i\,\delta F_i)\,dV + \iint_S u_i\,\delta\sigma_{ni}\,dS \\ = \iiint_V \left(\delta\sigma_{ji}\frac{\partial u_i}{\partial x_j}\right) dV - \iiint_V \frac{\partial}{\partial x_j}(u_i\,\delta\sigma_{ji})\,dV + \\ \iint_S u_i\,\delta\sigma_{ji}\,\ell_j\,dS\end{aligned} \tag{15.8.6}$$

and, using the divergence theorem, we get:

$$\begin{aligned}\iiint_V (u_i\,\delta F_i)\,dV + \iint_S u_i\,\delta\sigma_{ni}\,dS \\ = \iiint_V \left(\delta\sigma_{ji}\frac{\partial u_i}{\partial x_j}\right) dV \\ = \frac{1}{2}\iiint_V \delta\sigma_{ji}\left(\frac{\partial u_i}{\partial x_j} + \frac{\partial u_j}{\partial x_i}\right) dV.\end{aligned} \tag{15.8.7}$$

Therefore,

$$\iiint_V (u_i\,\delta F_i)\,dV + \iint_S u_i\,\delta\sigma_{ni}\,dS = \iiint_V e_{ij}\,\delta\sigma_{ij}\,dV. \tag{15.8.8}$$

Eq. (15.8.8) may be called the principle of virtual complementary work. This principle may be stated as follows: The virtual complementary work done under the actual state of strain and virtual stress change $\delta\sigma_{ij}$, which satisfies the differential equations of equilibrium, is equal to the complementary work done by the virtual external forces. Eq. (15.8.8) is applicable without restrictions regardless of the nature of the stress-strain relations governing the behavior of the body.

This principle is complementary to the principle of virtual work. Each represents an alternative approach to the solution of mechanics problems. While in the principle of virtual work the quantities to be varied are the displacements, in the principle of virtual complementary work the quantities to be varied are the internal stresses and the external forces.

15.9 Principle of Minimum Complementary Energy

This principle may be looked upon as a special case of the principle of virtual complementary work. Let us assume the existence of a function of the stresses called the complementary strain energy density function $U^*(\sigma_{ij})$ with the property that [1, 2]:

$$\frac{\partial U^*}{\partial \sigma_{ij}} = e_{ij}. \tag{15.9.1}$$

The principle of virtual complementary work may be written as:

$$\begin{aligned} &\iiint_V (u_i \delta F_i)\, dV + \iint_S u_i \delta\sigma_{ni}\, dS \\ &= \iiint_V \frac{\partial U^*}{\partial \sigma_{ij}} \delta\sigma_{ij}\, dV = \delta \iiint_V U^*\, dV. \end{aligned} \tag{15.9.2}$$

Since the volume is fixed and u_i are not varied, Eq. (15.9.2) can be written as:

$$\begin{aligned} \delta\Pi^* &= \delta\Big[\iiint_V U^*\, dV - \iiint_V u_i F_i\, dV - \iint_S u_i \sigma_{ni}\, dS\Big] \\ &= \delta(U_t^* - W) = 0. \end{aligned} \tag{15.9.3}$$

The quantity Π^* is a function of the stresses σ_{ij}, of the surface forces and of the body forces: It is called the complementary energy and is defined by:

$$\Pi^* = \iiint_V U^* \, dV - \iiint_V u_i F_i \, dV - \iint_S u_i \sigma_{ni} \, dS \\ = U_t^* - W. \tag{15.9.4}$$

We, therefore, have the following theorem: Of all the states of stress which satisfy the equations of equilibrium and boundary conditions where stresses are prescribed, the actual one is that which makes the complementary energy stationary. The proof that Π^* is a minimum is analogous to that in Sec. 15.6:

$$\Pi^*(\sigma_{ij} + \delta\sigma_{ij}) - \Pi^*(\sigma_{ij}) = \iiint_V [U^*(\sigma_{ij} + \delta\sigma_{ij}) - U^*(\sigma_{ij})] \, dV \\ - \iint_S [\sigma_{ni}(\sigma_{ij} + \delta\sigma_{ij}) - \sigma_{ni}(\sigma_{ij})] u_i \, dS \\ - \iiint_V u_i \, \delta F_i \, dV. \tag{15.9.5}$$

Expanding $U^*(\sigma_{ij} + \delta\sigma_{ij})$ in power series, and neglecting the terms of an order higher than the second, we get:

$$\Pi^*(\sigma_{ij} + \delta\sigma_{ij}) - \Pi^*(\sigma_{ij}) \\ = \iiint_V e_{ij} \, \delta\sigma_{ij} \, dV + \iiint_V \frac{1}{2} \frac{\partial^2 U^*}{\partial \sigma_{ij} \, \partial \sigma_{k\ell}} \delta\sigma_{ij} \, \delta\sigma_{k\ell} \\ - \iint_S [\delta\sigma_{ij} \ell_j] u_i \, dS - \iiint_V u_i \, \delta F_i \, dV. \tag{15.9.6}$$

The quantity

$$\iiint_V \frac{1}{2} \frac{\partial^2 U^*}{\partial \sigma_{ij} \, \partial \sigma_{k\ell}} \delta\sigma_{ij} \, \delta\sigma_{k\ell} = U^*(\delta\sigma_{ij}) \tag{15.9.7}$$

and the other terms in the right-hand side of Eq. (15.9.7), are equal to zero because of Eq. (15.9.3). Hence,

$$\Pi^*(\sigma_{ij} + \delta\sigma_{ij}) - \Pi^*(\sigma_{ij}) = U^*(\delta\sigma_{ij}). \tag{15.9.8}$$

If $U^*(\delta\sigma_{ij})$ is positive definite, we have the following theorem: Of all states which satisfy the equations of equilibrium and boundary conditions where stresses are prescribed, the actual one is that which makes the complementary energy minimum. It is to be noticed that at no time was linearity of the stress-strain relations of the body invoked. All that was required was the existence of the complementary strain energy density function. Thus, the principle is valid for any elastic body, linear or nonlinear.

If the body is linearly elastic, then

$$e_{ij} = \frac{1}{E}[(1 + \nu)\sigma_{ij} - \nu\delta_{ij}\sigma_{nn}] \tag{15.9.9}$$

and

$$U_t^* = U_t = \iiint_V \frac{1}{2E}[(1 + \nu)\sigma_{ij}\sigma_{ij} - \nu(\sigma_{nn})^2]\,dV. \tag{15.9.10}$$

Using the Maxwell and Morera stress functions (see Sec. 9.8), it can be shown [2] that if $\delta\Pi^* = 0$ for all variations of stresses $\delta\sigma_{ij}$ which satisfy the equations of equilibrium in the body and on the boundary where stresses are prescribed, then σ_{ij} also satisfies the equations of compatibility.

15.10 Castigliano's Second Theorem

Let us consider a linearly elastic body subjected to N generalized forces Q_m, and assume that the internal stresses have been expressed in terms of the generalized forces. The complementary energy Π^* can be written as:

$$\Pi^* = U_t(Q_1, Q_2 - - - Q_N) - \sum_{m=1}^{N} Q_m q_m, \tag{15.10.1}$$

where q_m is the generalized displacement corresponding to Q_m. According to the principle of minimum complementary energy,

$$\delta\Pi^* = 0 = \sum_{m=1}^{N} \frac{\partial\Pi^*}{\partial Q_m}\delta Q_m. \tag{15.10.2}$$

Since each variation δQ_m is independent of the others, then Eq. (15.10.2) yields N equations:

$$\frac{\partial\Pi^*}{\partial Q_m} = 0 \qquad m = 1, 2, - - - - N. \tag{15.10.3}$$

Applying Eq. (15.10.3) to Eq. (15.10.1), we get:

$$q_m = \frac{\partial U_t}{\partial Q_m} \qquad m = 1, 2, ----N. \tag{15.10.4}$$

Eq. (15.10.4) is referred to as Castigliano's second theorem, which states: In a linearly elastic structure, the partial derivative of the strain energy with respect to an externally applied generalized force is equal to the displacement corresponding to that force. This theorem is often used in the computation of the deflection of structures.(*)

15.11 Theorem of Least Work

The strain energy U_t for a statically indeterminate linearly elastic structure cannot be written in terms of the applied external forces Q_m alone. It can, however, be expressed in terms of the external forces and a number of redundant internal forces or reactions. If the structure contains N independent redundant forces $(X_1, X_2 --- X_N)$, internal or at the boundary, the expression for the complementary energy is:

$$\Pi^* = U_t(Q_1, Q_2 --- Q_m, X_1, X_2 -- X_N) - Q_m q_m. \tag{15.11.1}$$

The principle of minimum complementary energy yields two equations:

$$\frac{\partial U_t}{\partial Q_m} = q_m \tag{15.11.2}$$

and

$$\frac{\partial U_t}{\partial X_n} = 0. \tag{15.11.3}$$

Eq. (15.11.2) is Castigliano's second theorem, while Eq. (15.11.3) is referred to as the theorem of least work. This theorem states: For a statically indeterminate linearly elastic structure, the derivative of the strain energy with respect to any redundant reaction must be zero.

15.12 Summary of Energy Theorems

Before illustrating the use of the theorems derived in the previous sections, it is worthwhile to present them in a way that brings out the duality that exists between strain and complementary energies:

---

* See page A-70.

Principle of virtual work	Principle of complementary virtual work
Any Elastic Stress-Strain Relations	
Strain energy density function U exists: ↓ Principle of minimum potential energy	**Complementary strain energy density function U* exists:** ↓ Principle of minimum complementary energy
↓ Castigliano's first theorem	
Linear Elastic Stress-Strain Relations	
Reciprocal law of Betti and Maxwell	
	Castigliano's second theorem
	Theorem of least work

The theorems based on the principle of virtual work and those based on the principle of virtual complementary work offer two different approaches to the solution of elasticity problems. The application of these theorems to the same problems will yield the same answers if the problem can be solved exactly, but if approximate answers are desired the different approaches will, in many cases, yield slightly different answers. The reader is referred to Washizu's treatise [3] for a detailed examination of the interrelations among the various theorems.

"Remark": The existence of a strain energy density function U expressed in terms of the strain components and the existence of a complementary strain energy density function U^* expressed in terms of the stress components go hand in hand [2]. Also notice that,

$$U + U^* = \sigma_{ij} e_{ij} ,$$

and $$\frac{\partial U^*}{\partial \sigma_{kl}} = -\frac{\partial U}{\partial \sigma_{kl}} + \sigma_{ij} \frac{\partial e_{ij}}{\partial \sigma_{kl}} + e_{ij} \delta_{ik} \delta_{jl} .$$

Because of eq. (8.2.9, also 15.6.1),

$$\frac{\partial U^*}{\partial \sigma_{kl}} = \left(\sigma_{ij} - \frac{\partial U}{\partial e_{ij}} \right) \frac{\partial e_{ij}}{\partial \sigma_{kl}} + e_{kl} = e_{kl} .$$

Therefore,

$$\frac{\partial U^*}{\partial \sigma_{ij}} = e_{ij} .$$

15.13 Working Form of the Strain Energy for Linearly Elastic Slender Members

The name, slender member, is given to a solid generated by a plane section of contour C and area A whose centroid G describes a curve S in space: The radius of curvature of this curve is large compared to the dimensions of A, and the solid is generated in a way such that the principal axes of inertia of the cross section through its centroid maintain a constant angle with the principal normal and the binormal of S. In this manner, any point of the cross section will describe a fiber of the solid parallel to S. In Fig. 15.10, the OX_1 axis is taken tangent to S, and OX_2 and OX_3 are the principal axes of inertia through the

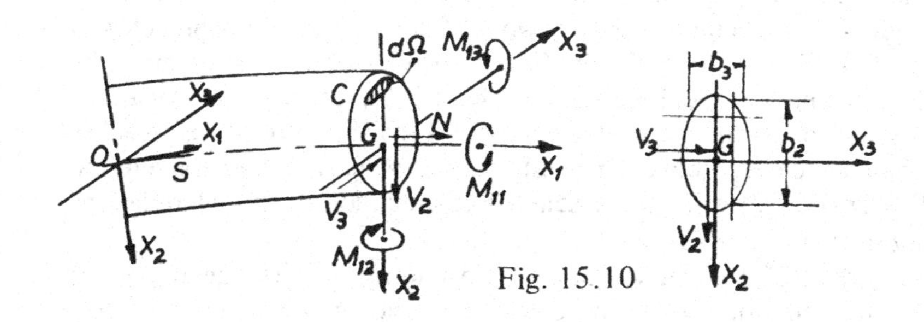

Fig. 15.10

centroid G. S is called the mean line of the member.

In this section, we shall compute the expression of the elastic strain energy in terms of the normal and shearing forces, as well as in terms of the bending and twisting moments. The only stresses to be considered are those acting on the surface A: namely, σ_{11}, σ_{12}, and σ_{13}; σ_{22}, σ_{33}, and σ_{23} are assumed equal to zero. We shall first examine the part of the strain energy connected with the normal stresses, then examine the part connected with the shearing stresses.

a. Parts of the strain energy due to σ_{11}

For a member of length L, and cross section A, Eq. (8.7.13) gives:

$$U_t = \int_0^L \left[\iint_A \frac{(\sigma_{11})^2}{2E} \, d\Omega \right] dS \tag{15.13.1}$$

If the member is subject to a normal force N, and to bending moments M_{12} and M_{13} at a distance S from the origin O (see Fig. 15.10), the normal stress σ_{11} is given by:

$$\sigma_{11} = \frac{N}{A} \pm \frac{M_{12}x_3}{I_2} \pm \frac{M_{13}x_2}{I_3}, \tag{15.13.2}$$

where I_2 and I_3 are the moments of inertia about the two principal axes GX_2 and GX_3. Substituting Eq. (15.13.2) into Eq. (15.13.1), and noting that

$$\iint_A \frac{2NM_{12}}{AI_2} x_3\, d\Omega = \iint_A \frac{2NM_{13}}{AI_3} x_2\, d\Omega = 0 \tag{15.13.3}$$

and that

$$\iint_A \left(\frac{N}{A}\right)^2 d\Omega = \frac{N^2}{A} \tag{15.13.4}$$

$$\iint_A \left(\frac{M_{12}}{I_2}\right)^2 x_3^2\, d\Omega = \frac{M_{12}^2}{I_2} \tag{15.13.5}$$

$$\iint_A \left(\frac{M_{13}}{I_3}\right)^2 x_2^2\, d\Omega = \frac{M_{13}^2}{I_3}, \tag{15.13.6}$$

we get:

$$U_t = \frac{1}{2E} \int_0^L \left(\frac{N^2}{A} + \frac{M_{12}^2}{I_2} + \frac{M_{13}^2}{I_3}\right) dS. \tag{15.13.7}$$

b. Part of the strain energy due to σ_{12} and σ_{13}

From Eq. (8.7.13), we have:

$$U_t = \int_0^L \left[\iint \frac{1}{2G}(\sigma_{12}^2 + \sigma_{13}^2)\, d\Omega\right] dS. \tag{15.13.8}$$

σ_{12} and σ_{13} are produced by applied couples and by shearing forces which may or may not be acting at the shear center. Recalling Eqs. (10.5.3), (10.5.4), and (12.2.5), we have (Fig. 15.10):

$$\sigma_{12} = \frac{V_2 Q_3}{I_3 b_3} + \frac{\partial \phi}{\partial x_3} \tag{15.13.9}$$

$$\sigma_{13} = \frac{V_3 Q_2}{I_2 b_2} - \frac{\partial \phi}{\partial x_2}, \tag{15.13.10}$$

where ϕ is Prandtl's stress function, and Q_2 and Q_3 are static moments about the OX_2 and OX_3 axes, respectively. A substitution of Eqs. (15.13.9) and (15.13.10) into Eq. (15.13.8) yields:

$$U_t = \frac{1}{2G} \int_0^L \left\{ \iint_A \left[\left(\frac{\partial \phi}{\partial x_3} \right)^2 + \left(\frac{\partial \phi}{\partial x_2} \right)^2 + \left(\frac{V_3 Q_2}{I_2 b_2} \right)^2 + \left(\frac{V_2 Q_3}{I_3 b_3} \right)^2 + 2 \frac{\partial \phi}{\partial x_3} \frac{V_2 Q_3}{I_3 b_3} - 2 \frac{\partial \phi}{\partial x_2} \frac{V_3 Q_2}{I_2 b_2} \right] d\Omega \right\} dS. \tag{15.13.11}$$

Let us now consider the various terms in the right-hand side of Eq. (15.13.11):

$$\iint_A \left(\frac{V_2 Q_3}{I_3 b_3} \right)^2 d\Omega = V_2^2 \iint_A \left(\frac{Q_3}{I_3 b_3} \right)^2 d\Omega = \frac{V_2^2}{C_2}, \tag{15.13.12}$$

where

$$\frac{1}{C_2} = \iint_A \left(\frac{Q_3}{I_3 b_3} \right)^2 d\Omega \tag{15.13.13}$$

is a property of the cross section.

$$\iint_A \left(\frac{V_3 Q_2}{I_2 b_2} \right)^2 d\Omega = \frac{V_3^2}{C_3}, \tag{15.13.14}$$

where

$$\frac{1}{C_3} = \iint_A \left(\frac{Q_2}{I_2 b_2} \right)^2 d\Omega \tag{15.13.15}$$

is a property of the cross section.

$$2 \iint_A \left(\frac{\partial \phi}{\partial x_3} \frac{V_2 Q_3}{I_3 b_3} - \frac{\partial \phi}{\partial x_2} \frac{V_3 Q_2}{I_2 b_2} \right) d\Omega$$
$$= 2 \oint_C \left\{ \left[\frac{V_2 Q_3}{I_3 b_3} \phi \right] dx_2 + \left[\frac{V_3 Q_2}{I_2 b_2} \phi \right] dx_3 \right\}, \tag{15.13.16}$$

according to the Green-Riemann formula (see Appendix A-10.1). But $\phi = 0$ on the contour C, so that the sum in the right-hand side of Eq. (15.13.16) vanishes.

Eq. (15.13.11) can now be written as follows:

$$U_t = \frac{1}{2G} \int_0^L \left[\frac{V_2^2}{C_2} + \frac{V_3^2}{C_3} \right] dS$$
$$+ \frac{1}{2G} \int_0^L \left\{ \iint \left[\left(\frac{\partial \phi}{\partial x_3} \right)^2 + \left(\frac{\partial \phi}{\partial x_2} \right)^2 \right] d\Omega \right\} dS. \tag{15.13.17}$$

Recalling Chapter 10, we can write:

$$U_t = \frac{1}{2G} \int_0^L \left[\frac{V_2^2}{C_2} + \frac{V_3^2}{C_3} + \frac{M_{11}^2}{J} \right] dS. \tag{15.13.18}$$

where GJ is the torsional rigidity and M_{11} is the twisting moment about the OX_1 axis. Adding Eqs. (15.13.7) and (15.13.18), we get the expression of the strain energy in a slender member:

$$U_t = \frac{1}{2} \int_0^L \left[\frac{N^2}{AE} + \frac{V_2^2}{GC_2} + \frac{V_3^2}{GC_3} + \frac{M_{11}^2}{GJ} + \frac{M_{12}^2}{EI_2} + \frac{M_{13}^2}{EI_3} \right] dS. \tag{15.13.19}$$

Temperature effects can be included in Eq. (15.13.19) by substituting for N^2/AE the quantity $N[N/AE + \alpha(\Delta T)]$, where α is the coefficient of thermal expansion and ΔT is the change in temperature.

For a rectangular cross section, the constants C_2 and C_3 are equal to $5A\ /\ 6$. For a circular and an elliptic cross section, they are equal to $0.9A$.

15.14 Strain Energy of a Linearly Elastic Slender Member in Terms of the Unit Displacements of the Centroid *G* and of the Unit Rotations

Under the effects of a normal force N acting at a distance S from the origin (see Fig. 15.10), the mean line suffers a change in length per unit length e_{1G}, where

$$e_{1G} = \frac{N}{AE}. \tag{15.14.1}$$

Therefore, in terms of e_{1G},

$$\frac{1}{2}\frac{N^2}{AE} = \frac{1}{2}(e_{1G})^2 AE. \tag{15.14.2}$$

From the elementary theory of beams, we know that the angle of rotation per unit length due to bending, β, is given by:

$$\beta = \frac{M}{EI}. \tag{15.14.3}$$

Therefore,

$$\frac{1}{2}\frac{M_{12}^2}{EI_2} = \frac{1}{2}(\beta_2)^2 EI_2, \quad \frac{1}{2}\frac{M_{13}^2}{EI_3} = \frac{1}{2}(\beta_3)^2 EI_3. \tag{15.14.4}$$

In Chapter 10, it was shown that the angle of twist per unit length is given by:

$$\alpha = \frac{M_{11}}{GJ}. \tag{15.14.5}$$

Therefore,

$$\frac{1}{2}\frac{M_{11}^2}{GJ} = \frac{1}{2}(\alpha)^2 GJ. \tag{15.14.6}$$

Due to a shearing force V_i ($i = 2$ or 3), the centroid G of A moves a distance e_{iG} per unit length of S along a direction normal to S (Fig. 15.11). Using Eqs. (15.13.13) and (15.13.14), the work per unit length done by V_2 and V_3 is

$$\frac{1}{2}V_2 e_{2G} = \frac{1}{2G}\frac{V_2^2}{C_2}, \quad \text{and} \quad \frac{1}{2}V_3 e_{3G} = \frac{1}{2G}\frac{V_3^2}{C_3}. \tag{15.14.7}$$

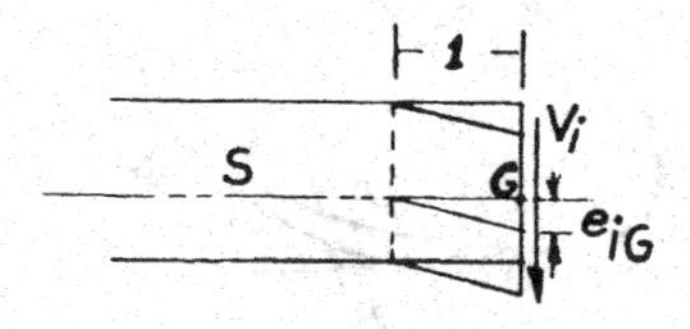

Fig. 15.11

Therefore,

$$e_{2G} = \frac{V_2}{GC_2}, \quad e_{3G} = \frac{V_3}{GC_3} \tag{15.14.8}$$

and

$$\frac{1}{2G}\frac{V_2^2}{C_2} = \frac{1}{2}GC_2(e_{2G})^2, \quad \frac{1}{2G}\frac{V_3^2}{C_3} = \frac{1}{2}GC_3(e_{3G})^2. \tag{15.14.9}$$

Substituting the previous equations into Eq. (15.13.19), we get:

$$U_t = \frac{1}{2}\int_0^L [EA(e_{1G})^2 + GC_2(e_{2G})^2 + GC_3(e_{3G})^2 + GJ(\alpha)^2 + EI_2(\beta_2)^2 + EI_3(\beta_3)^2]\,dS. \tag{15.14.10}$$

Temperature effects can be included by substituting for $EA(e_{1G})^2$ the quantity $EA\,e_{1G}[e_{1G} + \alpha(\Delta T)]$, where α is the coefficient of thermal expansion and ΔT is the change in temperature.

15.15 A Working Form of the Principles of Virtual Work and of Virtual Complementary Work for a Linearly Elastic Slender Member

Imagine the body shown in Fig. 15.12 to be subjected to a generalized force $\underline{F}$ that is imaginary and not a part of any force system on the body: $\underline{F}$ is a virtual force. Now suppose that the body undergoes a small change in shape as the result of some action other than $\underline{F}$. This change in shape may be due to a system of real applied loads, changes in

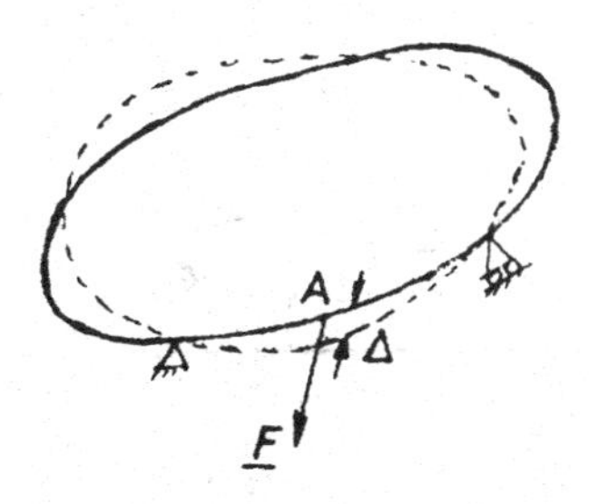

Fig. 15.12

temperature, or other causes. Consistent with the change in shape, internal distortions will result such as elongations, rotations, and shearing distortions. Let γ be the real distortion of a differential element and let Δ be the actual displacement of a point A in the direction of the virtual force $\underline{F}$. Also let $\underline{f}$ be the normal stress, moment, torque, or shear stress acting on the differential element as a result of the application of $\underline{F}$ at A. According to the principle of virtual complementary work,

$$\underline{F}\Delta = \sum \underline{f}\gamma. \tag{15.15.1}$$

$\underline{F}\Delta$ is the work done by the imaginary force $\underline{F}$ moving a distance Δ along its line of action. If $\underline{F} = 1$,

$$\Delta = \sum \underline{f}\gamma. \tag{15.15.2}$$

Therefore, to determine a displacement of a point A in any direction, apply a load $\underline{F}$ at the point in the desired direction. To determine the absolute direction of movement as well as the magnitude, determine separately the movement in two orthogonal directions. If it is desired to determine the rotation at A, apply a virtual couple. Let us now apply Eq. (15.15.1) to slender members:

For axial forces in beams or members of a truss:

$$\underline{F}\Delta = \int_0^L \underline{N} e_{1G}\, dS = \int_0^L \underline{N}\left[\frac{N}{AE} + \alpha(\Delta T)\right] dS, \tag{15.15.3}$$

where $\underline{N}$ is the normal force due to $\underline{F}$ and e_{1G} is the elongation per unit length of the real system at a distance S from the origin (see Fig. 15.11).

For bending moments in beams, frames, and arches:

$$\underline{F}\Delta = \int_0^L \underline{M}\frac{M\,dS}{EI}, \tag{15.15.4}$$

where $\underline{M}$ is the bending moment due to the application of $\underline{F}$, and M/EI is the angle of rotation per unit length of the real system.

For a torque M_{11}:

$$\underline{F}\Delta = \int_0^L \underline{M}_{11}\frac{M_{11}\,dS}{GJ}, \tag{15.15.5}$$

where $\underline{M}_{11}$ is the twisting moment resulting from the application of $\underline{F}$, and M_{11}/GJ is the angle of twist per unit length of the real system.

For shear:

$$\underline{F}\Delta = \int_0^L \underline{V}\frac{V\,dS}{GC}, \tag{15.15.6}$$

where $\underline{V}$ is the shearing force due to the application of $\underline{F}$, and V/GC is the displacement per unit length of the centroid G along V.

Adding Eqs. (15.15.3) to (15.15.6), we get the general working relation:

$$\underline{F}\Delta = \int_0^L \left\{\underline{N}\left[\frac{N}{EA} + \alpha(\Delta T)\right] + \underline{V}_2\frac{V_2}{GC_2} + \underline{V}_3\frac{V_3}{GC_3} + \underline{M}_{11}\frac{M_{11}}{GJ} + \underline{M}_{12}\frac{M_{12}}{EI_2} + \underline{M}_{13}\frac{M_{13}}{EI_3}\right\} dS. \tag{15.15.7}$$

It is to be noticed that the displacements caused by the virtual force $\underline{F}$ are assumed to be very small, so that the work done by the real system due to the displacements induced by $\underline{F}$ is negligible.

Although we have applied an imaginary force $\underline{F}$ to our elastic system and used the principle of virtual complementary work, we can consider $\underline{F}$ as being the only force of the real system and consider the real displacements of the body as imaginary displacements: In this case, the principle of virtual work yields Eq. (15.15.7). In the following sections, the use of Eq. (15.15.7) will be illustrated by various examples.

15.16 Examples of Application of the Theorems of Virtual Work and Virtual Complementary Work

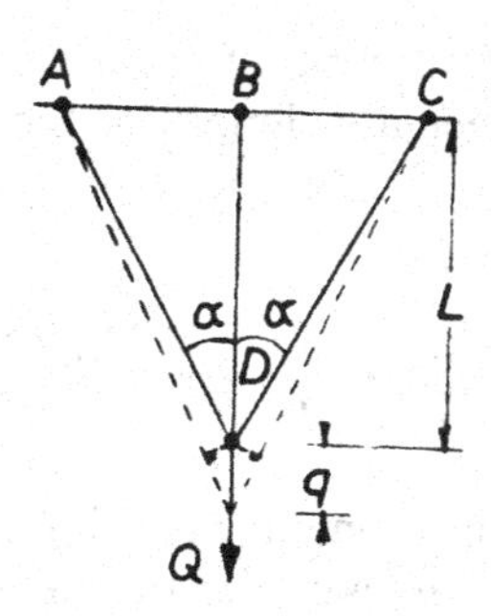

Fig. 15.13

1. *Statically indeterminate truss.* Let us consider the problem of a plane truss with three members acted on by a load Q as shown in Fig. 15.13. The tensions in the three members can be computed using the principle of virtual work and the condition of geometrical compatibility as follows:

$$T_{BD} = q\frac{AE}{L} \tag{15.16.1}$$

$$T_{AD} = T_{CD} = q\frac{AE\cos^2\alpha}{L}, \tag{15.16.2}$$

where A is the cross section of the bars and E is Young's modulus. Let us now impose a virtual displacement δq at D. According to the principle of virtual work,

$$Q\delta q = q\frac{AE}{L}\delta q + 2q\frac{AE\cos^3\alpha\,\delta q}{L}. \tag{15.16.3}$$

Therefore,

$$q = \frac{LQ}{AE(1 + 2\cos^3\alpha)} \tag{15.16.4}$$

and

$$T_{AD} = T_{CD} = \frac{Q \cos^2 \alpha}{1 + 2 \cos^3 \alpha}. \tag{15.16.5}$$

It is to be noticed that the equilibrium equations were not introduced in the solution since they are implied in the principle of virtual work.

2. *Deflection of a flexible elastic string.* Let us consider a string AB stretched by forces S between fixed points A and B (Fig. 15.14) and loaded by a distributed vertical load of intensity f, where f is a function of x_1. We shall assume that the initial tension of the string is so large

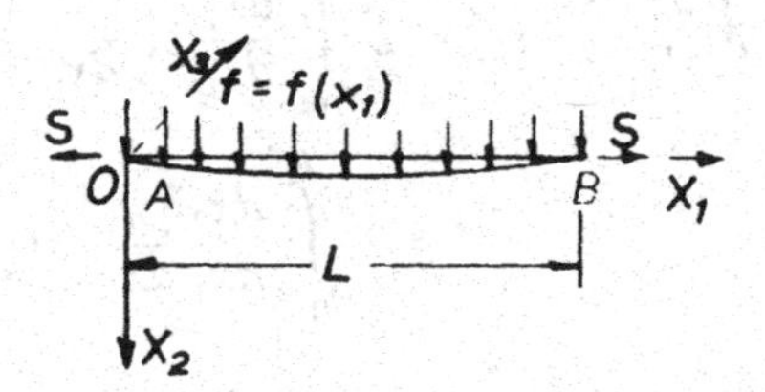

Fig. 15.14

that the increase in tensile force due to additional stretching during the deflection can be neglected. The deflection itself is also assumed to be small. The length of the deflected string is given by:

$$\int_0^L \sqrt{(dx_1)^2 + (dx_2)^2} = \int_0^L dx_1 \sqrt{1 + \left(\frac{dx_2}{dx_1}\right)^2}$$
$$\approx \int_0^L dx_1 \left[1 + \frac{1}{2}\left(\frac{dx_2}{dx_1}\right)^2\right]. \tag{15.16.6}$$

The stretching of the string is:

$$\int_0^L \left[dx_1 + \frac{1}{2}\left(\frac{dx_2}{dx_1}\right)^2 dx_1\right] - L = \frac{1}{2}\int_0^L \left(\frac{dx_2}{dx_1}\right)^2 dx_1, \tag{15.16.7}$$

and the increase in the strain energy is given by:

$$\frac{S}{2}\int_0^L \left(\frac{dx_2}{dx_1}\right)^2 dx_1 . \tag{15.16.8}$$

The principle of virtual work gives the following equation:

$$\int_0^L f\, dx_1\, \delta x_2 = \frac{S}{2}\int_0^L \delta\left(\frac{dx_2}{dx_1}\right)^2 dx_1 . \tag{15.16.9}$$

Calculating the variation in the right-hand side of Eq. (15.16.9), we find:

$$\begin{aligned}\int_0^L \delta\left(\frac{dx_2}{dx_1}\right)^2 dx_1 &= 2\int_0^L \frac{dx_2}{dx_1}\,\delta\left(\frac{dx_2}{dx_1}\right) dx_1 \\ &= 2\int_0^L \frac{dx_2}{dx_1}\frac{d}{dx_1}(\delta x_2)\, dx_1 .\end{aligned} \tag{15.16.10}$$

Integrating by parts, and remembering that at the ends of the string $\delta x_2 = 0$, we get:

$$\begin{aligned}2\int_0^L \frac{dx_2}{dx_1}\frac{d}{dx_1}(\delta x_2)\, dx_1 &= 2\left\{\left[\frac{dx_2}{dx_1}\delta x_2\right]_0^L - \int_0^L \frac{d^2 x_2}{dx_1^2}\delta x_2\, dx_1\right\} \\ &= -2\int_0^L \frac{d^2 x_2}{dx_1^2}\delta x_2\, dx_1 .\end{aligned} \tag{15.16.11}$$

Substituting Eq. (15.16.11) into Eq. (15.16.9), we obtain:

$$\int_0^L \left(S\frac{d^2 x_2}{dx_1^2} + f\right)\delta x_2\, dx_1 = 0. \tag{15.16.12}$$

This equation will be satisfied for any virtual displacement δx_2 only if

$$S\frac{d^2 x_2}{dx_1^2} + f(x_1) = 0. \tag{15.16.13}$$

This is the differential equation of the vertically loaded string.

3. *Indeterminate reaction in a slender member.* Consider the uniformly loaded beam of Fig. 15.15 with end A fixed and end B simply supported. It is desired to determine the reaction at B. Two equations of equilibrium can be written for such a beam, namely

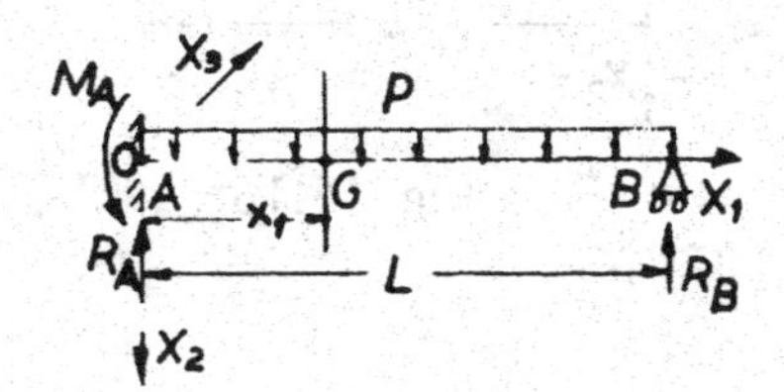

Fig. 15.15

$$R_A + R_B = pL \tag{15.16.14}$$

$$\frac{pL^2}{2} - R_B L = M_A . \tag{15.16.15}$$

These two equations are not sufficient to determine the three unknowns R_A, R_B, and M_A. To apply Eq. (15.15.7), let us consider any section at a distance x_1 from the origin O. At such a section, we have:

$$\frac{N}{AE} = 0, \quad \frac{V_2}{GC_2} = \frac{p(L - x_1) - R_B}{GC_2}, \quad \frac{V_3}{GC_3} = 0 \tag{15.16.16}$$

$$\frac{M_{11}}{GJ} = 0, \quad \frac{M_{12}}{EI_2} = 0, \quad \frac{M_{13}}{EI_3} = \frac{\left[R_B(L - x_1) - \dfrac{p(L - x_1)^2}{2}\right]}{EI_3} . \tag{15.16.17}$$

These can be considered as the imaginary displacements of the auxiliary system shown in Fig. 15.16. This system consists of a fixed end beam subjected to a force $\underline{F}$ at B. Therefore,

$$\underline{N} = 0, \quad \underline{V}_2 = -\underline{F}, \quad \underline{V}_3 = 0 \tag{15.16.18}$$

$$\underline{M}_{11} = 0, \quad \underline{M}_{12} = 0, \quad \underline{M}_{13} = +\underline{F}(L - x_1), \tag{15.16.19}$$

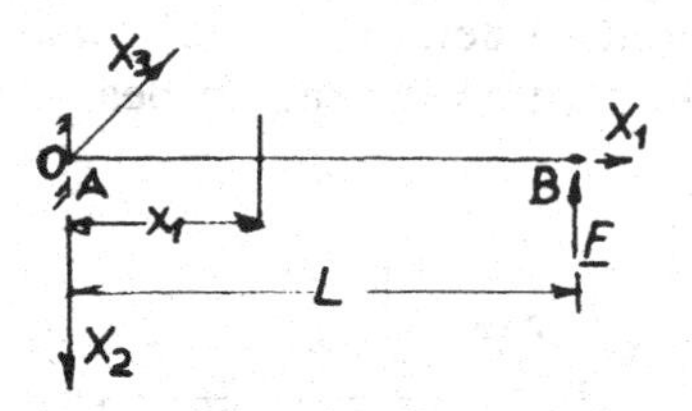

Fig. 15.16

with the boundary condition that B does not move ($\Delta = 0$ at B). Substituting Eqs. (15.16.16) to (15.16.19) into Eqs. (15.15.7), we get:

$$\underline{F}\Delta = \int_0^L \left\{ \frac{-\underline{F}\left[p(L - x_1) - R_B\right]}{GC_2} + \frac{\underline{F}(L - x_1)\left[R_B(L - x_1) - \dfrac{p(L - x_1)^2}{2}\right]}{EI_3} \right\} dx_1 = 0. \tag{15.16.20}$$

Integrating Eq. (15.16.20), we obtain:

$$R_B = \frac{P\left[\dfrac{L^4}{8EI_3} + \dfrac{L^2}{2GC_2}\right]}{\dfrac{L}{GC_2} + \dfrac{L^3}{3EI_3}}. \tag{15.16.21}$$

If we neglect the shear terms in Eq. (15.15.7), then:

$$R_B = \frac{3pL}{8}. \tag{15.16.22}$$

4. *Displacements in a slender member.* The displacement of the end point of a cantilever beam under the effect of a concentrated load P (Fig. 15.17), can be obtained by substituting in Eq. (15.15.7):

$$\frac{V_2}{GC_2} = \frac{P}{GC_2}, \quad \frac{M_{13}}{EI_3} = \frac{-P(L - x_1)}{EI_3}, \tag{15.16.23}$$

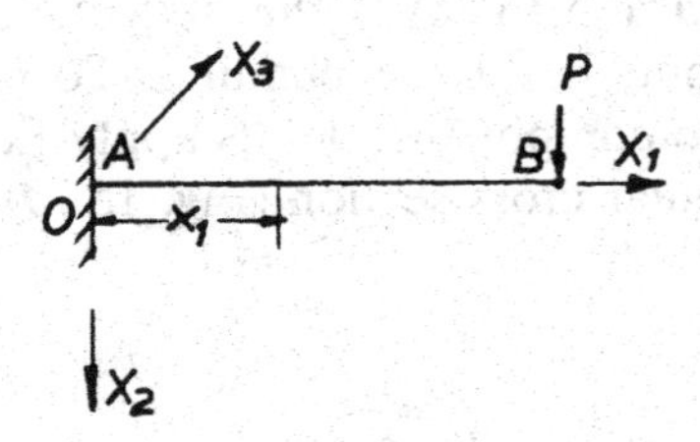

Fig. 15.17

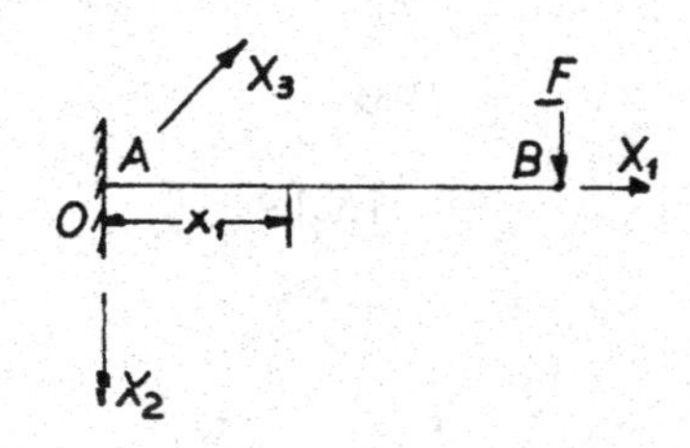

Fig. 15.18

and (Fig. 15.18):

$$\underline{V}_2 = \underline{F}, \quad \underline{M}_{13} = -\underline{F}(L - x_1), \tag{15.16.24}$$

with all the other values equal to zero. Thus,

$$\underline{F}\Delta = \int_0^L \left[\frac{\underline{F}P}{GC_2} + \frac{\underline{F}P(L - x_1)^2}{EI_3} \right] dx_1. \tag{15.16.25}$$

Therefore, the displacement of B is given by: .

$$\Delta = \frac{PL}{GC_2} + \frac{PL^3}{3EI_3}. \tag{15.16.26}$$

The first term of Eq. (15.16.26) is the deflection due to the shearing forces, and the second is due to bending. To compare the order of magnitude of the two components, let us apply Eq. (15.16.26) to a steel beam with a rectangular cross section (Fig. 15.19). For such a beam:

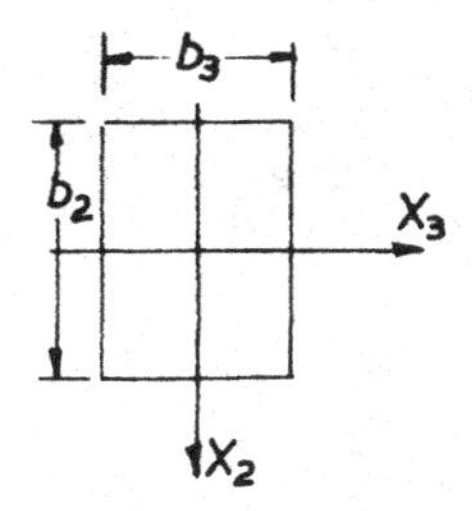

Fig. 15.19

$$C_2 = \frac{5}{6} b_2 b_3 = \frac{5}{6} A$$

$$\frac{G}{E} = \frac{2}{5}$$

and

$$\Delta = \frac{PL^3}{3EI_3}\left[1 + \frac{3}{4}\frac{b_2^2}{L^2}\right].$$

For $b/L = 1/10$, for example, $3b_2^2/4L^2 = 0.0075$, which is negligible compared to unity. Therefore, the approximation which is involved in neglecting the deflections due to shear in slender members is quite justified.

15.17 Examples of Application of Castigliano's First and Second Theorems

1. *Deflection of a wire.* Castigliano's first theorem can be used to determine the relation between the force Q acting at the center of an

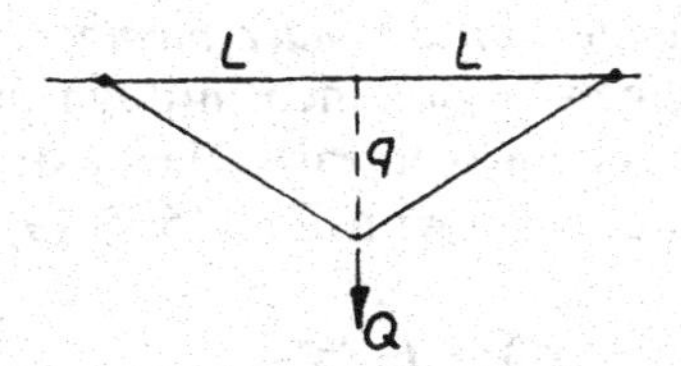

Fig. 15.20

elastic wire and the displacement q (Fig. 15.20). Such a relation is elastic but non-linear. The strain energy stored in the stretched wire is given by [see Eq. (15.14.2)]:

$$U_t = 2\left(\frac{1}{2}\frac{EA}{L}\Delta^2\right), \tag{15.17.1}$$

where A is the cross section of the wire, and Δ is the elongation of half the initial length $2L$. We have:

$$\Delta = \sqrt{L^2 + q^2} - L \tag{15.17.2}$$

and, in case of small deflections,

$$\Delta = L\left[1 + \frac{1}{2}\left(\frac{q}{L}\right)^2 + \cdots\right] - L; \tag{15.17.3}$$

or, approximately,

$$\Delta = \frac{q^2}{2L}. \tag{15.17.4}$$

Therefore, the strain energy can be written as:

$$U_t = \frac{EAq^4}{4L^3}, \tag{15.17.5}$$

and from the Castigliano's first theorem, we have:

$$Q = \frac{\partial U_t}{\partial q} = \frac{EAq^3}{L^3}. \tag{15.17.6}$$

The elastic nonlinearity of this problem results from geometry and not from material properties.

2. *Indeterminate reaction in a slender member.* In Fig. 15.15 of Sec. 15.16, the reaction R_B can be obtained through the use of Castigliano's second theorem. The system is linearly elastic and Eqs. (15.16.16) and (15.16.17) used in conjunction with Eq. (15.13.19), give:

$$U_t = \frac{1}{2}\int_0^L \left\{\frac{[P(L-x_1) - R_B]^2}{GC_2} + \frac{\left[R_B(L-x_1) - \dfrac{P(L-x_1)^2}{2}\right]^2}{EI_3}\right\} dx_1 . \tag{15.17.7}$$

According to Castigliano's second theorem,

$$\frac{\partial U_t}{\partial R_B} = q_b = 0 = \int_0^L \left\{\frac{-\left[P(L-x_1) - R_B\right.}{GC_2} + \frac{(L-x_1)\left[R_B(L-x_1) - \dfrac{P(L-x_1)^2}{2}\right]}{EI_3}\right\} dx_1 , \tag{15.17.8}$$

which is the same as Eq. (15.16.20).

3. *Deflection of curved bars.* Let us assume that the bar shown in Fig. 15.21 has a circular profile, and that its radius is large enough so that the terms due to shear in the energy equation can be neglected. It is

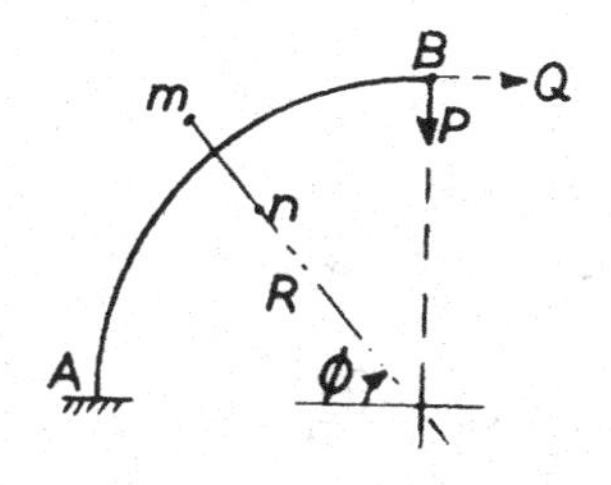

Fig. 15.21

desired to find the vertical and horizontal displacements of B under the effect of the force P. The strain energy is given by:

$$U_t = \frac{1}{2} \int_0^L \frac{M_{13}^2}{EI_3} \, dS, \tag{15.17.9}$$

and the bending moment M_{13} at any section mn is:

$$M_{13} = -PR \cos \phi. \tag{15.17.10}$$

Therefore, the vertical displacement Δ_V of B is:

$$\Delta_V = \frac{\partial}{\partial P} \int_0^{\frac{\Pi}{2}} \frac{P^2 R^3 \cos^2 \phi}{2EI_3} \, d\phi = \frac{\Pi PR^3}{4EI_3}. \tag{15.17.11}$$

To find the horizontal displacement of B, we can assume the existence of a horizontal force Q acting at B in addition to P, and obtain the value of $(\partial U_t / \partial Q)_{Q=0}$:

$$M_{13} = -\{PR \cos \phi + QR(1 - \sin \phi)\} \tag{15.17.12}$$

and

$$\Delta_H = \left(\frac{\partial U_t}{\partial Q}\right)_{Q=0} = \left[\frac{\partial}{\partial Q} \int_0^{\frac{\Pi}{2}} \frac{M_{13}^2 R}{2EI_3} \, d\phi\right]_{Q=0} = \frac{PR^3}{2EI_3}. \tag{15.17.13}$$

Remark

For curved bars whose cross section is large compared to their radius, the strain energy due to normal and shearing forces cannot be neglected. Recalling the notation used in Sec. 13.2, it can be shown [4] that the strain energy in such cases is given by:

$$U_t = \int_0^L \left[\frac{M_{13}^2}{2AEer_c} + \frac{N^2}{2AE} - \frac{M_{13} N}{AEr_c} + \frac{V_2^2}{2GC_2}\right] dS. \tag{15.17.14}$$

In the case of the circular arc previously examined (Fig. 15.21):

$$M_{13} = -PR \cos \phi \tag{15.17.15}$$

$$N = -P \cos \phi \tag{15.17.16}$$

$$V_2 = P \sin \phi \tag{15.17.17}$$

and

$$\Delta_V = \frac{\Pi PR}{4E}\left(\frac{R}{Ae} + \frac{E}{GC_2} - \frac{1}{A}\right), \tag{15.17.18}$$

where $R = r_c$ is the radius of the center line, and e is the distance between the center line and the neutral axis.

15.18 Examples of Application of the Principles of Minimum Potential Energy and Minimum Complementary Energy

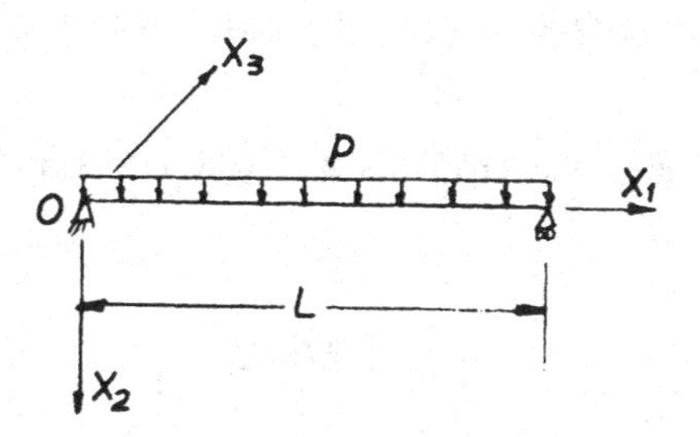

Fig. 15.22

1. *Deflection of the mean line of a beam.* The energy stored in the uniformly loaded beam shown in Fig. 15.22, is given by [see Eq. (15.14.4)]:

$$U_t = \int_0^L \frac{EI_3}{2}(\beta_3)^2 \, dx_1, \tag{15.18.1}$$

where the part due to shear has been neglected. Recalling Eqs. (12.2.2), we have:

$$U_t = \int_0^L \frac{EI_3}{2}\left(\frac{d^2u_2}{dx_1^2}\right)^2 dx_1. \tag{15.18.2}$$

Neglecting the body forces, the work done by the external forces to bring the beam to its final position is [see (Eq. 15.6.4)]:

$$W = \int_0^L p u_2 \, dx_1 . \tag{15.18.3}$$

The potential energy of the beam is:

$$\Pi_p = U_t - W = \int_0^L \frac{EI_3}{2}\left(\frac{d^2 u_2}{dx_1^2}\right)^2 dx_1 - \int_0^L p u_2 \, dx_1 . \tag{15.18.4}$$

Taking the variation of Π_p, we get:

$$\delta\Pi_p = EI_3 \int_0^L \frac{d^2 u_2}{dx_1^2}\frac{d^2(\delta u_2)}{dx_1^2} dx_1 - \int_0^L p(\delta u_2) dx_1 . \tag{15.18.5}$$

Integrating the first integral by parts:

$$\int_0^L \frac{d^2 u_2}{dx_1^2}\frac{d^2(\delta u_2)}{dx_1^2} dx_1 = \left[\frac{d^2 u_2}{dx_1^2}\frac{d(\delta u_2)}{dx_1}\right]_0^L - \int_0^L \frac{d(\delta u_2)}{dx_1}\frac{d^3 u_2}{dx_1^3} dx_1 \tag{15.18.6}$$

$$= \left[\frac{d^2 u_2}{dx_1^2}\frac{d(\delta u_2)}{dx_1} - \delta u_2 \frac{d^3 u_2}{dx_1^3}\right]_0^L + \int_0^L \delta u_2 \frac{d^4 u_2}{dx_1^4} dx_1 . \tag{15.18.7}$$

The condition $\delta\Pi_p = 0$ becomes:

$$\int_0^L \left(EI_3 \frac{d^4 u_2}{dx_1^4} - p\right)\delta u_2 \, dx_1 + \left[EI_3 \frac{d^2 u_2}{dx_1^2}\frac{d(\delta u_2)}{dx_1} - EI_3 \frac{d^3 u_2}{dx_1^3}\delta u_2\right]_0^L = 0. \tag{15.18.8}$$

Since $M_{13} = -EI_3 d^2 u_2/dx_1^2 = 0$ and $u_2 = 0$ at $x_1 = 0$ and $x_1 = L$, $\delta u_2 = 0$ at $x_1 = 0$ and $x_1 = L$ and the terms inside the bracket vanish on the boundary $x_1 = 0$ and $x_1 = L$. In between the boundaries, δu_2 is arbitrary and the condition

$$\int_0^L \left(EI_3 \frac{d^4 u_2}{dx_1^4} - p\right)\delta u_2 \, dx_1 = 0 \tag{15.18.9}$$

is possible only when

$$EI_3 \frac{d^4 u_2}{dx_1^4} = p, \tag{15.18.10}$$

which is the governing differential equation in this case (see Sec. 12.2).

2. *Torsion of prismatic bars.* In Chapter 10, it was shown that there exists a stress function ϕ such that (Fig. 15.23):

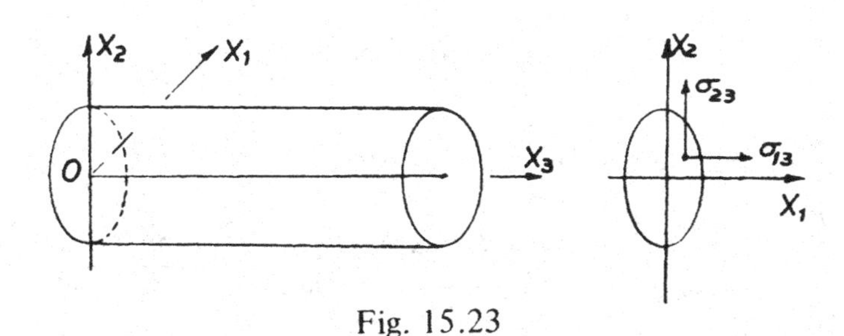

Fig. 15.23

$$\sigma_{13} = \frac{\partial \phi}{\partial x_2}, \quad \sigma_{23} = -\frac{\partial \phi}{\partial x_1}, \tag{15.18.11}$$

the other stress components being equal to zero. The variation of the stresses is therefore equivalent to the variation of the stress function. The strain energy stored in a twisted bar is equal to:

$$\begin{aligned} U_t &= \frac{1}{2G} \iiint_V [(\sigma_{13})^2 + (\sigma_{23})^2]\,dV \\ &= \frac{1}{2G} \int_0^L \left\{ \iint_R \left[\left(\frac{\partial \phi}{\partial x_2} \right)^2 + \left(\frac{\partial \phi}{\partial x_1} \right)^2 \right] d\Omega \right\} dS \\ &= \frac{L}{2G} \iint_R \left[\left(\frac{\partial \phi}{\partial x_2} \right)^2 + \left(\frac{\partial \phi}{\partial x_1} \right)^2 \right] dx_1\, dx_2, \end{aligned} \tag{15.18.12}$$

where R denotes the cross section of the bar. The lateral faces of the bar are free of forces. If at the end $x_3 = 0$, the displacements are equal to zero, the work done by the external forces to bring the bar to its final position is:

$$W = \left[\iint_R (\sigma_{13} u_1 + \sigma_{23} u_2) dx_1\, dx_2\right]_{x_3=L}. \qquad (15.18.13)$$

Following Saint-Venant's assumptions, the displacements u_1 and u_2 at $x_3 = L$ are

$$u_1 = -\alpha x_2 L, \quad u_2 = \alpha x_1 L. \qquad (15.18.14)$$

Hence,

$$W = \alpha L \iint_R \left(-x_2 \frac{\partial \phi}{\partial x_2} - x_1 \frac{\partial \phi}{\partial x_1}\right) dx_1\, dx_2. \qquad (15.18.15)$$

Eq. (15.18.15) can be transformed by the Green-Riemann formula, to give (see Sec. 10.5):

$$W = 2\alpha L \iint_R \phi\, dx_1\, dx_2 - \alpha L \oint_C (\phi x_1\, dx_2 - \phi x_2\, dx_1). \qquad (15.18.16)$$

The total complementary energy of the system is, therefore,

$$\Pi^* = U_t - W = \frac{L}{2G} \iint_R \left[\left(\frac{\partial \phi}{\partial x_2}\right)^2 + \left(\frac{\partial \phi}{\partial x_1}\right)^2 - 4G\alpha\phi\right] dx_1\, dx_2$$

$$+\alpha L \oint_C \phi(x_1\, dx_2 - x_2\, dx_1) \qquad (15.18.17)$$

and

$$\delta\Pi^* = \frac{L}{2G} \iint_R \left[2 \frac{\partial \phi}{\partial x_2} \frac{\partial(\delta\phi)}{\partial x_2} + 2 \frac{\partial \phi}{\partial x_1} \frac{\partial(\delta\phi)}{\partial x_1} - 4G\alpha(\delta\phi)\right] dx_1\, dx_2$$

$$+\alpha L \oint_C \delta\phi(x_1\, dx_2 - x_2\, dx_1). \qquad (15.18.18)$$

Since

$$\frac{\partial\phi}{\partial x_2}\frac{\partial(\delta\phi)}{\partial x_2}+\frac{\partial\phi}{\partial x_1}\frac{\partial(\delta\phi)}{\partial x_1}=\frac{\partial}{\partial x_1}\left(\delta\phi\frac{\partial\phi}{\partial x_1}\right)+\frac{\partial}{\partial x_2}\left(\delta\phi\frac{\partial\phi}{\partial x_2}\right)-\left(\frac{\partial^2\phi}{\partial x_1^2}+\frac{\partial^2\phi}{\partial x_2^2}\right)\delta\phi \tag{15.18.19}$$

$$\delta\Pi^*=-\frac{L}{G}\iint_R\left(\frac{\partial^2\phi}{\partial x_1^2}+\frac{\partial^2\phi}{\partial x_2^2}+2G\alpha\right)\delta\phi\,dx_1\,dx_2+\frac{L}{G}\oint_C\left[\left(\frac{\partial\phi}{\partial x_1}+G\alpha x_1\right)dx_2-\left(\frac{\partial\phi}{\partial x_2}+G\alpha x_2\right)dx_1\right]\delta\phi. \tag{15.18.20}$$

On the boundary C, the surface forces are prescribed which means that ϕ is prescribed so that:

$$\delta\phi=0 \qquad \text{on } C. \tag{15.18.21}$$

The line integral of Eq. (15.18.20) thus vanishes. Since $\delta\phi$ is arbitrary in R, the only way to make $\delta\Pi^*=0$, is to have:

$$\frac{\partial^2\phi}{\partial x_1^2}+\frac{\partial^2\phi}{\partial x_2^2}=-2G\alpha \qquad \text{in } R. \tag{15.18.22}$$

As shown in Chapter 10, the torsion problem is solved once a function ϕ satisfying Eq. (15.18.22) in R, and equal to a constant on the boundary, is found. This constant can be chosen equal to zero, and in such a case the expression for the total complementary energy is:

$$\Pi^*=\frac{L}{2G}\iint_R\left[\left(\frac{\partial\phi}{\partial x_2}\right)^2+\left(\frac{\partial\phi}{\partial x_1}\right)^2-4G\alpha\phi\right]dx_1\,dx_2. \tag{15.18.23}$$

Eq. (15.18.23) provides us with another avenue to approach the torsion problem. This problem can be considered as the one seeking the function ϕ which will minimize the total complementary energy Π^* and which satisfies the condition $\phi=0$ on the boundary.

15.19 Example of Application of the Theorem of Least Work

Consider the beam shown in Fig. 15.24. There are four unknown reactions which cannot be directly determined, since we only have two

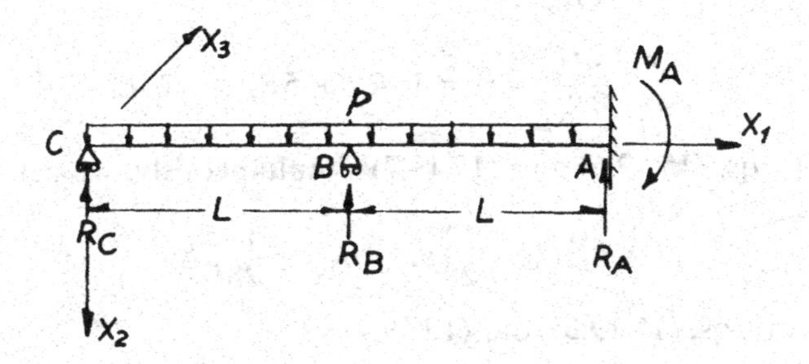

Fig. 15.24

equations of static equilibrium. We can consider R_B and R_C as the redundant forces. The equations of static equilibrium give:

$$R_A + R_B + R_C = 2pL \tag{15.19.1}$$

$$(R_B + 2R_C)L + M_A = 2pL^2. \tag{15.19.2}$$

These two equations can be solved to give M_A and R_A in terms of the two redundant forces R_B and R_C:

$$R_A = 2pL - R_B - R_C \tag{15.19.3}$$

$$M_A = 2pL^2 - L(R_B + 2R_C). \tag{15.19.4}$$

If we take into account the bending terms alone, the total strain energy of the beam is given by:

$$U_t = \int_O^L \frac{\left(R_C x_1 - \frac{px_1^2}{2}\right)^2}{2EI_3} dx_1 + \int_L^{2L} \frac{\left[R_C x_1 - \frac{px_1^2}{2} + R_B(x_1 - L)\right]^2}{2EI_3} dx_1 . \tag{15.19.5}$$

According to the theorem of least work,

$$\frac{\partial U_t}{\partial R_B} = 0 = 8R_B + 20R_C - 17pL \tag{15.19.6}$$

$$\frac{\partial U_t}{\partial R_C} = 0 = 16R_C + 5R_B - 12pL. \tag{15.19.7}$$

Solving Eqs. (15.19.6) and (15.19.7) simultaneously, we get:

$$R_B = \frac{32}{28}pL, \quad R_C = \frac{11}{28}pL; \tag{15.19.8}$$

and from Eqs. (15.19.3) and (15.19.4):

$$R_A = \frac{13}{28}pL, \quad M_A = \frac{2pL^2}{28}. \tag{15.19.9}$$

15.20 The Rayleigh-Ritz Method

The Rayleigh-Ritz method is a general procedure for obtaining approximate solutions of problems expressed in variational form. The procedure consists of assuming that the desired stationary function $y(x)$ of a given problem (see Sec. 15.4) is approximated by a combination of suitably chosen functions satisfying the boundary conditions but with undetermined parameters c_i. The relevant quantity I is then expressed as a function of the c_i's, which are so determined that the resultant expression is stationary. Therefore, instead of using the calculus of variations in attempting to determine that function which renders I stationary with reference to all admissible slightly varied functions, we consider only the family of all functions of the type assumed; we then use ordinary differential calculus to seek the member of that family for which I is stationary with reference to slightly modified functions belonging to the same family. The efficiency of this procedure thus depends on the choice of the functions combined to provide the stationary function y (x).

Application to the deflection of the mean line of a beam. In Sec. 15.18, we have seen that the solution of the problem of the uniformly loaded simply supported beam could be obtained by minimizing the potential energy Π_p; this resulted in Eq. (15.18.10), which can be solved to give the deflection $u_2 = u_2(x_1)$. Using the Rayleigh-Ritz method, one can directly obtain an approximate solution to the problem. For that, let us assume a deflection curve in the form of a trigonometric series:

$$u_2 = a_1 \sin\frac{\Pi x_1}{L} + a_2 \sin\frac{2\Pi x_1}{L} + \ldots + a_n \sin\frac{n\Pi x_1}{L} + \ldots \qquad (15.20.1)$$

$$= \sum_{n=1}^{n=\infty} a_n \sin\frac{n\Pi x_1}{L},$$

in which $a_1, a_2 - - - a_n$ are undetermined parameters. Thus, the deflection curve is obtained by superposition of sinusoidal curves (Fig. 15.25) each of which satisfies the boundary conditions of the problem. From Eq. (15.18.4), we have:

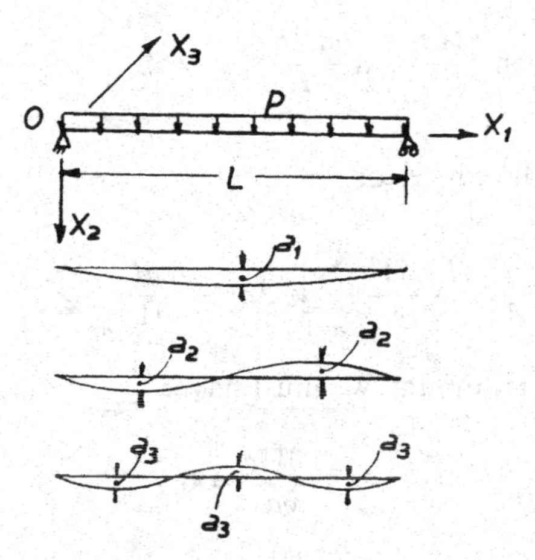

Fig. 15.25

$$\Pi_p = \int_0^L \frac{EI_3}{2}\left(\frac{d^2 u_2}{dx_1^2}\right)^2 dx_1 - \int_0^L p u_2\, dx_1 . \qquad (15.20.2)$$

Eq. (15.20.1) gives:

$$\frac{d^2 u_2}{dx_1^2} = -a_1 \frac{\Pi^2}{L^2} \sin\frac{\Pi x_1}{L} - 2^2 a_2 \frac{\Pi^2}{L^2} \sin\frac{2\Pi x_1}{L} \qquad (15.20.3)$$

$$- 3^2 a_3 \frac{\Pi^2}{L^2} \sin\frac{3\Pi x_1}{L} \cdots .$$

When squared, the right-hand side of Eq. (15.20.3) involves terms of two kinds, namely:

$$a_n^2 \frac{n^4 \Pi^4}{L^4} \sin^2 \frac{n\Pi x_1}{L} \text{ and } 2a_n a_m \frac{m^2 n^2 \Pi^4}{L^4} \sin \frac{n\Pi x_1}{L} \sin \frac{m\Pi x_1}{L}. \quad (15.20.4)$$

However, since,

$$\int_0^L \sin^2 \frac{n\Pi x_1}{L} dx_1 = \frac{L}{2} \text{ and}$$
$$\int_0^L \sin \frac{n\Pi x_1}{L} \sin \frac{m\Pi x_1}{L} dx_1 = 0 \text{ if } n \neq m, \quad (15.20.5)$$

Eq. (15.20.2) will reduce to:

$$\Pi_p = \frac{EI_3 \Pi^4}{4L^3} \sum_{n=1}^{\infty} n^4 a_n^2 - \frac{2pL}{\Pi} \sum_{n=1,3,5--}^{\infty} \frac{a_n}{n}. \quad (15.20.6)$$

For Π_p to be a minimum, we must have:

$$\frac{\partial \Pi p}{\partial a_n} = 0 \quad (15.20.7)$$

so that

$$\frac{EI_3 \Pi^4}{4L^3} 2n^4 a_n - \frac{2pL}{\Pi} \frac{1}{n} = 0 \text{ for } n \text{ odd} \quad (15.20.8)$$

and

$$\frac{EI_3 \Pi^4}{4L^3} 2n^4 a_n = 0 \text{ for } n \text{ even}. \quad (15.20.9)$$

Hence,

$$a_n = \frac{4pL^4}{EI_3 \Pi^5 n^5} \text{ for } n \text{ odd} \quad (15.20.10)$$

and

$$a_n = 0 \text{ for } n \text{ even .} \tag{15.20.11}$$

Therefore, the deflection curve can be written as:

$$u_2 = \frac{4pL^4}{EI_3\Pi^5} \sum_{n=1,3,5--}^{\infty} \frac{1}{n^5} \sin \frac{n\Pi x_1}{L}. \tag{15.20.12}$$

The series is rapidly convergent and only the first few terms are necessary to give a satisfactory approximation. For $x_1 = L/2$, we have:

$$(u_2)_{\max} = \frac{4pL^4}{EI_3\Pi^5}\left(1 - \frac{1}{3^5} + \frac{1}{5^5} - \ldots\right). \tag{15.20.13}$$

If we take the first term of this series, we obtain:

$$(u_2)_{\max} = \frac{pL^4}{76.6EI_3}. \tag{15.20.14}$$

The exact answer is:

$$(u_2)_{\max} = \frac{pL^4}{76.8EI_3}, \tag{15.20.15}$$

so that the error involved is only 0.26 percent.

Other examples can be found in [5,6,7].

PROBLEMS

1. Determine the curve between two points A and B, which by revolution about the OX_1 axis generates the surface of least area.
2. What is the Euler equation of the problem

$$\delta \iint_R F(x, y, u, u_x, u_y)\, dx\, dy = 0,$$

in which x and y are independent variables and $u\ (x,y)$ is prescribed along the closed boundary S of the region R?

3. A uniform elastic beam is fixed at both ends and carries a linearly increasing distributed load that varies from zero at one end to q_o at the other. Obtain the equation of the deflection of the beam using the energy terms due to bending alone and Euler's equation

(15.4.18). Eliminate the constants of integration using the boundary conditions.

4. What is the expression of the strain energy per unit length of a linearly elastic thick cylinder with inner and outer radii a and b, with free ends, and which is subjected to an internal pressure P_i? (See Sec. 11.2.)
5. A short thick cylindrical hub is shrunk-fit on a short shaft, so that the radial pressure at the interface is p. If the inner and outer radii of the hub are a and b, and if the shear modulus is G, find the strain energy density at any radius r of the hub.

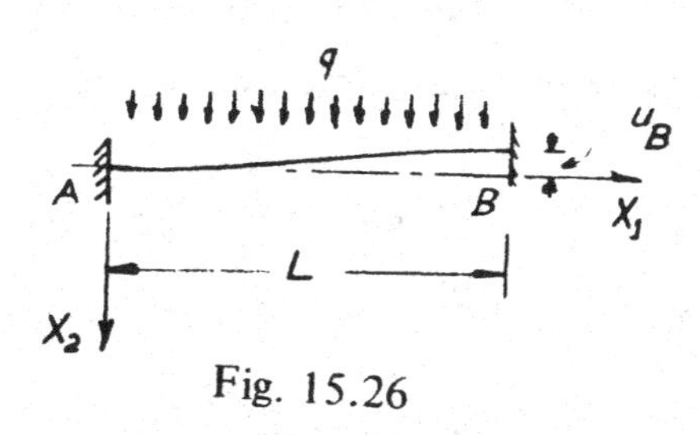

Fig. 15.26

6. Use the method of virtual work or of virtual complementary work to find the reactions and fixed-end moments for a beam AB, of length L, fixed at both ends, subjected to a uniformly distributed load q, and whose end B is given a small vertical displacement u_B (Fig. 15.26). Discuss the influence of the shearing forces on the reactions and the fixed-end moments.
7. A steel tube, 2 in. internal diameter and $\frac{1}{8}$ in. thick, stands vertically from a rigid base. At 3 ft. from the base, the tube is bent into a quadrant of a circle of 2 ft. radius, and at the end is suspended a load of 500 lbs. (Fig. 15.27). Considering the strain energy due to bending alone, use Castigliano's second theorem to calculate the vertical deflection of the load ($E = 30 \times 10^6$ psi).
8. Using the energy due to bending alone and the theorem of minimum potential energy, find the expression of the deflection of a cantilever beam subjected to an end load P (Fig. 15.17). Write down the forced and the natural boundary conditions.

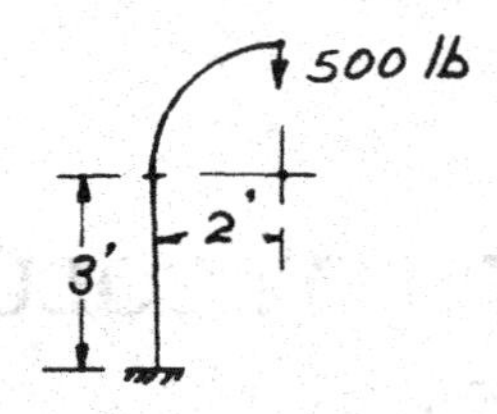

Fig. 15.27

9. Assuming that the deflection in Problem 8 is given by $u_2 = C(1 - \cos \Pi x_1 / 2L)$, where C is a constant and x_1 is the distance from the fixed end, obtain the deflection at the free end using the principle of minimum potential energy. Compare this deflection to that obtained in Problem 8.

10. Use Castigliano's 2nd theorem to solve problem 6.

11. Following steps similar to those of Sec. 15.20, use the Rayleigh-Ritz method to obtain the solution of the simply supported beam subjected to a concentrated load P at midspan.(*)

REFERENCES

[1] H. L. Langhaar, *Energy Methods in Applied Mechanics*, John Wiley & Son, New York, N. Y., 1962.
[2] Y. C. Fung, *Foundations of Solid Mechanics*, Prentice-Hall, Englewood Cliffs, N. J., 1965.
[3] K. Washizu, *Variational Methods in Elasticity and Plasticity*, Pergamon Press, New York, N. Y., 1968.
[4] S. Timoshenko, *Strength of Materials*, Vol. 1, Van Nostrand, Princeton, N. J., 1955.
[5] S. Timoshenko, *Strength of Materials*, Vol. 2, Van Nostrand, Princeton, N. J., 1955.
[6] S. Timoshenko and J. N. Goodier, *Theory of Elasticity*, McGraw-Hill, New York, N. Y., 1970.
[7] C. T. Wang, *Applied Elasticity*, McGraw-Hill, New York, N. Y., 1953.

* See pages A-70 to A-72 for additional problems.

CHAPTER 16

ELASTIC STABILITY: COLUMNS AND BEAM COLUMNS

16.1 Introduction

In the previous chapters, we discussed problems in which the stress-deformation relationships were generally linear with deformations quite small compared to the smallest dimension of the body. The forces were in equilibrium and, for linearily elastic bodies, stress and deformation patterns could be superimposed to produce complex configurations. In this chapter, we shall examine what happens when a body in equilibrium is slightly disturbed from its configuration: Does it tend to return to its equilibrium position or does it tend to depart from it? For example, a slender rod behaves normally when loaded in tension and can also carry a small amount of compression; however, as the compression load is increased, the rod becomes unstable and undergoes large deflections. The question of stability of a compressed bar can be investigated by using methods analagous to those used in investigating the stability of equilibrium configurations of rigid bodies. Consider, for example, the small weight on the frictionless surface of Fig. 16.1. In Fig. 16.1.a, the load is slightly displaced from its equilibrium position. The weight W and the reaction N are no longer in balance, but the resultant is a restoring force which accelerates the particle to its equilibrium position. Such an equilibrium is called stable. In Fig. 16.lc, the resultant unbalance is an upsetting force which accelerates the particle away from the equilibrium position. Such an equilibrium is called unstable. In Fig. 16.lb, there is no tendency to return to the original position or to go further. Such an equilibrium is called neutral.

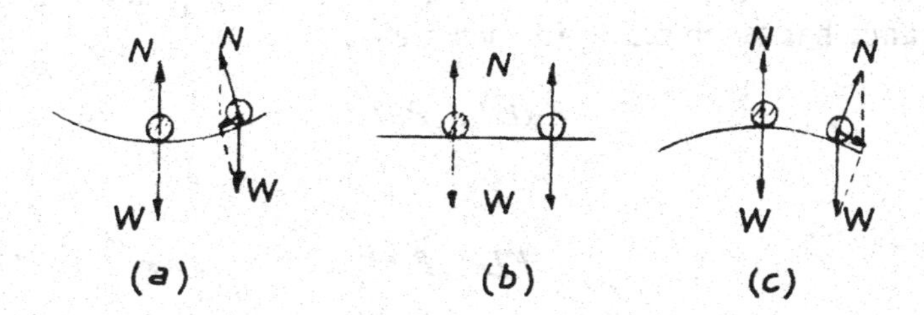

Fig. 16.1

Generalizing, a load carrying structure is said to be in a state of stable equilibrium if, for all admissible small displacements from the equilibrium position, restoring forces arise which tend to accelerate the structure back towards its equilibrium position. A classical example used in illustrating the problem of stability is that represented by Fig. 16.2. The load P acts on the infinitely rigid bar AB, which is hinged at

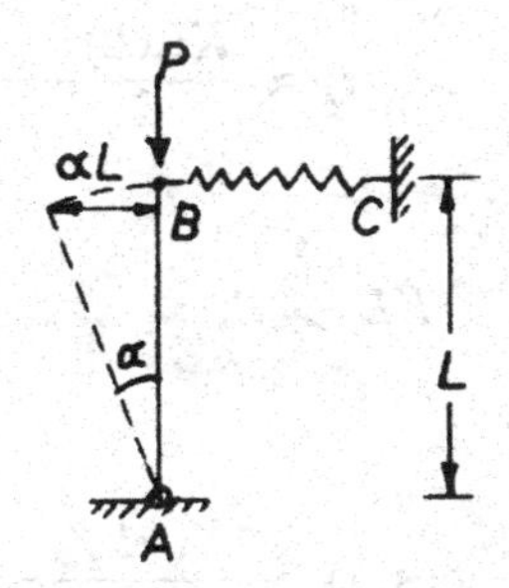

Fig. 16.2

A. For small values of P, the vertical position of the bar is stable.If a disturbing force produces a lateral displacement at B, the spring BC will return the bar to its vertical position. In this case, the moment about A of the spring force is higher than that of P, i.e.,

$$K\alpha L^2 > P\alpha L \tag{16.1.1}$$

or

$$KL > P. \tag{16.1.2}$$

The force P, however, may increase to such a point that the spring force is not sufficient to restore the bar to its original position after the disturbance has taken place. In such a case,

$$K\alpha L^2 < P\alpha L \tag{16.1.3}$$

or

$$KL < P, \tag{16.1.4}$$

and B is accelerated further away from its equilibrium position. The value of P for neutral equilibrium is, therefore,

$$P = KL. \tag{16.1.5}$$

One can use an energy method to arrive at Eqs.(16.1.2), (16.1.4), and (16.1.5). The system of Fig.16.2 is stable if, due to the disturbance, the change in potential energy is positive; and unstable if, due to the disturbance, the change in potential energy is negative. In the first case,

$$\Delta\Pi_p = \frac{K(\alpha L)^2}{2} - PL(1 - \cos\alpha) \approx \frac{K(\alpha L)^2}{2} - \frac{PL\alpha^2}{2} > 0 \tag{16.1.6}$$

or

$$KL > P; \tag{16.1.7}$$

and in the second case,

$$\Delta\Pi_p = \frac{K(\alpha L)^2}{2} - PL(1 - \cos\alpha) \approx \frac{K(\alpha L)^2}{2} - \frac{PL\alpha^2}{2} < 0 \tag{16.1.8}$$

or

$$KL < P. \tag{16.1.9}$$

The value of P for neutral equilibrium is obtained by writing:

$$\Delta\Pi_p = 0. \tag{16.1.10}$$

In the following sections, we shall examine the problems of prismatic bars subjected to axial compression and to a combination of axial compression and bending. In the first case, the bars will be called columns and in the second case they will be called beam columns.

16.2 Differential Equations of Columns and Beam-Columns

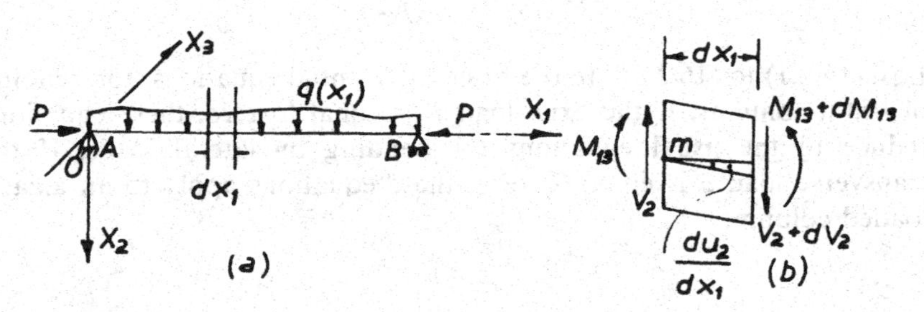

Fig. 16.3

Fig. 16.3a shows a beam subjected to a longitudinal load P as well as to a transverse load $q(x_1)$, and Fig.16.3b shows a portion of such a beam of length dx_1 between two cross sections normal to the undeflected axis.

The relations among load, shearing force, and bending moment are obtained by considering the equilibrium of the element in Fig.16.3b. Summing the forces in the OX_2 direction gives :

$$q = -\frac{dV_2}{dx_1}. \tag{16.2.1}$$

Taking moments about m, and assuming that the deflection of the beam is small, we get:

$$V_2 = \frac{dM_{13}}{dx_1} - P\frac{du_2}{dx_1}, \tag{16.2.2}$$

in which the terms of the second order have been neglected. If we neglect the effects of shear on the deformation, the expression for the curvature of the axis of the beam is:

$$\frac{d^2u_2}{dx_1^2} = -\frac{M_{13}}{EI_3}. \tag{16.2.3}$$

EI_3 is the flexural rigidity in the plane of bending, which is assumed to be a plane of symmetry. Combining Eqs. (16.2.1) to (16.2.3), we get :

$$EI_3\frac{d^3u_2}{dx_1^3} + P\frac{du_2}{dx_1} = -V_2 \tag{16.2.4}$$

and

$$EI_3 \frac{d^4 u_2}{dx_1^4} + P \frac{d^2 u_2}{dx_1^2} = q. \tag{16.2.5}$$

Eqs. (16.2.1) to (16.2.5) are the basic differential equations, for bending of beam-columns. If the axial load P is equal to zero, these equations reduce to the usual equations for bending by lateral loads. If the transverse load q is equal to zero, these equations apply to an axially loaded column.

16.3 Simple Columns

Let us consider the fixed end column of Fig. 16.4, and assume that it has a uniform bending stiffness EI_3 and that buckling occurs in the plane OX_1, OX_2. If the column is accidentally displaced from its straight position along the OX_1 axis, the force P produces a moment about O which tends to bend the column even further; on the other hand, the elastic forces in the column tend to restore it to its original position. For small values of P, the straight position is stable and the column is subjected to uniform compression. For large values of P, the straight position is unstable and the column buckles. If P is gradually increased, a condition is reached where an accidentally produced displacement does not disappear upon removal of the disturbing agent. This value of P is the critical load which can be defined [1] as the axial force which is sufficient to keep the column in a slightly bent shape. The equation governing the deflection of the column in Fig. 16.4 is:

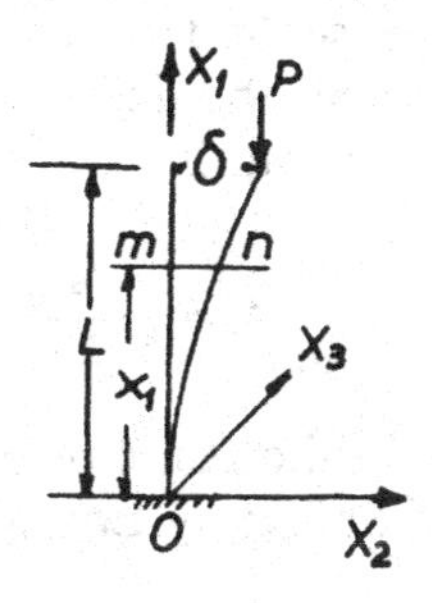

Fig. 16.4

$$EI_3 \frac{d^2 u_2}{dx_1^2} = -M_{13}. \tag{16.3.1}$$

The boundary conditions to be satisfied are :

$$\text{At} \quad x_1 = 0,\ u_2 = 0 \text{ and } \frac{du_2}{dx_1} = 0. \tag{16.3.2}$$

$$\text{At} \quad x_1 = L,\ u_2 = \delta. \tag{16.3.3}$$

At any section *mn*,

$$M_{13} = -P(\delta - u_2), \tag{16.3.4}$$

so that Eq. (16.3.1) becomes :

$$\frac{d^2 u_2}{dx_1^2} + K^2 u_2 = K^2 \delta, \tag{16.3.5}$$

where

$$K^2 = \frac{P}{EI_3}. \tag{16.3.6}$$

The general solution of Eq. (16.3.5) is :

$$u_2 = A \cos Kx_1 + B \sin Kx_1 + \delta, \tag{16.3.7}$$

in which A and B are constants of integration. The boundary conditions (16.3.2) give:

$$A = -\delta \text{ and } B = 0. \tag{16.3.8}$$

Therefore,

$$u_2 = \delta(1 - \cos Kx_1). \tag{16.3.9}$$

The boundary condition (16.3.3) yields:

$$\delta = \delta - \delta \cos KL, \tag{16.3.10}$$

which requires that either $\delta = 0$ or $\cos KL = 0$. If $\delta = 0$, there is no deflection and therefore no buckling. If $\cos KL = 0$, we must have:

$$KL = (2n - 1)\frac{\Pi}{2}, \tag{16.3.11}$$

where $n = 1, 2, \ldots\ldots$ Eq. (16.3.11) determines the values of K at which a buckled shape can exist. δ remains indeterminate and can take any value within the scope of small deflection theory [remember that Eq. (16.3.1) is based on such a theory]. One has to notice that in this problem, as P increases, the moment increases, thus increasing the deflection which in turn increases the moment: The problem is no longer linear and the results based on linear theory , such as Eq. (16.3.7), do not contain enough boundary conditions to permit a solution for the exact value of δ. For values of P slightly higher than the critical value, the deflections becomes so high that the linear theories do not apply. The smallest value of KL satisfying Eq.(16.3.11) corresponds to $n = 1$.
Thus

$$KL = \frac{\Pi}{2} \tag{16.3.12}$$

or

$$P = P_{cr} = \frac{\Pi^2 EI_3}{4L^2}. \tag{16.3.13}$$

This is the smallest axial force which can maintain a slightly bent shape: The column is in a neutral equilibrium position. Other values of n correspond to the deflection patterns shown in Fig. 16.5. Those patterns

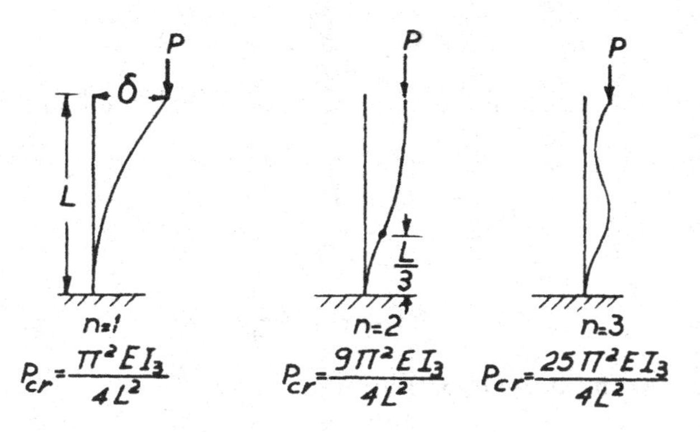

Fig. 16.5

are called buckling modes and correspond to values of P higher than P_{cr}. A value of P corresponding to $n = 2$ can only be attained if the mode of buckling corresponding to $n = 1$ is prevented. The high buckling modes are, however, mathematically important. The fact that $n = 1$

corresponds to the smallest axial force that can maintain a slightly bent shape, will be exploited to obtain an energy solution to the buckling problem.

The previous results could have been obtained by starting from Eq.(16.2.5) in which q is set equal to zero. The solution of this equation is:

$$u_2 = C_1 + C_2 x_1 + C_3 \sin Kx_1 + C_4 \cos Kx_1, \tag{16.3.14}$$

where C_1, C_2, C_3, and C_4 are constants of integration to be obtained from the following boundary conditions:

$$\text{At} \quad x_1 = 0,\ u_2 = 0 \text{ and } \frac{du_2}{dx_1} = 0 \tag{16.3.15}$$

$$\text{At} \quad x_1 = L,\ M_{13} = -EI_3 \frac{d^2 u_2}{dx_1^2} = 0 \text{ and} \tag{16.3.16}$$

$$V_2 = -P\frac{du_2}{dx_1} - EI_3 \frac{d^3 u_2}{dx_1^3} = 0.$$

Substituting these four boundary conditions into Eq. (16.3.14), we obtain the following simultaneous equations for the constants of integration:

$$\begin{aligned} C_1 + 0 + 0 + C_4 &= 0 \\ 0 + C_2 + C_3 K + 0 &= 0 \\ 0 + 0 - C_3 K \sin KL - C_4 K \cos KL &= 0 \\ 0 + C_2 P + 0 + 0 &= 0. \end{aligned} \tag{16.3.17}$$

This is a set of four homogeneous equations with four unknowns, a non-trivial solution of which exists only when the determinant formed by the coefficients is equal to zero. This type of problem, the eigenvalue problem, has been examined at length in Chapter 3. Setting the determinant equal to zero yields:

$$\cos KL = 0 \tag{16.3.18}$$

or

$$KL = (2n - 1)\frac{\Pi}{2}, \tag{16.3.19}$$

which is the same as Eq. (16.3.11) . To every value of n, there corresponds a value of K and consequently a value of P, which provides us with a possible solution. Those values of P are the eigenvalues of the problem, each of which corresponds to a form of u_2, in other words, to a buckling mode: Indeed, substituting Eq. (16.3.2) into Eq. (16.3.14), we get:

$$u_2 = C_1(1 - \cos Kx_1). \tag{16.3.20}$$

If we set $u_2 = \delta$ at $x_1 = L$, then

$$u_2 = \delta(1 - \cos Kx_1), \tag{16.3.21}$$

which is the same as Eq. (16.3.9). There are as many expressions of u_2 as there are values of K and, consequently, of P (Fig.16.5).

The indeterminacy in the value of δ can be lifted if the load P is slightly eccentric, thus inducing a couple M_0 at the free end (Fig. 16.6). In such a case, Eq. (16.3.5) becomes:

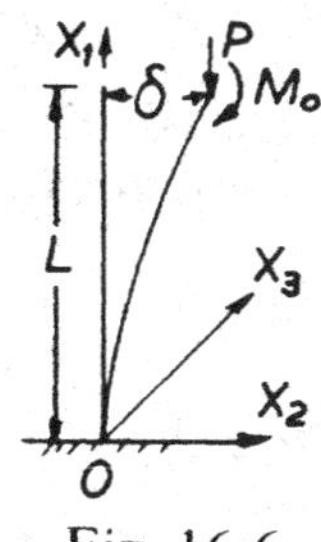

Fig. 16.6

$$\frac{d^2u_2}{dx_1^2} + K^2u_2 = K^2\delta + \frac{M_0}{EI_3}. \tag{16.3.22}$$

The general solution of this equation is :

$$u_2 = A\cos Kx_1 + B\sin Kx_1 + \delta + \frac{M_0}{K^2EI_3}. \tag{16.3.23}$$

The boundary conditions (16.3.2), give:

$$A = -\delta - \frac{M_0}{K^2EI_3}, \quad B = 0. \tag{16.3.24}$$

Therefore,

$$u_2 = \left(\delta + \frac{M_0}{K^2 EI_3}\right)(1 - \cos Kx_1). \tag{16.3.25}$$

The boundary condition (16.3.3) yields :

$$\delta = \frac{M_0}{P}\left(\frac{1 - \cos KL}{\cos KL}\right). \tag{16.3.26}$$

The expression for u_2 now becomes:

$$u_2 = \frac{M_0}{P}\left(\frac{1 - \cos Kx_1}{\cos KL}\right). \tag{16.3.27}$$

The buckling load of a column is quite sensitive to the nature of the supports of the ends of the columns :
a) *For a bar with hinged ends* (Fig.16.7), the boundary conditions are:

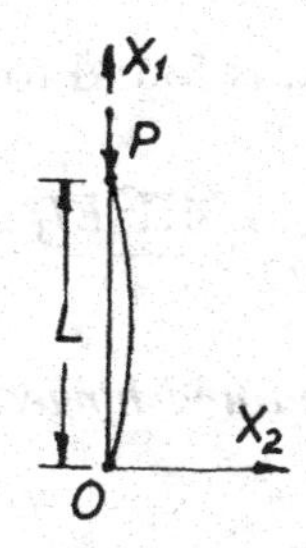

Fig.16.7

$$u_2 = 0 \quad \text{at } x_1 = 0 \quad \text{and } x_1 = L$$

$$M_{13} = -EI_3 \frac{d^2 u_2}{dx_1^2} = 0 \quad \text{at } x_1 = 0 \quad \text{and } x_1 = L.$$

The value of the buckling load is found to be :

$$P_{cr} = \frac{\Pi^2 EI_3}{L^2} \tag{16.3.28}$$

This case is called the fundamental case of buckling of a prismatic bar.
b) *For a bar with both ends built in* (Fig.16.8), the boundary conditions are:

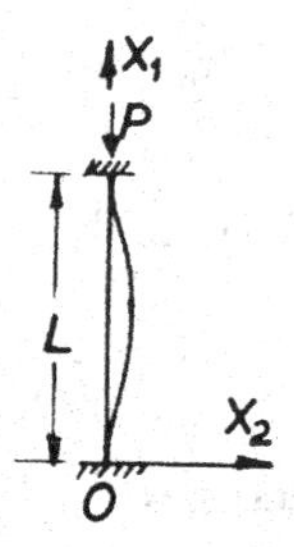

Fig. 16.8

$$u_2 = 0 \text{ at } x_1 = 0 \text{ and } x_1 = L$$

$$\frac{du_2}{dx_1} = 0 \text{ at } x_1 = 0 \text{ and } x_1 = L$$

The value of the buckling load is found to be :

$$P_{cr} = \frac{4\Pi^2 EI_3}{L^2}. \tag{16.3.29}$$

c) *For a bar built in at one end and hinged at the other* (Fig. 16.9), the boundary conditions are:

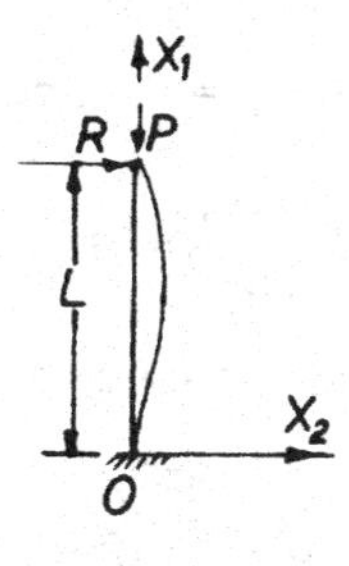

Fig. 16.9

$$u_2 = 0 \text{ at } x_1 = 0 \text{ and } x_1 = L$$

$$\frac{du_2}{dx_1} = 0 \text{ at } x_1 = 0$$

$$M_{13} = -EI_3 \frac{d^2 u_2}{dx_1^2} = 0 \text{ at } x_1 = L.$$

In this case, a reactive force R is developed at the pinned end. The critical load is found to be:

$$P_{cr} = \frac{\Pi^2 EI_3}{(0.7L)^2}. \tag{16.3.30}$$

16.4 Energy Solution of the Buckling Problem

Let us consider the equilibrium of a fixed end column when the compression load is equal to the buckling load. Under such a load, ,the column could be straight or, if disturbed, it could assume a bent form and keep it (Fig.16.10). During this change of form, the column is still

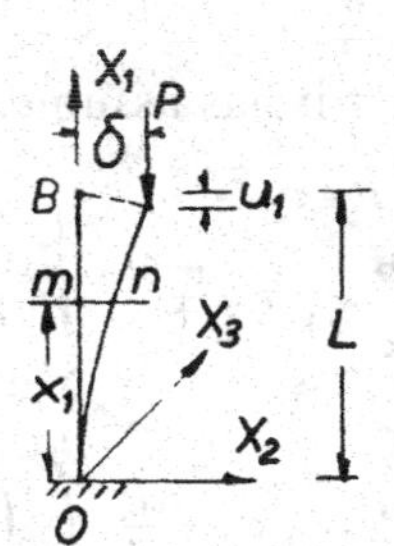

Fig. 16.10

in equilibrium, but its strain energy is increased since the energy of bending of the column will be added to the energy of compression. Thus, the total strain energy in the column is given by:

$$U_t = \frac{A}{2} \int_0^L \frac{P}{A} e_{11}\, dx_1 + \frac{1}{2} \int_0^L EI_3 \left(\frac{d^2 u_2}{dx_1^2}\right)^2 dx_1, \tag{16.4.1}$$

where A is the cross section of the column. The displacement u_1 is given by:

$$u_1 = \int_0^L \left(ds - dx_1\right) = \int_0^L \left[dx_1\sqrt{1 + \left(\frac{du_2}{dx_1}\right)^2} - dx_1\right] \approx \int_0^L \frac{1}{2}\left(\frac{du_2}{dx_1}\right)^2 dx_1. \tag{16.4.2}$$

The work done by the external force P during axial compression and bending is:

$$W = \int_0^L Pe_{11}\,dx_1 + \frac{P}{2}\int_0^L \left(\frac{du_2}{dx_1}\right)^2 dx_1. \tag{16.4.3}$$

The potential energy of the system prior to its assuming a bent form is:

$$(\Pi_p)_1 = \frac{A}{2}\int_0^L \frac{P}{A}e_{11}\,dx_1 - \int_0^L Pe_{11}\,dx_1, \tag{16.4.4}$$

and its potential energy after it has assumed a bent form:

$$(\Pi_p)_2 = \frac{A}{2}\int_0^L \frac{P}{A}e_{11}\,dx_1 + \frac{1}{2}\int_0^L EI_3\left(\frac{d^2u_2}{dx_1^2}\right)^2 dx_1 - \int_0^L Pe_{11}\,dx_1 - \frac{P}{2}\int_0^L \left(\frac{du_2}{dx_1}\right)^2 dx_1. \tag{16.4.5}$$

The change in potential energy is, therefore,

$$\Delta\Pi_p = \Delta U_t - \Delta W = \frac{1}{2}\int_0^L EI_3\left(\frac{d^2u_2}{dx_1^2}\right)^2 dx_1 - \frac{P}{2}\int_0^L \left(\frac{du_2}{dx_1}\right)^2 dx_1. \tag{16.4.6}$$

and the value of P for neutral equilibrium is obtained from writing $\Delta\Pi_p = 0$, in other words from:

$$P = \frac{\dfrac{1}{2}\displaystyle\int_0^L EI_3\left(\frac{d^2u_2}{dx_1^2}\right)^2 dx_1}{\dfrac{1}{2}\displaystyle\int_0^L \left(\frac{du_2}{dx_1}\right)^2 dx_1} = \frac{\Delta U_t}{u_1}. \tag{16.4.7}$$

Now, in Sec. 16.3, it was shown that there exists an infinite number of values of P which are possible solutions of the eigenvalue problem. Each value corresponds to a buckling mode; in other words, to a certain $u_2 = u_2(x_1)$. The one value of P we are interested in is P_{cr}, which is the smallest one and which corresponds to the first mode. Therefore, the buckling load will be given by the minimum value of Eq. (16.4.7);i.e.,by

$$P = \left[\frac{\displaystyle\int_0^L EI_3\left(\frac{d^2u_2}{dx_1^2}\right)^2 dx_1}{\displaystyle\int_0^L \left(\frac{du_2}{dx_1}\right)^2 dx_1}\right]_{\min} \tag{16.4.8}$$

The numerator in Eq. (16.4.7) is the increase in the strain energy of the column when, under P, it assumes the bent form; the denominator is the displacement of P along its line of action due to this bending. Eq. (16.4.8) is called Rayleigh's Formula.

The energy method is used to obtain an approximate solution to problems which cannot be solved exactly. For that, an expression for $u_2(x_1)$ satisfying the boundary conditions is assumed, and P is deduced from Eq. (16.4.7). This value will always be larger than P_{cr}. When using the energy method, Timoshenko [2] noticed that if the expression of the strain energy due to bending was written in terms of the moment, a better approximation would result. From Eq. (16.3.1) and for the case of the column in Fig.16.10, we have:

$$EI_3\left(\frac{d^2u_2}{dx_1^2}\right)^2 = \frac{M_{13}^2}{EI_3} = \frac{P^2(\delta - u_2)^2}{EI_3}. \tag{16.4.9}$$

Substituting Eq. (16.4.9) into Eq. (16.4.6) and setting $\Delta\Pi_p = 0$, we obtain Timoshenko's formula:

$$P = \frac{\displaystyle\int_0^L \left(\frac{du_2}{dx_1}\right)^2 dx_1}{\displaystyle\int_0^L \frac{(\delta - u_2)^2}{EI_3}\, dx_1} \tag{16.4.10}$$

and

$$P_{cr} = \left[\frac{\int_0^L \left(\frac{du_2}{dx_1}\right)^2 dx_1}{\int_0^L \frac{(\delta - u_2)^2}{EI_3} dx_1} \right]_{min} \quad (16.4.11)$$

Eq. (16.4.10) gives a better approximation than Eq. (16.4.7). Wang [1] gives an analysis of the errors in the buckling loads calculated by the energy method.

When a polynomial or a trigonometric series is introduced to express the value of $u_2(x_1)$, the Rayleigh- Ritz method is often used to minimize P and get as close as possible to P_{cr}.

16.5 Examples of Calculation of Buckling Loads by the Energy Method

In this section, we shall illustrate the use of the energy method by means of two examples whose correct answer is known. This will give us an idea about the approximations involved in this method .

1. *Column with one end built in and the other end free.* Let us assume that the deflection curve of the buckled bar (Fig. 16.10) is given by an equation of the form:

$$u_2 = \delta\left(1 - \cos\frac{\Pi x_1}{2L}\right). \quad (16.5.1)$$

Applying Eq. (16.4.7), we get:

$$P = \frac{\frac{\delta^2 \Pi^4}{64L^3}}{\frac{\delta^2 \Pi^2}{16L}} EI_3 = \frac{\Pi^2 EI_3}{4L^2}. \quad (16.5.2)$$

This is the correct answer $P = P_{cr}$, since we started by assuming a buckled shape which happened to be the correct one.

Let us now assume that u_2 is given by the following equation:

$$u_2 = \frac{\delta x_1^2}{2L^3}(3L - x_1). \quad (16.5.3)$$

This is the equation of the deflected column under the effect of a horizontal load acting at B. Applying Eq.(16.4.7), we get:

$$P = \frac{\dfrac{3EI_3\delta^2}{2L^3}}{\dfrac{3}{5}\dfrac{\delta^2}{L}} = \frac{2.5EI_3}{L^2}, \tag{16.5.4}$$

When compared to Eq. (16.5.2), it is found that this result is in error by 1.3 percent. If we now apply Timoshenko's Eq. (16.4.10), we get:

$$P = \frac{\dfrac{3}{5}\dfrac{\delta^2}{L}}{\dfrac{17}{70}\dfrac{\delta^2 L}{EI_3}} = \frac{2.4706EI_3}{L^2}, \tag{16.5.5}$$

which is in error by only 0.13 percent.

Let us now assume a very poor representation of $u_2(x_1)$. For that, a parabola whose equation is

$$u_2 = \frac{\delta x_1^2}{L^2} \tag{16.5.6}$$

is chosen. This representation does not satisfy the end conditions since it results in a constant curvature along the column. Yet the resulting approximate solution for P is quite satisfactory. Indeed, substituting Eq.(16.5.6) into Eq.(16.4.10), we get:

$$P = \frac{2.5EI_3}{L^2}. \tag{16.5.7}$$

The error is only 1.3 percent.

Finally, we can assume that

$$u_2 = c_0\left(\frac{x_1}{L}\right)^2 + c_1\left(\frac{x_1}{L}\right)^3, \tag{16.5.8}$$

where c_0 and c_1 are undetermined parameters. This form satisfies the boundary conditions $u_2 = 0$ and $du_2/dx_1 = 0$ at $x_1 = 0$. Substituting Eq.(16.5.8) into Eq.(16.4.7), we get :

$$P = \frac{4EI_3}{L^2}\,\frac{1 + 3\dfrac{c_1}{c_0} + 3\left(\dfrac{c_1}{c_0}\right)^2}{\dfrac{4}{3} + 3\dfrac{c_1}{c_0} + \dfrac{9}{5}\left(\dfrac{c_1}{c_0}\right)^2}. \tag{16.5.9}$$

The minimum value of P is obtained by differentiating with respect to c_1/c_0 and setting the result equal to zero (Rayleigh -Ritz method). Thus

$$\frac{dP}{d\left(\frac{c_1}{c_0}\right)} = 0 = 18\left(\frac{c_1}{c_0}\right)^2 + 22\frac{c_1}{c_0} + 5, \tag{16.5.10}$$

which gives $c_1/c_0 = -0.3$ and -0.92. Substituting these values into Eq.(16.5.9), we find that $c_1/c_0 = -0.3$ gives the smaller P, so that

$$P = \frac{2.49EI_3}{L^2}. \tag{16.5.11}$$

The error is 0.92 percent. Better accuracy can be obtained by taking more than two undetermined parameters.

2. *Prismatic column with hinged ends.* The boundary conditions in this case are (Fig.16.7):

$$u_2 = 0 \text{ and } d^2u_2/dx_1^2 = 0 \text{ at } x_1 = 0 \text{ and } x_1 = L.$$

These conditions are satisfied by assuming that the shape of the deflected column is represented by a trigonometric series:

$$u_2 = \sum_{n=1}^{\infty} c_n \sin\frac{n\Pi x_1}{L}. \tag{16.5.12}$$

We have:

$$\Delta U_t = \frac{1}{2}\int_0^L EI_3\left(\frac{d^2u_2}{dx_1^2}\right)^2 dx_1 = \frac{\Pi^4 EI_3}{4L^3}\sum_{n=1}^{\infty} n^4 c_n^2 \tag{16.5.13}$$

and

$$u_1 = \frac{1}{2}\int_0^L \left(\frac{du_2}{dx_1}\right)^2 dx_1 = \frac{\Pi^2}{4L}\sum_{n=1}^{\infty} n^2 c_n^2. \tag{16.5.14}$$

Substituting Eqs. (16.5.13) and (16.5.14) into Eq.(16.4.7),we get:

$$P = \frac{\Delta U_t}{u_1} = \frac{\Pi^2 EI_3}{L^2}\,\frac{1 + 16\left(\frac{c_2}{c_1}\right)^2 + 81\left(\frac{c_3}{c_1}\right)^2 + \cdots}{1 + 4\left(\frac{c_2}{c_1}\right)^2 + 9\left(\frac{c_3}{c_1}\right)^2 + \cdots} \tag{16.5.15}$$

The minimum P is obtained by adjusting the undetermined parameters c_2/c_1, c_3/c_1, etc. Using the Rayleigh- Ritz method, we must write

$$\frac{\partial P}{\partial\left(\frac{c_2}{c_1}\right)} = \frac{\partial P}{\partial\left(\frac{c_3}{c_1}\right)} = \cdots = 0.$$

Carrying out the calculation, we find that these conditions require that

$$\frac{c_2}{c_1} = \frac{c_3}{c_1} = \cdots = 0.$$

Hence

$$P_{\min} = \frac{\Pi^2 EI_3}{L^2},$$

which is the value given by the exact solution. This is the case because the assumed form of the deflection curve happened to include the exact solution.

Remark

For a wide variety of examples using the energy method, the reader is referred to Timoshenko and Gere's classic treatise [2].

16.6 Combined Compression and Bending

1. *Beam-column with a concentrated lateral load.* Let us consider the problem of a strut AB with hinged ends loaded by an axial force P and a force Q at a distance a from B (Fig.16.11). The differential equations of the two portions of the strut are:

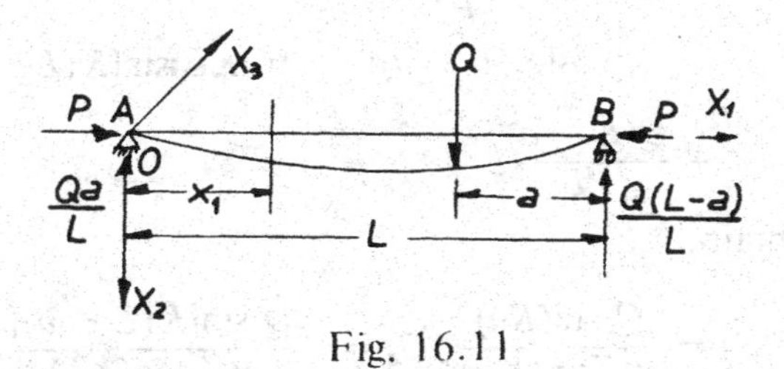

Fig. 16.11

$$\text{Left of } Q : EI_3 \frac{d^2 u_2}{dx_1^2} = -Pu_2 - \frac{Qa}{L} x_1. \tag{16.6.1}$$

$$\text{Right of } Q : EI_3 \frac{d^2 u_2}{dx_1^2} = -Pu_2 - \frac{Q(L-a)(L-x_1)}{L}. \tag{16.6.2}$$

Setting

$$K^2 = \frac{P}{EI_3}, \tag{16.6.3}$$

the solutions of these equations are:

$$\text{Left of } Q : u_2 = C_1 \cos(Kx_1) + C_2 \sin(Kx_1) - \frac{Qax_1}{PL}. \tag{16.6.4}$$

$$\text{Right of } Q : u_2 = C_3 \cos(Kx_1) + C_4 \sin(Kx_1) - \frac{Q(L-a)(L-x_1)}{PL}, \tag{16.6.5}$$

where C_1, C_2, C_3, and C_4 are constants of integration to be determined from the boundary conditions. At $x_1 = 0$ and $x_1 = L$, $u_2 = 0$. Therefore,

$$C_1 = 0 \text{ and } C_3 = -C_4 \tan KL. \tag{16.6.6}$$

At the point of application of the load Q, $x_1 = L - a$, the two portions of the deflection curve as given by Eqs. (16.6.4) and (16.6.5) must give the same deflection and slope. Then,

$$C_2 \sin[K(L-a)] = C_4\{\sin[K(L-a)] - \tan(KL)\cos[K(L-a)]\} \tag{16.6.7}$$

$$C_2 K \cos[K(L-a)] - \frac{Qa}{PL} = C_4 K\{\cos[K(L-a)] + \tan(KL)\sin[K(L-a)]\} + \frac{Q(L-a)}{PL}. \tag{16.6.8}$$

Solving, we obtain:

$$C_2 = \frac{Q \sin(Ka)}{PK \sin(KL)}, \quad C_4 = -\frac{Q \sin[K(L-a)]}{PK \tan(KL)}. \tag{16.6.9}$$

Substituting Eqs. (16.6.6) and (16.6.9) into Eqs. (16.6.1) and (16.6.2), we obtain for the portion of the strut to the left of Q:

$$u_2 = \frac{Q\sin(Ka)}{PK\sin(KL)}\sin(Kx_1) - \frac{Qa}{PL}x_1 \tag{16.6.10}$$

$$\frac{du_2}{dx_1} = \frac{Q\sin(Ka)}{P\sin(KL)}\cos(Kx_1) - \frac{Qa}{PL} \tag{16.6.11}$$

$$\frac{d^2u_2}{dx_1^2} = -\frac{QK\sin(Ka)}{P\sin(KL)}\sin(Kx_1), \tag{16.6.12}$$

and for the portion of the strut to the right of Q:

$$u_2 = \frac{Q\sin[K(L-a)]}{PK\sin(KL)}\sin[K(L-x_1)] - \frac{Q(L-a)(L-x_1)}{PL} \tag{16.6.13}$$

$$\frac{du_2}{dx_1} = -\frac{Q\sin[K(L-a)]}{P\sin(KL)}\cos[K(L-x_1)] + \frac{Q(L-a)}{PL} \tag{16.6.14}$$

$$\frac{d^2u_2}{dx_1^2} = -\frac{QK\sin[K(L-a)]}{P\sin(KL)}\sin[K(L-x_1)]. \tag{16.6.15}$$

In the particular case of a load Q applied at the center of the beam, the deflection curve is symmetrical, and for $x_1 = a = L/2$

$$u_2 = (u_2)_{max} = \frac{Q}{2PK}\left[\tan\left(\frac{KL}{2}\right) - \frac{KL}{2}\right] = \frac{QL^3}{48EI_3}\frac{3\left[\tan\left(\frac{KL}{2}\right) - \frac{KL}{2}\right]}{\left(\frac{KL}{2}\right)^3} \tag{16.6.16}$$

or

$$(u_2)_{max} = \frac{QL^3}{48EI_3}\chi. \tag{16.6.17}$$

Numerical values of χ for various $\frac{KL}{2}$ can be found in [2]. When $\frac{KL}{2}$ approaches $\Pi/2$,the deflection becomes infinite and

$$P = P_{cr} = \frac{\Pi^2 EI_3}{L^2}, \tag{16.6.18}$$

which is the buckling load. To find the slope of the deflection curve at the end of the beam column when Q is at the center, set $a = \frac{L}{2}$ and $x_1 = 0$ in Eq. (16.6.11). This gives:

$$\left(\frac{du_2}{dx_1}\right)_{x_1=0} = \frac{QL^2}{16EI_3}\,\frac{2\left[1-\cos\left(\frac{KL}{2}\right)\right]}{\left(\frac{KL}{2}\right)^2\cos\left(\frac{KL}{2}\right)} = \frac{QL^2}{16EI_3}\lambda, \quad (16.6.19)$$

where λ is a function of $\frac{KL}{2}$. The same way the maximum bending moment is obtained from Eq.(16.6.12) as

$$(M_{13})_{max} = -EI_3\left(\frac{d^2u_2}{dx_1^2}\right)_{x_1=\frac{L}{2}} = \frac{QKEI_3}{2P}\tan\left(\frac{KL}{2}\right)$$
$$= \frac{QL}{4}\,\frac{\tan\left(\frac{KL}{2}\right)}{\frac{KL}{2}}. \quad (16.6.20)$$

Numerical values of λ and $\tan \frac{KL}{2}/\frac{KL}{2}$ can be found in [2].

2. *Beam—column with several concentrated loads.* Eqs.(16.6.10) and (16.6.13) show that, for a given P, the deflections are linear functions of Q. On the other hand, P intervenes in these equations in a rather complicated way. For more than one lateral load, the principle of superposition can be applied but in a slightly modified form to take into account the effect of P. To prove this statement, let us consider a beam-column subjected to two lateral loads Q_1 and Q_2 at distances a_1 and a_2 from B (Fig. 16 .12). To the left of Q_1, the differential equation of the deflection curve is:

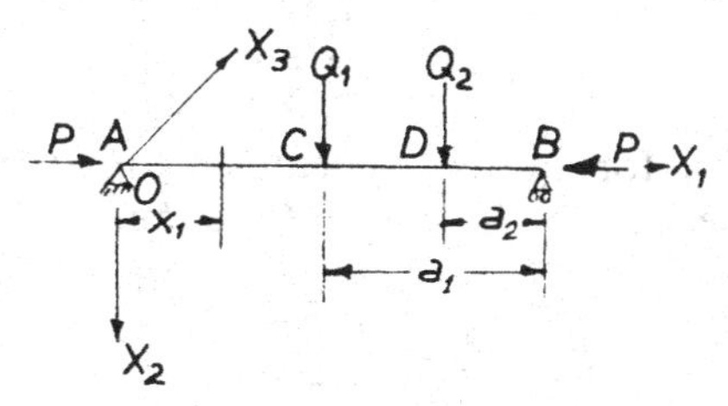

Fig. 16.12

$$EI_3 \frac{d^2 u_2}{dx_1^2} = -\frac{Q_1 a_1}{L} x_1 - \frac{Q_2 a_2}{L} x_1 - P u_2 . \tag{16.6.21}$$

Consider now Q_1 and Q_2 acting separately on the axially compressed strut and denote by $(u_2)_1$ the deflection caused by Q_1 , and by $(u_2)_2$ the deflection caused by Q_2. For the portion of the beam left of C and for Q_1 acting alone, we have:

$$EI_3 \frac{d^2 (u_2)_1}{dx_1^2} = -\frac{Q_1 a_1}{L} x_1 - P(u_2)_1 . \tag{16.6.22}$$

For Q_2 acting alone, we have:

$$EI_3 \frac{d^2 (u_2)_2}{dx_1^2} = -\frac{Q_2 a_2}{L} x_1 - P(u_2)_2 . \tag{16.6.23}$$

By adding these two equations, we find:

$$EI_3 \frac{d^2[(u_2)_1 + (u_2)_2]}{dx_1^2} = -\frac{Q_1 a_1}{L} x_1 - \frac{Q_2 a_2}{L} x_2 - P[(u_2)_1 + (u_2)_2]. \tag{16.6.24}$$

Eq.(16.6.24), for the sum of the deflections $(u_2)_1$ and $(u_2)_2$, is the same as Eq. (16.6.21). The same conclusion holds for points between Q_1 and Q_2 and points to the right of Q_2. Therefore, when there are several loads acting on a beam column, the resultant deflection can be obtained by superposition of the deflections produced separately by each lateral load acting in combination with the compressive force P.

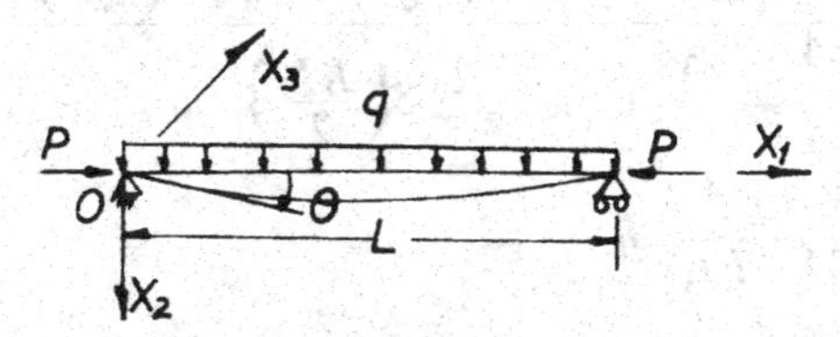

Fig. 16.13

3.*Beam column with a uniformly distributed load.* Consider the beam shown in Fig. 16.13, subjected to a uniformly distributed transverse load q and the axial force P. The differential equation of the deflection curve is:

$$EI_3 \frac{d^2 u_2}{dx_1^2} = -\frac{qL}{2} x_1 + \frac{q x_1^2}{2} - P u_2. \tag{16.6.25}$$

Setting $K^2 = \frac{P}{EI_3}$, the solution of this equation is:

$$u_2 = C_1 \sin(Kx_1) + C_2 \cos(Kx_1) + \frac{q}{2P}\left(x_1^2 - Lx_1 - \frac{2}{K^2}\right). \tag{16.6.26}$$

The boundary conditions that $u_2 = 0$ at $x_1 = 0$ and $x_1 = L$ determine C_1 and C_2 to be:

$$C_1 = \frac{q}{PK^2}\left[\frac{1 - \cos(KL)}{\sin(KL)}\right], \; C_2 = \frac{q}{PK^2}.$$

Therefore,

$$u_2 = \frac{q}{2PK^2}\left\{2\left[\frac{1 - \cos(KL)}{\sin(KL)} \sin(Kx_1) + \cos(Kx_1)\right] + x_1^2 K^2 - L x_1 K^2 - 2\right\}. \tag{16.6.27}$$

The deflection at the middle of the beam column is:

$$(u_2)_{x_1 = \frac{L}{2}} = \frac{5}{384} \frac{qL^4}{EI_3} \frac{12\left[2Sec\left(\frac{KL}{2}\right) - 2 - \left(\frac{KL}{2}\right)^2\right]}{5\left(\frac{KL}{2}\right)^4} = \frac{5}{384} \frac{qL^4}{EI_3} \eta. \tag{16.6.28}$$

Numerical values of η for various $\frac{KL}{2}$ can be found in [2]. As $\frac{KL}{2}$ approaches $\pi/2$, P approaches its critical value given by Eq.(16.3.28).

The slope at the end of the beam-column is obtained by differentiating Eq.(16.6.27). Thus,

$$\theta = \left(\frac{du_2}{dx_1}\right)_{x_1=0} = \frac{qL^3}{24EI_3}\,\frac{3\left[\tan\left(\frac{KL}{2}\right) - \frac{KL}{2}\right]}{\left(\frac{KL}{2}\right)^3} = \frac{qL^3}{24EI_3}\chi. \quad (16.6.29)$$

The maximum bending moment is obtained by differentiating Eq. (16. 6.27) twice. Thus,

$$(M_{13})_{\max} = -EI_3\left(\frac{d^2u_2}{dx_1^2}\right)_{x_1=\frac{L}{2}} = \frac{qL^2}{8}\,\frac{2\left[1 - \cos\left(\frac{KL}{2}\right)\right]}{\left(\frac{KL}{2}\right)^2\cos\left(\frac{KL}{2}\right)} \quad (16.6.30)$$

$$= \frac{qL^2}{8}\lambda.$$

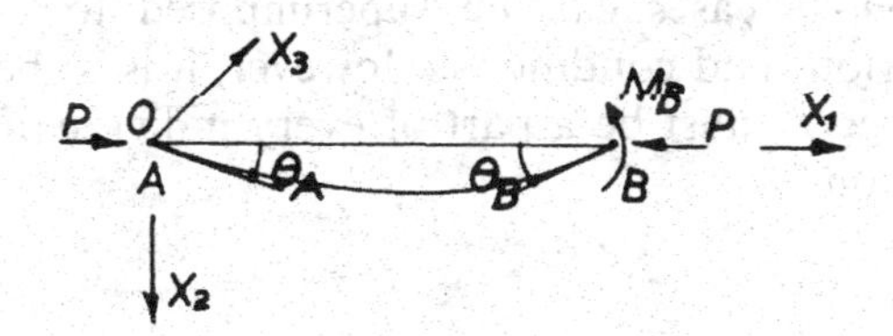

Fig. 16.14

4. *Beam-column with end couple.* This problem can be solved [2] by assuming in Fig. 16.11 that the distance a tends to zero while Q increases such that the product (Qa) remains finite and equal to M_B. In Eq.(16.6.10), (Ka) is substituted to sin(Ka) and M_B to Qa. Thus (Fig.16.14),

$$u_2 = \frac{M_B}{P}\left[\frac{\sin(Kx_1)}{\sin(KL)} - \frac{x_1}{L}\right] \quad (16.6.31)$$

and

$$\theta_A = \left(\frac{du_2}{dx_1}\right)_{x_1=0} = \frac{M_B L}{6EI_3}\frac{3}{\left(\frac{KL}{2}\right)}\left[\frac{1}{\sin 2\left(\frac{KL}{2}\right)} - \frac{1}{2\left(\frac{KL}{2}\right)}\right]$$

$$= \frac{M_B L}{6EI_3}\phi \tag{16.6.32}$$

$$\theta_B = -\left(\frac{du_2}{dx_1}\right)_{x_1=L} = \frac{M_B L}{3EI_3}\frac{3}{2\left(\frac{KL}{2}\right)}\left[\frac{1}{2\left(\frac{KL}{2}\right)} - \frac{1}{\tan 2\left(\frac{KL}{2}\right)}\right] \tag{16.6.33}$$

$$= \frac{M_B L}{3EI_3}\psi$$

Numerical values for ϕ and ψ for various $\frac{KL}{2}$ are given in [2].

Remark

The four previous cases can be superimposed to analyze beam-columns with various end conditions. However, it is to be remembered that the axial force P must be a part of every individual solution used in the superposition.

16.7 Lateral Buckling of Thin Rectangular Beams

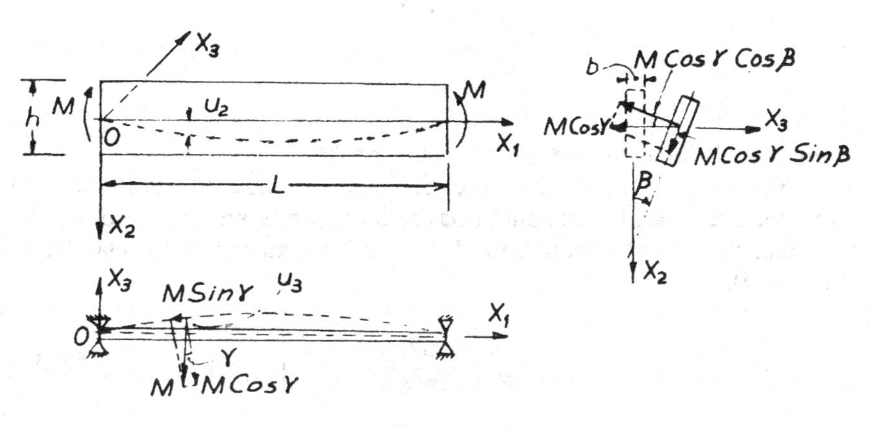

Fig. 16.15

If a beam is very stiff against bending in one plane and very flexible in a perpendicular plane (like a ruler), and if the beam is loaded in the stiff plane, it may become unstable at some critical value of the load and buckle sidewise as shown in Fig. 16.15. This bending in the less flexible direction is always associated with a twist. In this section, we shall consider two simple cases of lateral buckling.

1. *Simply supported beam bent by couples*. Consider a simple beam with a narrow rectangular cross section which is subjected to pure bending in the OX_1, OX_2 plane (Fig.16.15). Let us assume that the beam is disturbed so as to have a small lateral deflection u_3. At any point x_1, the bending moment vector $\overline{M}_{13}$ is directed along the $-OX_3$ axis (using the right-hand rule): Its magnitude is M and it can be decomposed into three components: namely (see Fig. 16.15), $M \sin \gamma$, which causes a twisting of the beam; and $M \cos\gamma \cos\beta$ and $M \cos\gamma \sin\beta$, which causes a bending of the beam in two orthogonal directions. Recalling the equations of the deflection and of the twisting of simple beams, we have:

$$-M \cos \gamma \cos \beta = EI_3 \frac{d^2 u_2}{dx_1^2} \tag{16.7.1}$$

$$-M \cos \gamma \sin \beta = EI_2 \frac{d^2 u_3}{dx_1^2} \tag{16.7.2}$$

$$M \sin \gamma = GJ\alpha = GJ \frac{d\beta}{dx_1}, \tag{16.7.3}$$

where I_2 and I_3 are the moments of inertia with respect to the OX_2 and OX_3 axes, GJ is the torsional rigidity, and $\alpha = d\beta/dx_1$ is the angle of rotation per unit length of the beam following the sense of the twisting moment.

For rectangular cross sections, we have:

$$I_2 = \frac{hb^3}{12}, I_3 = \frac{bh^3}{12}, J = \frac{hb^3}{16} K, \tag{16.7.4}$$

where K is given by the table in Sec. 10.7. Since γ and β as well as the displacements are small, then

$$\sin \gamma \approx \frac{du_3}{dx_1}, \cos \gamma \approx 1, \sin \beta \approx \beta, \cos \beta \approx 1. \tag{16.7.5}$$

Eq.(16.7.1) gives the vertical displacement of the beam and is of no interest in this case. Differentiating Eq. (16.7.3) with respect to x_1, and eliminating d^2u_3/dx_1^2 by using Eq.(16.7.2), we get:

$$GJ\frac{d^2\beta}{dx_1^2} + \frac{M^2}{EI_2}\beta = 0. \tag{16.7.6}$$

If the beam has a constant cross section, the solution of Eq. (16.7.6) is:

$$\beta = C_1 \sin\sqrt{\frac{M^2}{EGJI_2}}\,x_1 + C_2 \cos\sqrt{\frac{M^2}{EGJI_2}}\,x_1 \tag{16.7.7}$$

where C_1 and C_2 are constants to be determined from the boundary conditions: $\beta = 0$ at $x_1 = 0$ and $x_1 = L$. Therefore,

$$C_2 = 0, \quad C_1 \sin\sqrt{\frac{M^2}{EGJI_2}}\,L = 0. \tag{16.7.8}$$

If $C_1 = 0$, we have the trivial solution corresponding to the unbuckled form. Therefore the buckled form is only possible when

$$\sin\sqrt{\frac{M^2}{EGJI_2}}\,L = 0. \tag{16.7.9}$$

Therefore, the critical bending moment necessary to keep the beam in its laterally deformed position is given by :

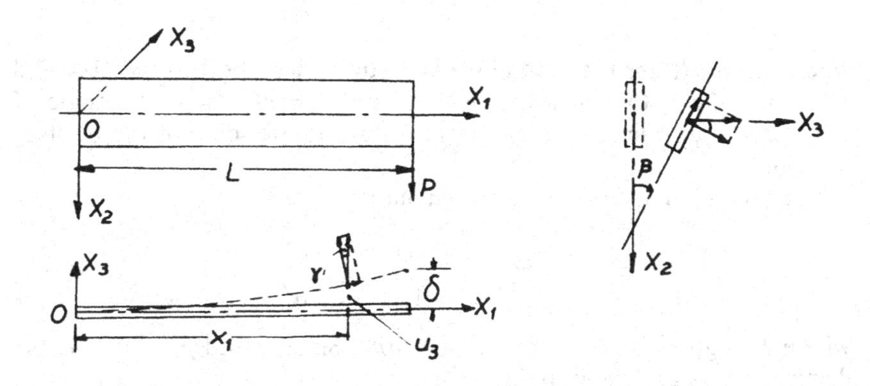

Fig. 16.16

$$(M)_{cr} = \frac{\Pi\sqrt{EGJI_2}}{L} \tag{16.7.10}$$

2. *Cantilever beam bent by an end load.* Let us consider the case of a cantilever beam bent by an end load P which is applied at the centroid of the cross section (Fig. 16.16) and in the OX_1, OX_2 plane. Let us assume that the beam is disturbed so as to have a small lateral displacement δ. At any distance x_1 from the origin, we have a bending moment and a torsional moment respectively equal in magnitude to $P(L - x_1)$ and $P(\delta - u_3)$. Following the same steps as in the previous derivation, we get:

$$EI_3 \frac{d^2 u_2}{dx_1^2} = P(L - x_1) \tag{16.7.11}$$

$$EI_2 \frac{d^2 u_3}{dx_1^2} = P(L - x_1)\beta \tag{16.7.12}$$

$$GJ\frac{d\beta}{dx_1} = -P(L - x_1)\frac{du_3}{dx_1} + P(\delta - u_3). \tag{16.7.13}$$

Eliminating u_3 from Eqs. (16.7.12) and (16.7.13), we get:

$$\frac{d^2\beta}{dx_1^2} + \frac{P^2 L^2}{GJEI_2}\left(1 - \frac{x_1}{L}\right)^2 \beta = 0. \tag{16.7.14}$$

Let $\xi = (1 - x_1/L)$ and $t^2 = P^2 L^2/GJEI_2$. Eq.(16.7.14) becomes :

$$\frac{d^2\beta}{d\xi^2} + t^2\xi^2\beta = 0. \tag{16.7.15}$$

The solution of this equation is:

$$\beta = \sqrt{(\xi)}\left[C_1 J_{\frac{1}{4}}\left(\frac{t\xi^2}{2}\right) + C_2 J_{-\frac{1}{4}}\left(\frac{t\xi^2}{2}\right)\right], \tag{16.7.16}$$

where $J_{\frac{1}{4}}$ and $J_{-\frac{1}{4}}$ are Bessel functions of the first kind of order $\frac{1}{4}$ and $-\frac{1}{4}$, respectively: C_1 and C_2 are constants to be determined from the conditions:

$$\beta = 0 \text{ at } x_1 = 0 \text{ and } \frac{d\beta}{dx_1} = 0 \text{ at } x_1 = L,$$

since the torsional moment is zero there. In terms of ξ, these boundary conditions become:

$$\beta = 0 \text{ for } \xi = 1 \text{ and } \frac{d\beta}{dx_1} = 0 \text{ for } \xi = 0.$$

This last condition requires that $C_1 = 0$. To obtain a non-trival solution, the second condition requires that

$$J_{-\frac{1}{4}}\left(\frac{t}{2}\right) = 0. \tag{16.7.17}$$

From a table of zeros of Bessel functions of the first kind and of order $-\frac{1}{4}$, we find that the smallest t is given by:

$$t = 4.013.$$

Therefore, the critical P necessary to keep the beam in its laterally deformed position is:

$$P_{cr} = \frac{4.013\sqrt{GJEI_2}}{L^2}. \tag{16.7.18}$$

Remark

Reference [2] gives solutions for various types of beams under different loading conditions.

PROBLEMS

1. Solve the problem of the column with one end built in and the other end free (Sec. 16.5) by assuming that the deflection curve of the buckled bar, $u_2 = u_2(x_1)$, is the same as that of a uniformly loaded beam built in at one end and free at the other. Use both Eqs. (16.4.7) and (16.4.10), and compare the axial load obtained in each case to the critical one.
2. Solve Problem 1 for the case of the column hinged at both ends.
3. A coupling rod is 10 ft. long, 2 in. wide, and 4 in. deep. It carries an axial compressive load of 10 tons and a uniformly distributed load of 100 lb./ft. run. Calculate the maximum stress in the rod ($E = 30 \times 10^6$ psi).

4. A long slender steel strut, originally straight and built in at one end and free at the other end, is loaded at the free end with an eccentric load whose line of action is parallel to the original axis of the strut. Determine the deviation of the free end from its original position and the greatest compressive stress, if the length of the strut is 10 ft., its cross section is circular with 2 in. external diameter and 1 in. internal diameter, the load is 800 lb., and the original eccentricity is 3 in. ($E = 30 \times 10^6$ psi).
5. Obtain the expressions for the bending moments at the ends and center of a uniform beam of length L, built in at both ends, and subjected to a uniform lateral load of intensity q, and to end thrusts of intensity P. Show, without elaborate analysis, from the expressions derived, which of the two bending moments is numerically greater than the other.
6. A simply supported beam of length L (and originally straight) is subjected to end thrusts P, together with a lateral load which increases uniformly in intensity from zero at one end to q per unit length at the other end. Deduce an expression from which the bending moment at any section of the beam may be calculated.
7. A straight vertical column is built in at the base and free at the top. It carries a vertical load P at the top and a horizontal side loading which varies uniformly in intensity from zero at the top to q per unit length at the bottom. Derive expressions for the maximum bending moment and maximum deflection.(*)

REFERENCES

[1] C.T.Wang, *Applied Elasticity*,McGraw-Hill, New York, N.Y., 1953.
[2] S.Timoshenko and J.Gere. *Theory of Elastic Stability*, McGraw-Hill, New York, N.Y., 1961.

---

* See pages A-73 to A-75 for additional problems.

CHAPTER 17

BENDING OF THIN FLAT PLATES

17.1 Introduction and Basic Assumptions. Strains and Stresses

A flat plate is a body bounded by two flat parallel surfaces, the distance between these surfaces (called the thickness) being very small in comparison with the dimensions of the surfaces. The plane parallel to the two faces of the plate, and bisecting the thickness h, is called the middle plane. The coordinate axes are such that the OX_1 and OX_2 axes are in the middle plane and the OX_3 axis is perpendicular to it (Fig. 17.1). In this chapter, the small deflection theory of thin plates is presented. The assumptions on which this theory is based and their implications are discussed, with the geometry of deformation being given special attention. The inverse method is used to solve a few simple problems gradually leading to Navier's solution of the simply supported rectangular plate. This solution is also obtained using the principle of minimum potential energy.

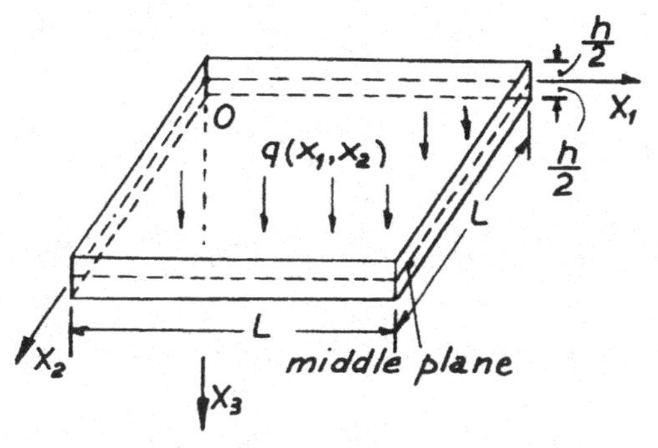

Fig. 17.1

If, when subjected to a load $q = q(x_1, x_2)$, the deflection of the thin plate is small compared to its thickness, the following assumptions attributed to Kirchhoff may be made:

1. The middle plane remains unstrained. This assumption will make it unnecessary to consider the equilibrium of the forces acting on an element of the plate in the OX_1 and OX_2 directions.

2. The normal strain e_{33} in the OX_3 direction is small enough to be neglected, and the normal stresses σ_{33} is small compared to σ_{11} and σ_{22} so that it can be neglected in the stress-strain relations. Therefore,

$$\begin{aligned} e_{33} &= \frac{\partial u_3}{\partial x_3} = 0, \quad e_{11} = \frac{\partial u_1}{\partial x_1} = \frac{1}{E}(\sigma_{11} - \nu\sigma_{22}), \\ e_{22} &= \frac{\partial u_2}{\partial x_2} = \frac{1}{E}(\sigma_{22} - \nu\sigma_{11}). \end{aligned} \tag{17.1.1}$$

3. The normals to the middle plane before bending remain normal to this plane after bending. This means that the out-of-plane shear strains are small enough to be neglected. Therefore,

$$e_{13} = \frac{1}{2}\left(\frac{\partial u_1}{\partial x_3} + \frac{\partial u_3}{\partial x_1}\right) \approx 0, \quad e_{23} = \frac{1}{2}\left(\frac{\partial u_2}{\partial x_3} + \frac{\partial u_3}{\partial x_2}\right) \approx 0. \tag{17.1.2}$$

The only shearing strain left is:

$$e_{12} = \frac{1}{2}\left(\frac{\partial u_1}{\partial x_2} + \frac{\partial u_2}{\partial x_1}\right). \tag{17.1.3}$$

The previous equations represent a generalization of the equations of the simplified theory of beams. They are such that all the strains and, consequently, all the stresses can be written in terms of the deflection u_3 of the middle plane. For example, Eq. (17.1.1) expresses the fact that u_3 is only a function of x_1 and x_2, i.e.,

$$u_3 = u_3(x_1, x_2). \tag{17.1.4}$$

Integrating Eqs. (17.1.2), we get:

$$u_1 = -x_3\frac{\partial u_3}{\partial x_1} + f_1(x_1, x_2) \tag{17.1.5}$$

and

$$u_2 = -x_3 \frac{\partial u_3}{\partial x_2} + f_2(x_1, x_2). \tag{17.1.6}$$

Now, f_1 and f_2 are two functions which represent displacements in the middle plane and, according to our first assumption, these displacements are negligible. Thus (Fig. 17.2):

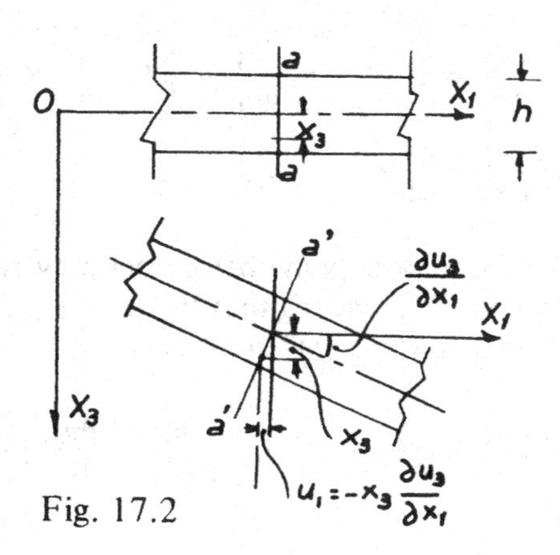

Fig. 17.2

$$u_1 = -x_3 \frac{\partial u_3}{\partial x_1}, \quad u_2 = -x_3 \frac{\partial u_3}{\partial x_2}. \tag{17.1.7}$$

Substituting Eqs. (17.1.7) into Eqs. (17.1.1) and (17.1.3), we get:

$$e_{11} = -x_3 \frac{\partial^2 u_3}{\partial x_1^2} = \frac{1}{E}(\sigma_{11} - \nu\sigma_{22}) \tag{17.1.8}$$

$$e_{22} = -x_3 \frac{\partial^2 u_3}{\partial x_2^2} = \frac{1}{E}(\sigma_{22} - \nu\sigma_{11}) \tag{17.1.9}$$

$$e_{12} = -x_3 \frac{\partial^2 u_3}{\partial x_1 \, \partial x_2} = \frac{1}{2G}\sigma_{12}. \tag{17.1.10}$$

Solving for the stresses, the following equations are obtained:

$$\sigma_{11} = \frac{E}{1-\nu^2}(e_{11} + \nu e_{11}) = -\frac{E}{1-\nu^2}x_3\left(\frac{\partial^2 u_3}{\partial x_1^2} + \nu\frac{\partial^2 u_3}{\partial x_2^2}\right) \qquad (17.1.11)$$

$$\sigma_{22} = \frac{E}{1-\nu^2}(e_{22} + \nu e_{11}) = -\frac{E}{1-\nu^2}x_3\left(\frac{\partial^2 u_3}{\partial x_2^2} + \nu\frac{\partial^2 u_3}{\partial x_1^2}\right) \qquad (17.1.12)$$

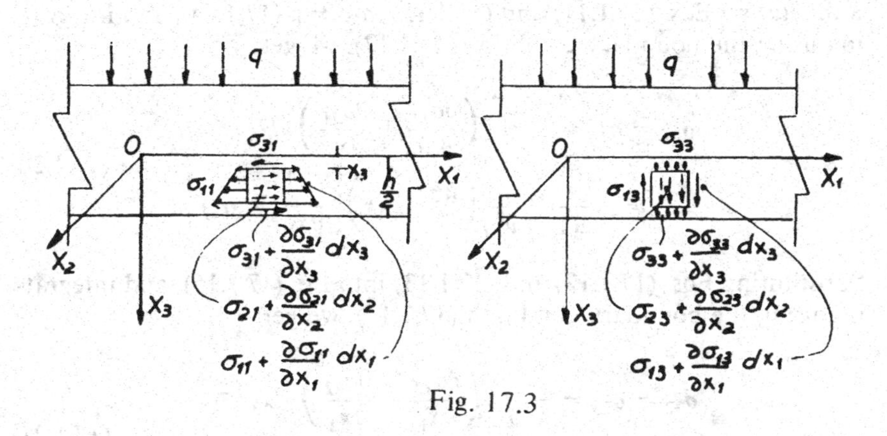

Fig. 17.3

$$\sigma_{12} = 2Ge_{12} = -\frac{E(1-\nu)}{1-\nu^2}x_3\frac{\partial^2 u_3}{\partial x_1\,\partial x_2}. \qquad (17.1.13)$$

Thus, at a given point, σ_{11}, σ_{22}, and σ_{12} vary linearly with x_3 (Fig. 17.3).

The fact that σ_{33} was considered small enough not to affect the stress-strain relations, and that e_{13} and e_{23} were neglected makes it impossible to determine σ_{33}, σ_{13}, and σ_{23} from Hooke's law. Those three quantities can, however, be determined from the differential equations of equilibrium (7.4.6). Neglecting the body forces in Eqs. (7.4.6), we get:

$$\frac{\partial\sigma_{11}}{\partial x_1} + \frac{\partial\sigma_{21}}{\partial x_2} + \frac{\partial\sigma_{31}}{\partial x_3} = 0 \qquad (17.1.14)$$

$$\frac{\partial\sigma_{12}}{\partial x_1} + \frac{\partial\sigma_{22}}{\partial x_2} + \frac{\partial\sigma_{32}}{\partial x_3} = 0 \qquad (17.1.15)$$

$$\frac{\partial\sigma_{13}}{\partial x_1} + \frac{\partial\sigma_{23}}{\partial x_2} + \frac{\partial\sigma_{33}}{\partial x_3} = 0. \qquad (17.1.16)$$

The boundary conditions for σ_{13}, σ_{23}, and σ_{33} are (Fig. 17.3):

$$\text{at } x_3 = \pm\frac{h}{2},\ \sigma_{13} = \sigma_{23} = 0 \tag{17.1.17}$$

$$\text{at } x_3 = +\frac{h}{2},\ \sigma_{33} = 0 \tag{17.1.18}$$

$$\text{at } x_3 = -\frac{h}{2},\ \sigma_{33} = -q. \tag{17.1.19}$$

Substituting Eqs. (17.1.11) and (17.1.13) into Eq. (17.1.14), and integrating using the boundary condition (17.1.17), we get:

$$\begin{aligned}\sigma_{31} = \sigma_{13} &= -\int\left(\frac{\partial\sigma_{11}}{\partial x_1} + \frac{\partial\sigma_{21}}{\partial x_2}\right) dx_3 \\ &= -\frac{E}{2(1-\nu^2)}\left(\frac{h^2}{4} - x_3^2\right)\left[\frac{\partial}{\partial x_1}(\nabla^2 u_3)\right].\end{aligned} \tag{17.1.20}$$

Substituting Eqs. (17.1.12) and (17.1.13) into Eq. (17.1.15), and integrating using the boundary condition (17.1.17), we get:

$$\begin{aligned}\sigma_{32} = \sigma_{23} &= -\int\left(\frac{\partial\sigma_{12}}{\partial x_1} + \frac{\partial\sigma_{22}}{\partial x_2}\right) dx_3 \\ &= -\frac{E}{2(1-\nu^2)}\left(\frac{h^2}{4} - x_3^2\right)\left[\frac{\partial}{\partial x_2}(\nabla^2 u_3)\right].\end{aligned} \tag{17.1.21}$$

From Eqs. (17.1.20) and (17.1.21), the variation of σ_{13} and σ_{23} with x_3 is seen to be parabolic. Substituting Eqs. (17.1.20) and (17.1.21) into Eq. (17.1.16), and integrating using the boundary condition (17.1.18), we get:

$$\begin{aligned}\sigma_{33} &= -\int\left(\frac{\partial\sigma_{13}}{\partial x_1} + \frac{\partial\sigma_{23}}{\partial x_2}\right) dx_3 \\ &= -\frac{E}{2(1-\nu^2)}\left(\frac{h^3}{12} - \frac{h^2 x_3}{4} + \frac{x_3^3}{3}\right)\nabla^4 u_3.\end{aligned} \tag{17.1.22}$$

This is the equation of a cubic parabola (Fig. 17.4).

At the upper surface of the plate, the boundary condition (17.1.19) gives:

$$q = \frac{Eh^3}{12(1-\nu^2)}\nabla^4 u_3. \tag{17.1.23}$$

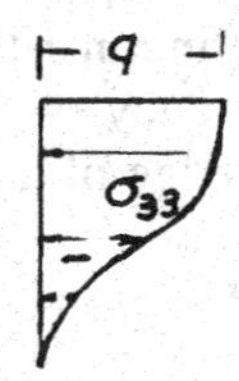

Fig. 17.4

Eq. (17.1.23) is a condition of equilibrium of the plate. Its counterpart in the theory of beams is $q = EI_3 d^4 u_3 / dx_1^4$. The quantity

$$\frac{Eh^3}{12(1 - \nu^2)} = D \qquad (17.1.24)$$

is defined as the flexural rigidity of the plate. Eq. (17.1.23) is called *Lagrange's equation*; in expanded form, it is written:

$$q = D\left(\frac{\partial^4 u_3}{\partial x_1^4} + 2\frac{\partial^4 u_3}{\partial x_1^2 \partial x_2^2} + \frac{\partial^4 u_3}{\partial x_2^4}\right) \qquad (17.1.25)$$

Lagrange's equation relates the vertical deformation u_3 to the applied load q. It will be derived in a different way in Sec. 17.4.

Looking back at the previous assumptions and derivations, we see that the stresses can be grouped into three classes: the stresses parallel to the middle plane of the plate σ_{11}, σ_{22}, and σ_{12}; the transverse normal stress σ_{33}; and the transverse shear stresses σ_{13} and σ_{23}. σ_{33} is of the order of magnitude of q, which rarely reaches values higher than 50 psi. It usually varies between 1 and 10 psi. This is negligible compared to the hundreds of psi reached by σ_{11} and σ_{22}. The total transverse load (Fig. 17.1) on the plate is of the order of qL^2. For equilibrium, this load must be balanced by transverse shearing forces of the order of $\sigma_{13} Lh$ or $\sigma_{23} Lh$. Therefore, σ_{13} and σ_{23} are of the order of $q(L/h)$. If we consider the bending of a strip of the plate of unit width, the bending moment is of the order of qL^2 and the resisting moment is of the order of $\sigma_{11} h^2$ or $\sigma_{22} h^2$ (see Sec. 17.3). Therefore, σ_{11}, σ_{22} (and, it is assumed, σ_{12} also), are of the order of $q(L/h)^2$. Thus, since L / h is relatively large for thin plates, then σ_{11}, σ_{22}, .and σ_{12} are greater than σ_{13} and σ_{23}, and much greater than σ_{33}. Since σ_{13}, σ_{23}, and σ_{33} are relatively small, our neglecting of their effects on the displacement u_3 was quite justified. This, however, will lead to some inconsistencies in the development of the theory. One such inconsistency will appear when writing the boundary conditions at the free edge of a plate in Sec. 17.5.

17.2 Geometry of Surfaces with Small Curvatures

Due to bending, the middle plane of a flat plate becomes slightly curved. The resulting surface is described by the equation (Fig. 17.5):

$$u_3 = u_3(x_1, x_2). \tag{17.2.1}$$

Let us first consider the slopes: At a point P of the surface, and when proceeding in the OX_1 direction, the slope is given by $\partial u_3/\partial x_1$; when proceeding in the OX_2 direction, the slope is given by $\partial u_3/\partial x_2$. Recalling the definition of the gradient of a function, we see that the two quantities $\partial u_3/\partial x_1$ and $\partial u_3/\partial x_2$ are the components of the vector

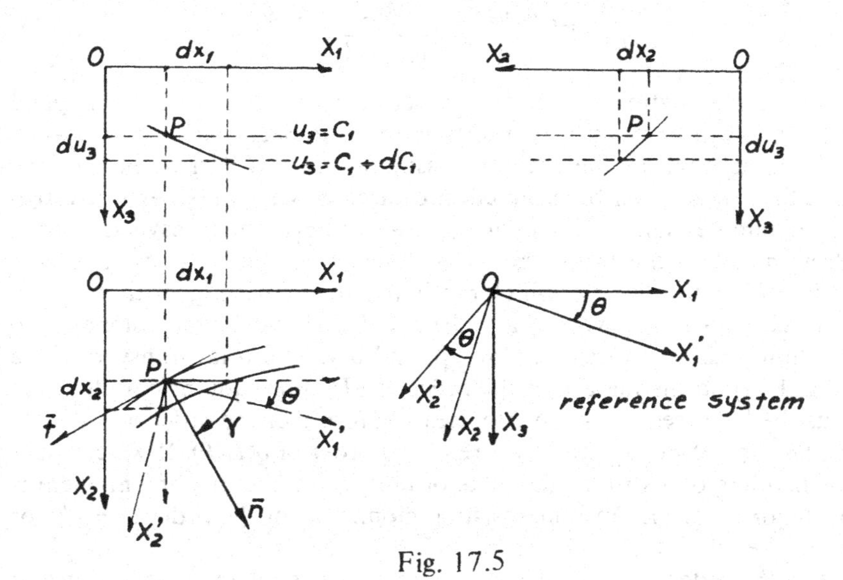

Fig. 17.5

$$\bar{\nabla} u_3 = \text{grad } u_3 = \bar{i}_1 \frac{\partial u_3}{\partial x_1} + \bar{i}_2 \frac{\partial u_3}{\partial x_2}, \tag{17.2.2}$$

whose magnitude is

$$\frac{du_3}{dn} = \sqrt{\left(\frac{\partial u_3}{\partial x_1}\right)^2 + \left(\frac{\partial u_3}{\partial x_2}\right)^2}, \tag{17.2.3}$$

and whose direction is along the normal $\bar{n}$ to the contour lines (Fig. 17.5). The angle made by $\bar{n}$ with the OX_1 axis is given by:

$$\tan \gamma = \frac{\partial u_3}{\partial x_2} / \frac{\partial u_3}{\partial x_1}. \tag{17.2.4}$$

The maximum slope of the surface at a point $P(x_1, x_2)$ is given by the magnitude of the gradient vector at that point; the direction along which this maximum slope occurs is that of $\bar{n}$ and is given by Eq. (17.2.4). The minimum slope of the surface at a point $P(x_1, x_2)$ is equal to zero, and it occurs in a direction $\bar{t}$ normal to $\bar{n}$ since it has to be tangent to the contour line. The gradient of u_3 being a vector quantity, it can be represented in any system of axes OX'_1, OX'_2 by means of the equation (see Chapter 3):

$$\frac{\partial u_3}{\partial x'_i} = \ell_{ij} \frac{\partial u_3}{\partial x_j} \qquad i, j = 1, 2, \tag{17.2.5}$$

where ℓ_{ij} are the direction cosines of the new system with respect to the old one. In terms of the angle of rotation θ (Fig. 17.5), Eq. (17.2.5) becomes:

$$\begin{bmatrix} \dfrac{\partial u_3}{\partial x'_1} \\ \dfrac{\partial u_3}{\partial x'_2} \end{bmatrix} = \begin{bmatrix} \cos \theta & \sin \theta \\ -\sin \theta & \cos \theta \end{bmatrix} \begin{bmatrix} \dfrac{\partial u_3}{\partial x_1} \\ \dfrac{\partial u_3}{\partial x_2} \end{bmatrix}. \tag{17.2.6}$$

Eq. (17.2.6) gives the slopes when proceeding in the directions of OX'_1 and OX'_2, respectively.

Let us now consider the curvatures: In the system of axes of Fig. 17.1, the curvature of the surface at a point P and in a plane parallel to the OX_1, OX_3 plane is approximately given by [see Eq. (12.2.1)]:

$$C_{11} = \frac{\partial}{\partial x_1}\left(\frac{\partial u_3}{\partial x_1}\right) = \frac{\partial^2 u_3}{\partial x_1^2}. \tag{17.2.7}$$

In a plane parallel to the OX_2, OX_3 plane, it is given by:

$$C_{22} = \frac{\partial}{\partial x_2}\left(\frac{\partial u_3}{\partial x_2}\right) = \frac{\partial^2 u_3}{\partial x_2^2}. \tag{17.2.8}$$

C_{11} and C_{22} represent the rate at which the slope of $u_3 = u_3(x_1, x_2)$ changes when proceeding in the OX_1 and OX_2 directions, respectively. In addition to the two second derivatives of Eqs. (17.2.7) and (17.2.8), the following two mixed second derivatives will be needed:

$$C_{12} = \frac{\partial}{\partial x_1}\left(\frac{\partial u_3}{\partial x_2}\right) = \frac{\partial^2 u_3}{\partial x_1\,\partial x_2} \tag{17.2.9}$$

and

$$C_{21} = \frac{\partial}{\partial x_2}\left(\frac{\partial u_3}{\partial x_1}\right) = \frac{\partial^2 u_3}{\partial x_2\,\partial x_1}. \tag{17.2.10}$$

C_{12} represents the rate at which the slope $\partial u_3/\partial x_2$ changes when moving in the OX_1 direction, and C_{21} represents the rate at which the slope $\partial u_3/\partial x_1$ changes when moving in the OX_2 direction. The geometrical interpretation of C_{12} and C_{21} can be obtained as follows: Consider the square element *abcd* whose sides are chosen equal to unity. Under the effect of local twisting couples, it will take the shape $a'b'c'd'$, shown in Fig. 17.6. On this figure, we notice that:

$$u_{3b} - u_{3a} = \frac{\partial u_3}{\partial x_1} \tag{17.2.11}$$

$$u_{3c} - u_{3d} = \frac{\partial u_3}{\partial x_1} + \frac{\partial}{\partial x_2}\left(\frac{\partial u_3}{\partial x_1}\right) = \frac{\partial u_3}{\partial x_1} + \frac{\partial^2 u_3}{\partial x_2\,\partial x_1} \tag{17.2.12}$$

$$u_{3d} - u_{3a} = \frac{\partial u_3}{\partial x_2} \tag{17.2.13}$$

$$u_{3c} - u_{3b} = \frac{\partial u_3}{\partial x_2} + \frac{\partial}{\partial x_1}\left(\frac{\partial u_3}{\partial x_2}\right) = \frac{\partial u_3}{\partial x_2} + \frac{\partial^2 u_3}{\partial x_1\,\partial x_2}. \tag{17.2.14}$$

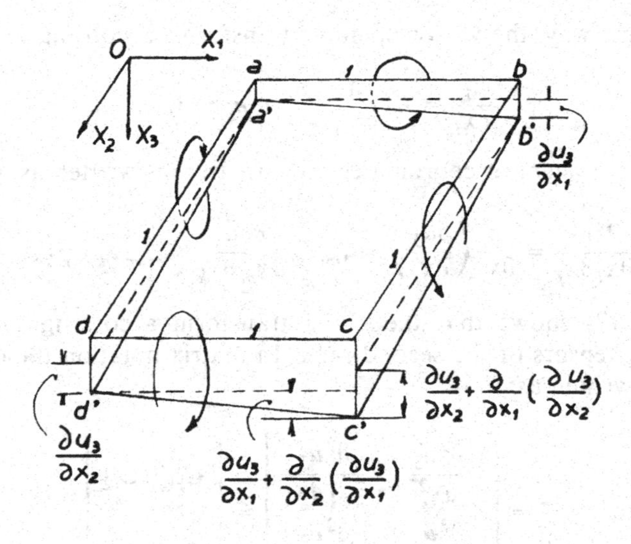

Fig. 17.6

Subtracting Eq. (17.2.11) from Eq. (17.2.12), and Eq. (17.2.13) from Eq. (17.2.14), we obtain the two following equations:

$$(u_{3c} + u_{3a}) - (u_{3b} + u_{3d}) = \frac{\partial^2 u_3}{\partial x_2\, \partial x_1} = C_{21} \qquad (17.2.15)$$

$$(u_{3c} + u_{3a}) - (u_{3b} + u_{3d}) = \frac{\partial^2 u_3}{\partial x_1\, \partial x_2} = C_{12}. \qquad (17.2.16)$$

Eqs. (17.2.15) and (12.7.16) show that $C_{12} = C_{21}$. $\partial^2 u_3/\partial x_1\, \partial x_2$ is called the twist. Therefore, at a point P, and with respect to a system of axes OX_1, OX_2, OX_3, we have defined three quantities—namely, two curvatures and one twist. We can prove that C_{11}, C_{22}, C_{12}, and C_{21} are the components of a tensor of the second rank called the curvature tensor. Indeed, let us consider the vector operator $\overline{\nabla}(\bar{i}_1\, \partial/\partial x_1, \bar{i}_2\, \partial/\partial x_2)$ and the vector gradient $\overline{\nabla} u_3(\bar{i}_1\, \partial u_3/\partial x_1, \bar{i}_2\, \partial u_3/\partial x_2)$. In a rotation of coordinates around the OX_3 axis, the vector gradient transforms according to Eq. (17.2.5); namely, according to

$$\frac{\partial u_3}{\partial x'_j} = \ell_{jk} \frac{\partial u_3}{\partial x_k} \qquad j, k = 1, 2.$$

In the same way, the vector operator transforms according to

$$\frac{\partial}{\partial x'_i} = \ell_{im}\frac{\partial}{\partial x_m} \qquad i, m = 1, 2.$$

In this new system of coordinates the curvature is written as:

$$C'_{ij} = \frac{\partial^2 u_3}{\partial x'_i \partial x'_j} = \frac{\partial}{\partial x'_i}\left(\frac{\partial u_3}{\partial x'_j}\right) = \ell_{im}\ell_{jk}\frac{\partial^2 u_3}{\partial x_m \partial x_k} = \ell_{im}\ell_{jk}C_{mk}. \qquad (17.2.17)$$

Eq. (17.2.17) shows that the C_{ij} ' s transform according to the law governing tensors of the second rank. In matrix notation the curvature tensor is written as:

$$C = \begin{bmatrix} \dfrac{\partial^2 u_3}{\partial x_1^2} & \dfrac{\partial^2 u_3}{\partial x_1 \partial x_2} \\ \dfrac{\partial^2 u_3}{\partial x_2 \partial x_1} & \dfrac{\partial^2 u_3}{\partial x_2^2} \end{bmatrix} = \begin{bmatrix} C_{11} & C_{12} \\ C_{21} & C_{22} \end{bmatrix}. \qquad (17.2.18)$$

The curvature tensor is symmetric and, like the strain and stress tensors, it can be diagonalized; in other words, a system of reference axes can be found in which the off-diagonal terms C_{12} and C_{21} disappear. It is also susceptible to a representation by means of Mohr's circle and enjoys the other properties presented in Chapter 3 on linear symmetric transformations.

For a rotation θ around the OX_3 axis (Fig. 17.5), Eq. (17.2.17) becomes:

$$C'_{11} = C_{11}\cos^2\theta + C_{22}\sin^2\theta + 2C_{12}\sin\theta\cos\theta \qquad (17.2.19)$$

$$C'_{22} = C_{11}\sin^2\theta + C_{22}\cos^2\theta - 2C_{12}\sin\theta\cos\theta \qquad (17.2.20)$$

$$C'_{12} = C'_{21} = -(C_{11} - C_{22})\sin\theta\cos\theta + C_{12}\cos 2\theta. \qquad (17.2.21)$$

The previous equations are similar to those written for two dimensional states of stress and strain. The principal curvatures are given by:

$$C_1 = \frac{C_{11} + C_{22}}{2} + \sqrt{\left(\frac{C_{11} - C_{22}}{2}\right)^2 + C_{12}^2} \qquad (17.2.22)$$

$$C_2 = \frac{C_{11} + C_{22}}{2} - \sqrt{\left(\frac{C_{11} - C_{22}}{2}\right)^2 + C_{12}^2}, \qquad (17.2.23)$$

and correspond to the maximum and minimum curvatures. They fall in the principal planes of curvature whose directions are given by an angle $\theta = \phi$, such that

$$\tan 2\phi = \frac{2C_{12}}{C_{11} - C_{22}}. \tag{17.2.24}$$

The invariants of the curvature tensor are:

$$C_{11} + C_{22} = C'_{11} + C'_{22} \tag{17.2.25}$$

$$C_{11}C_{22} - (C_{12})^2 = C'_{11}C'_{22} - (C'_{12})^2. \tag{17.2.26}$$

The representation on a Mohr diagram follows the same conventions established for the stresses in Sec. 7.11. Positive values of C_{11} and C_{22} are plotted on the positive side of the $o'C_n$ axis (Fig. 17.7), and negative values on the negative side. If C_{12} is positive, it is plotted below the $o'C_n$ axis for the curvature corresponding to the more clockwise of the two planes in which C_{ii}($i = 1, 2$, no sum) is computed; if C_{12} is negative, it is plotted above the $o'C_n$ axis. For the less clockwise of the two planes, the location of the point representing C_{12} is reversed. Fig. 17.7 shows the representation for C_{11}, C_{22}, and C_{12} positive. A plane (OX'_1,

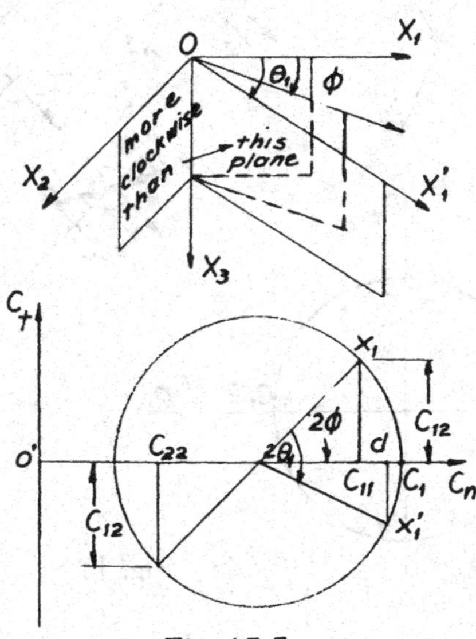

Fig. 17.7

OX_3) making an angle $\theta = \theta_1$ with (OX_1, OX_3) will have in it a positive curvature equal to $o'd$ and a negative twist equal to $x_1' d$. The plane making ϕ with (OX_1, OX_3) is the major principal plane of curvature. In this plane, the twist is equal to zero.

A surface which is convex downward will have values of C_{11} and C_{22} which are negative, while a surface which is convex upward will have values of C_{11} and C_{22} which are positive. Fig. 17.6 shows the direction of a positive twist.

If the two principal curvatures C_1 and C_2 are the same, Mohr's circle shrinks to a point, the curvature is the same in all directions, and there is no twist in any direction: The surface is purely spherical at this point. If the two principal curvatures are equal in magnitude and opposite in sign, the result is a saddle point: There are no curvatures in planes making 45° with the principal planes, just twists (Fig. 17.8).

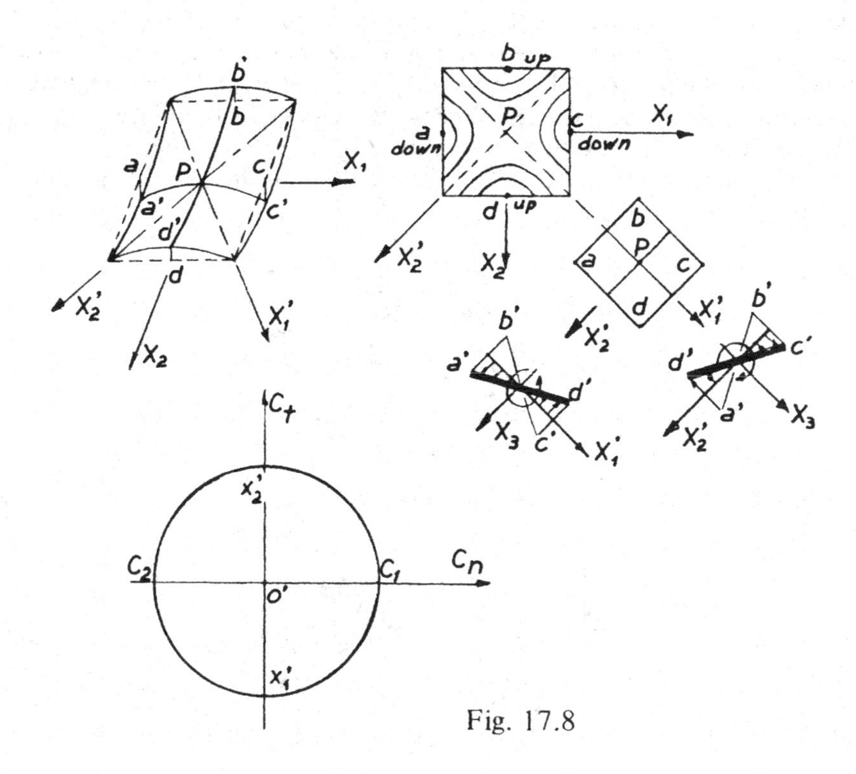

Fig. 17.8

Finally, in terms of the curvatures and twist, the stress-strain relations (17.1.11) to (17.1.13) can be written as follows:

$$\sigma_{ij} = -\frac{E}{1+\nu}x_3 C_{ij} - \delta_{ij}\frac{E\nu}{1-\nu^2}x_3 C_{nn}(i,j,n = 1,2). \quad (17.2.27)$$

In matrix notation, this equation can be written in either of the two forms:

$$\begin{bmatrix} \sigma_{11} & \sigma_{12} \\ \sigma_{21} & \sigma_{22} \end{bmatrix} = -\frac{Ex_3}{1+\nu}\begin{bmatrix} C_{11} & C_{12} \\ C_{21} & C_{22} \end{bmatrix} - \frac{E\nu x_3}{1-\nu^2}(C_{11} + C_{22})\begin{bmatrix} 1 & 0 \\ 0 & 1 \end{bmatrix} \quad (17.2.28)$$

$$\begin{bmatrix} \sigma_{11} \\ \sigma_{22} \\ \sigma_{12} \end{bmatrix} = -\frac{Ex_3}{1-\nu^2}\begin{bmatrix} 1 & \nu & 0 \\ \nu & 1 & 0 \\ 0 & 0 & 1-\nu \end{bmatrix}\begin{bmatrix} C_{11} \\ C_{22} \\ C_{12} \end{bmatrix}$$

$$\begin{bmatrix} \sigma_{11} \\ \sigma_{22} \\ \sigma_{12} \end{bmatrix} = -\frac{Ex_3}{1-\nu^2}\begin{bmatrix} 1 & \nu & 0 \\ \nu & 1 & 0 \\ 0 & 0 & 1-\nu \end{bmatrix}\begin{bmatrix} \dfrac{\partial^2 u_3}{\partial x_1^2} \\ \dfrac{\partial^2 u_3}{\partial x_2^2} \\ \dfrac{\partial^2 u_3}{\partial x_1 \partial x_2} \end{bmatrix}. \quad (17.2.29)$$

17.3 Stress Resultants and Stress Couples

Let us consider an element of a plate under the action of a normal distributed load $q = q(x_1, x_2)$. In addition to bending and twisting moments acting on the sides of the element, there will be shearing forces due to the shearing stresses σ_{13} and σ_{23}. Both moments and shearing forces are expressed per unit length of plate in the OX_1 and the OX_2 directions.

The directions for positive and negative stresses established in Sec. 7.2 always hold and apply to normal and shear forces on the plate's cross sections. *The convention for moments is that a positive bending moment produces a positive normal stress in the positive half of the plate and a positive twisting moment causes a positive shear stress on the positive half of the plate.* Both moments and shear forces carry the subscripts corresponding to the stresses they cause. If one wishes to use the right-hand rule for the positive moments, the thumb has to point in the direction of the double arrows in Fig. 17.9.

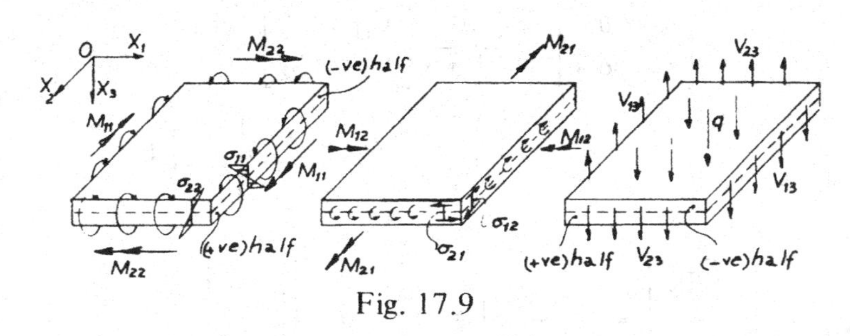

Fig. 17.9

The bending moment per unit length along the OX_1 direction is given by:

$$M_{11} = \int_{-\frac{h}{2}}^{+\frac{h}{2}} \sigma_{11} x_3 \, dx_3 = -\int_{-\frac{h}{2}}^{+\frac{h}{2}} \frac{Ex_3^2}{1-\nu^2}(C_{11} + \nu C_{22})\, dx_3 \tag{17.3.1}$$

or

$$M_{11} = -D(C_{11} + \nu C_{22}). \tag{17.3.2}$$

Similarly, the bending moment per unit length along the OX_2 direction is given by:

$$M_{22} = \int_{-\frac{h}{2}}^{+\frac{h}{2}} \sigma_{22} x_3 \, dx_3 = -D(C_{22} + \nu C_{11}). \tag{17.3.3}$$

The twisting moment per unit length along the OX_1 direction is given by:

$$M_{12} = \int_{-\frac{h}{2}}^{+\frac{h}{2}} \sigma_{12} x_3 \, dx_3 = -D(1-\nu)C_{12}. \tag{17.3.4}$$

Since $\sigma_{12} = \sigma_{21}$,

$$M_{21} = \int_{-\frac{h}{2}}^{+\frac{h}{2}} \sigma_{21} x_3 \, dx_3 = -D(1-\nu)C_{21} = M_{12}. \tag{17.3.5}$$

Notice that the effects of σ_{13} and σ_{23} were neglected in the expressions for M_{12} and M_{21}, respectively. This will result in an inconsistency in the writing of the boundary conditions for a free edge in Sec. 17.5. In index notation, Eqs. (17.3.2) to (17.3.5) are written as follows:

$$M_{ij} = -\frac{Eh^3}{12(1+\nu)}C_{ij} - \delta_{ij}\frac{E\nu h^3}{12(1-\nu^2)}C_{nn} \qquad (1, j, n = 1, 2). \tag{17.3.6}$$

In matrix notation, we have:

$$\begin{bmatrix} M_{11} & M_{12} \\ M_{12} & M_{22} \end{bmatrix} = -\frac{Eh^3}{12(1+\nu)}\begin{bmatrix} C_{11} & C_{12} \\ C_{12} & C_{22} \end{bmatrix} - \frac{E\nu h^3}{12(1-\nu^2)}(C_{11}+C_{22})\begin{bmatrix} 1 & 0 \\ 0 & 1 \end{bmatrix} \tag{17.3.7}$$

or

$$\begin{bmatrix} M_{11} \\ M_{22} \\ M_{12} \end{bmatrix} = -D\begin{bmatrix} 1 & \nu & 0 \\ \nu & 1 & 0 \\ 0 & 0 & 1-\nu \end{bmatrix}\begin{bmatrix} C_{11} \\ C_{22} \\ C_{12} \end{bmatrix}$$

$$\begin{bmatrix} M_{11} \\ M_{22} \\ M_{12} \end{bmatrix} = -D\begin{bmatrix} 1 & \nu & 0 \\ \nu & 1 & 0 \\ 0 & 0 & 1-\nu \end{bmatrix}\begin{bmatrix} \dfrac{\partial^2 u_3}{\partial x_1^2} \\ \dfrac{\partial^2 u_3}{\partial x_2^2} \\ \dfrac{\partial^2 u_3}{\partial x_1 \, \partial x_2} \end{bmatrix}. \tag{17.3.8}$$

The previous equations show that the four quantities $M_{ij}(i, j = 1, 2)$ are the components of a symmetric tensor of the second rank whose properties are similar to those of the curvature tensor. The tensor is called the moment tensor M. If Eq. (17.3.8) is inverted, we get:

$$\begin{bmatrix} C_{11} \\ C_{22} \\ C_{12} \end{bmatrix} = -\frac{1}{D(1-\nu^2)} \begin{bmatrix} 1 & -\nu & 0 \\ -\nu & 1 & 0 \\ 0 & 0 & 1+\nu \end{bmatrix} \begin{bmatrix} M_{11} \\ M_{22} \\ M_{12} \end{bmatrix}. \qquad (17.3.9)$$

Eqs. (17.3.8) and (17.3.9) represent the stress-strain relations for thin plates in terms of moments and curvatures.

All the equations written for the curvature tensor apply to the moment tensor. Thus Eqs. (17.2.17) to (17.2.26) can be rewritten here with M replacing C. The representation on Mohr's diagram follows the conventions established for stresses. Positive values of M_{11} and M_{22} (as shown in Fig. 17.9) are plotted on the positive side of the $o'M_n$ axis, and negative values on the negative side. If M_{12} is positive, it is plotted below the $o'M_n$ axis for the plane whose outward normal is parallel to the more clockwise of the two axes (here OX_2); if M_{12} is negative, it is plotted above the $o'M_n$ axis. Fig. 17.10a shows the representation on Mohr's diagram for the positive values of M_{11}, M_{22}, and M_{12} in Fig. 17.9. A plane whose outward normal makes an angle $\theta = \theta_1$ with the OX_1 axis (Fig. 17.10b), will have a positive bending moment M'_{11} acting on it equal to $o'd$ and a negative twisting moment M'_{12} equal to $x'_1 d$. The plane whose normal makes ϕ with OX_1 is subjected to the major principal bending moment and does not suffer any twisting moment.

The quantities in Fig. 17.10 are related to the quantities in Fig. 17.7

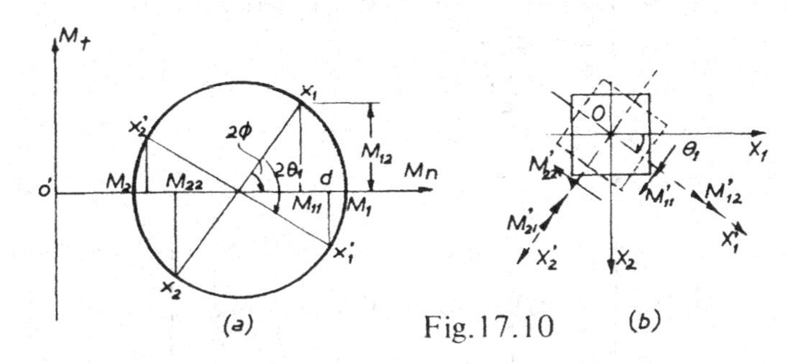

Fig. 17.10

by means of the coefficients in Eq. (17.3.8). The negative sign in front of the matrix of coefficients in Eq. (17.3.8) must be kept in mind when trying to visualize moments and curvatures.

The shearing forces per unit length on planes normal to the OX_1 and OX_2 directions are given by:

$$V_{13} = \int_{-\frac{h}{2}}^{+\frac{h}{2}} \sigma_{13}\, dx_3, \quad V_{23} = \int_{-\frac{h}{2}}^{+\frac{h}{2}} \sigma_{23}\, dx_3. \tag{17.3.10}$$

Substituting Eq. (17.1.20) into Eq. (17.3.10) and integrating, we obtain the following expression for V_{13}:

$$\begin{aligned} V_{13} &= -\int_{-\frac{h}{2}}^{+\frac{h}{2}} \frac{E}{2(1-\nu^2)}\left(\frac{h^2}{4} - x_3^2\right)\left[\frac{\partial}{\partial x_1}(\nabla^2 u_3)\right] dx_3 \\ &= -D\frac{\partial}{\partial x_1}(\nabla^2 u_3) \\ &= \frac{\partial M_{11}}{\partial x_1} + \frac{\partial M_{12}}{\partial x_2}. \end{aligned} \tag{17.3.11}$$

Similarly,

$$V_{23} = -D\frac{\partial}{\partial x_2}(\nabla^2 u_3) = \frac{\partial M_{12}}{\partial x_1} + \frac{\partial M_{22}}{\partial x_2}. \tag{17.3.12}$$

Eqs. (17.3.11) and (17.3.12) show that V_{13} and V_{23} are the two components of the vector $[-D\, \mathrm{grad}(\nabla^2 u_3)]$. Thus, in matrix notation:

$$\begin{bmatrix} V_{13} \\ \\ V_{23} \end{bmatrix} = -D\begin{bmatrix} \dfrac{\partial}{\partial x_1}(\nabla^2 u_3) \\ \dfrac{\partial}{\partial x_2}(\nabla^2 u_3) \end{bmatrix}. \tag{17.3.13}$$

Eqs. (17.3.11) and (17.3.12) will be derived in a different way in Sec. 17.4. Knowing V_{13} and V_{23} at a point, the vertical shearing forces per unit length on a pair of orthogonal planes whose normals have direction cosines $\ell_{ij}(i,j = 1,2)$ with OX_1 and OX_2 can be obtained from the transformation formula (see Chapter 3):

$$V'_{i3} = \ell_{ij} V_{j3}. \tag{17.3.14}$$

Therefore, for the rotation shown in Fig. 17.10b, we have:

$$\begin{bmatrix} V'_{13} \\ V'_{23} \end{bmatrix} = \begin{bmatrix} \cos\theta_1 & \sin\theta_1 \\ -\sin\theta_1 & \cos\theta_1 \end{bmatrix}\begin{bmatrix} V_{13} \\ V_{23} \end{bmatrix} \tag{17.3.15}$$

Once the moments and the shearing forces are known, the normal and the shearing stresses can be computed: Substituting Eq. (17.3.9) into Eq. (17.2.29), we obtain σ_{11}, σ_{22}, and σ_{12} in terms of the bending and twisting moments:

$$\begin{bmatrix} \sigma_{11} \\ \sigma_{22} \\ \sigma_{12} \end{bmatrix} = \frac{12x_3}{h^3} \begin{bmatrix} M_{11} \\ M_{22} \\ M_{12} \end{bmatrix}. \tag{17.3.16}$$

From Eqs. (17.1.20), (17.1.21), (17.3.11), and (17.3.12), we obtain σ_{13} and σ_{23} in terms of the shearing forces:

$$\begin{bmatrix} \sigma_{13} \\ \sigma_{23} \end{bmatrix} = \frac{3}{2h}\left(1 - 4\frac{x_3^2}{h^2}\right) \begin{bmatrix} V_{13} \\ V_{23} \end{bmatrix}. \tag{17.3.17}$$

The similarities between the expressions for the stresses in plates and in beams are clearly seen in the two previous equations.

17.4 Equations of Equilibrium of Laterally Loaded Thin Plates

To study the equilibrium of a small element of plate, one must consider the variations in the moments and shears along the OX_1 and OX_2 directions. Such variations are shown in Fig. 17.11. Summing the vertical forces leads to

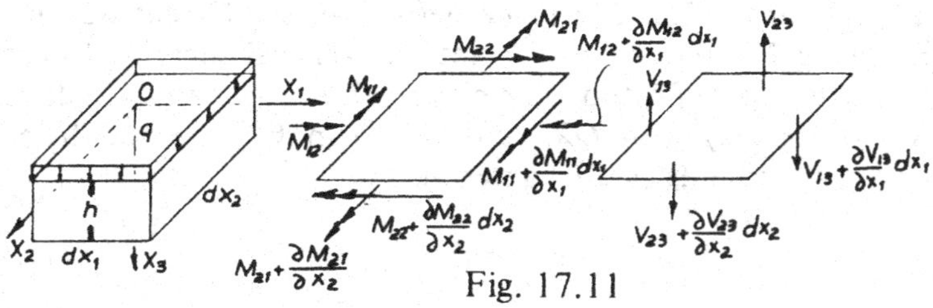

Fig. 17.11

$$\frac{\partial V_{13}}{\partial x_1} + \frac{\partial V_{23}}{\partial x_2} + q = 0. \tag{17.4.1}$$

Taking moments about the OX_2 axis and neglecting higher order differentials leads to

$$\frac{\partial M_{11}}{\partial x_1} + \frac{\partial M_{21}}{\partial x_2} - V_{13} = 0, \tag{17.4.2}$$

which is the same as Eq. (17.3.11). Taking moments about the OX_1 axis and neglecting higher order differentials leads to

$$\frac{\partial M_{12}}{\partial x_1} + \frac{\partial M_{22}}{\partial x_2} - V_{23} = 0, \tag{17.4.3}$$

which is the same as Eq. (17.3.12). If we now substitute the values of V_{13} and V_{23} from Eqs. (17.4.2) and (17.4.3) into Eq. (17.4.1), we get:

$$\frac{\partial^2 M_{11}}{\partial x_1^2} + 2\frac{\partial^2 M_{12}}{\partial x_1\, \partial x_2} + \frac{\partial^2 M_{22}}{\partial x_2^2} = -q. \tag{17.4.4}$$

In terms of the curvatures, this equation becomes:

$$\frac{\partial^2 C_{11}}{\partial x_1^2} + 2\frac{\partial^2 C_{12}}{\partial x_1\, \partial x^2} + \frac{\partial^2 C_{22}}{\partial x_2^2} = \frac{q}{D}. \tag{17.4.5}$$

In terms of the displacement u_3, we obtain the following equation:

$$\frac{\partial^4 u_3}{\partial x_1^4} + 2\frac{\partial^4 u_3}{\partial x_1^2\, \partial x_2^2} + \frac{\partial^4 u_3}{\partial x_2^4} = \nabla^4 u_3 = \frac{q}{D}. \tag{17.4.6}$$

This is Lagrange's equation, which was derived in a different way in Sec. 17.1. It is the basic plate equation and any $u_3(x_1, x_2)$ satisfying it is a solution of a plate problem. One must remember, however, that in Eq. (17.4.6) the effects of V_{13} and V_{23} on u_3 have been neglected since the relations between moments and curvatures did not account for σ_{13} and σ_{23}: It is recalled that this is the result of the third assumption in Sec. 17.1.

17.5 Boundary Conditions

The edges of a plate may be (1) built in, (2) simply supported, that is, free to rotate around the edge but not free to deflect there, or (3) free, that is unsupported. Let us consider each case separately.

1) *Built in or clamped edge* (Fig. 17.12). At the built in edge, we have:

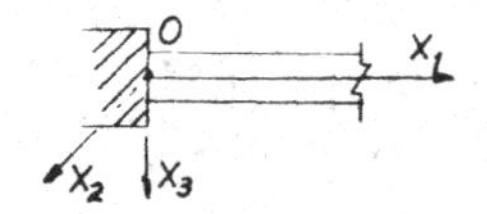

Fig. 17.12

$$(u_3)_{x_1=0} = 0, \quad \left(\frac{\partial u_3}{\partial x_1}\right)_{x_1=0} = 0, \tag{17.5.1}$$

where OX_1 is the normal to the clamped edge.

2) *Simply supported edge* (Fig. 17.13). At a simply supported edge, we have:

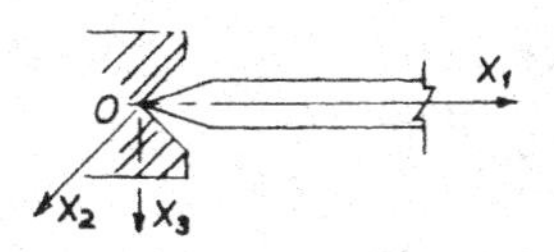

Fig. 17.13

$$(u_3)_{x_1=0} = 0, \quad (M_{11})_{x_1=0} = -D(C_{11} + \nu C_{22}) = 0, \tag{17.5.2}$$

where OX_1 is the normal to the simply supported edge. However, since at $x_1 = 0$, $u_3 = 0$, and $\partial u_3/\partial x_2 = \partial^2 u_3/\partial x_2^2 = 0$, the boundary conditions for this case become:

$$(u_3)_{x_1=0} = 0, \quad \left(\frac{\partial^2 u_3}{\partial x_1^2}\right)_{x_1=0} = 0. \tag{17.5.3}$$

3. *Free edge* (*Fig. 17.14*). At a free edge, there must be no bending or twisting moments as well as no shearing forces. One, therefore, could write:

$$(M_{11})_{x_1=a} = 0, \quad (V_{13})_{x_1=a} = 0, \quad (M_{12})_{x_1=a} = 0. \tag{17.5.4}$$

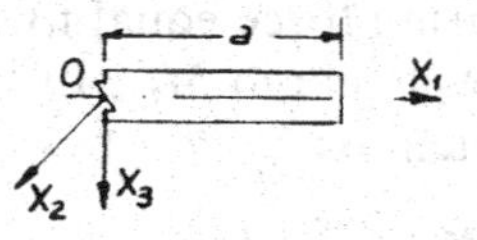

Fig. 17.14

It was shown, however, by Kirchhoff that two conditions are sufficient for the complete determination of u_3 satisfying Eq. (17.4.6). This inconsistency is due to the assumption that $e_{13} = e_{23} = 0$. This assumption, which resulted in neglecting the effects of σ_{13} and σ_{23} on the deflection u_3, was made so that all the strains, and consequently the stresses, could be expressed in terms of one dependent variable u_3. The twisting moment M_{12} at $x_1 = a$ cannot be specified independently of V_{13}. That one cannot specify three conditions along the boundary can be seen by examining the solution of the biharmonic equation [1]:

$$\nabla^4 u_3 = 0. \tag{17.5.5}$$

This solution involves the use of complex variable theory and we are only interested here in the final answer: If

$$z = x_1 + ix_2 \text{ and } \bar{z} = x_1 - ix_2, \tag{17.5.6}$$

where $i = \sqrt{-1}$, the solution of Eq. (17.5.5) is:

$$u_3 = \text{Real part of } [\bar{z}F_1(z) + F_2(z)]. \tag{17.5.7}$$

The functions F_1 and F_2 are analytic functions and are independent of each other. By means of F_1 and F_2, we can satisfy two and only two independent conditions at the boundary of a plate. Using a variational method, Kirchhoff [2] showed that the boundary conditions for the free edge are:

$$(M_{11})_{x_1=a} = 0, \quad R_{13} = \left(V_{13} + \frac{\partial M_{12}}{\partial x_2}\right)_{x_1=a} = 0, \tag{17.5.8}$$

where R_{13} is the vertical reaction at the edge. In effect, the second equation (17.5.8) states that a distribution of twisting moments M_{12} along an edge is equivalent to a distribution of vertical shearing forces. This equivalence is illustrated in Fig. (17.15): The twisting moment corresponding to a distribution of shearing stresses σ_{12} in (a) is replaced by a statically equivalent system in (b). This statically equivalent system results in a distributed shearing force equal to $\partial M_{12}/\partial x_2$ per unit length (c and d) and two end forces equal to M_{12}. In terms of u_3, the boundary conditions (17.5.8) are written as:

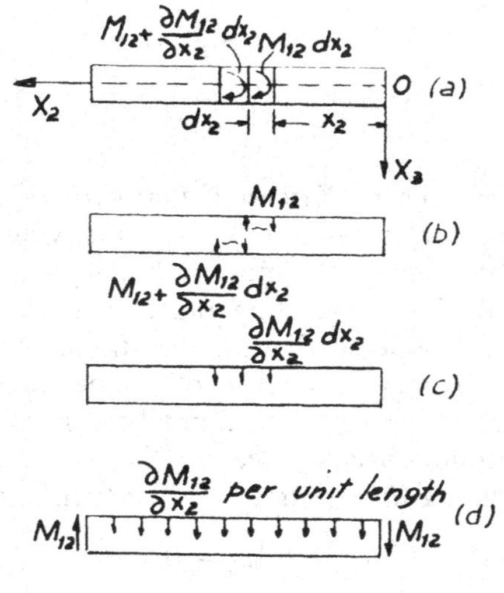

Fig. 17.15

$$\left[\frac{\partial^2 u_3}{\partial x_1^2} + \nu \frac{\partial^2 u_3}{\partial x_2^2}\right]_{x_1=a} = 0, \quad \left[\frac{\partial^3 u_3}{\partial x_1^3} + (2-\nu)\frac{\partial^3 u_3}{\partial x_1\,\partial x_2^2}\right]_{x_1=a} = 0. \tag{17.5.9}$$

Remark

In the case of plates with curvilinear boundaries, the expressions of the slopes, curvatures, moments, and shearing forces must first be obtained in a system of axes formed by the normal and the tangent to the boundary. The formulas for such transformations of axes were given in the previous sections.

17.6 Some Simple Solutions of Lagrange's Equation

Using the inverse method, a number of simple problems related to thin plates have been solved. An expression for $u_3 = u_3(x_1, x_2)$, satisfying Lagrange's equation, is examined and the type of boundary conditions to which it applies is obtained. Lagrange's equation being linear, superposition can be used to generate solutions to new problems. Let us consider the following cases:

1. $q = 0$, $u_3 = -Bx_1^2$, where B is a constant. This expression for u_3 satisfies the biharmonic equation $\nabla^4 u_3 = 0$. It is, therefore, a possible solution to the plate problem. The shape of the plate (Fig. 17.16) is a flat parabolic trough. The curvature tensor is:

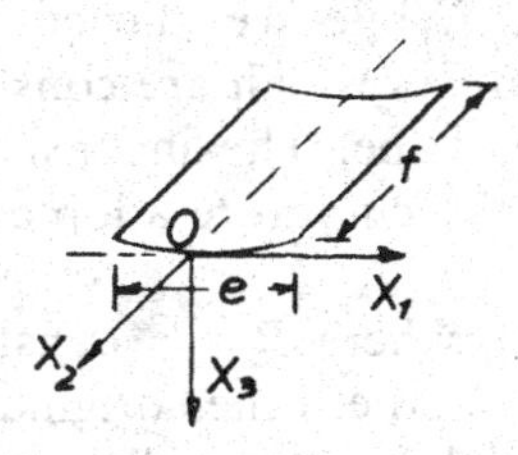

Fig. 17.16

$$C = \begin{bmatrix} -2B & 0 \\ 0 & 0 \end{bmatrix} \tag{17.6.1}$$

and is the same everywhere. The moment tensor is:

$$M = \begin{bmatrix} 2BD & 0 \\ 0 & 2BD\nu \end{bmatrix}. \tag{17.6.2}$$

The shearing forces V_{13} and V_{23} are both equal to zero. Therefore, if we cut from an infinite plate a rectangle of sides e and f (Fig. 17.6), the outside loading would consist of a uniformly distributed bending moment $2BD$ along the edge f and of a uniformly distributed bending moment $2BD\nu$ along the edge e. In spite of the latter, the edge f remains straight: Indeed $2BD\nu$ is the moment necessary to compensate for the Poisson effect, which would have caused an anticlastic surface if M_{11} was the only moment acting on the finite plate.

Similar results are obtained if we set $u_3 = -Bx_2^2$.

2. $q = 0$, $u_3 = -B(x_1^2 + x_2^2) = -Br^2$, where B is a constant. The shape of the deformed plate is that of a paraboloid of revolution. u_3

satisfies the biharmonic equation $\nabla^4 u = 0$. The curvature tensor is:

$$C = \begin{bmatrix} -2B & 0 \\ 0 & -2B \end{bmatrix}. \tag{17.6.3}$$

The moment tensor is:

$$M = \begin{bmatrix} 2BD(1+\nu) & 0 \\ 0 & 2BD(1+\nu) \end{bmatrix}. \tag{17.6.4}$$

The shearing forces V_{13} and V_{23} are both equal to zero. We have here a case of spherical bending. While we started with a paraboloid of revolution, we obtain curvatures which are constant at each point; this corresponds to a spherical shape. The inconsistency comes from the approximate expression of the curvature adopted in the theory of thin plates.

3. $q = 0$, $u_3 = B(x_1^2 - x_2^2)$, where B is a constant. The shape of the deformed plate is that of a saddle. Lines originally parallel to the OX_1 axis curve downwards, and lines originally parallel to the OX_2 axis curve upwards (Fig. 17.8). The curvature tensor is the same everywhere and its components are:

$$C = \begin{bmatrix} 2B & 0 \\ 0 & -2B \end{bmatrix}. \tag{17.6.5}$$

The representation of C on Mohr's diagram is shown in Fig. 17.8. The moment tensor is:

$$M = \begin{bmatrix} -2BD(1-\nu) & 0 \\ 0 & 2BD(1-\nu) \end{bmatrix}. \tag{17.6.6}$$

Lines making 45° with OX_1 and OX_2 remain straight after bending; they are subjected only to twists. Therefore, if as shown in Fig. 17.8, a plate *abcd* is cut from the original plate, its sides will be subjected to twisting moments only.

4. $q = 0$, $u_3 = Bx_1x_2$, where B is a constant. Within a rotation of coordinate axes, this case is equivalent to the previous one. Indeed, the curvature tensor is the same everywhere and its components are:

$$C = \begin{bmatrix} 0 & B \\ B & 0 \end{bmatrix}. \tag{17.6.7}$$

The moment tensor is given by:

$$M = \begin{bmatrix} 0 & -BD \\ -BD & 0 \end{bmatrix}. \tag{17.6.8}$$

After deformation, the plate takes the shape shown in Fig. 17.17a. The

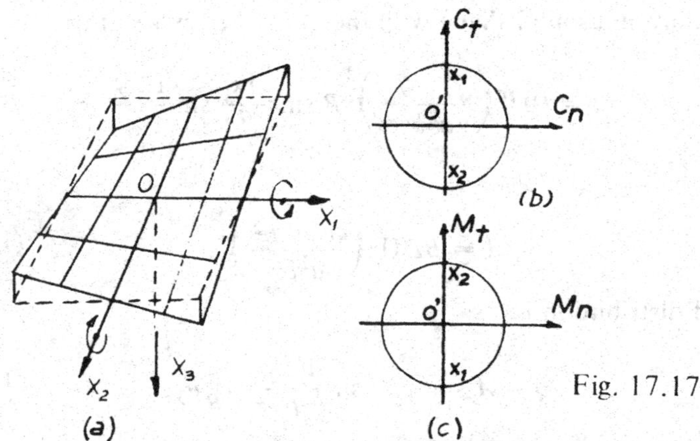

Fig. 17.17

representation on Mohr's diagram for curvatures and moments is shown in Fig. 17.17b and 17.17c, respectively. Because of the sign conventions adopted in the study of flat plates, we see that corresponding points on the two Mohr diagrams fall on opposite sides of the horizontal. From these two diagrams, we conclude that lines making 45° with OX_1 and OX_2 are subjected to bending moments alone. Therefore, if a plate whose sides make 45° with the axes is cut from the original one, it will be subjected to pure bending on its boundaries.

5. $q = ?$, $u_3 = B \sin(\Pi x_1/a) \sin(\Pi x_2/b)$, where B, a, and b are constants. The plate is infinite in extent and is bent in a double sinusoidal shape (Fig. 17.18).

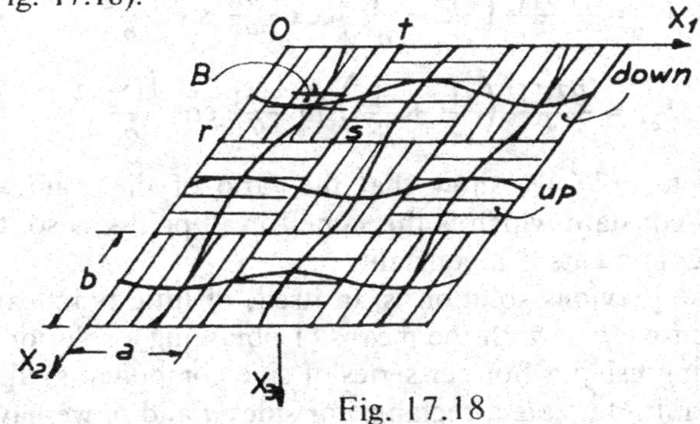

Fig. 17.18

Substituting the expression of u_3 into Lagrange's equation, we get:

$$\nabla^4 u_3 = \Pi^4\left(\frac{a^2+b^2}{a^2b^2}\right)^2 u_3 = \frac{q}{D}. \tag{17.6.9}$$

The doubly sinusoidal shape will, therefore, be obtained if:

$$q = D\Pi^4\left(\frac{a^2+b^2}{a^2b^2}\right)^2 B \sin\frac{\Pi x_1}{a}\sin\frac{\Pi x_2}{b}. \tag{17.6.10}$$

Setting:

$$A = BD\Pi^4\left(\frac{a^2+b^2}{a^2b^2}\right)^2, \tag{17.6.11}$$

the load distribution is

$$q = A\sin\frac{\Pi x_1}{a}\sin\frac{\Pi x_2}{b} = \frac{A}{B}u_3. \tag{17.6.12}$$

From Eqs. (17.3.8), the moments are given by:

$$M_{11} = \Pi^2 D\left(\frac{1}{a^2}+\frac{\nu}{b^2}\right)u_3 \tag{17.6.13}$$

$$M_{22} = \Pi^2 D\left(\frac{1}{b^2}+\frac{\nu}{a^2}\right)u_3 \tag{17.6.14}$$

$$M_{12} = -\frac{\Pi^2}{ab}BD(1-\nu)\cos\frac{\Pi x_1}{a}\cos\frac{\Pi x_2}{b}. \tag{17.6.15}$$

The shearing forces are given by:

$$V_{13} = \frac{BD\Pi^3}{a}\left(\frac{1}{a^2}+\frac{1}{b^2}\right)\cos\frac{\Pi x_1}{a}\sin\frac{\Pi x_2}{b} \tag{17.6.16}$$

$$V_{23} = \frac{BD\Pi^3}{b}\left(\frac{1}{a^2}+\frac{1}{b^2}\right)\sin\frac{\Pi x_1}{a}\cos\frac{\Pi x_2}{b}. \tag{17.6.17}$$

Eqs. (17.6.12) to (17.6.17) show that the ratio of the loading to the deflection is a constant which is the same for all points; also, the ratio of the bending moments is a constant.

Although the previous solution is, in itself, of little practical importance, it does provide us with the means of obtaining a solution for any alternate loading using a Fourier series of sine components; also, if we cut out of the infinite plate a rectangle of sides a and b, we have:

$$u_3 = 0 \text{ and } M_{11} = 0 \qquad \text{for } x_1 = 0 \text{ and } x_1 = a \quad (17.6.18)$$

$$u_3 = 0 \text{ and } M_{22} = 0 \qquad \text{for } x_2 = 0 \text{ and } x_2 = b. \quad (17.6.19)$$

These are the boundary conditions for a simply supported rectangular plate of sides a and b loaded according to Eq. (17.6.12). The previous solution is therefore valid for a simply supported rectangular plate, if we can provide along the sides the twisting moments and shearing forces given by Eqs. (17.6.15) to (17.6.17) (Fig. 17.19). Now, the twisting

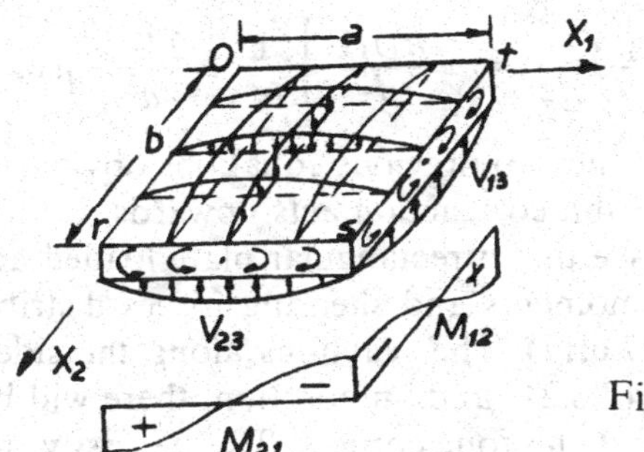

Fig. 17.19

moments acting on the sides of the rectangle parallel to the OX_1 axis are statically equivalent to a continuously distributed shear loading equal to $\partial M_{21}/\partial x_1$ and to concentrated forces acting at the corners, each equal in magnitude to M_{21} at these points (see Sec. 17.5). In the same way, the twisting moments acting on the sides parallel to the OX_2 axis are equivalent to a continuously distributed shear loading equal to

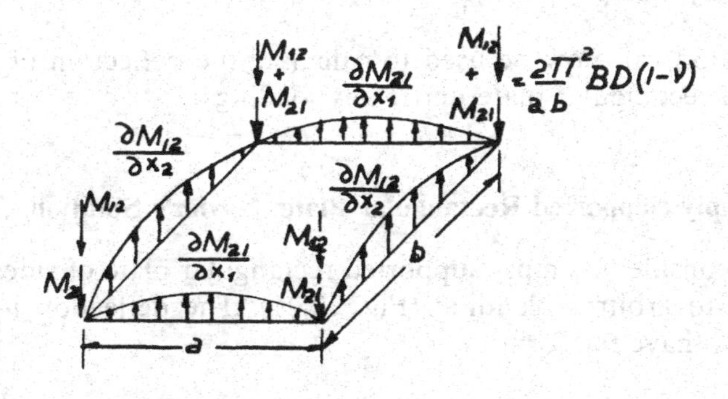

Fig. 17.20

$\partial M_{12}/\partial x_2$ and to concentrated forces acting at the corners, each equal in magnitude to M_{12} at these points (Fig. 17.20). At the four corners of the plate, the magnitude of the concentrated force is $2(\Pi^2/ab)BD(1 - \nu)$. The reaction along side b is given by:

$$R_{13} = \left(V_{13} + \frac{\partial M_{12}}{\partial x_2}\right)_{x_1=a} = -\frac{BD\Pi^3}{a}\left[\frac{1}{a^2} + \frac{2-\nu}{b^2}\right]\sin\frac{\Pi x_2}{b}. \quad (17.6.20)$$

The reaction along side a is given by:

$$R_{23} = \left(V_{23} + \frac{\partial M_{21}}{\partial x_1}\right)_{x_2=b} = -\frac{BD\Pi^3}{b}\left[\frac{1}{b^2} + \frac{2-\nu}{a^2}\right]\sin\frac{\Pi x_1}{a}. \quad (17.6.21)$$

The negative sign in the two previous equations obviously indicates that the sinusoidally distributed reaction acts upwards.

In summary, we see that a rectangular plate loaded according to Eq. (17.6.12) will have moments and shearing forces distribution given by Eqs. (17.6.13) to (17.6.17). The reactions along the sides are given by Eqs. (17.6.20) and (17.6.21) and, in addition, there will be four concentrated forces acting at the four corners. The necessity of these concentrated forces is easy to visualize: When loaded, the plate takes a dish-like shape and the corners tend to rise; they have to be pressed down. It is interesting to note that the sum of the concentrated forces is equal to the sum of the distributed shear loading due to the twisting moments. Indeed:

$$2\int_0^a \left(\frac{\partial M_{21}}{\partial x_1}\right)_{x_2=0} dx_1 + 2\int_0^b \left(\frac{\partial M_{12}}{\partial x_2}\right)_{x_1=0} = \frac{8\Pi^2}{ab} B\,D(1-\nu). \quad (17.6.22)$$

This solution can now be used to calculate the deflection of a simply supported rectangular plate under any loading.

17.7 Simply Supported Rectangular Plate. Navier's Solution

Let us consider a simply supported rectangular plate of sides a and b subjected to arbitrary loading (Fig. 17.21). The deflection u_3 will be assumed to have the form:

$$u_3 = \sum_{m=1}^{\infty}\sum_{n=1}^{\infty} B_{mn}\sin\frac{m\Pi x_1}{a}\sin\frac{n\Pi x_2}{b}. \quad (17.7.1)$$

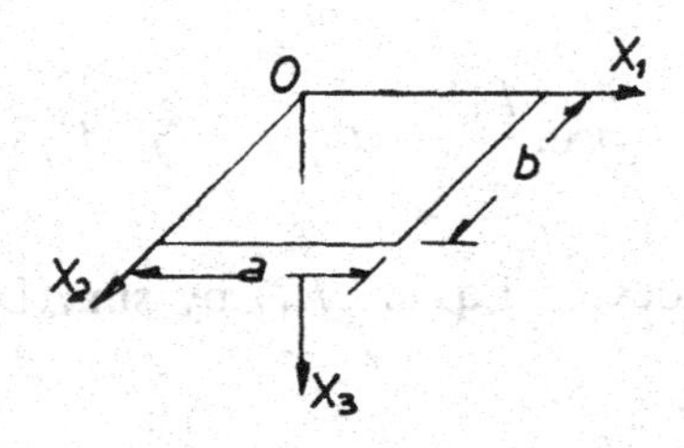

Fig. 17. 21

This assumption satisfies the boundary conditions (17.6.18) and (17.6.19), and is dictated by the results obtained in the previous section. Substituting Eq. (17.7.1) into Lagrange's equation, we get:

$$q = D\Pi^4 \sum_{m=1}^{\infty} \sum_{n=1}^{\infty} B_{mn} \left(\frac{m^2}{a^2} + \frac{n^2}{b^2} \right)^2 \sin \frac{m\Pi x_1}{a} \sin \frac{n\Pi x_2}{b}. \tag{17.7.2}$$

Setting:

$$A_{mn} = D\Pi^4 \left(\frac{m^2}{a^2} + \frac{n^2}{b^2} \right)^2 B_{mn}, \tag{17.7.3}$$

the load distribution is:

$$q = \sum_{m=1}^{\infty} \sum_{n=1}^{\infty} A_{mn} \sin \frac{m\Pi x_1}{a} \sin \frac{n\Pi x_2}{b}. \tag{17.7.4}$$

By means of Eq. (17.7.4), one can approximate any load distribution $q = q(x_1, x_2)$ and easily obtain the corresponding displacements $u_3 = u_3(x_1, x_2)$ from Eqs. (17.7.3) and (17.7.1).

The first step is to find the value of the coefficients A_{mn} of Eq. (17.7.4) as a function of q. For that, we make use of the two following identities:

$$\int_0^a \sin \frac{n\Pi x_1}{a} \sin \frac{k\Pi x_1}{a} \, dx_1 = 0 \text{ when } n \neq k \tag{17.7.5}$$

$$\int_0^a \sin \frac{n\Pi x_1}{a} \sin \frac{k\Pi x_1}{a} \, dx_1 = \frac{a}{2} \text{ when } n = k. \tag{17.7.6}$$

Multiplying both sides of Eq. (17.7.4) by $\sin k\Pi x_1/a$ and integrating from 0 to a, we get:

$$\int_0^a q(x_1, x_2)\sin\frac{k\Pi x_1}{a}\,dx_1 = \frac{a}{2}\sum_{n=1}^{\infty} A_{kn}\sin\frac{n\Pi x_2}{b}. \tag{17.7.7}$$

Multiplying both sides of Eq. (17.7.7) by $\sin j\Pi x_2/b$ and integrating from 0 to b, we get:

$$\frac{4}{ab}\int_0^b\int_0^a q(x_1, x_2)\sin\frac{k\Pi x_1}{a}\sin\frac{j\Pi x_2}{b}\,dx_1\,dx_2 = A_{kj}. \tag{17.7.8}$$

Therefore, given the function $q(x_1, x_2)$, we can find any coefficient A_{kj}. Substituting Eq. (17.7.8) into Eq. (17.7.3) and using the subscripts m and n instead of k and j, we get:

$$B_{mn} = \frac{4}{D\Pi^4\left(\frac{m^2}{a^2}+\frac{n^2}{b^2}\right)^2 ab}$$

$$\int_0^b\int_0^a q(x_1, x_2)\sin\frac{m\Pi x_1}{a}\sin\frac{n\Pi x_2}{b}\,dx_1\,dx_2. \tag{17.7.9}$$

The deflection u_3, corresponding to the loading $q(x_1, x_2)$, can now be obtained using Eqs. (17.7.1) and (17.7.9). Let us now consider two cases:

1. $q(x_1, x_2) = q_0$ *is uniformly distributed over the area of the plate.* In this case, the double integral in Eq. (17.7.9) can be split into two single integrals which are quite elementary to evaluate. B_{mn} is found to be given by

$$B_{mn} = \frac{16q_0}{D\Pi^6 mn\left(\frac{m^2}{a^2}+\frac{n^2}{b^2}\right)^2}, \tag{17.7.10}$$

where m and n take only odd values 1,3,5---∞. The expression of u_3 is, therefore,

$$u_3 = \frac{16q_0}{\Pi^6 D} \sum_{m=1}^{\infty} \sum_{n=1}^{\infty} \frac{\sin \frac{m\Pi x_1}{a} \sin \frac{n\Pi x_2}{b}}{mn\left(\frac{m^2}{a^2} + \frac{n^2}{b^2}\right)^2}, \tag{17.7.11}$$

with m = 1,3,5---∞ and n = 1,3,5---∞. This series converges quite rapidly. The maximum deflection occurs at the center of the plate, and is:

$$(u_3)_{\max} = \frac{16q_0}{\Pi^6 D} \sum_{m=1}^{\infty} \sum_{n=1}^{\infty} \frac{(-1)^{\frac{m+n}{2}-1}}{mn\left(\frac{m^2}{a^2} + \frac{n^2}{b^2}\right)^2}. \tag{17.7.12}$$

Using the expression of the deflection given by Eq. (17.7.11), we can find by differentiation the moments and the shearing forces at any point of the plate.

2. *$q(x_1, x_2)$ is a concentrated force P acting at any point $x_1 = e$, $x_2 = f$* (Fig. 17.22). In thise case, we can replace the force P by a continuous load q_0 distributed over an infinitisimal area dx_1, dx_2 such that:

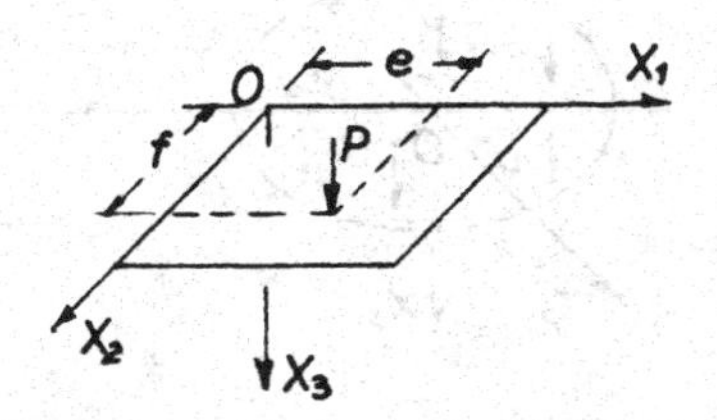

Fig. 17.22

$$q_0 = \frac{P}{dx_1\, dx_2}. \tag{17.7.13}$$

The function $q(x_1, x_2)$ in Eq. (17.7.9) is zero everywhere except at the point $x_1 = e$, $x_2 = f$ where it is equal to q_0. The double integral in Eq. (17.7.9) becomes $P \sin(m\Pi e/a)\sin(m\Pi f/b)$ and

$$B_{mn} = \frac{4P \sin \frac{m\Pi e}{a} \sin \frac{n\Pi f}{b}}{D\Pi^4 \left(\frac{m^2}{a^2} + \frac{n^2}{b^2}\right)^2 ab}. \tag{17.7.14}$$

The deflection of the middle plane is, therefore,

$$u_3 = \frac{4P}{D\Pi^4 ab} \sum_{m=1}^{\infty} \sum_{n=1}^{\infty} \frac{\sin \dfrac{m\Pi e}{a} \sin \dfrac{n\Pi f}{b}}{\left(\dfrac{m^2}{a^2} + \dfrac{n^2}{b^2}\right)^2} \sin \frac{m\Pi x_1}{a} \sin \frac{n\Pi x_2}{b}. \quad (17.7.15)$$

The moments and the shearing forces can now be obtained at any point of the plate by differentiation.

Remark

A large number of problems related to simply supported rectangular plates can be found in [2]. Care should be exercised when using the equations for the moments and the shears, since sign conventions differ from text to text.

17.8 Elliptic Plate with Clamped Edges under Uniform Load (Fig. 17.23)

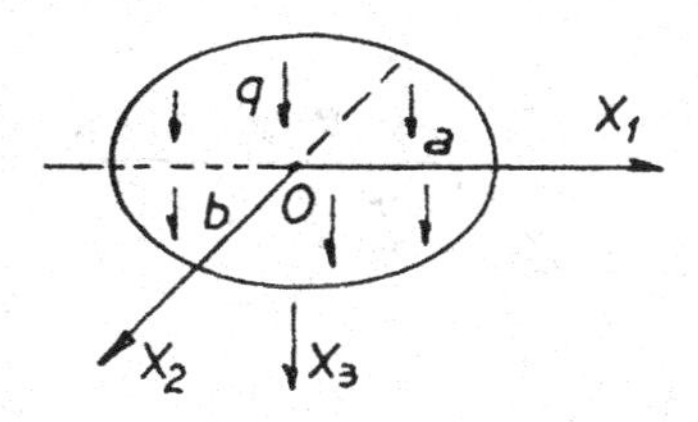

Fig. 17.23

Let us assume that the deflection u_3 is given by:

$$u_3 = A\left(\frac{x_1^2}{a^2} + \frac{x_2^2}{b^2} - 1\right)^2, \quad (17.8.1)$$

where the equation of the boundary of the ellipse is

$$\frac{x_1^2}{a^2} + \frac{x_2^2}{b^2} = 1 \quad (17.8.2)$$

and A is a constant. u_3 is equal to zero on the boundary and the two components of the gradient vector,

$$\frac{\partial u_3}{\partial x_1} = 2A\left(\frac{x_1^2}{a^2} + \frac{x_2^2}{b^2} - 1\right)\left(\frac{2x_1}{a^2}\right) \tag{17.8.3}$$

and

$$\frac{\partial u_3}{\partial x_2} = 2A\left(\frac{x_1^2}{a^2} + \frac{x_2^2}{b^2} - 1\right)\left(\frac{2x_2}{b^2}\right), \tag{17.8.4}$$

also vanish on the boundary. Therefore, the assumed deflection satisfies the boundary conditions of a clamped plate. Substituting Eq. (17.8.1) into Lagrange's equation, we obtain:

$$8A\left[\frac{3}{a^4} + \frac{2}{a^2b^2} + \frac{3}{b^4}\right] = \frac{q}{D}, \tag{17.8.5}$$

which shows that $q(x_1, x_2)$ is a constant q_0. From Eqs. (17.8.1) and (17.8.5), we deduce that

$$u_3 = \frac{q_0}{8D\left[\frac{3}{a^4} + \frac{2}{a^2b^2} + \frac{3}{b^4}\right]}\left(\frac{x_1^2}{a^2} + \frac{x_2^2}{b^2} - 1\right)^2 \tag{17.8.6}$$

is the solution to the problem of the elliptic plate with clamped edges subjected to a uniformly distributed load q_0.

The moments at any point are obtained from Eq. (17.3.8). From these values, the bending and twisting moments around any two directions can be obtained by means of a transformation of coordinates (or Mohr's circle).

The shearing forces at any point are obtained from Eq. (17.3.13). The transformation expressed by Eqs. (17.3.15) allows one to find the shearing forces on any pair of orthogonal planes whose normals have known direction cosines. For the case of the ellipse, transformations of coordinates are necessary to obtain moments, shearing forces, and reactions at the boundaries.

17.9 Bending of Circular Plates

In the discussion of bending of circular plates, it is convenient to use cylindrical coordinates. The coordinates r and θ will be taken as shown in Fig. 17.24. Eqs. (6.2.17) to (6.2.24), coupled with the results of Sec. 6.4, allow us to write quite easily all the equations of the bending of thin plates in cylindrical coordinates. From Eq. (6.4.28), the expression of the Laplacian of $u_3 = u_z$ is:

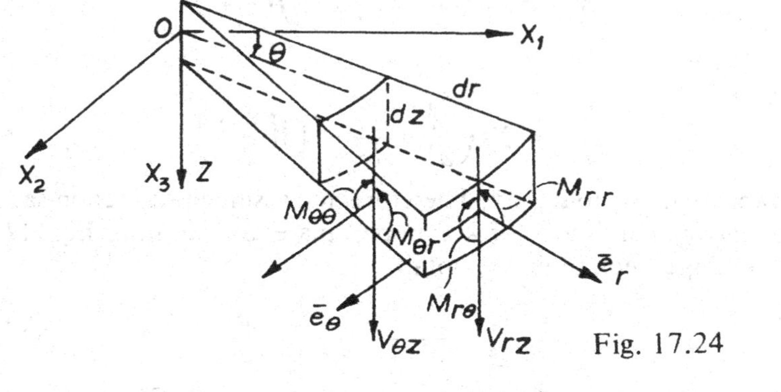

Fig. 17.24

$$\nabla^2 u_z = \frac{\partial u_z}{\partial r^2} + \frac{1}{r}\frac{\partial u_z}{\partial r} + \frac{1}{r^2}\frac{\partial^2 u_z}{\partial \theta^2}, \tag{17.9.1}$$

so that Lagrange's equation in cylindrical coordinates is:

$$\begin{aligned}\nabla^4 u_z &= \left(\frac{\partial}{\partial r^2} + \frac{1}{r}\frac{\partial}{\partial r} + \frac{1}{r^2}\frac{\partial^2}{\partial \theta^2}\right)\left(\frac{\partial^2 u_z}{\partial r^2} + \frac{1}{r}\frac{\partial u_z}{\partial r} + \frac{1}{r^2}\frac{\partial^2 u_z}{\partial \theta^2}\right) \\ &= \frac{q}{D}.\end{aligned} \tag{17.9.2}$$

Positive directions for bending moments, twisting moments, and shearing forces are shown in Fig. 17.24.

To obtain the expressions for M_{rr}, $M_{\theta\theta}$, $M_{r\theta}$, V_{rz}, and $V_{\theta z}$ in terms of $u_3 = u_z$, let us consider the element shown in Fig. 17.24 and assume that the OX_1 axis coincides with the radial direction $\bar{e}_r$. Therefore, M_{rr}, $M_{\theta\theta}$, $M_{r\theta}$, V_{rz}, and $V_{\theta z}$ have the same values as M_{11}, M_{22}, M_{12}, V_{13}, and V_{23} at the same point. Setting $\theta = 0$ in Eqs. (6.2.20), (6.2.21), and (6.2.22), we get:

$$\begin{aligned}M_{rr} &= -D\left(\frac{\partial^2 u_z}{\partial x_1^2} + \nu\frac{\partial^2 u_z}{\partial x_2^2}\right)_{\theta=0} \\ &= -D\left[\frac{\partial^2 u_z}{\partial r^2} + \nu\left(\frac{1}{r}\frac{\partial u_z}{\partial r} + \frac{1}{r^2}\frac{\partial^2 u_z}{\partial \theta^2}\right)\right]\end{aligned} \tag{17.9.3}$$

$$\begin{aligned}M_{\theta\theta} &= -D\left(\frac{\partial^2 u_z}{\partial x_2^2} + \nu\frac{\partial^2 u_z}{\partial x_1^2}\right)_{\theta=0} \\ &= -D\left[\frac{1}{r}\frac{\partial u_z}{\partial r} + \frac{1}{r^2}\frac{\partial^2 u_z}{\partial \theta^2} + \nu\frac{\partial^2 u_z}{\partial r^2}\right]\end{aligned} \tag{17.9.4}$$

$$M_{r\theta} = -D(1-\nu)\left(\frac{\partial^2 u_z}{\partial x_1\,\partial x_2}\right)_{\theta=0}$$
$$= -D(1-\nu)\left(\frac{1}{r}\frac{\partial^2 u_z}{\partial r\partial\theta} - \frac{1}{r^2}\frac{\partial u_z}{\partial\theta}\right) \tag{17.9.5}$$

$$V_{rz} = -D\frac{\partial}{\partial x_1}(\nabla^2 u_z)_{\theta=0} = -D\frac{\partial}{\partial r}\left(\frac{\partial^2 u_z}{\partial r^2} + \frac{1}{r}\frac{\partial u_z}{\partial r} + \frac{1}{r^2}\frac{\partial^2 u_z}{\partial\theta^2}\right) \tag{17.9.6}$$

$$V_{\theta z} - D\frac{\partial}{\partial x_2}(\nabla^2 u_z)_{\theta=0} = -\frac{D}{r}\frac{\partial}{\partial\theta}\left(\frac{\partial^2 u_z}{\partial r^2} + \frac{1}{r}\frac{\partial u_z}{\partial r} + \frac{1}{r^2}\frac{\partial^2 u_z}{\partial\theta^2}\right). \tag{17.9.7}$$

The boundary conditions at the edge of a circular plate of radius a are as follows:

For a simply supported edge:

$$u_z = 0 \text{ and } M_{rr} = 0. \tag{17.9.8}$$

For a clamped edge:

$$u_z = 0 \text{ and } \frac{\partial u_z}{\partial r} = 0. \tag{17.9.9}$$

For a free edge:

$$M_{rr} = 0 \text{ and } R_{rz} = V_{rz} + \frac{1}{r}\frac{\partial M_{r\theta}}{\partial\theta} = 0. \tag{17.9.10}$$

If the load q is symmetrically distributed about the Z axis, u_z is independent of θ and Lagrange's equation becomes:

$$\frac{1}{r}\frac{d}{dr}\left\{r\frac{d}{dr}\left[\frac{1}{r}\frac{d}{dr}\left(r\frac{du_z}{dr}\right)\right]\right\} = \frac{q}{D}. \tag{17.9.11}$$

This equation can easily be integrated when $q = q\,(r)$ is given:

1. $q(r) = q_0 =$ *constant*. Eq. (17.9.11) yields:

$$u_z = \frac{q_0 r^4}{64D} + \frac{C_1 r^2}{4}(\ell n\, r - 1) + C_2\frac{r^2}{4} + C_3\,\ell n\, r + C_4, \tag{17.9.12}$$

where C_1, C_2, C_3, and C_4 are constants of integration to be determined from the boundary conditions. From Eq. (17.9.6):

$$V_{rz} = -D\frac{d}{dr}\left(\frac{d^2 u_z}{dr^2} + \frac{1}{r}\frac{du_z}{dr}\right) = -D\left(\frac{q_0 r}{2D} + \frac{C_1}{r}\right). \tag{17.9.13}$$

For a circular plate without a central hole, we see that V_{rz} becomes infinite for $r = 0$. Since this is impossible, C_1 must be equal to zero. Since u_z is finite for $r = 0$, C_3 must also be equal to zero. Thus, for a uniformly loaded circular plate without a center hole:

$$u_z = \frac{q_0 r^4}{64D} + C_2 \frac{r^2}{4} + C_4. \tag{17.9.14}$$

If we assume that the edge is clamped, then according to Eq. (17.9.9), we have:

$$\frac{q_0 a^4}{64D} + \frac{C_2 a^2}{4} + C_4 = 0, \quad \frac{q_0 a^3}{16D} + \frac{C_2 a}{2} = 0. \tag{17.9.15}$$

Solving, we get:

$$C_2 = -\frac{q_0 a^2}{8D}, \quad C_4 = \frac{q_0 a^4}{64D}. \tag{17.9.16}$$

Therefore, the deflection u_z of a uniformly loaded circular plate clamped at the edge is given by:

$$u_z = \frac{q_0}{64D}(a^2 - r^2)^2. \tag{17.9.17}$$

Substituting Eq. (17.9.17) into Eqs. (17.9.3) and (17.9.4), we get:

$$M_{rr} = \frac{q_0}{16}[a^2(1 + \nu) - r^2(3 + \nu)] \tag{17.9.18}$$

$$M_{\theta\theta} = \frac{q_0}{16}[a^2(1 + \nu) - r^2(1 + 3\nu)]. \tag{17.9.19}$$

Because of the symmetry, $M_{r\theta} = 0$. Substituting $r = a$ in the two previous expressions, we find for the bending moments at the boundary:

$$(M_{rr})_{r=a} = -\frac{q_0 a^2}{8}, \quad (M_{\theta\theta})_{r=a} = -\frac{\nu q_0 a^2}{8}. \tag{17.9.20}$$

At the center:

$$(M_{rr})_{r=0} = (M_{\theta\theta})_{r=0} = \frac{q_0 a^2}{16}(1 + \nu). \tag{17.9.21}$$

The stresses σ_{rr} and $\sigma_{\theta\theta}$ are equal at the center and, from Eq. (17.3.16), are given by:

$$\sigma_{rr} = \sigma_{\theta\theta} = \frac{3za^2 q_0}{4h^3}(1 + \nu). \tag{17.9.22}$$

At the lower face $z = h/2$, and Eq. (17.9.22) becomes:

$$\sigma_{rr} = \sigma_{\theta\theta} = \frac{3a^2 q_0}{8h^2}(1 + \nu). \tag{17.9.23}$$

2. *The plate is loaded by a single concentrated force P at the center.* Eq. (17.9.12), in which q_0 is set equal to zero, applies to this case except at the center where q_0 is infinite. We know from experience that a solution exists with a finite deflection u_z, and that the slope du_z/dr at the center is equal to zero. From Eq. (17.9.12), we get:

$$\frac{du_z}{dr} = 0 + \frac{C_1}{2}\left(r\ell nr - \frac{r}{2}\right) + \frac{C_2 r}{2} + \frac{C_3}{r}. \qquad (17.9.24)$$

For du_z/dr to be equal to zero for $r = 0$, the constant C_3 must be equal to zero. Let us now isolate a small vertical cylinder of radius r around P. Equilibrium requires that $V_{rz} = -P/2\Pi r$. From Eq. (17.9.13) $V_{rz} = -DC_1/r$ so that $C_1 = P/2\Pi D$. Eq. (17.9.12) becomes:

$$u_z = \frac{Pr^2}{8\Pi D}(\ell nr - 1) + C_2\frac{r^2}{4} + C_4. \qquad (17.9.25)$$

If we assume that the edge is clamped, then from Eq. (17.9.9), we have:

$$C_2 = -\frac{P}{4\Pi D}(2\ell na - 1), \quad C_4 = \frac{Pa^2}{16\Pi D}, \qquad (17.9.26)$$

and

$$u_z = \frac{P}{16\Pi D}\left[2r^2\,\ell n\frac{r}{a} + a^2 - r^2\right].$$

Moments and stresses can be obtained using Eqs. (17.9.3) and (17.9.4), together with Eq. (17.3.16).

A wide variety of problems related to circular plates can be founa in [2].

17.10 Strain Energy and Potential Energy of a Thin Plate in Bending

From Eqs. (8.7.7) and (8.7.11), the expression for the strain energy stored in an elastic body is given by:

$$U_t = \frac{1}{2}\iiint_V \sigma_{ij}\, e_{ij}\, dV \qquad (17.10.1)$$

In the case of a thin plate in which σ_{33}, e_{13}, and e_{23} are neglected, this expression is reduced to:

$$U_t = \iiint_V \left[\frac{1}{2E}(\sigma_{11}^2 + \sigma_{22}^2 - 2\nu\sigma_{11}\sigma_{22}) + \frac{1+\nu}{E}\sigma_{12}^2\right] dx_1\, dx_2\, dx_3. \qquad (17.10.2)$$

Substituting Eq. (17.2.29) in the above equation, we obtain:

$$U_t = \frac{D}{2}\iint_A \left\{\left(\frac{\partial^2 u_3}{\partial x_1^2} + \frac{\partial^2 u_3}{\partial x_2^2}\right)^2 - 2(1-\nu)\left[\frac{\partial^2 u_3}{\partial x_1^2}\frac{\partial^2 u_3}{\partial x_2^2} - \left(\frac{\partial^2 u_3}{\partial x_1\,\partial x_2}\right)^2\right]\right\} dx_1\,dx_2, \tag{17.10.3}$$

where A is the area of the plate. This is the expression of the strain energy of a thin plate in bending. It can be put in a simpler form for plates with clamped edges and for rectangular plates with $u_3 = 0$ along the edges. Indeed, integrating twice by parts the last term of Eq. (17.10.3), we obtain:

$$\begin{aligned}\iint_A \frac{\partial^2 u_3}{\partial x_1\,\partial x_2}\frac{\partial^2 u_3}{\partial x_1\,\partial x_2}\,dx_1\,dx_2 &= \oint_C \frac{\partial^2 u_3}{\partial x_1\,\partial x_2}\frac{\partial u_3}{\partial x_1}\,dx_1 - \iint_A \frac{\partial u_3}{\partial x_1}\frac{\partial^3 u_3}{\partial x_1\,\partial x_1\,\partial x_2^2}\,dx_1\,dx_2 \\ &= \oint_C \frac{\partial^2 u_3}{\partial x_1\,\partial x_2}\frac{\partial u_3}{\partial x_1}\,dx_1 - \oint_C \frac{\partial^2 u_3}{\partial x_2^2}\frac{\partial u_3}{\partial x_1}\,dx_2 \\ &\quad + \iint_A \frac{\partial^2 u_3}{\partial x_1^2}\frac{\partial^2 u_3}{\partial x_2^2}\,dx_1\,dx_2,\end{aligned} \tag{17.10.4}$$

where $\oint$ is the integral taken around the contour C of the plate. For plates with clamped edges, the two components of the gradient vector $\partial u_3/\partial x_1$ and $\partial u_3/\partial x_2$ along the edges vanish. For a rectangular plate with $u_3 = 0$ on the boundary, $\partial u_3/\partial x_1 = 0$ along the edge parallel to the OX_1 axis, and $\partial u_3/\partial x_2 = \partial^2 u_3/\partial x_2^2 = 0$ along the edge parallel to the OX_2 axis. Therefore, the two first integrals in the right-hand side of Eq. (17.10.4) vanish in these two cases. With this result, the expression of the strain energy U_t becomes:

$$U_t = \frac{D}{2}\iint_A \left(\frac{\partial^2 u_3}{\partial x_1^2} + \frac{\partial^2 u_3}{\partial x_2^2}\right)^2 dx_1\,dx_2 \tag{17.10.5}$$

If a plate has clamped edges or is simply supported and is subjected to a load $q = q(x_1, x_2)$, the work done by the external forces is:

$$W = \int\int q u_3 \, dx_1 \, dx_2, \tag{17.10.6}$$

and the potential energy of the plate is:

$$\Pi_p = U_t - W = \int\int \left\{ \frac{D}{2} \left(\frac{\partial^2 u_3}{\partial x_1^2} + \frac{\partial^2 u_3}{\partial x_2^2} \right)^2 - q u_3 \right\} dx_1 \, dx_2 . \tag{17.10.7}$$

The minimization of the potential energy gives, in theory, the solution of the plate problem. Such a minimization is not easy to achieve and an approximate solution can be obtained using the Rayleigh-Ritz method (Sec. 15.20).

17.11 Application of the Principle of Minimum Potential Energy to Simply Supported Rectangular Plates

In Sec. 17.7 it was seen that the deflection of a simply supported rectangular plate can be represented in the form of a double trigonometric series:

$$u_3 = \sum_{m=1}^{\infty} \sum_{n=1}^{\infty} B_{mn} \sin \frac{m\Pi x_1}{a} \sin \frac{n\Pi x_2}{b} . \tag{17.11.1}$$

Substituting Eq. (17.11.1) into Eq. (17.10.5), we get:

$$U_t = \frac{\Pi^4 Dab}{8} \sum_{m=1}^{\infty} \sum_{n=1}^{\infty} B_{mn}^2 \left(\frac{m^2}{a^2} + \frac{n^2}{b^2} \right)^2 . \tag{17.11.2}$$

The coefficients B_{mn}, which determine the shape of the plate, can be obtained by means of the principle of minimum potential energy:

1. $q(x_1, x_2) = q_0$ *is uniformly distributed over the area of the plate.* The work done by the applied loads is:

$$W = \int_0^a \int_o^b q_0 \sum_{m=1}^{\infty} \sum_{n=1}^{\infty} B_{mn} \sin \frac{m\Pi x_1}{a} \sin \frac{n\Pi x_2}{b} \, dx_1 \, dx_2$$
$$= \frac{4 q_0 ab}{\Pi^2} \sum_{m=1}^{\infty} \sum_{n=1}^{\infty} \frac{B_{mn}}{mn} . \tag{17.11.3}$$

The potential energy Π_p is to be minimized with respect to the amplitude term B_{mn}, so that

$$\delta\Pi p = \left[\frac{\Pi^4 Dab}{4} B_{mn}\left(\frac{m^2}{a^2} + \frac{n^2}{b^2}\right)^2 - \frac{4q_0 ab}{\Pi^2 mn}\right]\delta B_{mn} = 0. \quad (17.11.4)$$

Thus,

$$B_{mn} = \frac{16q_0}{D\Pi^6 mn\left(\frac{m^2}{a^2} + \frac{n^2}{b^2}\right)^2}, \quad (17.11.5)$$

which is the same as Eq. (17.7.10).

2. *$q(x_1, x_2)$ is a concentrated force P acting at any point $x_1 = e$, $x_2 = f$.* The work done by the applied load is:

$$W = P \sum_{m=1}^{\infty} \sum_{n=1}^{\infty} B_{mn} \sin \frac{m\Pi e}{a} \sin \frac{n\Pi f}{b} \quad (17.11.6)$$

and

$$\delta\Pi_p = \left[\frac{\Pi^4 Dab}{4} B_{mn}\left(\frac{m^2}{a^2} + \frac{n^2}{b^2}\right)^2 - P \sin \frac{m\Pi e}{a} \sin \frac{n\Pi f}{b}\right]\delta B_{mn} \quad (17.11.7)$$
$$= 0.$$

Thus,

$$B_{mn} = \frac{4P \sin \frac{m\Pi e}{a} \sin \frac{n\Pi f}{b}}{D\Pi^4 ab\left(\frac{m^2}{a^2} + \frac{n^2}{b^2}\right)^2}, \quad (17.11.8)$$

which is the same as Eq. (17.7.14).

PROBLEMS

1. A thin rectangular plate is subjected to the following uniform edge moments: M_{11} = 15 in.-lb/in., M_{22} = − 10 in.-lb/in., and M_{12} = 5 in.-lb/in. If the plate is 1 inch thick:
   (a) Find the components of the curvature tensor.
   (b) Find the directions of the planes corresponding to the principal curvatures.

(c) Finu the magnitude and directions of the principal stresses.
(d) Draw the Mohr circles representing the moment, the curvature, and the stress tensors. E and ν for the material of the plate are 10^7 psi and 0.3, respectively.

2. A thin rectangular plate 20 in. long, 10 in, wide, and 1 in. thick lies in the OX_1, OX_2 plane with its long side parallel to the OX_1 axis. Find the magnitudes of the radii of curvature if the edges parallel to the OX_2 axis are subjected to a moment $M_{11} = 20$ in.-lb/in. ($E = 30 \times 10^6$ psi, $\nu = 0.3$).

3. A simply supported rectangular plate of sides a and b (see Fig. 17.21) and thickness h is subjected to a hydrostatic pressure $q(x_1, x_2) = q_0 x_1 / a$. Find the expressions of the deflection, the moments, and the stresses in the plate.

4. Show that

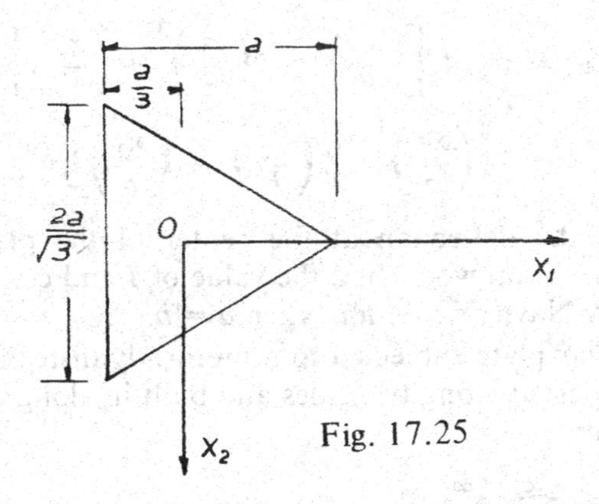

Fig. 17.25

$$u_3 = \frac{q_0}{64aD}\left[x_1^3 - 3x_2^2 x_1 - a(x_1^2 + x_2^2) + \frac{4}{27}a^3\right]\left[\frac{4}{9}a^2 - x_1^2 - x_2^2\right]$$

is the solution of the problem of the simply supported plate having the shape of an equilateral triangle, and being subjected to a uniformly distributed load q_0 (Fig. 17.25). Find the maximum values of M_{11} and M_{22}.

5. Find the expressions of the deflection and of the stresses σ_{rr}, $\sigma_{\theta\theta}$, and $\sigma_{r\theta}$ for a circular simply supported plate subjected to a uniformly distributed load $q(r, \theta) = q_0$.
6. Solve Problem 6 for a concentrated load P applied at the center of the plate.
7. A circular plate with radius a has a concentric circular hole with radius b. The plate is subjected to a uniformly distributed load q_0 and has its inner edge built in and its outer edge free. Find the maximum deflection, the maximum moment, and the maximum bending stresses, if $b / a = 6$.
8. A rectangular plate of sides a and b is subjected to a uniformly distributed load q_0, and is simply supported along its edges. Assume that the deflection is given by

$$u_3 = 10.24\ell\left[\left(\frac{x_1}{a}\right)^4 - 2\left(\frac{x_1}{a}\right)^3 + \left(\frac{x_1}{a}\right)\right]$$
$$\left[\left(\frac{x_2}{b}\right)^4 - 2\left(\frac{x_2}{b}\right)^3 + \left(\frac{x_2}{b}\right)\right],$$

where ℓ is the deflection at the center of the plate. Using the principle of virtual work, find the value of ℓ and compare it to that obtained by Navier's solution when $a = b$.

9. A rectangular plate subjected to a uniformly distributed load q_0 is simply supported along two sides and built in along the two others. The equation

$$u_3 = \sum_{m=1}^{\infty} \sum_{n=1}^{\infty} B_{mn}\left(1 - \cos\frac{2m\Pi x_1}{a}\right)\sin\frac{n\Pi x_2}{b}$$

satisfies the boundary conditions. Determine B_{mn} using the Rayleigh-Ritz method.

REFERENCES

[1] W. Kaplan, *Advanced Calculus*, Addison-Wesley, Reading, Mass., 1952.

[2] S. Timoshenko and S. Woinowski-Kreiger, *Theory of Plates and Shells*, McGraw-Hill, New York, N. Y., 1959.

CHAPTER 18

INTRODUCTION TO THE THEORY OF THIN SHELLS

18.1 Introduction

A shell structure may be defined as a body enclosed between two closely spaced and curved surfaces. If the thickness is small compared with the overall dimensions of the bounding surfaces, the shell is called a thin shell. Before studying the theory of thin shells, it is important that one acquires a clear understanding of the theory of surfaces and of the curves that are drawn on them. These topics belong to the subject of differential geometry [1].

In this chapter, we shall present and discuss those relations of differential geometry which are needed in the development of thin shell theory. The relations among forces, moments, and stresses will be given, and the equations of equilibrium will be derived. Both will then be applied to a few particular cases of thin shells. It is recommended that, prior to reading this chapter, a thorough understanding of Chapter 6 be acquired.

18.2 Space Curves

In vector analysis, a curve is defined as the locus of a point whose position vector $\bar{r}$, relative to some fixed origin O, is a function of one parameter ξ_1. If the curve is embedded in a tridimensional cartesian space, the coordinates of any point on the curve are given by (Fig. 18.1):

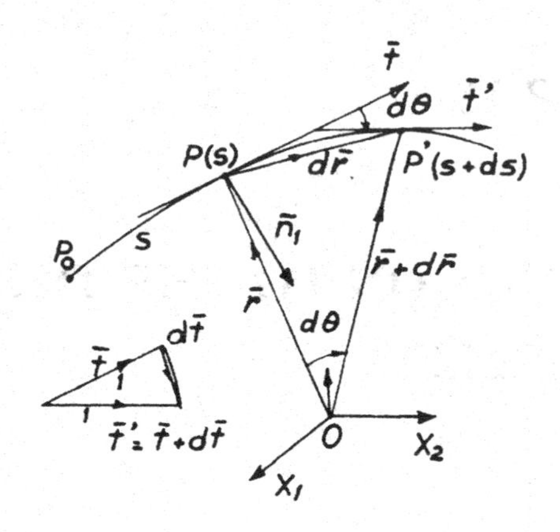

Fig. 18.1

$$x_1 = x_1(\xi_1), \quad x_2 = x_2(\xi_1), \quad x_3 = x_3(\xi_1). \tag{18.2.1}$$

Eqs. (18.2.1) are the parametric equations of a space curve. x_1, x_2, and x_3 are assumed to be single valued functions of ξ_1 to insure that each value of ξ_1 gives one single point on the space curve.

Let $\bar{r}$ be a vector giving the position of a point P on a space curve with respect to an arbitrary origin O (Fig. 18.1), and let s be a parameter of the curve representing the distance of P measured along the curve from a reference point P_o. The vector $\bar{r}$ can be considered as dependent on the parameter s. If $\bar{t}$ represents the unit vector at P in the direction of increasing s, then

$$\bar{t} = \frac{d\bar{r}}{ds}. \tag{18.2.2}$$

Let us define a unit vector,

$$\bar{n}_1 = A\frac{d\bar{t}}{ds}, \tag{18.2.3}$$

where $d\bar{t}$ denotes the increment in $\bar{t}$ in passing from the point $P(s)$ to $P'(s + ds)$, and A is a positive factor of proportionality. Since $\bar{t}$ is also a unit vector, $d\bar{t}$ will be perpendicular to $\bar{t}$. The two vectors $\bar{t}$ and $\bar{n}_1$ determine a plane called the osculating plane of the curve at P. The osculating plane is also defined as that plane containing three consecutive points or two consecutive tangents on a space curve.

Finally, let $\bar{n}_2$ be a unit vector at P perpendicular to the osculating plane and in a direction such that (Fig. 18.2):

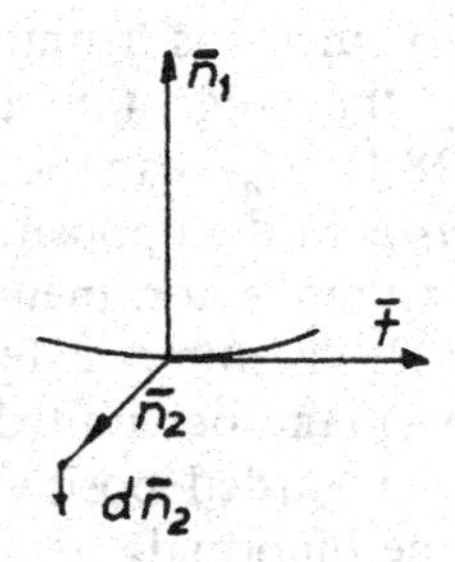

Fig. 18.2

$$\bar{n}_2 = \bar{t} \times \bar{n}_1 . \tag{18.2.4}$$

The three unit vectors $\bar{t}$, $\bar{n}_1$, and $\bar{n}_2$ constitute a unit right-handed system associated with P. $\bar{t}$ is called the (vector) tangent, $\bar{n}_1$ the (vector) principal normal, and $\bar{n}_2$ the (vector) binormal. Since

$$\bar{t} \cdot \bar{n}_1 = \bar{n}_1 \cdot \bar{n}_2 = \bar{n}_2 \cdot \bar{t} = 0, \tag{18.2.5}$$

then, by differentiation, we get:

$$\bar{t} \cdot d\bar{n}_1 = -\bar{n}_1 \cdot d\bar{t}, \quad \bar{n}_1 \cdot d\bar{n}_2 = -\bar{n}_2 \cdot d\bar{n}_1, \quad \bar{n}_2 \cdot d\bar{t} = -\bar{t} \cdot d\bar{n}_2 . \tag{18.2.6}$$

But from Eq. (18.2.3), $\bar{n}_1$ and $d\bar{t}$ are parallel; therefore, $\bar{n}_2$ and $d\bar{t}$ must be perpendicular to each other. And from Eq. (18.2.6), $\bar{t}$ and $d\bar{n}_2$ must be perpendicular to each other. Now, since $\bar{n}_2$ is a unit vector, it must also be perpendicular to $d\bar{n}_2$, so that $d\bar{n}_2$ and $\bar{n}_1$ must be collinear (Fig. 18.2).

Let us now define two characteristic quantities associated with the curve, at P, by the equations:

$$C_{n1} = \bar{n}_1 \cdot \frac{d\bar{t}}{ds} = \bar{n}_1 \cdot \frac{d^2\bar{r}}{ds^2} \tag{18.2.7}$$

$$C_{n2} = -\bar{n}_1 \cdot \frac{d\bar{n}_2}{ds} . \tag{18.2.8}$$

C_{n1} is seen to be equal to the reciprocal of the positive factor of proportionality A in Eq. (18.2.3), and is called the first curvature or flexure (or just the curvature) of the curve at P; C_{n2} is called the second curvature or torsion of the curve at P. C_{n1} is always positive and, since $\bar{n}_1$ is a unit vector in the same direction as $d\bar{t}$, then C_{n1} is equal in magnitude to $|d\bar{t}/ds|$; also since $\bar{t}$ is a unit vector, C_{n1} is equal to the angle turned through by the tangent to the curve per unit distance traveled along it (Fig. 18.1). C_{n2} can be either positive or negative depending on whether $d\bar{n}_2$ is in the opposite or in the same direction as $\bar{n}_1$. Moreover, since $\bar{n}_1$ is a unit vector, then $C_{n2} = \mp|d\bar{n}_2/ds|$; also, since $\bar{n}_2$ is a unit vector, then C_{n2} is numerically equal to the angle turned through by the osculating plane per unit distance traversed along the curve. The torsion C_{n2} is regarded positive when the rotation of the osculating plane (i.e., of the binormal), as s increases, follows the right-hand rule with the thumb in the direction of $\bar{t}$. The torsion of a plane curve is obviously equal to zero.

The inverses of C_{n1} and C_{n2} are called the radii of curvature of the curve at P by:

$$R_{n1} = \frac{1}{C_{n1}} \tag{18.2.9}$$

is called the radius of flexure, and

$$R_{n2} = \frac{1}{C_{n2}} \tag{18.2.10}$$

is called the radius of torsion. The point at a distance R_{n1} from P in the direction of $\bar{n}_1$ is called the first center of curvature or the center of flexure. The point at a distance R_{n2} from P in the direction of $d\bar{n}_2$ (i.e., in the direction of $\pm\bar{n}_1$) is called the second center of curvature or the center of torsion.

18.3 Elements of the Theory of Surfaces

In this section, we shall give a short presentation of the elements of the theory of surfaces which will enable us to deduce the results needed in the study of the theory of thin shells.

1) *Gaussian surface coordinates. First fundamental form*

In vector analysis, a surface is defined as the locus of a point whose position vector $\bar{r}$, relative to some fixed origin O, is a function of two independent parameters ξ_1 and ξ_2. If the surface is embedded in a tridimensional cartesian space, the coordinates of any point of the surface are given by:

$$x_1 = x_1(\xi_1, \xi_2), \quad x_2 = x_2(\xi_1, \xi_2), \quad x_3 = x_3(\xi_1, \xi_2). \qquad (18.3.1)$$

Eqs. (18.3.1) are the parametric equations of a surface. If we eliminate ξ_1 and ξ_2 from Eqs. (18.3.1), we obtain the equation of the surface in cartesian coordinates, namely:

$$F(x_1, x_2, x_3) = 0. \qquad (18.3.2)$$

It is understood that the surface is regular in the sense that the functions x_1, x_2, and x_3 of Eqs. (18.3.1), together with their derivatives with respect to the parameters ξ_1 and ξ_2 to the order required in the following discussions, are continuous. Any relation between ξ_1 and ξ_2, such as

$$g(\xi_1, \xi_2) = 0, \qquad (18.3.3)$$

represents, together with Eqs. (18.3.1) a curve on the surface. If either of the parameters (say, ξ_1) is held constant, then Eqs. (18.3.1) together with the additional equation ξ_1 = constant will represent a line on the surface (Fig. 18.3). Such lines are called coordinate lines or parametric

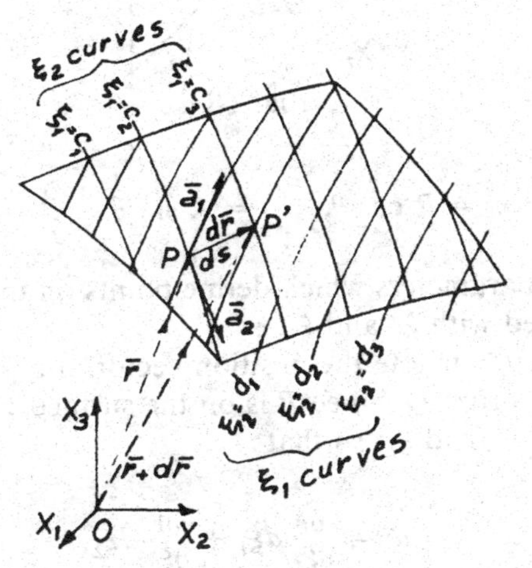

Fig. 18.3

curves. Through each point on the surface there will pass two coordinate lines, one corresponding to ξ_1 = constant and one corresponding to ξ_2 = constant. The parameters ξ_1 and ξ_2 thus constitute a system of curvilinear coordinates for points on the surface (see Sec. 6.2). The curve along which ξ_2 is constant and ξ_1 varies is called a ξ_1 curve; the other, along which ξ_1 is constant and ξ_2 varies, is called a ξ_2 curve. ξ_1 and ξ_2 may be considered as surface coordinates of the point P: They are called *Gaussian coordinates.* As an example, consider the vertical circular cylinder of Fig. 18.4. If R is the radius of the cylinder, the cartesian coordinates of a point on the cylinder are:

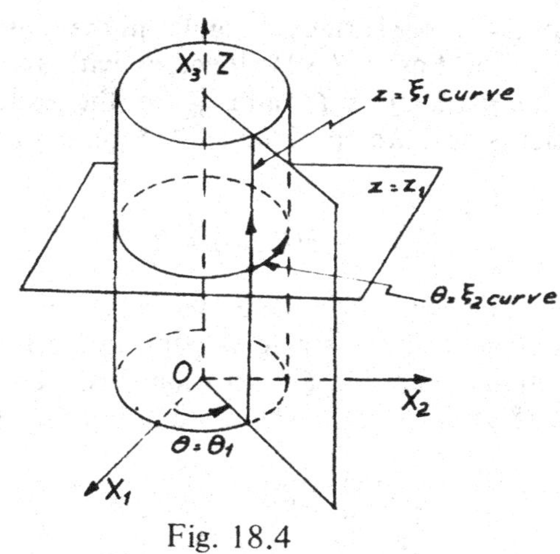

Fig. 18.4

$$x_1 = R\cos\theta, \quad x_2 = R\sin\theta, \quad x_3 = z. \tag{18.3.4}$$

z and θ are the parameters which define points on the surface, and they may be identified with ξ_1 and ξ_2.

In Fig. 18.3, let $\bar{r}$ denote the position vector of a point P with respect to any arbitrary origin O. Since P is on the surface, $\bar{r}$ can be considered as a function of ξ_1 and ξ_2, so that

$$d\bar{r} = \frac{\partial \bar{r}}{\partial \xi_1} d\xi_1 + \frac{\partial \bar{r}}{\partial \xi_2} d\xi_2, \tag{18.3.5}$$

where $d\bar{r}$ denotes the differential increment in $\bar{r}$ which occurs in passing from $P(\xi_1, \xi_2)$ to $P'(\xi_1 + d\xi_1, \xi_2 + d\xi_2)$. The vector $\partial\bar{r}/\partial\xi_1$ denotes the derivative of $\bar{r}$ with respect to ξ_1 when ξ_2 remains constant (compare to Sec. 6.3). Therefore, $\partial\bar{r}/\partial\xi_1$ is tangent to the ξ_1 curve. Similarly, $\partial\bar{r}/\partial\xi_2$ is tangent to the ξ_2 curve. These two vectors will be denoted by:

$$\bar{a}_1 = \frac{\partial \bar{r}}{\partial \xi_1} \tag{18.3.6}$$

$$\bar{a}_2 = \frac{\partial \bar{r}}{\partial \xi_2}, \tag{18.3.7}$$

and may be considered as constituting an $\bar{a}_1, \bar{a}_2$ base system at point P. Any vector $\bar{A}$ associated with point P, which can be represented by a line tangent to the surface at P, is called a surface vector and can be represented by [compare to Eq. (6.3.13)]:

$$\bar{A} = A_1\bar{a}_1 + A_2\bar{a}_2. \tag{18.3.8}$$

Thus, for the differential increment in $\bar{r}$, we have:

$$d\bar{r} = \bar{a}_1 d\xi_1 + \bar{a}_2 d\xi_2. \tag{18.3.9}$$

The square of the magnitude of $d\bar{r}$ is given by:

$$(ds)^2 = d\bar{r} \cdot d\bar{r} = \bar{a}_1 \cdot \bar{a}_1 (d\xi_1)^2 + 2\bar{a}_1 \cdot \bar{a}_2 \, d\xi_1 \, d\xi_2 + \bar{a}_2 \cdot \bar{a}_2 (d\xi_2)^2 \tag{18.3.10}$$

Setting

$$\bar{a}_1 \cdot \bar{a}_1 = a_{11} = E \tag{18.3.11}$$

$$\bar{a}_1 \cdot \bar{a}_2 = a_{12} = F \tag{18.3.12}$$

$$\bar{a}_2 \cdot \bar{a}_2 = a_{22} = G, \tag{18.3.13}$$

Eq. (18.3.10) can be written as:

$$(ds)^2 = a_{ij} d\xi_i d\xi_j = E(d\xi_1)^2 + 2F d\xi_1 d\xi_2 + G(d\xi_2)^2. \tag{18.3.14}$$

The differential quadratic form (18.3.14) is called the *first fundamental form of the surface* and E, F, G are called the *first fundamental magnitudes*. These magnitudes are the metric coefficients for the surface and, as can be seen from Eq. (18.3.14), they are the link between the length of an element and the differentials $d\xi_i$. If $d\bar{r}$ is taken along the ξ_1 curve, Eq. (18.3.14) gives:

$$ds_1 = \sqrt{E}\, d\xi_1 = \sqrt{a_{11}}\, d\xi_1. \tag{18.3.15}$$

If $d\bar{r}$ is taken along the ξ_2 curve, Eq. (18.3.14) gives:

$$ds_2 = \sqrt{G}\, d\xi_2 = \sqrt{a_{22}}\, d\xi_2 . \tag{18.3.16}$$

The cosine of the angle between $\bar{a}_1$ and $\bar{a}_2$ is given by:

$$\cos\theta = \frac{\bar{a}_1 \cdot \bar{a}_2}{|\bar{a}_1||\bar{a}_2|} = \frac{a_{12}}{\sqrt{a_{11}a_{22}}} = \frac{F}{\sqrt{EG}} . \tag{18.3.17}$$

Since $\cos\theta \leq 1$, then

$$EG \geq F^2, \tag{18.3.18}$$

and the quantity

$$H^2 = EG - F^2 \tag{18.3.19}$$

is always positive. The parametric curves ξ_1 and ξ_2 will form an orthogonal system of curvilinear coordinates if, all over the surface, $a_{12} = F = 0$. In this case, the first fundamental form becomes:

$$(ds)^2 = E(d\xi_1)^2 + G(d\xi_2)^2 . \tag{18.3.20}$$

From a given point (ξ_1, ξ_2) any direction on the surface is determined by the increments $d\xi_1$ and $d\xi_2$. Let ds and ds' be elements of arc lengths of two surface curves C and C' intersecting at P (Fig. 18.5). Then,

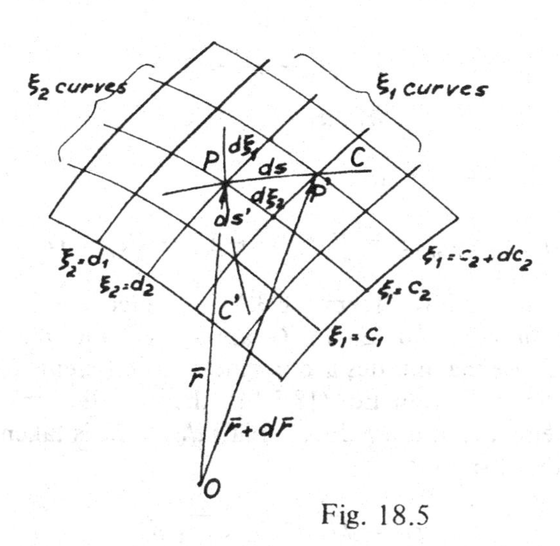

Fig. 18.5

$$\frac{d\bar{r}}{ds} = \frac{\partial \bar{r}}{\partial \xi_1}\frac{d\xi_1}{ds} + \frac{\partial \bar{r}}{\partial \xi_2}\frac{d\xi_2}{ds} \tag{18.3.21}$$

$$\frac{d\bar{r}}{ds'} = \frac{\partial \bar{r}}{\partial \xi_1}\frac{d\xi_1}{ds'} + \frac{\partial \bar{r}}{\partial \xi_2}\frac{d\xi_2}{ds'}. \tag{18.3.22}$$

The angle θ between C and C' is given by:

$$\cos\theta = \frac{d\bar{r}}{ds}\cdot\frac{d\bar{r}}{ds'} = E\frac{d\xi_1}{ds}\frac{d\xi_1}{ds'} + F\left(\frac{d\xi_1}{ds}\frac{d\xi_2}{ds'} + \frac{d\xi_1}{ds'}\frac{d\xi_2}{ds}\right) + G\frac{d\xi_2}{ds}\frac{d\xi_2}{ds'}. \tag{18.3.23}$$

For C and C' to be orthogonal, we must have:

$$E\frac{d\xi_1}{ds}\frac{d\xi_1}{ds'} + F\left(\frac{d\xi_1}{ds}\frac{d\xi_2}{ds'} + \frac{d\xi_1}{ds'}\frac{d\xi_2}{ds}\right) + G\frac{d\xi_2}{ds}\frac{d\xi_2}{ds'} = 0. \tag{18.3.24}$$

It is desirable to eliminate s and s' from Eq. (18.3.24). For that, we write:

$$\frac{d\xi_1}{ds} = \left(\frac{d\xi_1}{d\xi_2}\right)_C \frac{d\xi_2}{ds} \tag{18.3.25}$$

$$\frac{d\xi_1}{ds'} = \left(\frac{d\xi_1}{d\xi_2}\right)_{C'} \frac{d\xi_2}{ds'}, \tag{18.3.26}$$

where $(d\xi_1/d\xi_2)_C$ and $(d\xi_1/d\xi_2)_{C'}$ are to be computed along the curves C and C', respectively. Substituting Eqs. (18.3.25) and (18.3.26) into Eq. (18.3.24), we get:

$$E\left(\frac{d\xi_1}{d\xi_2}\right)_C\left(\frac{d\xi_1}{d\xi_2}\right)_{C'} + F\left[\left(\frac{d\xi_1}{d\xi_2}\right)_C + \left(\frac{d\xi_1}{d\xi_2}\right)_{C'}\right] + G = 0, \tag{18.3.27}$$

which is the condition for C and C' to be orthogonal.

2) *Second fundamental form*

A normal section of a surface at a point P is the section defined by a plane containing the normal to the surface at the point (Fig. 18.6). This section is a plane curve whose principal normal is collinear with the normal to the surface at that point. The curvature of a normal section, such as the curve AB (Fig. 18.6), is called the *normal curvature* of the surface at P in the direction of AB. The unit vector along the normal to the surface at P is given by:

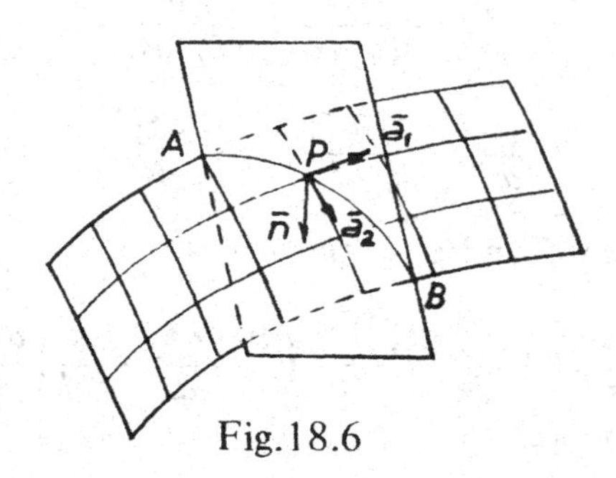

Fig. 18.6

$$\bar{n}(\xi_1, \xi_2) = \frac{\bar{a}_1 \times \bar{a}_2}{|\bar{a}_1 \times \bar{a}_2|}, \tag{18.3.28}$$

where the quantity in the denominator is the magnitude of the vector product $\bar{a}_1 \times \bar{a}_2$. But

$$|\bar{a}_1 \times \bar{a}_2| = |\bar{a}_1||\bar{a}_2|\sin\theta = \sqrt{EG}\sqrt{\frac{EG - F^2}{EG}} = H. \tag{18.3.29}$$

Therefore,

$$\bar{n} = \frac{\bar{a}_1 \times \bar{a}_2}{H}. \tag{18.3.30}$$

On Fig. 18.5, in passing along a curve C from $P(\xi_1, \xi_2)$ to the near point $P'(\xi_1 + d\xi_1, \xi_2 + d\xi_2)$, let us assume that $\bar{n}$ and $\bar{r}$ increase by $d\bar{n}$ and $d\bar{r}$. Then,

$$d\bar{n} = \frac{\partial \bar{n}}{\partial \xi_1} d\xi_1 + \frac{\partial \bar{n}}{\partial \xi_2} d\xi_2, \quad d\bar{r} = \frac{\partial \bar{r}}{\partial \xi_1} d\xi_1 + \frac{\partial \bar{r}}{\partial \xi_2} d\xi_2. \tag{18.3.31}$$

Forming the scalar product of $d\bar{n}$ and $d\bar{r}$, we get:

$$-d\bar{n} \cdot d\bar{r} = L(d\xi_1)^2 + 2M d\xi_1 d\xi_2 + N(d\xi_2)^2, \tag{18.3.32}$$

where

$$L = -\frac{\partial \bar{n}}{\partial \xi_1} \cdot \frac{\partial \bar{r}}{\partial \xi_1}, \quad M = -\frac{1}{2}\left[\frac{\partial \bar{n}}{\partial \xi_1} \cdot \frac{\partial \bar{r}}{\partial \xi_2} + \frac{\partial \bar{n}}{\partial \xi_2} \cdot \frac{\partial \bar{r}}{\partial \xi_1}\right]$$
$$N = -\frac{\partial \bar{n}}{\partial \xi_2} \cdot \frac{\partial \bar{r}}{\partial \xi_2}. \tag{18.3.33}$$

L, M, and N are called the *fundamental magnitudes of the second order*. It will be shown that they are intimately connected with the curvature properties of the surface. The expression in the right-hand side of Eq. (18.3.32) is called the *second fundamental form* for the surface. The coefficients L, M, and N of this form can be evaluated as follows: Since $\bar{n}$ is normal to the surface, and since $\partial\bar{r}/\partial\xi_1$ and $\partial\bar{r}/\partial\xi_2$ are tangent to the surface, therefore,

$$\bar{n} \cdot \frac{\partial \bar{r}}{\partial \xi_1} = \bar{n} \cdot \frac{\partial \bar{r}}{\partial \xi_2} = 0,$$

and upon partial differentiation of these expressions, the following relations are found:

$$\frac{\partial \bar{n}}{\partial \xi_1} \cdot \frac{\partial \bar{r}}{\partial \xi_1} = -\bar{n} \cdot \frac{\partial^2 \bar{r}}{\partial \xi_1^2} \tag{18.3.34}$$

$$\frac{\partial \bar{n}}{\partial \xi_2} \cdot \frac{\partial \bar{r}}{\partial \xi_2} = -\bar{n} \cdot \frac{\partial^2 \bar{r}}{\partial \xi_2^2} \tag{18.3.35}$$

$$\frac{\partial \bar{n}}{\partial \xi_2} \cdot \frac{\partial \bar{r}}{\partial \xi_1} = -\bar{n} \cdot \frac{\partial^2 \bar{r}}{\partial \xi_1 \, \partial \xi_2} \tag{18.3.36}$$

$$\frac{\partial \bar{n}}{\partial \xi_1} \cdot \frac{\partial \bar{r}}{\partial \xi_2} = -\bar{n} \cdot \frac{\partial^2 \bar{r}}{\partial \xi_1 \, \partial \xi_2}. \tag{18.3.37}$$

Hence, from Eqs. (18.3.33), we have:

$$L = \bar{n} \cdot \frac{\partial^2 \bar{r}}{\partial \xi_1^2} \tag{18.3.38}$$

$$M = \bar{n} \cdot \frac{\partial^2 \bar{r}}{\partial \xi_1 \, \partial \xi_2} \tag{18.3.39}$$

$$N = \bar{n} \cdot \frac{\partial^2 \bar{r}}{\partial \xi_2^2}. \tag{18.3.40}$$

Thus, L, M, and N are the projections of the second derivatives of $\bar{r}$ on the normal $\bar{n}$ to the surface.

Since $\bar{n}$ is a unit vector, $d\bar{n}$ is normal to $\bar{n}$ and, therefore, parallel to the surface. Consequently, $\partial\bar{n}/\partial\xi_1$ can be expressed in terms of its components along $\bar{a}_1$ and $\bar{a}_2$:

$$\frac{\partial \bar{n}}{\partial \xi_1} = p\bar{a}_1 + q\bar{a}_2 = p\frac{\partial \bar{r}}{\partial \xi_1} + q\frac{\partial \bar{r}}{\partial \xi_2}, \tag{18.3.41}$$

where p and q are unknowns to be determined. Forming the scalar product of each side of Eq. (18.3.41) by $\bar{a}_1$ and $\bar{a}_2$, respectively, and recalling Eqs. (18.3.11) to (18.3.13), we get:

$$-L = pE + qF \tag{18.3.42}$$

$$-M = pF + qG. \tag{18.3.43}$$

Solving for p and q, we obtain:

$$p = \frac{FM - LG}{H^2} \tag{18.3.44}$$

$$q = \frac{FL - EM}{H^2}. \tag{18.3.45}$$

Therefore,

$$\frac{\partial \bar{n}}{\partial \xi_1} = \frac{FM - LG}{H^2}\frac{\partial \bar{r}}{\partial \xi_1} + \frac{FL - EM}{H^2}\frac{\partial \bar{r}}{\partial \xi_2}. \tag{18.3.46}$$

Similarly,

$$\frac{\partial \bar{n}}{\partial \xi_2} = \frac{FN - MG}{H^2}\frac{\partial \bar{r}}{\partial \xi_1} + \frac{FM - EN}{H^2}\frac{\partial \bar{r}}{\partial \xi_2}. \tag{18.3.47}$$

From Eqs. (18.3.30), (18.3.46), and (18.3.47), the values of the following scalar triple products can be obtained:

$$\bar{n} \cdot \frac{\partial \bar{n}}{\partial \xi_1} \times \frac{\partial \bar{r}}{\partial \xi_1} = \frac{EM - FL}{H} \tag{18.3.48}$$

$$\bar{n} \cdot \frac{\partial \bar{n}}{\partial \xi_1} \times \frac{\partial \bar{r}}{\partial \xi_2} = \frac{FM - GL}{H} \tag{18.3.49}$$

$$\bar{n} \cdot \frac{\partial \bar{n}}{\partial \xi_2} \times \frac{\partial \bar{r}}{\partial \xi_1} = \frac{EN - FM}{H} \tag{18.3.50}$$

$$\bar{n} \cdot \frac{\partial \bar{n}}{\partial \xi_2} \times \frac{\partial \bar{r}}{\partial \xi_2} = \frac{FN - GM}{H}. \tag{18.3.51}$$

3) *Curvature of a normal section. Meunier's theorem*

It was previously pointed out that the fundamental magnitudes of the second order L, M, and N are connected with the curvature properties of the surface. Let us consider a normal section of a surface at point P (Fig. 18.6); that is to say, the section by a plane containing the normal to the surface at P. Such a section AB is a plane curve whose principal normal $\bar{n}_1$ (see Sec. 18.2) is collinear with the normal to the surface. *We shall adopt the convention that the surface coordinates are chosen in a way such that the normal $\bar{n}$ to the surface points in the same direction as the principal normal $\bar{n}_1$ of the section AB.* With this convention, the first center of curvature of the section AB [see Eq. (18.2.9)] falls on the positive side of $\bar{n}$. In what follows, the curvature of a normal section will be called C_n, and its radius of curvature will be called $R_n = 1/C_n$. From Eq. (18.2.7), we have:

$$C_n = \bar{n} \cdot \frac{d^2\bar{r}}{ds^2}, \tag{18.3.52}$$

where s is the distance measured along the normal section AB (Fig. 18.6). Using Eq. (18.3.21), we have:

$$\begin{aligned}\frac{d^2\bar{r}}{ds^2} = \frac{\partial\bar{r}}{\partial\xi_1}\frac{d^2\xi_1}{ds^2} + \frac{\partial\bar{r}}{\partial\xi_2}\frac{d^2\xi_2}{ds^2} + \frac{\partial^2\bar{r}}{\partial\xi_1^2}\left(\frac{d\xi_1}{ds}\right)^2 + 2\frac{\partial^2\bar{r}}{\partial\xi_1\,\partial\xi_2}\frac{d\xi_1}{ds}\frac{d\xi_2}{ds} \\ + \frac{\partial^2\bar{r}}{\partial\xi_2^2}\left(\frac{d\xi_2}{ds}\right)^2\end{aligned} \tag{18.3.53}$$

Recalling Eqs. (18.3.38) to (18.3.40), the expression for C_n becomes:

$$C_n = L\left(\frac{d\xi_1}{ds}\right)^2 + 2M\frac{d\xi_1}{ds}\frac{d\xi_2}{ds} + N\left(\frac{d\xi_2}{ds}\right)^2. \tag{18.3.54}$$

Substituting Eq. (18.3.14) into Eq. (18.3.54), we obtain the expression of C_n, the normal curvature, in terms of the first and second fundamental forms:

$$C_n = \frac{L(d\xi_1)^2 + 2M(d\xi_1)(d\xi_2) + N(d\xi_2)^2}{E(d\xi_1)^2 + 2F(d\xi_1)(d\xi_2) + G(d\xi_2)^2} \tag{18.3.55}$$

or

$$C_n = -\frac{d\bar{n}\cdot d\bar{r}}{d\bar{r}\cdot d\bar{r}}\,. \tag{18.3.56}$$

Dividing the numerator and denominator of Eq. (18.3.55) by $(d\xi_2)^2$, we get:

$$C_n = \frac{L\left(\dfrac{d\xi_1}{d\xi_2}\right)^2 + 2M\left(\dfrac{d\xi_1}{d\xi_2}\right) + N}{E\left(\dfrac{d\xi_1}{d\xi_2}\right)^2 + 2F\left(\dfrac{d\xi_1}{d\xi_2}\right) + G}. \tag{18.3.57}$$

In Eq. (18.3.57), the quantities L, M, N, E, F, G are functions of the coordinates ξ_1, and ξ_2, and have a given constant value for each point P on the surface. Therefore, at every point, the normal curvature depends only on the ratio $(d\xi_1/d\xi_2)$. Thus, it can be stated that all surface curves through a point P which are tangent to the same direction have the same normal curvature.

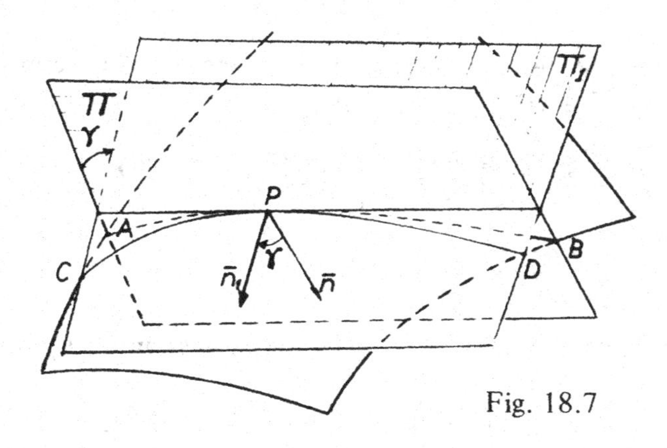

Fig. 18.7

Let us assume now that the section CD of a surface (Fig. 18.7) by a plane Π_1 at a point P is not a normal section. Then $\bar{n}_1$, the principal normal of the curve, is not parallel to $\bar{n}$, the normal to the surface. From Eqs. (18.2.3) and (18.2.7), we deduce that the principal normal can be written as

$$\bar{n}_1 = \frac{1}{C_{n1}} \frac{d^2\bar{r}}{ds^2}, \tag{18.3.58}$$

where C_{n1} is the first curvature of the section. Let γ be the inclination of the plane of the section to the normal plane Π which touches the curve at P. Then γ is the angle between $\bar{n}$ and $\bar{n}_1$. Hence,

$$\cos\gamma = \bar{n}\cdot\bar{n}_1 = \frac{\bar{n}}{C_{n1}}\cdot\frac{d^2\bar{r}}{ds^2}. \tag{18.3.59}$$

Therefore,

$$\cos\gamma = \frac{C_n}{C_{n1}} \tag{18.3.60}$$

or

$$C_n = C_{n1}\cos\gamma. \tag{18.3.61}$$

This is Meunier's theorem connecting the normal curvature in any direction with the curvature of any other section through the same tangent line.

4) *Principal directions and lines of curvature*

The normals at consecutive points of a surface do not intersect in general. However, at any point P of a surface, there are two directions at right angle to each other such that the normal at a consecutive point in either of these directions meets the normal at P. The two directions are called principal directions at P. To show this, let $\bar{r}$ be the position vector of P, and $\bar{n}$ be the unit normal to the surface there (Fig. 18.8).

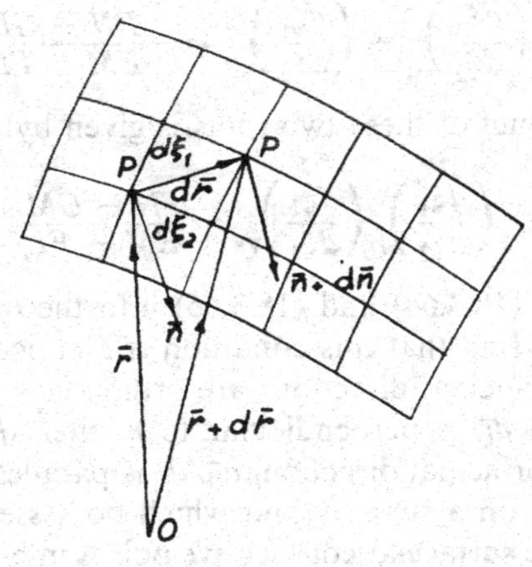

Fig. 18.8

Let $\bar{r} + d\bar{r}$ be an adjacent point in a direction defined by $d\xi_1$ and $d\xi_2$, and $\bar{n} + d\bar{n}$ be the unit normal at this point. The normals will intersect if $\bar{n}$, $\bar{n} + d\bar{n}$, and $d\bar{r}$ are coplanar; that is to say, if $\bar{n}$, $d\bar{n}$, and $d\bar{r}$ are coplanar (Fig. 18.8). The condition that three vectors be coplanar is that their scalar triple product vanishes, so that

$$\bar{n} \cdot d\bar{n} \times d\bar{r} = 0. \tag{18.3.62}$$

This condition can be expanded in terms of $d\xi_1$ and $d\xi_2$ using Eqs. (18.3.31) and Eqs. (18.3.48) to (18.3.51), to give:

$$(EM - FL)(d\xi_1)^2 + (EN - GL)d\xi_1\, d\xi_2 + (FN - GM)(d\xi_2)^2 = 0$$

or

$$(EM - FL)\left(\frac{d\xi_1}{d\xi_2}\right)^2 + (EN - GL)\left(\frac{d\xi_1}{d\xi_2}\right) + (FN - GM) = 0. \tag{18.3.63}$$

This equation gives two values of the ratio $(d\xi_1/d\xi_2)$ and, therefore, two directions on the surface for which the required property holds. Let these two directions be along two curves C and C' intersecting at P. From the theory of equations, we know that the sum of the two roots of Eq. (18.3.63) is given by:

$$\left(\frac{d\xi_1}{d\xi_2}\right)_C + \left(\frac{d\xi_1}{d\xi_2}\right)_{C'} = -\frac{EN - GL}{EM - FL}, \tag{18.3.64}$$

and that the product of these two roots is given by:

$$\left(\frac{d\xi_1}{d\xi_2}\right)_C \left(\frac{d\xi_1}{d\xi_2}\right)_{C'} = \frac{FN - GM}{EM - FL}. \tag{18.3.65}$$

Substituting Eqs. (18.3.64) and (18.3.65) into the orthogonality condition (18.3.27), we find that this condition is identically satisfied, which means that the principal directions are orthogonal. Since $d\bar{r}$ is perpendicular to $\bar{n}$, and $d\bar{n}$ is perpendicular to $\bar{n}$, then $d\bar{n}$ is parallel to $d\bar{r}$. Therefore, for a principal direction, $d\bar{n}/ds$ is parallel to $d\bar{r}/ds$.

A curve drawn on a surface, and which possesses the property that the normals to the surface at consecutive points intersect, is called a line of curvature. Therefore, the direction of a line of curvature at any point is a principal direction at that point. Through each point on the surface two lines of curvature pass, cutting each other at right angle. On the surface, there are two systems of lines of curvature whose differential equation is (18.3.63). It is interesting to remark that Eq. (18.3.63) gives the directions of the maximum and minimum normal curvatures at a point P. Indeed, if we differentiate C_n in Eq. (18.3.57) with respect to $(d\xi_1/d\xi_2)$ and equate the result to zero, we obtain Eq. (18.3.63). Thus, the principal directions at a point are the directions of greatest and least normal curvatures.

It is convenient in the development of the theory of thin shells to refer to its lines of curvature as parametric curves. If this is done, the differential equation (18.3.63) for the lines of curvature becomes identical with the differential equation of the parametric curves; that is,

$$d\xi_1 \, d\xi_2 = 0. \tag{18.3.66}$$

Hence, we must have:

$$EM - FL = 0 \tag{18.3.67}$$

$$FN - GM = 0 \tag{18.3.68}$$

and

$$EN - GL \neq 0. \tag{18.3.69}$$

Multiplying Eq. (18.3.67) by N, and Eq. (18.3.68) by L, and adding, we get:

$$(EN - GL)M = 0. \tag{18.3.70}$$

Multiplying Eq. (18.3.67) by G, and Eq. (18.3.68) by E, and adding, we get:

$$(EN - GL)F = 0. \tag{18.3.71}$$

In view of Eq. (18.3.69), the conditions that the parametric curves also be lines of curvature are:

$$F = M = 0. \tag{18.3.72}$$

5) *Principal curvatures, first and second curvatures*

The point of intersection of consecutive normals along a line of curvature at P is called a center of curvature of the surface: Its distance from P, measured in the direction of the unit normal $\bar{n}$, is called a principal radius of curvature of the surface at P. The reciprocal of a principal radius of curvature is called a principal curvature. Thus, at each point of a surface two principal curvatures exist and these are the normal curvatures of the surface in the direction of the lines of curvature. They must not be confused with the (first) curvatures of the line of curvature because the principal normal of a line of curvature is not, in general, the normal to the surface. In other words, the osculating plane of a line of curvature does not as a rule give a normal section of the surface; however, the curvature of a line of curvature is connected with the corresponding principal curvature by Meunier's theorem (18.3.61).

Those portions of the surface on which the two principal curvatures have the same sign are called synclastic: for example, the surface of a sphere is synclastic at all points. If the principal curvatures have opposite signs on any part of the surface, this part is said to be anticlastic: for example, the surface of a hyperbolic paraboloid is anticlastic at all points.

At any point of a surface, there are two centers of curvature—one for each principal direction. Both lie on the normal to the surface. Let the principal curvatures be denoted by $C(C_1$ or $C_2)$, and the principal radii of curvature by $R(R_1$ or $R_2)$. To determine the principal curvatures at any point, we proceed as follows: Let $\bar{r}$ be the position vector of the point P on the surface, $\bar{n}$ be the unit normal there, and $R(R_1$ or $R_2)$ be a principal radius of curvature (Fig. 18.9a). Then, the corresponding center of curvature is given by $\bar{\rho}$, where

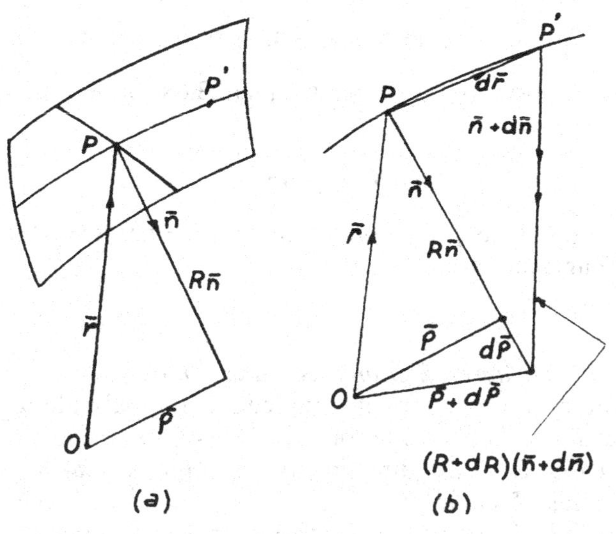

Fig. 18.9

$$\bar{\rho} = \bar{r} + R\bar{n}. \tag{18.3.73}$$

If P' is a point adjacent to P along a line of curvature of the surface (Fig. 18.9b), then

$$d\bar{\rho} = d\bar{r} + d(R\bar{n}) = d\bar{r} + R\,d\bar{n} + \bar{n}\,dR. \tag{18.3.74}$$

Now, the vector $(d\bar{r} + R\,d\bar{n})$ is tangential to the surface since both $d\bar{r}$ and $d\bar{n}$ are. The vector $d\bar{\rho}$ has the direction of $\bar{n}$ (see Fig. 18.9b); consequently, we must have:

$$d\bar{r} + Rd\bar{n} = 0.$$

If C (i.e., C_1 or C_2) is the corresponding principal curvature,

$$Cd\bar{r} + d\bar{n} = 0. \tag{18.3.75}$$

Eq. (18.3.75) is called Rodrigue's formula. Inserting Eqs. (18.3.31) into Eq. (18.3.75) and rearranging terms, we get:

$$\left(C\frac{\partial \bar{r}}{\partial \xi_1} + \frac{\partial \bar{n}}{\partial \xi_1}\right)d\xi_1 + \left(C\frac{\partial \bar{r}}{\partial \xi_2} + \frac{\partial \bar{n}}{\partial \xi_2}\right)d\xi_2 = 0. \tag{18.3.76}$$

Forming the scalar product of this equation with $\partial\bar{r}/\partial\xi_1$ and $\partial\bar{r}/\partial\xi_2$ successively, we obtain:

$$(CE - L)d\xi_1 + (CF - M)d\xi_2 = 0 \tag{18.3.77}$$

$$(CF - M)d\xi_1 + (CG - N)d\xi_2 = 0. \tag{18.3.78}$$

These two equations determine the principal curvatures and the directions of the lines of curvature. Eliminating $d\xi_1/d\xi_2$, we get:

$$H^2C^2 - (EN - 2FM + GL)C + (LN - M^2) = 0. \tag{18.3.79}$$

This is a quadratic in C, whose two roots are the principal curvatures C_1 and C_2. When the principal curvatures have been determined from Eq. (18.3.79), the direction of the lines of curvature is given by either Eq. (18.3.77) or (18.3.78). Thus, corresponding to C_1, the principal direction is given by:

$$\left(\frac{d\xi_1}{d\xi_2}\right)_1 = -\frac{C_1F - M}{C_1E - L} = -\frac{C_1G - N}{C_1F - M}, \tag{18.3.80}$$

and, corresponding to C_2, the principal direction is given by:

$$\left(\frac{d\xi_1}{d\xi_2}\right)_2 = -\frac{C_2F - M}{C_2E - L} = -\frac{C_2G - N}{C_2F - M}. \quad (18.3.81)$$

The directions of the lines of curvature may also be found by eliminating C from Eqs. (18.3.77) and (18.3.78). This elimination leads to:

$$(EM - FL)(d\xi_1)^2 + (EN - GL)d\xi_1\, d\xi_2 + (FN - GM)(d\xi_2)^2 = 0, \quad (18.3.82)$$

which was previously obtained in a different way. Eq. (18.3.82) fails to determine the principal directions if the coefficients vanish identically; that is to say, when

$$E : F : G = L : M : N.$$

In this case, the normal curvature as given by Eq. (18.3.55) is independent of the ratio $(d\xi_1/d\xi_2)$ and, consequently, has the same value for all directions through the point. Such a point is called an umbilic on the surface.

The first curvature of the surface at any point is defined as the sum of the principal curvatures. It is denoted by J. Thus,

$$J = C_1 + C_2 = \frac{1}{H^2}(EN - 2FM + GL). \quad (18.3.83)$$

The second curvature of the surface, also called the Gaussian or the specific curvature, is defined as the product of the principal curvatures. It is denoted by K. Thus,

$$K = C_1 C_2 = \frac{LN - M^2}{H^2}. \quad (18.3.84)$$

6) *Euler's theorem*

The condition that the parametric curves also be lines of curvature were found to be (18.3.72), $F = M = 0$. $F = 0$ is the condition of orthogonality of the parametric curves (18.3.17). Euler's theorem expresses the normal curvature at a point in any direction in terms of the principal curvatures. If, in the expression of the normal curvature (18.3.55),

$$C_n = \frac{L(d\xi_1)^2 + 2Md\xi_1\, d\xi_2 + N(d\xi_2)^2}{E(d\xi_1)^2 + 2Fd\xi_1\, d\xi_2 + Gd\xi_2^2}, \quad (18.3.85)$$

we set $M = F = 0$ and $d\xi_2 = 0$ (since along the ξ_1 curves ξ_2 is constant), we obtain:

$$C_1 = \frac{L}{E} = \frac{1}{R_1}. \tag{18.3.86}$$

Similarly, with $d\xi_1 = 0$, we get:

$$C_2 = \frac{N}{G} = \frac{1}{R_2}. \tag{18.3.87}$$

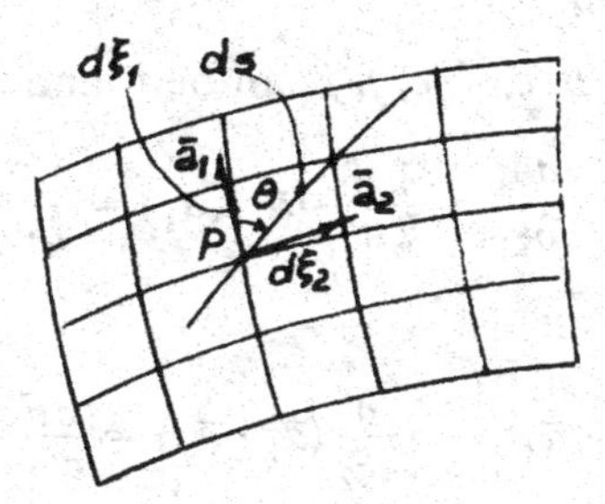

Fig. 18.10

Let us now consider a normal section of the surface in a direction making an angle θ with the ξ_1 curve at P (Fig. 18.10). Using Eqs. (18.3.15) and (18.3.16), we have:

$$\cos\theta = \sqrt{E}\,\frac{d\xi_1}{ds} \tag{18.3.88}$$

$$\sin\theta = \sqrt{G}\,\frac{d\xi_2}{ds}, \tag{18.3.89}$$

so that the normal curvature of this section is by Eq. (18.3.85):

$$C_n = \frac{L}{E}\cos^2\theta + \frac{N}{G}\sin^2\theta \tag{18.3.90}$$

so that

$$C_n = C_1\cos^2\theta + C_2\sin^2\theta. \tag{18.3.91}$$

This is known as Euler's theorem on normal curvature. From this theorem, it follows that the sum of normal curvatures in two directions at right angle is constant, and equal to the sum of the principal curvatures.

7) *Rate of change of the vectors* $\bar{a}_i$ *and the corresponding unit vectors along the parametric lines*

Recalling that the fundamental magnitudes L, M, and N are the projections on $\bar{n}$ of the second derivatives of $\bar{r}$, we can write:

$$\frac{\partial \bar{a}_1}{\partial \xi_1} = L\bar{n} + b\bar{a}_1 + e\bar{a}_2 \tag{18.3.92}$$

$$\frac{\partial \bar{a}_1}{\partial \xi_2} = \frac{\partial \bar{a}_2}{\partial \xi_1} = M\bar{n} + c\bar{a}_1 + f\bar{a}_2 \tag{18.3.93}$$

$$\frac{\partial \bar{a}_2}{\partial \xi_2} = N\bar{n} + d\bar{a}_1 + j\bar{a}_2, \tag{18.3.94}$$

where the coefficients b, c, d, e, f, j can be found as follows:

$$\bar{a}_1 \cdot \frac{\partial \bar{a}_1}{\partial \xi_1} = \frac{1}{2}\frac{\partial}{\partial \xi_1}(\bar{a}_1 \cdot \bar{a}_1) = \frac{1}{2}\frac{\partial E}{\partial \xi_1} \tag{18.3.95}$$

and

$$\bar{a}_2 \cdot \frac{\partial \bar{a}_1}{\partial \xi_1} = \frac{\partial}{\partial \xi_1}(\bar{a}_1 \cdot \bar{a}_2) - \frac{1}{2}\frac{\partial}{\partial \xi_2}(\bar{a}_1 \cdot \bar{a}_1) = \frac{\partial F}{\partial \xi_1} - \frac{1}{2}\frac{\partial E}{\partial \xi_2}. \tag{18.3.96}$$

By forming the scalar product of each side of Eq. (18.3.92) with $\bar{a}_1$ and $\bar{a}_2$, respectively, we get:

$$\frac{1}{2}\frac{\partial E}{\partial \xi_1} = bE + eF \tag{18.3.97}$$

$$\frac{\partial F}{\partial \xi_1} - \frac{1}{2}\frac{\partial E}{\partial \xi_2} = bF + eG. \tag{18.3.98}$$

Solving for b and e, we get:

$$b = \frac{1}{2H^2}\left(G\frac{\partial E}{\partial \xi_1} - 2F\frac{\partial F}{\partial \xi_1} + F\frac{\partial E}{\partial \xi_2}\right) \tag{18.3.99}$$

$$e = \frac{1}{2H^2}\left(2E\frac{\partial F}{\partial \xi_1} - E\frac{\partial E}{\partial \xi_2} - F\frac{\partial E}{\partial \xi_1}\right). \tag{18.3.100}$$

By forming the scalar product of each side of Eq. (18.3.93) with $\bar{a}_1$ and $\bar{a}_2$, respectively, the two following expressions for c and f are obtained:

$$c = \frac{1}{2H^2}\left(G\frac{\partial E}{\partial \xi_2} - F\frac{\partial G}{\partial \xi_1}\right) \tag{18.3.101}$$

$$f = \frac{1}{2H^2}\left(E\frac{\partial G}{\partial \xi_1} - F\frac{\partial E}{\partial \xi_2}\right). \tag{18.3.102}$$

By forming the scalar product of each side of Eq. (18.3.94) with $\bar{a}_1$ and $\bar{a}_2$, respectively, the two following expressions for d and j are obtained:

$$d = \frac{1}{2H^2}\left(2G\frac{\partial F}{\partial \xi_2} - G\frac{\partial G}{\partial \xi_1} - F\frac{\partial G}{\partial \xi_2}\right) \qquad (18.3.103)$$

$$j = \frac{1}{2H^2}\left(E\frac{\partial G}{\partial \xi_2} - 2F\frac{\partial F}{\partial \xi_2} + F\frac{\partial G}{\partial \xi_1}\right). \qquad (18.3.104)$$

Eqs. (18.3.92), (18.3.93), and (18.3.94), with the coefficients given by Eqs. (18.3.99) to (18.3.104), give the required derivatives of $\bar{a}_1$ and $\bar{a}_2$.

When the parametric curves are orthogonal, the values of the coefficients b, c, d, e, f, j are simplified, since $F = 0$ and $H^2 = EG$. The derivatives become:

$$\frac{\partial \bar{a}_1}{\partial \xi_1} = L\bar{n} + \frac{\frac{\partial E}{\partial \xi_1}}{2E}\bar{a}_1 - \frac{\frac{\partial E}{\partial \xi_2}}{2G}\bar{a}_2 \qquad (18.3.105)$$

$$\frac{\partial \bar{a}_2}{\partial \xi_1} = \frac{\partial \bar{a}_1}{\partial \xi_2} = M\bar{n} + \frac{\frac{\partial E}{\partial \xi_2}}{2E}\bar{a}_1 + \frac{\frac{\partial G}{\partial \xi_1}}{2G}\bar{a}_2 \qquad (18.3.106)$$

$$\frac{\partial \bar{a}_2}{\partial \xi_2} = N\bar{n} - \frac{\frac{\partial G}{\partial \xi_1}}{2E}\bar{a}_1 + \frac{\frac{\partial G}{\partial \xi_2}}{2G}\bar{a}_2. \qquad (18.3.107)$$

If $\bar{e}_1$ and $\bar{e}_2$ are the unit vectors parallel to $\bar{a}_1$ and $\bar{a}_2$, then, from Eqs. (18.3.11) and (18.3.13), we have:

$$\bar{e}_1 = \frac{\bar{a}_1}{\sqrt{a_{11}}} = \frac{\bar{a}_1}{\sqrt{E}} \qquad (18.3.108)$$

$$\bar{e}_2 = \frac{\bar{a}_2}{\sqrt{a_{22}}} = \frac{\bar{a}_2}{\sqrt{G}}. \qquad (18.3.109)$$

$\bar{e}_1$, $\bar{e}_2$, and $\bar{n}$ form a right-handed system of unit vectors mutually perpendicular. From Eqs. (18.3.105) to (18.3.109), we easily deduce that

$$\frac{\partial \bar{e}_1}{\partial \xi_1} = \frac{L}{\sqrt{E}}\bar{n} - \frac{\frac{\partial E}{\partial \xi_2}}{2H}\bar{e}_2 \qquad (18.3.110)$$

$$\frac{\partial \bar{e}_1}{\partial \xi_2} = \frac{M}{\sqrt{E}}\bar{n} + \frac{\frac{\partial G}{\partial \xi_1}}{2H}\bar{e}_2 \qquad (18.3.111)$$

$$\frac{\partial \bar{e}_2}{\partial \xi_1} = \frac{M}{\sqrt{G}}\bar{n} + \frac{\frac{\partial E}{\partial \xi_2}}{2H}\bar{e}_1 \qquad (18.3.112)$$

$$\frac{\partial \bar{e}_2}{\partial \xi_2} = \frac{N}{\sqrt{G}}\bar{n} - \frac{\frac{\partial G}{\partial \xi_1}}{2H}\bar{e}_1. \qquad (18.3.113)$$

From Eqs. (18.3.46) and (18.3.47), in which we set $F = 0$ and $H^2 = EG$, we get:

$$\frac{\partial \bar{n}}{\partial \xi_1} = -\frac{L}{\sqrt{E}}\bar{e}_1 - \frac{M}{\sqrt{G}}\bar{e}_2 \tag{18.3.114}$$

$$\frac{\partial \bar{n}}{\partial \xi_2} = -\frac{M}{\sqrt{E}}\bar{e}_1 - \frac{N}{\sqrt{G}}\bar{e}_2. \tag{18.3.115}$$

The previous derivatives can further be simplified if the parametric curves are lines of curvature. In such a case, M is set equal to zero and the derivatives of the unit vectors $\bar{e}_1$, $\bar{e}_2$, and $\bar{n}$ become:

$$\frac{\partial \bar{e}_1}{\partial \xi_1} = \frac{L}{\sqrt{E}}\bar{n} - \frac{\frac{\partial E}{\partial \xi_2}}{2H}\bar{e}_2 = \frac{\sqrt{E}}{R_1}\bar{n} - \frac{1}{\sqrt{G}}\frac{\partial \sqrt{E}}{\partial \xi_2}\bar{e}_2 \tag{18.3.116}$$

$$\frac{\partial \bar{e}_1}{\partial \xi_2} = \frac{\frac{\partial G}{\partial \xi_1}}{2H}\bar{e}_2 = \frac{1}{\sqrt{E}}\frac{\partial \sqrt{G}}{\partial \xi_1}\bar{e}_2 \tag{18.3.117}$$

$$\frac{\partial \bar{e}_2}{\partial \xi_1} = \frac{\frac{\partial E}{\partial \xi_2}}{2H}\bar{e}_1 = \frac{1}{\sqrt{G}}\frac{\partial \sqrt{E}}{\partial \xi_2}\bar{e}_1 \tag{18.3.118}$$

$$\frac{\partial \bar{e}_2}{\partial \xi_2} = \frac{N}{\sqrt{G}}\bar{n} - \frac{\frac{\partial G}{\partial \xi_1}}{2H}\bar{e}_1 = \frac{\sqrt{G}}{R_2}\bar{n} - \frac{1}{\sqrt{E}}\frac{\partial \sqrt{G}}{\partial \xi_1}\bar{e}_1 \tag{18.3.119}$$

$$\frac{\partial \bar{n}}{\partial \xi_1} = -\frac{L}{\sqrt{E}}\bar{e}_1 = -\frac{\sqrt{E}}{R_1}\bar{e}_1 \tag{18.3.120}$$

$$\frac{\partial \bar{n}}{\partial \xi_2} = -\frac{N}{\sqrt{G}}\bar{e}_2 = -\frac{\sqrt{G}}{R_2}\bar{e}_2. \tag{18.3.121}$$

Eqs. (18.3.116) to (18.3.121) could have been directly obtained from Eqs. (6.5.35) to (6.5.38). This is easily seen by making the following substitutions in the latter group of equations:

y_1, y_2, and y_3 are replaced by ξ_1, ξ_2, and ξ_3

$\bar{e}_3$ is replaced by $\bar{n}$

h_1 is replaced by $\sqrt{E}\left(1 - \frac{\xi_3}{R_1}\right)$

h_2 is replaced by $\sqrt{G}\left(1 - \frac{\xi_3}{R_2}\right)$

h_3 is set equal to unity.

The three last substitutions will be justified in Sec. 18.4.

8) *The Gauss-Codazzi conditions*

The six fundamental magnitudes E, F, G, L, M, N are not functionally independent but are connected by three differential relations. These relations are to be satisfied if the six magnitudes are to determine a surface uniquely, except for its position and orientation in space (compare to compatibility equations of strain). We shall restrict ourselves to the cases in which the parametric curves are lines of curvature. The fundamental magnitudes are thus reduced to four, since in this case $F = M = 0$. The three relations are derived by writing the equality of the mixed second derivatives of the unit vectors. Let us first consider the unit vector $\bar{n}$. From Eqs. (18.3.120) and (18.3.121), we have:

$$\frac{\partial}{\partial \xi_2}\left(-\frac{\sqrt{E}}{R_1}\bar{e}_1\right) = \frac{\partial}{\partial \xi_1}\left(-\frac{\sqrt{G}}{R_2}\bar{e}_2\right). \tag{18.3.122}$$

If we carry the differentiation, we get:

$$\bar{e}_1\left[\frac{\partial}{\partial \xi_2}\left(\frac{\sqrt{E}}{R_1}\right) - \frac{1}{R_2}\left(\frac{\partial\sqrt{E}}{\partial \xi_2}\right)\right] + \bar{e}_2\left[\frac{1}{R_1}\frac{\partial\sqrt{G}}{\partial \xi_1} - \frac{\partial}{\partial \xi_1}\left(\frac{\sqrt{G}}{R_2}\right)\right] = 0. \tag{18.3.123}$$

This equation is satisfied only if the coefficients of $\bar{e}_1$ and $\bar{e}_2$ vanish; hence,

$$\frac{1}{R_2}\frac{\partial\sqrt{E}}{\partial \xi_2} = \frac{\partial}{\partial \xi_2}\left(\frac{\sqrt{E}}{R_1}\right) \tag{18.3.124}$$

$$\frac{1}{R_1}\frac{\partial\sqrt{G}}{\partial \xi_1} = \frac{\partial}{\partial \xi_1}\left(\frac{\sqrt{G}}{R_2}\right). \tag{18.3.125}$$

Eqs. (18.3.124) and (18.3.125) are known as the Codazzi conditions.

If we repeat the previous steps with Eqs. (18.3.116) and (18.3.117), we obtain two conditions of which only one is new—namely,

$$\frac{\partial}{\partial \xi_1}\left(\frac{1}{\sqrt{E}}\frac{\partial\sqrt{G}}{\partial \xi_1}\right) + \frac{\partial}{\partial \xi_2}\left(\frac{1}{\sqrt{G}}\frac{\partial\sqrt{E}}{\partial \xi_2}\right) = -\frac{\sqrt{EG}}{R_1 R_2}. \tag{18.3.126}$$

This equation is known as the Gauss condition. It is useless to consider the two equations (18.3.118) and (18.3.119), since they do not lead to any new relations. The three conditions (18.3.124), (18.3.125), and (18.3.126) are known as the Gauss-Codazzi conditions. We now state a fundamental theorem of the theory of surfaces: If E, G, L, and N are given functions of the real curvilinear coordinates ξ_1 and ξ_2, are differentiable, and satisfy the Gauss-Codazzi conditions while $E > 0$ and $G > 0$, then a real surface exists which is uniquely determined except for its position in space and which has $[E(d\xi_1)^2 + G(d\xi_2)^2]$ and $[L(d\xi_1)^2 + N(d\xi_2)^2]$ as first and second fundamental forms. The Gauss-Codazzi conditions are referred to as the compatibility conditions of the theory of surfaces.

9) *Application to surfaces of revolution*

A surface of revolution is obtained by rotation of a plane curve about an axis lying in the plane of the curve. The curve is called the meridian, and its plane is the meridian plane. Let the axis of rotation be the OZ axis, and let R_o be the perpendicular from any point P on the surface to the OZ axis (Fig. 18.11). The equation of a meridian is:

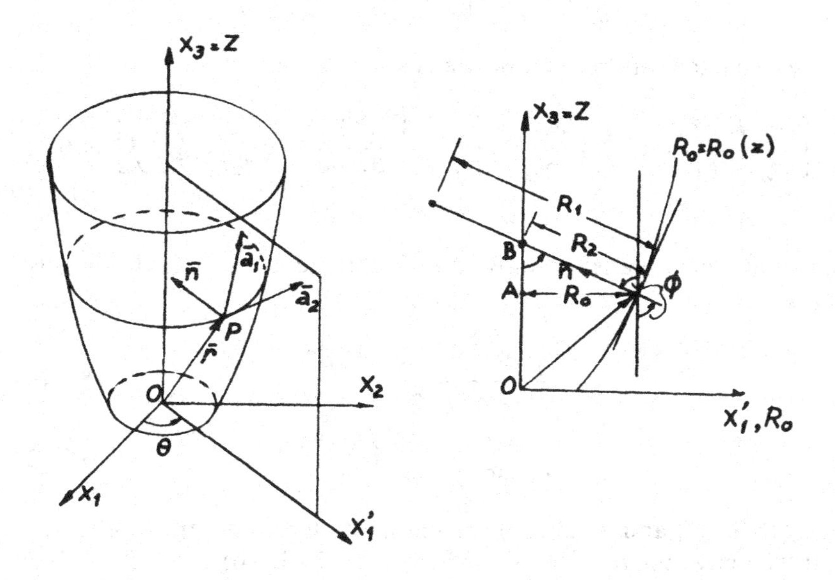

Fig. 18.11

$$R_o = R_o(z), \tag{18.3.127}$$

and its position is defined by the angle θ with the OX_1, OZ plane. The intersections of the surface with planes perpendicular to the OZ axis are circles called parallels. The position of a parallel is defined by the equation:

$$z = \text{constant}. \tag{18.3.128}$$

The cartesian coordinates of P are:

$$x_1 = R_o \cos \theta, \quad x_2 = R_o \sin \theta, \quad x_3 = z. \tag{18.3.129}$$

The position vector of P is:

$$\bar{r} = \bar{i}_1 R_o \cos \theta + \bar{i}_2 R_o \sin \theta + \bar{i}_3 z. \tag{18.3.130}$$

The parametric equations of the surface are:

$$x_1 = x_1(\theta, z) = R_o(z) \cos \theta \tag{18.3.131}$$

$$x_2 = x_2(\theta, z) = R_o(z) \sin \theta \tag{18.3.132}$$

$$x_3 = x_3(\theta, z) = z. \tag{18.3.133}$$

For each constant value of θ, there corresponds a meridian of the surface; and for each constant of z, there corresponds a parallel on the surface. Let us take the meridians and the parallels as our parametric curves and identify z with ξ_1, and θ with ξ_2. Then

$$\bar{a}_1 = \frac{\partial \bar{r}}{\partial \xi_1} = \frac{\partial \bar{r}}{\partial z} = \bar{i}_1 R'_o \cos \theta + \bar{i}_2 R'_o \sin \theta + \bar{i}_3 \tag{18.3.134}$$

$$\bar{a}_2 = \frac{\partial \bar{r}}{\partial \xi_2} = \frac{\partial \bar{r}}{\partial \theta} = -\bar{i}_1 R_o \sin \theta + \bar{i}_2 R_o \cos \theta, \tag{18.3.135}$$

where

$$R'_o = \frac{dR_o}{dz}. \tag{18.3.136}$$

The first fundamental magnitudes are:

$$E = \bar{a}_1 \cdot \bar{a}_1 = \frac{\partial \bar{r}}{\partial z} \cdot \frac{\partial \bar{r}}{\partial z} = 1 + (R'_o)^2 \tag{18.3.137}$$

$$F = \bar{a}_1 \cdot \bar{a}_2 = \frac{\partial \bar{r}}{\partial z} \cdot \frac{\partial \bar{r}}{\partial \theta} = 0 \tag{18.3.138}$$

$$G = \bar{a}_2 \cdot \bar{a}_2 = \frac{\partial \bar{r}}{\partial \theta} \cdot \frac{\partial \bar{r}}{\partial \theta} = R_o^2 \tag{18.3.139}$$

$$H = \sqrt{EG - F^2} = R_o \sqrt{1 + (R'_o)^2}\,. \tag{18.3.140}$$

The first fundamental form is:

$$(ds)^2 = [1 + (R_o')^2](dz)^2 + R_o^2(d\theta)^2. \tag{18.3.141}$$

The normal is given by:

$$\bar{n} = \frac{\bar{a}_1 \times \bar{a}_2}{H} = -\frac{1}{H}(\bar{i}_1 R_o \cos\theta + \bar{i}_2 R_o \sin\theta - \bar{i}_3 R_o R_o'), \tag{18.3.142}$$

and the second fundamental magnitudes are:

$$L = \bar{n} \cdot \frac{\partial^2 r}{\partial \xi_1^2} = -\frac{R_o R_o''}{H} \tag{18.3.143}$$

$$M = \bar{n} \cdot \frac{\partial \bar{r}}{\partial \xi_1 \partial \xi_2} = 0 \tag{18.3.144}$$

$$N = \bar{n} \cdot \frac{\partial^2 \bar{r}}{\partial \xi_2^2} = \frac{R_o^2}{H}, \tag{18.3.145}$$

where

$$R_o'' = \frac{d^2 R_o}{dz^2}.$$

Since both F and M are equal to zero, the parametric curves are also lines of curvature. The principal radii of curvature are calculated from Eqs. (18.3.86) and (18.3.87), with the result that

$$R_1 = \frac{E}{L} = -\frac{[1 + (R_o')^2]^{\frac{3}{2}}}{R_o''} \tag{18.3.146}$$

$$R_2 = \frac{G}{N} = R_o\sqrt{1 + (R_o')^2}\,. \tag{18.3.147}$$

We see (Fig. 18.11) that R_1 is the radius of curvature of the generating curve $R_o = R_o(z)$. R_2 is the length of the normal intercepted between P and the OZ axis. Both R_1 and R_2 are positive quantities, which means that they are measured in the positive direction of $\bar{n}$. Finally, it is easy to check that E, G, R_1, and R_2 define a valid surface by substituting their values in the Gauss-Codazzi equations which are satisfied.

An alternate and useful description of a surface of revolution is based on the independent variables ϕ and θ, where ϕ is the angle between the axis of revolution of the surface and the normal to the surface at that point. In this case, the first fundamental form is:

$$(ds)^2 = R_1^2(d\phi)^2 + R_o^2(d\theta)^2, \tag{18.3.148}$$

where the first term in the right-hand side represents the square of the differential length of arc along a meridian, and the second term represents the square of the differential length of arc along a parallel. In this case, ξ_1 is identified with ϕ, and ξ_2 with θ. The first fundamental magnitudes in this case are R_1^2 and R_o^2.

10) *Important remarks*

a) The analogy between the metric properties of a surface [as expressed through the metric coefficients a_{ij} of its first fundamental form] and the corresponding metric properties of a three dimensional Euclidean space [as expressed through the metric coefficients g_{ij} of the quadratic differential form (6.3.12) associated with any general system of coordinates], is evident. The expressions for g_{ij} and a_{ij} are the same, except that the indices of a take the values 1 and 2, while the indices of g take the values of 1, 2, and 3. The quantities h_1, h_2 of Chapter 6 can be identified with $\sqrt{E}$ and $\sqrt{G}$, respectively. A difference which is of importance exists, however, between the two cases:

In the case of a three dimensional Euclidean space, it is always possible, by a change of coordinates, to reduce the quadratic differential form to a sum of squares of differential coordinates; indeed, this is a characteristic of Euclidean spaces. In the case of a surface, unless the Gaussian curvature vanishes, it is not possible to reduce its first fundamental form to a sum of squares of the differentials of its Gaussian coordinates. Therefore, a surface for which the Gaussian curvature K does not vanish is called a two dimensional non-Euclidean space. The Gaussian curvature vanishes for a flat surface as well as for a developable surface, such as that of a cylinder or of a cone.

b) In a system of orthogonal curvilinear coordinates, the curves of intersection of the three coordinate surfaces S_1, S_2, and S_3 (Fig. 6.1) are lines of curvature of these surfaces (Dupin's Theorem [1]).

c) In some texts, the normal to the surface is chosen in a way such that it points away from the center of curvature. This results in some changes in sign in those expressions containing $\bar{n}$. Caution should be exercised when comparing the equations deduced in this text with those in which such a convention is adopted.

18.4 Basic Assumptions and Reference System of Coordinates

In developing the theory of thin elastic shells, the following assumptions (which are called Love's assumptions) are made:

1. The shell is thin. This means that the thickness of the shell h is small compared with the radii of curvature R_1 and R_2 of the middle surface, so that their ratio is small compared to unity.

2. The deflections of the shell are small, and the strains in the direction of the normal are small enough to be neglected. This assumption allows us to refer the analyses to the initial configuration of the shell.

3. The normal stresses acting on planes parallel to the middle surface are negligible compared with other stress components and may be neglected in the stress-strain relations. This assumption will generally be valid except in the vicinity of highly concentrated loads.

4. The components of the displacements (u_1 and u_2) are linearly distributed across the thickness.

5. The shear strains which cause the distortions of the normals to the middle surface (e_{13} and e_{23}) can be neglected. This, added to the previous assumption, allows us to conclude that the normals to the undeformed middle surface remain normal to it after deformation.

From the study of the theory of surfaces, we know that any point on a shell can be located by means of three parameters, two of which vary on the middle surface while the third one varies along the normal to the middle surface. The middle surface will be the reference surface and its lines of curvature will be chosen as parametric curves. Those lines, together with the normal, form our orthogonal system of reference. An arbitrary point in the space occupied by the shell can therefore be located by means of the position vector $\bar{\xi}(\xi_1, \xi_2, \xi_3)$:

$$\bar{\xi}(\xi_1, \xi_2, \xi_3) = \bar{r}(\xi_1, \xi_2) + \xi_3 \bar{n}(\xi_1, \xi_2), \tag{18.4.1}$$

where $\bar{r}$ is the position vector of a corresponding point on the middle surface, $\bar{n}$ is the unit normal vector, ξ_3 is the distance of the arbitrary point from the middle surface measured along $\bar{n}$ (Fig. 18.12). The magnitude of an element of length is given by:

$$(ds)^2 = d\bar{\xi} \cdot d\bar{\xi} = (d\bar{r} + \xi_3 \, d\bar{n} + \bar{n} \, d\xi_3) \cdot (d\bar{r} + \xi_3 \, d\bar{n} + \bar{n} \, d\xi_3). \tag{18.4.2}$$

If this scalar product is carried out, keeping in mind the orthogonality of the coordinates, we get:

$$(ds)^2 = E\left(1 - \frac{\xi_3}{R_1}\right)^2 (d\xi_1)^2 + G\left(1 - \frac{\xi_3}{R_2}\right)^2 (d\xi_2)^2 + (d\xi_3)^2. \tag{18.4.3}$$

This expression contains all the information needed to measure lengths, areas, and volumes in a shell. In other words, it contains all the metric properties of the shell. The first two terms in the right-hand side represent the first fundamental form of a surface at a distance ξ_3 from the middle surface (Fig. 18.12). The lengths of the edges of this element of surface are:

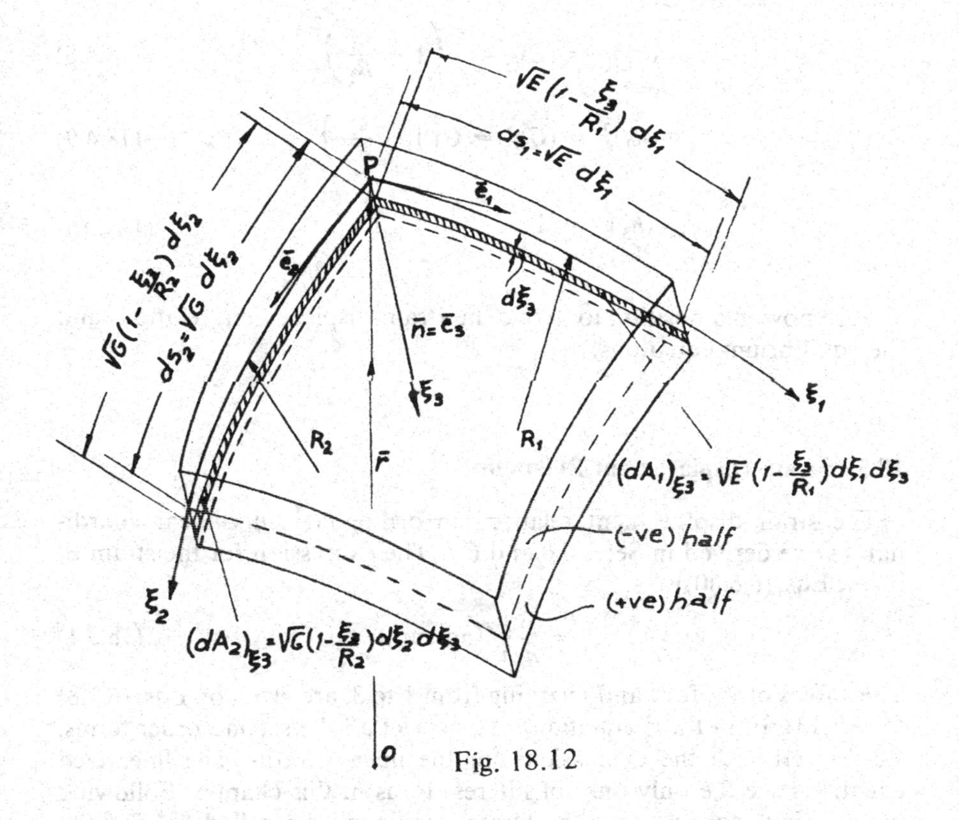

Fig. 18.12

$$(ds_1)_{\xi_3} = \sqrt{E}\left(1 - \frac{\xi_3}{R_1}\right)d\xi_1 \tag{18.4.4}$$

$$(ds_2)_{\xi_3} = \sqrt{G}\left(1 - \frac{\xi_3}{R_2}\right)d\xi_2, \tag{18.4.5}$$

and the differential areas of the edge faces are:

$$(dA_1)_{\xi_3} = \sqrt{E}\left(1 - \frac{\xi_3}{R_1}\right)d\xi_1\,d\xi_3 \tag{18.4.6}$$

$$(dA_2)_{\xi_3} = \sqrt{G}\left(1 - \frac{\xi_3}{R_2}\right)d\xi_2\,d\xi_3 . \tag{18.4.7}$$

Therefore, the metric coefficients of a surface at a distance ξ_3 from the middle surface are:

$$(h_1)^2_{\xi_3} = (E)_{\xi_3} = E\left(1 - \frac{\xi_3}{R_1}\right)^2 \tag{18.4.8}$$

$$(h_2)^2_{\xi_3} = (G)_{\xi_3} = G\left(1 - \frac{\xi_3}{R_2}\right)^2 . \tag{18.4.9}$$

$$(h_3)^2_{\xi_3} = 1 \tag{18.4.10}$$

We are now in a position to derive the strain-displacement relations and the equilibrium equations.

18.5 Strain-Displacement Relations

The strain-displacement relations in orthogonal curvilinear coordinates were derived in Secs. 6.6 and 6.7. The expression for the strain ε_{ij} is [see Eqs. (6.6.30)]

$$\varepsilon_{ij} = \frac{\gamma_{ij}}{h_i h_j} \text{ (no sum)} . \tag{18.5.1}$$

The values of γ_{ij}, for i and j varying from 1 to 3, are given by Eqs. (6.7.8) to (6.7.13). If, in these equations, we neglect all the second order terms, we are left with the expressions for the linear strains. The linearized equations are the only ones of interest to us in this chapter. Following our previous conventions, the linear strains will be called e_{ij}. For the application to thin shells, we make the following substitutions:

$$\begin{aligned} &y_1 \text{ is replaced by } \xi_1 \\ &y_2 \text{ is replaced by } \xi_2 \\ &y_3 \text{ is replaced by } \xi_3 . \end{aligned} \tag{18.5.2}$$

u_1, u_2, and u_3 are the components of the displacement of a point along the ξ_1, ξ_2, and ξ_3 curves. Hence, from Eqs. (18.5.1) and (6.7.8) to (6.7.13), we have:

$$e_{11} = \frac{1}{h_1}\frac{\partial u_1}{\partial \xi_1} + \frac{u_2}{h_1 h_2}\frac{\partial h_1}{\partial \xi_2} + \frac{u_3}{h_1 h_3}\frac{\partial h_1}{\partial \xi_3} \tag{18.5.3}$$

$$e_{22} = \frac{1}{h_2}\frac{\partial u_2}{\partial \xi_2} + \frac{u_3}{h_2 h_3}\frac{\partial h_2}{\partial \xi_3} + \frac{u_1}{h_2 h_1}\frac{\partial h_2}{\partial \xi_1} \tag{18.5.4}$$

$$e_{33} = \frac{1}{h_3}\frac{\partial u_3}{\partial \xi_3} + \frac{u_1}{h_1 h_3}\frac{\partial h_3}{\partial \xi_1} + \frac{u_2}{h_3 h_2}\frac{\partial h_3}{\partial \xi_2} \tag{18.5.5}$$

$$e_{12} = \frac{1}{2}\left(\frac{1}{h_2}\frac{\partial u_1}{\partial \xi_2} + \frac{1}{h_1}\frac{\partial u_2}{\partial \xi_1} - \frac{u_2}{h_1 h_2}\frac{\partial h_2}{\partial \xi_1} - \frac{u_1}{h_1 h_2}\frac{\partial h_1}{\partial \xi_2}\right) \tag{18.5.6}$$

$$e_{23} = \frac{1}{2}\left(\frac{1}{h_3}\frac{\partial u_2}{\partial \xi_3} + \frac{1}{h_2}\frac{\partial u_3}{\partial \xi_2} - \frac{u_3}{h_2 h_3}\frac{\partial h_3}{\partial \xi_2} - \frac{u_2}{h_2 h_3}\frac{\partial h_2}{\partial \xi_3}\right) \tag{18.5.7}$$

$$e_{13} = \frac{1}{2}\left(\frac{1}{h_1}\frac{\partial u_3}{\partial \xi_1} + \frac{1}{h_3}\frac{\partial u_1}{\partial \xi_3} - \frac{u_1}{h_3 h_1}\frac{\partial h_1}{\partial \xi_3} - \frac{u_3}{h_3 h_1}\frac{\partial h_3}{\partial \xi_1}\right). \tag{18.5.8}$$

In Eqs. (18.5.3) to (18.5.8), h_1^2, h_2^2, and h_3^2 are the metric coefficients of a surface at a distance ξ_3 from the middle surface. From Eqs. (18.4.8) to (18.4.10), h_1, h_2, and h_3 are given by:

$$h_1 = \sqrt{E}\left(1 - \frac{\xi_3}{R_1}\right) \tag{18.5.9}$$

$$h_2 = \sqrt{G}\left(1 - \frac{\xi_3}{R_2}\right) \tag{18.5.10}$$

$$h_3 = 1. \tag{18.5.11}$$

The above strain-displacement relations are general and do not reflect the assumptions of Sec. 18.4:

a) The second assumption, in conjunction with Eqs. (18.5.5) and (18.5.11), leads us to:

$$e_{33} = \frac{\partial u_3}{\partial \xi_3} = 0, \tag{18.5.12}$$

that is to say:

$$u_3 = u_3(\xi_1, \xi_2) = u_{30}, \tag{18.5.13}$$

where u_{30} is the displacement of the middle surface in the direction of the normal.

b) The fourth assumption allows us to represent the components of the displacement at each point as follows:

$$u_1 = u_{10} + \xi_3 \left(\frac{\partial u_1}{\partial \xi_3} \right)_{\xi_3=0} \tag{18.5.14}$$

$$u_2 = u_{20} + \xi_3 \left(\frac{\partial u_2}{\partial \xi_3} \right)_{\xi_3=0} . \tag{18.5.15}$$

u_{10} and u_{20} represent the components of the displacements of points on the middle surface $\xi_3 = 0$. $(\partial u_1 / \partial \xi_3)_{\xi_3=0}$ and $(\partial u_2 / \partial \xi_3)_{\xi_3=0}$ represent the angles of rotation (more precisely their tangent) of the normal to the parametric curves ξ_1 and ξ_2, which lie on the middle surface. In the following, we shall refer to these two angles as β_1 and β_2.

c) The fifth assumption results in $e_{23} = e_{13} = 0$. Taking into account Eqs. (18.5.9) to (18.5.11), we have:

$$e_{23} = \frac{1}{2} \left[\frac{\partial u_2}{\partial \xi_3} + \frac{1}{\sqrt{G}\left(1 - \frac{\xi_3}{R_2}\right)} \frac{\partial u_3}{\partial \xi_2} + \frac{u_2}{R_2\left(1 - \frac{\xi_3}{R_2}\right)} \right] = 0 \tag{18.5.16}$$

and setting $\xi_3 = 0$, we obtain:

$$\left(\frac{\partial u_2}{\partial \xi_3} \right)_{\xi_3=0} = \left(-\frac{1}{\sqrt{G}} \frac{\partial u_3}{\partial \xi_2} - \frac{u_{20}}{R_2} \right) = \beta_2 . \tag{18.5.17}$$

In the same way, $e_{13} = 0$ leads to:

$$\left(\frac{\partial u_1}{\partial \xi_3} \right)_{\xi_3=0} = \left(-\frac{1}{\sqrt{E}} \frac{\partial u_3}{\partial \xi_1} - \frac{u_{10}}{R_1} \right) = \beta_1 . \tag{18.5.18}$$

Hence,

$$u_1 = u_{10} - \xi_3 \left(\frac{1}{\sqrt{E}} \frac{\partial u_3}{\partial \xi_1} + \frac{u_{10}}{R_1} \right) = u_{10} + \xi_3 \beta_1 \tag{18.5.19}$$

$$u_2 = u_{20} - \xi_3 \left(\frac{1}{\sqrt{G}} \frac{\partial u_3}{\partial \xi_2} + \frac{u_{20}}{R_2} \right) = u_{20} + \xi_3 \beta_2 . \tag{18.5.20}$$

Substituting Eqs. (18.5.9), (18.5.10), (18.5.11), (18.5.19), and (18.5.20) into the strain-displacement relations (18.5.3) to (18.5.8), and neglecting in the final result ξ_3 / R_1 and ξ_3 / R_2 compared to unity (first assumption), we get:

$$e_{13} = e_{23} = e_{33} = 0 \tag{18.5.21}$$

$$e_{11} = e_{110} - \xi_3 K_{11} \tag{18.5.22}$$

$$e_{22} = e_{220} - \xi_3 K_{22} \tag{18.5.23}$$

$$e_{12} = e_{120} - \xi_3 K_{12}, \tag{18.5.24}$$

where the subscript 0 refers to the middle surface and

$$e_{110} = \frac{1}{\sqrt{E}}\frac{\partial u_{10}}{\partial \xi_1} + \frac{u_{20}}{\sqrt{EG}}\frac{\partial \sqrt{E}}{\partial \xi_2} - \frac{u_3}{R_1} \tag{18.5.25}$$

$$e_{220} = \frac{1}{\sqrt{G}}\frac{\partial u_{20}}{\partial \xi_2} + \frac{u_{10}}{\sqrt{EG}}\frac{\partial \sqrt{G}}{\partial \xi_1} - \frac{u_3}{R_2} \tag{18.5.26}$$

$$e_{120} = \frac{1}{2}\left[\sqrt{\frac{G}{E}}\frac{\partial}{\partial \xi_1}\left(\frac{u_{20}}{\sqrt{G}}\right) + \sqrt{\frac{E}{G}}\frac{\partial}{\partial \xi_2}\left(\frac{u_{10}}{\sqrt{E}}\right)\right] \tag{18.5.27}$$

$$K_{11} = \frac{1}{\sqrt{E}}\frac{\partial}{\partial \xi_1}\left(\frac{u_{10}}{R_1} + \frac{1}{\sqrt{E}}\frac{\partial u_3}{\partial \xi_1}\right) + \frac{1}{\sqrt{EG}}\left(\frac{u_{20}}{R_2} + \frac{1}{\sqrt{G}}\frac{\partial u_3}{\partial \xi_2}\right)\frac{\partial \sqrt{E}}{\partial \xi_2} \tag{18.5.28}$$

$$= -\frac{1}{\sqrt{E}}\frac{\partial}{\partial \xi_1}(\beta_1) - \frac{1}{\sqrt{EG}}(\beta_2)\frac{\partial \sqrt{E}}{\partial \xi_2}$$

$$K_{22} = \frac{1}{\sqrt{G}}\frac{\partial}{\partial \xi_2}\left(\frac{u_{20}}{R_2} + \frac{1}{\sqrt{G}}\frac{\partial u_3}{\partial \xi_2}\right) + \frac{1}{\sqrt{EG}}\left(\frac{u_{10}}{R_1} + \frac{1}{\sqrt{E}}\frac{\partial u_3}{\partial \xi_1}\right)\frac{\partial \sqrt{G}}{\partial \xi_1} \tag{18.5.29}$$

$$= -\frac{1}{\sqrt{G}}\frac{\partial}{\partial \xi_2}(\beta_2) - \frac{1}{\sqrt{EG}}(\beta_1)\frac{\partial \sqrt{G}}{\partial \xi_1}$$

$$K_{12} = \frac{1}{2}\left[\sqrt{\frac{G}{E}}\frac{\partial}{\partial \xi_1}\left(\frac{u_{20}}{\sqrt{G}\,R_2} + \frac{1}{G}\frac{\partial u_3}{\partial \xi_2}\right) + \sqrt{\frac{E}{G}}\frac{\partial}{\partial \xi_2}\left(\frac{u_{10}}{\sqrt{E}\,R_1} + \frac{1}{E}\frac{\partial u_3}{\partial \xi_1}\right)\right] \tag{18.5.30}$$

$$= -\frac{1}{2}\left[\sqrt{\frac{G}{E}}\frac{\partial}{\partial \xi_1}\left(\frac{\beta_2}{\sqrt{G}}\right) + \sqrt{\frac{E}{G}}\frac{\partial}{\partial \xi_2}\left(\frac{\beta_1}{\sqrt{E}}\right)\right].$$

In the above expressions, e_{110}, e_{220}, e_{120} may be interpreted physically as strains in the middle surface of the shell. K_{11} and K_{22} represent the changes in curvature of the middle surface during deformation. K_{12} represents the change in twist of the middle surface during deformation. Eqs. (18.5.25) to (18.5.30) are the strain-displacement relations for thin shells.

18.6 Stress Resultants and Stress Couples

In the previous sections, the strains (therefore, the stresses) have been shown to be linearly distributed across the thickness of the shell. It is convenient, as was done in the study of thin plates, to integrate the stress distribution through the thickness, and to replace the stresses by equivalent stress resultants and stress couples. The variations with respect to ξ_3 are thus completely eliminated. Let us consider an element of a shell subjected to a lateral load $q = q(\xi_1, \xi_2)$. In addition to bending and twisting moments, there will be normal and shearing forces acting on the sides of the element. Both moments and forces are expressed per unit length of shell along the ξ_1 and ξ_2 directions. The convention for

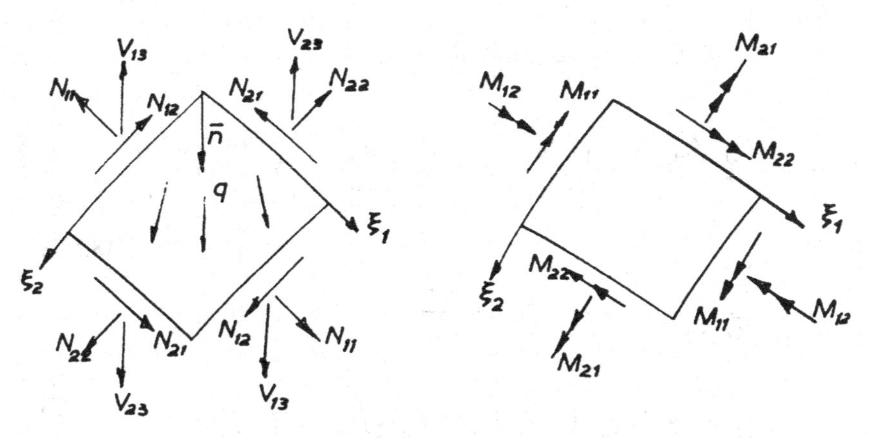

Fig. 18.13

moments is that positive moments give positive stresses on the positive half of the shell. Stresses and forces follow the conventions established in Sec. 7.2. Fig. 18.13 shows the directions for positive moments and forces. For moments, the right-hand rule applies with the thumb pointing in the direction of the double arrow. Forces and moments carry the subscript of the stresses they cause.

Recalling that the lengths of the edges of the element in Fig. 18.13 (for any distance ξ_3 from the middle surface) are given by Eqs. (18.4.4) and (18.4.5), we have:

$$N_{11}\sqrt{G}\,d\xi_2 = \int_{-\frac{h}{2}}^{+\frac{h}{2}} \sigma_{11}\sqrt{G}\left(1 - \frac{\xi_3}{R_2}\right) d\xi_2\,d\xi_3 \tag{18.6.1}$$

or

$$N_{11} = \int_{-\frac{h}{2}}^{+\frac{h}{2}} \sigma_{11}\left(1 - \frac{\xi_3}{R_2}\right) d\xi_3 . \tag{18.6.2}$$

Similarly,

$$N_{22} = \int_{-\frac{h}{2}}^{+\frac{h}{2}} \sigma_{22}\left(1 - \frac{\xi_3}{R_1}\right) d\xi_3 \tag{18.6.3}$$

$$\begin{aligned} N_{12} &= \int_{-\frac{h}{2}}^{+\frac{h}{2}} \sigma_{12}\left(1 - \frac{\xi_3}{R_2}\right) d\xi_3 , \\ N_{21} &= \int_{-\frac{h}{2}}^{+\frac{h}{2}} \sigma_{21}\left(1 - \frac{\xi_3}{R_1}\right) d\xi_3 \end{aligned} \tag{18.6.4}$$

$$\begin{aligned} V_{13} &= \int_{-\frac{h}{2}}^{+\frac{h}{2}} \sigma_{13}\left(1 - \frac{\xi_3}{R_2}\right) d\xi_3 , \\ V_{23} &= \int_{-\frac{h}{2}}^{+\frac{h}{2}} \sigma_{23}\left(1 - \frac{\xi_3}{R_1}\right) d\xi_3 \end{aligned} \tag{18.6.5}$$

$$\begin{aligned} M_{11} &= \int_{-\frac{h}{2}}^{+\frac{h}{2}} \sigma_{11}\left(1 - \frac{\xi_3}{R_2}\right)\xi_3\, d\xi_3 , \\ M_{22} &= \int_{-\frac{h}{2}}^{+\frac{h}{2}} \sigma_{22}\left(1 - \frac{\xi_3}{R_1}\right)\xi_3\, d\xi_3 \end{aligned} \tag{18.6.6}$$

$$M_{12} = \int_{-\frac{h}{2}}^{+\frac{h}{2}} \sigma_{12}\left(1 - \frac{\xi_3}{R_2}\right)\xi_3\, d\xi_3,$$
$$M_{21} = \int_{-\frac{h}{2}}^{+\frac{h}{2}} \sigma_{21}\left(1 - \frac{\xi_3}{R_1}\right)\xi_3\, d\xi_3. \tag{18.6.7}$$

From the above equations, we see that although $\sigma_{12} = \sigma_{21}$, N_{12} is not equal to N_{21}, and M_{12} is not equal to M_{21} because R_1 is not necessarily equal to R_2. However, taking into account the first assumption of Sec. 18.4, the quantities ξ_3/R_1 and ξ_3/R_2 are negligible compared to unity, so that $N_{12} = N_{21}$ and $M_{12} = M_{21}$; Eqs. (18.6.2) to (18.6.7) then become similar to those used in the theory of thin flat plates.

The expressions for forces and moments which involve σ_{11}, σ_{22}, and σ_{12} can be written in terms of the strains using the stress-strain relations:

$$\begin{aligned}\sigma_{11} &= \frac{E}{1-\nu^2}(e_{11} + \nu e_{22}) \\ &= \frac{E}{1-\nu^2}[e_{110} + \nu e_{220} - \xi_3(K_{11} + \nu K_{22})]\end{aligned} \tag{18.6.8}$$

$$\begin{aligned}\sigma_{22} &= \frac{E}{1-\nu^2}(e_{22} + \nu e_{11}) \\ &= \frac{E}{1-\nu^2}[e_{220} + \nu e_{110} - \xi_3(K_{22} + \nu K_{11})]\end{aligned} \tag{18.6.9}$$

$$\sigma_{12} = \frac{E}{1+\nu}e_{12} = \frac{E}{1+\nu}(e_{120} - \xi_3 K_{12}). \tag{18.6.10}$$

Because of the assumptions on e_{33}, e_{13}, and e_{23}, Hooke's law gives us only three stress-strain relations. Substituting Eqs. (18.6.8) to (18.6.10) into the expressions of N_{ij} and M_{ij}, and neglecting ξ_3/R_1 and ξ_3/R_2 compared to unity, we obtain:

$$N_{11} = \frac{Eh}{1-\nu^2}(e_{110} + \nu e_{220}), \quad N_{22} = \frac{Eh}{1-\nu^2}(e_{220} + \nu e_{110}) \tag{18.6.11}$$

$$N_{12} = N_{21} = \frac{Eh}{1+\nu}e_{120} \tag{18.6.12}$$

$$M_{11} = -D(K_{11} + \nu K_{22}), \quad M_{22} = -D(K_{22} + \nu K_{11}) \tag{18.6.13}$$

$$M_{12} = M_{21} = -D(1-\nu)K_{12}, \tag{18.6.14}$$

where

$$D = \frac{Eh^3}{12(1-\nu^2)}. \tag{18.6.15}$$

Eqs. (18.6.11) to (18.6.14) are the force-strain relations for thin shells. V_{13} and V_{23} cannot be written in terms of the strains. They can, however, be obtained from the general equations of equilibrium.

For the sake of clarity and easy comparison with the equations of thin flat plates, Eqs. (18.6.8) to (18.6.14) will be rewritten in matrix form:

$$\begin{bmatrix} \sigma_{11} \\ \sigma_{22} \\ \sigma_{12} \end{bmatrix} = \frac{E}{1-\nu^2}\begin{bmatrix} 1 & \nu & 0 \\ \nu & 1 & 0 \\ 0 & 0 & 1-\nu \end{bmatrix}\begin{bmatrix} e_{110} \\ e_{220} \\ e_{120} \end{bmatrix} - \frac{E\xi_3}{1-\nu^2}\begin{bmatrix} 1 & \nu & 0 \\ \nu & 1 & 0 \\ 0 & 0 & 1-\nu \end{bmatrix}\begin{bmatrix} K_{11} \\ K_{22} \\ K_{12} \end{bmatrix} \tag{18.6.16}$$

$$\begin{bmatrix} N_{11} \\ N_{22} \\ N_{12} \end{bmatrix} = \frac{Eh}{1-\nu^2}\begin{bmatrix} 1 & \nu & 0 \\ \nu & 1 & 0 \\ 0 & 0 & 1-\nu \end{bmatrix}\begin{bmatrix} e_{110} \\ e_{220} \\ e_{120} \end{bmatrix} \tag{18.6.17}$$

$$\begin{bmatrix} M_{11} \\ M_{22} \\ M_{12} \end{bmatrix} = -D\begin{bmatrix} 1 & \nu & 0 \\ \nu & 1 & 0 \\ 0 & 0 & 1-\nu \end{bmatrix}\begin{bmatrix} K_{11} \\ K_{22} \\ K_{12} \end{bmatrix}. \tag{18.6.18}$$

From Eq. (18.6.17), one can obtain the expression of the strains in terms of the forces:

$$\begin{bmatrix} e_{110} \\ e_{220} \\ e_{120} \end{bmatrix} = \frac{1}{Eh}\begin{bmatrix} 1 & -\nu & 0 \\ -\nu & 1 & 0 \\ 0 & 0 & 1+\nu \end{bmatrix}\begin{bmatrix} N_{11} \\ N_{22} \\ N_{12} \end{bmatrix}. \tag{18.6.19}$$

At this stage, it is appropriate to make a remark regarding Love's assumptions: When investigating the deformations of the surfaces parallel to the middle surface, e_{13} and e_{23} were equated to zero; in other words, the effects of σ_{13} and σ_{23} on the deformation of these surfaces were neglected. This does not mean that V_{13} and V_{23} can be neglected, since these shearing forces are essential to equilibrium.

18.7 Equations of Equilibrium of Loaded Thin Shells

To derive the equilibrium equations, let us consider a small element separated from a shell by four sections perpendicular to the middle surface. The external forces acting on the elements are the body forces, which will be neglected, and the surface forces. The internal forces are the stresses acting on the sides of the elements. Both external and internal forces are reduced to statically equivalent systems acting on the middle surface. As was done in Sec. 7.12 for the derivation of the equilibrium equations in orthogonal curvilinear coordinates, we shall express the equilibrium of forces and moments in vector form, then write down the component equations. Using the sign conventions of Fig. 18.13, the sides of the element of the middle surface (Fig. 18.14a) are acted upon by the following forces and moments (Figs. 18.14b, 18.14c):

On side OC, the force per unit length is

$$-\overline{N}_{\xi 1} = -(N_{11}\bar{e}_1 + N_{12}\bar{e}_2 + V_{13}\bar{n}), \tag{18.7.1}$$

and the moment per unit length is

$$-\overline{M}_{\xi 1} = -(M_{11}\bar{e}_2 - M_{12}\bar{e}_1). \tag{18.7.2}$$

On side OA, the force per unit length is

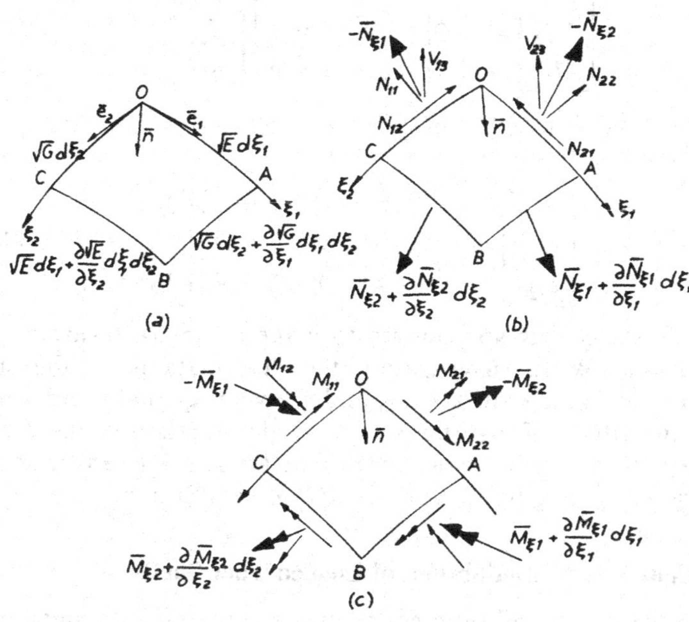

Fig. 18.14

$$-\bar{N}_{\xi 2} = -(N_{21}\bar{e}_1 + N_{22}\bar{e}_2 + V_{23}\bar{n}), \tag{18.7.3}$$

and the moment per unit length is

$$-\bar{M}_{\xi 2} = -(M_{21}\bar{e}_2 - M_{22}\bar{e}_1). \tag{18.7.4}$$

On the side OB, the force per unit length is

$$\bar{N}_{\xi 1} + \frac{\partial \bar{N}_{\xi 1}}{\partial \xi_1} d\xi_1, \tag{18.7.5}$$

and the moment per unit length is

$$\bar{M}_{\xi 1} + \frac{\partial \bar{M}_{\xi 1}}{\partial \xi_1} d\xi_1. \tag{18.7.6}$$

On side BC, the force per unit length is

$$\bar{N}_{\xi 2} + \frac{\partial \bar{N}_{\xi 2}}{\partial \xi_2} d\xi_2, \tag{18.7.7}$$

and the moment per unit length is

$$\bar{M}_{\xi 2} + \frac{\partial \bar{M}_{\xi 2}}{\partial \xi_2} d\xi_2. \tag{18.7.8}$$

In addition, the element is subjected to the external surface loading q per unit area:

$$\bar{q} = q_1\bar{e}_1 + q_2\bar{e}_2 + q_3\bar{n}. \tag{18.7.9}$$

The area of the element is given by $\sqrt{EG}\, d\xi_1\, d\xi_2$. The condition that the resultant vector of all the forces acting on the element of the middle surface is to vanish, can be written as follows:

$$\begin{aligned}
&\left(\bar{N}_{\xi 1} + \frac{\partial \bar{N}_{\xi 1}}{\partial \xi_1} d\xi_1\right)\left(\sqrt{G}\, d\xi_2 + \frac{\partial \sqrt{G}}{\partial \xi_1} d\xi_1\, d\xi_2\right) - \bar{N}_{\xi 1}\sqrt{G}\, d\xi_2 \\
&+ \left(\bar{N}_{\xi 2} + \frac{\partial \bar{N}_{\xi 2}}{\partial \xi_2} d\xi_2\right)\left(\sqrt{E}\, d\xi_1 + \frac{\partial \sqrt{E}}{\partial \xi_2} d\xi_1\, d\xi_2\right) - \bar{N}_{\xi 2}\sqrt{E}\, d\xi_1 \\
&+ \bar{q}\sqrt{EG}\, d\xi_1\, d\xi_2 = 0.
\end{aligned} \tag{18.7.10}$$

Neglecting the infinitesimals of the third order, the previous equation is reduced to:

$$\frac{\partial}{\partial \xi_1}(\sqrt{G}\,\overline{N}_{\xi 1}) + \frac{\partial}{\partial \xi_2}(\sqrt{E}\,\overline{N}_{\xi 2}) + \sqrt{EG}\,\overline{q} = 0. \tag{18.7.11}$$

Replacing $\overline{N}_{\xi 1}$, $\overline{N}_{\xi 2}$, and $\overline{q}$ by their expressions in terms of their components we have:

$$\begin{aligned} &\frac{\partial}{\partial \xi_1}[\sqrt{G}\,(N_{11}\overline{e}_1 + N_{12}\overline{e}_2 + V_{13}\overline{n})] \\ &+ \frac{\partial}{\partial \xi_2}[\sqrt{E}\,(N_{21}\overline{e}_1 + N_{22}\overline{e}_2 + V_{23}\overline{n})] \\ &+ \sqrt{EG}\,(q_1\overline{e}_1 + q_2\overline{e}_2 + q_3\overline{n}) = 0. \end{aligned} \tag{18.7.12}$$

Eqs. (18.3.116) to (18.3.121) are now used in taking the partial derivatives of the unit vectors in Eq. (18.7.12), which becomes:

$$\begin{aligned} &\left[\frac{\partial}{\partial \xi_1}(\sqrt{G}\,N_{11}) + \frac{\partial}{\partial \xi_2}(\sqrt{E}\,N_{21}) + N_{12}\frac{\partial\sqrt{E}}{\partial \xi_2} - N_{22}\frac{\partial\sqrt{G}}{\partial \xi_1}\right. \\ &\qquad \left. - V_{13}\frac{\sqrt{EG}}{R_1} + q_1\sqrt{EG}\right]\overline{e}_1 \\ &+ \left[\frac{\partial}{\partial \xi_1}(\sqrt{G}\,N_{12}) + \frac{\partial}{\partial \xi_2}(\sqrt{E}\,N_{22}) + N_{21}\frac{\partial\sqrt{G}}{\partial \xi_1} - N_{11}\frac{\partial\sqrt{E}}{\partial \xi_2}\right. \\ &\qquad \left. - V_{23}\frac{\sqrt{EG}}{R_2} + q_2\sqrt{EG}\right]\overline{e}_2 \\ &+ \left[\frac{\partial}{\partial \xi_1}(\sqrt{G}\,V_{13}) + \frac{\partial}{\partial \xi_2}(\sqrt{E}\,V_{23}) + \sqrt{EG}\left(\frac{N_{11}}{R_1} + \frac{N_{22}}{R_2}\right)\right. \\ &\qquad \left. + q_3\sqrt{EG}\right]\overline{n} = 0. \end{aligned} \tag{18.7.13}$$

In order that this vector equation be satisfied, the coefficients of $\overline{e}_1$, $\overline{e}_2$, and $\overline{n}$ must identically vanish. Hence, the following three differential equations of equilibrium of forces are obtained:

$$\begin{aligned} &\frac{\partial}{\partial \xi_1}(\sqrt{G}\,N_{11}) + \frac{\partial}{\partial \xi_2}(\sqrt{E}\,N_{21}) + N_{12}\frac{\partial\sqrt{E}}{\partial \xi_2} - N_{22}\frac{\partial\sqrt{G}}{\partial \xi_1} \\ &\qquad - V_{13}\frac{\sqrt{EG}}{R_1} + q_1\sqrt{EG} = 0 \end{aligned} \tag{18.7.14}$$

$$\frac{\partial}{\partial \xi_1}(\sqrt{G}\,N_{12}) + \frac{\partial}{\partial \xi_2}(\sqrt{E}\,N_{22}) + N_{21}\frac{\partial \sqrt{G}}{\partial \xi_1} - N_{11}\frac{\partial \sqrt{E}}{\partial \xi_2} - V_{23}\frac{\sqrt{EG}}{R_2} + q_2\sqrt{EG} = 0 \tag{18.7.15}$$

$$\frac{\partial}{\partial \xi_1}(\sqrt{G}\,V_{13}) + \frac{\partial}{\partial \xi_2}(\sqrt{E}\,V_{23}) + \sqrt{EG}\left(\frac{N_{11}}{R_1} + \frac{N_{22}}{R_2}\right) + q_3\sqrt{EG} = 0. \tag{18.7.16}$$

To find the equilibrium equations for the moments, we must vectorially add the internal moments shown in Fig. 18.14c to those moments due to the forces shown in Fig. 18.14b. The sum of the internal moments shown in Fig. 18.14c is:

$$\begin{aligned} &\left(\overline{M}_{\xi 1} + \frac{\partial \overline{M}_{\xi 1}}{\partial \xi_1} d\xi_1\right)\left(\sqrt{G}\,d\xi_2 + \frac{\partial \sqrt{G}}{\partial \xi_1} d\xi_1\,d\xi_2\right) - \overline{M}_{\xi 1}\sqrt{G}\,d\xi_2 \\ &+ \left(\overline{M}_{\xi 2} + \frac{\partial \overline{M}_{\xi 2}}{\partial \xi_2} d\xi_2\right)\left(\sqrt{E}\,d\xi_1 + \frac{\partial \sqrt{E}}{\partial \xi_2} d\xi_1\,d\xi_2\right) - \overline{M}_{\xi 2}\sqrt{E}\,d\xi_1 \\ &= \left[\frac{\partial}{\partial \xi_1}(\sqrt{G}\,\overline{M}_{\xi 1}) + \frac{\partial}{\partial \xi_2}(\sqrt{E}\,\overline{M}_{\xi 2})\right] d\xi_1\,d\xi_2. \end{aligned} \tag{18.7.17}$$

in which the infinitesimals of the third order have been neglected. In considering the moments about O of the forces in Fig. 18.14b, we shall neglect the infinitesimals of the third order. In that respect, we notice that:

a) The moment due to the external force $\overline{q}$ is computed by multiplying q first by the area of the element (a second order quantity), then by the moment arm (a first order quantity). The result is that q will be multiplied by an infinitesimal of the third order and, consequently, its moment can be neglected.

b) The components of the forces in the plane of the middle surface (i.e., the N_{ij}' s) give moments of significance only about the normal $\overline{n}$. On the other hand, the components of the forces normal to the middle surface (i.e., the V_{ij}' s) give moments of significance only about the ξ_1 and ξ_2 lines. When multiplying a force by its appropriate moment arm, we can neglect the differential forces since they lead to third order terms. Taking the previous remarks into account (in other words, retaining only the second order terms), we obtain the following expression for the moment of all the forces about 0:

$$[V_{23}\bar{e}_1 - V_{13}\bar{e}_2 + (N_{12} - N_{21})\bar{n}]\sqrt{EG}\, d\xi_1\, d\xi_2. \tag{18.7.18}$$

For equilibrium of moments, the sum of Eqs. (18.7.17) and (18.7.18) must be equal to zero. Thus,

$$\begin{aligned}\frac{\partial}{\partial \xi_1}(\sqrt{G}\,\bar{M}_{\xi 1}) + \frac{\partial}{\partial \xi_2}(\sqrt{E}\,\bar{M}_{\xi 2})\\ + \sqrt{EG}\,[V_{23}\bar{e}_1 - V_{13}\bar{e}_2 + (N_{12} - N_{21})\bar{n}] = 0.\end{aligned} \tag{18.7.19}$$

Replacing $\bar{M}_{\xi 1}$ and $\bar{M}_{\xi 2}$ by their expressions in terms of their components, we have:

$$\begin{aligned}\frac{\partial}{\partial \xi_1}[\sqrt{G}\,(M_{11}\bar{e}_2 - M_{12}\bar{e}_1)] + \frac{\partial}{\partial \xi_2}[\sqrt{E}\,(M_{21}\bar{e}_2 - M_{22}\bar{e}_1)]\\ + \sqrt{EG}\,[V_{23}\bar{e}_1 - V_{13}\bar{e}_2 + (N_{12} - N_{21})\bar{n}] = 0.\end{aligned} \tag{18.7.20}$$

Taking the derivatives, Eq. (18.7.20) becomes:

$$\begin{aligned}&\left[-\frac{\partial}{\partial \xi_1}(\sqrt{G}\,M_{12}) - \frac{\partial}{\partial \xi_2}(\sqrt{E}\,M_{22}) - M_{21}\frac{\partial\sqrt{G}}{\partial \xi_1} + M_{11}\frac{\partial\sqrt{E}}{\partial \xi_2}\right.\\ &\qquad\left. + \sqrt{EG}\,V_{23}\right]\bar{e}_1\\ &+ \left[\frac{\partial}{\partial \xi_1}(\sqrt{G}\,M_{11}) + \frac{\partial}{\partial \xi_2}(\sqrt{E}\,M_{21}) + M_{12}\frac{\partial\sqrt{E}}{\partial \xi_2} - M_{22}\frac{\partial\sqrt{G}}{\partial \xi_1}\right.\\ &\qquad\left. - \sqrt{EG}\,V_{13}\right]\bar{e}_2\\ &+ \sqrt{EG}\left[-\frac{M_{12}}{R_1} + \frac{M_{21}}{R_2} + N_{12} - N_{21}\right]\bar{n} = 0.\end{aligned} \tag{18.7.21}$$

In order that this vector equation be satisfied, the coefficients of $\bar{e}_1$, $\bar{e}_2$, and $\bar{n}$ must identically vanish. Hence, the following three differential equations of equilibrium of moments are obtained:

$$\frac{\partial}{\partial \xi_1}(\sqrt{G}\, M_{12}) + \frac{\partial}{\partial \xi_2}(\sqrt{E}\, M_{22}) + M_{21}\frac{\partial \sqrt{G}}{\partial \xi_1} - M_{11}\frac{\partial \sqrt{E}}{\partial \xi_2} - \sqrt{EG}\, V_{23} = 0 \tag{18.7.22}$$

$$\frac{\partial}{\partial \xi_1}(\sqrt{G}\, M_{11}) + \frac{\partial}{\partial \xi_2}(\sqrt{E}\, M_{21}) + M_{12}\frac{\partial \sqrt{E}}{\partial \xi_2} - M_{22}\frac{\partial \sqrt{G}}{\partial \xi_1} - \sqrt{EG}\, V_{13} = 0 \tag{18.7.23}$$

$$N_{12} - N_{21} + \frac{M_{21}}{R_2} - \frac{M_{12}}{R_1} = 0. \tag{18.7.24}$$

The six Eqs. (18.7.14) to (18.7.16) and (18.7.22) to (18.7.24) are the conditions for the equilibrium of a small element of the shell. Upon examination of these equations, we notice that Eq. (18.7.24) is an identity. Indeed, if we introduce in it the definitions given in Eqs. (18.6.4) and (18.6.7), we get:

$$N_{12} - N_{21} + \frac{M_{21}}{R_2} - \frac{M_{12}}{R_1} = \int_{-\frac{h}{2}}^{+\frac{h}{2}} \left(1 - \frac{\xi_3}{R_1}\right)\left(1 - \frac{\xi_3}{R_2}\right)(\sigma_{12} - \sigma_{21})\, d\xi_3 = 0, \tag{18.7.25}$$

since $\sigma_{12} = \sigma_{21}$. We also notice that V_{13} and V_{23} can be eliminated from the set of six equations by solving Eqs. (18.7.22) and (18.7.23) for these quantities and substituting the resulting expressions into the remaining equations. Thus the definition of V_{13} and V_{23} has no bearing on the analysis.

The derivation of the equations of equilibrium did not involve any equality between N_{12} and N_{21} or M_{12} and M_{21}. However, since these equations will be used in conjunction with the force-strain relations and the strain-displacement relations in which those equalities were assumed, the equilibrium equations may be rewritten setting $N_{12} = N_{21}$ and $M_{12} = M_{21}$.

In summary, we have five equations of equilibrium, six force-strain relations [Eqs. (18.6.11) to (18.6.14)], and six strain-displacement relations [Eqs. (18.5.25) to (18.5.30)] for a total of 17 equations in terms of 17 variables: N_{11}, N_{22}, $N_{12} = N_{21}$, V_{13}, V_{23}, M_{11}, M_{22}, $M_{12} = M_{21}$, e_{110}, e_{220}, e_{120}, K_{11}, K_{22}, K_{12}, u_{10}, u_{20}, u_{30}. In theory, the problem can be solved once the boundary conditions are specified.

18.8 Boundary Conditions

We shall only examine the case where the boundaries coincide with the lines of curvature of the middle surface, and assume for the purpose of this analysis that the boundary coincides with the ξ_1 line (Fig. 18.15). Acting on this boundary are the five quantities M_{21}, M_{22}, V_{23}, N_{22}, and N_{21}, and one would think that the number of conditions necessary to completely determine the solution must be five. In the following, we shall prove that the problem is completely defined by four, and not five, boundary conditions. The reasoning closely follows that made in the case of flat plates (see Sec. 17.5).

Let us consider a segment of the boundary near a point m_1, and approximate this segment by two equal chords mm_1 and m_1m_2. The value of the twisting moment per unit length at the middle of mm_1 is M_{21}, and that at the middle of m_1m_2 is $(M_{21} + \partial M_{21}/\partial\xi_1 \, d\xi_1)$. The total

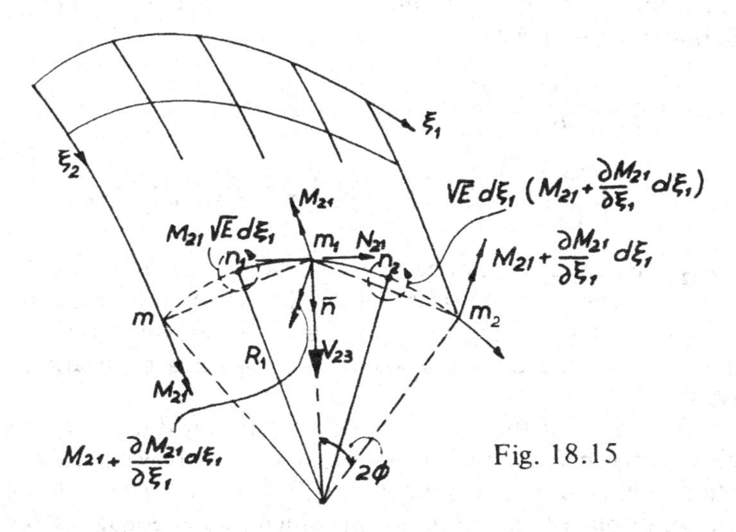

Fig. 18.15

twisting moments acting on mm_1 and m_1m_2 are $M_{21}\sqrt{E}\,d\xi_1$ and $(M_{21} + \partial M_{21}/\partial\xi_1 \, d\xi_1)(\sqrt{E}\,d\xi_1)$, respectively. Each of these moments can be replaced by two parallel forces equal in magnitude and opposite in direction at the ends of mm_1 and m_1m_2, as shown in Fig. 18.15. The force at m is parallel to the normal to the chord mm_1, and the force at m_2 is parallel to the normal to the chord m_1m_2. Projecting the forces at m_1 along the normal, we get:

$$\left(\frac{\partial M_{21}}{\partial \xi_1} d\xi_1 + M_{21} - M_{21}\right)\cos\phi \approx \frac{\partial M_{21}}{\partial \xi_1} d\xi_1, \tag{18.8.1}$$

and along the tangent, we get:

$$\begin{aligned}\left(M_{21} + M_{21} + \frac{\partial M_{21}}{\partial \xi_1} d\xi_1\right)\sin\phi &\approx \left(2M_{21} + \frac{\partial M_{21}}{\partial \xi_1} d\xi_1\right)\frac{\sqrt{E}\, d\xi_1}{2R_1} \\ &\approx \frac{M_{21}\sqrt{E}\, d\xi_1}{R_1}.\end{aligned} \tag{18.8.2}$$

Thus, along the edge of the shell, the twisting moment can be replaced by distributed shearing forces in the direction of the normal equal to:

$$\frac{1}{\sqrt{E}\, d\xi_1}\left(\frac{\partial M_{21}}{\partial \xi_1} d\xi_1\right) = \frac{1}{\sqrt{E}}\frac{\partial M_{21}}{\partial \xi_1}, \tag{18.8.3}$$

and by distributed shearing forces in the direction of the tangent equal to:

$$-\frac{1}{\sqrt{E}\, d\xi_1}\left(\frac{M_{21}\sqrt{E}\, d\xi_1}{R_1}\right) = -\frac{M_{21}}{R_1}. \tag{18.8.4}$$

Therefore, when the boundary is a ξ_1 line, the four quantities,

$$N_{22}, \quad N_{21} - \frac{M_{21}}{R_1}, \quad V_{23} + \frac{1}{\sqrt{E}}\frac{\partial M_{21}}{\partial \xi_1}, \quad M_{22}, \tag{18.8.5}$$

completely determine the state of stress at the edge of the shell. The same reasoning can be repeated when the ξ_2 line is the boundary line. Therefore, the number of boundary conditions at each edge must be equal to four. It is obvious that the boundary conditions are not always expressed in terms of forces and moments, since one often prescribes the displacements and the angles of rotation. The total number of conditions, however, cannot exceed four:

1) *Built-in or clamped edge*

At a built-in edge, we have:

$$u_1 = 0, \quad u_2 = 0, \quad u_3 = 0, \quad \beta_2 = -\frac{1}{\sqrt{G}}\frac{\partial u_3}{\partial \xi_2} - \frac{u_{20}}{R_2} = 0 \tag{18.8.6}$$

2) *Simply supported edge*

At a simply supported edge, we have:

$$u_1 = 0, \quad u_2 = 0, \quad u_3 = 0, \quad M_{22} = 0. \tag{18.8.7}$$

3) *Free edge*
At a free edge:

$$N_{22} = 0, \quad N_{21} - \frac{M_{21}}{R_1} = 0, \quad V_{23} + \frac{1}{\sqrt{E}}\frac{\partial M_{21}}{\partial \xi_1} = 0, \quad M_{22} = 0. \tag{18.8.8}$$

The concept of boundary conditions loses its meaning when the shell is closed. The coordinate lines ξ_1 and ξ_2 on the middle surface are closed curves and one periodically returns to the same point along a curve $\xi_1 =$ constant or $\xi_2 =$ constant. In such cases, one must impose that the solution be periodic functions of ξ_1 and ξ_2, and the boundary conditions are replaced by the periodicity conditions.

18.9 Membrane Theory of Shells

In many problems of thin shells, the loadings are such that the bending and twisting moments are zero, or so small that they can be neglected. The stresses in the shell are mainly due to the forces N_{11}, N_{22}, N_{12}, and N_{21}. The theory of thin shells, based on the assumption of zero stress couples, is called membrane theory. The conditions of equilibrium for this case can be obtained by setting:

$$M_{11} = M_{22} = M_{12} = M_{21} = 0 \tag{18.9.1}$$

first in the differential equations of equilibrium of moments, and second in the differential equations of equilibrium of forces. This gives:

$$V_{23} = V_{13} = 0 \tag{18.9.2}$$

$$N_{12} = N_{21} \tag{18.9.3}$$

$$\frac{\partial}{\partial \xi_1}(\sqrt{G}\, N_{11}) + \frac{\partial}{\partial \xi_2}(\sqrt{E}\, N_{21}) + N_{12}\frac{\partial \sqrt{E}}{\partial \xi_2} - N_{22}\frac{\partial \sqrt{G}}{\partial \xi_1} + q_1\sqrt{EG} = 0 \tag{18.9.4}$$

$$\frac{\partial}{\partial \xi_1}(\sqrt{G}\, N_{12}) + \frac{\partial}{\partial \xi_2}(\sqrt{E}\, N_{22}) + N_{21}\frac{\partial \sqrt{G}}{\partial \xi_1} - N_{11}\frac{\partial \sqrt{E}}{\partial \xi_2} + q_2\sqrt{EG} = 0 \tag{18.9.5}$$

$$\frac{N_{11}}{R_1} + \frac{N_{22}}{R_2} + q_3 = 0. \tag{18.9.6}$$

Eqs. (18.9.2) to (18.9.6) show that the applied loads are supported by internal forces in the plane of the shell. They are the equations of equilibrium for a shell in a membrane state of stress. The force-strain relations are given by Eq. (18.6.17). The strain-displacements relations are given by Eq. (18.5.25) to (18.5.27).

The three equations of equilibrium (18.9.4) to (18.9.6) contain three unknowns—N_{11}, N_{22}, and $N_{12} = N_{21}$. The problem is, therefore, statically determinate. Once those values are found, the strains can be obtained from Eq. (18.6.19). Knowing the strains, the displacements in the membrane shell are obtained by integration of Eqs. (18.5.25) to (18.5.27).

Finally, the boundary conditions along a line of curvature must be limited to two. Indeed, if the assumptions of the membrane theory are introduced in Eq. (18.8.5), the two remaining quantities are N_{22} and N_{21}, to be specified along a ξ_1 line. When the boundary conditions are specified in terms of displacements, the four quantities involved are u_{10}, u_{20}, u_3, and β_2. However, it is not possible to impose conditions on u_3 and β_2, since this would affect the values of V_{23} and M_{22}. If we specify, for example, that $u_3 = \beta_2 = 0$ on the boundary, the condition that $V_{23} = M_{22} = 0$ in membrane shells cannot be satisfied. It follows that, on the edge of the membrane one can only specify the components of the displacement tangential to the middle surface—namely, u_{10} and u_{20}.

18.10 Membrane Shells of Revolution

Thin shells of revolution are extensively used in various types of structures, such as containers, tanks, and domes. Let us take the ξ_1 lines along the meridians, and the ξ_2 lines along the circles in the planes perpendicular to the axis of revolution (Fig. 18.16).

It was shown in Sec. 18.3 that the radii of curvature at A of the ξ_1 and ξ_2 lines lie on the normal to the surface but have different lengths. The first fundamental form, as given by Eq. (18.3.148), is:

$$(ds)^2 = R_\phi^2(d\phi)^2 + R_o^2(d\theta)^2, \tag{18.10.1}$$

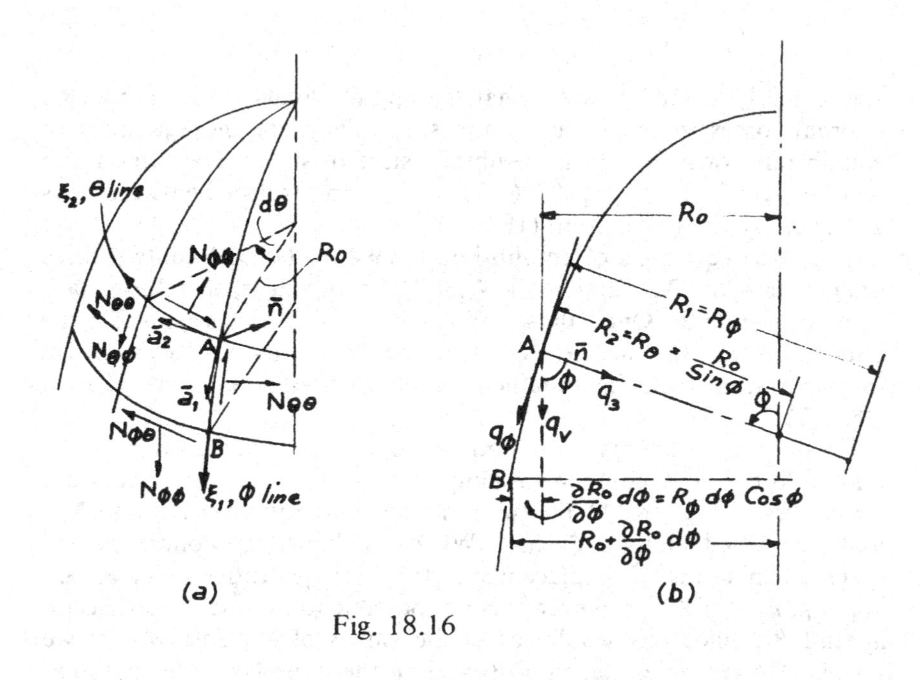

Fig. 18.16

where we have replaced the subscript 1 by ϕ. Remembering that R_ϕ and R_o, (the first fundamental magnitudes) are independent of θ, and that $\partial R_o/\partial\phi = R_\phi \cos\phi$ (Fig. 18.6), the equations of equilibrium (18.9.4) to (18.9.6) become:

$$\frac{\partial}{\partial\phi}(R_o N_{\phi\phi}) + R_\phi \frac{\partial N_{\theta\phi}}{\partial\theta} - N_{\theta\theta} R_\phi \cos\phi + q_\phi R_\phi R_o = 0 \tag{18.10.2}$$

$$\frac{\partial}{\partial\phi}(R_o N_{\phi\theta}) + R_\phi \frac{\partial N_{\theta\theta}}{\partial\theta} + N_{\theta\phi} R_\phi \cos\phi + q_\theta R_\phi R_o = 0 \tag{18.10.3}$$

$$\frac{N_{\phi\phi}}{R_\phi} + \frac{N_{\theta\theta}}{R_\theta} + q_3 = 0. \tag{18.10.4}$$

The force-strain relations are given by Eqs. (18.6.17), in which the subscripts 1 and 2 are changed to ϕ and θ.

The strain-displacement relations (18.5.25) to (18.5.27) become:

$$e_{\phi\phi 0} = \frac{1}{R_\phi}\left(\frac{\partial u_{\phi 0}}{\partial \phi} - u_3\right) \tag{18.10.5}$$

$$e_{\theta\theta 0} = \frac{1}{R_o}\left(\frac{\partial u_{\theta 0}}{\partial \theta} + u_{\phi 0}\cos\phi - u_3\sin\phi\right) \tag{18.10.6}$$

$$e_{\phi\theta 0} = \frac{1}{2}\left[\frac{1}{R_\phi}\frac{\partial u_{\theta 0}}{\partial \phi} - \frac{1}{R_o}\left(u_{\theta 0}\cos\phi - \frac{\partial u_{\phi 0}}{\partial \theta}\right)\right]. \tag{18.10.7}$$

To the previous equations, one must add the boundary conditions of the type presented in Sec. 18.9.

Let us now consider the special case in which the *shell is loaded symmetrically*. All the quantities in the previous equations do not depend on θ, and both $N_{\phi\theta}$ and q_θ must be equal to zero. Eq. (18.10.3) is thus identically satisfied. By solving the two Eqs. (18.10.2) and (18.10.4), the values of $N_{\phi\phi}$ and $N_{\theta\theta}$ can be calculated. Noticing that $R_o = R_\theta \sin\phi$, the value of $N_{\theta\theta}$ obtained from Eq. (18.10.4) is:

$$N_{\theta\theta} = -\frac{N_{\phi\phi}R_o}{R_\phi \sin\phi} - \frac{R_o q_3}{\sin\phi}. \tag{18.10.8}$$

Substituting Eq. (18.10.8) into Eq. (18.10.2) and multiplying the result by $\sin\phi$, we find that:

$$\frac{d}{d\phi}(R_o N_{\phi\phi}\sin\phi) + R_o R_\phi(q_\phi \sin\phi + q_3\cos\phi) = 0. \tag{18.10.9}$$

Integrating with respect to ϕ, we obtain:

$$N_{\phi\phi} = -\frac{1}{2\Pi R_o \sin\phi}\left[\int_0^\phi 2\Pi R_o R_\phi(q_\phi\sin\phi + q_3\cos\phi)\,d\phi\right]. \tag{18.10.10}$$

Since $(q_\phi \sin\phi + q_3\cos\phi)$ has a resultant q_v which acts on the annular area (Fig. 18.16) $2\Pi R_o R_\phi\, d\phi$, the integral in the right-hand side of Eq. (18.10.10) can be replaced by the resultant of the total load acting on that part of the shell corresponding to the angle ϕ. If we set:

$$\int_0^\phi 2\Pi R_o R_\phi(q_\phi\sin\phi + q_3\cos\phi)\,d\phi = F, \tag{18.10.11}$$

Eq. (18.10.10) gives:

$$N_{\phi\phi} = -\frac{F}{2\Pi R_o \sin\phi}. \tag{18.10.12}$$

Instead of solving Eqs. (18.10.2) and (18.10.4), it is more convenient to obtain $N_{\phi\phi}$ from Eq. (18.10.12), then obtain $N_{\theta\theta}$ from Eq. (18.10.8). The strains can now be obtained from the force-strain relations and the displacements by integrating the strain-displacement relations.

Instead of the equilibrium of an element, the equilibrium of the portion of the shell above the parallel circle defined by the angle ϕ may be considered. If the resultant of the total load on that portion of the shell is denoted by F (Fig. 18.17), equilibrium requires that

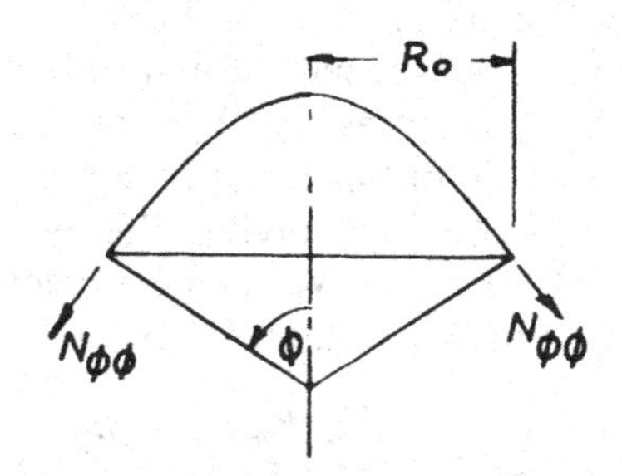

Fig. 18.17

$$2\Pi R_o N_{\phi\phi} \sin \phi + F = 0, \qquad (18.10.13)$$

which is the same as (18.10.12)

Example 1. Spherical Dome of Constant Thickness under Its Own Weight

Let γ be the unit weight of the material from which the shell is made. The gravitational force per unit area of the shell is γh. This force has the components

$$q_\phi = \gamma h \sin \phi, \quad q_\theta = 0, \quad q_3 = \gamma h \cos \phi. \qquad (18.10.14)$$

If the radius of the middle surface of the dome is a (Fig. 18.18),

$$R_\phi = R_\theta = a, \qquad (18.10.15)$$

Eq. (18.10.12) gives:

$$N_{\phi\phi} = -\frac{\gamma h a}{1 + \cos \phi}, \qquad (18.10.16)$$

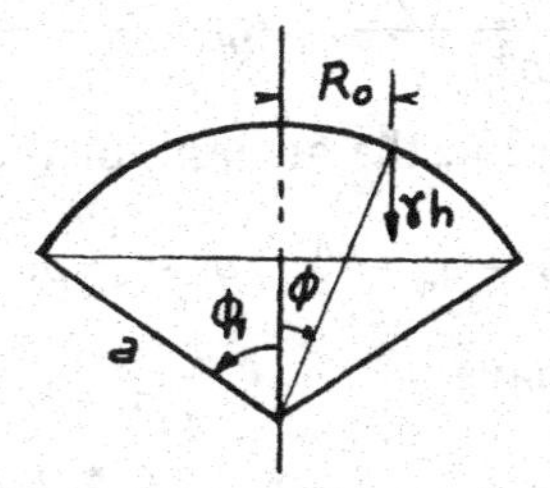

Fig. 18.18

where it was assumed that no external loads are acting on the dome. Substituting this value of $N_{\phi\phi}$ into (18.10.8), we get:

$$N_{\theta\theta} = -a\gamma h\left(\cos\phi - \frac{1}{1+\cos\phi}\right). \tag{18.10.17}$$

$N_{\phi\phi}$ is always compressive. $N_{\theta\theta}$ is compressive for small values of ϕ and becomes tensile for $\phi > 51° \; 50'$.

Example 2. Shell in the Form of an Ellipsoid of Revolution

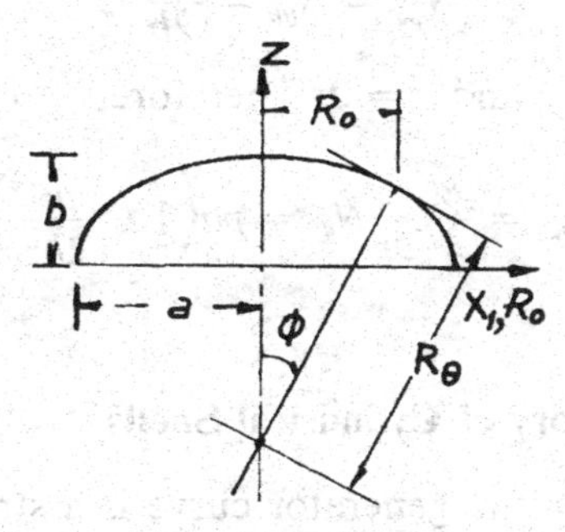

Fig. 18.19

Such shells are used in the construction of the ends of cylindrical boilers and air tanks (Fig. 18.19). The equation of the ellipse is:

$$\frac{R_o^2}{a^2} + \frac{z^2}{b^2} = 1. \tag{18.10.18}$$

The magnitudes of the principal radii of curvature can be computed from Eqs. (18.3.146) and (18.3.147):

$$R_\phi = \frac{(a^4 z^2 + b^4 R_o^2)^{3/2}}{a^4 b^4}, \quad R_\theta = \frac{(a^4 z^2 + b^4 R_o^2)^{1/2}}{b^2}. \tag{18.10.19}$$

If p is the uniform pressure in the air tank or the boiler, then

$$q_\theta = q_\phi = 0, \quad q_3 = -p. \tag{18.10.20}$$

From Eq. (18.10.12), we have:

$$N_{\phi\phi} = \frac{\Pi R_o^2 p}{2\Pi R_o \sin\phi} = \frac{pR_o}{2\sin\phi} = \frac{pR_\theta}{2} = \frac{p(a^4 z^2 + b^4 R_o^2)^{1/2}}{2b^2} \tag{18.10.21}$$

and

$$\begin{aligned} N_{\theta\theta} &= -\frac{R_\theta}{R_\phi} N_{\phi\phi} + R_\theta p \\ &= \frac{p(a^4 z^2 + b^4 R_o^2)^{1/2}}{2b^2}\left(2 - \frac{a^4 b^2}{a^4 z^2 + b^4 R_o^2}\right). \end{aligned} \tag{18.10.22}$$

At the top of the shell, $R_o = 0$ and $z = b$: Therefore,

$$N_{\phi\phi} = N_{\theta\theta} = \frac{pa^2}{2b}. \tag{18.10.23}$$

At the equator, $R_o = a$ and $z = 0$: Therefore,

$$N_{\phi\phi} = \frac{pa}{2}, \quad N_{\theta\theta} = pa\left(1 - \frac{a^2}{2b^2}\right). \tag{18.10.24}$$

18.11 Membrane Theory of Cylindrical Shells

In cylindrical shells, the generator curve is a straight line parallel to the axis of revolution. Therefore, the angle ϕ between the normal to the generator and the axis is $\Pi / 2$. In the OX_1, OX_2 plane the generator follows a path (Fig. 18.20):

$$R_o = R_o(\theta). \tag{18.11.1}$$

Since the generator is a straight line, $R_\phi = \infty$ and

$$\lim_{R_\phi \to \infty}(R_\phi\, d\phi) = dz. \tag{18.11.2}$$

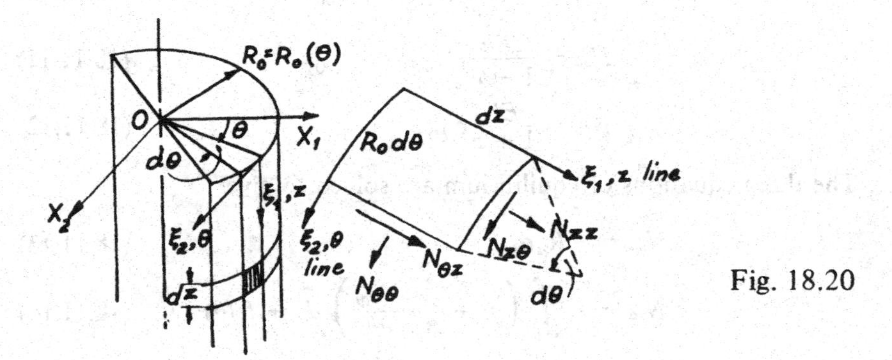

Fig. 18.20

We shall, therefore, associate ξ_1 with z and ξ_2 with θ.

The first fundamental form is:

$$(ds)^2 = (dz)^2 + R_o^2(d\theta)^2. \tag{18.11.3}$$

The three equations of equilibrium (18.9.4) to (18.9.6) become:

$$R_o \frac{\partial N_{zz}}{\partial z} + \frac{\partial N_{\theta z}}{\partial \theta} + R_o q_z = 0 \tag{18.11.4}$$

$$R_o \frac{\partial N_{z\theta}}{\partial z} + \frac{\partial N_{\theta\theta}}{\partial \theta} + R_o q_\theta = 0 \tag{18.11.5}$$

$$\frac{N_{\theta\theta}}{R_o} + q_3 = 0, \tag{18.11.6}$$

where we have replaced the subscripts 1 and 2 by z and θ, E by 1, and G by R_o^2.

The strain-displacement relations (18.5.25) to (18.5.27) become:

$$e_{zz0} = \frac{\partial u_{z0}}{\partial z} \tag{18.11.7}$$

$$e_{\theta\theta 0} = \frac{1}{R_o}\frac{\partial u_{\theta 0}}{\partial \theta} - \frac{u_3}{R_o} \tag{18.11.8}$$

$$e_{z\theta 0} = \frac{1}{2}\left(\frac{\partial u_{\theta 0}}{\partial z} + \frac{1}{R_o}\frac{\partial u_{z0}}{\partial \theta}\right). \tag{18.11.9}$$

The force-strain relations (18.6.17) become:

$$N_{zz} = \frac{Eh}{1 - \nu^2}(e_{zz0} + \nu e_{\theta\theta 0}) \tag{18.11.10}$$

$$N_{\theta\theta} = \frac{Eh}{1 - \nu^2}(e_{\theta\theta 0} + \nu e_{zz0}) \tag{18.11.11}$$

$$N_{z\theta} = \frac{Eh}{1 + \nu} e_{z\theta 0}. \tag{18.11.12}$$

The three equations of equilibrium are solved to give:

$$N_{\theta\theta} = -R_o q_3 \tag{18.11.13}$$

$$N_{z\theta} = -\int \left(q_\theta + \frac{1}{R_o}\frac{\partial N_{\theta\theta}}{\partial \theta} \right) dz + f_1(\theta) \tag{18.11.14}$$

$$N_{zz} = -\int \left(q_z + \frac{1}{R_o}\frac{\partial N_{\theta z}}{\partial \theta} \right) dz + f_2(\theta), \tag{18.11.15}$$

where $f_1(\theta)$ and $f_2(\theta)$ are two functions of θ.

The force-strain relations (18.6.19) become:

$$e_{zz0} = \frac{1}{Eh}(N_{zz} - \nu N_{\theta\theta}) \tag{18.11.16}$$

$$e_{\theta\theta 0} = \frac{1}{Eh}(N_{\theta\theta} - \nu N_{zz}) \tag{18.11.17}$$

$$e_{z\theta 0} = \frac{1 + \nu}{Eh} N_{z\theta}. \tag{18.11.18}$$

The strain-displacement relations result in:

$$u_{z0} = \int \frac{1}{Eh}(N_{zz} - \nu N_{\theta\theta})\, dz + f_3(\theta) \tag{18.11.19}$$

$$u_{\theta 0} = \int \frac{2(1 + \nu)}{Eh} N_{z\theta}\, dz - \int \frac{1}{R_o}\frac{\partial u_{z0}}{\partial \theta}\, dz + f_4(\theta) \tag{18.11.20}$$

$$u_3 = \frac{\partial u_{\theta 0}}{\partial \theta} - \frac{R_o}{Eh}(N_{\theta\theta} - \nu N_{zz}), \tag{18.11.21}$$

where $f_3(\theta)$ and $f_4(\theta)$ are two additional functions of θ. Eqs. (18.11.13) to (18.11.21) represent the complete solution of a cylindrical membrane shell. The four functions f_1, f_2, f_3, and f_4 require four boundary conditions. They are all functions of θ and can only be applied to edges along which θ varies; in other words, along which z is constant. Therefore, one cannot satisfy the conditions at edges along which θ is constant. This drawback can only be remedied by including the bending resistance of the shell. However, when the shell is closed these difficulties do not arise.

Example 1. Circular Tube Filled with Liquid and Supported at the Ends (Fig. 18.21)

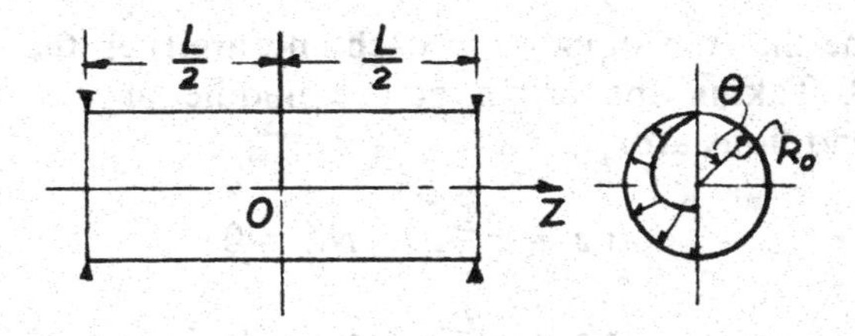

Fig. 18.21

The pressure at any point of the shell is in a direction opposite to that of the positive normal and equal to the weight of the unit column of the liquid at that point. If γ is the unit weight of the liquid,

$$q_z = q_\theta = 0 \quad q_3 = -\gamma R_o(1 - \cos\theta). \tag{18.11.22}$$

Substituting these values in Eqs. (18.11.13) to (18.11.15), we get:

$$N_{\theta\theta} = \gamma R_o^2(1 - \cos\theta) \tag{18.11.23}$$

$$N_{z\theta} = -\gamma R_o z \sin\theta + f_1(\theta) \tag{18.11.24}$$

$$N_{zz} = \frac{\gamma z^2 \cos\theta}{2} - \frac{z}{R_o}\frac{df_1}{d\theta} + f_2(\theta). \tag{18.11.25}$$

The strain-displacement relations reduce to:

$$u_{z0} = \int \frac{1}{Eh}\left[\frac{\gamma z^2 \cos\theta}{2} - \frac{z}{R_o}\frac{df_1}{d\theta} + f_2(\theta) - \nu\gamma R_o^2(1 - \cos\theta)\right] dz + f_3(\theta) \tag{18.11.26}$$

$$u_{\theta 0} = \int \frac{2(1+\nu)}{Eh}\left[-\gamma R_o z \sin\theta + f_1(\theta)\right] dz - \int \frac{1}{R_o}\frac{\partial u_{z0}}{\partial\theta}\, dz + f_4(\theta) \tag{18.11.27}$$

$$u_3 = \frac{\partial u_{\theta 0}}{\partial \theta} - \frac{R_o}{Eh}\left\{\gamma R_o^2(1 - \cos\theta) - \nu\left[\frac{\gamma z^2 \cos\theta}{2} - \frac{z}{R_o}\frac{df_1}{d\theta} + f_2(\theta)\right]\right\}. \tag{18.11.28}$$

Let us assume that the supports at each end are such that $u_{\theta 0} = 0$ and that $N_{zz} = 0$. Taking the origin at the middle of the cylinder, the boundary conditions are:

$$\text{at } z = \pm\frac{L}{2} \qquad N_{zz} = 0 \tag{18.11.29}$$

$$\text{at } z = \pm\frac{L}{2} \qquad u_{\theta 0} = 0. \tag{18.11.30}$$

The first two conditions in conjunction with Eqs. (18.11.24) and (18.11.25) lead to:

$$f_1 = 0, \quad f_2(\theta) = -\frac{\gamma L^2 \cos\theta}{8}. \tag{18.11.31}$$

Notice that those two conditions result in f_1 being a constant, but since no torque is applied to the cylinder this constant must be equal to zero. The third and fourth conditions lead to:

$$f_3 = 0 \tag{18.11.32}$$

$$f_4(\theta) = \frac{\gamma L^2 \sin\theta}{8EhR_o}\left[R_o^2(2 + \nu) + \frac{5L^2}{48}\right]. \tag{18.11.33}$$

Substituting f_1, f_2, f_3, and f_4 in the values of the forces and the displacements, we get:

$$N_{\theta\theta} = \gamma R_o^2(1 - \cos\theta) \tag{18.11.34}$$

$$N_{z\theta} = -\gamma R_o z \sin\theta \tag{18.11.35}$$

$$N_{zz} = -\frac{\gamma \cos\theta}{8}(L^2 - 4z^2) \tag{18.11.36}$$

$$u_{z0} = \frac{\gamma}{Eh}\left[\frac{z^3 \cos\theta}{6} - \frac{L^2 z \cos\theta}{8} - \nu R_o^2 z(1 - \cos\theta)\right] \tag{18.11.37}$$

$$u_{\theta 0} = \frac{\gamma \sin\theta}{8Eh}(L^2 - 4z^2)\left[(2 + \nu)R_o + \frac{5L^2 - 4z^2}{48R_o}\right] \tag{18.11.38}$$

$$u_3 = \frac{\gamma \cos\theta}{8Eh}(L^2 - 4z^2)\left(2R_o + \frac{5L^2 - 4z^2}{48R_o}\right) - \frac{\gamma R_o^3}{Eh}(1 - \cos\theta). \tag{18.11.39}$$

If the cylindrical tube is loaded only by an internal pressure $q_3 = -p_o$,

$$N_{\theta\theta} = p_o R_o, \quad N_{z\theta} = N_{zz} = 0 \tag{18.11.40}$$

$$u_{z0} = -\frac{\nu R_o p_o z}{Eh}, \quad u_{\theta 0} = 0, \quad u_3 = -\frac{p_o R_o^2}{Eh}. \tag{18.11.41}$$

Eqs. (18.11.34) to (18.11.41) are only valid if no force N_{zz} is applied at the ends of the tube.

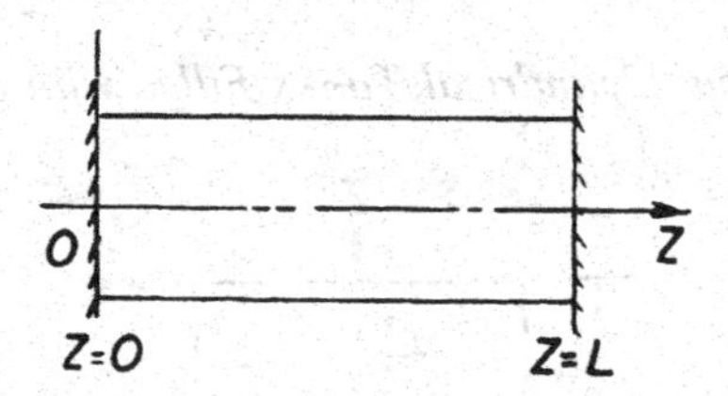

Fig. 18.22

Let us now consider the case in which the ends of the tube are built in and assume that at these ends $e_{\theta\theta 0} = 0$. Taking the origin at one end of the cylinder (Fig. 18.22), the boundary conditions necessary to determine $f_1(\theta)$ and $f_2(\theta)$ are:

$$\text{at } z = 0 \text{ and } z = L, \; N_{\theta\theta} = \nu N_{zz}. \tag{18.11.42}$$

Thus,

$$f_2(\theta) = \frac{\gamma R_o^2}{\nu}(1 - \cos\theta), \quad f_1(\theta) = \frac{\gamma R_o L}{2}\sin\theta + C, \tag{18.11.43}$$

where C is a constant of integration. If we substitute $f_1(\theta)$ into Eq. (18.11.24), we see that C represents a shearing force uniformly distributed around the tube. If there is no torque applied, such a force must be equal to zero. Thus, with the boundary conditions (18.11.42), the forces in the tube are:

$$N_{\theta\theta} = \gamma R_o^2(1 - \cos\theta) \tag{18.11.44}$$

$$N_{z\theta} = \gamma R_o \sin\theta\left(\frac{L}{2} - z\right) \tag{18.11.45}$$

$$N_{zz} = -\frac{\gamma z \cos\theta}{2}(L - z) + \frac{\gamma R_o^2}{\nu}(1 - \cos\theta). \tag{18.11.46}$$

If the supports are rigid and cannot move, there will be no change in the length of the generators, which means that u_{z0} must be equal to zero. However, it is apparent from Eq. (18.11.26) that this is not the case. Such a result indicates that bending will occur in the shell and that the membrane theory is not sufficient to describe the deformation.

If the cylinder is loaded only by an internal pressure $q_3 = -p_o$,

$$N_{\theta\theta} = p_o R_o, \quad N_{z\theta} = 0, \quad N_{zz} = \frac{p_o R_o}{2}. \tag{18.11.47}$$

Example 2. Vertical Cylindrical Tanks Filled with Liquid

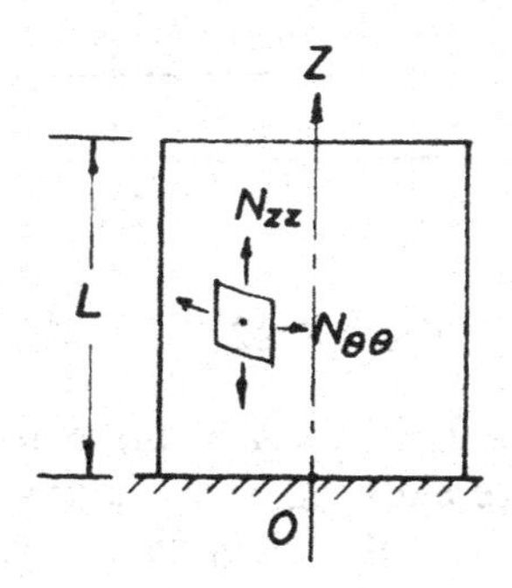

Fig. 18.23

Consider a cylindrical tank of radius R_o and height L (Fig. 18.23), which is filled with liquid of unit weight γ and rigidly built in at the base. The pressure at any point of the shell is in a direction opposite to that of the positive normal, and has the components:

$$q_z = q_\theta = 0 \quad q_3 = -\gamma(L - z). \tag{18.11.48}$$

Substituting these values into Eqs. (18.11.13) to (18.11.15), we get:

$$N_{\theta\theta} = R_o \gamma(L - z), \quad N_{z\theta} = f_1(\theta), \quad N_{zz} = -\frac{z}{R_o}\frac{\partial f_1}{\partial \theta} + F_2(\theta). \tag{18.11.49}$$

The boundary conditions necessary to determine $f_1(\theta)$ and $f_2(\theta)$ are:

$$\text{at } z = 0, \qquad N_{z\theta} = 0 \tag{18.11.50}$$

$$\text{at } z = L, \qquad N_{zz} = 0. \tag{18.11.51}$$

Thus,

$$f_1(\theta) = f_2(\theta) = 0, \tag{18.11.52}$$

so that

$$N_{\theta\theta} = R_o\gamma(L - z), \quad N_{z\theta} = 0, \quad N_{zz} = 0. \tag{18.11.53}$$

The displacements are given by Eqs. (18.11.19) to (18.11.21). Thus,

$$u_{z0} = -\frac{1}{Eh}\left(\nu R_o\gamma Lz - \frac{\nu R_o\gamma z^2}{2}\right) + f_3(\theta) \tag{18.11.54}$$

$$u_{\theta 0} = -\frac{z}{R_o}\frac{df_3}{d\theta} + f_4(\theta) \tag{18.11.55}$$

$$u_3 = -\frac{z}{R_o}\frac{d^2 f_3}{d\theta} + \frac{df_4}{d\theta} - \frac{R_o^2\gamma}{Eh}(L - z). \tag{18.11.56}$$

To determine $f_3(\theta)$ and $f_4(\theta)$, we have the two boundary conditions:

$$\text{at } z = 0, \quad u_{z0} = u_{\theta 0} = 0 \tag{18.11.57}$$

Thus,

$$f_3(\theta) = f_4(\theta) = 0, \tag{18.11.58}$$

so that

$$u_{z0} = -\frac{\gamma\nu R_o z}{2Eh}(2L - z), \quad u_{\theta 0} = 0, \quad u_3 = -\frac{R_o^2\gamma}{Eh}(L - z). \tag{18.11.59}$$

One notices that at $z = 0$, u_3 is not equal to zero, which cannot be true if the end is fixed. This result indicates that bending will occur in the shell and that the membrane theory does not completely describe the deformation.

Example 3. Cantilever Circular Cylindrical Shell under Its Own Weight Cylindrical shells are commonly used as roofing structures and, as such, they may either be supported at the ends or cantilevered as shown in Fig. 18.24. If the gravitational force per unit area is γh, this force has the components:

$$q_z = 0, \quad q_\theta = \gamma h \sin\theta, \quad q_3 = \gamma h \cos\theta. \tag{18.11.60}$$

Substituting these values into Eqs. (18.11.13) to (18.11.15), we get:

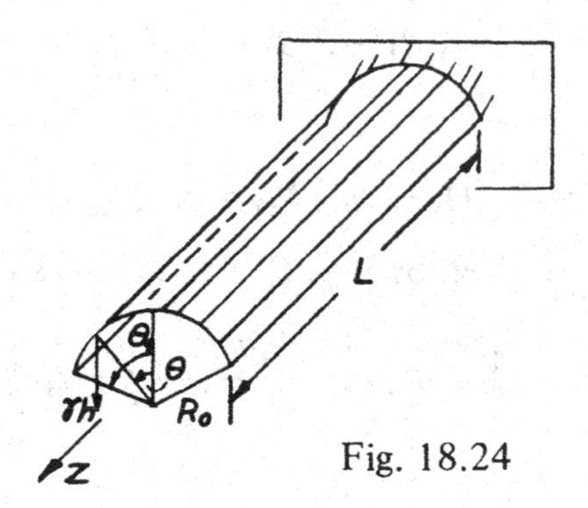

Fig. 18.24

$$N_{\theta\theta} = -R_o\gamma h \cos\theta, \quad N_{z\theta} = -2\gamma hz \sin\theta + f_1(\theta) \quad (18.11.61)$$

$$N_{zz} = \frac{\gamma hz^2 \cos\theta}{R_o} - \frac{z}{R_o}\frac{df_1}{d\theta} + f_2(\theta). \quad (18.11.62)$$

The boundary conditions for this problem are that

$$\text{at } z = L, \quad N_{zz} = N_{z\theta} = 0. \quad (18.11.63)$$

Thus,

$$f_1(\theta) = 2\gamma hL \sin\theta, \quad f_2(\theta) = \frac{\gamma hL^2 \cos\theta}{R_o}, \quad (18.11.64)$$

so that the forces are now

$$N_{\theta\theta} = -R_o\gamma h \cos\theta, \quad N_{z\theta} = 2\gamma h(L - z)\sin\theta,$$
$$N_{zz} = \frac{\gamma h \cos\theta}{R_o}(L - z)^2. \quad (18.11.65)$$

As mentioned at the beginning of this section, one cannot satisfy the conditions on the boundaries along which θ is constant for this open shell. Thus, at the edges $\theta = \pm\theta_1$,

$$N_{\theta\theta} = -R_o\gamma h \cos\theta_1 \neq 0 \quad (18.11.66)$$

$$N_{z\theta} = \pm 2\gamma h(L - z)\sin\theta_1 \neq 0, \quad (18.11.67)$$

and the conditions on these boundaries are violated. The membrane theory is unsatisfactory for this case and one has to use the general theory which incorporates bending.

18.12 General Theory of Circular Cylindrical Shells

The general equations of the theory of circular cylindrical shells are obtained by associating ξ_1 with z and ξ_2 with θ, and by setting $\sqrt{E} = 1$ and $\sqrt{G} = R_o$ in the equations of equilibrium (Sec. 18.7), the strain-displacement relations (Sec. 18.5), and the stress resultants and stress couples relations (Sec. 18.6). The subscripts 1 and 2 are replaced by z and θ, respectively: R_1 is infinite and $R_2 = R_o$. The first fundamental form is

$$(ds)^2 = (dz)^2 + R_o^2(d\theta)^2. \tag{18.12.1}$$

The differential equations of equilibrium of forces (18.7.14) to (18.7.16) become (Fig. 18.25):

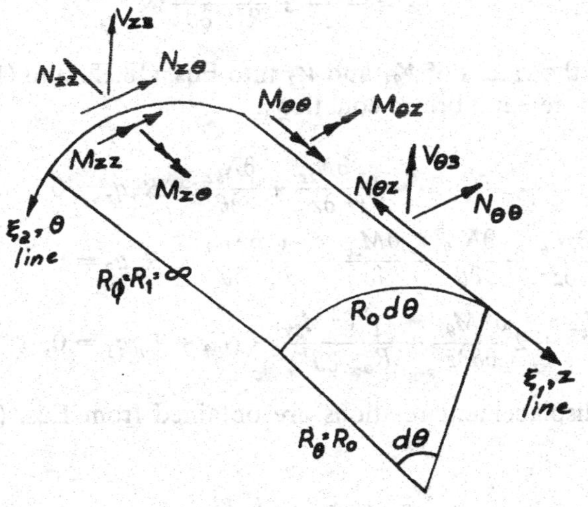

Fig. 18.25

$$R_o \frac{\partial N_{zz}}{\partial z} + \frac{\partial N_{\theta z}}{\partial \theta} + R_o q_z = 0 \tag{18.12.2}$$

$$R_o \frac{\partial N_{z\theta}}{\partial z} + \frac{\partial N_{\theta\theta}}{\partial \theta} - V_{\theta 3} + R_o q_\theta = 0 \tag{18.12.3}$$

$$R_o \frac{\partial V_{z3}}{\partial z} + \frac{\partial V_{\theta 3}}{\partial \theta} + N_{\theta\theta} + R_o q_3 = 0. \tag{18.12.4}$$

We shall assume that the ratio of the thickness to R_o is small compared to unity, so that $M_{z\theta} = M_{\theta z}$ and $N_{z\theta} = N_{\theta z}$. The differential equations of equilibrium of moments (18.7.22) and (18.7.23), become:

$$R_o \frac{\partial M_{z\theta}}{\partial z} + \frac{\partial M_{\theta\theta}}{\partial \theta} - R_o V_{\theta 3} = 0 \tag{18.12.5}$$

$$R_o \frac{\partial M_{zz}}{\partial z} + \frac{\partial M_{\theta z}}{\partial \theta} - R_o V_{z3} = 0. \tag{18.12.6}$$

From these two equations, we find that:

$$V_{\theta 3} = \frac{\partial M_{z\theta}}{\partial z} + \frac{1}{R_o}\frac{\partial M_{\theta\theta}}{\partial \theta} \tag{18.12.7}$$

$$V_{z3} = \frac{\partial M_{zz}}{\partial z} + \frac{1}{R_o}\frac{\partial M_{\theta z}}{\partial \theta}. \tag{18.12.8}$$

Substituting the values of $V_{\theta 3}$ and V_{z3} into Eqs. (18.12.2) to (18.12.4), we obtain the three equilibrium equations:

$$R_o \frac{\partial N_{zz}}{\partial z} + \frac{\partial N_{\theta z}}{\partial \theta} + R_o q_z = 0 \tag{18.12.9}$$

$$R_o \frac{\partial N_{z\theta}}{\partial z} + \frac{\partial N_{\theta\theta}}{\partial \theta} - \frac{\partial M_{z\theta}}{\partial z} - \frac{1}{R_o}\frac{\partial M_{\theta\theta}}{\partial \theta} + R_o q_\theta = 0 \tag{18.12.10}$$

$$R_o \frac{\partial^2 M_{zz}}{\partial z^2} + 2\frac{\partial^2 M_{\theta z}}{\partial \theta \partial z} + \frac{1}{R_o}\frac{\partial^2 M_{\theta\theta}}{\partial \theta^2} + N_{\theta\theta} + R_o q_3 = 0. \tag{18.12.11}$$

The strain-displacement relations are obtained from Eqs. (18.5.22) to (18.5.30):

$$e_{zz} = e_{zz0} - \xi_3 K_{zz} \tag{18.12.12}$$

$$e_{\theta\theta} = e_{\theta\theta 0} - \xi_3 K_{\theta\theta} \tag{18.12.13}$$

$$e_{z\theta} = e_{z\theta 0} - \xi_3 K_{z\theta}, \tag{18.12.14}$$

where

$$e_{zz0} = \frac{\partial u_{z0}}{\partial z} \tag{18.12.15}$$

$$e_{\theta\theta 0} = \frac{1}{R_o}\frac{\partial u_{\theta 0}}{\partial \theta} - \frac{u_3}{R_o} \tag{18.12.16}$$

$$e_{z\theta 0} = \frac{1}{2}\left(\frac{\partial u_{\theta 0}}{\partial z} + \frac{1}{R_o}\frac{\partial u_{z0}}{\partial \theta}\right) \tag{18.12.17}$$

$$K_{zz} = \frac{\partial^2 u_3}{\partial z_2} \tag{18.12.18}$$

$$K_{\theta\theta} = \frac{1}{R_o^2}\frac{\partial}{\partial \theta}\left(u_{\theta 0} + \frac{\partial u_3}{\partial \theta}\right) \tag{18.12.19}$$

$$K_{z\theta} = \frac{1}{R_o}\left(\frac{1}{2}\frac{\partial u_{\theta 0}}{\partial z} + \frac{\partial^2 u_3}{\partial \theta \partial z}\right). \tag{18.12.20}$$

Substituting the previous equations into Eqs. (18.6.11) to (18.6.15), we obtain:

$$N_{zz} = \frac{Eh}{1 - \nu^2}\left[\frac{\partial u_{z0}}{\partial z} + \nu\left(\frac{1}{R_o}\frac{\partial u_{\theta 0}}{\partial \theta} - \frac{u_3}{R_o}\right)\right] \tag{18.12.21}$$

$$N_{\theta\theta} = \frac{Eh}{1 - \nu^2}\left(\frac{1}{R_o}\frac{\partial u_{\theta 0}}{\partial \theta} - \frac{u_3}{R_o} + \nu\frac{\partial u_{z0}}{\partial z}\right) \tag{18.12.22}$$

$$N_{z\theta} = N_{\theta z} = \frac{Eh}{2(1 + \nu)}\left(\frac{\partial u_{\theta 0}}{\partial z} + \frac{1}{R_o}\frac{\partial u_{z0}}{\partial \theta}\right) \tag{18.12.23}$$

$$M_{zz} = -D\left[\frac{\partial^2 u_3}{\partial z^2} + \frac{\nu}{R_o^2}\left(\frac{\partial u_{\theta 0}}{\partial \theta} + \frac{\partial^2 u_3}{\partial \theta^2}\right)\right] \tag{18.12.24}$$

$$M_{\theta\theta} = -D\left[\frac{1}{R_o^2}\left(\frac{\partial u_{\theta 0}}{\partial \theta} + \frac{\partial^2 u_3}{\partial \theta^2}\right) + \nu\frac{\partial^2 u_3}{\partial z^2}\right] \tag{18.12.25}$$

$$M_{z\theta} = M_{\theta z} = -\frac{D(1 - \nu)}{R_o}\left(\frac{1}{2}\frac{\partial u_{\theta 0}}{\partial z} + \frac{\partial^2 u_3}{\partial \theta \partial z}\right). \tag{18.12.26}$$

The three equations of equilibrium (18.12.9) to (18.12.11) can now be written in terms of the displacements u_{z0}, $u_{\theta 0}$, and u_3 of the middle surface. Substituting Eqs. (18.12.21) to (18.12.26) into Eqs. (18.12.9) to (18.12.11), we get:

$$\frac{\partial^2 u_{z0}}{\partial z^2} + \frac{(1-\nu)}{2R_o^2}\frac{\partial^2 u_{z0}}{\partial \theta^2} + \frac{1+\nu}{2R_o}\frac{\partial^2 u_{\theta 0}}{\partial z \partial \theta} - \frac{\nu}{R_o}\frac{\partial u_3}{\partial z} + \frac{q_z(1-\nu^2)}{Eh} = 0 \qquad (18.12.27)$$

$$\frac{(1+\nu)}{2R_o}\frac{\partial^2 u_{z0}}{\partial z \partial \theta} + \frac{1-\nu}{2}\frac{\partial^2 u_{\theta 0}}{\partial z^2} + \frac{1}{R_o^2}\frac{\partial^2 u_{\theta 0}}{\partial \theta^2} - \frac{1}{R_o^2}\frac{\partial u_3}{\partial \theta}$$
$$+ \frac{h^2}{12R_o^2}\left(\frac{\partial^3 u_3}{\partial \theta \partial z^2} + \frac{1}{R_o^2}\frac{\partial^3 u_3}{\partial \theta^3}\right) \qquad (18.2.28)$$
$$+ \frac{h^2}{12R_o^2}\left(\frac{1-\nu}{2}\frac{\partial^2 u_{\theta 0}}{\partial z^2} + \frac{1}{R_o^2}\frac{\partial^2 u_{\theta 0}}{\partial \theta^2}\right) + \frac{(1-\nu^2)}{Eh}q_\theta = 0$$

$$\nu\frac{\partial u_{z0}}{\partial z} + \frac{1}{R_o}\frac{\partial u_{\theta 0}}{\partial \theta} - \frac{u_3}{R_o}$$
$$- \frac{h^2}{12}\left(R_o\frac{\partial^4 u_3}{\partial z^4} + \frac{2}{R_o}\frac{\partial^4 u_3}{\partial z^2 \partial \theta^2} + \frac{1}{R_o^3}\frac{\partial^4 u_3}{\partial \theta^4}\right) \qquad (18.12.29)$$
$$- \frac{h^2}{12}\left(\frac{1}{R_o}\frac{\partial^3 u_{\theta 0}}{\partial z^2 \partial \theta} + \frac{1}{R_o^3}\frac{\partial^3 u_{\theta 0}}{\partial \theta^3}\right) + \frac{1-\nu^2}{Eh}R_o q_3 = 0.$$

The problem of the circular cylindrical shell reduces in each particular case to the solution of this system of three partial differential equations.

18.13 Circular Cylindrical Shell Loaded Symmetrically with Respect to its Axis

Eqs. (18.12.27) to (18.12.29) are much simplified in the case of circular cylindrical shells loaded symmetrically with respect to their axes. Because of this symmetry, $q_\theta = 0$, $u_{\theta 0} = 0$, and u_{z0} and u_3 are functions of z only. Eq. (18.12.28) is identically satisfied, while Eqs. (18.12.27) and (18.12.29) become:

$$\frac{d^2 u_{z0}}{dz^2} - \frac{\nu}{R_o}\frac{du_3}{dz} + \frac{q_z(1-\nu^2)}{Eh} = 0 \qquad (18.13.1)$$

$$\nu\frac{du_{z0}}{dz} - \frac{u_3}{R_o} - \frac{h^2}{12}R_o\frac{d^4 u_3}{dz^4} + \frac{(1-\nu^2)}{Eh}R_o q_3 = 0. \qquad (18.13.2)$$

If the cylinder is subjected to radial pressure only, $q_z = 0$, then Eq. (18.13.1) can be integrated to give:

$$\frac{du_{z0}}{dz} - \frac{\nu u_3}{R_o} = C, \qquad (18.13.3)$$

where C is a constant of integration. But, from Eq. (18.12.21),

$$N_{zz} = \frac{Eh}{1-\nu^2}\left(\frac{du_{z0}}{dz} - \frac{\nu u_3}{R_o}\right). \qquad (18.13.4)$$

Therefore, the forces N_{zz} are constants. Eliminating du_{z0}/dz from Eq. (18.13.2), the following linear differential equation of the fourth order is obtained:

$$\frac{d^4 u_3}{dz^4} + 4\beta^4 u_3 = \frac{q_3}{D} + \frac{\nu N_{zz}}{DR_o}, \qquad (18.13.5)$$

where

$$\beta^4 = \frac{3(1-\nu^2)}{R_o^2 h^2}, \quad D = \frac{Eh^3}{12(1-\nu^2)}. \qquad (18.13.6)$$

The solution of this differential equation is

$$u_3 = e^{-\beta z}(C_1 \cos \beta z + C_2 \sin \beta z) + e^{\beta z}(C_3 \cos \beta z + C_4 \sin \beta z) + f(z), \qquad (18.13.7)$$

where C_1, C_2, C_3, and C_4 are constants of integration, and $f(z)$ is the particular integral. Once u_3 has been found, the forces and moments per unit length can be obtained from Eqs. (18.12.21) to (18.12.26). N_{zz} has already been shown to be a constant and, if this constant is equal to 0,

$$\frac{\partial u_{z0}}{\partial z} = \frac{\nu u_3}{R_o}. \qquad (18.13.8)$$

Substituting Eq. (18.13.8) into Eq. (18.12.22), we get:

$$N_{\theta\theta} = -\frac{Ehu_3}{R_o}. \qquad (18.13.9)$$

Also,

$$N_{z\theta} = 0, \quad M_{zz} = -D\frac{d^2u_3}{dz^2}, \quad M_{\theta\theta} = -D\nu\frac{d^2u_3}{dz^2}. \qquad (18.13.10)$$

From Eqs. (18.12.7) and (18.12.8):

$$V_{\theta 3} = 0, \quad V_{z3} = -D\frac{d^3u_3}{dz^3}. \qquad (18.13.11)$$

Example 1: Vertical Cylincrical Tank Filled with Liquid
This problem was analyzed in the previous section by means of the membrane theory, and the conclusion was reached that such a theory does not completely describe the deformation. It is interesting to compare the results obtained using the membrane theory to those obtained using the general theory. The components of the pressure at any point of the shell are (Fig. 18.23):

$$q_z = q_\theta = 0, \quad q_3 = -\gamma(L - z). \tag{18.13.12}$$

N_{zz} is a constant throughout the shell and, assuming that no load is applied to it in the vertical direction, N_{zz} is equal to zero. Eq. (18.13.5) becomes:

$$\frac{d^4 u_3}{dz^4} + 4\beta^4 u_3 = -\frac{\gamma(L - z)}{D}, \tag{18.13.13}$$

and its particular integral is:

$$f(z) = -\gamma(L - z)\frac{R_o^2}{Eh}. \tag{18.13.14}$$

Eq. (18.13.7) has four constants of integration to be determined from the boundary conditions. The constant β is inversely proportional to $\sqrt{R_o h}$; if the length L is large compared to $\sqrt{R_o h}$, the tank may be considered as infinitely long. In this case, the two constants C_3 and C_4 must be equal to zero if one is to have meaningful displacements at the top. At the bottom of the tank, we have the two conditions:

$$\text{at } z = 0 \quad u_3 = 0, \quad \frac{du_3}{dz} = 0. \tag{18.13.15}$$

Substituting Eq. (18.13.15) into Eq. (18.13.7), we find:

$$C_1 = \frac{\gamma R_o^2 L}{Eh}, \quad C_2 = \frac{\gamma R_o^2}{Eh}\left(L - \frac{1}{\beta}\right). \tag{18.13.16}$$

The deflection u_3 then becomes:

$$u_3 = -\frac{\gamma R_o^2}{Eh}\left\{L - z - e^{-\beta z}\left[L \cos \beta z + \left(L - \frac{1}{\beta}\right)\sin \beta z\right]\right\} \tag{18.13.17}$$

and

$$N_{zz} = N_{z\theta} = 0 \tag{18.13.18}$$

$$N_{\theta\theta} = -\frac{Ehu_3}{R_o} = \gamma R_o\left\{L - z - e^{-\beta z}\left[L\cos\beta z + \left(L - \frac{1}{\beta}\right)\sin\beta z\right]\right\} \tag{18.13.19}$$

$$M_{zz} = -D\frac{d^2u_3}{dz^2} = \frac{\gamma R_o Lh}{\sqrt{12(1-\nu^2)}}e^{-\beta z}\left[-\sin\beta z + \left(1 - \frac{1}{\beta L}\right)\cos\beta z\right] \tag{18.13.20}$$

$$M_{\theta\theta} = -D\nu\frac{d^2u_3}{dz^2} = \frac{\nu\gamma R_o Lh}{\sqrt{12(1-\nu^2)}}e^{-\beta z}\left[-\sin\beta z + \left(1 - \frac{1}{\beta L}\right)\cos\beta z\right] \tag{18.13.21}$$

$$V_{\theta 3} = 0 \tag{18.13.22}$$

$$V_{z3} = -D\frac{d^3u_3}{dz^3} = -\frac{\gamma R_o h}{\sqrt{12(1-\nu^2)}}e^{-\beta z}[(2\beta L - 1)\cos\beta z - \sin\beta z]. \tag{18.13.23}$$

Knowing the values of the forces and moments, the stresses can easily be calculated. The maximum moment and shearing force occur at the bottom of the tank, and are given by:

$$(M_{zz})_{\max} = \left(1 - \frac{1}{\beta L}\right)\frac{\gamma R_o Lh}{\sqrt{12(1-\nu^2)}},$$
$$(V_{z3})_{\max} = -(2\beta L - 1)\frac{\gamma R_o h}{\sqrt{12(1-\nu^2)}}. \tag{18.13.24}$$

One notices that the solution based on membrane theory is contained in the previous expressions for $N_{\theta\theta}$ and u_3. Close to the base, the results differ quite substantially; the differences decrease as the value of z increases. A more severe criticism of the membrane theory is that it does not give any indication of the nature or values of the bending moments M_{zz}, which give the most critical stress condition at the base of the tank. Tables and charts which considerably facilitate the numerical computations involved in this problem can be found in [2].

Example 2: Cylindrical Pressure Tank with Rigid End Plates

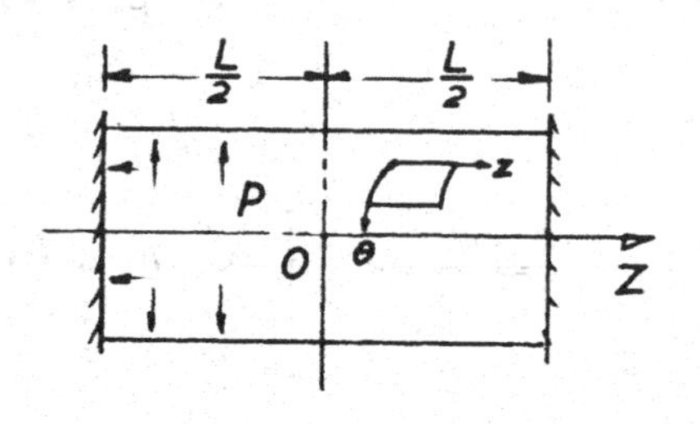

Fig. 18.26

Let us consider a cylindrical pressure tank of radius R_o and of length L (Fig. 18.26) subjected to an internal air pressure p, and let us choose the origin at the center of the tank. The components of the pressure at any point are:

$$q_z = q_\theta = 0, \quad q_3 = -p. \tag{18.13.25}$$

Eq. (18.13.5) becomes:

$$\frac{d^4 u_3}{dz^4} + 4\beta^4 u_3 = -\frac{p}{D}. \tag{18.13.26}$$

The solution of Eq. (18.13.26) is:

$$u_3 = e^{-\beta z}(C_1 \cos \beta z + C_2 \sin \beta z) + e^{\beta z}(C_3 \cos \beta z + C_4 \sin \beta z) - \frac{pR_o^2}{Eh}. \tag{18.13.27}$$

It is convenient to replace the exponential functions by hyperbolic functions according to the well known formulas:

$$e^{\beta z} = \cosh \beta z + \sinh \beta z, \quad e^{-\beta z} = \cosh \beta z - \sinh \beta z. \tag{18.13.28}$$

Thus,

$$\begin{aligned} u_3 = {} & A_1 \sin \beta z \sinh \beta z + A_2 \sin \beta z \cosh \beta z \\ & + A_3 \cos \beta z \sinh \beta z + A_4 \cos \beta z \cosh \beta z - \frac{pR_o^2}{Eh}, \end{aligned} \tag{18.13.29}$$

where A_1, A_2, A_3, and A_4 are constants to be obtained from the boundary conditions. Because of symmetry, the displacement u_3 must be the same at equal distances on both sides of the origin. The two terms in Eq. (18.13.29) that are symmetrical with respect to the origin are $A_1 \sin \beta z \sinh \beta z$ and $A_4 \cos \beta z \cosh \beta z$, so that A_2 and A_3 must be equal to zero and

$$u_3 = A_1 \sin \beta z \sinh \beta z + A_4 \cos \beta z \cosh \beta z - \frac{pR_o^2}{Eh}. \qquad (18.13.30)$$

Since the ends of the shell are rigid, the boundary conditions are:

$$\text{at } z \pm \frac{L}{2} \quad u_3 = \frac{du_3}{dz} = 0. \qquad (18.13.31)$$

Substituting Eq. (18.13.31) into Eq. (18.13.30), and setting:

$$\alpha = \frac{\beta L}{2}, \qquad (18.13.32)$$

we get:

$$A_1 = \frac{2pR_o^2}{Eh}\left[\frac{\sin \alpha \cosh \alpha - \cos \alpha \sinh \alpha}{\sin 2\alpha + \sinh 2\alpha}\right] \qquad (18.13.33)$$

$$A_4 = \frac{2pR_o^2}{Eh}\left[\frac{\cos \alpha \sinh \alpha + \sin \alpha \cosh \alpha}{\sin 2\alpha + \sinh 2\alpha}\right]. \qquad (18.13.34)$$

The expressions of the stress resultants and stress couples can now easily be obtained by substitution of u_3 into Eqs. (18.13.9) to (18.13.11):

$$N_{\theta\theta} = pR_o - \frac{Eh}{R_o}[A_1 \sin \beta z \sinh \beta z + A_4 \cos \beta z \cosh \beta z] \qquad (18.13.35)$$

$$N_{z\theta} = 0 \qquad (18.13.36)$$

$$M_{zz} = -2D\beta^2[A_1 \cos \beta z \cosh \beta z - A_4 \sin \beta z \sinh \beta z] \qquad (18.13.37)$$

$$M_{\theta\theta} = \nu M_{zz} \qquad (18.13.38)$$

$$V_{\theta 3} = 0 \qquad (18.13.39)$$

$$V_{z3} = -2D\beta^3[-(A_1 + A_4)\sin \beta z \cosh \beta z + (A_1 - A_4)\cos \beta z \sinh \beta z], \qquad (18.13.40)$$

where A_1 and A_4 are given by Eqs. (18.13.33) and (18.13.34). Additional examples can be found in [2] and [3].

PROBLEMS

1. Find the principal directions and the principal curvatures on the surface

$$x_1 = a(\xi_1 + \xi_2), \quad x_2 = b(\xi_1 - \xi_2), \quad x_3 = \xi_1 \xi_2 .$$

What are the expressions for the first and second curvatures?

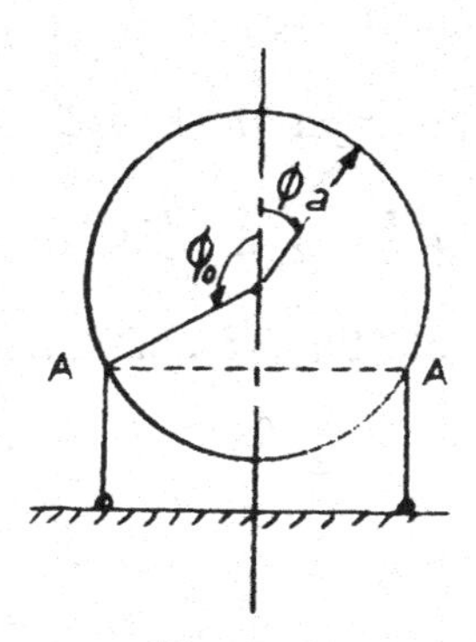

Fig. 18.27

2. Derive the first fundamental form and the second fundamental magnitudes of the following surfaces of revolution: 1) a flat circular sheet 2) a cone 3) a sphere 4) a paraboloid 5) an ellipsoid.

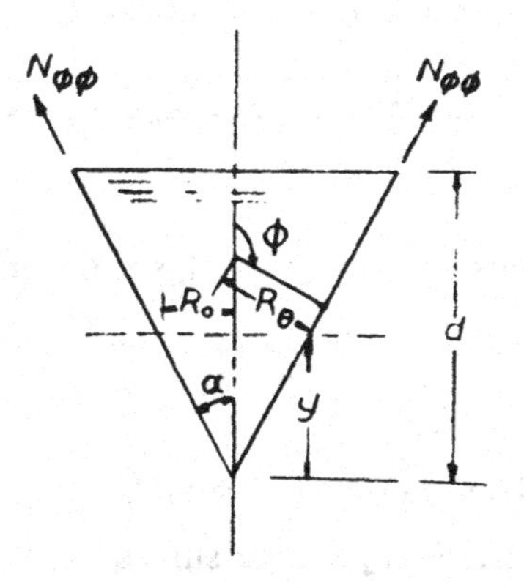

Fig. 18.28

3. A spherical tank, supported along a parallel circle *AA* (Fig. 18.27), is filled with a liquid of unit weight γ. Using the membrane theory of shells of revolution, find $N_{\phi\phi}$ and $N_{\theta\theta}$ in terms of ϕ, a, and γ for $\phi < \phi_0$ and $\phi > \phi_0$.

4. A conical shell, filled with a liquid of unit weight γ is supported by forces in the direction of the generatrices, as shown in Fig. 18.28. Using the membrane theory of shells, find $N_{\phi\phi}$ and $N_{\theta\theta}$ in terms of α, γ, d and y. d is the depth of the liquid and y is an arbitrary distance from the apex of the cone.

5. A horizontal thin circular cylinder has its ends built in and is under its own weight. Find the membrane forces.

6. Draw the distribution of the moments and forces with respect to depth for a tank similar to the one shown in Fig. 18.23, full of water, and having the following dimensions: $R_o = 30$ ft., $L = 30$ ft., $h = 1$ ft. ($\nu = 0.3, E = 3 \times 10^6$ psi, $\gamma = 62.5$ lb / ft^3)

REFERENCES

[1] D. J. Struik, *Differential Geometry*, Addison-Wesley, Reading, Pa., 1950.

[2] S. Timoshenko and S. Woinowsky-Kreiger, *Theory of Plates and Shells*, McGraw-Hill, New York, N. Y., 1959.

[3] J. E. Gibson, *Linear Elastic Theory of Thin Shells*, Pergamon Press, New York, N. Y., 1965.

CHAPTER 19

SOLUTIONS OF ELASTICITY PROBLEMS BY MEANS OF COMPLEX VARIABLES

19.1 Introduction

Functions of a complex variable can be used to great advantage in constructing stress functions. They were introduced in the solution of plane elastic problems by Kolosov [1] and later used systematically by Muskhelishvili [2] to solve a variety of problems, in particular two dimensional ones.

In this chapter, a brief summary of complex variables theory is given. This theory is then used to solve a few problems, some expressed in terms of curvilinear coordinates. The goal is to show the power of the method, as well as, to lay a good foundation for more advanced studies; while obtaining results that are used in topics such as fracture and crack propagation, and stresses in semi-infinite media. In addition to the references mentioned above, other accounts of complex variable techniques have been given by Sokolnikoff [3], Green and Zerna [4] and Milne-Thomson [5]. For detailed descriptions on the functions of a complex variable the reader is referred to the treatises of Churchill [6] and McLachlan [7].

19.2 Complex Variables and Complex Functions: A Short Review

A complex variable z is formed by two real variables x_1 and x_2 so that

$$z = x_1 + ix_2 \tag{19.2.1}$$

where $i = \sqrt{-1}$. x_1 is the real part of the complex variable z and x_2 is the imaginary part. By definition, when two complex variables are equal, their real parts are equal and their imaginary parts are equal. Similarly, when adding or subtracting complex variables, the real parts and imaginary parts are added or subtracted separately. Multiplications and divisions are carried on just as for real numbers. Thus, for example:

$$\begin{aligned} z^2 &= (x_1 + ix_2)^2 \\ &= x_1^2 + 2ix_1x_2 + (ix_2)^2 \\ &= (x_1^2 - x_2^2) + i(2x_1x_2) \end{aligned} \tag{19.2.2}$$

Complex variables can be geometrically represented in an X_1, X_2 plane called the complex plane, on a diagram called Argand's diagram. Fig. 19.1 shows such a diagram in which the X_1 axis is called the real axis and the X_2 axis is called the imaginary axis. In polar coordinates we have

$$z = x_1 + ix_2 = r(\cos\theta + i\sin\theta) = re^{i\theta} \tag{19.2.3}$$

z represents the vector OA. The vector OA' is represented by

$$\bar{z} = x_1 - ix_2 = r(\cos\theta - i\sin\theta) = re^{-i\theta} \tag{19.2.4}$$

$\bar{z}$ is called the conjugate of z. The product $z\bar{z}$ is equal to r^2 where $r = \sqrt{x_1^2 + x_2^2}$ is called the modulus of z. The angle θ is called the argument of z. The argument of the product $z_1\, z_2$ is the sum of the two arguments θ_1 and θ_2. When a complex number z is multiplied by i, the resulting vector iz is the one obtained by rotating the vector z through a right angle, counterclockwise, without changing the length.

A function of a complex variable is called a complex function. It has a real part and an imaginary part. Thus,

$$\zeta = F(z) = F(x_1 + ix_2) = \phi(x_1, x_2) + i\psi(x_1, x_2) \tag{19.2.5}$$

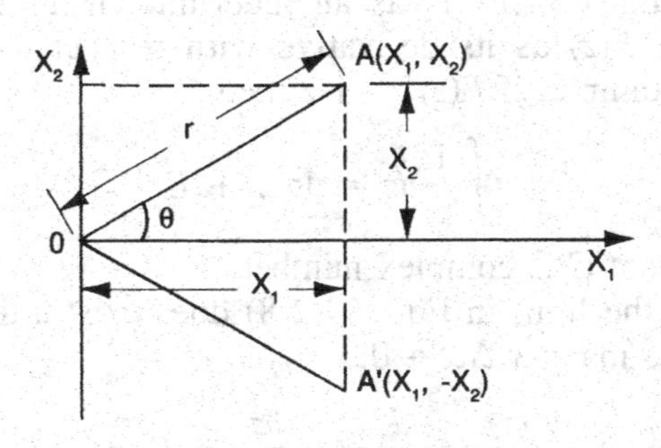

Fig. 19.1

where ϕ and ψ are real functions of x_1 and x_2 only. Thus, in Eq. (19.2.2), $\phi = x_1^2 - x_2^2$ and $\psi = 2x_1x_2$. ψ is called the conjugate of ϕ.

The conjugate function of the complex function $F(z)$ is $\overline{F}(\overline{z}) \equiv \overline{F(z)}$, where

$$\overline{F}(\overline{z}) = \phi(x_1, x_2) - i\psi(x_1, x_2) \tag{19.2.6}$$

If the derivative of a complex function $F(z)$ is unique at every point of a region R and depends only on z, the function is said to be analytic or regular. It is called holomorphic if it is also single valued. By definition if $\zeta = F(z)$

$$\frac{d\zeta}{dz} = \frac{dF(z)}{dz} = F'(z) = \lim_{\Delta z \to 0} \frac{F(z + \Delta z) - F(z)}{\Delta z} = \lim_{\Delta z \to 0} \frac{\Delta \zeta}{\Delta z} \tag{19.2.7}$$

Δz is a vector in the complex plane. For the derivative to be unique, it must be independent of the manner Δz approaches zero. For example, $\zeta = x_1 - ix_2 = \overline{z}$ is not analytic anywhere since its derivative is not unique. Indeed,

$$\begin{aligned}\lim_{\Delta z \to 0} \frac{\Delta \zeta}{\Delta z} &= \lim_{\substack{\Delta x_1 \to 0 \\ \Delta x_2 \to 0}} \frac{x_1 + \Delta x_1 - i(x_2 + \Delta x_2) - x_1 + ix_2}{\Delta x_1 + i\Delta x_2} \\ &= \lim_{\substack{\Delta x_1 \to 0 \\ \Delta x_2 \to 0}} \frac{\Delta x_1 - i\Delta x_2}{\Delta x_1 + i\Delta x_2}\end{aligned} \tag{19.2.8}$$

If Δz approaches zero along a line parallel to the X_1 axis so that $\Delta x_2 = 0$ throughout the limiting process, the limit in Eq. (19.2.8) is equal to $+1$; whereas, if Δz approaches zero along a line parallel to the imaginary X_2 axis so that $\Delta x_1 = 0$ throughout the process, the limit is equal to -1.

An analytic function $F(z)$ has an indefinite integral defined as the function having $F(z)$ as its derivative with respect to z; it is written $\int F(z)dz$. For example, if $F(z) = 1/z$ then,

$$\int \frac{1}{z}\, dz = \ln z + C \tag{19.2.9}$$

where the constant C is complex number.

Suppose that the limit in Eq. (19.2.8) does exist uniquely and is independent of the manner $\Delta z \to 0$.

$$\frac{\partial F(z)}{\partial x_1} = \frac{d}{dz} F(z) \frac{\partial z}{\partial x_1} = F'(z) \tag{19.2.10}$$

$$\frac{\partial F(z)}{\partial x_2} = \frac{d}{dz} F(z) \frac{\partial z}{\partial x_2} = iF'(z) \tag{19.2.11}$$

If we set as in Eq. (19.2.5)

$$F(z) = \psi + i\Gamma \ ,$$

$$\frac{\partial F}{\partial x_1} = \frac{\partial \psi}{\partial x_1} + i \frac{\partial \Gamma}{\partial x_1} = F'(z)$$

and

$$\frac{\partial F}{\partial x_2} = \frac{\partial \psi}{\partial x_2} + i \frac{\partial \Gamma}{\partial x_2} = iF'(z)$$

Therefore, if $F'(z)$ is to be unique,

$$\frac{\partial \psi}{\partial x_1} + i \frac{\partial \Gamma}{\partial x_1} = \frac{1}{i} \frac{\partial \psi}{\partial x_2} + \frac{\partial \Gamma}{\partial x_2} = -i \frac{\partial \psi}{\partial x_2} + \frac{\partial \Gamma}{\partial x_2}$$

Equating real and imaginary parts, we get

$$\frac{\partial \psi}{\partial x_1} = \frac{\partial \Gamma}{\partial x_2}, \frac{\partial \psi}{\partial x_2} = -\frac{\partial \Gamma}{\partial x_1} \tag{19.2.12}$$

These equations are called the "Cauchy-Riemann" equations. They are the necessary and sufficient conditions for $F(z)$ to be analytic and relate the real and imaginary parts to each other. It is clear that $\zeta = x_1 - ix_2$ does not satisfy Eq. (19.2.12). Eliminating Γ from Eq. (19.2.12) by differentiating the first equation with respect to x_1 and the second with respect to x_2 one obtains

$$\nabla^2 \psi = \frac{\partial^2 \psi}{\partial x_1^2} + \frac{\partial^2 \psi}{\partial x_2^2} = 0 \tag{19.2.13}$$

Similarly eliminating ψ, we find

$$\nabla^2 \Gamma = \frac{\partial^2 \Gamma}{\partial x_1^2} + \frac{\partial^2 \Gamma}{\partial x_2^2} = 0 \tag{19.2.14}$$

Eqs. (19.2.13) and (19.2.14) show that the real and imaginary parts of any analytic function of a complex variable are solutions of Laplace's equation. ψ and Γ are, therefore, harmonic functions.

In the X_1, X_2 plane, two families of curves, $\psi(x_1, x_2) = c_1$ and

$\Gamma(x_1, x_2) = c_2$ intersect at right angles. Indeed, at the point of intersection of two curves, the slopes of the tangent lines are respectively given by,

$$\left(\frac{dx_2}{dx_1}\right)_{\psi=c_1} = -\frac{\partial\psi/\partial x_1}{\partial\psi/\partial x_2}$$

$$\left(\frac{dx_2}{dx_1}\right)_{\Gamma=c_2} = -\frac{\partial\Gamma/\partial x_1}{\partial\Gamma/\partial x_2}$$

But, as a result of Eq. (19.2.12),

$$\left(\frac{dx_2}{dx_1}\right)_{\psi=c_1} = -1/\left(\frac{dx_2}{dx_1}\right)_{\Gamma=c_2} \tag{19.2.15}$$

Hence, the two slopes are negative reciprocals, and therefore, the two curves intersect at right angle; that is every curve of one family intersects every curve of the other family at right angle. This is expressed by saying that the families of curves corresponding to two conjugate functions form an orthogonal system.

For every harmonic function ψ there exists a conjugate harmonic function Γ that can be calculated as follows:

$$d\Gamma = \frac{\partial\Gamma}{\partial x_1}dx_1 + \frac{\partial\Gamma}{\partial x_2}dx_2$$

$$d\Gamma = -\frac{\partial\psi}{\partial x_2}dx_1 + \frac{\partial\psi}{\partial x_1}dx_2$$

$$\Gamma = \int\left(-\frac{\partial\psi}{\partial x_2}dx_1 + \frac{\partial\psi}{\partial x_1}dx_2\right) + \text{constant}$$

If ψ is single valued, Γ may not be. Consider, for example, the function

$$\psi = \ln r = \frac{1}{2}\ln(x_1^2 + x_2^2)\ .$$

$$\frac{\partial\psi}{\partial x_1} = \frac{x_1}{r^2}\,, \qquad \frac{\partial\psi}{\partial x_2} = \frac{x_2}{r^2}\ .$$

$$\Gamma = \int\left(-\frac{x_2}{r^2}dx_1 + \frac{x_1}{r^2}dx_2\right) = \int(\sin^2\theta + \cos^2\theta)\,d\theta + \text{constant}$$

$$\Gamma = \theta + \text{constant}$$

Also, Γ is determined within an arbitrary additive constant.

Finally, it is well known that a real function $F(\theta)$ given in an interval $0 \leq \theta \leq 2\pi$ may be represented in the form of a Fourier series

$$F(\theta) = 1/2\, \alpha_0 + \sum_{k=1}^{\infty} (\alpha_k \cos k\theta + \beta_k \sin k\theta) \quad (19.2.16)$$

Sines and cosines can be replaced by imaginary exponentials to lead to the following complex form

$$F(\theta) = \sum_{-\infty}^{+\infty} a_k e^{ik\theta} \quad (19.2.17)$$

where the summation extends over all integers from $-\infty$ to $+\infty$. The coefficients a_k for $k = n$ are given by

$$a_n = \frac{1}{2\pi} \int_0^{2\pi} F(\theta) e^{-in\theta}\, d\theta \quad (19.2.18)$$

The quantities a_n and a_{-n} are conjugate complex numbers.

Consider now an expression of the form $F_1(\theta) + iF_2(\theta)$ where F_1 and F_2 are real functions which may be represented in the interval $(0, 2\pi)$ by Fourier series of the form Eq. (19.2.17). Adding the series one gets

$$F_1(\theta) + iF_2(\theta) = \sum_{-\infty}^{+\infty} a_k e^{ik\theta} \quad (19.2.19)$$

with

$$a_n = \frac{1}{2\pi} \int_0^{2\pi} (F_1 + iF_2) e^{-in\theta}\, d\theta \quad (19.2.20)$$

The only difference from the preceding case is that the quantities a_n and a_{-n} will not, in general, be conjugate.

19.3 Line Integrals of Complex Functions. Cauchy's Integral Theorem.

If C is a piecewise smooth curve in the complex plane between z_0 and z_1, the line integral of a function $F(z)$ along C is defined by,

$$\begin{aligned} \int_C F(z)\, dz &= \int_C (\psi + i\Gamma)(dx_1 + i\, dx_2) \\ &= \int_C (\psi\, dx_1 - \Gamma\, dx_2) + i \int_C (\Gamma\, dx_1 + \psi\, dx_2) \end{aligned} \quad (19.3.1)$$

The conditions that the two integrands are total differentials are the requirement that the Cauchy-Riemann equations be satisfied. Thus, the integral is independent of the path joining z_0 and z_1 if C is enclosed in a simple contour inside which $F(z)$ is analytic. Therefore,

$$\int_{z_0}^{z_1} F(z)\,dz = f(z_1) - f(z_0)$$

where $f(z)$ is a function whose derivative is $F(z)$. If z_0 and z_1 coincide, then $\oint F(z)\,dz = 0$. This result is known as Cauchy's integral theorem which states that, if $F(z)$ is analytic in a simply connected region R then for every closed contour C within R, (Fig. 19.2)

$$\oint_C F(z)\,dz = 0 \tag{19.3.2}$$

The region R is said to be simply connected if every simple closed curve within it encloses only points of R. The positive direction around a closed contour is shown in Fig. 19.2.

If ψ and Γ in Eq. (19.3.1) are written in polar coordinates we get

$$\int_C F(z)\,dz = \int_C (\psi + i\Gamma)(e^{i\theta}\,dr + ir\,e^{i\theta}\,d\theta) \tag{19.3.3}$$

Therefore,

$$\begin{aligned}\int_C F(z)\,dz = &\int_C [(\psi \operatorname{Cos}\theta - \Gamma \sin\theta)\,dr - (\Gamma \operatorname{Cos}\theta + \psi \sin\theta)r d\theta] \\ &+ i\int [(\Gamma \operatorname{Cos}\theta + \psi \sin\theta)\,dr + (\psi \operatorname{Cos}\theta - \Gamma \sin\theta)r d\theta]\end{aligned} \tag{19.3.4}$$

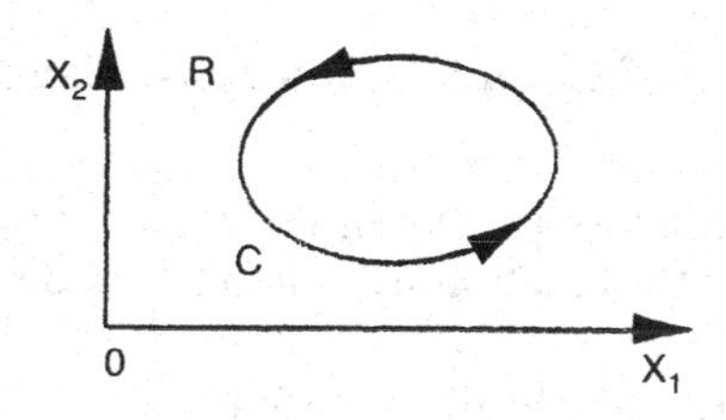

Fig. 19.2

The requirement that the quantities in brackets be exact differentials can be reduced to

$$\frac{\partial \psi}{\partial r} = \frac{1}{r}\frac{\partial \Gamma}{\partial \theta}, \qquad \frac{1}{r}\frac{\partial \psi}{\partial \theta} = -\frac{\partial \Gamma}{\partial r} \tag{19.3.5}$$

These are the Cauchy-Riemann equations in cylindrical coordinates.

Cauchy's integral theorem has been deduced with the assumption that the closed curve C (Fig. 19.2) is the boundary of a simply connected region. The theorem can be extended to the case of a multiply connected region (Fig. 19.3). By introducing the cuts AB and DE, the region may be transformed into a simply connected region. We thus have

$$\oint_S F(z)\, dz = 0$$

where the curve S includes the outer curve C traversed in the positive (counterclockwise) direction, the curves C_1 and C_2 traversed in the negative direction and the cuts AB and DE. Thus,

$$\oint_S F(z)\, dz = \oint_C F(z)\, dz + \oint_{C_1} F(z)\, dz + \oint_{C_2} F(z)\, dz$$

$$+ \int_A^B F(z)\, dz + \int_B^A F(z)\, dz + \int_D^E F(z)\, dz + \int_E^D F(z)\, dz = 0$$

The integrals along the cuts cancel in pairs. Transposing, we obtain

$$\oint_C F(z)\, dz = \oint_{C_1} F(z)\, dz + \oint_{C_2} F(z)\, dz \tag{19.3.6}$$

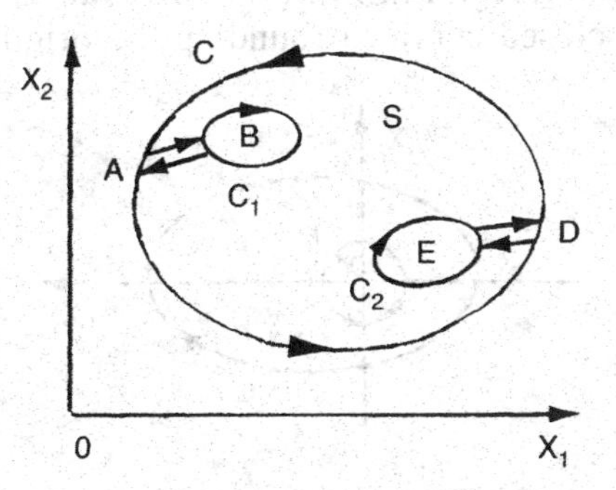

Fig. 19.3

As an example of the use of Cauchy's integral theorem, let us compute the integral (Fig. 19.4)

$$I = \oint_C \frac{dz}{z} \tag{19.3.7}$$

where C is a simple closed curve. The function $F(z) = 1/z$ is analytic for any value of z except for $z = 0$. Let us draw a small circle C_1 of radius r with center at the origin as shown in Fig. (19.4). Since the function $1/z$ is analytic in the region between C and C_1, we have

$$\oint_C \frac{dz}{z} = \oint_{C_1} \frac{dz}{z}$$

On the circle C_1, r is constant and we have

$$z = re^{i\theta}, \qquad dz = ire^{i\theta}\, d\theta .$$

Hence,

$$\oint_{C_1} \frac{dz}{z} = \oint_0^{2\pi} i\, d\theta = 2\pi i \tag{19.3.8}$$

we thus have

$$\oint_C \frac{dz}{z} = \begin{cases} 0 \text{ if } C \text{ does not surround the origin} \\ 2\pi i \text{ if } C \text{ surrounds the origin} \end{cases} \tag{19.3.9}$$

If $F(z) = z^n$ where n is an integer, and C_1 is a unit circle $|z| = 1$,

$$\oint_{C_1} z^n\, dz = \int_0^{2\pi} e^{ni\theta}(ie^{i\theta}\, d\theta) = 0 \; (n \neq -1) \tag{19.3.10}$$

More generally the integral in Eq. (19.3.10) vanishes under the same conditions for any closed curve surrounding the origin.

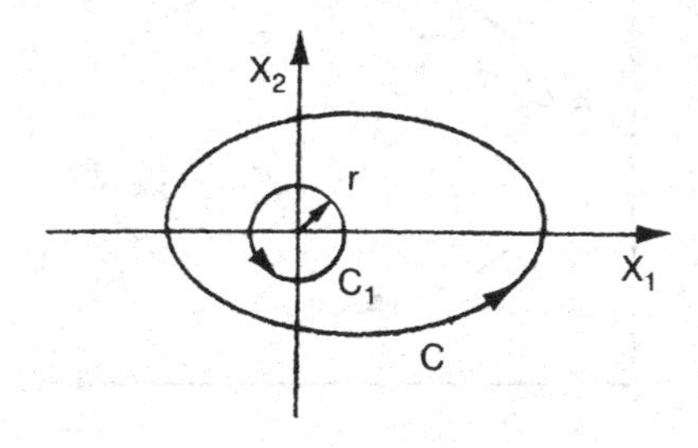

Fig. 19.4

If we replace z by $z - a$ where a is a complex number, we can deduce the additional results:

$$\oint_C (z - a)^n \, dz = 0 \qquad (n \neq -1) \tag{19.3.11}$$

$$\oint_C \frac{dz}{z - a} = 2\pi i \tag{19.3.12}$$

where C is a closed curve enclosing $z = a$ in the positive direction and n is an integer (Fig. 19.5).

If the point a is outside the contour C, then Eq. (19.3.11) is true whatever the value of n.

As an example, consider the integral

$$\frac{1}{2\pi i} \oint_C \frac{d\alpha}{\alpha(\alpha - z)}$$

where the contour C surrounds the origin $z = 0$. In this integral z is held constant while α traverses the contour C. We have

$$\frac{1}{\alpha(\alpha - z)} = \frac{1}{z(\alpha - z)} - \frac{1}{\alpha z} \tag{19.3.13}$$

and

$$\frac{1}{2\pi i} \oint_C \frac{d\alpha}{\alpha(\alpha - z)} = \frac{1}{2\pi i z} \oint_C \frac{d\alpha}{\alpha - z} - \frac{1}{2\pi i z} \oint_C \frac{d\alpha}{\alpha}$$

From Eqs. (19.3.8) and (19.3.12)

$$\frac{1}{2\pi i} \oint_C \frac{d\alpha}{\alpha(\alpha - z)} = \frac{2\pi i}{2\pi i z} - \frac{2\pi i}{2\pi i z} = 0 \tag{19.3.14}$$

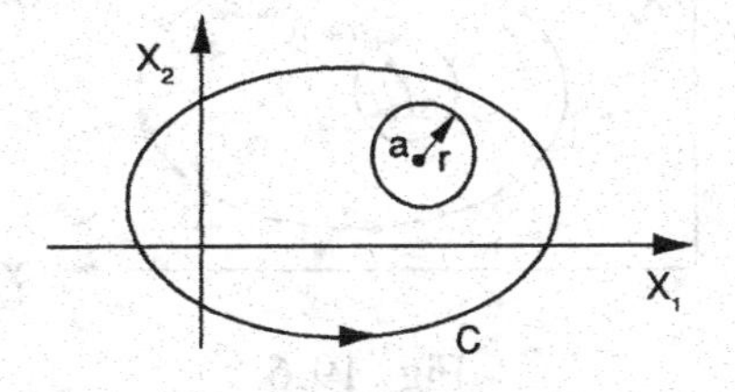

Fig. 19.5

In general,

$$\frac{1}{2\pi i}\oint_C \frac{d\alpha}{\alpha^n(\alpha - z)} = 0 \text{ if } n > 0 \qquad (19.3.15)$$
$$= 1 \text{ if } n = 0$$

19.4 Cauchy's Integral Formula

C is a closed contour inside which and along which $F(z)$ is analytic. α is a point of the region R bounded by the contour C, and C_1 is a small circle of radius r with center at α (Fig. 19.6).

The function $F(z)/(z - \alpha)$ is analytic between C and C_1, therefore

$$\oint_C \frac{F(z)}{z - \alpha}\, dz = \oint_{C_1} \frac{F(z)}{z - \alpha}\, dz \qquad (19.4.1)$$

for all positive values of r. Eq. (19.4.1) must be true in the limit as $r \to 0$. For a point on C_1 we have

$$z = \alpha + re^{i\theta}, \quad dz = ire^{i\theta}\, d\theta \quad \text{for } 0 \leqslant \theta < 2\pi$$

Hence,

$$\oint_{C_1} \frac{F(z)}{z - \alpha}\, dz = \int_0^{2\pi} F(\alpha + re^{i\theta})i\, d\theta$$

As $r \to 0$ we obtain

$$\oint_C \frac{F(z)}{z - \alpha}\, dz = \lim_{r \to 0} \int_0^{2\pi} F(\alpha + re^{i\theta})i\, d\theta = 2\pi i\, F(\alpha)$$

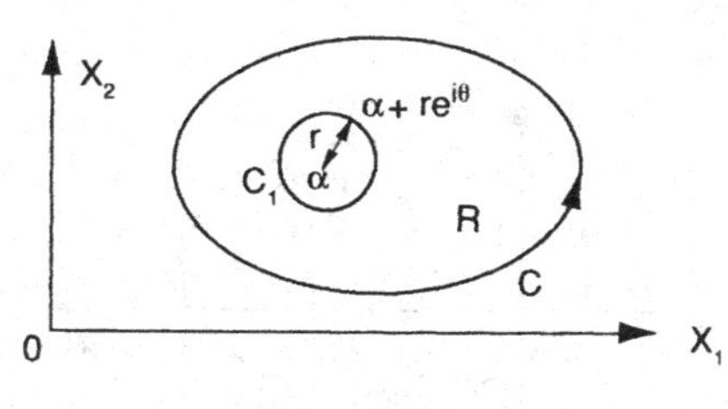

Fig. 19.6

Thus,

$$F(\alpha) = \frac{1}{2\pi i} \oint_C \frac{F(z)}{z - \alpha} dz \tag{19.4.2}$$

This is Cauchy's integral formula. It is remarkable in that it enables one to compute the value of $F(z)$ inside a region in which it is analytic, from the values of the function on the boundary.

Another form of Cauchy's integral formula is obtained by interchanging α and z in Eq. (19.4.2).

$$F(z) = \frac{1}{2\pi i} \oint_C \frac{F(\alpha)}{\alpha - z} d\alpha \tag{19.4.3}$$

where C is now a closed contour enclosing the point z which is held fixed in the integration while α traverses the curve C. It can be shown that the integral in Eq. (19.4.3) may be differentiated under the integral sign and so may the results. Thus,

$$\frac{dw}{dz} = F'(z) = \frac{1}{2\pi i} \oint_C \frac{F(\alpha)}{(\alpha - z)^2} d\alpha$$

$$F''(z) = \frac{2!}{2\pi i} \oint_C \frac{F(\alpha)}{(\alpha - z)^3} d\alpha$$

$$F^{(n)}(z) = \frac{n!}{2\pi i} \oint_C \frac{F(\alpha)}{(\alpha - z)^{n+1}} d\alpha \tag{19.4.4}$$

Note that the function $F(\alpha)$ represents the values of an analytic function $F(z)$ on the boundary C of the region R.

19.5 Taylor Series

The integral formula of Cauchy can be used to show that if $F(z)$ is analytic in the neighborhood of a point $z = a$, then $F(z)$ can be expanded in a Taylor series of powers of $(z - a)$. We notice that the function $1/(\alpha - z)$ can be expanded in the geometric series (Fig 19.7):

$$\frac{1}{\alpha - z} = \frac{1}{(\alpha - a) - (z - a)} = \frac{1}{\alpha - a} \frac{1}{1 - (z - a)/(\alpha - a)}$$

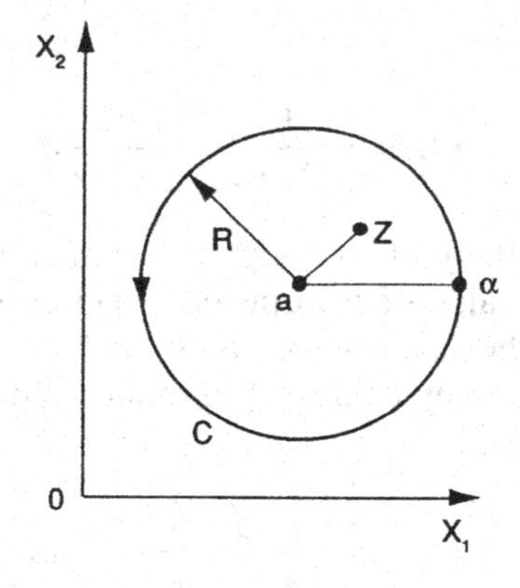

Fig. 19.7

Thus,

$$\frac{1}{\alpha - z} = \frac{1}{\alpha - a} \sum_{n=0}^{\infty} \left(\frac{z - a}{\alpha - a} \right)^n \quad \text{with } (|z - a| < |\alpha - a|) \tag{19.5.1}$$

Take C as any circle with center $z = a$ inside which and on which $F(z)$ is analytic (Fig. 19.7). For points inside C we have $|z - a| < |\alpha - a| = R$ where R is the radius of the circle. For such points the series in Eq. (19.5.1) converges and can be multiplied by $F(\alpha)$ and integrated around C term by term. Eq. (19.4.3) can be combined with Eq. (19.5.1) to give

$$F(z) = \sum_{n=0}^{\infty} \frac{1}{2\pi i} \left[\oint_C \frac{F(\alpha)\, d\alpha}{(\alpha - a)^{n+1}} \right] (z - a)^n \quad \text{with } (|z - a| < R) \tag{19.5.2}$$

or

$$F(z) = \sum_{n=0}^{\infty} A_n (z - a)^n \quad \text{with } (|z - a| < R) \tag{19.5.3}$$

where

$$A_n = \frac{1}{2\pi i} \oint_C \frac{F(\alpha)\, d\alpha}{(\alpha - a)^{n+1}} \tag{19.5.4}$$

Therefore, if $F(z)$ is analytic throughout the circle $|z - a| < R$, it can be expanded in Taylor series converging in this circle with coefficients given by Eq. (19.5.4).

By differentiating Eq. (19.5.3) n times, setting $z = a$ in the result and using Eq. (19.5.4) we obtain

$$F^{(n)}(a) = n! A_n = \frac{n!}{2\pi i} \oint_C \frac{F(\alpha)\, d\alpha}{(\alpha - a)^{n+1}} \tag{19.5.5}$$

If we change the notation by substituting z to a, we get

$$F^{(n)}(z) = \frac{n!}{2\pi i} \oint_C \frac{F(\alpha)\, d\alpha}{(\alpha - z)^{n+1}} \tag{19.5.6}$$

where C is now a circle with radius R about z (Fig. 19.8). Notice that Eq. (19.5.6) is the same as Eq. (19.4.4). From the above, we conclude that if $F(z)$ is analytic in the neighborhood of a point $z = a$, $F(z)$ can be expanded in the Taylor series

$$F(z) = F(a) + F'(a)(z - a) \ldots \frac{F^n(a)}{n!}(z - a)^n$$

$$F(z) = \sum_{n=0}^{\infty} \frac{F^n(a)}{n!}(z - a)^n\,, \tag{19.5.7}$$

with the circle of convergence coinciding with the largest circle with center at $z = a$ in which $F(z)$ is analytic.

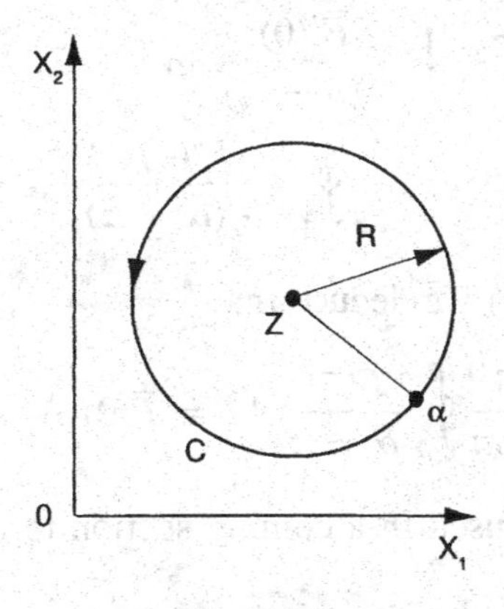

Fig. 19.8

When $a = 0$, Eq. (19.5.7) reduces to Maclaurin's series

$$F(z) = F(0) + \sum_{n=1}^{\infty} \frac{F^n(0)}{n!} z^n \tag{19.5.8}$$

which gives the expansion of $F(z)$ in a power series in the neighborhood of the origin.

Often a function is expanded in a Maclaurin's series followed by a term-by-term integration. Consider, for example, the integral

$$\frac{1}{2\pi i} \oint_\gamma \frac{\overline{F}(\overline{\alpha})}{(\alpha - z)} d\alpha \tag{19.5.9}$$

where γ is a unit circle $|z| = 1$. The function $F(\alpha)$ when decomposed in a Maclaurin's series gives

$$F(\alpha) = F(0) + F'(0)\alpha + \frac{1}{2} F''(0)\alpha^2 + \ldots \tag{19.5.10}$$

$F(\alpha)$ is analytic in the region bounded by the unit circle. Now $\overline{\alpha} = 1/\alpha$ and

$$\overline{F}(\overline{\alpha}) = \overline{F}(0) + \overline{F}'(0)\frac{1}{\alpha} + \frac{1}{2}\overline{F}''(0)\frac{1}{\alpha^2} \ldots \tag{19.5.11}$$

The integral (19.5.9) becomes

$$\frac{1}{2\pi i}\left[\oint_\gamma \frac{\overline{F}(0)}{(\alpha - z)} d\alpha + \oint_\gamma \frac{\overline{F}'(0)}{\alpha(\alpha - z)} d\alpha + \oint_\gamma \frac{\overline{F}''(0)}{2\alpha^2(\alpha - z)} d\alpha + \ldots\right] \tag{19.5.12}$$

Using Eq. (19.3.15) we deduce that

$$\frac{1}{2\pi i} \oint_\gamma \frac{\overline{F}(\overline{\alpha})}{\alpha - z} d\alpha = \overline{F}(0) \ldots \tag{19.5.13}$$

This result will be used in a coming section to obtain the formula of Schwarz.

19.6 Laurent Series, Residues and Cauchy's Residue Theorem

The Taylor series expansion can be generalized to involve both positive and negative integral powers of $z - a$. This is the case when there is a circular annulus S for which $R_1 < |z - a| < R_2$, such that $F(z)$ is analytic and single valued in S and on its inner and outer circular boundaries C_1 and C_2 (Fig. 19.9a).

Introduce a cross cut between C_1 and C_2 and apply Cauchy's integral formula in Eq. (19.4.3) to the resultant simply connected region S' (Fig. 19.9b) bounded by C'. By taking the limit as the width of the cut tends to zero we have

$$F(z) = \frac{1}{2\pi i}\oint_{C_2} \frac{F(\alpha)}{\alpha - z}\,d\alpha - \frac{1}{2\pi i}\oint_{C_1} \frac{F(\alpha)}{\alpha - z}\,d\alpha \qquad (19.6.1)$$

The first integral is dealt with as in Section 19.5 by expanding $1/(\alpha - z)$ in powers of $(z - a)/(\alpha - a)$ and integrating term-by-term to yield the relation

$$\frac{1}{2\pi i}\oint_{C_2} \frac{F(\alpha)}{\alpha - z}\,d\alpha = \sum_{n=0}^{\infty} A_n(z - a)^n \qquad \text{with} \quad (|z - a| < R_2) \qquad (19.6.2)$$

where

$$A_n = \frac{1}{2\pi i}\oint_{C_2} \frac{F(\alpha)\,d\alpha}{(\alpha - a)^{n+1}} \quad (n = 0, 1, 2, \ldots) \qquad (19.6.3)$$

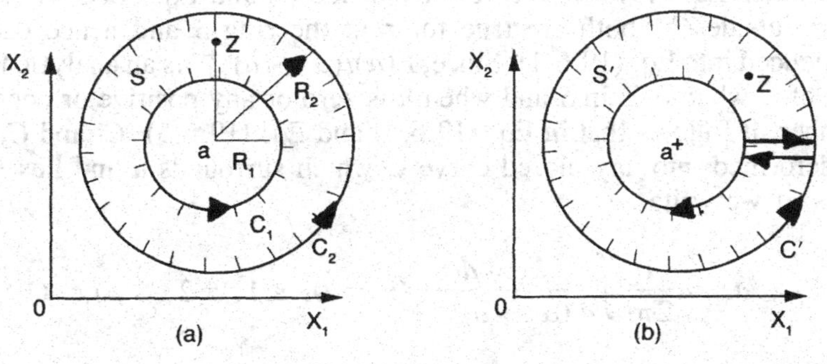

Fig. 19.9

In the second integral of Eq. (19.6.1) we expand $1/(\alpha - z)$ in powers of $(\alpha - a)/(z - a)$,

$$-\frac{1}{\alpha - z} = \frac{1}{z - a} \frac{1}{1 - \dfrac{\alpha - a}{z - a}} = \sum_{n=1}^{\infty} \frac{(\alpha - a)^{n-1}}{(z - a)^n}$$

with $(|z - a|) > R_1$,

so that, after multiplying this convergent power series by $F(\alpha)/2\pi i$ and integrating term by term around C_1 we get

$$\begin{aligned} -\frac{1}{2\pi i} \oint_{C_1} \frac{F(\alpha)}{\alpha - z} d\alpha &= \sum_{n=1}^{\infty} B_{-n}(z - a)^{-n} \\ &= \sum_{n=-\infty}^{-1} B_n(z - a)^n \end{aligned} \tag{19.6.4}$$

with

$$(|z - a|) > R_1$$

and

$$B_n = \frac{1}{2\pi i} \oint_{C_1} \frac{F(\alpha)\, d\alpha}{(\alpha - a)^{n+1}} \quad (n = -1, -2, \ldots) \tag{19.6.5}$$

Now since Eq. (19.6.2) is true for z inside C_2 and Eq. (19.6.4) is true for z outside C_1, both are true for z in the ring S and hence can be introduced into Eq. (19.6.1). Since, $F(\alpha)/(\alpha - a)^{n+1}$ is an analytic function of α when α is in S and when n is zero or any positive or negative number, it follows that in Eq. (19.6.3) and Eq. (19.6.5), C_1 and C_2 can be deformed into any closed curve C which surrounds a and lies in S. Thus, if we write

$$a_n = \frac{1}{2\pi i} \oint_{C} \frac{F(\alpha)\, d\alpha}{(\alpha - a)^{n+1}} \quad (n = 0, \pm 1, \pm 2 \ldots), \tag{19.6.6}$$

it follows that $A_n = a_n$ when $n \geqslant 0$ and $B_n = a_n$ when $n \leqslant -1$. Eq. (19.6.1) can now be written

$$F(z) = \sum_{n=-\infty}^{\infty} a_n(z - a)^n \,, \qquad (R_1 < |z - a| < R_2) \quad (19.6.7)$$

This expansion is known as the Laurent series expansion of $F(z)$ in the specified annulus and involves both positive and negative powers of $(z - a)$. The expansion is unique in the sense that if an expansion of the form

$$F(z) = \sum_{n=-\infty}^{\infty} b_n(z - a)^n$$

can be obtained in any way and is valid in the annulus

$$R_1 < |z - a| < R_2$$

then $b_n = a_n$. This allows one to obtain a Laurent series by elementary methods.

Consider, for example, the function

$$F(z) = \frac{1}{z(z - 1)}$$

This function is analytic at all points of the z plane except at $z = 0$ and $z = 1$. Therefore, it is possible to expand the function in a Laurent series in an annular region about $z = 0$ or $z = 1$. To expand about $z = 0$ we have

$$\frac{1}{z(z - 1)} = -\frac{1}{z}(1 - z)^{-1} = -\frac{1}{z}(1 + z + z^2 + z^3 + \ldots) \qquad 0 < |z| < 1$$

To expand about $z = 1$ we write

$$\frac{1}{z(z - 1)} = \frac{1}{z - 1}\,\frac{1}{(z - 1) + 1}$$

$$= \frac{1}{z - 1}(1 - (z - 1) + (z - 1)^2 \ldots)$$

$$0 < |z - 1| < 1$$

If $F(z)$ also is analytic inside and on C_1, then $B_n = 0$, i.e., $a_n = 0$ for $n < 0$ and Eq. (19.6.7) becomes the Taylor series

$$F(z) = \sum_{n=0}^{\infty} A_n(z - a)^n$$

valid everywhere inside C_2. On the other hand, when $F(z)$ also is analytic on and outside C_2, the series involves only a constant term and negative powers of $(z - a)$. It can then be considered as an ordinary power series in $1/(z - a)$ valid everywhere outside C_1. Here $F(z)$ also must be analytic at infinity.

To study the behavior of a function F(z) *for large values of* $|z|$ we make the substitution

$$z = \frac{1}{t} \qquad (19.6.8)$$

$F(z)$ is transformed to a new function $G(t)$,

$$G(t) = F\left(\frac{1}{t}\right) \qquad (19.6.9)$$

As z moves indefinitely far from the origin in any direction $|z| \to \infty$, the corresponding point t approaches $t = 0$. We say $F(z)$ is analytic at z_∞ if $G(t) = F(1/t)$ is analytic at $t = 0$, and similarly, if $G(t)$ has a particular type of singularity at $t = 0$, then $F(z)$ has the same type of singularity at z_∞. If $G(t) = F(1/t)$ is analytic at $t = 0$, then it can be expanded in a series

$$F\left(\frac{1}{t}\right) = A_0 + A_1 t + A_2 t^2 + \ldots = \sum_{n=0}^{\infty} A_n t^n \qquad (19.6.10)$$

in a circle about $t = 0$ and conversely. Hence, replacing t by $1/z$ it follows that $F(z)$ is analytic at z_∞, if and only if, it can be represented by a series of the form

$$F(z) = A_0 + \frac{A_1}{z} + \frac{A_2}{z^2} + \ldots = \sum_{n=0}^{\infty} \frac{A_n}{z^n} \qquad (19.6.11)$$

for values of z outside a sufficiently large circle with center at $z = 0$. In particular, $F(z)$ must approach a limit $F(\infty)$ as $z \to z_\infty$, that is as $|z| \to \infty$ in any way and Eq. (19.6.11) shows that $A_0 = F(\infty)$. It is appropriate to mention here that a function is called holomorphic at the point $z = \infty$

if in the neighborhood of that point (i.e. for sufficiently large $|z|$) it may be represented by a series of the form Eq. (19.6.11).

In the above, we notice that a small circle surrounding the point $t = 0$ corresponds to a large circle with center at $z = 0$ in the complex plane. Just as we may consider a circle of radius equal to zero as enclosing the origin, we may consider the exterior of a circle C_∞ of infinite radius as consisting only of the point z_∞. Consequently, the exterior of a closed contour C may be considered as composed of the region between C and the infinite circle C_∞ together with an added exterior point z_∞ at infinity.

Consider now the coefficient a_{-1} *of* $(z - a)^{-1}$ *in the Laurent series (19.6.7)*

$$a_{-1} = \frac{1}{2\pi i}\oint_C F(\alpha)\,d\alpha \tag{19.6.12}$$

where C is a curve surrounding a. a_{-1} is a complex number in general and is called the residue of $F(z)$ at the point $z = a$. If we consider, for example, the function

$$F(z) = \frac{1}{z},$$

the Laurent series about the origin has only one term $1/z$ with $a_{-1} = 1$. Eq. (19.6.12) then gives

$$2\pi i = \oint_C \frac{1}{z}\,dz$$

where the integral is taken around the curve surrounding the origin.

Let us now consider a function $F(z)$ which is analytic inside a region S at all points except at points $z_1, z_2, z_3 \ldots z_n$. Let $F(z)$ also be analytic at all points on the boundary C of the region S (Fig. 19.10). Let us surround $z_1, z_2 \ldots z_n$ by the closed curves $C_1, C_2 \ldots C_n$. By Cauchy's integral theorem we have

$$\int_C F(z)\,dz = \int_{C_1} F(z)\,dz + \int_{C_2} F(z)\,dz + \ldots \tag{19.6.13}$$

However, from Eq. (19.6.12) we have

$$\int_{C_r} F(z)\,dz = 2\pi i \quad \text{Res}\,F(z) \quad \text{for } z = z_r \tag{19.6.14}$$

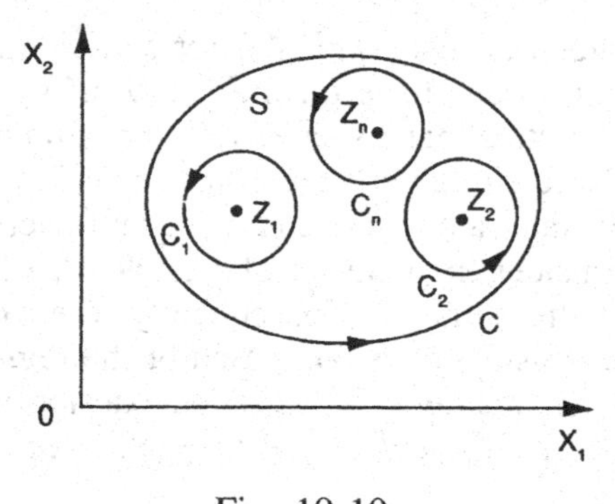

Fig. 19.10

Substituting in Eq. (19.6.13) we get

$$\int_C F(z)\, dz = 2\pi i \sum_{r=1}^{r=n} \text{Res } F(z_r) \tag{19.6.15}$$

This is Cauchy's residue theorem.

19.7 Singular Points of an Analytic Function

If a function is analytic in every neighborhood of a point a except at a itself, then a is called an "isolated singular point" of the function. The function $1/z$ furnishes a simple example of a function which is analytic everywhere except at $z = 0$. Thus, the origin is an isolated singular point of $1/z$. On the other hand, the function $\ln z$ has a singular point at the origin that is not isolated because each neighborhood of the origin includes points on the negative real axis x where $\ln z$ is not analytic.

At an isolated singular point of $F(z)$ a Laurent series of the form

$$F(z) = \sum_{n=-\infty}^{\infty} a_n(z - a)^n \qquad (0 < |z - a| < R) \tag{19.7.1}$$

is valid. If $F(z)$ were to remain finite as $z \to a$, then all a's with negative subscripts in Eq. (19.7.1) would be zero. Thus, everywhere inside the circle C of radius R, except at $z = a$, $F(z)$ could be expressed in the form

$$F(z) = a_0 + a_1(z - a) + a_2(z - 2)^2 + \ldots \tag{19.7.2}$$

If it were also true that $F(a) = a_0$ for $z = a$, then $F(z)$ would be analytic everywhere inside C. That is $F(z)$ would become analytic at $z = a$, if it were suitably defined there. Such a point is called a "removable singular point". We will assume that removable singular points have been removed and it follows that $F(z)$ cannot remain finite at an isolated singular point.

Let us concern ourselves only with single valued functions of the complex variable $F(z)$. Such functions may then have two types of singularities:

1. Poles, or non-essential singular points
2. Essential singular points

Let a be a singular point of $F(z)$. The expansion of $F(z)$ in a Laurent series contains powers of $(z - a)$ with negative exponents. Hence,

$$F(z) = \frac{a_{-m}}{(z - a)^m} + \frac{a_{-m+1}}{(z - a)^{m-1}} + \ldots \frac{a_{-1}}{z - a} + a_0 + a_1(z - a) + \ldots \tag{19.7.3}$$

There are two possibilities:

1. The expansion Eq. (19.7.3) has only a finite number of powers of $(z - a)$ with negative exponents. In this case, $F(z)$ is said to have a pole at $z = a$. If m is the largest of the negative exponents and if the function

$$\phi(z) = (z - a)^m F(z) \qquad (m \text{ is a positive integer})$$

is analytic and is not zero at the point a, m is called the order of the pole a and $F(z)$ is said to have a pole of the mth order at the point a. The sum of the terms with negative exponents

$$\frac{a_{-1}}{(z - a)} + \frac{a_{-2}}{(z - a)^2} + \ldots \frac{a_{-m}}{(z - a)^m} \tag{19.7.4}$$

is called the "principal part" of $F(z)$ at $z = a$.

2. The Laurent expansion of $F(z)$ about a has an infinite number of negative powers of $z - a$ and is of the form

$$F(z) = \sum_{n=-\infty}^{\infty} a_n(z - a)^n \tag{19.7.5}$$

In this case, the point $z = a$ is said to be an essential singular point and $F(z)$ is said to have an essential singularity at $z = a$.

For example, the function

$$F(z) = \frac{1}{(z - 5)^2(z - 1)} \tag{19.7.6}$$

has a pole of the second order at $z = 5$ and a pole of the first order or a simple pole at $z = 1$. On the other hand, the function

$$F(z) = e^{1/z}$$

has a Laurent series whose expansion contains an infinite number of negative powers of z and has an essential singularity at $z = 0$. This series is

$$e^{1/z} = 1 + \frac{1}{z} + \frac{1}{2!z^2} + \frac{1}{3!z^3} \ldots \tag{19.7.7}$$

Consider now the function ln z,

$$w = F(z) = \ln z = \ln |z| + i\theta \tag{19.7.8}$$

where θ is determinate only within an integral multiple of 2π. We speak of that value of θ in the range $0 \leqslant \theta \leqslant 2\pi$ as the principal value of $\theta = \theta_p$. Then any other permissible value of θ is of the form $\theta = \theta_p + 2k\pi$ where k is integral and Eq. (19.7.8) becomes

$$\begin{aligned} w = F(z) &= \ln z \\ &= \ln |z| + i(\theta_p + 2k\pi) \quad (k = 0, \pm 1, \pm 2 \ldots) \end{aligned} \tag{19.7.9}$$

Thus, each time one goes once around the origin and returns to the same point in the z plane, the imaginary part of w varies by 2π.

When working with multi-valued functions, it is conventional to specify one particular branch of the function and artificially prevent the possibility of transition from the branch to another branch. In the case of the function $w = \ln z$, it is conventional to imagine the complex plane to be cut along the entire positive real axis, so that there is no transition across the part of the axis (Fig. 19.11). In the "cut plane" any convenient branch of ln z, say the branch for $k = 0$,

$$\ln z = \ln |z| + i\theta_p = \ln r + i\theta_p (0 \leqslant \theta_p < 2\pi)$$

is a single-valued function of z which is analytic everywhere except on the cut. $z = 0$ is a branch point. For large values of z, following Eq. (19.6.8), we set $z = 1/t$, so that $F(1/t) = \ln(1/t) = -\ln t$. Thus,

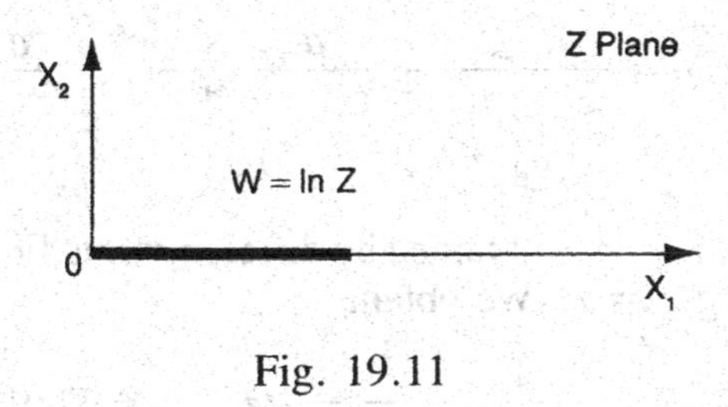

Fig. 19.11

$F(z) = \ln z$ has a branch point at z_∞, as well as, $z = 0$ and the cut introduced above can be considered as joining the two branch points.

19.8 Evaluation of Residues

The evaluation of the residues of a function $F(z)$ at its poles may be done in several ways. By definition, the residue of $F(z)$ at a simple pole $z = a$ is the coefficient a_{-1} in the Laurent series expansion

$$F(z) = \frac{a_{-1}}{z - a} + a_0 + a_1(z - a) + a_2(z - a)^2 + \ldots \qquad (19.8.1)$$

where a is a simple pole. Multiply both sides of Eq. (19.8.1) by $z - a$ and take the limit as $z \to a$:

$$\lim_{z \to a} (z - a)\, F(z) = a_{-1} = \text{Res } F(z) \qquad (19.8.2)$$

For example, the function

$$F(z) = \frac{1}{z^2 + 1} = \frac{1}{(z+i)(z-i)}$$

has two simple poles at $z = i$ and $z = -i$. To evaluate the residue at $z = i$, we form the limit

$$\lim_{z \to i} (z - i)\, F(z) = \lim_{z \to i} \frac{1}{z + i} = \frac{1}{2i} = \text{Res } (i)$$

Similarly, the limit at $z = -i$ is $-1/2i = \text{Res } (-i)$.

If the function $F(z)$ has a multiple pole at $z = a$ of order m then the Laurent series expansion of $F(z)$ is

$$F(z) = \frac{a_{-m}}{(z - a)^m} + \frac{a_{-m+1}}{(z - a)^{m-1}} \cdots \frac{a_{-1}}{z - a} + a_0 + a_1(z - a) + \ldots \qquad (19.8.3)$$

The residue at $z = a$ is obtained by differentiating Eq. (19.8.3) $m - 1$ times and setting $z = a$. We obtain

$$a_{-1} = \frac{1}{(m - 1)!} \frac{d^{m-1}}{dz^{m-1}} [(z - a)^m F(z)]_{z=a} \qquad (19.8.4)$$

For example, the function

$$F(z) = \frac{1}{z^2(z - 1)}$$

has a simple pole at $z = 1$ and a double pole at $z = 0$. At $z = 1$, from Eq. (19.8.2) we obtain

$$\text{Res }(1) = \lim_{z \to 1} \frac{1}{z^2} = 1$$

At $z = 0$, from Eq. (19.8.4) we obtain

$$\text{Res }(0) = \left[\frac{d}{dz}\left(\frac{1}{z - 1}\right)\right]_{z=0} = -1$$

Residues can be used to evaluate definite integrals. For example, let us determine the value of

$$I = \int_0^{2\pi} \frac{1}{5 + 2 \text{ Cos } \theta} d\theta$$

Along a unit circle γ, $z = e^{i\theta}$ and $dz = ie^{i\theta}\, d\theta = iz\, d\theta$. Thus, $d\theta = \frac{dz}{iz}$. On the other hand,

$$\text{Cos } \theta = (e^{i\theta} + e^{-i\theta})/2 = \left(z + \frac{1}{z}\right)/2 = (1 + z^2)/2z$$

Therefore,

$$I = \oint_\gamma \frac{1}{5 + (1 + z^2)/z} \frac{dz}{iz} = \frac{1}{i} \oint_\gamma \frac{dz}{z^2 + 5z + 1}$$

Now,

$$z^2 + 5z + 1 = (z - z_1)(z - z_2)$$

where

$$z_1 = \frac{-5 + \sqrt{25 - 4}}{2} = -.208$$

and

$$z_2 = \frac{-5 - \sqrt{25 - 4}}{2} = -4.79$$

It is seen that $-.208$ is the only possible pole within the unit circle. Thus, from Eq. (19.8.2)

$$\text{Res } F(z) = \lim_{z \to z_1} \frac{(z - z_1)}{(z - z_1)(z - z_2)} = \frac{1}{(z_1 - z_2)} = \frac{1}{\sqrt{21}}$$

and from Eq. (19.6.15)

$$I = 2\pi i \frac{1}{i} \frac{1}{\sqrt{21}} = \frac{2\pi}{\sqrt{21}}$$

Frequently, it is required to evaluate the residues of a function $w(z)$ that has the form

$$w(z) = \frac{F(z)}{G(z)} \tag{19.8.5}$$

where $G(z)$ has simple zeros and hence $w(z)$ has simple poles. If $z = a$ is a simple pole of $w(z)$ then by Eq. (19.8.2)

$$\text{Res } w(z)_{z=a} = \lim_{z \to a} [(z - a)w(z)] = \lim_{z \to a} \left[(z - a) \frac{F(z)}{G(z)} \right] \tag{19.8.6}$$

Since $z = a$ is a simple pole of $w(z)$ we must have $G(a) = 0$, so that the expression Eq. (19.8.6) becomes 0/0. To evaluate it we use l'Hospital's rule and obtain

$$\text{Res } w(z)_{z=a} = \lim_{z=a} \frac{F(z) + (z - a)\, F'(z)}{G'(z)} = \frac{F(a)}{G'(a)} \tag{19.8.7}$$

For example, let us compute the residue of

$$w(z) = \frac{e^{iz}}{z^2 + a^2}$$

at the simple pole $z = ia$. Using Eq. (19.8.7) we have

$$\text{Res}\left(\frac{e^{iz}}{z^2 + a^2}\right)_{z=ia} = \frac{e^{-a}}{2ia}$$

19.9 Conformal Representation or Conformal Mapping

The properties of a function of a real variable $f(x_1)$ are exhibited geometrically by the graph of the function. $x_2 = f(x_1)$ establishes a correspondence between points x_1 on the X_1 axis and points x_2 on the X_2 axis. One can say that $f(x_1)$ maps each point on the X_1 axis into a point in the X_1X_2 plane at a distance x_2 above or below that point. The result of this mapping is a curve. In the case of complex variables we have to use two planes, the Z and the ζ plane (Fig. 19.12).

Consider the relation $\zeta = F(z)$ where $\zeta = \xi_1 + i\xi_2$. If $F(z)$ is single valued, then corresponding to each point z in the Z plane there is only one corresponding value in the ζ plane. The reverse is true if $F(z)$ is analytic and the Jacobian

$$\begin{vmatrix} \dfrac{\partial \xi_1}{\partial x_1} & \dfrac{\partial \xi_1}{\partial x_2} \\ \dfrac{\partial \xi_2}{\partial x_1} & \dfrac{\partial \xi_2}{\partial x_2} \end{vmatrix} = \left(\frac{\partial \xi_1}{\partial x_1}\right)^2 + \left(\frac{\partial \xi_2}{\partial x_1}\right)^2 = |F'(z)|^2 \neq 0 \quad (19.9.1)$$

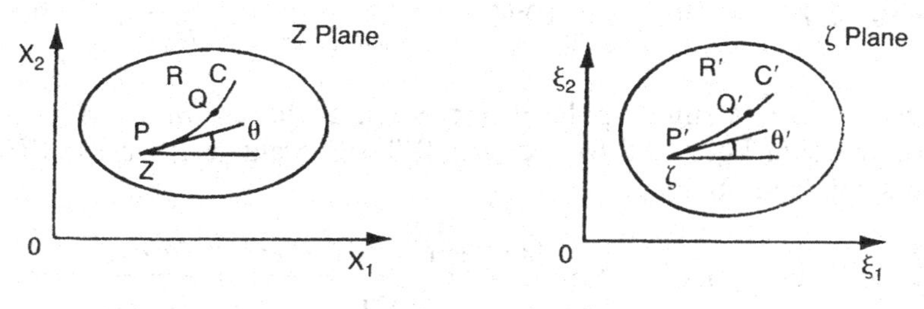

Fig. 19.12

at points z. Under such conditions there is a one-to-one correspondence between points in the two planes, and one can solve $\zeta = F(z)$ to obtain $z = f(\zeta)$. When P describes the curve C in the region R, P' describes the curve C' in the region R'. The correspondence is called mapping. If we now consider a second point such as Q on C and its image Q' on C', we can write

$$\frac{\Delta\zeta}{\Delta z} = \frac{\Delta F}{\Delta z} \tag{19.9.2}$$

This ratio is a complex number whose modulus is the ratio of the chords $P'C'$ and PC and whose argument is the angle between the directions of these chords. In the limit, as Δz and $\Delta\zeta$ approach zero, the limiting argument is the angle between the directions of C and C' at corresponding points and the limiting modulus $|F'(z)|$ is a local magnification factor in the neighborhood of P. Since for analytic functions, the ratio in Eq. (19.9.2) tends to $F'(z)$ independently of the direction of the chord PC, all curves passing through a point such as P are rotated through the same angle with respect to the mapped curves, and are magnified in the same ratio $|F'(z)|$ taken in the neighborhood of P. Thus, relative angles and shapes are preserved in a mapping where $F(z)$ is analytic and it is said to be conformal.

By means of a conformal mapping function $\zeta = F(z)$, an analytic function of the complex variable z becomes another analytic function of the complex variable ζ and its real and imaginary parts will satisfy the Laplace equation. In addition, the boundary curve which in the domain R may not be convenient to work with, may be mapped into the domain R' which allow the boundary conditions to be satisfied more easily. Indeed, a theorem due to Caratheodory and Riemann states that if we have two arbitrary simply connected regions R and R', there always exists a mapping function $\zeta = F(z)$ which will map the region R onto R' conformally in such a way that the mapping is continuous up and including the boundaries C and C'; furthermore, this mapping function is uniquely determined if we specify the correspondence of two arbitrarily chosen points z_0 and ζ_0 and the directions of arbitrarily chosen linear elements passing through these points.

The theorem of Caratheodory is of little help in the actual construction of the mapping functions. There are some formulas that permit one to construct functions for certain classes of problems. One such formula is the one known as the Schwarz-Christoffel Transformation. The Schwarz-Christoffel theorem states (Fig. 19.13).

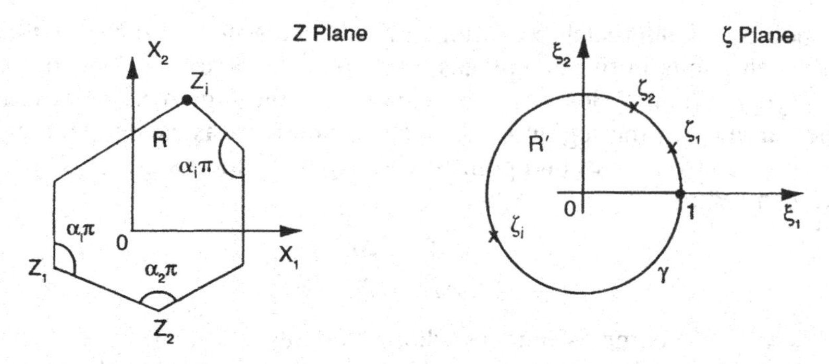

Fig. 19.13

If the region R is that bounded by a rectilinear polygon of n sides, the function $f(\zeta)$ that maps the interior of the polygon onto the unit circle $|\zeta| \leq 1$ has the form

$$z = f(\zeta) = A \int_0^{\zeta} (\zeta - \zeta_1)^{\alpha_1 - 1}(\zeta - \zeta_2)^{\alpha_2 - 1} \ldots$$

$$(\zeta - \zeta_n)^{\alpha_n - 1}\, d\zeta + B \tag{19.9.3}$$

where ζ_i are the points on the boundary γ of the unit circle that correspond to the vertices of the polygon in the z plane and the numbers $\alpha_i\pi$ are interior angles at the vertices of the polygon.

19.10 Examples of Mapping by Elementary Functions

Example 1.

$$\text{The function } \zeta = F(z) = \frac{1}{z} \tag{19.10.1}$$

Eq. (19.10.1) can be thought of as a transformation which brings the point z to the point ζ in the Z plane (Fig. 19.14). In polar coordinates

$$\zeta = \frac{1}{r} e^{-i\theta}$$

which shows that the transformation (or the new point) is an inversion with respect to the unit circle $r = 1$, followed by a reflection with respect to the real axis. Thus, points outside the unit circle are transformed to,

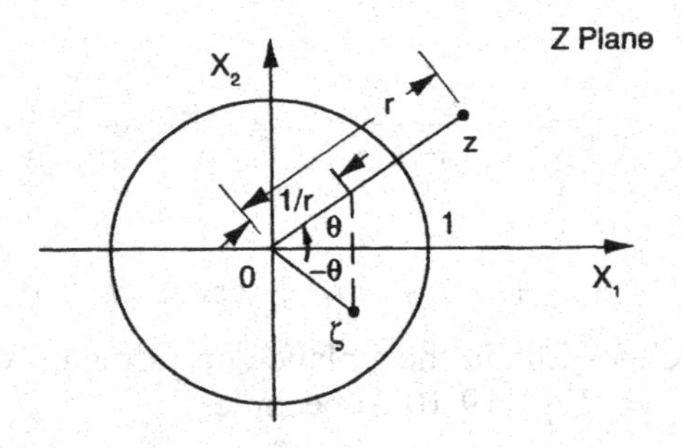

Fig. 19.14

or mapped into points inside the circle, and conversely. Points on the unit circle are mapped into points on the circle. The point $\zeta = 0$ is not the image of any point in the finite z plane. But we notice that as r increases, ζ gets closer to the origin. By making $r = R$ sufficiently large, the images of all points z falling outside a circle of radius R will fall within an arbitrarily small neighborhood of the point $\zeta = 0$. It is convenient to use the concept of the point at infinity, or the infinite point $z = \infty$. Formally, this point is the image of the point $\zeta = 0$ under the transformation $\zeta = 1/z$; so that we can talk interchangeably about the behavior of a function at $z = \infty$ and its behavior at $\zeta = 0$. Infinity is conveniently regarded as a single point and the behavior of a function $F(z)$ at infinity is considered by making the substitution $z = 1/t$ and examining $F(1/t)$ at $t = 0$.

Let us now study the mapping from the Z plane onto the ζ plane and vice versa by means of the functions

$$\zeta = \frac{1}{z} \quad \text{or} \quad z = \frac{1}{\zeta} \tag{19.10.2}$$

Eq. (19.10.2) can be written as

$$\zeta = \xi_1 + i\xi_2 = \frac{1}{x_1 + ix_2} \tag{19.10.3}$$

This equation leads to

$$\xi_1 = \frac{x_1}{x_1^2 + x_2^2}, \qquad \xi_2 = \frac{-x_2}{x_1^2 + x_2^2} \tag{19.10.4}$$

as well as to

$$x_1 = \frac{\xi_1}{\xi_1^2 + \xi_2^2}, \qquad x_2 = \frac{-\xi_2}{\xi_1^2 + \xi_2^2} \tag{19.10.5}$$

The equation

$$a(x_1^2 + x_2^2) + bx_1 + cx_2 + d = 0 \tag{19.10.6}$$

represents a circle or line in the z plane depending on whether $a = 0$ or not. With $\zeta = 1/z$, Eq. (19.10.6) becomes

$$d(\xi_1^2 + \xi_2^2) + b\xi_1 - c\xi_2 + a = 0 . \tag{19.10.7}$$

Whenever ξ_1 and ξ_2 satisfy Eq. (19.10.7), x_1 and x_2 satisfy Eq. (19.10.6). If a and d are both different from zero, both the curve and its image are circles. That is, circles not passing through $z = 0$ map into circles not passing through $\zeta = 0$. Similarly, every circle through the origin $z = 0$ transforms into a straight line in the ζ plane. Lines in the Z plane transform into circles through the origin $\zeta = 0$ unless the line passes through $z = 0$, in which case the image is a line through the origin $\zeta = 0$. The lines $x_1 = c_1$ are mapped onto circles

$$\xi_1^2 + \xi_2^2 - \frac{\xi_1}{c_1} = 0 \tag{19.10.8}$$

tangent to the ξ_2 axis at the origin and the lines $x_2 = c_2$ onto circles

$$\xi_1^2 + \xi_2^2 + \frac{\xi_2}{c_2} = 0 , \tag{19.10.9}$$

if $c_1 \neq 0$ and $c_2 \neq 0$ (Fig. 19.15).

The half plane $x_1 > c_1$ has for its image the region

$$\frac{\xi_1}{\xi_1^2 + \xi_2^2} > c_1 . \tag{19.10.10}$$

When $c_1 > 0$ we have

$$\left(\xi_1 - \frac{1}{2c_1}\right)^2 + \xi_2^2 < \left(\frac{1}{2c_1}\right)^2 ; \tag{19.10.11}$$

which means that the point ζ is inside a circle tangent to the ξ_2 axis at the origin. Consequently, every point inside the circle is the image of

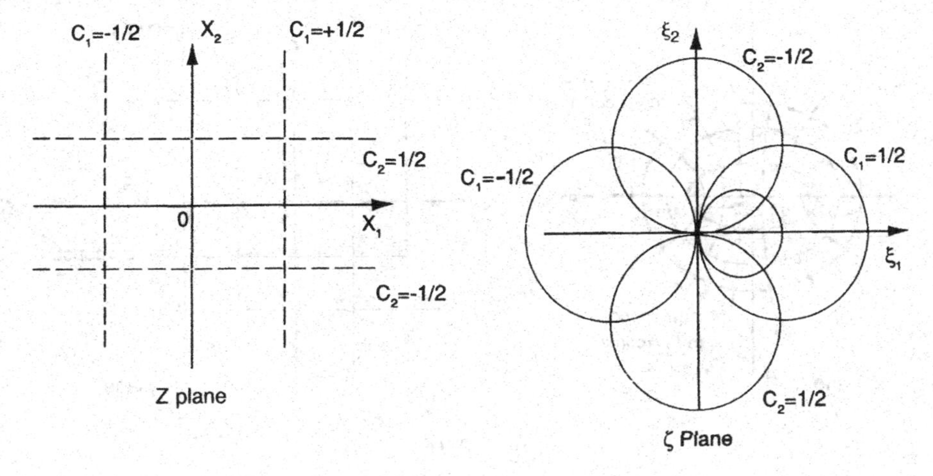

Fig. 19.15

some point in the half plane. Thus, the image of the half plane is the entire circular region (Eq. 19.10.11).

Example 2.

$$\text{The function } z = c \cosh \zeta \ . \tag{19.10.12}$$

c is a constant. Separating the real and imaginary parts we get

$$x_1 + ix_2 = c(\cosh \xi_1 \cos \xi_2 + i \sinh \xi_1 \sin \xi_2)$$
$$x_1 = c \cosh \xi_1 \cos \xi_2 \ , \qquad x_2 = \sinh \xi_1 \sin \xi_2 \tag{19.10.13}$$

If we eliminate ξ_2 from Eq. (19.10.13) we obtain

$$\frac{x_1^2}{c^2 \cosh^2 \xi_1} + \frac{x_2^2}{c^2 \sinh^2 \xi_1} = 1 \tag{19.10.14}$$

For a constant value of ξ_1, this equation becomes that of an ellipse drawn in the Z plane with foci at $x_1 = \pm c$ and with semi axis $c \cosh \xi_1$ and $c \sinh \xi_1$. For different ξ_1 we obtain different ellipses with foci at $x_1 = \pm c$, i.e., confocal ellipses (Fig. 19.16).

If we now eliminate ξ_1 from Eq. (19.10.13) we obtain

$$\frac{x_1^2}{c^2 \cos^2 \xi_2} - \frac{x_2^2}{c^2 \sin^2 \xi_2} = 1 \tag{19.10.15}$$

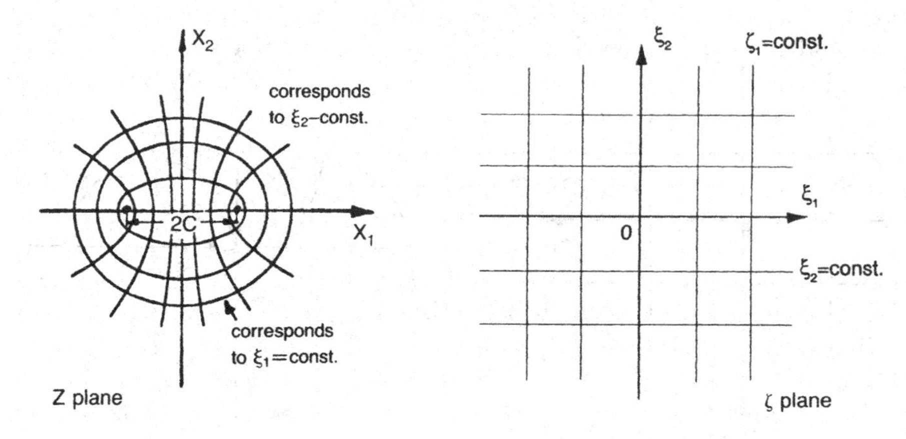

Fig. 19.16

For a constant ξ_2, this equation represents a hyperbola having the same foci as the ellipses. For different ξ_2 we obtain different confocal hyperbolas which intersect with the ellipses at right angle. Such a system can be used as a system of orthogonal curvilinear coordinates in the $X_1 0 X_2$ plane (see problem 1, Chapter 6).

Thus, the transformation given by Eq. (19.10.12) transforms sets of straight-lines parallel to the axes in the ζ plane into confocal ellipses and hyperbolas in the Z plane.

Example 3.

$$\text{The function } z = c\left(\zeta + \frac{m}{\zeta}\right) \tag{19.10.16}$$

$c > 0$ is a constant and $0 \leqslant m \leqslant 1$. Separating real and imaginary parts we get

$$x_1 = c\left(\rho + \frac{m}{\rho}\right)\cos\theta \qquad x_2 = c\left(\rho - \frac{m}{\rho}\right)\sin\theta \tag{19.10.17}$$

where $\rho = |\zeta|$. These are the parametric equations of an ellipse in the Z plane whose semi axes are:

$$a = c\left(\rho + \frac{m}{\rho}\right), \quad \text{and} \quad b = c\left(\rho - \frac{m}{\rho}\right) \tag{19.10.18}$$

provided $\rho^2 \geqslant m$. Eq. (19.10.16) maps the region outside this ellipse in the Z plane onto the region outside a circle in the ζ plane. Points on the ellipse correspond to points on the circle. When a point ζ moves on a circle of radius ρ in the ζ plane counterclockwise, the point z moves around an ellipse whose major and minor axes are given by Eq. (19.10.18), counterclockwise, in the Z plane. The unit circle γ in the ζ plane maps in the Z plane into an ellipse whose semi axes are

$$a = c(1 + m), \quad \text{and} \quad b = c(1 - m)$$

so that,

$$c = \frac{a + b}{2}, \quad \text{and} \quad m = \frac{a - b}{a + b}$$

If $m = 0$, the ellipse becomes a circle. If $m = 1$ the point in the Z plane traces out the segment of the x_1 axis between $x_1 = 2c$ and $x_1 = -2c$, twice, as the point ζ describes once the boundary $|\zeta| = 1$ in the ζ plane. In this case Eq. (19.10.16) maps the Z plane slit along the line joining the points $(2c,0)$ and $(-2c,0)$ onto $|\zeta| = 1$ (Fig. 19.17).

Any circle of radius ρ_1 maps onto an ellipse C_1, and a circle of radius ρ_2 maps onto an ellipse C_2 conformally. If ρ_2 is increased indefinitely, Eq. (19.10.16) maps the region exterior to C_1 onto the region exterior to the circe $\rho = \rho_1$. If $m = \rho_1^2$ the ellipse C_1 degenerates into a segment of the real axis (already seen for $\rho = m = 1$).

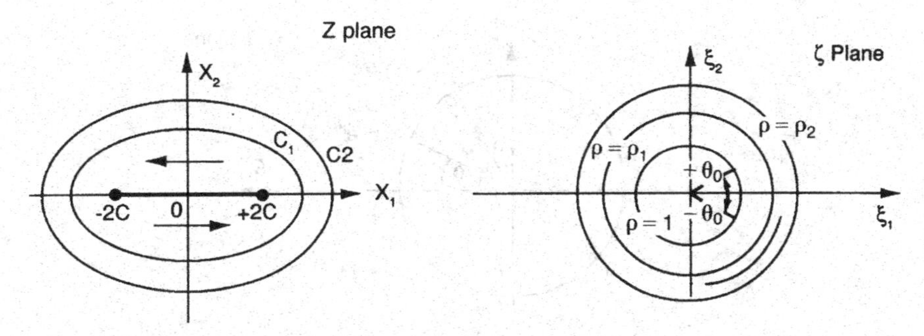

Fig. 19.17

19.11 The Theorem of Harnack and the Formulas of Schwarz and Poisson

When applying conformal mapping to the solution of elasticity problems we often deal with regions bounded by a unit circle, i.e. regions where $|z| \leq 1$ in the complex Z plane or regions where $|\zeta| \leq 1$ in the complex ζ plane. In the Z plane $z = x_1 + ix_2$ and in the ζ plane $\zeta = \xi_1 + i\xi_2$. Since the unit circle will generally be in the ζ plane, the formulas that follow will be developed using the notation specific to that plane. Following Muskhelishvili [2] the contour of the unit circle will be denoted by γ and points on this boundary by σ. Thus, on this contour $\sigma = e^{i\theta}$, where θ is the argument. All the functions of θ will be assumed to be periodic so that $f(\theta + 2\pi) = f(\theta)$ (Fig. 19.18).

The theorem of Harnack states that [3], if $f(\theta)$ and $\phi(\theta)$ are continuous real functions of the argument θ defined on the unit circle γ and if

$$\frac{1}{2\pi i}\oint_\gamma \frac{f(\theta)\,d\sigma}{\sigma - \zeta} = \frac{1}{2\pi i}\oint_\gamma \frac{\phi(\theta)\,d\sigma}{\sigma - \zeta} \tag{19.11.1}$$

for all values of ζ inside γ, then

$$f(\theta) = \phi(\theta) \tag{19.11.2}$$

If the point ζ is outside γ and if Eq. (19.11.1) holds for all values of ζ then

$$f(\theta) = \phi(\theta) + \text{constant} \tag{19.11.3}$$

The constant in Eq. (19.11.3) can be removed if in addition to Eq. (19.11.1) we have the equality

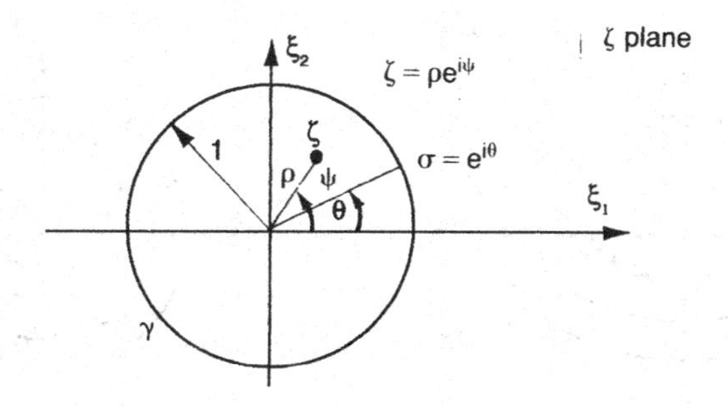

Fig. 19.18

$$\frac{1}{2\pi i}\oint_\gamma f(\theta)\,\frac{d\sigma}{\sigma} = \frac{1}{2\pi i}\oint_\gamma \phi(\theta)\,\frac{d\sigma}{\sigma} \tag{19.11.4}$$

A corollary that follows from the above theorem states that, if we have four real continuous functions f_1, f_2, ϕ_1, ϕ_2 and the following simultaneous equalities hold for all values of ζ:

$$\frac{1}{2\pi i}\oint_\gamma \frac{f_1 + if_2}{\sigma - \zeta}\,d\sigma = \frac{1}{2\pi i}\oint_\gamma \frac{\phi_1 + i\phi_2}{\sigma - \zeta}\,d\sigma\,,$$

$$\frac{1}{2\pi i}\oint_\gamma \frac{f_1 - if_2}{\sigma - \zeta}\,d\sigma = \frac{1}{2\pi i}\oint_\gamma \frac{\phi_1 - i\phi_2}{\sigma - \zeta}\,d\sigma$$

then

$$\phi_1 = f_1\,, \quad \text{and} \quad \phi_2 = f_2\,, \quad \text{if} \quad |\zeta| < 1$$

and

$$\phi_1 = f_1 + \text{const}\,, \quad \text{and} \quad \phi_2 = f_2 + \text{const}, \quad \text{if}\ |\zeta| > 1$$

The proof of the above theorem and its corollary can be found in references [2] and [3].

The formulas of Schwarz and Poisson are useful in solving boundary value problems for a circular region. Consider the region inside a unit circle centered at the origin. It is required to determine a harmonic function $u(\xi_1,\xi_2)$ which on the boundary of the circle γ assumes the values $u = f(\theta)$ where $f(\theta)$ is a real function of θ. The conjugate function of u being v, the function

$$F(\zeta) = u(\xi_1,\xi_2) + iv(\xi_1,\xi_2)$$

is analytic for all values inside $|\zeta| = 1$. Assuming $F(\zeta)$ to be continuous in the closed region $|\zeta| \leq 1$, we can write the boundary condition $u = f(\theta)$ as

$$F(\sigma) + \overline{F}(\overline{\sigma}) = 2f(\theta) \text{ on } \gamma \tag{19.11.5}$$

Multiplying both members by $1/2\pi i \oint_\gamma d\sigma/(\sigma - \zeta)$ where ζ is any point interior to γ we obtain the formula

$$\frac{1}{2\pi i}\oint_\gamma \frac{F(\sigma)}{\sigma - \zeta}\,d\sigma + \frac{1}{2\pi i}\oint_\gamma \frac{\overline{F}(\overline{\sigma})}{\sigma - \zeta}\,d\sigma$$

$$= \frac{1}{\pi i}\int \frac{f(\theta)}{\sigma - \zeta}\,d\sigma \tag{19.11.6}$$

By Harnack's theorem this equation if equivalent to Eq. (19.11.5). The first integral is equal to $F(\zeta)$ [see Eq. (19.4.3)]. The second integral is equal to $\bar{F}(0)$ [see Eq. (19.5.13)]. Setting $\bar{F}(0) = a_0 - ib_0$, Eq. (19.11.6) becomes

$$F(\zeta) = \frac{1}{\pi i} \oint_\gamma \frac{f(\theta)}{\sigma - \zeta} \, d\sigma - a_0 + ib_0 \tag{19.11.7}$$

For $\zeta = 0$,

$$a_0 + ib_0 = \frac{1}{\pi i} \oint_\gamma \frac{f(\theta)}{\sigma} \, d\sigma - a_0 + ib_0$$

Therefore,

$$2a_0 = \frac{1}{\pi i} \oint_\gamma \frac{f(\theta)}{\sigma} \, d\sigma = \frac{1}{\pi} \int_0^{2\pi} f(\theta) \, d\theta$$

Substituting into Eq. (19.11.7) we get

$$F(\zeta) = \frac{1}{\pi i} \oint_\gamma \frac{f(\theta)}{\sigma - \zeta} \, d\sigma - \frac{1}{2\pi i} \oint_\gamma \frac{f(\theta) \, d\sigma}{\sigma} + ib_0 \tag{19.11.8}$$

or

$$F(\zeta) = \frac{1}{2\pi i} \oint_\gamma f(\theta) \frac{\sigma + \zeta}{\sigma - \zeta} \frac{d\sigma}{\sigma} + ib_0 \tag{19.11.9}$$

This is the formula of Schwarz in which b_0 is undetermined since $v(\xi_1, \xi_2)$ is determined within an arbitrary real constant.

If we now substitute $\zeta = \rho e^{i\psi}$ and $\sigma = e^{i\theta}$ in Eq. (19.11.9) and separate real and imaginary parts we get, for the real part,

$$\text{Re } F(\zeta) = u(\xi_1, \xi_2) = \frac{1}{2\pi} \int_0^{2\pi} \frac{(1 - \rho^2) f(\theta) \, d\theta}{1 - 2\rho \cos(\theta - \psi) + \rho^2} \tag{19.11.10}$$

This is Poisson's integral formula for the harmonic function u in the circle. The integral of Poisson gives the solution of the first boundary value problem of potential theory. This problem, called the Dirichlet problem, is one of finding a function that is harmonic in a specific region and that satisfies prescribed conditions on the boundary. Here, the boundary is a circle. It is possible to generalize the formulas obtained above so as to make them apply to any simply connected region. This is often done by using mapping functions.

If instead of the function itself being prescribed on the boundary, its normal derivative is, the boundary value problem at hand is called the second boundary value problem of potential theory or the Neumann problem. This problem has been encountered in Section 10.3 while studying the torsion of non-circular prismatic bars.

19.12 Torsion of Prismatic Bars Using Complex Variables

In Chapter 10 it was found that the solution of the torsion problem amounted to finding a warping function ψ satisfying Laplace's equation

$$\nabla^2 \psi = 0 \tag{19.12.1}$$

throughout the section and

$$\left(\frac{\partial \psi}{\partial x_1} - x_2\right)\frac{dx_2}{ds} - \left(\frac{\partial \psi}{\partial x_2} + x_1\right)\frac{dx_1}{ds} = 0 \tag{19.12.2}$$

on the boundary C. Also, as indicated in Chapter 10, Eq. (19.12.2) can be replaced by Eq. (10.3.24) to cast the problem as a Neumann boundary value one.

If complex variables are used the problem may be reduced to the determination of an analytic function $F(z) = \psi + i\Gamma$, whose real part satisfies the boundary condition (Eq. 19.12.2). This boundary condition is not convenient to work with, and it may be recalled that it was simplified through the use of Prandtl's stress function ϕ in Section 10.5. Let us, therefore, examine how it can be recast in a more convenient form. The Cauchy-Riemann conditions are written

$$\frac{\partial \Gamma}{\partial x_1} = -\frac{\partial \psi}{\partial x_2}, \quad \frac{\partial \Gamma}{\partial x_2} = \frac{\partial \psi}{\partial x_1} \tag{19.12.3}$$

Substituting into Eq. (19.12.2) we get

$$\left(\frac{\partial \Gamma}{\partial x_2} - x_2\right)\frac{dx_2}{ds} - \left(-\frac{\partial \Gamma}{\partial x_1} + x_1\right)\frac{dx_1}{ds} = 0, \text{ on the boundary } C$$

or

$$\frac{\partial \Gamma}{\partial x_2}\frac{dx_2}{ds} + \frac{\partial \Gamma}{\partial x_1}\frac{dx_1}{ds} = x_2\frac{dx_2}{ds} + x_1\frac{dx_1}{ds}$$

Therefore,

$$\frac{d\Gamma}{ds} = \frac{d}{ds}\frac{x_1^2 + x_2^2}{2}$$

and

$$\Gamma = \left(\frac{x_1^2 + x_2^2}{2}\right) + K \qquad \text{on } C.$$

Γ is a harmonic function and K is a constant.

The constant K will not influence the stresses and can be left undetermined. Since $z\bar{z} = x_1^2 + x_2^2$,

$$\Gamma = \frac{1}{2} z\bar{z} \qquad \text{on } C$$

Thus, the solution of the torsion problem amounts to finding a harmonic function Γ (the imaginary part of $F(z)$), such that it has the value 1/2 $z\bar{z}$ on the boundary C.

In other words we must have:

$$\frac{\partial^2\Gamma}{\partial x_1^2} + \frac{\partial^2\Gamma}{\partial x_2^2} = 0 \qquad \text{through the section} \tag{19.12.4}$$

and

$$\Gamma = \frac{1}{2}(x_1^2 + x_2^2) = \frac{1}{2} z\bar{z} \qquad \text{on the boundary} \tag{19.12.5}$$

This is the Dirichlet problem whose solution is unique.

It is of interest to point out the relation between Γ and Prandtl's stress function ϕ defined and used in Sec. 10.5. In 1903, Prandtl [8] defined a function $\Psi(x_1, x_2)$ as follows

$$\Psi = \Gamma - \frac{1}{2}(x_1^2 + x_2^2)\,. \tag{19.12.6}$$

Recalling Eqs. (10.5.3), (10.5.4) and (19.12.3) one finds that

$$\sigma_{13} = G\alpha\frac{\partial\Psi}{\partial x_2}\,, \qquad \sigma_{23} = -G\alpha\frac{\partial\Psi}{\partial x_1} \tag{19.12.7}$$

and

$$\frac{\partial^2\Psi}{\partial x_1^2} + \frac{\partial^2\Psi}{\partial x_2^2} = -2\,. \tag{19.12.8}$$

The two functions ϕ of Sec. 10.5 and Ψ of Eq. (19.12.6) are essentially the same; within the multiplying factor $G\alpha$. Ψ is also called Prandtl's stress function. There should not be any confusion since those functions will be referred to by their Greek symbol. Let us go back to Eqs. (19.12.4) and (19.12.5) and examine, for example, the case of a circular region of radius equal to unity. On the boundary $\Gamma = 1/2$. The formula of Schwarz can be written in the Z plane (Fig. 19.19) as

$$\Phi(z) = \frac{1}{2\pi i}\oint_C \frac{d\alpha}{\alpha - z} - \frac{1}{4\pi i}\oint_C \frac{d\alpha}{\alpha} \tag{19.12.9}$$

where the constant ib_0 is ignored. This equation is easily integrated to give (see Sec. 19.3)

$$\Phi(z) = 1 - \frac{1}{2} = \frac{1}{2}$$

Therefore, $\Gamma = 1/2$ for a unit circle. Thus,

$$\Psi = -\frac{1}{2}(x_1^2 + x_2^2 - 1)$$

and the Prandtl stress function ϕ is given by

$$\phi = -\frac{G\alpha}{2}(x_1^2 + x_2^2 - 1)$$

This is indeed the answer that one gets from Eq. (10.6.1) if $a = b = 1$. In general, one has to use the residues theorem to integrate Schwarz's or Poisson's formulas.

For general cross sections the use of conformal mapping allows one

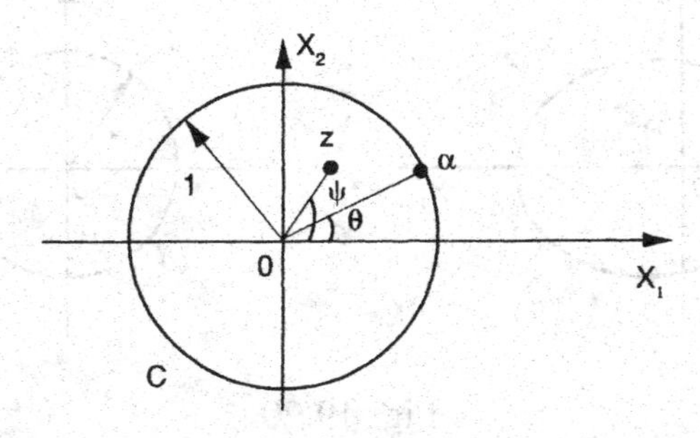

Fig. 19.19

to reduce the problem to that of the torsion of a circular prismatic bar (Fig. 19.20). Consider the mapping function

$$z = f(\zeta) \tag{19.12.10}$$

The function $F(z) = \psi + i\Gamma$ becomes

$$F(z) = F[f(\zeta)] = F_1(\zeta) \tag{19.12.11}$$

and the boundary condition Eq. (19.12.5) becomes

$$\Gamma = \frac{1}{2} f(\sigma)\bar{f}(\bar{\sigma}) \qquad \text{on } \gamma \tag{19.12.12}$$

where γ is the unit circle $|\zeta| = 1$. Now

$$\frac{1}{i} F_1(\zeta) = \Gamma - i\psi \tag{19.12.13}$$

Thus, the torsion problem will be solved if we succeed in determining the real part of the analytic function $(1/i)\ F_1(\zeta)$ which on the boundary γ assumes the value given by Eq. (19.12.12). Again, this problem was examined in the previous section and led to the formulas of Schwarz and Poisson. From Eq. (19.11.7).

$$\frac{1}{i} F_1(\zeta) = \frac{1}{\pi i} \oint_\gamma \frac{\frac{1}{2} f(\sigma)\bar{f}(\bar{\sigma})}{\sigma - \zeta}\, d\sigma - a_0 + ib_0$$

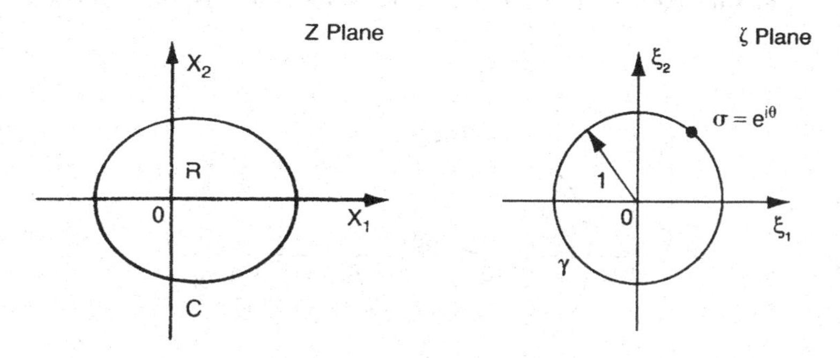

Fig. 19.20

Remembering that on the unit circle $\zeta = \sigma = e^{i\theta}$ and $\overline{\zeta} = e^{-i\theta} = 1/\sigma$,

$$F_1(\zeta) = \psi + i\Gamma = \frac{1}{2\pi}\oint_\gamma \frac{f(\sigma)\overline{f}\left(\frac{1}{\sigma}\right)}{\sigma - \zeta}\, d\sigma + \text{const.} \qquad (19.12.14)$$

This formula gives the warping function and its conjugate so that the solution of the torsion problem is reduced to finding the integral in Eq. (19.12.14).

a) Determination of the stresses in terms of $F_1(\zeta)$: From Eqs. (10.5.3) and (10.5.4) we have

$$\sigma_{13} - i\sigma_{23} = G\alpha\left(\frac{\partial\psi}{\partial x_1} - x_2 - i\frac{\partial\psi}{\partial x_2} - ix_1\right).$$

Using Eq. (19.12.3), we get

$$\sigma_{13} - i\sigma_{23} = G\alpha\left(\frac{\partial\psi}{\partial x_1} + i\frac{\partial\Gamma}{\partial x_1} - i(x_1 - ix_2)\right)$$

but,

$$\frac{\partial\psi}{\partial x_1} + i\frac{\partial\Gamma}{\partial x_1} = \frac{\partial F}{\partial x_1} = \frac{dF}{dz} = F'(z)$$

and

$$\sigma_{13} - i\sigma_{23} = G\alpha(F'(z) - i\overline{z}) \qquad (19.12.15)$$

Using Eq. (19.12.11) we have

$$F'(z) = F_1'(\zeta)\frac{d\zeta}{dz} = F_1'(\zeta)\frac{1}{f'(\zeta)}$$

and Eq. (19.12.15) becomes

$$\sigma_{13} - i\sigma_{23} = G\alpha\left[\frac{F_1'(\zeta)}{f'(\zeta)} - i\overline{f}(\overline{\zeta})\right] \qquad (19.12.16)$$

Eq. (19.12.16) is used to calculate the stress.

b) Determination of the torsional rigidity in terms of $F_1(\zeta)$: Eq. (10.3.17) gives for J,

$$J = \iint_R \left(x_1^2 + x_2^2 + x_1 \frac{\partial \psi}{\partial x_2} - x_2 \frac{\partial \psi}{\partial x_1} \right) dx_1\, dx_2$$

$$J = \iint_R \left[\frac{\partial}{\partial x_2} (x_1^2 x_2) + \frac{\partial}{\partial x_1} (x_1 x_2^2) \right] dx_1\, dx_2$$

$$+ \iint_R \left[\frac{\partial}{\partial x_2} (x_1 \psi) - \frac{\partial}{\partial x_1} (x_2 \psi) \right] dx_1\, dx_2$$

$$J = -\oint_C x_1 x_2 (x_1\, dx_1 - x_2\, dx_2) - \oint_C \psi (x_1\, dx_1 + x_2\, dx_2)$$

where the Green-Riemann formula was used to transform the surface integrals. Since,

$$x_1 = \frac{z + \bar{z}}{2}, \qquad x_2 = \frac{z - \bar{z}}{2i}$$

we find

$$\oint_C x_1 x_2 (x_1\, dx_1 - x_2\, dx_2) = \frac{1}{8i} \oint_C (z^2 - (\bar{z})^2)(z\, dz + \bar{z}\, d\bar{z})$$

$$= \frac{1}{8i} \oint_C (z^3\, dz + z^2 \bar{z}\, d\bar{z} - (\bar{z})^2 z\, dz - (\bar{z})^3\, d\bar{z})$$

But from Cauchy's integral theorem

$$\oint_C z^3\, dz = 0, \qquad \oint_C (\bar{z})^3\, d\bar{z} = 0.$$

Also,

$$\oint_C z^2 \bar{z}\, d\bar{z} = \oint_C z^2\, d\left(\frac{(\bar{z})^2}{2}\right) = -\oint_C (\bar{z})^2 z\, dz.$$

Therefore,

$$J = \frac{1}{4i} \oint_C (\bar{z})^2 z\, dz - \oint_C \psi\, d\left(\frac{z\bar{z}}{2}\right)$$

Now

$$\psi = \frac{1}{2}[F_1(\zeta) + \bar{F}_1(\bar{\zeta})]$$

and on the contour of the unit circle $\zeta = \sigma = e^{i\theta}$. Thus,

$$J = \frac{1}{4i}\oint_\gamma [\bar{f}(\bar{\sigma})]^2 f(\sigma)\, df(\sigma) - \frac{1}{4}\oint_\gamma [F_1(\sigma) + \bar{F}_1(\bar{\sigma})]\, d\,[f(\sigma)\bar{f}(\bar{\sigma})] \tag{19.12.17}$$

This expression can be written as

$$J = I_0 + D_0$$

where

$$I_0 = -\frac{i}{4}\oint_\gamma [\bar{f}(\bar{\sigma})]^2 f(\sigma)\, df(\sigma) \tag{19.12.18}$$

and

$$D_0 = -\frac{1}{4}\oint_\gamma [F_1(\sigma) + \bar{F}_1(\bar{\sigma})]\, d\,[f(\sigma)\bar{f}(\bar{\sigma})] \tag{19.12.19}$$

If the mapping function $z = f(\zeta)$ *is given*, on the unit circle it becomes a function of θ only and can be expanded into a complex Fourier series:

$$f(\sigma) = f(e^{i\theta}) = \sum_{-\infty}^{+\infty} A_n\, e^{in\theta} \tag{19.12.20}$$

where the Fourier coefficients are given by the formula

$$A_n = \frac{1}{2\pi}\int_0^{2\pi} f(\sigma)e^{-in\theta}\, d\theta$$

Since $F_1(\zeta)$ is analytic in the interior of the unit circle γ, it can be expanded into the power series

$$F_1(\zeta) = \sum_{n=0}^{\infty} B_n\, \zeta^n \tag{19.12.21}$$

On the unit circle we have

$$F_1(\sigma) = \sum_{n=0}^{\infty} B_n e^{in\theta} \quad \text{and} \quad \overline{F}_1(\overline{\sigma}) = \sum_{n=0}^{\infty} \overline{B}_n e^{-in\theta} \tag{19.12.22}$$

where $\overline{B}_n$ are the conjugates of B_n.

From the definition of conjugate functions we have

$$2i\Gamma = F_1(\zeta) - \overline{F}_1(\overline{\zeta}) \tag{19.12.23}$$

and the boundary condition Eq. (19.12.12) becomes

$$F_1(\sigma) - \overline{F}_1(\overline{\sigma}) = if(\sigma)\overline{f}(\overline{\sigma}) \tag{19.12.24}$$

Substituting Eq. (19.12.22) into Eq. (19.12.24) we find

$$\begin{aligned} &\sum_0^{\infty} B_n e^{in\theta} - \sum_0^{\infty} \overline{B}_n e^{-in\theta} = \\ &\qquad i\left(\sum_{-\infty}^{+\infty} A_n e^{in\theta}\right)\left(\sum_{-\infty}^{\infty} \overline{A}_n e^{-in\theta}\right) = i\sum_{-\infty}^{\infty} C_n e^{in\theta} \end{aligned} \tag{19.12.25}$$

where

$$C_n = \sum_{m=-\infty}^{\infty} A_{n+m}\overline{A}_m \quad \text{and} \quad C_{-n} = \overline{C}_n,\ n = 0, 1, 2, \ldots \infty$$

Equating the coefficients of $e^{in\theta}$ on both sides of the equation except $n = 0$, we find

$$B_n = iC_n\ ,\ \overline{B}_n = -i\overline{C}_n \tag{19.12.26}$$

Thus,

$$f(\sigma)\overline{f}(\overline{\sigma}) = \sum_{-\infty}^{+\infty} C_n e^{in\theta} \tag{19.12.27}$$

and

$$F_1(\zeta) = i\sum_{n=1}^{\infty} C_n \zeta^n + \text{constant} \tag{19.12.28}$$

Separating the real and imaginary parts, Eq. (19.12.28) gives us the warping function ψ and its conjugate Γ in terms of the coefficients C_n yielded by the mapping function.

Finally, it must be pointed out that it may often be more advantageous to use other techniques to solve torsion problems. The reader is referred to Sokolnikoff's treatise [3] for a short discussion of this subject.

19.13 Torsion of Prismatic Bars with Various Shapes

Example 1. The bar with a cardioid cross section.

The equation of the cardioid in the z plane is (Fig. 19.21a)

$$r = 2K(1 + \text{Cos}\ \theta) \tag{19.13.1}$$

where K is a constant. The cardioid section in the Z plane can be mapped onto a unit circle in the ζ plane by the mapping function

$$z = f(\zeta) = K(1 - \zeta)^2 \tag{19.13.1}$$

On the unit circle $\zeta = e^{i\theta}$ we have

$$f(\sigma) = K(1 - 2e^{i\theta} + e^{2i\theta}) \quad \text{and}$$
$$\bar{f}(\bar{\sigma}) = K(1 - 2e^{-i\theta} + e^{-2i\theta})$$

Hence, following Eq. (19.12.27) we have

$$f(\sigma)\bar{f}(\bar{\sigma}) = K^2(e^{2i\theta} - 4e^{i\theta} - 4e^{-i\theta} + e^{-2i\theta} + 6)$$

from which we find

$$C_1 = -4K^2, \qquad C_2 = K^2$$

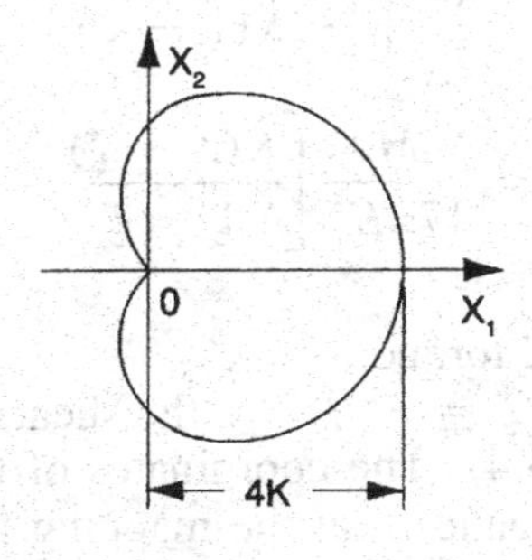

Fig. 19.21a

Therefore,

$$F_1(\zeta) = iK^2(\zeta^2 - 4\zeta) + \text{constant} \tag{19.13.2}$$

For the torsional rigidity the constant J can be computed from Eq. (19.12.17):

$$J = \frac{K^4}{4i}\oint_\gamma \left(1 - \frac{2}{\sigma} + \frac{1}{\sigma^2}\right)^2 (1 - 2\sigma + \sigma^2)(-2 + 2\sigma)\, d\sigma$$

$$\frac{-K^4}{4}\oint_\gamma \left[i(\sigma^2 - 4\sigma) - i\left(\frac{1}{\sigma^2} - \frac{4}{\sigma}\right)\right]\left(2\sigma - 4 + \frac{4}{\sigma^2} - \frac{2}{\sigma^3}\right) d\sigma$$

$$= \frac{K^4}{2i}\oint_\gamma \left(\sigma^3 - 7\sigma^2 + 21\sigma - 35 + \frac{35}{\sigma} - \frac{21}{\sigma^2} + \frac{7}{\sigma^3} - \frac{1}{\sigma^4}\right) d\sigma$$

$$- \frac{K^4 i}{2}\oint_\gamma \left(\sigma^3 - 6\sigma^2 + 8\sigma + 6 - \frac{18}{\sigma} + \frac{6}{\sigma^2} + \frac{8}{\sigma^3} - \frac{6}{\sigma^4} + \frac{1}{\sigma^5}\right) d\sigma$$

Using the residues theorem,

$$J = 2\pi i\left[\frac{K^4}{2i}(35) - \frac{K^4 i}{2}(-18)\right] = 17\pi K^4 \tag{19.13.3}$$

The undetermined constant in Eq. (19.13.2) does not affect J.

The shearing stresses can be computed from Eq. (19.12.16)

$$\sigma_{13} - i\sigma_{23} = G\alpha\left[\frac{K^2 i(2\zeta - 4)}{-2K(1 - \zeta)} - i(1 - \bar{\zeta})^2\right]$$

$$= \frac{iM_t}{17\pi K^4}\left[\frac{K(2 - \zeta)}{1 - \zeta} - (1 - \bar{\zeta})^2\right] \tag{19.13.4}$$

where M_t is the applied torque.

For various values of $\zeta = \xi_1 + i\xi_2$, the shearing stresses can be computed using Eq. (19.13.4). The coordinates of the corresponding point in the Z plane can be found from the mapping function (Eq. 19.13.1).

Example 2. The bar with a cross section bounded by the inverse of an ellipse with respect to its center [3].

Consider the ellipse

$$\frac{x_1^2}{a^2} + \frac{x_2^2}{b^2} = 1 \tag{19.13.5}$$

when its points are inverted with respect to the origin 0, a point (x_1, x_2) is carried to (x_1', x_2') such that

$$r^2(r')^2 = (x_1^2 + x_2^2)[(x_1')^2 + (x_2')] = 1 \tag{19.13.6}$$

The resulting curve is shown on Fig. (19.21b). The parametric equations of Eq. (19.13.5) with u_1 as a parameter can be written as:

$$x_1 = \frac{1}{c}\cosh k \cos u_1, \qquad x_2 = \frac{1}{c}\sinh k \sin u_1 \tag{19.13.7}$$

with a different ellipse for each value of c and k. Here $a = (1/c)\cosh k$ and $b = (1/c)\sinh k$, so that $c = 1/\sqrt{a^2 - b^2}$ and $\tanh k = b/a$. In terms of the parameter u_1 the equations of the inverse of the ellipse are

$$\frac{x_1}{x_1^2 + x_2^2} = \frac{1}{c}\cosh k \cos u_1 \tag{19.13.8}$$

$$\frac{x_2}{x_1^2 + x_2^2} = \frac{1}{c}\sinh k \sin u_1 \tag{19.13.9}$$

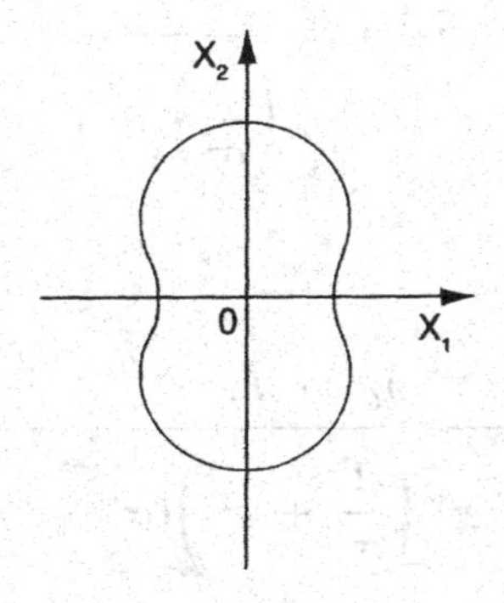

Fig. 19.21b

Thus,

$$\frac{x_1 - ix_2}{(x_1 + ix_2)(x_1 - ix_2)} = \frac{1}{c}\cosh k \cos u_1 - \frac{i}{c}\sinh k \sin u_1$$
$$\frac{1}{x_1 + ix_2} = \frac{1}{c}\cos ik \cos u_1 - \frac{1}{c}\sin ik \sin u_1 \tag{19.13.10}$$

and

$$x_1 + ix_2 = c \sec (u_1 + ik) \tag{19.13.11}$$

This equation can be written as

$$z = c \sec (w + ik), \qquad u_2 = 0 \tag{19.13.12}$$

where $z = x + iy$ and $w = u_1 + iu_2$. If we put $\zeta = e^{iw}$ in Eq. (19.13.12) we see that the resulting function

$$z = f(\zeta) = \frac{2ce^k\zeta}{\zeta^2 + e^{2k}}, \quad c > 0, \quad k > 0 \tag{19.13.13}$$

maps the cross section of the cylinder upon the interior of the unit circle $|\zeta| \leqslant 1$ [3]. On the unit circle $\zeta = \sigma = e^{i\theta}$ we have

$$f(\sigma) = \frac{2ce^k\sigma}{\sigma^2 + e^{2k}} \tag{19.13.14}$$

and

$$\overline{f}(\overline{\sigma}) = \frac{2ce^k\left(\dfrac{1}{\sigma}\right)}{\left(\dfrac{1}{\sigma^2}\right) + e^{2k}} \tag{19.13.15}$$

Eq. (19.12.14) gives

$$F_1(\zeta) = \psi + i\Gamma$$
$$= \frac{1}{2\pi}\oint_\gamma \frac{4c^2e^{2k}\, d\sigma}{(\sigma^2 + e^{2k})\left(\dfrac{1}{\sigma^2} + e^{2k}\right)(\sigma - \zeta)} + \text{const.}$$
$$= \frac{2c^2}{\pi}\oint_\gamma \frac{\sigma^2\, d\sigma}{(\sigma^2 + e^{2k})(\sigma^2 + e^{-2k})(\sigma - \zeta)} + \text{const.} \tag{19.13.16}$$

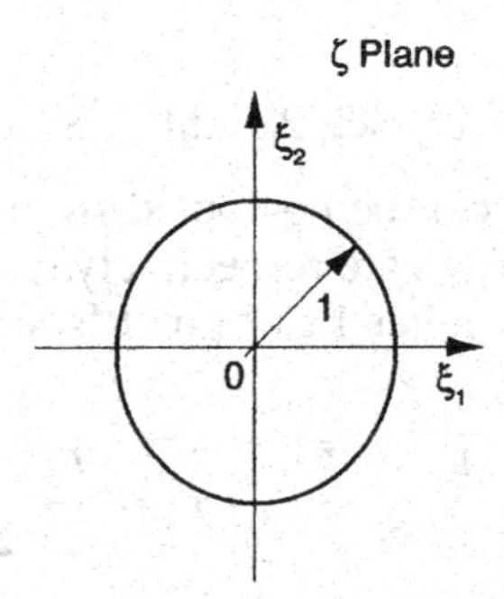

Fig. 19.21c

This integral can be obtained by the method of residues. From Cauchy's residue theorem,

$$F_1(\zeta) = \psi + i\Gamma = -4ic^2(R_1 + R_2) + \text{constant} \qquad (19.13.17)$$

where R_1 and R_2 are the residues of the integrand in Eq. (19.13.16) at $\sigma = ie^k$ and $\sigma = -ie^k$ respectively. R_1 and R_2 can be obtained from Eq. (19.8.6) as

$$R_1 = \lim_{\sigma = ie^k} \left[\frac{\sigma^2}{(\sigma + ie^k)(\sigma^2 + e^{-2k})(\sigma - \zeta)} \right] = \frac{ie^k}{4(\zeta - ie^k)\sinh 2k}$$

$$R_2 = \lim_{\sigma = -ie^k} \left[\frac{\sigma^2}{(\sigma - ie^k)(\sigma^2 + e^{-2k})(\sigma - \zeta)} \right] = \frac{-ie^k}{4(\zeta + ie^k)\sinh 2k}$$

Let the constant in Eq. (19.13.17) be chosen as equal to $-ic^2 \operatorname{csch} 2k$ so that this equation now becomes

$$\begin{aligned} F_1(\zeta) = \psi + i\Gamma &= c^2 \operatorname{csch} 2k \left[-i \frac{\zeta^2 - e^{2k}}{\zeta^2 + e^{2k}} \right] \qquad (19.13.18) \\ &= c^2 \operatorname{csch} 2k \left[-i \frac{e^{2iw} - e^{2k}}{e^{2iw} + e^{2k}} \right] \\ &= c^2 \operatorname{csch} 2k \left[-i \frac{e^{iw-k} - e^{-iw+k}}{e^{iw-k} + e^{-iw+k}} \right] \end{aligned}$$

Thus,

$$F_1(\zeta) = \psi + i\Gamma = c^2 \operatorname{csch} 2k \tan(w + ik) \qquad (19.13.19)$$

The shearing stresses can be computed from Eq. (19.12.7). The real and imaginary parts of $F_1(\zeta)$ are respectively the warping function ψ and the function Γ which is related to Prandtl's stress function. Using the relationships

$$\psi = \frac{F_1 + \bar{F}_1}{2}, \quad \Gamma = \frac{F_1 - \bar{F}_1}{2i},$$

we have

$$\begin{aligned} \psi &= \frac{1}{2} c^2 \operatorname{csch} 2k[\tan(w + ik) + \tan(\bar{w} - ik)] \\ &= c^2 \operatorname{csch} 2k \frac{\tan u_1[1 - \tanh^2(u^2 + k)]}{1 + \tan^2 u_1 \tanh^2(u_2 + k)} \\ &= c^2 \operatorname{csch} 2k \frac{\sin 2u_1}{\cos 2u_1 + \cosh 2(u_2 + k)} \end{aligned} \qquad (19.13.20)$$

and

$$\begin{aligned} \Gamma &= \frac{1}{2i} c^2 \operatorname{csch} 2k[\tan(w + ik) - \tan(\bar{w} - ik)] \\ &= c^2 \operatorname{csch} 2k \frac{\tanh(u_2 + k)[1 + \tan^2 u_1]}{1 + \tan^2 u_1 \tanh^2(u_2 + k)} \\ &= c^2 \operatorname{csch} 2k \frac{\sinh 2(u_2 + k)}{\cos 2u_1 + \cosh 2(u_2 + k)}, \end{aligned} \qquad (19.13.21)$$

so that

$$\begin{aligned} F_1(\zeta) &= \psi + i\Gamma \\ &= c^2 \operatorname{csch} 2k \frac{\sin 2u_1 + i \sinh 2(u_2 + k)}{\cos 2u_1 + \cosh 2(u_2 + k)} \end{aligned} \qquad (19.13.22)$$

From Eq. (19.12.6)

$$\Psi = \frac{c^2 \operatorname{csch} 2k \sinh 2(u_2 + k)}{\cos 2u_1 + \cosh 2(u_2 + k)} - \frac{x_1^2 + x_2^2}{2} \qquad (19.13.23)$$

In the expression $z = c \sec(w + ik)$, with specific magnitudes of c and k, different values of u_1 and a values of $u_2 = 0$ lead to points located on the curve shown in Fig. (19.21b). Points located inside this boundary can be obtained by varying the value of u_2. Thus, we can write

$$x_1^2 + x_2^2 = z\bar{z} = c^2 \sec(w + ik) \sec(\bar{w} - ik) \tag{19.13.24}$$

$$= \frac{2c^2}{\cos 2u_1 + \cosh 2(u_2 + k)} \tag{19.13.25}$$

Thus,

$$\Psi = \frac{-c^2[1 - \operatorname{csch} 2k \sinh 2(u_2 + k)]}{\cos 2u_1 + \cosh 2(u_2 + k)} \tag{19.13.26}$$

To use Eq. (19.12.7) one must take the derivatives of Ψ with respect to x_1 and x_2. Now

$$\frac{\partial \Psi}{\partial x_1} = \frac{\partial \Psi}{\partial u_1}\frac{\partial u_1}{\partial x_1} + \frac{\partial \Psi}{\partial u_2}\frac{\partial u_2}{\partial x_1} \tag{19.13.27}$$

$$\frac{\partial \Psi}{\partial x_2} = \frac{\partial \Psi}{\partial u_1}\frac{\partial u_1}{\partial x_2} + \frac{\partial \Psi}{\partial u_2}\frac{\partial u_2}{\partial x_2} \tag{19.13.28}$$

The equation

$$z = x_1 + ix_2 = c \sec(w + ik) = c \sec[u_1 + i(u_2 + k)]$$

allows one to obtain separate expressions for x_1 and x_2 in terms of u_1 and u_2. Those expressions are

$$x_1 = \frac{2c \cos u_1 \cosh(u_2 + k)}{\cos 2u_1 + \cosh 2(u_2 + k)} \tag{19.13.29}$$

$$x_2 = \frac{-2c \sin u_1 \sinh(u_2 + k)}{\cos 2u_1 + \cosh 2(u_2 + k)} \tag{19.13.30}$$

They can be used in Eqs. (19.13.27) and (19.13.28) after the partial derivatives $\partial u_i/\partial x_j$ are written in terms of $\partial x_i/\partial u_j$:

$$\frac{\partial \Psi}{\partial x_1} = \left[\frac{\partial \Psi}{\partial u_1}\frac{\partial x_1}{\partial u_1} + \frac{\partial \Psi}{\partial u_2}\frac{\partial x_1}{\partial u_2}\right] \Big/ \left[\left(\frac{\partial x_1}{\partial u_1}\right)^2 + \left(\frac{\partial x_1}{\partial u_2}\right)^2\right], \tag{19.13.31}$$

$$\frac{\partial \Psi}{\partial x_2} = \left[-\frac{\partial \Psi}{\partial u_1}\frac{\partial x_1}{\partial u_2} + \frac{\partial \Psi}{\partial u_2}\frac{\partial x_1}{\partial u_1}\right] \Big/ \left[\left(\frac{\partial x_1}{\partial u_1}\right)^2 + \left(\frac{\partial x_1}{\partial u_2}\right)^2\right] \tag{19.13.32}$$

The calculations involved in the use of Eq. (19.12.7) are long but simple. They yield

$$\sigma_{13} = G\alpha \frac{\partial \Psi}{\partial x_2}$$

$$= -2\,G\alpha c \sin u_1 \left[\frac{\operatorname{csch} 2k \cosh(u_2 + k)}{\cos 2u_1 - \cosh 2(u_2 + k)} + \frac{\sinh(u_2 + k)}{\cos 2u_1 + \cosh 2(u_2 + k)} \right] \tag{19.13.33}$$

$$\sigma_{23} = -G\alpha \frac{\partial \Psi}{\partial x_1}$$

$$= -2G\alpha c \cos u_1 \left[\frac{\operatorname{csch} 2k \sinh(u_2 + k)}{\cos 2u_1 - \cosh 2(u_2 + k)} - \frac{\cosh(u_2 + k)}{\cos 2u_1 + \cosh 2(u_2 + k)} \right] \tag{19.13.34}$$

To obtain the torsional rigidity J one can use Eqs. (19.12.17) to (19.12.19) together with Eqs. (19.13.14) and (19.13.15):

$$I_0 = 4c^4 i \oint_\gamma \frac{\sigma^3(\sigma^2 - e^{2k})}{(\sigma^2 + e^{2k})^3(\sigma^2 + e^{-2k})^2}\, d\sigma \tag{19.13.35}$$

The integral can be obtained by the method of residues so that

$$I_0 = -8\pi c^4(R_3 + R_4)$$

where R_3 and R_4 are the residues of the integrand at $\sigma = ie^{-k}$ and $-ie^{-k}$ respectively. The residues are given by

$$R_3 = \frac{d}{d\sigma}\left[\frac{\sigma^3(\sigma^2 - e^{2k})}{(\sigma^2 + e^{2k})^3(\sigma + ie^{-k})^2}\right]_{\sigma = ie^{-k}}$$

$$= \frac{-\operatorname{csch}^4 2k(2 + \cosh 4k)}{16} = R_4$$

so that $I_0 = \pi c^4(2 + \cosh 4k)\operatorname{csch}^4 2k$

Similarly,

$$D_0 = -i4c^4 \operatorname{csch} 2k \int_\gamma \frac{\sigma(1 - \sigma^4)^2}{(\sigma^2 + e^{2k})^3(\sigma^2 + e^{-2k})^3}\, d\sigma$$

and

$$D_0 = 8\pi c^4 \operatorname{csch} 2k(R_5 + R_6)$$

in which R_5 and R_6 are the residues of the integrand at $\sigma = ie^{-k}$ and $\sigma = -ie^{-k}$.

$$R_5 = \frac{1}{2}\frac{d^2}{d\sigma^2}\left[\frac{\sigma(1-\sigma^4)^2}{(\sigma^2+e^{2k})^3(\sigma+ie^{-k})^3}\right]_{\sigma=ie^{-k}}$$

$$R_5 = -\frac{1}{8}\operatorname{csch}^3 2k = R_6$$

and

$$D_0 = -2\pi c^4 \operatorname{csch}^4 2k$$

Therefore, the torsional rigidity J is given by

$$J = I_0 + D_0 = \pi c^4(2\operatorname{csch}^2 2k + \operatorname{csch}^4 2k) \quad (19.13.37)$$

and the twisting moment $M_t = M_{33}$ *is given by*

$$M_t = JG\alpha = \pi G\alpha c^4(2\operatorname{csch}^2 2k + \operatorname{csch}^4 2k)\,. \quad (19.13.38)$$

Eq. (19.13.38) can be written in terms of the semi minor and semi major axes of the cross section:

$$d = \frac{1}{a} = c\operatorname{sech} k, \quad \text{and} \quad e = \frac{1}{b} = c\operatorname{csch} k$$

$$M_t = \pi G\alpha(d^4 + e^4 + 6d^2e^2)/16 \quad (19.13.39)$$

If $d = e$ the original ellipse is a circle, the inverse of which, with respect to the center is also a circle. Eq. (19.13.38) then yields

$$M_t = G\alpha\frac{\pi d^4}{2}$$

which is the known expression of the twisting moment for a circular cross section of radius d.

19.14 The Plane Stress and Strain Problems and the Solution to the Biharmonic Equation.

In Sections 9.9 and 9.10 it was shown that the solution to a plane strain or a plane stress problem with no body (or gravitational) forces can be reduced to the determination of the stress function ϕ which satisfies the biharmonic equation

$$\nabla^4\phi = 0 \tag{19.14.1}$$

This equation can be solved using complex variables. We first start by writing Eq. (19.14.1) in the form

$$\nabla^2(\nabla^2\phi) = 0$$

Thus, the function

$$P_1 = \nabla^2\phi \tag{19.14.2}$$

satisfies Laplace's equation $\nabla^2 P_1 = 0$. If P_2 is the conjugate of P_1, then,

$$F(z) = P_1 + iP_2 \tag{19.14.3}$$

is an analytic function and its integral with respect to z is also analytic. Let

$$\psi(z) = Q_1 + iQ_2 = \frac{1}{4}\int F(z)\,dz \tag{19.14.4}$$

Then,

$$\begin{aligned}\psi'(z) &= \frac{\partial Q_1}{\partial x_1} + i\frac{\partial Q_2}{\partial x_1} = \frac{\partial Q_2}{\partial x_2} - i\frac{\partial Q_1}{\partial x_2} \\ &= \frac{1}{4}F(z) = \frac{1}{4}(P_1 + iP_2)\end{aligned} \tag{19.14.5}$$

Thus,

$$\frac{\partial Q_2}{\partial x_2} = \frac{\partial Q_1}{\partial x_1} = \frac{1}{4}P_1 \tag{19.14.6}$$

From the above we conclude that

$$\nabla^2(x_1Q_1 + x_2Q_2) = 2\frac{\partial Q_1}{\partial x_1} + 2\frac{\partial Q_2}{\partial x_2} = P_1 \tag{19.14.7}$$

Substituting Eq. (19.14.2) into Eq. (19.14.7) we get

$$\nabla^2(\phi - x_1Q_1 - x_2Q_2) = \nabla^2\phi - 2\frac{\partial Q_1}{\partial x_1} - 2\frac{\partial Q_2}{\partial x_2} = 0 \tag{19.14.8}$$

Thus, the function $(\phi - x_1Q_1 - x_2Q_2)$ is a harmonic function which we will call U_1 and the stress function ϕ can be written as

$$\phi = U_1 + x_1 Q_1 + x_2 Q_2 \tag{19.14.9}$$

which shows that any stress function can be made of suitably chosen conjugate harmonic functions Q_1 and Q_2 and a harmonic function U_1. The use of Q_1 and Q_2, however, is not necessary. Indeed, instead of Eq. (19.14.8), Timoshenko [9] writes

$$\nabla^2(\phi - 2x_1 Q_1) = \nabla^2\phi - 4\frac{\partial Q_1}{\partial x_1} = 0 \tag{19.14.10}$$

indicating that $(\phi - 2x_1 Q_1)$ is harmonic equal to H_1 say, so that any stress function can be expressed in the form

$$\phi = 2x_1 Q_1 + H_1 \tag{19.14.11}$$

The same kind of equation can be rewritten in terms of Q_2, i.e.

$$\phi = 2x_2 Q_2 + H_2 \tag{19.14.12}$$

where H_1 and H_2 are suitably chosen harmonic functions. Thus, only two harmonic functions are necessary. Eqs. (19.14.11) and (19.14.12) will be referred to as Timoshenko's equation. We will return to them in Sec. 19.15.

The stress function ϕ can be written in several different forms. We can construct the function

$$\chi(z) = U_1 + iU_2 \tag{19.14.13}$$

where U_2 is the conjugate of U_1. Noting that

$$\bar{z}\psi(z) = (x_1 - ix_2)(Q_1 + iQ_2) = (x_1 Q_1 + x_2 Q_2) + i(x_1 Q_2 - x_2 Q_1)$$

we can write

$$\phi = \text{Re}[\bar{z}\psi(z) + \chi(z)] \tag{19.14.14}$$

Thus, Airy's stress function can be written in terms of two analytic functions ψ and χ. Eq. (19.14.14) can be written as

$$\phi = \frac{1}{2}[\bar{z}\psi(z) + z\bar{\psi}(\bar{z}) + \chi(z) + \bar{\chi}(\bar{z})] \tag{19.14.15}$$

From the above we see that we have a general solution in terms of two analytic functions of z as given by Eq. (19.14.14) or in terms of two harmonic functions (also called logarithmic potential functions) given by either Eq. (19.14.11) or (19.14.12). Eq. (19.14.14) enables us to use the powerful tools of complex variable theory to obtain solutions, whereas Eqs. (19.14.11) and (19.14.12) are more restricted in application.

1. *Displacements and stresses*

Let us first consider displacements and stresses and write their expressions in terms of Airy's stress function.

For plane stress problems without body forces (see Sec. 8.17)

$$e_{11} = \frac{\partial u_1}{\partial x_1} = \frac{1}{E}(\sigma_{11} - \nu\sigma_{22}) = \frac{1}{E}\left(\frac{\partial^2\phi}{\partial x_2^2} - \nu\frac{\partial^2\phi}{\partial x_1^2}\right) \tag{19.14.16}$$

$$e_{22} = \frac{\partial u_2}{\partial x_2} = \frac{1}{E}(\sigma_{22} - \nu\sigma_{11}) = \frac{1}{E}\left(\frac{\partial^2\phi}{\partial x_1^2} - \nu\frac{\partial^2\phi}{\partial x_2^2}\right) \tag{19.14.17}$$

$$e_{12} = \frac{1}{2}\left(\frac{\partial u_1}{\partial x_2} + \frac{\partial u_2}{\partial x_1}\right) = \frac{2(1 + \nu)}{E}\sigma_{12} \tag{19.14.18}$$

$$= -\frac{2(1 + \nu)}{E}\frac{\partial^2\phi}{\partial x_1\, dx_2}$$

Recalling Eqs. (19.14.2) and (19.14.6), we obtain

$$\frac{\partial u_1}{\partial x_1} = \frac{1}{E}\left[P_1 - \frac{\partial^2\phi}{\partial x_1^2} - \nu\frac{\partial^2\phi}{\partial x_1^2}\right] = \frac{1}{E}\left[4\frac{\partial Q_1}{\partial x_1} - (1 + \nu)\frac{\partial^2\phi}{\partial x_1^2}\right] \tag{19.14.19}$$

and

$$\frac{\partial u_2}{\partial x_2} = \frac{1}{E}\left[4\frac{\partial Q_2}{\partial x_2} - (1 + \nu)\frac{\partial^2\phi}{\partial x_2^2}\right] \tag{19.14.20}$$

Integrating we find

$$u_1 = \frac{1}{E}\left[4Q_1 - (1 + \nu)\frac{\partial\phi}{\partial x_1} + g_1(x_2)\right]$$

$$u_2 = \frac{1}{E}\left[4Q_2 - (1 + \nu)\frac{\partial\phi}{\partial x_2} + g_2(x_1)\right]$$

where $g_1(x_2)$ and $g_2(x_1)$ are arbitrary functions of x_2 and x_1 respectively. Substituting these expressions into Eq. (19.14.18) we get

$$4\left(\frac{\partial Q_1}{\partial x_2} + \frac{\partial Q_2}{\partial x_1}\right) - 2(1 + \nu)\frac{\partial^2 \phi}{\partial x_1 \partial x_2} + \frac{dg_1}{dx_2} + \frac{dg_2}{dx_1} = -2(1 + \nu)\frac{\partial^2 \phi}{\partial x_1 \partial x_2}$$

But,

$$\frac{\partial Q_1}{\partial x_2} = -\frac{\partial Q_2}{\partial x_1},$$

thus,

$$\frac{dg_1}{dx_2} + \frac{dg_2}{dx_1} = 0$$

Therefore,

$$\frac{dg_1}{dx_2} = -\frac{dg_2}{dx_1} = C$$

where C is a constant, so that

$$g_1 = Cx_2 + C_1, \qquad g_2 = -Cx_1 + C_2 .$$

Those components of the displacement will not produce any strain and represent rigid body displacements. They can be ignored, and

$$u_1 = \frac{1}{E}\left[4Q_1 - (1 + \nu)\frac{\partial \phi}{\partial x_1}\right] \tag{19.14.21}$$

$$u_2 = \frac{1}{E}\left[4Q_2 - (1 + \nu)\frac{\partial \phi}{\partial x_2}\right] \tag{19.14.22}$$

$$u_1 + iu_2 = \frac{1}{E}\left[4(Q_1 + iQ_2) - (1 + \nu)\left(\frac{\partial \phi}{\partial x_1} + i\frac{\partial \phi}{\partial x_2}\right)\right] \tag{19.14.23}$$

Noting that

$$\frac{\partial z}{\partial x_1} = \frac{\partial \bar{z}}{\partial x_1} = 1, \quad \frac{\partial z}{\partial x_2} = i, \quad \frac{\partial \bar{z}}{\partial x_2} = -i$$

Eq. (19.14.15) gives

$$\frac{\partial\phi}{\partial x_1} = \frac{1}{2}\,[\bar{z}\psi'(z) + \psi(z) + z\bar{\psi}'(\bar{z}) + \bar{\psi}(\bar{z}) + \chi'(z) + \bar{\chi}'(\bar{z})]$$

$$\frac{\partial\phi}{\partial x_2} = \frac{i}{2}\,[\bar{z}\psi'(z) - \psi(z) - z\bar{\psi}'(\bar{z}) + \bar{\psi}(\bar{z}) + \chi'(z) - \bar{\chi}'(\bar{z})]$$

Therefore,

$$\left.\begin{aligned} \frac{\partial\phi}{\partial x_1} + i\,\frac{\partial\phi}{\partial x_2} &= \psi(z) + z\bar{\psi}'(\bar{z}) + \bar{\chi}'(\bar{z}) \\ \frac{\partial\phi}{\partial x_1} - i\,\frac{\partial\phi}{\partial x_2} &= \bar{\psi}(\bar{z}) + \bar{z}\psi'(z) + \chi'(z) \\ x_1\,\frac{\partial\phi}{\partial x_1} + x_2\,\frac{\partial\phi}{\partial x_2} &= \mathrm{Re}\left[z\left(\frac{\partial\phi}{\partial x_1} - i\,\frac{\partial\phi}{\partial x_2}\right)\right] \end{aligned}\right\} \qquad (19.14.24)$$

Recalling Eq. (19.14.4),

$$u_1 + iu_2 = \frac{3-\nu}{E}\,\psi(z) - \frac{1+\nu}{E}\,[z\bar{\psi}'(\bar{z}) + \bar{\chi}'(\bar{z})] \qquad (19.14.25)$$

or

$$2G(u_1 + iu_2) = \frac{3-\nu}{1+\nu}\,\psi(z) - z\bar{\psi}'(\bar{z}) - \bar{\chi}'(\bar{z}) \qquad (19.14.26)$$

The expressions for u_1 and u_2 can be obtained by taking one half of the sum and one half of the difference of the right hand side of Eq. (19.14.25) and of its conjugate. Thus,

$$4Gu_1 = \frac{3-\nu}{1+\nu}\,[\psi(z) + \bar{\psi}(\bar{z})] - z\bar{\psi}'(\bar{z}) - \bar{z}\psi'(z) - \chi'(z) - \bar{\chi}'(\bar{z})$$

$$\begin{aligned} 4Gu_2 = {} & -i\,\frac{3-\nu}{1+\nu}\,[\psi(z) - \bar{\psi}(\bar{z})] \\ & + i[z\bar{\psi}'(\bar{z}) - \bar{z}\psi'(z) - \chi'(z) + \bar{\chi}'(\bar{z})] \end{aligned}$$

Sometimes the rotation ω_{21} is used. Its expression can be obtained from Eq. (1.2.3) as

$$\omega_{21} = \frac{1}{2}\left(\frac{\partial u_2}{\partial x_1} - \frac{\partial u_1}{\partial x_2}\right) = \frac{4 \text{ Im } \psi'(z)}{E} = -\frac{4i}{2E}[\psi'(z) - \bar{\psi}'(\bar{z})] \qquad (19.14.27)$$

Eq. (19.14.26) allows one to calculate the components of the displacement for plane stress problems when the "complex potentials" $\psi(z)$ and $\chi(z)$ are given.

For plane strain $\nu/(1 - \nu)$ is substituted for ν in the right hand side of Eq. (19.14.26). Therefore, for plane strain

$$2G(u_1 + iu_2) = (3 - 4\nu)\psi(z) - z\bar{\psi}'(\bar{z}) - \bar{\chi}'(\bar{z})$$

or in terms of λ and μ,

$$2\mu(u_1 + iu_2) = \frac{\lambda + 3\mu}{\lambda + \mu}\psi(z) - z\bar{\psi}'(\bar{z}) - \bar{\chi}'(\bar{z})$$

It is advantageous to write the expressions for the displacements as

$$2G(u_1 + iu_2) = \kappa\psi(z) - z\bar{\psi}'(\bar{z}) - \bar{\chi}'(\bar{z}) \qquad (19.14.28)$$

where

$$\kappa = \begin{cases} \dfrac{3 - \nu}{1 + \nu} \text{ for plane stress} \\ 3 - 4\nu \text{ for plane strain} \end{cases}$$

Now let us examine the stress components. Differentiating Eq. (19.14.24) with respect to x_1 and x_2 we find that

$$\frac{\partial^2 \phi}{\partial x_1^2} + i\frac{\partial^2 \phi}{\partial x_1 \partial x_2} = \psi'(z) + z\bar{\psi}''(\bar{z}) + \bar{\psi}'(\bar{z}) + \bar{\chi}''(\bar{z})$$

$$\frac{\partial^2 \phi}{\partial x_1 \partial x_2} + i\frac{\partial^2 \phi}{\partial x_2^2} = i[\psi'(z) - z\bar{\psi}''(\bar{z}) + \bar{\psi}'(\bar{z}) - \bar{\chi}''(\bar{z})]$$

Multiplying the second equation by i and subtracting it from the first we get, using Eq. (9.10.4)

$$\sigma_{11} + \sigma_{22} = 2\psi'(z) + 2\bar{\psi}'(\bar{z}) = 4 \text{ Re } \psi'(z) \qquad (19.14.29)$$

Similarly, by adding we get

$$\sigma_{22} - \sigma_{11} - 2i\sigma_{12} = 2[z\bar{\psi}''(\bar{z}) + \bar{\chi}''(\bar{z})] ;$$

or by changing i to $-i$ on both sides of the equation,

$$\sigma_{22} - \sigma_{11} + 2i\sigma_{12} = 2[\bar{z}\psi''(z) + \chi''(z)] \qquad (19.14.30)$$

Eqs. (19.14.29) and (19.14.30) give the stresses in terms of the complex potentials $\psi(z)$ and $\chi(z)$. By choosing specific functions for $\psi(z)$ and $\chi(z)$ we can get possible states of stress from these equations. Such stresses provide the solution to an elasticity problem.

Eqs. (19.14.29) and (19.14.30) can be rewritten in a form that can be more convenient to special kinds of problems. Thus,

$$\sigma_{11} = \frac{1}{2}\{2[\psi'(z) + \bar{\psi}'(\bar{z})] - \bar{z}\psi''(z) - z\bar{\psi}''(\bar{z}) - \chi''(z) - \bar{\chi}''(\bar{z})\} \qquad (19.14.31)$$

$$\sigma_{22} = \frac{1}{2}\{2[\psi'(z) + \bar{\psi}'(\bar{z})] + \bar{z}\psi''(z) + z\bar{\psi}''(\bar{z}) + \chi''(z) + \bar{\chi}''(\bar{z})] \qquad (19.14.32)$$

$$\sigma_{12} = -\frac{i}{2}[\bar{z}\psi''(z) - z\bar{\psi}''(\bar{z}) + \chi''(z) - \bar{\chi}''(\bar{z})] \qquad (19.14.33)$$

Using the relations

$$\bar{z}\psi''(z) + z\bar{\psi}''(\bar{z}) = 2x_1 \text{ Re } \psi''(z) + 2x_2 \text{ Im } \psi''(z) = 2 \text{ Re } \bar{z}\psi''(z) \qquad (19.14.34)$$

$$\bar{z}\psi''(z) - z\bar{\psi}''(\bar{z}) = 2i[x_1 \text{ Im } \psi''(z) - x_2 \text{ Re } \psi''(z)] \qquad (19.14.35)$$

we get

$$\sigma_{11} = \text{Re}[2\psi'(z)] - \text{Re } \chi''(z) - x_1 \text{ Re } \psi''(z) - x_2 \text{ Im } \psi''(z) \qquad (19.14.36)$$

$$\sigma_{22} = \text{Re}[2\psi'(z)] + \text{Re } \chi''(z) + x_1 \text{ Re}\psi''(z) + x_2 \text{ Im } \psi''(z) \qquad (19.14.37)$$

$$\sigma_{12} = -x_2 \text{ Re}\psi''(z) + x_1 \text{ Im } \psi''(z) + \text{Im } \chi''(z) \qquad (19.14.38)$$

Now looking at Eq. (19.14.26) and the expressions for u_1 and u_2 we have

$$2Gu_1 = \kappa \text{ Re } \psi(z) - x_1 \text{ Re } \psi'(z) - x_2 \text{ Im } \psi'(z) - \text{Re } \chi'(z) \qquad (19.14.39)$$

$$2Gu_2 = \kappa \operatorname{Im} \psi(z) + x_1 \operatorname{Im} \psi'(z) - x_2 \operatorname{Re} \psi'(z) + \operatorname{Im} \chi'(z) \tag{19.14.40}$$

Eqs. (19.14.36) to (19.14.40) can now be easily specialized to cases where the stress components are symmetric or anti-symmetric relative to the x_1 axis.

2. *Boundary conditions*

Let us now look at the boundary conditions and think of a plate of unit thickness. (Fig. 19.22)

Recalling Sec. 7.11 and Fig. 10.10, we have

$$\sigma_{n1} = \sigma_{11}l_1 + \sigma_{12}l_2 = \sigma_{11}\cos\varepsilon - \sigma_{12}\sin\varepsilon$$
$$\sigma_{n2} = \sigma_{12}l_1 + \sigma_{22}l_2 = \sigma_{12}\cos\varepsilon - \sigma_{22}\sin\varepsilon$$

With these equations,

$$\sigma_{n1} = \frac{\partial^2\phi}{\partial x_2^2}\frac{dx_2}{ds} + \frac{\partial^2\phi}{\partial x_1 \partial x_2}\frac{dx_1}{ds} = \frac{d}{ds}\left(\frac{\partial\phi}{\partial x_2}\right) \tag{19.14.41}$$
$$\sigma_{n2} = -\frac{\partial^2\phi}{\partial x_1^2}\frac{dx_1}{ds} - \frac{\partial^2\phi}{\partial x_1 \partial x_2}\frac{dx_2}{ds} = -\frac{d}{ds}\left(\frac{\partial\phi}{\partial x_1}\right)$$

$$\sigma_{n1} + i\sigma_{n2} = -i\frac{d}{ds}\left(\frac{\partial\phi}{\partial x_1} + i\frac{\partial\phi}{\partial x_2}\right)$$

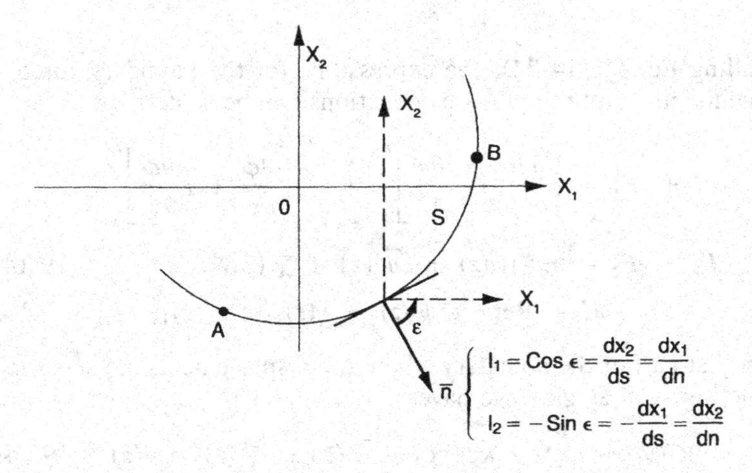

Fig. 19.22

Recalling Eq. (19.14.24) we get

$$\sigma_{n1} + i\sigma_{n2} = -i\frac{d}{ds}[\psi(z) + z\overline{\psi}'(\overline{z}) + \overline{\chi}'(\overline{z})]$$

and

$$\psi(z) + z\overline{\psi}'(\overline{z}) + \overline{\chi}'(\overline{z}) = \int_S (-\sigma_{n2} + i\sigma_{n1})\, ds \quad (19.14.42)$$

The components of the resultant force acting on AB are

$$F_1 = \int_A^B \sigma_{n1}\, ds = \int_A^B \frac{d}{ds}\left(\frac{\partial \phi}{dx_2}\right) ds = \left[\frac{\partial \phi}{\partial x_2}\right]_A^B$$

$$F_2 = \int_A^B \sigma_{n2}\, ds = -\int_A^B \frac{d}{ds}\left(\frac{\partial \phi}{\partial x_1}\right) ds = -\left[\frac{\partial \phi}{\partial x_1}\right]_A^B \quad (19.14.43)$$

The moment of the forces acting on AB about the origin is

$$M = \int_A^B (x_1\sigma_{n2} - x_2\sigma_{n1})\, ds = -\int_A^B \left[x_1\, d\left(\frac{\partial \phi}{\partial x_1}\right) + x_2\, d\left(\frac{\partial \phi}{\partial x_2}\right)\right]$$

Integration by part gives

$$M = -\left[x_1\frac{\partial \phi}{\partial x_1} + x_2\frac{\partial \phi}{\partial x_2}\right]_A^B + [\phi]_A^B$$

Recalling Eq. (19.14.24), the expressions for the boundary forces and moments given by the previous equations can be written as,

$$F_1 + iF_2 = \left[\frac{\partial \phi}{\partial x_2} - i\frac{\partial \phi}{\partial x_1}\right]_A^B = -i\left[\frac{\partial \phi}{\partial x_1} + i\frac{\partial \phi}{\partial x_2}\right]_A^B$$

$$F_1 + iF_2 = -i\,[\psi(z) + z\overline{\psi}'(\overline{z}) + \overline{\chi}'(\overline{z})]_A^B \quad (19.14.44)$$

$$M = \text{Re}[-z\overline{z}\psi'(z) + \chi(z) - z\chi'(z)]_A^B \quad (19.14.45)$$

On that part of the boundary where the displacements are specified by a function such as $g(z)$, we have

$$2G(u_1 + iu_2) = \kappa\psi(z) - z\overline{\psi}'(\overline{z}) - \overline{\chi}'(\overline{z}) = g(z) \quad (19.14.46)$$

3. *The structure of the functions $\psi(z)$ and $\chi(z)$ in simply connected regions.*

Let us ask the question: How arbitrary are the functions $\psi(z)$ and $\chi(z)$? Are they unique and single valued? The answer can be obtained by considering separately the cases where the stresses are given and those where the displacements are specified. $\chi(z)$ occurs in both stresses and displacements only through its first and second derivatives. On the other hand, $\psi(z)$ and its first derivative appears in the displacements while its second derivative appears in the stresses. It is convenient in the following discussion to set $\chi'(z) = \gamma(z)$. Let us first consider simply connected regions:

For a given stress distribution the function $\psi(z)$ is uniquely determined to within an arbitrary linear function of z, and the function $\gamma(z)$ is determined to within an arbitrary complex constant. To prove this statement, assume that there are two sets of functions (ψ, γ) and (ψ_0, γ_0) giving the same stress distribution. Since

$$\sigma_{11} + \sigma_{22} = 4 \operatorname{Re} \psi'(z) = 4 \operatorname{Re} \psi_0'(z) ,$$

then

$$\operatorname{Re} \psi'(z) = \operatorname{Re} \psi_0'(z) ,$$

so that

$$\psi_0'(z) = \psi'(z) + ic ,$$

and

$$\psi_0(z) = \psi(z) + icz + \alpha \tag{19.14.47}$$

which proves the first part of the statement. Now, from Eq. (19.14.30)

$$\sigma_{22} - \sigma_{11} + 2i\sigma_{12} = 2[\bar{z}\psi''(z) + \gamma'(z)] = 2[\bar{z}\psi_0''(z) + \gamma_0'(z)] .$$

Therefore,

$$\gamma'(z) = \gamma_0'(z)$$

and

$$\gamma_0(z) = \gamma(z) + \beta \tag{19.14.48}$$

which proves the second part of the statement. Note that c is a real constant, while α and β may be complex constant.

Looking now at the displacements, we can state that for a given set of displacements, the function $\psi(z)$ is determined to within an arbitrary constant and the function $\gamma(z)$ is then uniquely determined if this arbitrary constant is specified. Indeed, from Eq. (19.14.28)

$$\begin{aligned} 2G(u_1 + iu_2) &= \kappa\psi(z) - z\overline{\psi}'(\overline{z}) - \overline{\chi}'(\overline{z}) \\ &= \kappa\psi_0(z) - z\overline{\psi}_0'(\overline{z}) - \overline{\gamma}_0(\overline{z}) \\ &= \kappa[\psi(z) + icz + \alpha] - z[\overline{\psi}'(\overline{z}) - ic] - \overline{\gamma}(\overline{z}) - \overline{\beta} \end{aligned}$$

Therefore,

$$icz(\kappa + 1) + \kappa\alpha - \overline{\beta} = 0$$

Since z is arbitrary,

$$c = 0, \quad \kappa\alpha = \overline{\beta} \qquad (19.14.49)$$

which proves the statement.

If the origin of coordinates is taken within the simply connected region the functions $\psi(z)$ and $\gamma(z)$ will be determined uniquely for a given state of stress if c, α and β are chosen, so that

$$\psi(0) = 0, \quad \text{Im}\, \psi'(0) = 0, \quad \gamma(0) = 0$$

For a given state of displacement c is necessarily zero and α can be chosen so that $\psi(0) = 0$. This choice fixes the value of β. It is to be remembered that $\psi(z)$ and $\gamma(z) = \chi'(z)$ are single-valued analytic functions of z, i.e., holomorphic functions, and as such can be represented by the power series

$$\psi(z) = \sum_{n=0}^{\infty} a_n z^n, \text{ and } \gamma(z) = \sum_{n=0}^{\infty} b_n z^n$$

in a simply connected domain

4. *The structure of the functions $\psi(z)$ and $\chi(z)$ in finite, multiply connected regions.*

If a region S is finite but not simply connected, ψ and γ need not be single valued but their structure can be determined if both stresses and displacements are assumed to be single valued [2]. Indeed, from Eqs. (19.14.29) and (19.14.3) to (19.14.5),

$$\sigma_{11} + \sigma_{22} = 4\, \text{Re}\, \psi'(z) = \text{Re}\, F(z) = \text{Re}(P_1 + iP_2)\,,$$

and P_1 is single valued. However, this does not mean that its imaginary part P_2 is single valued. Indeed, for one circuit like L_K' surrounding one of the contours L_k (Fig. 19.23), P_2 may acquire a constant increment. Denoting this increment by $8\pi A_k$, the function $\psi'(z)$ acquires an increment $2\pi i A_k$, where A_k is a real constant. Now consider the function $A_k \ln(z - z_k)$ where z_k is a point inside the contour L_k. This function acquires an increment $2\pi i A_k$ in going once around this contour (see Eq. 19.7.9). Hence, we can write

$$\Phi(z) = \psi'(z) - \sum_{k=1}^{m} A_k \ln(z - z_k) \tag{19.14.50}$$

and $\Phi(z)$ is a single valued function, analytic in S and which returns to its initial value for a circuit around any closed contour.

Integrating, we get

$$\psi(z) = \sum_{k=1}^{m} A_k[(z - z_k) \ln(z - z_k) - (z - z_k)] + \int_{z_0}^{z} \Phi(z)\, dz + \text{const.} \tag{19.14.51}$$

where z_0 is an arbitrary fixed point in S. Now $\int_{z_0}^{z} \Phi(z)dz$ is a function of a complex variable which may undergo an increase $2\pi i c_k$ for a circuit around one of the contours L_k. Thus,

$$\int_{z_0}^{z} \Phi(z)\, dz = \sum_{k=1}^{m} c_k \ln(z - z_k) + \Omega(z) \tag{19.14.52}$$

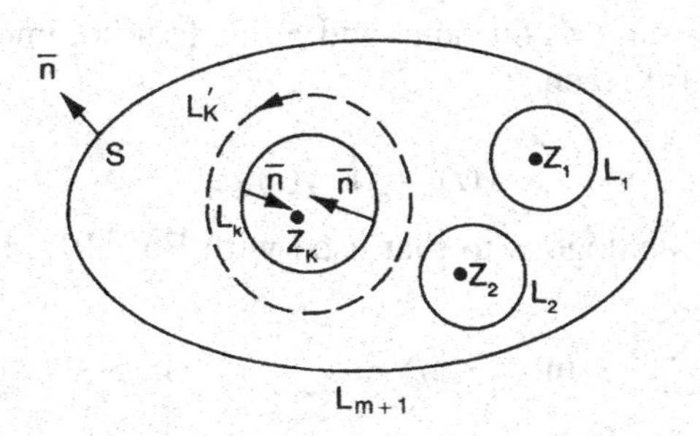

Fig. 19.23

where $\Omega(z)$ is analytic and single valued (holomorphic) in S. Note the c_k maybe complex since both the real and imaginary parts of the integral may be multiple valued. Thus,

$$\psi(z) = z\sum_{k=1}^{m} A_k \ln(z - z_k) + \sum_{k=1}^{m} (c_k - A_k z_k) \ln(z - z_k) - \sum_{k=1}^{m} A_k(z - z_k) + \Omega(z) \qquad (19.14.53)$$

$$= z\sum_{k=1}^{m} A_k \ln(z - z_k) + \sum_{k=1}^{m} \varepsilon_k \ln(z - z_k) + \phi^*(z) \qquad (19.14.54)$$

where

$$\varepsilon_k = c_k - A_k z_k \qquad (19.14.55)$$

and

$$\phi^*(z) = \Omega - \sum_{k=1}^{m} A_k(z - z_k) \qquad (19.14.56)$$

$\phi^*(z)$ is holomorphic in S and ε_k are general complex constants with A_k being real. Now, from Eq. (19.14.30), $\gamma'(z)$ must be single valued since $\psi''(z)$ is single valued. Its integral however can be mutiple valued and we can write

$$\gamma(z) = \sum_{k=1}^{m} \epsilon_k^* \ln(z - z_k) + \gamma^*(z) \qquad 19.14.57)$$

where $\varepsilon^* k$ are complex constants and $\gamma^*(z)$ is a holomorphic function. Since $\chi'(z) = \gamma(z)$, then

$$\chi(z) = \int \gamma(z)\, dz$$

and in a manner analogous to that used with Eq. (19.14.51) we get,

$$\chi(z) = z\sum_{k=1}^{m} \varepsilon_k^* \ln(z - z_k) + \sum_{k=1}^{m} \varepsilon_k^{**} \ln(z - z_k) + \chi^*(z) \qquad (19.14.58)$$

where ε_k^{**} are generally complex constants and $\chi^*(z)$ is holomorphic.

Let us now make use of the single valuedness of the displacements. From Eq. (19.14.28) we have

$$2G(u_1 + iu_2) = \kappa\psi(z) - z\overline{\psi}'(\overline{z}) - \overline{\gamma}(\overline{z})$$

$$= \kappa\left[z \sum_{k=1}^{m} A_k \ln(z - z_k) + \sum_{k=1}^{m} \varepsilon_k \ln(z - z_k) + \phi^*(z)\right]$$

$$- z\left[\overline{\Phi}(\overline{z}) + \sum_{k=1}^{m} A_k \ln(\overline{z} - \overline{z}_k)\right]$$

$$- \sum_{k=1}^{m} \overline{\varepsilon}_k^* \ln(\overline{z} - \overline{z}_k) + \overline{\gamma}^*(\overline{z}) \qquad (19.14.59)$$

The increment that is gained by the displacements in describing a contour L_k' (Fig. 19.23) is zero. Thus,

$$\kappa(z\, 2\pi i A_k + 2\pi i \varepsilon_k) + z\, 2\pi i A_k + 2\pi i \overline{\varepsilon}_k^* = 0$$

$$2\pi i[(\kappa + 1)A_k z + \kappa\varepsilon_k + \overline{\varepsilon}_k^*] = 0$$

or

$$A_k = 0, \text{ and } \kappa\varepsilon_k + \overline{\varepsilon}_k^* = 0 \text{ for } k = 1, 2 \ldots m. \qquad (19.14.60)$$

which is similar to Eq. (19.14.49).

The quantities ε_k and ε_k^* can be expressed in terms of F_{1k} and F_{2k}, the components of the resultant force acting on the contour $L_k (k = 1, 2, \ldots m)$. From Eq. (19.14.44)

$$F_{1k} + iF_{2k} = -i[\psi(z) + z\overline{\psi}'(\overline{z}) + \overline{\chi}'(\overline{z})]_{L_k} \qquad (19.14.61)$$

Notice that this formula refers to Fig. 19.22 where the normal $\overline{n}$ points out, with the direction of integration in the counterclockwise direction. In the present case (Fig. 19.23) the normal $\overline{n}$ is such that the contour must be traversed in the clockwise direction. Hence, we can reverse the sign in the right hand side of Eq. (19.14.61) and integrate in the counterclockwise direction. Therefore,

$$F_{1k} + iF_{2k} = i[\psi(z) + z\overline{\psi}'(\overline{z}) + \overline{\chi}'(\overline{z})]_{L_k} \qquad (19.14.62)$$

Now

$$\psi(z)]_{L_k} = 2\pi i \varepsilon_k$$
$$z\overline{\psi}'(\overline{z})]_{L_k} = 0 \quad \text{since} \quad A_k = 0$$
$$\overline{\chi}'(\overline{z})]_{L_k} = -2\pi i \overline{\varepsilon}_k^*$$

Thus,

$$F_{1k} + iF_{2k} = -2\pi(\varepsilon_k - \overline{\varepsilon}_k^*) \tag{19.14.63}$$

Using Eq. (19.14.60) we get

$$\varepsilon_k = -\frac{F_{1k} + iF_{2k}}{2\pi(1+\kappa)}, \qquad \overline{\varepsilon}_k^* = \frac{\kappa(F_{1k} + iF_{2k})}{2\pi(1+\kappa)} \tag{19.14.64}$$

This leads to the following expressions for $\psi(z)$ and $\gamma(z)$:

$$\psi(z) = -\frac{1}{2\pi(1+\kappa)} \sum_{k=1}^{m} (F_{1k} + iF_{2k}) \ln(z - z_k) + \phi^*(z) \tag{19.14.65}$$

$$\gamma(z) = -\frac{\kappa}{2\pi(1+\kappa)} \sum_{k=1}^{m} (F_{1k} - iF_{2k}) \ln(z - z_k) + \gamma^*(z) \tag{19.14.66}$$

5. *The structure of the functions* $\psi(z)$ *and* $\chi(z)$ *in infinite multiply connected regions.*

Let us now consider the case when the region S consists of the entire Z plane from which finite parts bounded by simple contours have been removed. This is the case of an infinite plate with holes for example. In Fig. 19.23 this corresponds to the contour L_{m+1} moved to infinity. The formulas established in the previous subsection hold, but we must now consider the behavior of $\psi(z)$ and $\gamma(z)$ in the neighborhood of the point at infinity in the Z plane.

Draw about the origin as center a circle L_R of radius R sufficiently large so that all the contours L_k lie inside L_R. Let z be a point outside L_R and at a sufficiently large distance, so that,

$$|z| > |z_k|$$

Therefore,

$$\ln(z - z_k) = \ln z + \ln\left(1 - \frac{z_k}{z}\right) = \ln z - \left[\frac{z_k}{z} + \frac{1}{2}\left(\frac{z_k}{z}\right)^2 + \ldots\right]$$
$$= \ln z + \text{a function holomorphic outside } L_R.$$

From Eqs. (19.14.65) and (19.14.66)

$$\psi(z) = -\frac{1}{2\pi(1+\kappa)}\sum_{k=1}^{m}(F_{1k} + iF_{2k})\ln z + \phi^{**}(z) \qquad (19.14.67)$$

$$\gamma(z) = \frac{K}{2\pi(1+\kappa)}\sum_{k=1}^{m}(F_{1k} - iF_{2k})\ln z + \gamma^{**}(z) \qquad (19.14.68)$$

$\gamma^{**}(z)$ and $\phi^{**}(z)$ are functions holomorphic outside L_R with the possible exception of the point at infinity, since $\gamma^*(z)$ and $\phi^*(z)$ may not be analytic at infinity.

By the theorem of Laurent, the two functions $\phi^{**}(z)$ and $\gamma^{**}(z)$ can be represented outside L_R by the series

$$\phi^{**}(z) = \sum_{-\infty}^{+\infty} a_n z^n, \qquad \gamma^{**}(z) = \sum_{-\infty}^{+\infty} a'_n z^n$$

which will uniformly converge for every finite region outside L_R. In Sec. 19.6 it was shown that expansions in Laurent series referred to an annulus. Here the inner ring may be shrunk to a point and the outer one made infinitely large.

To go beyond this stage, one has to introduce additional conditions related to the distribution of stresses in the neighborhood of the point at infinity of the plane. Let us assume that the stresses are bounded throughout the region S. Then, from Eq. (19.14.29),

$$\sigma_{11} + \sigma_{22} = 2\left[-\frac{\sum_{k=1}^{m} F_{1k} + iF_{2k}}{2\pi(1+\kappa)z} - \frac{\sum_{k=1}^{m} F_{1k} - iF_{2k}}{2\pi(1+\kappa)\bar{z}} + \sum_{n=-\infty}^{+\infty} n(a_n z^{n-1} + \bar{a}_n(\bar{z})^{n-1})\right] \qquad (19.14.69)$$

The terms which grow beyond bound with $|z|$ are

$$\sum_{n=2}^{\infty} n(a_n z^{n-1} + \bar{a}_n(\bar{z})^{n-1}) = \sum_{n=2}^{\infty} nr^{n-1}(a_n e^{(n-1)i\theta} + \bar{a}_n e^{-(n-1)i\theta})$$

If $\sigma_{11} + \sigma_{22}$ is to remain finite as $r \to \infty$, then

$$a_n = \bar{a}_n = 0 \quad \text{for} \quad n \geqslant 2$$

Therefore,

$$\psi(z) = -\frac{\sum_{k=1}^{m} (F_{1k} + iF_{2k})}{2\pi(1 + \kappa)} \ln z + a_1 z + \sum_{n=0}^{\infty} a_n z^{-n} \quad (19.14.70)$$

From Eq. (19.14.30)

$$\sigma_{22} - \sigma_{11} + 2i\sigma_{12} = 2\left[\bar{z} \frac{\sum_{k=1}^{m} (F_{1k} + iF_{2k})}{2\pi(1 + \kappa)} \frac{1}{z^2} \right.$$

$$\left. + \bar{z} \sum_{-\infty}^{-1} n(n - 1)a_n z^{n-2} + \kappa \frac{\sum_{k=1}^{m} (F_{1k} - iF_{2k})}{2\pi(1 + \kappa)} \frac{1}{z} + \sum_{-\infty}^{+\infty} n\, a_n' z^{n-1} \right]$$

(19.14.71)

For the right hand side to remain finite, we must have

$$\sum_{-\infty}^{+\infty} n\, a_n' z^{n-1} = \sum_{-\infty}^{+\infty} n\, a_n' r^{n-1} e^{i(n-1)\theta} \text{ be bounded.}$$

This means that

$$a_n' = 0 \quad \text{for} \quad n \geqslant 2$$

Therefore,

$$\gamma(z) = \frac{\kappa \sum_{k=1}^{m} (F_{1k} - iF_{2k})}{2\pi(1 + \kappa)} \ln z + a_1' z + \sum_{0}^{\infty} a_n' z^{-n} \quad (19.14.72)$$

Let us set in Eqs. (19.14.70) and (19.14.72),

$$a_1 = B + iC\,, \qquad a_1' = B' + iC'\,, \qquad (19.14.73)$$

$$\psi_0(z) = \sum_{n=0}^{\infty} a_n z^{-n}\,, \qquad \gamma_0(z) = \sum_{n=0}^{\infty} a_n' z^{-n} \qquad (19.14.74)$$

Thus, $\psi_0(z)$ and $\gamma_0(z)$ are both holomorphic outside L_R including the point at infinity. Since it was shown that the stresses are not altered if $\psi(z)$ and $\gamma(z)$ are changed by the addition of a constant, we can assume that $a_0 = a_0' = 0$ without affecting them. This gives

$$\psi_0(\infty) = \gamma_0(\infty) = 0$$

Since it was also shown that $\psi(z)$ is arbitrary to the extent of a term icz (see Eq. (19.14.47), where c is real and that $c = 0$ (see Eq. 19.14. 49) for specified displacements, the functions $\psi(z)$ and $\gamma(z)$ can now be written as

$$\psi(z) = -\frac{\sum_{k=1}^{m} (F_{1k} + iF_{2k})}{2\pi(1 + \kappa)} \ln z + (B + iC)z + \psi_0(z) \qquad (19.14.75)$$

$$\gamma(z) = \frac{\kappa \sum_{k=1}^{m} (F_{1k} - iF_{2k})}{2\pi(1 + \kappa)} \ln z + (B' + iC')z + \gamma_0(z) \qquad (19.14.76)$$

with the summations in Eqs. (19.14.74) taken between $n = 1$ and $n = \infty$

Now $\quad \lim_{z\to\infty} (\sigma_{11} + \sigma_{22}) = \lim_{z\to\infty} 2[\psi'(z) + \bar{\psi}'(\bar{z})] = 4B$

and

$$\lim_{z\to\infty} (\sigma_{22} - \sigma_{11} + 2i\sigma_{12}) = \lim_{z\to\infty} 2[\bar{z}\psi''(z) + \gamma'(z)] = 2(B' + iC')$$

Therefore,

$$\sigma_{11}(\infty) + \sigma_{22}(\infty) = 4B$$
$$-\sigma_{11}(\infty) + \sigma_{22}(\infty) = 2B'$$

so that the constants are given by

$$\left.\begin{aligned} B &= \frac{\sigma_{11}(\infty) + \sigma_{22}(\infty)}{4} \\ B' &= \frac{\sigma_{22}(\infty) - \sigma_{11}(\infty)}{2} \\ C' &= \sigma_{12}(\infty) \end{aligned}\right\} \qquad (19.14.77)$$

Thus, in the neighborhood of the point at infinity the stresses are uniformly distributed.

Let us now look at the conditions for the displacements to remain bounded for large values of $|z|$. Taking the limits in the expression of the displacements:

$$\lim_{z\to\infty} 2G(u_1 + iu_2) = \lim_{z\to\infty} \kappa\left[-\frac{\sum_{k=1}^{m}(F_{1k} + iF_{2k})}{2\pi(1+\kappa)}\ln z + Bz + \psi_0(z)\right]$$

$$- z\left[-\frac{\sum_{k=1}^{m}(F_{1k} - iF_{2k})}{2\pi(1+\kappa)}\frac{1}{\bar{z}} + B + \bar{\psi}_0'(\bar{z})\right]$$

$$-\left[\frac{\kappa\sum_{k=1}^{m}(F_{1k} + iF_{2k})}{2\pi(1+\kappa)}\ln\bar{z} + (B' - iC')\bar{z} + \bar{\gamma}_0(\bar{z})\right]$$

Now, $\psi_0(\infty) = \gamma_0(\infty) = 0$, and

$$2G(u_1 + iu_2)]_{z=\infty} = -\frac{\kappa\sum_{k=1}^{m}(F_{1k} + iF_{2k})}{2\pi(1+\kappa)}\ln(z\bar{z})$$

$$+ (\kappa - 1)Bz - (B' - iC')\bar{z} + \frac{\sum_{k=1}^{m}(F_{1k} - iF_{2k})}{2\pi(1+\kappa)}\frac{z}{\bar{z}} \qquad (19.14.78)$$

which shows that for $(u_1 + iu_2)$ to be bounded

$$\sum_{k=1}^{m} F_{1k} = \sum_{k=1}^{m} F_{2k} = 0 \quad \text{and} \quad B = B' = C' = 0\,, \qquad (19.14.79)$$

i.e. the stresses at infinity are zero.

Finally, the constant C which does not affect the stresses can be related to the rotation of an infinitely remote part of the plane. Using Eq. (19.14.27) it can directly be shown that [2],

$$C = \frac{2G\omega_{21}(\infty)}{1+\kappa} \qquad (19.14.80)$$

In the analysis of stress C can always be set equal to zero.

6. *The first and second boundary value problems in plane elasticity.* Problems I and II of the general theory of elasticity were defined in Sec. 8.8. The situation is the same here, but in addition we shall make the following assumptions:

a) For the first boundary value problem for an infinite region, the stresses are finite and specified at infinity according to Eq. (19.14.77) and $C = 0$.

b) For the second boundary value problem for an infinite region it will be assumed that the quantities B, C, B', C', $\sum_{k=1}^{m} F_{1k}$, $\sum_{k=1}^{m} F_{2k}$ are given; i.e., that not only the value of the stresses as well as of the rotation at infinity are given, but also the resultant vectors of all external forces applied to the boundary. Without the latter condition it can be shown [2] that the solution is not unique.

The following theorems can readily be proven [2]:

a) The state of stress for the first boundary value problem in a simply connected body depends only on its shape and not on the material (i.e., it is independent of the elastic constants).

b) For a multiply connected region the state of stress does not depend on the material's properties, if and only if, the resultant vectors of the external forces applied to each of the contours separately are zero.

Another important theorem states that, the first and second boundary value problems of plane elasticity reduce to finding two analytic functions $\psi(z)$ and $\gamma(z)$ which on the boundary of the region S satisfy the condition

$$\begin{aligned} \psi(z) + z\overline{\psi}'(\overline{z}) + \overline{\gamma}(\overline{z}) &= \int_0^{s_j} (-\sigma_{n2} + i\sigma_{n1})\, ds + \text{const.} \\ &= f_{1j}(s) + if_{2j}(s) + \text{const.} \qquad (19.14.81) \end{aligned}$$

for the first boundary value problem, and

$$\begin{aligned} -\kappa\psi(z) + z\overline{\psi}'(\overline{z}) + \overline{\gamma}(\overline{z}) &= -2G(g_{1j} + ig_{2j}) \qquad (19.14.82) \\ &= -2G(u_1(s_j) + u_2(s_j)] \end{aligned}$$

for the second boundary value problem. The subscript j refers to the jth contour and s_j is the arc length along the jth contour. j varies from 1 to m. The proof follows directly from Eqs. (19.14.42) and (19.14.46). The constant in Eq. (19.14.81) can be related to the arbitrary constants of Eqs. (19.14.47) and (19.14.48). Thus, for the first boundary value problem, since $\psi(z)$ is arbitrary to the extent of $icz + \alpha$ and $\gamma(z)$ is arbitrary to the extent of β, we can write

$$\begin{aligned}\psi(z) + z\overline{\psi}'(\overline{z}) + \overline{\gamma}(\overline{z}) &= \psi(z) + icz + \alpha + z\overline{\psi}'(\overline{z}) \\ &\quad - icz + \overline{\gamma}(\overline{z}) + \overline{\beta} \\ &= \psi(z) + z\overline{\psi}'(\overline{z}) + \overline{\gamma}(\overline{z}) \\ &\quad + \alpha + \overline{\beta}\end{aligned} \tag{19.14.83}$$

The left hand side of Eq. (19.14.81) can be replaced by $\psi(z) + z\overline{\psi}'(\overline{z}) + \overline{\gamma}(\overline{z}) + \alpha + \overline{\beta}$. By suitably choosing the value of $\alpha + \overline{\beta}$, we automatically fix the constant in Eq. (19.14.81); if we fix the constant we can only fix one of α or β. It is customary to fix the constants α, β and c by assuming that

$$\psi(0) = 0, \quad \gamma(0) = 0 \quad \text{and} \quad \text{Im}\, \psi'(0) = 0 \tag{19.14.84}$$

or

$$\psi(0) = 0, \quad \text{and the constant in Eq. (19.14.81)} = 0 \tag{19.14.85}$$

For the second boundary value problem, as stated in subsection (19.14.3), α can be chosen so that $\psi(0) = 0$. c is necessarily equal to zero.

For a multiply connected region, the constant may be fixed for only one contour. On the remaining contours these constants are determined from the conditions that the displacements and stresses be single valued.

In conclusion, if we find a pair of analytic functions $\psi(z)$ and $\gamma(z)$ which satisfy the above conditions on the boundary, we have a solution to the plane problem, which, by the uniqueness theorem is unique [2].

7. *Displacements and stresses in curvilinear coordinates.*

The expressions for the displacements and the stresses can easily be written in terms of *cylindrical (or polar) coordinates* as follows (Fig. 19.24a):

$$\begin{aligned}u_1 + iu_2 &= (u_r \cos\theta - u_\theta \sin\theta) + i(u_r \sin\theta + u_\theta \cos\theta) \\ &= (u_r + iu_\theta)e^{i\theta}\end{aligned}$$

Thus,

$$u_r + iu_\theta = e^{-i\theta}(u_1 + iu_2) \tag{19.14.86}$$

or

$$2G(u_r + iu_\theta) = e^{-i\theta}\left[\frac{3 - \nu}{1 + \nu}\psi(z) - z\overline{\psi}'(\overline{z}) - \overline{\chi}'(\overline{z})\right] \tag{19.14.87}$$

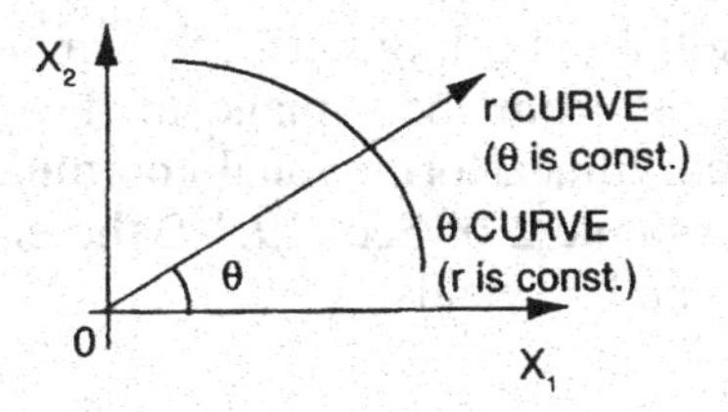

Fig. 19.24a

for plane stress problems. Similarly, the expressions for the stresses can be obtained from a rotation of axes (see Sec. 7.11). Recalling that

$$\sigma_{r\theta} = (\sigma_{22} - \sigma_{11})\sin\theta\cos\theta + \sigma_{12}(\cos^2\theta - \sin^2\theta)$$

we have

$$\sigma_{rr} + \sigma_{\theta\theta} = \sigma_{11} + \sigma_{22}$$

$$\sigma_{\theta\theta} - \sigma_{rr} + 2i\sigma_{r\theta} = (\sigma_{22} - \sigma_{11} + 2i\sigma_{12})e^{2i\theta}$$

$$\sigma_{rr} - i\sigma_{r\theta} = \frac{1}{2}(\sigma_{11} + \sigma_{22}) - \frac{1}{2}(\sigma_{22} - \sigma_{11} + 2i\sigma_{12})e^{2i\theta}$$

Thus,

$$\sigma_{rr} + \sigma_{\theta\theta} = 2\psi'(z) + 2\overline{\psi}'(\overline{z}) = 4\,\text{Re}\,\psi'(z) \tag{19.14.88}$$

$$\sigma_{\theta\theta} - \sigma_{rr} + 2i\sigma_{r\theta} = 2[\overline{z}\psi''(z) + \chi''(z)]e^{2i\theta} \tag{19.14.89}$$

$$\sigma_{rr} - i\sigma_{r\theta} = \psi'(z) + \overline{\psi}'(\overline{z}) - [\overline{z}\psi''(z) + \chi''(z)]e^{2i\theta} \tag{19.14.90}$$

Similar equations can be written in terms of *elliptical coordinates* (see problem 1, Chapter 6). The relation between the system of axes x_1, x_2 and the curvilinear system y_1, y_2 are given by

$$x_1 = a\cosh y_1 \cos y_2\,, \qquad x_2 = a\sinh y_1 \sin y_2$$

The notation used in this section suggests using η_1 and η_2 in place of y_1 and y_2 and c in place of a so that

$$x_1 = c\cosh\eta_1 \cos\eta_2\,, \qquad x_2 = a\sinh\eta_1 \sin\eta_2$$

In terms of complex variables these relations can be written

$$z = c\cosh\zeta, \quad \text{with} \quad \zeta = \eta_1 + i\eta_2$$

Notice that we are not here dealing with conformal mapping and we are not transforming points from the z plane to the ζ plane; just using a different system of curvilinear orthogonal coordinates $\zeta\,(\eta_1, \eta_2)$. However, as we did in Example 2 of Sec. 19.10, the elimination of η_2 from the expressions of x_1 and x_2 yields

$$\frac{x_1^2}{c^2 \cosh^2 \eta_1} + \frac{x_2^2}{c^2 \sinh^2 \eta_1} = 1$$

This is the equation of a family of confocal ellipses each corresponding to a value of η_1. Similarly, the elimination of η_1 yield the equation

$$\frac{x_1^2}{c^2 \cos^2 \eta_2} - \frac{x_2^2}{c^2 \sin^2 \eta_2} = 1$$

This is the equation of a family of confocal hyperbolas which intersect the previous ellipses at right angles and with which they form a system of orthogonal curvilinear coordinates. As was done in Chapter 6, the curves along which η_1 varies are called η_1 curves and those along which η_2 varies are called η_2 curves. Thus, the ellipses are η_2 curves and the hyperbolas are η_1 curves (Fig. 19.24b). Eqs. (19.14.86) to (19.14.90) can be repeated here with η_1 replacing r and η_2 replacing θ. This system of coordinate is useful in solving problems related to elliptical holes in plates.

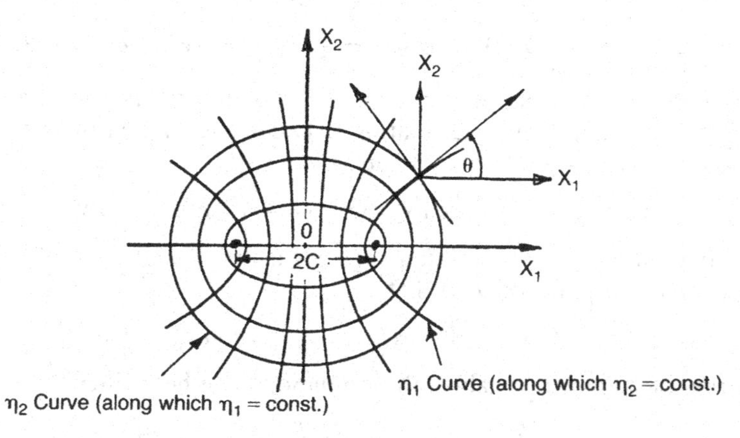

Fig. 19.24b

$$(u_{\eta 1} + iu_{\eta 2}) = (u_1 + iu_2)e^{-i\theta} \tag{19.14.91}$$

$$\sigma_{\eta 1} + \sigma_{\eta 2} = 4 \operatorname{Re} \psi'(z) \tag{19.14.92}$$

$$\sigma_{\eta 2} - \sigma_{\eta 1} + 2i\sigma_{\eta 12} = 2[\bar{z}\psi''(z) + \chi''(z)]e^{2i\theta} \tag{19.14.93}$$

$$\sigma_{\eta 1} - i\sigma_{\eta 12} = \psi'(z) + \bar{\psi}'(\bar{z}) - [\bar{z}\psi''(z) + \chi''(z)]e^{2i\theta} \tag{19.14.94}$$

8. *Conformal mapping for plane problems.*

The expressions for the displacements, stresses and boundary conditions can be written using conformal mapping with the aid of the mapping function $z = f(\zeta)$. In what follows we will assume that the region of interest R is simply connected and map it conformally on the unit circle $|\zeta| \leq 1$ by the analytic function $z = f(\zeta)$. $f'(\zeta)$ is not equal to zero anywhere in the region and the boundary L of R has a continuously changing curvature. Now, let the region R be finite and the origin $z = 0$ taken in the interior of the region. By making the point $z = 0$ correspond to $\zeta = 0$ we can represent the mapping function by the power series

$$z = f(\zeta) = \sum_{n=1}^{\infty} k_n \zeta^n \qquad |\zeta| \leq 1 \tag{19.14.95}$$

In the ζ plane we will use polar coordinates ρ and α so that $\zeta = \rho e^{i\alpha}$. In Figs. 19.25 the circles $\rho =$ const. and the radii $\alpha =$ const. in the ζ plane, correspond to curves in the z plane denoted also by $\rho =$ const. and $\alpha =$ const. (see Section 19.10). The quantities ρ and α may be

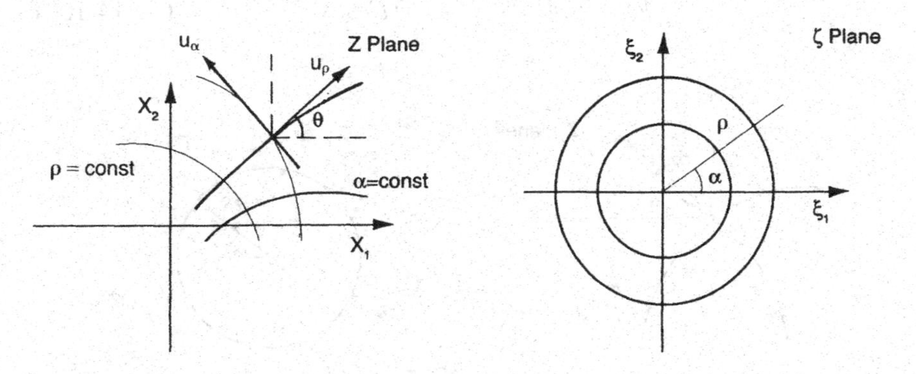

Fig. 19.25

considered as curvilinear coordinates of the point (x_1, x_2) of the z plane. They are related to x_1, x_2 by the equation

$$z = x_1 + ix_2 = f\zeta) = f(\rho e^{i\alpha})$$

Now let the point z move a distance dz in the α = constant direction (Fig. 19.26). The corresponding point ζ undergoes a displacement $d\rho$ in the radial direction. Thus, $dz = e^{i\theta}|dz|$ and $d\zeta = e^{i\alpha}|d\zeta|$. Since $dz/d\zeta = f'(\zeta)$, then $dz = f'(\zeta)\, d\zeta$ and,

$$e^{i\theta} = \frac{f'(\zeta)\, d\zeta}{|f'(\zeta)|\, |d\zeta|} = \frac{f'(\zeta)e^{i\alpha}}{|f'(\zeta)|} = \frac{\zeta f'(\zeta)}{\rho|f'(\zeta)|} \tag{19.14.96}$$

Therefore,

$$e^{-i\theta} = \frac{\overline{\zeta}\,\overline{f}'(\overline{\zeta})}{\rho|f'(\zeta)|} = \frac{\overline{f}'(\overline{\zeta})}{|f'(\zeta)|}\, e^{-i\alpha} \tag{19.14.97}$$

and

$$e^{2i\theta} = \frac{\zeta^2 f'(\zeta)}{\rho^2 \overline{f}'(\overline{\zeta})} \tag{19.14.98}$$

Thus, the components u_ρ, u_α of the displacement vector in the z plane are related to the cartesian components u_1 and u_2 by

$$u_\rho + iu_\alpha = e^{-i\theta}(u_1 + iu_2) = \frac{\overline{f}'(\overline{\zeta})}{|f'(\zeta)|}\, e^{-i\alpha}(u_1 + iu_2) \tag{19.14.99}$$

Remembering that $\chi'(z) = \gamma(z)$, we have,

$$\psi(z) = \psi[f(\zeta)] = \psi_1(\zeta) \tag{19.14.100}$$

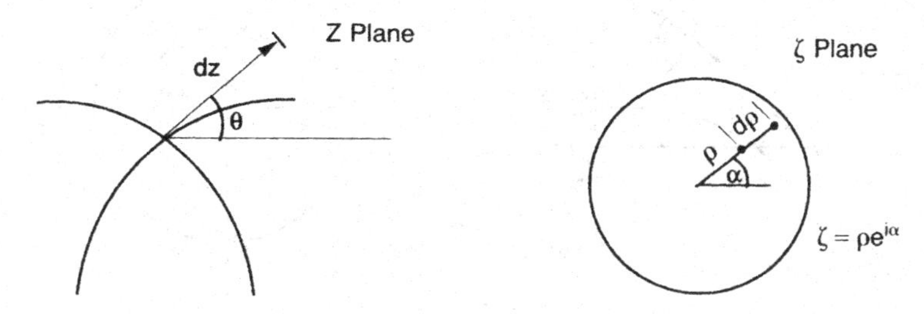

Fig. 19.26

$$\psi'(z) = \frac{d\psi}{dz} = \frac{d\psi_1}{d\zeta}\frac{d\zeta}{dz} = \frac{\psi_1'(\zeta)}{f'(\zeta)} \tag{19.14.101}$$

$$\begin{aligned}\psi''(z) &= \frac{d}{d\zeta}\left[\frac{\psi_1'(\zeta)}{f'(\zeta)}\right]\frac{d\zeta}{dz} \\ &= \frac{\psi_1''(\zeta)f'(\zeta) - \psi_1'(\zeta)f''(\zeta)}{[f'(\zeta)]^3}\end{aligned} \tag{19.14.102}$$

$$\chi(z) = \chi[f(\zeta)] = \chi_1(\zeta)$$

$$\chi'(z) = \chi'[f(\zeta)] = \gamma_1(\zeta) = \frac{\chi_1'(\zeta)}{f'(\zeta)} \tag{19.14.103}$$

$$\begin{aligned}\chi''(z) = \gamma'(z) &= \frac{d\gamma_1}{d\zeta}\frac{d\zeta}{dz} = \frac{\gamma_1'(\zeta)}{f'(\zeta)} \\ &= \frac{\chi_1''(\zeta)f'(\zeta) - \chi_1'(\zeta)f''(\zeta)}{[f'(\zeta)]^3}\end{aligned} \tag{19.14.104}$$

Inserting the equations above in Eqs. (19.14.28), (19.14.29), (19.14.30), (19.14.44) and (19.14.45) we get,

$$2G(u_1 + iu_2) = \kappa\psi_1(\zeta) - \frac{f(\zeta)\overline{\psi}_1'(\overline{\zeta})}{\overline{f}'(\overline{\zeta})} - \overline{\gamma}_1(\overline{\zeta}) \tag{19.14.105}$$

$$\begin{aligned}2G(u_1 + iu_2)e^{-i\theta} &= 2G(u_\rho + iu_\alpha) \\ &= \frac{\overline{f}'(\overline{\zeta})}{|f'(\zeta)|}e^{-i\alpha}\left[\kappa\psi_1(\zeta) - \frac{f(\zeta)\overline{\psi}_1'(\overline{\zeta})}{\overline{f}'(\overline{\zeta})} - \overline{\gamma}_1(\overline{\zeta})\right]\end{aligned} \tag{19.14.106}$$

$$\sigma_{11} + \sigma_{22} = 4\,\text{Re}\,\frac{\psi_1'(\zeta)}{f'(\zeta)} = 2\left[\frac{\psi_1'(\zeta)}{f'(\zeta)} + \frac{\overline{\psi}_1'(\overline{\zeta})}{\overline{f}'(\overline{\zeta})}\right] \tag{19.14.107}$$

$$\begin{aligned}&\sigma_{22} - \sigma_{11} + 2i\sigma_{12} = \\ &2\left[\frac{\overline{f}(\overline{\zeta})\psi_1''(\zeta)f'(\zeta) - \overline{f}(\overline{\zeta})\psi_1'(\zeta)f''(\zeta)}{[f'(\zeta)]^3} + \frac{\gamma_1'(\zeta)}{f'(\zeta)}\right]\end{aligned} \tag{19.14.108}$$

$$F_1 + iF_2 = -i\left[\psi_1(\zeta) + \frac{f(\zeta)\overline{\psi}_1'(\overline{\zeta})}{\overline{f}'(\overline{\zeta})} + \overline{\gamma}_1(\overline{\zeta})\right]_A^B \tag{19.14.109}$$

$$M = \text{Re}\left[-f(\zeta)\overline{f}(\overline{\zeta})\frac{\psi_1'(\zeta)}{f'(\zeta)} + \chi_1(\zeta) - f(\zeta)\gamma_1(\zeta)\right] \tag{19.14.110}$$

Eqs. (19.14.81) and (19.14.82) can be written

$$\psi_1(\zeta) + \frac{f(\zeta)}{\bar{f}'(\bar{\zeta})}\bar{\psi}_1'(\bar{\zeta}) + \bar{\gamma}_1(\bar{\zeta}) = F(\delta) \qquad (19.14.111)$$

$$\kappa\psi_1(\zeta) - \frac{f(\zeta)}{\bar{f}'(\bar{\zeta})}\bar{\psi}_1'(\bar{\zeta}) - \bar{\gamma}_1(\bar{\zeta}) = G(\delta) \qquad (19.14.112)$$

where $F(\delta)$ and $G(\delta)$ are known functions determined from the known values, $f_{1j} + if_{2j}$ + const. and $2G(g_{1j} + ig_{2j})$ specified on the contour L_j. If the boundary in the Z plane is mapped onto a unit circle, then (see Sec. 19.11) $\zeta = \sigma = e^{i\alpha}$.

If in Eqs. (19.14.29), (19.14.30), (19.14.107), and (19.14.108) the stresses are to remain finite at infinity then the functions ψ', χ'' and ψ_1', γ_1' (also χ_1'') must also remain finite. If they are to be represented by a power series they must have the form

$$\psi'(z) = \sum_{n=0}^{\infty} A_n z^{-n} \qquad \chi''(z) = \sum_{n=0}^{\infty} B_n z^{-n} \qquad (19.14.113)$$

$$\psi_1'(\zeta) = \sum_{n=0}^{\infty} C_n \zeta^{-n} \qquad \gamma_1'(\zeta) = \sum_{n=0}^{\infty} D_n \zeta^{-n} \qquad (19.14.114)$$

where A_n, B_n, C_n and D_n are complex constants.

9. *Solution by means of power series for simply connected regions.* The functions $\psi_1(\zeta)$ and $\gamma_1(\zeta)$ are analytic and can be expressed as power series

$$\psi_1(\zeta) = \sum_{k=1}^{\infty} a_k \zeta^k \qquad \gamma_1(\zeta) = \sum_{k=0}^{\infty} b_k \zeta_k \qquad (19.14.115)$$

where we have taken $\psi_1(0) = 0$ and therefore $a_0 = 0$. For points on the unit circle, σ replaces ζ and the boundary conditions Eqs. (19.14.111) and (10.14.112) can be written

$$\beta\psi_1(\sigma) + \frac{f(\sigma)}{\bar{f}'(\bar{\sigma})}\bar{\psi}_1'(\bar{\sigma}) + \bar{\gamma}_1(\bar{\sigma}) = H(\sigma) \qquad (19.14.116)$$

where $\sigma = e^{i\alpha}$. $\beta = 1$ and $H(\sigma) = F(\delta)$ for the first boundary value problem and $\beta = -\kappa$ and $H(\sigma) = -G(\delta)$ for the second boundary value problem.

We now expand $H(\sigma)$ in a Fourier series

$$H(\sigma) = \sum_{k=-\infty}^{\infty} t_k e^{ik\theta} = \sum_{k=-\infty}^{\infty} t_k \sigma^k \qquad (19.14.117)$$

with

$$t_k = \frac{1}{2\pi}\int_0^{2\pi} H(\sigma)e^{-ik\theta}\, d\theta$$

Also,

$$\frac{f(\sigma)}{\bar{f}\,'(\bar{\sigma})} = \sum_{k=-\infty}^{\infty} c_k e^{ik\alpha} = \sum_{-\infty}^{\infty} c_k \sigma^k \qquad (19.14.118)$$

Substituting into the boundary conditions and noting that

$$\psi_1'(\sigma) = \sum_{m=1}^{\infty} m a_m \sigma^{m-1}, \quad \bar{\psi}_1'(\bar{\sigma}) = \sum_{m=1}^{\infty} m\bar{a}_m \sigma^{1-m}$$

we get

$$\beta \sum_{k=1}^{\infty} a_k \sigma^k + \sum_{k=-\infty}^{\infty} c_k \sigma^k \sum_{m=1}^{\infty} m\bar{a}_m \sigma^{1-m} + \sum_{k=0}^{\infty} \bar{b}_k \sigma^{-k} = \sum_{k=-\infty}^{\infty} t_k \sigma^k$$

Now

$$\sum_{k=-\infty}^{\infty} c_k \sigma^k \sum_{m=1}^{\infty} m\bar{a}_m \sigma^{1-m} = \sum_{k=-\infty}^{\infty}\sum_{m=1}^{\infty} c_k m\bar{a}_m \sigma^{1-m+k}$$

Let $1 - m + k = l$, $k = l + m - 1$ with $-\infty \leqslant l \leqslant \infty$. Therefore,

$$\sum_{k=-\infty}^{\infty} c_k \sigma^k \sum_{m=1}^{\infty} m\bar{a}_m \sigma^{1-m} =$$

$$\sum_{l=1}^{\infty}\left(\sum_{m=1}^{\infty} m c_{l+m-1}\bar{a}_m\right)\sigma^l + \sum_{l=0}^{\infty}\left(\sum_{m=1}^{\infty} m c_{-l+m-1}\bar{a}_m\right)\sigma^{-l}$$

and

$$\beta \sum_{l=1}^{\infty} a_l \sigma^l + \sum_{l=1}^{\infty}\left(\sum_{m=1}^{\infty} mc_{l+m-1}\bar{a}_m\right)\sigma^l + \sum_{l=0}^{\infty}\left(\sum_{m=1}^{\infty} mc_{-l+m-1}\bar{a}_m\right)\sigma^{-l} + \sum_{l=0}^{\infty} \bar{b}_l \sigma^{-l}$$
$$= \sum_{l=1}^{\infty} t_l \sigma^l + \sum_{l=0}^{\infty} t_{-l}\sigma^{-l} \tag{19.14.119}$$

Comparing coefficients with the same power of σ, we get

$$\beta a_l + \sum_{m=1}^{\infty} mc_{l+m-1}\bar{a}_m = t_l \text{ for } l = 1, 2, 3 \ldots \infty \tag{19.14.120}$$

and

$$\bar{b}_l + \sum_{m=1}^{\infty} mc_{-l+m-1}\bar{a}_m = t_{-l} \text{ for } l = 0, 1, 2 \ldots \infty \tag{19.14.121}$$

We can solve the first set of equations for the a_l's and then the second set to get the b_l's:

$$\left.\begin{array}{l}
\beta a_1 + \bar{a}_1 c_1 + 2\bar{a}_2 c_2 + 3\bar{a}_3 c_3 + \ldots = t_1 \\
\beta a_2 + \bar{a}_1 c_2 + 2\bar{a}_2 c_3 + 3\bar{a}_3 c_4 + \ldots = t_2 \\
\beta a_3 + \bar{a}_1 c_3 + 2\bar{a}_2 c_4 + 3\bar{a}_3 c_5 + \ldots = t_3 \\
\text{—} \\
\text{—} \\
\text{—} \\
\bar{b}_0 + \bar{a}_1 c_0 + 2\bar{a}_2 c_1 + 3\bar{a}_3 c_3 + \ldots = t_0 \\
\bar{b}_1 + \bar{a}_1 c_{-1} + 2\bar{a}_2 c_0 + 3\bar{a}_3 c_1 + \ldots = t_{-1}
\end{array}\right\} \tag{19.14.122}$$

If the mapping function $f(\zeta)$ is a polynominal of degree n, the a_k's can be obtained as the solution of n linear equations in n unknowns [3]. Thus,

$$\begin{array}{l}
\beta a_1 + \bar{a}_1 c_1 + 2\bar{a}_2 c_2 + \ldots n\bar{a}_n c_n = t_1 \\
\beta a_2 + \bar{a}_1 c_2 + 2\bar{a}_2 c_3 + \ldots (n-1)\bar{a}_{n-1} c_n = t_2 \\
\text{—} \\
\text{—} \\
\beta a_n + \bar{a}_1 c_n + \ldots = t_n
\end{array} \tag{19.14.123}$$

For the rest of the terms

$$a_k = t_k \text{ for } k \geqslant n + 1$$

The b_k's are then obtained from

$$\bar{b}_k = -\sum_{m=1}^{n+k+1} m\bar{a}_m c_{m-k-1} + t_{-k} \tag{19.14.124}$$

Consider for example the *case of a circular region of radius R* in the z plane, which is mapped into a unit circle. The mapping function is

$$z = f(\zeta) = R\zeta \tag{19.14.125}$$

$$f'(\zeta) = R = \bar{f}'(\bar{\zeta}) \tag{19.14.126}$$

$$\frac{f(\zeta)}{\bar{f}'(\bar{\zeta})} = \zeta = \sum_{n=-\infty}^{+\infty} c_n\zeta^n \tag{19.14.127}$$

$$c_1 = 1$$

$$c_n = 0 \text{ for } n \neq 1$$

Then

$$a_1 + \bar{a}_1 c_1 = t_1 = a_1 + \bar{a}_1 \tag{19.14.128}$$

$$a_n = t_n \text{ for } n \geqslant 2$$

$$b_n = \bar{t}_{-n} - (2 + n)a_{2+n}$$

or

$$\text{Re } a_1 = \frac{t_1}{2} \text{ and Im } t_1 = 0 \tag{19.14.129}$$

Now

$$\psi_1(\zeta) = a_1\zeta + \sum_{n=2}^{\infty} t_n\zeta^n \tag{19.14.130}$$

and to determine $\psi_1(\zeta)$ uniquely we chose

$$\text{Im } \psi'(0) = 0, \text{ i.e., Im } a_1 = 0$$

so that $a_1 = t_1/2$. Therefore,

$$\psi_1(\zeta) = \frac{t_1}{2}\zeta + \sum_{n=2}^{\infty} t_n\zeta^n \tag{19.14.132}$$

$$\gamma_1(\zeta) = \sum_{n=0}^{\infty} [\bar{t}_{-n} - (2 + n)t_{n+2}]\zeta^n \qquad (19.14.133)$$

Setting $\zeta = z/R$ we get those two functions in terms of z:

$$\psi_1(z) = \frac{t_1}{2}\frac{z}{R} + \sum_{n=2}^{\infty} t_n\left(\frac{z}{R}\right)^n \qquad (19.14.134)$$

$$\gamma_1(z) = \sum_{n=0}^{\infty} [t_{-n} - (2 + n)t_{n+2}]\left(\frac{z}{R}\right)^n \qquad (19.14.135)$$

10. *Mapping of infinite regions.*

For an infinite region R, the mapping into a unit circle is done by assuming that $z = 0$ is and exterior point and letting $z = \infty$ become $\zeta = 0$. The mapping function can then be written [compare to Eq. (19.14.95)].

$$z = \frac{c}{\zeta} + \sum_{n=0}^{\infty} k_n\zeta^n = \frac{1}{\zeta}\left[c + \sum_{n=0}^{\infty} k_n\zeta^{n+1}\right] \qquad (19.14.136)$$

If we now substitute this relation into Eqs. (19.14.75) and (19.14.76), we get

$$\psi_1(\zeta) = \frac{\sum_{k=1}^{m}(F_{1k} + iF_{2k})}{2\pi(1 + \kappa)}\ln\zeta + \frac{Bc}{\zeta} + \psi^0(\zeta) \qquad (91.14.137)$$

$$\gamma_1(\zeta) = \frac{-\kappa\sum_{k=1}^{m}(F_{1k} - iF_{2k})}{2\pi(1 + \kappa)}\ln\zeta + (B' + iC')\frac{c}{\zeta} + \gamma^0(\zeta) \qquad (19.14.138)$$

where $\psi^0(\zeta)$ and $\gamma^0(\zeta)$ are holomorphic functions for $|\zeta| < 1$,

$$\psi^0(\zeta) = \psi_0(\zeta) - \frac{\sum_{k=1}^{m}(F_{1k} + iF_{2k})}{2\pi(1 + \kappa)}\ln\left(c + \sum_{n=0}^{\infty} k_n\zeta^{n+1}\right) + B\sum_{n=0}^{\infty} k_n\zeta^n \qquad (19.14.139)$$

and

$$\gamma^0(\zeta) = \gamma_0(\zeta) + \frac{\kappa\left(\sum_{k=1}^{m} F_{1k} - iF_{2k}\right)}{2\pi(1 + \kappa)} \ln\left(c + \sum_{n=0}^{\infty} k_n \zeta^{n+1}\right) \tag{19.14.140}$$

with

$$B = \frac{\sigma_{11}(\infty) + \sigma_{22}(\infty)}{4}$$

$$B' = \frac{\sigma_{22}(\infty) - \sigma_{11}(\infty)}{2}$$

$$C' = \sigma_{12}(\infty)$$

To obtain the boundary conditions for $\psi^0(\zeta)$ and $\gamma^0(\zeta)$ in the first and second boundary value problems we substitute Eqs. (19.14.137) and (19.14.138) into conditions (19.14.111) and (19.14.112) in which ζ and δ are replaced by σ(since we are on the unit circle). We get

$$\psi^0(\sigma) + \frac{f(\sigma)}{\bar{f}'(\bar{\sigma})}(\bar{\psi}^0)'(\bar{\sigma}) + \bar{\gamma}^0(\bar{\sigma}) = F^0(\sigma) \tag{19.14.141}$$

and

$$-\kappa\psi^0(\sigma) + \frac{f(\sigma)}{\bar{f}'(\bar{\sigma})}(\bar{\psi}^0)'(\bar{\sigma}) + \bar{\gamma}^0(\bar{\sigma}) = G^0(\sigma) \tag{19.14.142}$$

where

$$F^0(\sigma) = F(\sigma) - \frac{\sum_{k=1}^{m} (F_{1k} + iF_{2k})}{2\pi} \ln\sigma - \frac{Bc}{\sigma} - (B' - iC')\bar{c}\sigma$$

$$- \frac{f(\sigma)}{\bar{f}'(\bar{\sigma})}\left[\frac{\sum_{k=1}^{m} (F_{1k} - iF_{2k})}{2\pi(1 + \kappa)}\sigma - B\bar{c}\sigma^2\right] \tag{19.14.143}$$

and

$$G^0(\sigma) = G(\sigma) + \frac{\kappa Bc}{\sigma} - \frac{f(\sigma)}{\overline{f}'(\overline{\sigma})}\left[\frac{\sum_{k=1}^{m}(F_{1k} - iF_{2k})}{2\pi(1+\kappa)}\sigma - B\overline{c}\sigma^2\right] + (B' - iC')\overline{c}\sigma \quad (19.14.144)$$

in which

$$F(\sigma) = f_{1j} + if_{2j} \quad \text{and} \quad G(\sigma) = -2G(g_{1j} + g_{2j}) \quad (19.14.145)$$

Thus, the solution of our two boundary value problems amounts to finding two functions $\psi^0(\zeta)$ and $\gamma^0(\zeta)$ analytic in the circle $|\zeta| < 1$ and which satisfy the boundary conditions Eq. (19.14.141) or Eq. (19.14.142).

19.15 Solutions Using Timoshenko's Equations. Westergaard's Stress Function.

Equations (19.14.11) or (19.14.12) can be used to solve a certain class of problems involving the half plane. Starting with Eq. (19.14.12) we have

$$\sigma_{11} = \frac{\partial^2\phi}{\partial x_2^2} = \frac{\partial^2 H_2}{\partial x_2^2} + 2x_2\frac{\partial^2 Q_2}{\partial x_2^2} + 4\frac{\partial Q_2}{\partial x_2} \quad (19.15.1)$$

$$\sigma_{22} = \frac{\partial^2\phi}{\partial x_1^2} = \frac{\partial^2 H_2}{\partial x_1^2} + 2x_2\frac{\partial^2 Q_2}{\partial x_1^2} \quad (19.15.2)$$

$$\sigma_{12} = -\frac{\partial^2\phi}{\partial x_1\partial x_2} = -\frac{\partial^2 H_2}{\partial x_1\partial x_2} - 2x_2\frac{\partial^2 Q_2}{\partial x_1\partial x_2} - 2\frac{\partial Q_2}{\partial x_1}. \quad (19.15.3)$$

We can simplify these equations by writing them in terms of three harmonic functions instead of two and they will contain only first order derivatives. For that let us introduce two analytic functions Z and W,

$$Z = T + i\lambda\,, \qquad W = \beta + i\Omega \quad (19.15.4)$$

Because of the Cauchy-Riemann conditions, if $T = 2\partial Q_2/\partial x_2$, then $\lambda = 2\partial Q_2/\partial x_1$. Also, if $\beta = -(\partial H_2/\partial x_1)$, then $\Omega = \partial H_2/\partial x_2$. Therefore, the equations of stress become

$$\sigma_{11} = \frac{\partial\Omega}{\partial x_2} + x_2\frac{\partial T}{\partial x_2} + 2T \quad (19.15.5)$$

$$\sigma_{22} = -\frac{\partial \beta}{\partial x_1} + x_2 \frac{\partial \lambda}{\partial x_1} \tag{19.15.6}$$

$$\sigma_{12} = \frac{\partial \beta}{\partial x_2} - x_2 \frac{\partial T}{\partial x_1} - \lambda \tag{19.15.7}$$

Making use of the Cauchy-Riemann equations we can write

$$\sigma_{11} = \frac{\partial \beta}{\partial x_1} + x_2 \frac{\partial T}{\partial x_2} + 2T \tag{19.15.8}$$

$$\sigma_{22} = -\frac{\partial \beta}{\partial x_1} - x_2 \frac{\partial T}{\partial x_2} \tag{19.15.9}$$

$$\sigma_{12} = \frac{\partial \beta}{\partial x_2} - x_2 \frac{\partial T}{\partial x_1} - \lambda \tag{19.15.10}$$

The stresses are thus given in terms of three harmonic functions, two of which T and λ are conjugates.

There is a special class of problems, such as that of the half plane, where the shear stresses are equal to zero for $x_2 = 0$. In this case Eq. (19.15.10) yields

$$\frac{\partial \beta}{\partial x_2} = \lambda \tag{19.15.11}$$

at $x_2 = 0$. This condition leads to interesting results. Indeed, since T and λ are conjugates.

$$\frac{\partial T}{\partial x_2} = -\frac{\partial \lambda}{\partial x_1} = -\frac{\partial^2 \beta}{\partial x_1 \partial x_2}$$

so that

$$T = -\frac{\partial \beta}{\partial x_1} \tag{19.15.12}$$

and $\partial\beta/\partial x_2$ and $-(\partial\beta/\partial x_1)$ are conjugate. Now,

$$\lambda = \frac{\partial \beta}{\partial x_2} = \frac{\partial(2Q_2)}{\partial x_1} = \frac{\partial}{\partial x_2}\left(-\frac{\partial H_2}{\partial x_1}\right),$$

therefore,

$$\frac{\partial}{\partial x_1}\left(2Q_2 + \frac{\partial H_2}{\partial x_2}\right) = 0 .$$

$$\text{Also, } T = -\frac{\partial \beta}{\partial x_1} = \frac{\partial(2Q_2)}{\partial x_2} = \frac{\partial^2 H_2}{\partial x_1^2} = -\frac{\partial \Omega}{\partial x_2} = -\frac{\partial^2 H_2}{\partial x_2^2}$$

Therefore,

$$\frac{\partial}{\partial x_2}\left(2Q_2 + \frac{\partial H_2}{\partial x_2}\right) = 0$$

and

$$2Q_2 = -\frac{\partial H_2}{\partial x_2} \tag{19.15.13}$$

The stress Eqs. (19.15.8) to (19.15.10) then become

$$\sigma_{11} = T + x_2 \frac{\partial T}{\partial x_2} \tag{19.15.14}$$

$$\sigma_{22} = T - x_2 \frac{\partial T}{\partial x_2} \tag{19.15.15}$$

$$\sigma_{12} = -x_2 \frac{\partial T}{\partial x_1} \tag{19.15.16}$$

and are all expressed in terms of the single harmonic function T. They can be written in terms of complex variables as follows:

$$Z = T + i\lambda, \qquad T = \text{Re } Z, \qquad \lambda = \text{Im } Z$$

$$\frac{\partial T}{\partial x_1} = \frac{\partial}{\partial x_1}(\text{Re } Z) = \text{Re } \frac{\partial Z}{\partial x_1} = \text{Re } \frac{dZ}{dz}\frac{\partial z}{\partial x_1} = \text{Re } Z'$$

Also,

$$\frac{\partial T}{\partial x_2} = -\frac{\partial \lambda}{\partial x_1} = -\frac{\partial}{\partial x_1}\text{Im } Z = -\text{Im } Z' .$$

Therefore,

$$\sigma_{11} = \text{Re } Z - x_2 \text{ Im } Z' \tag{19.15.17}$$

$$\sigma_{22} = \text{Re } Z + x_2 \text{ Im } Z' \tag{19.15.18}$$

$$\sigma_{12} = -x_2 \text{ Re } Z' \tag{19.15.19}$$

This is the solution given by Westergaard [10]. It implies that

$$\sigma_{12} = 0 \quad \text{at} \quad x_2 = 0 \tag{19.15.20}$$

and

$$\sigma_{11} = \sigma_{22} \quad \text{at} \quad x_2 = 0 , \tag{19.15.21}$$

and it will only be correct if these conditions are consistent with the boundary conditions for the specific problem under consideration. The analytic function Z is known as Westergaard's stress function.

The connection between Z and Airy's stress function ϕ can easily be established. From Eqs. (19.15.4), (19.15.11) and (19.15.12) we have,

$$Z = -\frac{\partial \beta}{\partial x_1} + i\frac{\partial \beta}{\partial x_2} \tag{19.15.22}$$

$$= -\left(\frac{\partial \beta}{\partial x_1} + i\frac{\partial \Omega}{\partial x_1}\right) = -\frac{d}{dz}(\beta + i\Omega) = -\frac{dW}{dz}$$

Thus,

$$\beta + i\Omega = -\int Z\,dz \equiv -\tilde{Z} = -\frac{\partial H_2}{\partial x_1} + i\frac{\partial H_2}{\partial x_2} \tag{19.15.23}$$

Let K_2 be the conjugate of H_2. Then, $\partial H_2/\partial x_2 = -\partial K_2/\partial x_1$, and

$$\tilde{Z} = \frac{\partial H_2}{\partial x_1} + i\frac{\partial K_2}{\partial x_1} \tag{19.15.24}$$

or

$$\int\!\!\int Z\,dz \equiv \tilde{\tilde{Z}} = H_2 + iK_2 \tag{19.15.25}$$

Now, from Eqs. (19.14.12) and (19.15.13),

$$\phi = 2x_2Q_2 + H_2 = H_2 - x_2\frac{\partial H_2}{\partial x_2} = \mathrm{Re}\,\tilde{\tilde{Z}} + x_2\,\mathrm{Im}\,\tilde{Z} \tag{19.15.26}$$

where $\sim$ represents integration; Eq. (19.15.26) gives the relation between Airy's stress function ϕ and Westergaard's stress function Z.

As an example, consider the two-dimensional Boussinesq problem examined in Secs. 14.5 and 14.6. Using the coordinate system suggested by Westergaard, (Fig. 19.27) the complex potential

$$Z = \frac{q}{i\pi z} = -i\frac{q}{\pi r}e^{-i\theta} = -\frac{q}{\pi r}(\sin\theta + i\cos\theta) \tag{19.15.27}$$

gives a solution to the problem.

$$Z' = \frac{-q}{i\pi z^2} = \frac{iq}{\pi r^2}e^{-i2\theta} = \frac{q}{\pi r^2}(\sin 2\theta + i\cos 2\theta) \tag{19.15.28}$$

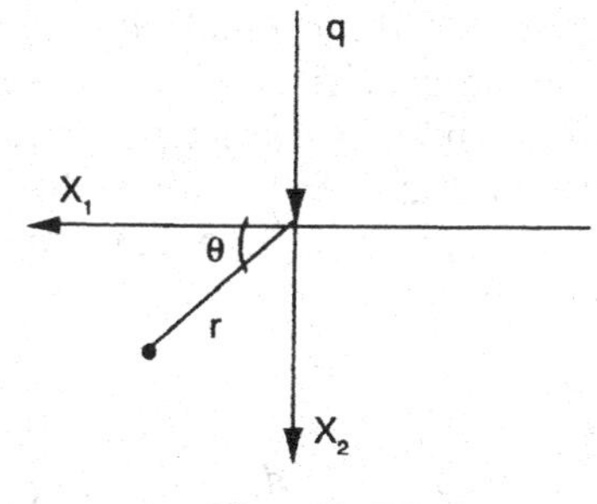

Fig. 19.27

From Eqs. (19.14.48) to (19.14.50) we have

$$\sigma_{rr} + \sigma_{\theta\theta} = 2 \text{ Re } Z = -\frac{2q}{\pi r} \sin \theta \tag{19.15.29}$$

$$\begin{aligned} \sigma_{\theta\theta} - \sigma_{rr} + 2i\sigma_{r\theta} &= (\sigma_{22} - \sigma_{11} + 2i\sigma_{12})e^{2i\theta} \\ &= (2x_2 \text{ Im } Z' - 2ix_2 \text{ Re } Z')e^{2i\theta} \\ &= \frac{2x_2q}{\pi r^2}(\cos 2\theta - i \sin 2\theta) \\ &\quad \times (\cos 2\theta + i \sin 2\theta) \\ &= \frac{2x_2q}{\pi r^2} = \frac{2rq \sin \theta}{\pi r^2} = \frac{2q \sin \theta}{\pi r} \end{aligned} \tag{19.15.30}$$

Thus,

$$\sigma_{r\theta} = 0$$

$$\sigma_{\theta\theta} - \sigma_{rr} = \frac{2q \sin \theta}{\pi r}$$

This, together with Eq. (19.15.29) leads to

$$\sigma_{r\theta} = 0, \qquad \sigma_{\theta\theta} = 0, \qquad \sigma_{rr} = -\frac{2q \sin \theta}{\pi r} \tag{19.15.31}$$

This is the result obtained in Sec. 14.5 if one takes into account the reference from which θ is measured. In terms of the stress function ϕ, we have from Eq. (19.15.26)

$$\phi = \text{Re } \tilde{\tilde{Z}} + x_2 \text{ Im } \tilde{Z}, \qquad x_2 = r \sin \theta$$

Now

$$\tilde{Z} = \frac{q}{i\pi} \ln z = \frac{q}{i\pi} (\ln r + i\theta) = -\frac{iq}{\pi} \ln r + \frac{q\theta}{\pi}$$

$$\tilde{\tilde{Z}} = \frac{q}{i\pi} (z \ln z - z) = \frac{qre^{i\theta}}{i\pi} (\ln r + i\theta - 1)$$

$$= \frac{qr}{\pi} \{\theta \cos \theta + (\ln r - 1)\sin \theta$$
$$+ i[-(\ln r - 1)\cos \theta + \theta \sin \theta]\}$$

Then,

$$\phi = \frac{qr}{\pi} [\theta \cos \theta + (\ln r - 1)\sin \theta + \sin \theta(-\ln r)] \tag{19.15.32}$$
$$= \frac{qr}{\pi} (\theta \cos \theta - \sin \theta)$$

Using Eqs. (9.9.11) one gets (19.15.31).

Let us now return to Eq. (19.14.30) or to Eq. (19.14.38). *In problems possessing* x_1 *axis symmetry for both loading and geometry* $\sigma_{12} = 0$ along $x_2 = 0$. Thus,

$$\text{Im}[\bar{z}\psi''(z) + \chi''(z)]_{x_2=0} = 0 \tag{19.15.33}$$

Remembering that $\bar{z} = z$ along the x_1 axis this equation can be satisfied most generally by requiring that

$$z\psi''(z) + \chi''(z) - A = 0 \tag{19.15.34}$$

at all points of the complex Z plane. The constant A is real and is determined by the stress boundary conditions appropriate to a problem in which the applied loads are such as to produce symmetric stresses. Integration of Eq. (19.15.34) yields

$$\chi'(z) = \psi(z) - z\psi'(z) + Az + B \tag{19.15.35}$$

where B is an arbitrary complex constant which can at most represent a rigid body displacement and may be set to zero.

Equation (19.15.34) is equivalent to the pair of equations,

$$x_1 \text{ Re } \psi''(z) - x_2 \text{ Im } \psi''(z) + \text{Re } \chi''(z) - A = 0 \tag{19.15.36}$$
$$x_1 \text{ Im } \psi''(z) + x_2 \text{ Re } \psi''(z) + \text{Im } \chi''(z) = 0 \tag{19.15.37}$$

and Eq. (19.15.35) is equivalent to the pair of equations:

$$x_1 \operatorname{Re} \psi'(z) - x_2 \operatorname{Im} \psi'(z) - \operatorname{Re} \psi(z) + \operatorname{Re} \chi'(z) - Ax_1 = 0 \qquad (19.15.38)$$

$$x_2 \operatorname{Re} \psi'(z) + x_1 \operatorname{Im} \psi'(z) - \operatorname{Im} \psi(z) + \operatorname{Im} \chi'(z) - Ax_2 = 0 \qquad (19.15.39)$$

Equations (19.14.36) to (19.14.40) now become:

$$\sigma_{11} = \operatorname{Re}[2\psi'(z)] - x_2 \operatorname{Im}[2\psi''(z)] - A \qquad (19.15.40)$$

$$\sigma_{22} = \operatorname{Re}[2\psi'(z)] + x_2 \operatorname{Im}[2\psi''(z)] + A \qquad (19.15.41)$$

$$\sigma_{12} = -x_2 \operatorname{Re}[2\psi''(z)] \qquad (19.15.42)$$

$$2Gu_1 = (\kappa - 1)\operatorname{Re} \psi(z) - x_2 \operatorname{Im}[2\psi'(z)] - Ax_1 \qquad (19.15.43)$$

$$2Gu_2 = (\kappa + 1)\operatorname{Im} \psi(z) - x_2 \operatorname{Re}[2\psi'(z)] + Ax_2 \qquad (19.15.44)$$

The symmetric problem is thus reduced to the determination of a single analytic function $\psi(z)$. By setting

$$Z(z) = 2\psi'(z) \qquad (19.15.45)$$

and $A = 0$ one obtains Westergaard's equations Eq. (19.15.17) to Eq. (19.15.19). The constant A however is in general, not equal to zero and it would be incorrect to assume it to be so unless the conditions of the problem permit it [12, 13]. Eqs. (19.15.40) to (19.15.44) can be referred to as the modified Westergaard field equations for symmetric plane problems. To avoid confusion we may set, if needed,

$$Z_1(z) = 2\psi'(z) \qquad (19.15.46)$$

in Eqs. (19.15.40) to (19.15.44) and call $Z_1(z)$ the modified Westergaard stress function for problems symmetric w.r.t. the $0X_1$ axis.

19.16 Simple Examples Using Complex Potentials.

Example 1: Find the state of stress corresponding to the complex potentials

$$\psi(z) = \frac{1}{4} S(1 + \lambda)z , \qquad \chi(z) = S(1 - \lambda)\frac{z^2}{4} \qquad (19.16.1)$$

The components of the state of stress are given by Eqs. (19.14.29) and (19.14.30).

$$\psi'(z) = \frac{1}{4} S(1 + \lambda), \qquad \psi''(z) = 0 \tag{19.16.2}$$

$$\chi'(z) = \frac{1}{2} S(1 + \lambda)z, \qquad \chi''(z) = \frac{1}{2} S(1 - \lambda)$$

$$\sigma_{11} + \sigma_{22} = S(1 + \lambda) \tag{19.16.3}$$

$$\sigma_{22} - \sigma_{11} + 2i\sigma_{12} = 2\left[\frac{1}{2} S(1 - \lambda)\right] \tag{19.16.4}$$

Thus,

$$\sigma_{11} = S\lambda, \qquad \sigma_{22} = S, \qquad \sigma_{12} = 0 \tag{19.16.5}$$

which represents a biaxial state of stress applied to an infinite sheet (Fig. 19.28)

Example 2: The two complex potentials

$$\psi(z) = \frac{1}{2} M \ln z, \qquad \chi(z) = -\frac{M}{2} z \ln z \tag{19.16.6}$$

are suggested for the solution of the problem of stresses in wedges (see Sec. 14.13) subjected to point loads. Find the stress as well as the Airy stress function corresponding to the problem.

Working in polar coordinates we have:

$$\psi'(z) = \frac{M}{2z} = \frac{Me^{-i\theta}}{2r}, \qquad \psi''(z) = -\frac{M}{2}\frac{e^{-2i\theta}}{r^2}, \tag{19.16.7}$$

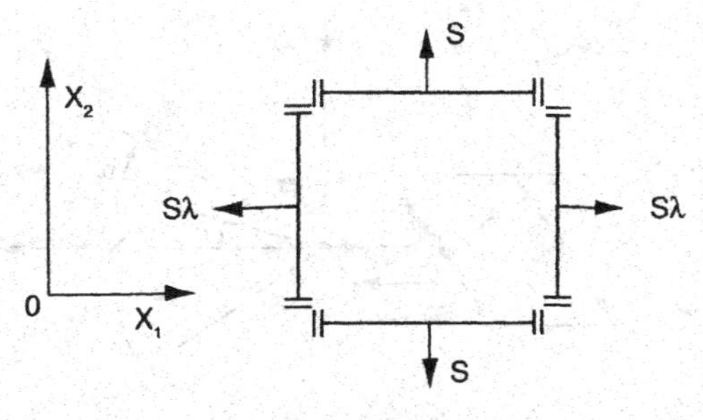

Fig. 19.28

$$\chi'(z) = -\frac{M}{2}(\ln z + 1), \quad \chi''(z) = -\frac{M}{2z} = -\frac{Me^{-i\theta}}{2r}, \tag{19.16.8}$$
$$\bar{z} = re^{-i\theta}$$

Thus, from Eqs. (19.14.88) and (19.14.89):

$$\sigma_{rr} + \sigma_{\theta\theta} = \frac{2M}{r}\cos\theta$$

$$\sigma_{\theta\theta} - \sigma_{rr} + 2i\sigma_{r\theta} = 2\left[-re^{-i\theta}\frac{M}{2r^2}e^{-2i\theta} - \frac{M}{2r}e^{-i\theta}\right]e^{2i\theta}$$
$$= -\frac{M}{r}(e^{-3i\theta} + e^{-i\theta})e^{2i\theta} = -\frac{2M}{r}\cos\theta$$

Therefore,

$$\sigma_{rr} = \frac{2M\cos\theta}{r}, \qquad \sigma_{\theta\theta} = 0, \qquad \sigma_{r\theta} = 0 \tag{19.16.9}$$

This is a purely radial state of stress (Fig. 19.29)

To find the total force acting on the boundary ABC, due to the force per unit length q, use is made of Eq. (19.14.44):

$$F_1 + iF_2 = -i[\psi(z) + z\overline{\psi}'(\bar{z}) + \overline{\chi}'(\bar{z})]_C^A \tag{19.16.10}$$

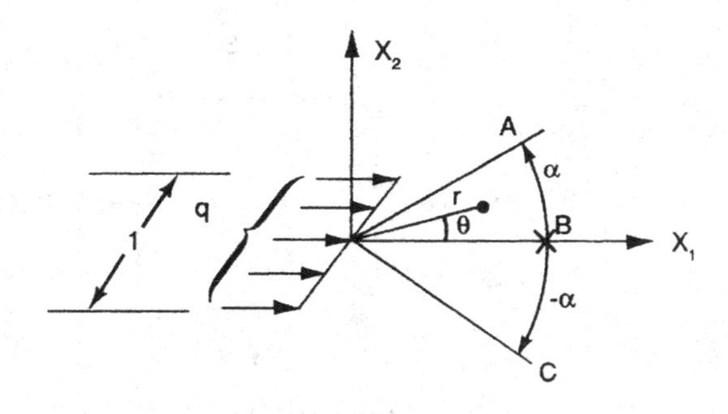

Fig. 19.29

Point A has coordinates $re^{i\alpha}$ and point C has coordinates $re^{-i\alpha}$. Thus,

$$F_1 + iF_2 = -i\left[\frac{M}{2}(\ln r + i\alpha) + \frac{M}{2}e^{2i\alpha} - \frac{M}{2}(\ln r - i\alpha) - \frac{M}{2}\right.$$

$$\left. - \frac{M}{2}(\ln r - i\alpha) - \frac{M}{2}e^{-2i\alpha} + \frac{M}{2}(\ln r + i\alpha) + \frac{M}{2}\right]$$

$$= M(2\alpha + \sin 2\alpha)$$

Therefore,

$$F_1 = M(2\alpha + \sin 2\alpha), \qquad F_2 = 0 \tag{19.16.11}$$

If a force q is distributed over the unit thickness of the wedge, then (compare to Eq. 14.13.5)

$$M = \frac{q}{2\alpha + \sin 2\alpha} \tag{19.16.12}$$

The Airy stress function is given by Eq. (19.14.14)

$$\phi = \text{Re}[\bar{z}\psi(z) + \chi(z)]$$

$$\phi = \text{Re}\left[re^{-i\theta}\left(\frac{M}{2}\ln r + \frac{M}{2}i\theta\right)\right.$$

$$\left. - re^{i\theta}\left(\frac{M}{2}(\ln r + \frac{M}{2}i\theta\right)\right]$$

$$\phi = Mr\theta \sin\theta \tag{19.16.13}$$

This is the Airy stress function presented in Sec. 14.13 for the solution of a point-loaded wedge.

Example 3: Find the state of stress and the displacements in a disk of radius R subjected to a uniform pressure p (Fig. 19.30).

$$\sigma_{n1} = -p\cos\theta, \qquad \sigma_{n2} = -p\sin\theta$$

From Eq. (19.14.81), in which the subscript j is not needed here, we get,

$$f_1(s) + if_2(s) = \int(-\sigma_{n2} + i\sigma_{n1})\,ds$$

$$= -ip\int(\cos\theta + i\sin\theta)R\,d\theta = -ipe^{i\theta}R\,d\theta$$

$$= -p\,\text{R}e^{i\theta}$$

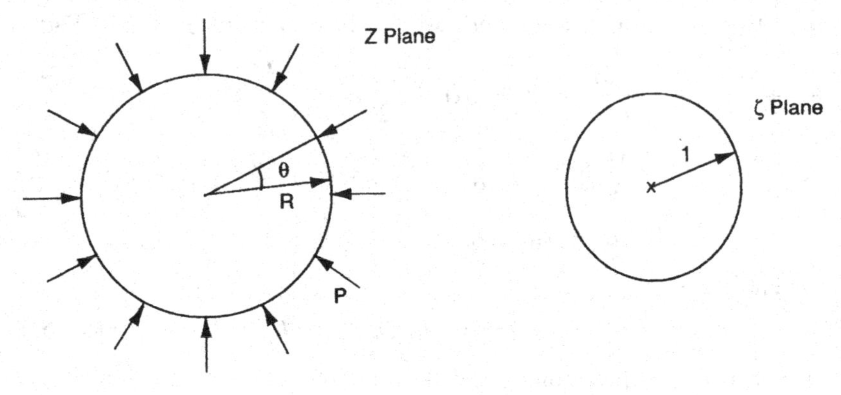

Fig. 19.30

Now

$$z = f(\zeta) = R\zeta$$

and from Eq. (19.14.96)

$$e^{i\theta} = e^{i\alpha} = \sigma$$

so that (see Eq. 19.14.117)

$$H(\sigma) = -pR\sigma = \sum_{k=-\infty}^{+\infty} t_k \sigma^k$$

Therefore,

$$t_1 = -pR$$
$$t_n = 0 \qquad \text{for } n \neq 1$$

and from Eqs. (19.14.132) and (19.14.133)

$$\psi_1(\zeta) = -\frac{pR\zeta}{2} \tag{19.16.14}$$

$$\gamma_1(\zeta) = 0 \tag{19.16.15}$$

so that

$$\psi(z) = -\frac{pz}{2}$$

$$\gamma(z) = 0$$

From Eqs. (19.14.88) to (19.14.90) we get

$$\sigma_{rr} = \sigma_{\theta\theta} = -p \tag{19.16.16}$$

$$\sigma_{r\theta} = 0 \tag{19.16.17}$$

From Eq. (19.14.87) we get

$$u_r = \frac{1-\kappa}{4G} pr \tag{19.16.18}$$

$$u_\theta = 0 \tag{19.16.19}$$

Example 4: Find the complex potentials appropriate to the problem of the concentrated forces acting on a circular boundary (Fig. 19.31).

The concentrated forces with the components shown in Fig. 19.31 can be considered as the limit of a uniformly distributed stress p applied on a small segment ΔL of the circle's boundary. As ΔL decreases and tends to zero, $p\Delta L$ tends to P. From Eq. (19.14.81) the function

$$f_1(s) + if_2(s) = -\int \sigma_{n2}\, ds$$

will be constant along part of the boundary where no load is applied but will suffer a discontinuity of magnitude P as one passes over the point of application of the load. From Eq. (19.14.81)

$$\begin{aligned} f_1 + if_2 &= 0 && \text{for } 0 \leqslant \theta < \beta \\ &= P && \text{for } \beta < \theta < 2\pi - \beta \\ &= 0 && 2\pi - \beta < \theta \leqslant 2\pi \end{aligned} \tag{19.16.20}$$

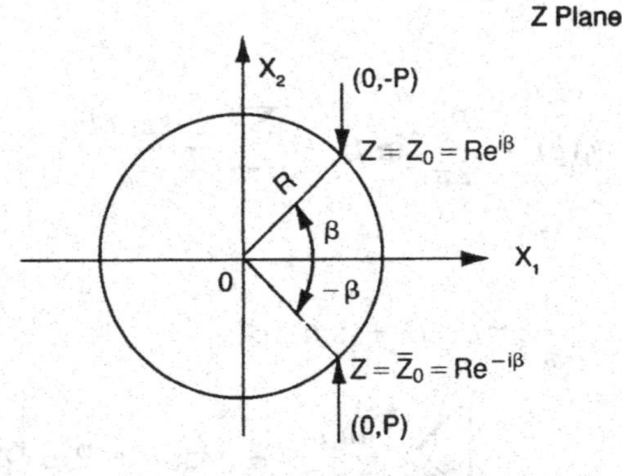

Fig. 19.31

Hence, from Eq. (19.14.127),

$$H(\sigma) = \sum_{k=-\infty}^{\infty} t_k \sigma^k \qquad (19.16.21)$$

where

$$t_k = \frac{1}{2\pi} \int_0^{2\pi} (f_1 + if_2) e^{-ik\theta} \, d\theta = \frac{P}{2\pi} \int_\beta^{2\pi - \beta} e^{-ik\theta} \, d\theta$$

$$= \frac{iP}{2\pi k} [e^{-ik(2\pi - \beta)} - e^{-ik\beta}]$$

$$= \frac{iP}{2\pi k} [e^{ik\beta} - e^{-ik\beta}] \quad \text{for } k \neq 0 \quad \text{since } e^{-2\pi ik} = 1 \qquad (19.16.22)$$

$$= \frac{P}{\pi} (\pi - \beta) \quad \text{for} \quad k = 0 \, .$$

Thus,

$$t_0 = \frac{P}{\pi} (\pi - \beta) \quad \text{and} \quad t_k = -\frac{P}{\pi k} \sin k\beta \quad \text{for} \quad k \neq 0$$

and from Eqs. (19.14.132) and (19.14.133) we have

$$\psi_1(\zeta) = -\frac{P\zeta}{2\pi} \sin \beta + \sum_{n=2}^{\infty} -\frac{P}{\pi} \frac{\sin n\beta}{n} \zeta^n$$

or

$$\psi_1(\zeta) = \frac{P\zeta}{2\pi} \sin \beta - \sum_{n=1}^{\infty} \frac{P}{\pi} \frac{\sin n\beta}{n} \zeta^n \qquad (19.16.23)$$

and

$$\gamma_1(\zeta) = \frac{P}{\pi} [\pi - \beta + \sin 2\beta] - \frac{P}{\pi} \left[\sum_{n=1}^{\infty} \frac{\sin n\beta}{n} - \sin(n + 2)\beta \right] \zeta^n \qquad (19.16.24)$$

If we revert to the exponential form we can have a closed form solution rather than a series. Indeed,

$$t_0 = \frac{P}{\pi}(\pi - \beta), \qquad t_n = \frac{iP}{2\pi n}(e^{in\beta} - e^{-in\beta}), \qquad n \neq 0$$

$$\psi_1(\zeta) = \frac{t_1}{2}\zeta + \sum_{n=2}^{\infty} t_n\zeta^n = -\frac{t_1}{2}\zeta + \sum_{n=1}^{\infty} t_n\zeta^n$$

$$= -\frac{iP}{4\pi}(e^{i\beta} - e^{-i\beta})\zeta + \sum_{n=1}^{\infty} \frac{Pi}{2\pi n}(e^{in\beta} - e^{-in\beta})\zeta^n$$

$$\gamma_1(\zeta) = \frac{P}{\pi}\left[\pi - \beta - \frac{i}{2}(e^{i2\beta} - e^{-i2\beta})\right] + \sum_{n=1}^{\infty} [t_{-n} - (2 + n)t_{2+n}]\zeta^n$$

$$= \frac{P}{\pi}\left[\pi - \beta - \frac{i}{2}(e^{i2\beta} - e^{-i2\beta})\right] + \frac{iP}{2\pi}\sum_{n=1}^{\infty}\left[\frac{e^{in\beta} - e^{-in\beta}}{n} - (e^{i(2+n)\beta} - e^{-i(2+n)\beta})\right]\zeta^n$$

Now

$$\sum_{n=1}^{\infty} \frac{(e^{i\beta}\zeta)^n}{n} = -\ln(1 - e^{i\beta}\zeta)$$

$$\sum_{n=1}^{\infty} (e^{i\beta}\zeta)^n = \frac{1}{1 - e^{i\beta}\zeta}$$

so that

$$\psi_1(\zeta) = -\frac{iP}{4\pi}(e^{i\beta} - e^{-i\beta})\zeta + \frac{iP}{2\pi}\ln\left(\frac{1 - e^{-i\beta}\zeta}{1 - e^{i\beta}\zeta}\right)$$

$$= \frac{iP}{2\pi}\left[\ln\left(\frac{e^{i\beta} - \zeta}{e^{-i\beta} - \zeta}e^{-2i\beta}\right) - \left(\frac{e^{i\beta} - e^{-i\beta}}{2}\right)\zeta\right]$$

$$\begin{aligned}\psi_1(\zeta) &= \frac{P}{\pi}\beta + \frac{iP}{2\pi}\left[\ln\frac{e^{i\beta}-\zeta}{e^{-i\beta}-\zeta} - \frac{e^{i\beta}-e^{-i\beta}}{2}\zeta\right]\\ &= \frac{P}{\pi}\beta + \frac{P\zeta}{2\pi}\sin\beta + \frac{iP}{2\pi}\ln\frac{e^{i\beta}-\zeta}{e^{-i\beta}-\zeta}\end{aligned} \qquad (19.16.25)$$

Similarly, we can show that

$$\begin{aligned}\gamma_1(\zeta) &= \frac{P\beta}{\pi} + \frac{P}{\pi}(\pi - \beta + \sin 2\beta)\\ &\quad + \frac{iP}{2\pi}\left[\ln\frac{e^{i\beta}-\zeta}{e^{-i\beta}-\zeta} - \frac{e^{i\beta}}{e^{-i\beta}-\zeta} + \frac{e^{-i\beta}}{e^{i\beta}-\zeta}\right]\end{aligned} \qquad (19.16.26)$$

Since $\zeta = z/R$, $e^{i\beta} = (1/R)z_0$, $e^{-i\beta} = (1/R)\overline{z}_0$,

$$\psi(z) = \frac{P\beta}{\pi} + \frac{iP}{2\pi}\ln\frac{z_0 - z}{\overline{z}_0 - z} + \frac{P\sin\beta}{2\pi R}z \qquad (19.16.27)$$

$$\begin{aligned}\gamma(z) &= \frac{P}{\pi}(\pi - \beta + \sin 2\beta) + \frac{P\beta}{\pi}\\ &\quad + \frac{iP}{2\pi}\left[\ln\frac{z_0 - z}{\overline{z}_0 - z} - \frac{z_0}{\overline{z}_0 - z} + \frac{\overline{z}_0}{z_0 - z}\right]\end{aligned} \qquad (19.16.28)$$

In the two equations above the constants can be ignored since they do not affect the stresses.

Thus,

$$\psi(z) = \frac{iP}{2\pi}\left(\ln\frac{z_0 - z}{\overline{z}_0 - z} - \frac{z_0 - \overline{z}_0}{2R^2}z\right) \qquad (19.16.29)$$

$$\gamma(z) = \frac{iP}{2\pi}\left(\ln\frac{z_0 - z}{\overline{z}_0 - z} + \frac{\overline{z}_0}{z_0 - z} - \frac{z_0}{\overline{z}_0 - z}\right) \qquad (19.16.30)$$

Stresses and displacements can be found without difficulty from Eqs. (19.14.28) to (19.14.30) [2].

19.17 The Infinite Plate with a Circular Hole.

Let the infinite plate have a hole of radius a, and let the origin be taken at the center of the circle. If the boundary conditions are given in terms of prescribed stresses over the boundary of the hole, then σ_{rr} and $\sigma_{r\theta}$ are known at $z = ae^{i\theta}$. In Eqs. (19.14.88) and (19.14.89) the analytic functions $\psi'(z)$ and $\chi''(z)$ can be expanded into power series. Since the stresses must remain finite when $r \to \infty$, we find from Eq. (19.14.88) to Eq. (19.14.90) that these functions must remain finite at $r = \infty$. Therefore, they must have the form

$$\psi'(z) = \sum_{n=0}^{\infty} A_n z^{-n}, \qquad \chi''(z) = \sum_{n=0}^{\infty} B_n z^{-n} \tag{19.17.1}$$

where A_n and B_n are complex constants. From Eqs. (19.14.88) to (19.14.90) we see that the stresses at infinity are given by the real part of A_0 and B_0. The imaginary part of the complex constant A_0 does not affect the state of stress. Integrating Eq. (19.17.1) with respect to z we get

$$\psi(z) = A_0 z + A_1 \ln z - \sum_{n=2}^{\infty} \frac{A_n z^{-n+1}}{n-1} + C_1 \tag{19.17.2}$$

$$\chi'(z) = B_0 z + B_1 \ln z - \sum_{n=2}^{\infty} \frac{B_n z^{-n+1}}{n-1} + C_2 \tag{19.17.3}$$

where C_1 and C_2 are complex constants. Since

$$\overline{\psi}'(\overline{z}) = \sum_{n=0}^{\infty} \overline{A}_n (\overline{z})^{-n}$$

and

$$\overline{\chi}'(\overline{z}) = \overline{B}_0 \overline{z} + \overline{B}_1 \ln \overline{z} - \sum_{n=2}^{\infty} \frac{\overline{B}_n (\overline{z})^{-n+1}}{n-1} + \overline{C}_2$$

the displacements in Eq. (19.14.87) become

$$\begin{aligned} u_r + i u_\theta = e^{-i\theta} \Bigg[& \frac{3-\nu}{E} \left(A_0 z + A_1 \ln z - \sum_{n=2}^{\infty} \frac{A_n z^{-n+1}}{n-1} + C_1 \right) \\ & - \frac{1+\nu}{E} \left(\overline{A}_0 z + r^2 \sum_{n=1}^{\infty} \overline{A}_n (\overline{z})^{-(n+1)} \right) \\ & - \frac{1+\nu}{E} \left(\overline{B}_0 \overline{z} + \overline{B}_1 \ln \overline{z} - \sum_{n=2}^{\infty} \frac{\overline{B}_n (\overline{z})^{-n+1}}{n-1} + \overline{C}_2 \right) \Bigg] \end{aligned} \tag{19.17.4}$$

As established in Sec. 19.7, ln z is not single valued because if we trace a path around the hole the value of θ increases from some value θ_1 to $\theta_1 + 2\pi$. The increment of $u_r + iu_\theta$ in going around the hole is therefore,

$$2\pi i e^{-i\theta}\left(\frac{3-\nu}{E}A_1 + \frac{1+\nu}{E}\overline{B}_1\right).$$

For $u_r + iu_\theta$ to be single valued, we must require that

$$(3-\nu)A_1 + (1+\nu)\overline{B}_1 = 0 \qquad \text{(see Eq. 19.14.49)}$$

Thus,

$$A_1 = -\frac{1+\nu}{3-\nu}\overline{B}_1 \tag{19.17.5}$$

Since σ_{rr} and $\sigma_{r\theta}$ are given at $r = a$, we can expand $(\sigma_{rr} - i\sigma_{r\theta})_{r=a}$ in a complex Fourier series; thus, (see Eq. 19.2.19)

$$(\sigma_{rr} - i\sigma_{r\theta})_{r=a} = \sum_{n=-\infty}^{\infty} C_n e^{in\theta} \tag{19.17.6}$$

where C_n are given by the formula (see Eq. 19.2.20)

$$C_n = \frac{1}{2\pi}\int_0^{2\pi}[\sigma_{rr}(\theta) - \sigma_{r\theta}(\theta)]_{r=a}\, e^{-in\theta}\, d\theta$$

$$n = 0, \pm 1, \pm 2, \ldots \tag{19.17.7}$$

Let us now substitute the series in Eqs. (19.17.1) and (19.17.6) in Eq. (19.14.90) and set $r = a$ since we are on the boundary of the hole:

$$\sum_{n=-\infty}^{\infty} C_n e^{in\theta} = \sum_{n=0}^{\infty}\frac{A_n}{a^n}e^{-in\theta} + \sum_{n=0}^{\infty}\frac{\overline{A}_n}{a^n}e^{in\theta}$$

$$+ \sum_{n=0}^{\infty}\frac{nA_n}{a^n}e^{-in\theta} - \sum_{n=0}^{\infty}\frac{B_n}{a^n}e^{-i(n-2)\theta}$$

$$\sum_{n=-\infty}^{\infty} C_n e^{in\theta} = \sum_{n=0}^{\infty}\left[(1+n)A_n - \frac{B_{n+2}}{a^2}\right]\frac{e^{in\theta}}{a^n}$$

$$- \frac{B_1}{a}e^{i\theta} - B_0 e^{i2\theta} + \sum_{n=0}^{\infty}\frac{\overline{A}_n}{a^n}e^{in\theta}$$

Comparing the coefficients of $e^{in\theta}$ on both sides of the above equation, we get

$$A_0 + \overline{A}_o - \frac{B_2}{a^2} = C_0 \qquad (19.17.8)$$

$$\frac{\overline{A}_1}{a} - \frac{B_1}{a} = C_1 \qquad (19.17.9)$$

$$\frac{\overline{A}_2}{a^2} - B_0 = C_2 \qquad (19.17.10)$$

$$\frac{\overline{A}_n}{a^n} = C_n \quad \text{if } n \geqslant 3 \qquad (19.17.11)$$

$$\frac{1+n}{a^n} A_n - \frac{B_{n+2}}{a^{n+2}} = C_{-n} \quad \text{if } n \geqslant 1 \qquad (19.17.12)$$

Since $A_0 + \overline{A}_0$ and B_0 describe the behavior of stresses at infinity, they are assumed to be known. Also, as previously indicated, the stresses do not depend on the magnitude of the imaginary part of the constant A_o. Now, if we examine Eq. (19.17.4) we notice that the imaginary part of A_o contributes only a rigid body motion to the displacement. Therefore, the magnitude of the imaginary part of A_0 is irrelevant to the problem and can be assigned a zero value. Then A_0 becomes a real constant and $A_0 + \overline{A}_0 = 2A_0$. Now, from Eq. (19.17.5),

$$\overline{A}_1 = -\frac{1+\nu}{3-\nu} B_1 \qquad (19.17.13)$$

and substituting into Eq. (19.17.9) we get

$$B_1 = -\frac{(3-\nu)}{4} C_1 a \qquad (19.17.14)$$

From Eq. (19.17.5) it follows that

$$A_1 = \frac{(1+\nu)}{4} \overline{C}_1 a ,$$

and from Eq. (19.17.10),

$$A_2 = \overline{B}_0 a^2 + \overline{C}_2 a^2 . \qquad (19.17.15)$$

Since it was established that A_0 is real, Eq. (19.17.8) yields

$$B_2 = 2A_0 a^2 - C_0 a^2 . \qquad (19.17.16)$$

Eqs. (19.17.11) and (19.17.12) give

$$A_n = \bar{C}_n a^n \quad \text{for} \quad n \geq 3 \tag{19.17.17}$$

and

$$B_n = (n - 1)a^2 A_{n-2} - a^n C_{-n+2} \quad \text{for} \quad n \geq 3 \tag{19.17.18}$$

Thus, the coefficients are all explicitly determined and the problem is completely solved if the stress distributions are given on the circular boundaries of the hole.

Examples

a) *The plate has a hole of radius* a, *located in the middle, and is subjected to a uniform tension in the* $0X_1$ *direction* (Fig. 19.32). The boundary conditions are:

$$1) \text{ At } r = \infty, \qquad \sigma_{11} = S, \qquad \sigma_{22} = \sigma_{12} = 0 \tag{19.17.19}$$

$$2) \text{ At r} = a, \qquad \sigma_{rr} = \sigma_{r\theta} = 0 \tag{19.17.20}$$

From formulas (19.14.29) and (19.14.30), we find for $r = \infty$,

$$\sigma_{11} = S = 4 \text{ Re } \psi'(z) = 4A_0, \quad \text{and} \quad -S = 2\bar{B}_0$$

Therefore,

$$A_0 = \frac{S}{4}, \qquad \bar{B}_0 = -\frac{S}{2}$$

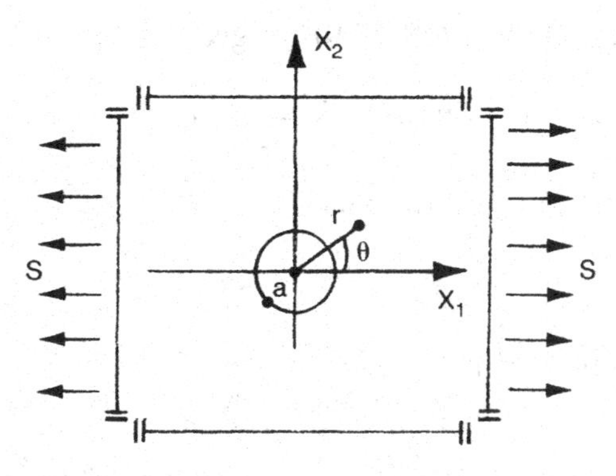

Fig. 19.32

Moreover, from Eqs. (19.17.6) and (19.17.20) we conclude that all the Fourier coefficients C_n vanish. In this case Eqs. (19.17.14) to (19.17.18) give

$$A_1 = 0, \quad B_1 = 0, \quad A_2 = -\frac{Sa^2}{2}, \quad B_2 = \frac{Sa^2}{2}$$

$$A_n = 0 \quad \text{if } n \geqslant 3$$

$$B_3 = 0, \quad B_4 = -\frac{3Sa^4}{2}$$

$$B_n = 0 \quad \text{if } n \geqslant 5$$

Accordingly the functions $\psi'(z)$ and $\chi''(z)$ are given by

$$\psi'(z) = \frac{S}{4}\left(1 - \frac{2a^2}{z^2}\right), \qquad \chi''(z) = -\frac{S}{2}\left(1 - \frac{a^2}{z^2} + \frac{3a^4}{z^4}\right)$$

Substituting these expressions into Eqs. (19.14.88) and (19.14.89) we get

$$\sigma_{rr} + \sigma_{\theta\theta} = S\left(1 - 2\frac{a^2}{r^2}\cos 2\theta\right)$$

$$\sigma_{\theta\theta} - \sigma_{rr} + 2i\sigma_{r\theta} = S\left(\frac{2a^2}{r^2}e^{-i2\theta} - e^{i2\theta} + \frac{a^2}{r^2} - \frac{3a^4}{r^4}e^{-i2\theta}\right)$$

Therefore,

$$\sigma_{rr} = \frac{S}{2}\left(1 - \frac{a^2}{r^2}\right) + \frac{S}{2}\left(1 + 3\frac{a^4}{r^4} - 4\frac{a^2}{r^2}\right)\cos 2\theta \tag{19.17.19}$$

$$\sigma_{\theta\theta} = \frac{S}{2}\left(1 + \frac{a^2}{r^2}\right) - \frac{S}{2}\left(1 + 3\frac{a^4}{r^4}\right)\cos 2\theta \tag{19.17.20}$$

$$\sigma_{r\theta} = -\frac{S}{2}\left(1 + 2\frac{a^2}{r^2} - 3\frac{a^4}{r^4}\right)\sin 2\theta \tag{19.17.21}$$

Dropping the constants C_1 and $\overline{C}_2$ from Eq. (19.17.4) and separating real and imaginary we get

$$u_r = \frac{S(1 + \nu)}{2Er}\left[\frac{1 - \nu}{1 + \nu} r^2 + a^2 + \left(\frac{4a^2}{1 + \nu} + r^2 - \frac{a^4}{r^2}\right)\cos 2\theta\right] \tag{19.17.22}$$

$$u_\theta = -\frac{S(1 + \nu)}{2Er}\left(\frac{1 - \nu}{1 + \nu} 2a^2 + r^2 + \frac{a^4}{r^2}\right)\sin 2\theta \tag{19.17.23}$$

b) *The plate has a hole or radius* a, *located in the middle and is uniformly stressed* (Fig. 19.33). The boundary conditions are:

1) At r $= \infty$ $\quad \sigma_{11} = \sigma_{22} = S, \quad \sigma_{12} = 0$ (19.17.24)

2) At $r = a$ $\quad \sigma_{rr} = \sigma_{r\theta} = 0$ (19.17.25)

From formulas (19.14.29) and (19.14.30) we find for $r = \infty$

$$\sigma_{11} + \sigma_{22} = 4 \text{ Re } \psi'(z)_{r=\infty} = 4A_0 = 2S, \text{ and } \overline{B}_0 = 0$$

Therefore,

$$A_0 = \frac{S}{2}, \qquad B_0 = 0$$

Here, too, from Eqs. (19.17.6) and (19.17.25) we conclude that all the Fourier coefficients C_n vanish and Eqs. (19.17.14) to (19.17.18) give

$$A_1 = 0, \quad B_1 = 0, \quad A_2 = 0, \quad B_2 = Sa^2$$
$$A_n = 0 \quad \text{if } n \geq 3$$

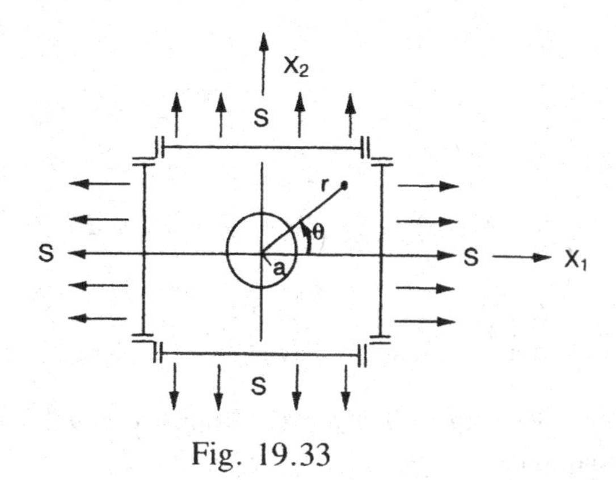

Fig. 19.33

$$B_3 = 0 \quad B_4 = 0$$
$$B_n = 0 \quad \text{if } n \geq 5$$

The functions $\psi'(z)$ and $\chi''(z)$ are given by

$$\psi'(z) = \frac{S}{2}, \qquad \chi''(z) = \frac{Sa^2}{z^2}$$

Substituting these expressions into Eqs. (19.14.88) and (19.14.89) we get

$$\sigma_{rr} + \sigma_{\theta\theta} = 2S$$
$$\sigma_{\theta\theta} - \sigma_{rr} + 2i\sigma_{r\theta} = 2e^{2i\theta}\frac{Sa^2}{z^2} = \frac{2Sa^2}{r^2}$$

Therefore,

$$\sigma_{rr} = S\left(1 - \frac{a^2}{r^2}\right) \tag{19.17.26}$$

$$\sigma_{\theta\theta} = S\left(1 + \frac{a^2}{r^2}\right) \tag{19.17.27}$$

$$\sigma_{r\theta} = 0 \tag{19.17.28}$$

Eq. (19.17.4) now yields, after dropping C_1 and $\overline{C}_2$,

$$u_r = \frac{S(1+\nu)}{Er}\left[\frac{1-\nu}{1+\nu}r^2 + a^2\right], \qquad u_\theta = 0 \tag{19.17.29}$$

It is interesting to see how the same results can be obtained using conformal mapping and the results of subsection (19.14.10). Consider the mapping function:

$$z = f(\zeta) = \frac{c}{\zeta} = \frac{a}{\zeta} \tag{19.17.30}$$

$$f'(\zeta) = -\frac{a}{\zeta^2} \tag{19.17.31}$$

$$\frac{f(\sigma)}{\overline{f}'(\overline{\sigma})} = \frac{a}{\sigma(-a\sigma^2)} = -\frac{1}{\sigma^3} \tag{19.17.32}$$

$$\sum_{k=1}^{m} (F_{1k} + iF_{2k}) = 0 \tag{19.17.33}$$

$$f_{1j} + if_{2j} = 0 \tag{19.17.34}$$

$$B = \frac{\sigma_{11}(\infty) + \sigma_{22}(\infty)}{4} = \frac{S}{2} \tag{19.17.35}$$

$$B' = \frac{\sigma_{22}(\infty) - \sigma_{11}(\infty)}{2} = 0 \tag{19.17.36}$$

$$C' = \sigma_{12}(\infty) = 0 \tag{19.17.37}$$

$$c = \bar{c} = a \tag{19.17.38}$$

Therefore,

$$F^0(\sigma) = -\frac{Sa}{2\sigma} + \frac{1}{\sigma^3}\left(-\frac{Sa}{2}\sigma^2\right) = -\frac{Sa}{\sigma} \tag{19.17.39}$$

If we now set $\psi^0(\zeta) = \sum_{k=1}^{\infty} a_k \zeta^k$, we have on the circle

$$\psi^0(\sigma) = \sum_{k=1}^{\infty} a_k \sigma^k = a_1 \sigma + a_2 \sigma^2 + \ldots \tag{19.17.40}$$

$$(\bar{\psi}^0)'(\bar{\sigma}) = \bar{a}_1 + \frac{2\bar{a}_2}{\sigma} + \frac{3\bar{a}_3}{\sigma_2} + \ldots \tag{19.17.41}$$

If we also set $\gamma^0(\zeta) = \sum_{k=0}^{\infty} b_k \zeta^k$, we have on the circle

$$\gamma^0(\sigma) = \sum_{k=0}^{\infty} b_k \sigma^k \tag{19.17.42}$$

$$\bar{\gamma}^0(\bar{\sigma}) = \bar{b}_0 + \frac{\bar{b}_1}{\sigma} + \frac{\bar{b}_2}{\sigma^2} + \ldots \tag{19.17.43}$$

From Eq. (19.14.141) we get

$$\begin{aligned} &a_1\sigma + a_2\sigma^2 + a_3\sigma^3 + \ldots \\ &- \frac{1}{\sigma^3}\left(\bar{a}_1 + \frac{2\bar{a}_2}{\sigma} + \frac{3\bar{a}_3}{\sigma^2} + \ldots\right) \\ &+ \bar{b}_0 + \frac{\bar{b}_1}{\sigma} + \frac{\bar{b}_2}{\sigma^2} + \ldots = -\frac{Sa}{\sigma} \end{aligned} \tag{19.17.44}$$

Comparison of like powers of σ yields

$$\begin{aligned} a_n &= 0 \quad \text{and therefore } \bar{a}_n = 0 \\ \bar{b}_1 &= -Sa \\ \bar{b}_n &= 0 \quad \text{for } n \neq 1 \\ b_1 &= -Sa \\ b_n &= 0 \quad \text{for } n \neq 1 \end{aligned}$$

The expressions for $\psi^0(\zeta)$ and $\gamma^0(\zeta)$ now become

$$\begin{aligned} \psi^0(\zeta) &= 0 \\ \gamma^0(\zeta) &= -Sa\zeta \end{aligned}$$

so that Eqs. (19.14.137) and (19.14.138) give

$$\begin{aligned} \psi_1(\zeta) &= \frac{Bc}{\zeta} = \frac{Sa}{2\zeta} \\ \gamma_1(\zeta) &= -Sa\zeta \end{aligned}$$

Therefore,

$$\begin{aligned} \psi(z) &= \frac{S}{2} z \\ \gamma(z) &= -\frac{Sa^2}{z} \end{aligned}$$

Those two complex potentials yield the results given by Eqs. (19.17.26) to (19.17.29).

c) *The plate has a hole or radius* a, *located in the middle, and there is a uniform pressure p applied along the edge of the hole* (Fig. 19.34). The boundary conditions are

1. At $r = \infty \qquad \sigma_{11} = \sigma_{22} = \sigma_{12} = 0$
2. At $r = a \qquad \sigma_{rr} = -p, \ \sigma_{r\theta} = 0$

From formulas (19.14.29) and (19.14.30) we find for $r = \infty$

$$\sigma_{11} + \sigma_{22} = 4 \operatorname{Re} \psi'(z)_{r=\infty} = 4A_0 = 0 \quad \text{and } \bar{B}_0 = 0$$

From Eq. (19.17.7)

$$-p = C_0$$

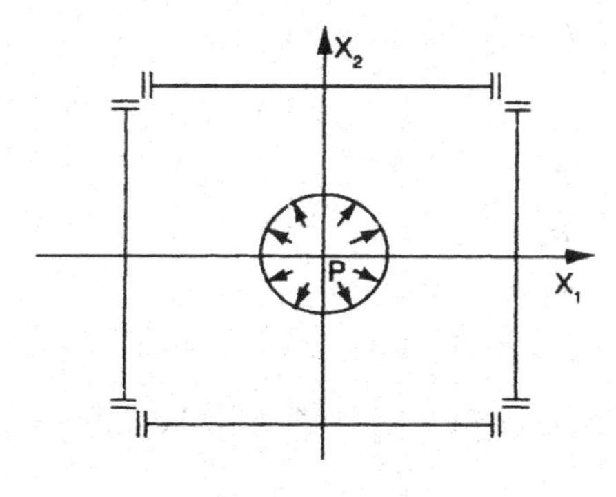

Fig. 19.34

and all other C's are equal to zero. Thus,

$$A_1 = B_1 = A_2 = 0$$
$$B_2 = +pa^2$$
$$A_n = 0 \quad \text{for } n \geqslant 3$$
$$B_n = 0 \quad \text{for } n \geqslant 3$$

Accordingly the functions $\psi'(z)$ and $\chi''(z)$ are given by

$$\psi'(z) = 0$$
$$\chi''(z) = \frac{pa^2}{z^2}$$

Substituting these expressions into Eqs. (19.14.88) and (19.14.89) we get

$$\sigma_{rr} + \sigma_{\theta\theta} = 0$$
$$\sigma_{\theta\theta} - \sigma_{rr} = \frac{2pa^2}{r^2}$$

Therefore,

$$\sigma_{rr} = -\frac{pa^2}{r^2}, \qquad \sigma_{\theta\theta} = \frac{pa^2}{r^2}, \qquad \sigma_{r\theta} = 0$$

Eq. (19.17.4) yield, after dropping C_1 and $\overline{C}_2$

$$u_r = \frac{pa^2(1 + \nu)}{Er}, \qquad u_\theta = 0$$

19.18 The Infinite Plate under the Action of a Concentrated Force and Moment.

Let us consider an infinite plate of thickness h with a circular hole of radius a (Fig. 19.35). Assume that the external force acting on the boundary of the hole has components along $0X_1$ and $0X_2$ equal to P_1 and P_2. The stress distribution can be obtained by analysing the effect of the constant stress distribution

$$T_1 = \frac{P_1}{2\pi ah}, \qquad T_2 = \frac{P_2}{2\pi ah}, \tag{19.18.1}$$

acting on the boundary $|z| = a$. The stresses at infinity are assumed to equal zero. The components of the stress vector on the tangent plane at A are

$$(\sigma_{rr})_{r=a} = -\frac{1}{2\pi ah}(P_1 \cos\theta + P_2 \sin\theta)$$

$$(\sigma_{r\theta})_{r=a} = -\frac{1}{2\pi ah}(-P_1 \sin\theta + P_2 \cos\theta)$$

Therefore, on the circle

$$(\sigma_{rr} - i\sigma_{r\theta})_{r=a} = -\frac{1}{2\pi ah}(P_1 - iP_2)e^{i\theta} \tag{19.18.2}$$

The only coefficient that does not vanish in Eq. (19.17.6) is

$$C_1 = -\frac{1}{2\pi ah}(P_1 - iP_2) \tag{19.18.3}$$

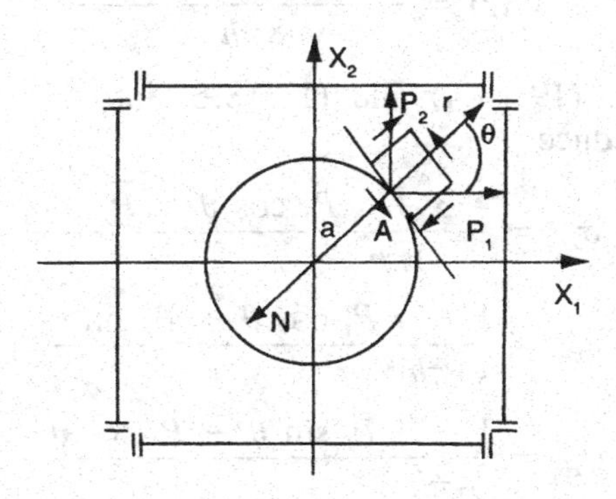

Fig. 19.35

Since the stresses at infinity are assumed equal to zero, then (see Sec. 19.17)

$$A_0 = B_0 = 0$$

From Eq. (19.17.14) to (19.17.18) we find that

$$A_1 = \frac{-(1 + \nu)(P_1 + iP_2)}{8\pi h}, \qquad A_n = 0 \quad \text{if} \quad n \geqslant 2$$

$$B_1 = \frac{(3 - \nu)(P_1 - iP_2)}{8\pi h}, \qquad B_2 = 0$$

$$B_3 = \frac{-a^2(1 + \nu)(P_1 + iP_2)}{4\pi h}, \qquad B_n = 0 \quad \text{if} \quad n \geqslant 4$$

Thus,

$$\psi'(z) = \frac{-(1 + \nu)(P_1 + iP_2)}{8\pi h}\frac{1}{z} \tag{19.18.4}$$

and

$$\chi''(z) = \frac{(3 - \nu)(P_1 - iP_2)}{8\pi h}\frac{1}{z} - \frac{(1 + \nu)(P_1 + iP_2)}{4\pi h}\frac{a^2}{z^3} \tag{19.18.5}$$

If we now let the radius a tend to zero, while allowing T_1 and T_2 to increase in such a way that the resultant of P_1 and P_2 has always the same magnitude and direction, we find that the second term of $\chi''(z)$ approaches zero. Thus,

$$\chi''(z) = \frac{(3 - \nu)(P_1 - iP_2)}{8\pi h}\frac{1}{z} \tag{19.18.6}$$

Substituting Eqs. (19.18.4) and (19.18.6) into Eqs. (19.14.88) and (19.14.89) we deduce

$$\sigma_{rr} = -\frac{3 + \nu}{4\pi h}\frac{P_1 \cos\theta + P_2 \sin\theta}{r} \tag{19.18.6}$$

$$\sigma_{\theta\theta} = \frac{1 - \nu}{4\pi h}\frac{P_1 \cos\theta + P_2 \sin\theta}{r} \tag{19.18.7}$$

$$\sigma_{r\theta} = \frac{1 - \nu}{4\pi h}\frac{P_1 \sin\theta - P_2 \cos\theta}{r} \tag{19.18.8}$$

Those three formulas give the distribution of stress due to a concentrated force applied at the origin of coordinates of an infinite plate.

To find the stress distribution due to a couple applied at the origin we consider that a uniform tangential stress of magnitude τ is applied along the edge of the circular hole of radius a. Then, at $r = a$,

$$\sigma_{rr} = 0, \qquad \sigma_{r\theta} = \tau$$

Here, too, we will assume that the stresses vanish at infinity so that, $A_0 = B_0 = 0$. Since

$$(\sigma_{rr} - i\sigma_{r\theta})_{r=a} = -i\tau$$

the coefficients in the Fourier expansion in Eq. (19.17.6) all vanish except the constant term

$$C_0 = -i\tau$$

Also, from Eq. (19.17.14) to Eq. (19.17.18), we find

$$A_n = 0 \quad \text{for all values of } n$$

$$B_n = 0 \quad \text{for all values of } n \text{ except } n = 2$$

$$B_2 = -C_0 a^2 = i\tau a^2$$

Now the moment of the external forces applied to the boundary of the hole is

$$M = -2\pi a^2 h\tau$$

so that

$$B_2 = -\frac{iM}{2\pi h}, \qquad \psi'(z) = 0, \qquad \chi''(z) = -\frac{iM}{2\pi h}\frac{1}{z^2}$$

The above leads to

$$\sigma_{rr} = \sigma_{\theta\theta} = 0 \tag{19.18.9}$$

$$\sigma_{r\theta} = -\frac{M}{2\pi h r^2} \tag{19.18.10}$$

Formulas (19.18.9) and (19.18.10) give the distribution in an infinite plate due to a couple M acting uniformly about a hole of radius a. These formulas do not change if we reduce the size of the hole to zero and increase τ so that M remains constant. In this case the formulas give the distribution in an infinite plate due to a concentrated couple of moment M acting at the origin.

19.19 The Infinite Plate with an Elliptic Hole Subjected to a Tensile Stress Normal to the Major Principal Axis of the Ellipse. (Fig. 19.36)

Referring to the representation in terms of elliptical coordinates in Sec. 19.14 we have

$$z = c \cosh \zeta, \quad x_1 = c \cosh \eta_1 \cos \eta_2, \quad x_2 = c \sinh \eta_1 \sin \eta_2 \tag{19.19.1}$$

with the principal radii of the ellipse given by

$$a = c \cosh \eta_1, \qquad b = c \sinh \eta_1 \tag{19.19.2}$$

Different values of c and η_1 yield different values of a and b and thus, different ellipses. (The reader is cautioned about the way c appears in the various sections.) Notice that as $\eta_1 \rightarrow 0$ the ellipse becomes an opening (a crack) of length $2c = 2a$. As the angle η_2 varies from 0 to 2π a point (x_1, x_2) goes all around the ellipse

$$\frac{x_1^2}{a^2} + \frac{x_2^2}{b^2} = 1. \tag{19.19.3}$$

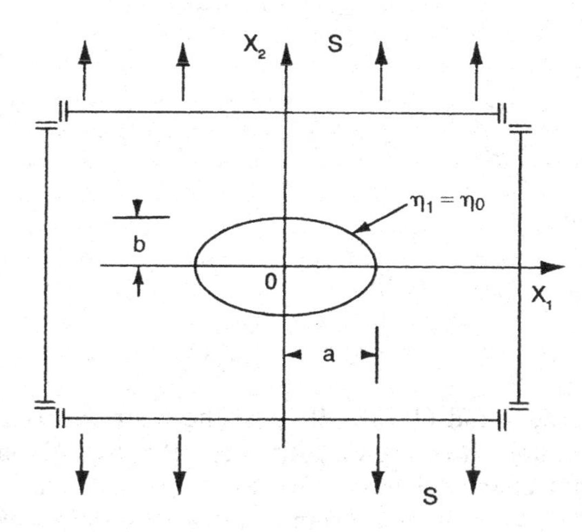

Fig. 19.36

Therefore, the continuity in the stress and displacements demands that their expression be periodic in η_2 with period 2π so that they have the same value at 0 and at 2π. We shall use here the elliptical coordinates defined in Sec. 19.14 and Fig. 19.24b. The stress components $\sigma_{\eta 1}$, $\sigma_{\eta 2}$ and $\sigma_{\eta 12}$ will refer to directions defined by the η_1 and η_2 curves (Fig. 19.37).

The boundary conditions are (from Eqs. 19.14.29 and 19.14.30)

$$\left.\begin{aligned} 4 \operatorname{Re} \psi'(z) &= S \\ 2[\bar{z}\psi''(z) + \chi''(z)] &= S \end{aligned}\right\} \text{at infinity;} \qquad (19.19.4)$$

and on the edges of the elliptical hole, $\eta_1 = \eta_0$ say,

$$\sigma_{\eta 1} = \sigma_{\eta 12} = 0. \qquad (19.19.5)$$

Inglis [11] found complex potentials which satisfy both these boundary conditions and which are periodic in η_2 with period 2π, as follows:

$$4\psi(z) = Sc[(1 + e^{2\eta_0})\sinh \zeta - e^{2\eta_0} \cosh \zeta] \qquad (19.19.6)$$

$$4\chi(z) = -Sc^2 \left[(\cosh 2\eta_0 - \cosh \pi)\zeta + \frac{1}{2} e^{2\eta_0} \cosh 2\left(\zeta - \eta_0 - i\frac{\pi}{2}\right) \right] \qquad (19.19.7)$$

Since $\sigma_{\eta 1}$ is zero on the surface of the hole we can obtain $\sigma_{\eta 2}$ on this surface from Eq. (19.14.92):

$$(\sigma_{\eta 2})_{\eta_1 = \eta_0} = S \left[\frac{\sinh 2\eta_0 - 1 + e^{2\eta_0} \cos 2\eta_2}{\cosh 2\eta_0 - \cos 2\eta_2} \right] \qquad (19.19.8)$$

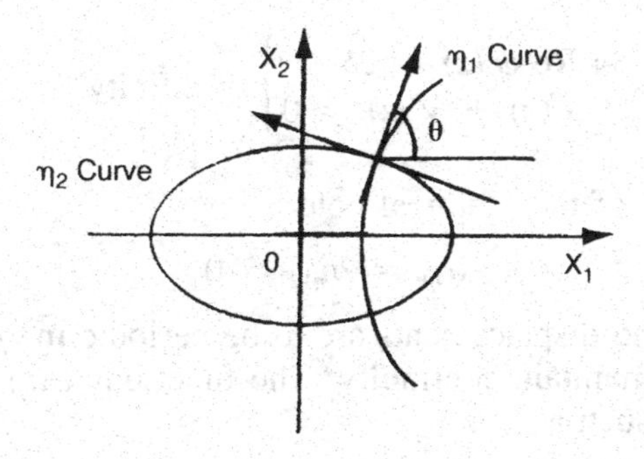

Fig. 19.37

The maximum values of $\sigma_{\eta 2}$ correspond to those at the ends of the major axis where $\eta = 0$ or π. Here $\cos 2\eta_2 = 1$ and we obtain

$$(\sigma_{\eta 2})_{\eta_2 = 0,\ \pi} \equiv \sigma_{22} = S\left[\frac{\sinh 2\eta_0 - 1 + e^{2\eta_0}}{\cosh 2\eta_0 - 1}\right] \tag{19.19.9}$$

Since $c^2 = a^2 - b^2$ we have

$$\sinh 2\eta_0 = \frac{2ab}{c^2}, \qquad \cosh 2\eta_0 = \frac{a^2 + b^2}{c^2} \tag{19.19.10}$$

Thus, at the tips of the hole (crack tips),

$$(\sigma_{\eta 2})_{\eta_2 = 0,\ \pi} \equiv \sigma_{22} = S\left(1 + 2\frac{a}{b}\right) \tag{19.19.11}$$

This value increases without limit as the elliptic hole becomes longer and more slender; and when $a = b$, it becomes identical to the result given by Eq. (19.17.20) when we set $r = a$ and $\theta = \pi/2$. At the tip of the ellipse the radius of curvature $\rho = b^2/a$ so that

$$(\sigma_{\eta 2})_{\eta_2 = 0,\ \pi} = S\left(1 + 2\sqrt{\frac{a}{\rho}}\right) \tag{19.19.12}$$

This problem will be examined again using conformal mapping.

19.20 Infinite Plate with an Elliptic Hole Subjected to a Uniform all around Tension *S*. (Fig. 19.38)

The boundary conditions are

$$\left.\begin{aligned} 4\,\mathrm{Re}\,\psi'(z) &= 2S \\ \bar{z}\psi''(z) + \chi''(z) &= 0 \end{aligned}\right\} \text{ at infinity}, \tag{19.20.1}$$

and on the edges of the elliptical hole

$$\sigma_{\eta 1} = \sigma_{\eta 12} = 0 \tag{19.20.2}$$

The stresses and displacements are to be periodic in η_2 with a period equal to 2π, to maintain continuity. The functions $\psi(z)$ and $\chi(z)$ could thus take forms such as

$$\sinh n\zeta = \sinh n\eta_1 \cos n\eta_2 + i \cosh n\eta_1 \sin n\eta_2$$

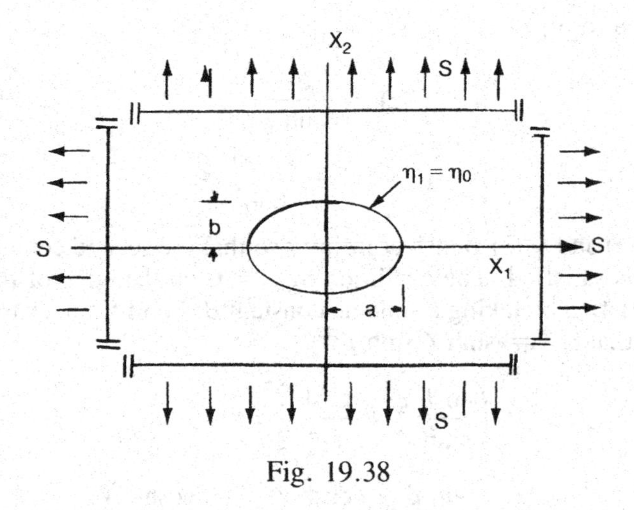

Fig. 19.38

and

$$\cosh n\zeta = \cosh n\eta_1 \cos n\eta_2 + i \sinh n\eta_1 \sin n\eta_2$$

where n is an integer. The function $\chi(z) = Bc^2\zeta$ with B being a constant is also an acceptable function for this problem. Let us take

$$\psi(z) = Ac \sinh \zeta \tag{19.20.3}$$

and,

$$\chi(z) = Bc^2\zeta \tag{19.20.4}$$

A is a constant. Now,

$$\psi'(z) = Ac \cosh \zeta \frac{d\zeta}{dz} = \frac{Ac \cosh \zeta}{c \sinh \zeta} = A \coth \zeta \tag{19.20.5}$$

since $d\zeta/dz = 1/(dz/d\zeta)$. At infinity η_1 is infinite and $\coth \zeta = 1$. Thus, Eq. (19.20.1) gives, $2A = S$.

$$\psi''(z) = -\frac{A}{c}\frac{1}{\sinh^3 \zeta} \tag{19.20.6}$$

and

$$\bar{z}\psi''(z) = -\frac{A \cosh \bar{\zeta}}{\sinh^3 \zeta} \tag{19.20.7}$$

From Eq. (19.20.4),

$$\chi'(z) = \frac{Bc}{\sinh \zeta} \tag{19.20.8}$$

$$\chi''(z) = -\frac{B \cosh \zeta}{\sinh^3 \zeta} \tag{19.20.9}$$

Both $\bar{z}\psi''(z)$ and $\chi''(z)$ vanish at infinity so that the second condition (Eq. 19.20.1) is satisfied. The condition $\sigma_{\eta 12} = 0$ on the edge of the ellipse can be satisfied by taking a suitable constant B: From Eq. (19.14.94) and recalling that $e^{2i\theta} = \sinh \zeta / \sinh \bar{\zeta}$;

$$\begin{aligned}\sigma_{\eta 1} - i\sigma_{\eta 12} &= A\left(\frac{\cosh \zeta}{\sinh \zeta} + \frac{\cosh \bar{\zeta}}{\sinh \bar{\zeta}}\right) \\ &\quad + \frac{\sinh \zeta}{\sinh \bar{\zeta}}\left(A\frac{\cosh \bar{\zeta}}{\sinh^3 \zeta} + B\frac{\cosh \zeta}{\sinh^3 \zeta}\right) \\ &= \frac{1}{\sinh^2 \zeta \sinh \bar{\zeta}}\{A[\sinh \zeta \sinh(\zeta + \bar{\zeta}) + \cosh \bar{\zeta}] \\ &\quad + B \cosh \zeta\}\end{aligned} \tag{19.20.10}$$

At the boundary of the hole $\eta_1 = \eta_0$ and $(\zeta + \bar{\zeta}) = 2\eta_0$, thus, Eq. (19.20.10) reduces to

$$\sigma_{\eta 1} - i\sigma_{\eta 12} = \frac{1}{\sinh^2 \zeta \sinh \bar{\zeta}}(A \cosh 2\eta_0 + B)\cosh \zeta$$

Thus, the condition that $\sigma_{\eta 12} = 0$ on the edge of the ellipse is satisfied if

$$B = -A \cosh 2\eta_0 = -\frac{S}{2} \cosh 2\eta_0 \tag{19.20.11}$$

The two potentials $\psi(z)$ and $\chi(z)$ are now given by

$$\psi(z) = \frac{1}{2} Sc \sinh \zeta \tag{19.20.12}$$

$$\chi(z) = -\frac{1}{2}(Sc^2 \cosh 2\eta_0)\zeta \tag{19.20.13}$$

These potentials satisfy all the boundary conditions and they also result in no discontinuities in the displacements; indeed, Eq. (19.14.26) gives

$$2G(u_1 + iu_2) = \frac{3 - \nu}{1 + \nu}\frac{Sc}{2}\sinh\zeta - \frac{Sc}{2}\cosh\zeta\coth\bar{\zeta} + \frac{Sc}{2}\frac{\cosh 2\eta_0}{\sinh\bar{\zeta}} \tag{19.20.14}$$

Both the real and imaginary parts of the hyperbolic functions are periodic in η_2. Thus, a cycle around any ellipse η_1 = constant will bring u_1 and u_2 back to their initial values. Thus, the complex potentials Eqs. (19.20.12) and (19.20.13) give the solution to the problem.

The stress component $\sigma_{\eta 2}$ at the rim of the hole is obtained by setting $\sigma_{\eta 1} = 0$ and $\eta_1 = \eta_0$ in Eqs. (19.14.92) and (19.20.5):

$$(\sigma_{\eta 2})_{\eta_1 = \eta_0} = \frac{2S\sinh 2\eta_0}{\cosh 2\eta_0 - \cos 2\eta_2} \tag{19.20.15}$$

The maximum value in Eq. (19.20.15) occurs at the ends of the major axes where $\eta_2 = 0$ and π and $\cos 2\eta_2 = 1$. Thus,

$$(\sigma_{\eta 2})_{\max} = \frac{2S\sinh 2\eta_0}{\cosh 2\eta_0 - 1} = \frac{2Sa}{b}, \tag{19.20.16}$$

since $c^2 = a^2 - b^2$, $\sinh 2\eta_0 = 2ab/c^2$ and $\cosh 2\eta_0 = (a^2 + b^2)/c^2$. As the ellipse becomes more slender this value becomes larger and larger. The minimum value in Eq. (19.20.15) occurs at the ends of the minor axes where $\cos 2\eta_2 = -1$. Thus,

$$(\sigma_{\eta 2})_{\min} = \frac{2Sb}{a}. \tag{19.20.17}$$

When the ellipse becomes a circle, $a = b$, and we get results in agreement with those obtained for a circular hole, i.e., $(\sigma_{\eta 2})_{\max} = (\sigma_{\eta 2})_{\min} = 2S$.

19.21 Conformal Mapping Applied to the Problem of the Elliptic hole.

Consider again the case of an infinite region bounded by an ellipse

$$\frac{x_1^2}{a^2} + \frac{x_2^2}{b^2} = 1 \tag{19.21.1}$$

The mapping function

$$z = f(\zeta) = c\left(\frac{1}{\zeta} + m\zeta\right), \quad c > 0, \quad 0 \leqslant m \leqslant 1 \quad (19.21.2)$$

with (Fig. 19.39)

$$c = \frac{a + b}{2}, \qquad m = \frac{a - b}{a + b} \qquad (19.21.3)$$

transforms the region exterior to the ellipse (19.21.1) into a circle $|\zeta| \leqslant 1$. This can be shown as follows: At $\zeta = 0$, $z = \infty$ and,

$$z = x_1 + ix_2 = c\left(\frac{1}{\zeta} + m\zeta\right) = c\left[\frac{1}{\xi_1 + i\xi_2} + m(\xi_1 + i\xi_2)\right]$$
$$= c\left[\frac{\xi_1 - i\xi_2}{\xi_1^2 + \xi_2^2} + m\xi_1 + im\xi_2\right].$$

But $\xi_1^2 + \xi_2^2 = 1$ on the unit circle, therefore,

$$x_1 = c(1 + m)\xi_1, \qquad x_2 = -c(1 - m)\xi_2 \qquad (19.21.4)$$

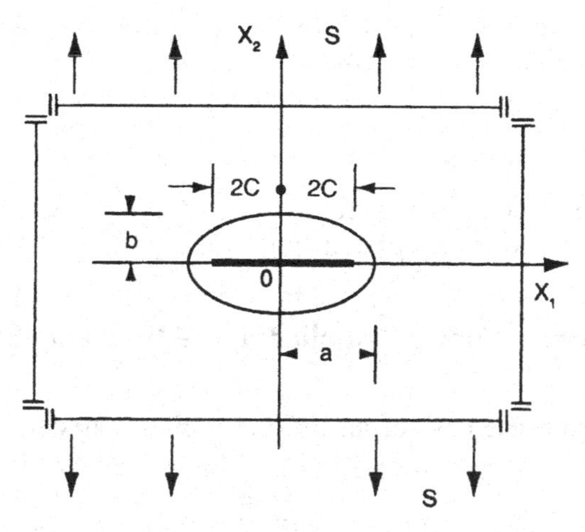

Fig. 19.39

Substituting in the left hand side of Eq. (19.21.1) we get

$$\frac{c^2}{a^2}(1 + m)^2\xi_1^2 + \frac{c^2}{b^2}(1 - m)^2\xi_2^2 = 1 , \qquad (19.21.5)$$

since

$$\frac{c^2}{a^2}(1 + m)^2 = \frac{c^2}{b^2}(1 - m)^2 = 1 . \qquad (19.21.6)$$

As the point $\zeta = e^{i\alpha}$ describes the circle $|\zeta| = 1$ in the positive or counter clockwise direction, the corresponding point z traces the ellipse in the clockwise direction. Indeed, Eqs. (19.21.4) are the parametric equations of the ellipse and as ξ_2 increases positively, x_2 increases negatively.

If $m = 0$ the ellipse becomes a circle and if $m = 1$ the ellipse reduces to a slit along the X_1 axis. The point in the Z plane traces out the segment of the $0X_1$ axis between $x_1 = 2c$ and $x_1 = -2c$ twice as the point ζ describes once the boundary $|\zeta| = 1$. Thus, in this case the function (19.21.2) maps the Z plane slit along the line joining the points $(+2c, 0)$ and $(-2c, 0)$ onto $|\zeta| \leqslant 1$.

It was shown in Subsection 10, Eq. (19.14.141) that the solution of the first boundary value problem for an infinite simply connected domain can be reduced to the determination of two functions $\psi^0(\zeta)$ and $\gamma^0(\zeta)$ analytic in the circle $|\zeta| < 1$, and which satisfy the boundary condition.

$$\psi^0(\sigma) + \frac{f(\sigma)}{\bar{f}'(\bar{\sigma})}(\bar{\psi}^0)'(\bar{\sigma}) + \bar{\gamma}^0(\bar{\sigma}) = F^0(\sigma) \qquad (19.21.7)$$

Let us follow with this equation the same steps followed in Subsection 9, Section 14 with Eq. (19.14.116).

$$\psi^0 = \sum_{k=1}^{\infty} a_k\zeta^k \qquad (19.21.8)$$

$$\gamma^0 = \sum_{k=0}^{\infty} b_k\zeta^k \qquad (19.21.9)$$

$$F^0(\sigma) = \sum_{k=-\infty}^{\infty} t_k e^{ik\theta} = \sum_{k=-\infty}^{\infty} t_k\sigma^k \qquad (19.21.10)$$

with

$$t_k = \frac{1}{2\pi}\int_0^{2\pi} F^0(\sigma)e^{-ik\theta}\,d\theta \tag{19.21.11}$$

Now, in our problem

$$f(\sigma) = \frac{c}{\sigma}(1 + m\sigma^2) \tag{19.21.12}$$

$$f'(\sigma) = c(-\frac{1}{\sigma^2} + m) \tag{19.21.13}$$

$$\overline{f}'(\overline{\sigma}) = c(-\sigma^2 + m) \tag{19.21.14}$$

so that

$$\frac{f(\sigma)}{\overline{f}'(\overline{\sigma})} = \frac{1}{\sigma}\frac{(1 + m\sigma^2)}{(m - \sigma^2)} \tag{19.21.15}$$

$$= -\frac{1 + m\sigma^2}{\sigma^3}\left(1 + \frac{m}{\sigma^2} + \frac{m^2}{\sigma^4} + \ldots\right) \tag{19.21.16}$$

Also [see Eq. (19.14.118)]

$$\frac{f(\sigma)}{\overline{f}'(\overline{\sigma})} = \sum_{k=-\infty}^{\infty} c_k e^{ik\alpha} = \sum_{-\infty}^{\infty} c_k\sigma^k \tag{19.21.17}$$

Therefore,

$$c_k = 0 \quad \text{for } k \geqslant 0$$

From Eq. (19.14.120)

$$a_k + \sum_{m=1}^{\infty} mc_{k+m-1}\overline{a}_m = t_k$$

But since $c_{k+m-1} = 0$ for all $k + m - 1 \geqslant 0$, then

$$a_k = t_k \quad k \geqslant 1 \tag{19.21.18}$$

Therefore,

$$\psi^0(\zeta) = \sum_{k=1}^{\infty} t_k\zeta^k \tag{19.21.19}$$

where

$$t_k = \frac{1}{2\pi}\int F^0(\sigma)e^{-i\kappa\theta}\,d\theta \tag{19.21.20}$$

and $F^0(\sigma)$ is given by Eq. (19.14.143)

The functions $\psi^0(\zeta)$ [as well as $\gamma^0(\zeta)$] can also be obtained through an integral representation. Indeed, from Eqs. (19.21.19) and (19.21.20)

$$\psi^0(\zeta) = \sum_{k=1}^{\infty}\frac{1}{2\pi}\left[\left[\int_0^{2\pi} F^0(\sigma)e^{-ik\theta}\,d\theta\right]\zeta^k\right. \tag{19.21.21}$$

Setting

$$\sigma = e^{i\theta} \quad\text{and}\quad \frac{1}{\sigma} = e^{-i\theta}$$

$$d\sigma = ie^{i\theta}\,d\theta \quad\text{and}\quad d\theta = \frac{1}{i}e^{-i\theta}\,d\sigma = \frac{1}{i\sigma}\,d\sigma\,.$$

Therefore,

$$\psi^0(\zeta) = \frac{1}{2\pi}\int_\gamma \frac{F^0(\sigma)}{i\sigma}\sum_{k=1}^{\infty}\frac{\zeta^k}{\sigma^k}\,d\sigma \tag{19.21.22}$$

Since

$$\sum_{k=0}^{\infty}\frac{\zeta^k}{\sigma^k} = \frac{1}{1-\dfrac{\zeta}{\sigma}}$$

Then

$$\psi^0(\zeta) = \frac{1}{2\pi i}\int_\gamma \frac{F^0(\sigma)}{\sigma}\left[\frac{1}{1-\dfrac{\zeta}{\sigma}} - 1\right]d\sigma$$

$$\psi^0(\zeta) = \frac{\zeta}{2\pi i}\int_\gamma \frac{F^0(\sigma)}{\sigma}\,\frac{1}{\sigma-\zeta}\,d\sigma \tag{19.21.23}$$

The same way, instead of using a series representation to find $\gamma^0(\zeta)$ based on the calculation of the coefficients b_k one can obtain an integral representation: Taking the conjugate of Eq. (19.21.7),

$$\bar{\psi}^0(\bar{\sigma}) + \frac{\bar{f}(\bar{\sigma})}{f'(\sigma)}(\psi^0)'(\sigma) + \gamma^0(\sigma) = \bar{F}^0(\bar{\sigma}), \quad (19.21.24)$$

multiplying all the terms by $(1/2\pi i)(d\sigma/\sigma - \zeta)$, and integrating, we get

$$\frac{1}{2\pi i}\int_\gamma \frac{\bar{\psi}^0(\bar{\sigma})}{\sigma - \zeta} d\sigma + \frac{1}{2\pi i}\int_\gamma \frac{\bar{f}(\bar{\sigma})}{f'(\sigma)} \frac{(\psi^0)'(\sigma)}{\sigma - \zeta} d\sigma + \frac{1}{2\pi i}\int_\gamma \frac{\gamma^0(\sigma)}{\sigma - \zeta} d\sigma$$
$$= \frac{1}{2\pi i}\int_\gamma \frac{\bar{F}^0(\bar{\sigma})}{\sigma - \zeta} d\sigma \quad (19.21.25)$$

Referring to Sec. 19.11, the first integral is equal to $\bar{\psi}^0(0) = 0$. Also, by Cauchy's integral formula,

$$\frac{1}{2\pi i}\int_\gamma \frac{\gamma^0(\sigma)}{\sigma - \zeta} d\sigma = \gamma^0(\zeta) \quad (19.21.26)$$

Therefore,

$$\gamma^0(\zeta) = \frac{1}{2\pi i}\int_\gamma \frac{\bar{F}^0(\bar{\sigma})}{\sigma - \zeta} d\sigma - \frac{1}{2\pi i}\int_\gamma \frac{\bar{f}(\bar{\sigma})}{f'(\sigma)} \frac{(\psi^0)'(\sigma)}{\sigma - \zeta} d\sigma$$

But from Eq. (19.21.15) we have (for $m < 1$)

$$\frac{\bar{f}(\bar{\sigma})}{f'(\sigma)} = \sigma \frac{1 + \frac{m}{\sigma^2}}{m - \frac{1}{\sigma^2}} = \frac{\sigma(\sigma^2 + m)}{m\sigma^2 - 1} \quad (19.21.28)$$

so that

$$\frac{1}{2\pi i}\int_\gamma \frac{\bar{f}(\bar{\sigma})}{f'(\sigma)} \frac{(\psi^0)'(\sigma)}{\sigma - \zeta} d\sigma = \frac{1}{2\pi i}\int_\gamma \frac{\sigma(\sigma^2 + m)}{m\sigma^2 - 1} \frac{(\psi^0)'(\sigma)}{\sigma - \zeta} d\sigma .$$

By Cauchy's integral theorem,

$$\frac{1}{2\pi i}\int_\gamma \frac{\sigma(\sigma^2 + m)}{m\sigma^2 - 1} \frac{(\psi^0)'(\sigma)}{\sigma - \zeta} d\sigma = \frac{\zeta(\zeta^2 + m)}{m\zeta^2 - 1}(\psi^0)'(\zeta) .$$

Finally,

$$\gamma^0(\zeta) = \frac{1}{2\pi i}\int_\gamma \frac{\bar{F}^0(\bar{\sigma})}{\sigma - \zeta} d\sigma + \frac{\zeta(\zeta^2 + m)}{1 - m\zeta^2}(\psi^0)'(\zeta) \quad (19.21.29)$$

Let us go back to Fig. 19.39.

$$\sigma_{11}(\infty) = 0, \quad \sigma_{22}(\infty) = S, \quad \sigma_{12}(\infty) = 0.$$

Thus the constants, B, B', C' in Eq. (19.14.143) are determined by

$$B = \frac{S}{4}, \quad B' = +\frac{S}{2}, \quad C' = 0. \qquad (19.21.30)$$

Compare now Eq. (19.14.136) and Eq. (19.21.2). c is a real quantity and the summation in Eq. (19.14.136) has been replaced by $cm\zeta$. Eq. (19.14.143) becomes in our case,

$$F^0(\sigma) = -\frac{S}{4}\frac{c}{\sigma} - (+\frac{S}{2}c\sigma) - \frac{1 + m\sigma^2}{\sigma(m - \sigma^2)}(-\frac{S}{4}c\sigma^2)$$

$$= -\frac{Sc}{4}\left[\frac{1}{\sigma} + 2\sigma + \frac{\sigma(1 + m\sigma^2)}{\sigma^2 - m}\right] \qquad (19.21.31)$$

and

$$\overline{F}^0(\overline{\sigma}) = -\frac{Sc}{4}\left[\sigma + \frac{2}{\sigma} + \frac{\sigma^2 + m}{\sigma(1 - m\sigma^2)}\right] \qquad (19.21.32)$$

Substitution of these two equations in Eq. (19.21.23) and (19.21.29) yields $\psi^0(\zeta)$ and $\gamma^0(\zeta)$. Those two functions when introduced in Eqs. (19.14.137) and (19.14.138) give $\psi_1(\zeta)$ and $\gamma_1(\zeta)$:

$$\psi_1(\zeta) = \frac{Sc\zeta}{4}(-2-m + \frac{1}{\zeta^2}) \qquad (19.21.33)$$

$$\gamma_1(\zeta) = -\frac{Sc}{2}\left[-\frac{1}{\zeta} - \frac{\zeta}{m} + \frac{(1 + m^2)(1 + m)}{m}\frac{\zeta}{1 - m\zeta^2}\right] \qquad (19.21.34)$$

from which the displacements and stresses can be computed [2,3].

19.22 Infinite Plate with an Elliptic Hole Subjected to a Uniform Pressure *P*. (Fig. 19.40)

The magnitude of the pressure acting on the elliptical contour is p. Thus,

$$\sigma_{n1} = -p\,\frac{dx_2}{ds} \qquad \sigma_{n2} = p\,\frac{dx_1}{ds} \tag{19.22.1}$$

Thus,

$$\begin{aligned} H(\sigma) &= i\int_s (\sigma_{n1} + i\sigma_{n2})\,ds \\ &= -p\int_s \left(\frac{dx_1}{ds} + i\,\frac{dx_2}{ds}\right) ds = -pz \end{aligned} \tag{19.22.2}$$

Assuming $\sigma_{ij}(\infty) = 0$, $F^0(\sigma) = H(\sigma) = -pz = -pf(\zeta)$. Therefore,

$$F^0(\sigma) = -pc\left(\frac{1}{\sigma} + m\sigma\right) = -pc\,\frac{1 + m\sigma^2}{\sigma} \tag{19.22.3}$$

and

$$\overline{F}^0(\overline{\sigma}) = -pc\left(\sigma + \frac{m}{\sigma}\right) = -pc\left(\frac{\sigma^2 + m}{\sigma}\right) \tag{19.22.4}$$

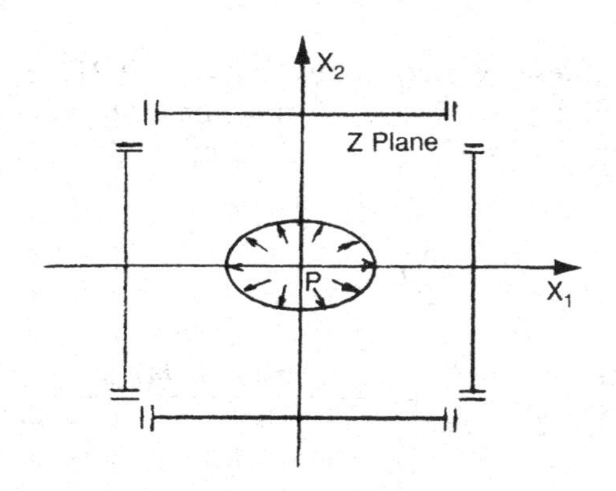

Fig. 19.40

We now have

$$\psi^0(\zeta) = \frac{\zeta}{2\pi i}\int_\gamma \frac{F^0(\sigma)}{\sigma(\sigma - \zeta)}\,d\sigma = -\frac{pc\zeta}{2\pi i}\int_\gamma \frac{1 + m\sigma^2}{\sigma^2(\sigma - \zeta)}\,d\sigma$$

$$= -\frac{pc\zeta}{2\pi i}\int_\gamma \frac{d\sigma}{\sigma^2(\sigma - \zeta)} - \frac{pc\zeta m}{2\pi i}\int_\gamma \frac{d\sigma}{(\sigma - \zeta)} \tag{19.22.5}$$

The first integral in Eq. (19.22.5) is equal to zero, and the second is equal to $2\pi i$. Thus,

$$\psi^0(\zeta) = -pc\zeta m \tag{19.22.6}$$

Now, turning to $\gamma^0(\zeta)$ we have,

$$(\psi^0)'(\zeta) = -pcm$$

and

$$\frac{1}{2\pi i}\int_\gamma \frac{\overline{F}^0(\overline{\sigma})}{\sigma - \zeta}\,d\sigma = \frac{-pc}{2\pi i}\int_\gamma \frac{\sigma\left(1 + \dfrac{m}{\sigma^2}\right)}{\sigma - \zeta}\,d\sigma$$

$$= \frac{-pc}{2\pi i}\int_\gamma \frac{\sigma}{\sigma - \zeta}\,d\sigma - \frac{pc}{2\pi i}\int_\gamma \frac{m}{\sigma^2(\sigma - \zeta)}\,d\sigma$$

$$= -pc\zeta$$

since the first integral is equal to $2\pi i\zeta$ and the second one is equal to zero. Thus,

$$\gamma^0(\zeta) = -pc\zeta - \frac{\zeta(\zeta^2 + m)}{1 - m\zeta^2}\,pcm$$

or

$$\gamma^0(\zeta) = -pc\zeta\left(\frac{1 + m^2}{1 - m\zeta^2}\right) \tag{19.22.7}$$

Returning to Eqs. (19.14.137) and (19.14.138) and assuming that $\sigma_{ij}(\infty) = 0$ and setting $F_{1k} = F_{2k} = 0$. We get

$$\left.\begin{aligned} \psi_1(\zeta) &= \psi^0(\zeta) \\ &\text{and} \\ \gamma_1(\zeta) &= \gamma^0(\zeta) \end{aligned}\right\} \tag{19.22.8}$$

Equations (19.22.8) yield the stresses, strains and displacements within the body.

19.23 Application to Fracture Mechanics

The complex potentials for a crack in an infinite sheet subjected to a tensile stress normal to the direction of the crack (Fig. 19.41) are

$$\psi(z) = \frac{S}{4}\left[2(z^2 - a^2)^{1/2} - z\right] \tag{19.23.1}$$

$$\chi'(z) = \gamma(z) = \frac{S}{2}\left(z - \frac{a^2}{(z^2 - a^2)^{1/2}}\right) \tag{19.23.2}$$

These expressions satisfy the boundary conditions of the problem. Indeed, using Eqs. (19.14.31) to (19.14.33) we find that

$$\lim_{z\to\infty} \sigma_{11} = 0, \qquad \lim_{z\to\infty} \sigma_{22} = S, \qquad \lim_{z\to\infty} \sigma_{12} = 0\,,$$

and

$$\sigma_{22} = \sigma_{12} = 0 \quad \text{for } -a \leqslant x_1 \leqslant a$$

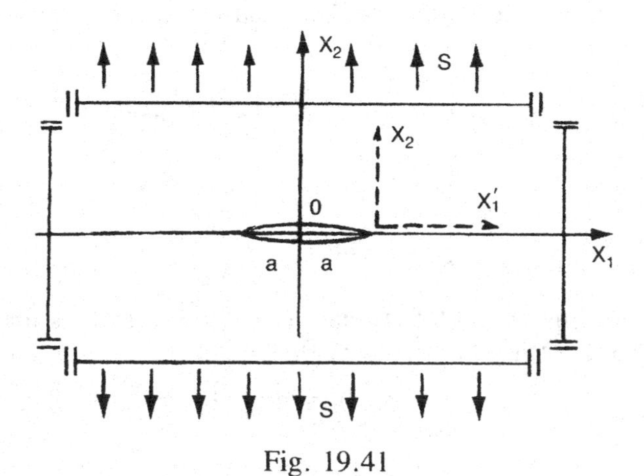

Fig. 19.41

The expression of σ_{22} for $x_2 = 0$ is

$$\sigma_{22} = \frac{S}{\sqrt{1 - a^2/x_1^2}}, \quad x_1 \geqslant a \tag{19.23.3}$$

At the tip the stress is infinite. The modified Westergaard stress function that solves this problem is given by

$$Z_1(z) = 2\psi'(z) = \left[\frac{Sz}{\sqrt{z^2 - a^2}} - A\right], \quad A = \frac{S}{2} \tag{19.23.4}$$

For large values of z Eqs. (19.15.40) to (19.15.42) yield

$$\sigma_{11} = S - A - A = 0, \qquad \text{thus } A = \frac{S}{2}$$

$$\sigma_{22} = S - A + A = S, \qquad \text{for all } A$$

$$\sigma_{12} = 0$$

Thus, by varying the real constant A we can decide on the degree of biaxiality of the stress applied to the infinite plate. $A = S/2$ results in an uniaxial field and $A = 0$ results in a biaxial field with

$$Z_1(z) = \frac{Sz}{\sqrt{z^2 - a^2}} \tag{19.23.5}$$

REFERENCES

[1] G. V. Kolosov, "On an Application of Complex Function Theory to a Plane Problem of the Mathematical Theory of Elasticity," *A Dissertation* at Dorpat University, 1909.

[2] N. I. Muskhelishvili, *Some Basic Problems of the Mathematical Theory of Elasticity*, Translation by J. R. M. Radok, P. Noordhoff, N. V., Groningen, Netherlands, 1953.

[3] I. S. Sokolnikoff, *Mathematical Theory of Elasticity*, McGraw-Hill, New York, N.Y., 1956.

[4] A. E. Green and W. Zerna, *Theoretical Elasticity*, Oxford University Press, Fair Lawn, N.J., 1954.

[5] L. M. Milne-Thomson, *Antiplane Elastic Systems*, Academic Press, New York, N.Y., 1962.

[6] R. V. Churchill, *Complex Variables and Applications*, McGraw-Hill, New York, N.Y., 1960.

[7] N. W. McLachlan, *Complex Variable Theory and Transform Calculus with Technical Applications*, 2nd Edition, Cambridge, University Press, London, 1953.

[8] L. Prandtl, *Physikalische Zeitschrift*, Vol. 4, pp. 758–770, 1903.

[9] S. Timoshenko and J. N. Goodier, *Theory of Elasticity*, McGraw-Hill, New York, N.Y., 1970.

[10] H. M. Westergaard, "Bearing Pressures and Cracks," *J. Appl. Mech.*, Vol. 6, pp. A49–53, 1937.

[11] C. E. Inglis, *Transactions of the Institute of Naval Architects*, Vol. 55, pp. 219–230, London, 1913.

[12] G. C. Sih, "On the Westergaard Method of Crack Analysis," *Journal of Fracture Mechanics*, Vol. 2, pp. 628–630, 1966.

[13] J. Eftis and H. Liebowitz, "On the Modified Westergaard Equations for Certain Plane Crack Problems," *International Journal of Fracture Mechanics*, Vol. 8, No. 4, pp. 383–392, 1972.

ADDENDUM

to

ELASTICITY, THEORY AND APPLICATIONS – 2ND EDITION

- Comments and Detailed Explanations.
- Additional Solved Examples
- Additional Problems

CHAPTER 1

[At the bottom of Page 8]

It is worthwhile at this stage to write Eqs. (1.1.6), (1.1.7), (1.2.4) and (1.2.5) using a matrix notation:

$$\begin{bmatrix} d\xi_1 \\ d\xi_2 \\ d\xi_3 \end{bmatrix} = \begin{bmatrix} \left(1+\frac{\partial u_1}{\partial x_1}\right) & \frac{\partial u_1}{\partial x_2} & \frac{\partial u_1}{\partial x_3} \\ \frac{\partial u_2}{\partial x_1} & \left(1+\frac{\partial u_2}{\partial x_2}\right) & \frac{\partial u_2}{\partial x_3} \\ \frac{\partial u_3}{\partial x_1} & \frac{\partial u_3}{\partial x_2} & \left(1+\frac{\partial u_3}{\partial x_3}\right) \end{bmatrix} \begin{bmatrix} dx_1 \\ dx_2 \\ dx_3 \end{bmatrix} \quad (1.1.6)$$

$$\begin{bmatrix} du_1 \\ du_2 \\ du_3 \end{bmatrix} = \begin{bmatrix} \frac{\partial u_1}{\partial x_1} & \frac{\partial u_1}{\partial x_2} & \frac{\partial u_1}{\partial x_3} \\ \frac{\partial u_2}{\partial x_1} & \frac{\partial u_2}{\partial x_2} & \frac{\partial u_2}{\partial x_3} \\ \frac{\partial u_3}{\partial x_1} & \frac{\partial u_3}{\partial x_2} & \frac{\partial u_3}{\partial x_3} \end{bmatrix} \begin{bmatrix} dx_1 \\ dx_2 \\ dx_3 \end{bmatrix} \quad (1.1.7)$$

The matrix in (1.1.7) is called the *displacement gradient matrix.*

$$\begin{bmatrix} d\xi_1 \\ d\xi_2 \\ d\xi_3 \end{bmatrix} = \begin{bmatrix} (1+e_{11}) & (e_{12}-\omega_{21}) & (e_{13}+\omega_{13}) \\ (e_{12}+\omega_{21}) & (1+e_{22}) & (e_{23}-\omega_{32}) \\ (e_{13}-\omega_{13}) & (e_{23}+\omega_{32}) & (1+e_{33}) \end{bmatrix} \begin{bmatrix} dx_1 \\ dx_2 \\ dx_3 \end{bmatrix} \quad (1.2.4)$$

$$\begin{bmatrix} du_1 \\ du_2 \\ du_3 \end{bmatrix} = \begin{bmatrix} e_{11} & e_{12}-\omega_{21} & e_{13}+\omega_{13} \\ e_{12}+\omega_{21} & e_{22} & e_{23}-\omega_{32} \\ e_{13}-\omega_{13} & e_{23}+\omega_{32} & e_{33} \end{bmatrix} \begin{bmatrix} dx_1 \\ dx_2 \\ dx_3 \end{bmatrix} \quad (1.2.5)$$

Notice that all the terms within the matrices above are numbers. Indeed, they are nothing but partial derivatives of the functions u_1, u_2 and u_3 taken at a specific point M whose coordinates are x_1, x_2 and x_3. For example if, $u_1 = c(2x_1 + x_2^2)$, $u_2 = c(x_1^2 - 3x_2^2)$, $u_3 = c(x_1 x_3)$, where c is a constant, then the values of the partial derivatives at a point M (1,2,4) are:

$$\frac{\partial u_1}{\partial x_1} = 2c, \quad \frac{\partial u_1}{\partial x_2} = 2cx_2 = 4c, \quad \frac{\partial u_1}{\partial x_3} = 0$$

$$\frac{\partial u_2}{\partial x_1} = 2cx_1 = 2c, \quad \frac{\partial u_2}{\partial x_2} = -6cx_2 = -12c, \quad \frac{\partial u_2}{\partial x_3} = 0$$

$$\frac{\partial u_3}{\partial x_1} = cx_3 = 4c, \quad \frac{\partial u_3}{\partial x_2} = 0 \quad \frac{\partial u_3}{\partial x_3} cx_1 = c.$$

If now the constant c is equal to 10^{-2} and the element MN has a unit length, Eq. (1.2.5 becomes:

$$\begin{bmatrix} du_1 \\ du_2 \\ du_3 \end{bmatrix} = 10^{-2} \begin{bmatrix} 2 & 4 & 0 \\ 2 & -12 & 0 \\ 4 & 0 & 1 \end{bmatrix} \begin{bmatrix} \ell_1 \\ \ell_2 \\ \ell_3 \end{bmatrix}$$

where ℓ_1, ℓ_2 and ℓ_3 are the direction cosines of MN. With the notation presented in Section 1.2 we have at the point M:

$$e_{11} = 2 \times 10^{-2}, \; e_{22} = -12 \times 10^{-2}, \; e_{33} = 10^{-2},$$
$$e_{12} = 3 \times 10^{-2}, \; e_{13} = 2 \times 10^{-2}, \; e_{23} = 0,$$
$$-\omega_{21} = 1 \times 10^{-2}, \; \omega_{13} = -2 \times 10^{-2}, \; -\omega_{23} = 0.$$

[Problems]

1. The deformation of a body is defined by the displacements of its points as follows:

$u_1 = k(3x_1^2 + x_2^2)$ ft, $u_2 = k(2x_2^2 + x_3)$ ft, $u_3 = k(4x_3^2 + x_1)$ ft, where k is a constant. At the point M whose coordinates are (1, 1, 1) and for a value of k=10^{-3}:

(a) Write down the displacement gradient matrix;
(b) Find the values of the e_{ij}'s and ω_{ij}'s;
(c) Write down Eq. (1.2.5) at the point M.

2. Repeat problem 1 with M(1, 0, 2) and,

$$u_1 = 10^{-3}(3x_1^2 x_2 + 6) \; ft$$
$$u_2 = 10^{-3}(x_2^2 + 6x_1 x_3) \; ft$$
$$u_3 = 10^{-3}(6x_3^2 + 12x_2 x_3 + 60) \; ft.$$

CHAPTER 2

[Additional Problems]

9. Given the two matrices

$$[a]=\begin{bmatrix}1 & 2 & 3\\ 1 & 3 & 3\\ 1 & 2 & 4\end{bmatrix}, \text{ and } [b]=\begin{bmatrix}1 & -2 & 3\\ 2 & 1 & 4\\ -3 & -4 & 1\end{bmatrix},$$

show that, if

$$[c]=[a]^{-1}[b][a],$$

the traces and the determinants of $[c]$ and $[b]$ are equal

10. In Problem 2 show that the matrix $[a]$ is singular.

11. Write in detail the three sets of linear equations represented by

$$[a][M]=[M][D],$$

where,

$$[a]=\begin{bmatrix}-2 & -1 & 4\\ 2 & 1 & -2\\ -1 & -1 & 3\end{bmatrix}, [M]=\begin{bmatrix}x_{11} & x_{12} & x_{13}\\ x_{21} & x_{22} & x_{23}\\ x_{31} & x_{32} & x_{33}\end{bmatrix}, [D]=\begin{bmatrix}2 & 0 & 0\\ 0 & 1 & 0\\ 0 & 0 & -1\end{bmatrix}.$$

Show that in each set only two of the equations are independent

12. Using the matrix [a] of problem 9 show that the transpose of its inverse is equal to the inverse of its transpose.

13. Show that the inverse of the symmetric matrix

$$[b]=\begin{bmatrix}4 & 3 & 2\\ 3 & 2 & 1\\ 2 & 1 & 1\end{bmatrix}$$

is also a symmetric matrix.

CHAPTER 3

[At the end of Section 3.3]

The process can of course be reversed. If one starts from a sphere centered at the origin, it becomes an ellipsoid with the same principal directions; provided the matrix [a] of the transformation can be inverted, in other words it is non-singular. The radii of the sphere along the principal directions become the semi axes of the ellipsoid. Depending on the nature of [a] they may or may not rotate but they will keep their orthogonality after transformation. The largest change in length occurs along the major principal axis and the smallest along the minor principal axis of the ellipsoid.

As an example let us find the principal directions of the transformation:

$$\begin{bmatrix} \xi_1 \\ \xi_2 \\ \xi_3 \end{bmatrix} = \begin{bmatrix} 2 & 1 & 1 \\ -1 & 2 & 1 \\ 1 & -1 & 2 \end{bmatrix} \begin{bmatrix} x_1 \\ x_2 \\ x_3 \end{bmatrix} \qquad (3.3.3)$$

The equation of the characteristic ellipsoid is

$$6x_1^2 + 6x_2^2 + 6x_3^2 - 2x_1x_2 + 6x_1x_3 + 2x_2x_3 = R^2 .$$

From analytic geometry, the general equation of a quadric with its center at the origin is

$$ax_1^2 + bx_2^2 + cx_3^2 + 2fx_2x_3 + 2gx_1x_3 + 2hx_1x_2 + d = 0 .$$

In this case a = 6, b = 6, c = 6, f = 1, g = 3, h = -1.

The discriminating cubic is,

$$\begin{vmatrix} 6-\beta & -1 & 3 \\ -1 & 6-\beta & 1 \\ 3 & 1 & 6-\beta \end{vmatrix} = 0$$

The roots of this equation and the principal directions are given by the β's and their corresponding cosines (see Section 3.6 on Characteristic Equations and Eigenvalues; or use an eigenvalue software):

$$\beta_3 = 9.00000 \qquad \beta_2 = 6.56155 \qquad \beta_1 = 2.43845$$

$$\begin{Bmatrix} 0.70711 \\ 0.00000 \\ 0.70711 \end{Bmatrix} \qquad \begin{Bmatrix} -0.26096 \\ 0.92941 \\ 0.26096 \end{Bmatrix} \qquad \begin{Bmatrix} -0.65719 \\ -0.36905 \\ 0.65719 \end{Bmatrix} \tag{3.3.4}$$

The equation of the characteristic ellipsoid in a trirectangular system of axes along the principal directions is

$$2.43845\ x_1^2 + 6.56155\ x_2^2 + 9x_3^2 = R^2 \tag{3.3.5}$$

If we assume that R = 1, Eq. (3.3.5) becomes

$$\frac{x_1^2}{0.4101} + \frac{x_2^2}{0.1524} + \frac{x_3^2}{0.1111} = 1$$

If we now apply the transformation (3.3.3) to points located on the principal axes such points will move but, together with the origin, they will determine directions that are still normal to each other. Thus, the principal system will rotate while keeping the orthogonality of its axes. Let us apply (3.3.3) to the three directions specified by (3.3.4). The numbers here are nothing but the coordinates of points on the principal axes at a unit distance from the origin. The operations are readily done in matrix notation:

$$\begin{bmatrix} 2 & 1 & 1 \\ -1 & 2 & 1 \\ 1 & -1 & 2 \end{bmatrix} \begin{bmatrix} -0.65719 & -0.26096 & 0.70711 \\ -0.36905 & 0.92941 & 0.00000 \\ 0.65719 & 0.26096 & 0.70711 \end{bmatrix} = \begin{bmatrix} -1.02624 & .66845 & 2.12133 \\ .57628 & 2.38074 & .0000 \\ 1.02624 & -.66845 & 2.12133 \end{bmatrix}$$

The vectors represented by the columns of the resulting matrix are normal to each other as indicated by their scalar product.

A good visualization can be had if one stays in two dimensions. Consider the transformation

$$\begin{bmatrix} \xi_1 \\ \xi_2 \end{bmatrix} = \begin{bmatrix} 2 & 2 \\ -1 & 2 \end{bmatrix} \begin{bmatrix} x_1 \\ x_2 \end{bmatrix} \tag{3.3.6}$$

The quadric here is an elliptical cylinder whose equation is

$$5x_1^2 + 8x_2^2 + 4x_1x_2 = R^2$$

The β's and their corresponding principal directions are

$$\beta_1 = 4.00000 \qquad \beta_2 = 9.00000$$

$$\begin{Bmatrix} 0.89443 \\ -0.44721 \end{Bmatrix} \qquad \begin{Bmatrix} 0.44721 \\ 0.89443 \end{Bmatrix} \tag{3.3.7}$$

The principal direction O1 makes an angle

$$\phi = \cos^{-1} 0.89443 = 26.565^{\circ}$$

with the OX_1 axis.

The equation of the ellipse for R = 1 is (see Fig. 3.8a)

$$\frac{x_1^2}{0.25} + \frac{x_2^2}{0.11111} = 1. \tag{3.3.8}$$

When the transformation (3.3.6) is applied to the unit vectors of (3.3.7) we get the direction ratios of O1* and O2*:

$$\begin{bmatrix} 2 & 2 \\ -1 & 2 \end{bmatrix} \begin{bmatrix} 0.89443 & 0.44721 \\ -0.44721 & 0.89443 \end{bmatrix} = \begin{bmatrix} 0.89443 & 2.68328 \\ -1.78885 & 1.34165 \end{bmatrix} \tag{3.3.9}$$

Notice again, that the columns of the resulting matrix, are normal to each other. Under this last transformation, point A on the ellipse moves to C on the circle whose radius R is 1. Point B moves to D. In the plane (OX_1, OX_2) the principal directions O1 and O2 rotate to O1* and O2* while remaining orthogonal.

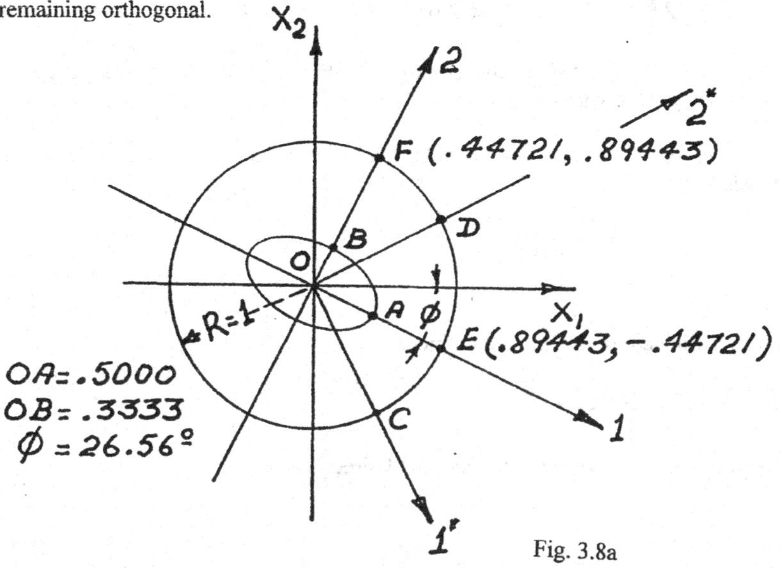

Fig. 3.8a

[At the end of Section 3.4]

The equations of transformation in this example are best derived in two dimensions. Consider Fig. 3.10a:

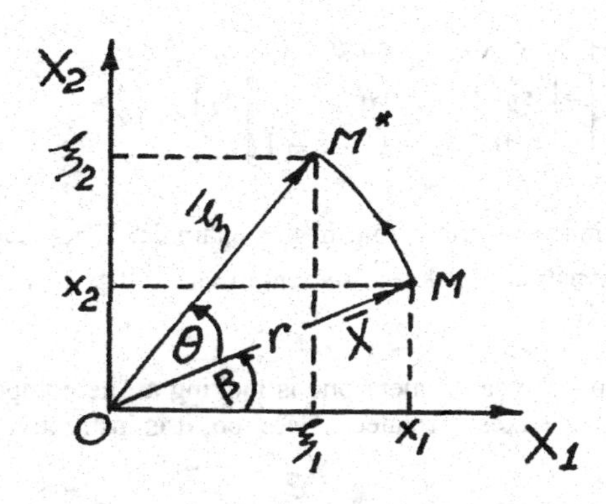

Fig. 3.10a

The vector $\bar{x}$ of length r is transformed to $\bar{\xi}$ through a rotation of an angle θ. Thus,

$$\xi_1 = r\cos(\beta+\theta) = r\cos\beta\cos\theta - r\sin\beta\sin\theta$$
$$\xi_2 = r\sin(\beta+\theta) = r\sin\beta\cos\theta + r\cos\beta\sin\theta.$$

Since,

$$x_1 = r\cos\beta\text{, and } x_2 = r\sin\beta,$$

then

$$\xi_1 = x_1\cos\theta - x_2\sin\theta$$
$$\xi_2 = x_1\sin\theta + x_2\cos\theta.$$

or

$$\begin{bmatrix}\xi_1\\ \xi_2\end{bmatrix} = \begin{bmatrix}\cos\theta & -\sin\theta\\ \sin\theta & \cos\theta\end{bmatrix}\begin{bmatrix}x_1\\ x_2\end{bmatrix}$$

Here $a_{11} = \cos\theta$, $a_{12} = -\sin\theta$, $a_{21} = \sin\theta$ and $a_{22} = \cos\theta$. The elements of this matrix satisfy Eq. (3.4.6); its inverse is equal to its transpose and its rows and columns form vectors that are normal to each other.

In three dimensions one would get

$$\begin{bmatrix} \xi_1 \\ \xi_2 \\ \xi_3 \end{bmatrix} = \begin{bmatrix} \cos\theta & -\sin\theta & 0 \\ \sin\theta & \cos\theta & 0 \\ 0 & 0 & +1 \end{bmatrix} \begin{bmatrix} x_1 \\ x_2 \\ x_3 \end{bmatrix} \tag{3.4.11}$$

which is the equation we started from in this example. The vector $\overline{\mathrm{OP}}$ is here subjected to an orthogonal transformation that keeps its length and brings it to $\overline{O\Pi}$, in the original system of axes OX_1, OX_2, OX_3.

Consider now the case where one is looking at the components of a single vector when the system of reference axes is rotated. Here too, it is instructive to start in two dimensions. In Fig. 3.10b:

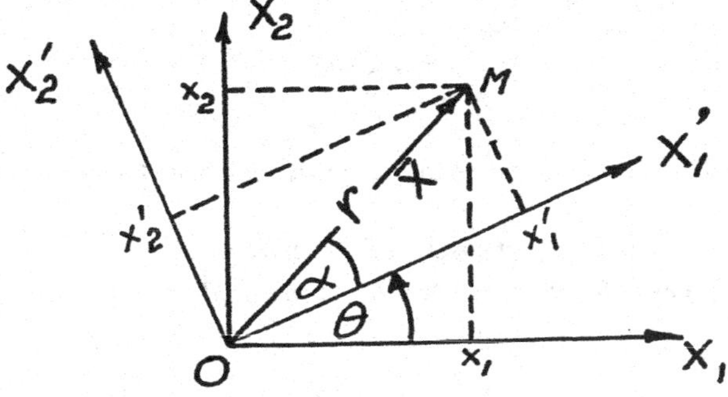

Fig. 3.10b

The components of the vector OM, of length r, in the original system of axes are x_1 and x_2; and in the rotated system x_1' and x_2'. We have,

$$x_1 = r\cos(\alpha+\theta) = r\cos\alpha\cos\theta - r\sin\alpha\sin\theta$$
$$x_2 = r\sin(\alpha+\theta) = r\sin\alpha\cos\theta + r\cos\alpha\sin\beta$$

Since, $x_1' = r\cos\alpha$, and $x_2' = r\sin\alpha$, then

$$x_1 = x_1'\cos\theta - x_2'\sin\theta$$
$$x_2 = x_1'\sin\theta + x_2'\cos\theta$$

or

$$\begin{bmatrix} x_1 \\ x_2 \end{bmatrix} = \begin{bmatrix} \cos\theta & -\sin\theta \\ \sin\theta & \cos\theta \end{bmatrix} \begin{bmatrix} x_1' \\ x_2' \end{bmatrix}. \tag{3.4.12}$$

Solving, we get,

$$x_1' = x_1 \cos\theta + x_2 \sin\theta$$
$$x_2' = -x_1 \sin\theta + x_2 \cos\theta$$

or

$$\begin{bmatrix} x_1' \\ x_2' \end{bmatrix} = \begin{bmatrix} \cos\theta & \sin\theta \\ -\sin\theta & \cos\theta \end{bmatrix} \begin{bmatrix} x_1 \\ x_2 \end{bmatrix} \tag{3.4.13}$$

The matrix in (3.4.13), which is the inverse of the one in (3.4.12), is a matrix of direction cosines of the axes of the rotated system with respect to the original one. Indeed the direction OX_1' has direction cosines $\ell_{11} = \cos\theta$ and $\ell_{12} = \sin\theta$; and the direction OX_2' has direction cosines $\ell_{21} = -\sin\theta$ and $\ell_{22} = \cos\theta$. So,

$$\begin{bmatrix} x_1' \\ x_2' \end{bmatrix} = \begin{bmatrix} \ell_{11} & \ell_{12} \\ \ell_{21} & \ell_{22} \end{bmatrix} \begin{bmatrix} x_1 \\ x_2 \end{bmatrix} \tag{3.4.13a}$$

In three dimensions (see Fig. 3.11) one would get,

$$\begin{bmatrix} x_1' \\ x_2' \\ x_3' \end{bmatrix} = \begin{bmatrix} \ell_{11} & \ell_{12} & \ell_{13} \\ \ell_{21} & \ell_{22} & \ell_{23} \\ \ell_{31} & \ell_{32} & \ell_{33} \end{bmatrix} \begin{bmatrix} x_1 \\ x_2 \\ x_3 \end{bmatrix} \tag{3.4.14}$$

Just as in the case of Eq. (3.4.11) the matrix of (3.4.14) has elements that satisfy Eq. (3.4.6). It is an orthogonal matrix with its inverse equal to its transpose; its rows and columns form vectors that are normal to each other.

There is however a fundamental difference between the two matrices in Eqs. (3.4.11) and (3.4.14):

The matrix in (3.4.11) is the matrix of a linear transformation that transforms a vector $\bar{x}$ to a vector $\bar{\xi}$ in one system of coordinates OX_1, OX_2, OX_3. On the other hand the matrix in (3.4.14) is a matrix of direction cosines of new axes OX_1', OX_2', OX_3' with respect to old ones. It straddles two systems of axes and is used to obtain the components (projections) of a vector $\bar{x}$ in a new, rotated system of axes OX_1', OX_2', OX_3' with respect to its components in the original system OX_1, OX_2, OX_3.

By definition, (3.4.14) expresses an orthogonal transformation of coordinates where the components of a vector in the new system with respect to the old one are given by

$$x_i' = \ell_{ij} x_j \tag{3.4.14a}$$

Here too, we have (Eqs. 3.4.6 to 3.4.10)

$$\ell_{ij}\ell_{ik} = \delta_{jk}$$
$$[\ell]^{-1} = [\ell]'$$
$$\ell_{ji}\ell_{ki} = \delta_{jk}$$
$$|[\ell]|^2 = 1$$

If $|[\ell]| = 1$ the orthogonal transformation of coordinates is called proper and amounts to a rotation around the origin; if $|[\ell]| = -1$ it is called improper and turns the coordinate system from right handed to left handed. In this text we will only deal with proper orthogonal transformations of coordinates, in other words, rotations around an origin.

[At the end of Section 3.8]

Example

(a) Let us now give a numerical example covering the content of the previous sections. Consider the linear transformation

$\xi_i = a_{ij} x_j$ where the matrix [a] is

$$\begin{bmatrix} -2 & -1 & 4 \\ 2 & 1 & -2 \\ -1 & -1 & 3 \end{bmatrix}$$

The invariant directions are given by Eq. (3.6.1):

$$\begin{bmatrix} \xi_1 \\ \xi_2 \\ \xi_3 \end{bmatrix} = \begin{bmatrix} -2 & -1 & 4 \\ 2 & 1 & -2 \\ -1 & -1 & 3 \end{bmatrix} \begin{bmatrix} x_1 \\ x_2 \\ x_3 \end{bmatrix} = \begin{bmatrix} \lambda x_1 \\ \lambda x_2 \\ \lambda x_3 \end{bmatrix} \tag{3.8.6}$$

therefore,

$$(-2-\lambda)x_1 - x_2 + 4x_3 = 0$$
$$2x_1 + (1-\lambda)x_2 - 2x_3 = 0 \qquad (3.8.7)$$
$$-x_1 - x_2 + (3-\lambda)x_3 = 0.$$

Solving for x_1, x_2 and x_3 using determinants we get

$$x_1 = \frac{0}{|D|},\ x_2 = \frac{0}{|D|},\ x_3 = \frac{0}{|D|} \qquad (3.8.8)$$

where,

$$|D| = \begin{vmatrix} -2-\lambda & -1 & 4 \\ 2 & 1-\lambda & -2 \\ -1 & -1 & 3-\lambda \end{vmatrix}$$

We notice that there is no solution unless the denominator $|D|$ is equal to zero; in which case we have an indeterminate number of solutions. Setting $|D| = 0$ we get the cubic equation

$$\lambda^3 - 2\lambda^2 - \lambda + 2 = 0 \qquad (3.8.9)$$

This equation has three real roots,

$$\lambda_1 = 2, \qquad \lambda_2 = 1 \qquad \lambda_3 = -1$$

Each of these values when substituted in (3.8.7) should, in theory, give at set (x_1, x_2, x_3), which, with the origin, determines a direction that remains parallel to itself (invariant direction) after transformation. The set related to λ_1 is written x_{11}, x_{21}, x_{31}, the one related to λ_2 is written x_{12}, x_{22}, x_{32} and the one related to λ_3 is written x_{13}, x_{23}, x_{33}. Initially the question was asked if one could find a direction that remained parallel to itself, and it appears that we found three!

Substituting $\lambda_1 = 2$ in (3.8.7) we obtain the three homogeneous equations

$$-4x_{11} \quad -x_{21} \quad +4x_{31} = 0$$
$$2x_{11} \quad -x_{21} \quad -2x_{31} = 0 \qquad (3.8.10)$$
$$-x_{11} \quad -x_{21} \quad +x_{31} = 0$$

We are faced again with solving three homogeneous equations to get a point, which, with the origin, will determine that direction that does not change. However, those 3 equations are not independent and what we are really asking for is a direction, which can be determined by 2 intersecting planes. So one can ignore one of the three equations and find the direction ratios(or direction cosines) of the line of intersection of two planes. This can be done by choosing any convenient value for x_{11} (say $x_{11} = 1$) and substituting it in the first two equations of (3.8.10):

$$-x_{21} + 4x_{31} = 4$$
$$x_{21} + 2x_{31} = 2$$

This leads to $x_{21} = 0$ and $x_{31} = 1$. Thus the eigenvector correspondent to $\lambda_1 = 2$ is

$$\{\bar{x}_1\} = \begin{bmatrix} 1 \\ 0 \\ 1 \end{bmatrix}$$

The vector is usually normalized by dividing its component by its length, thus

$$\{\bar{x}_1\} = \begin{bmatrix} \frac{1}{\sqrt{2}} \\ 0 \\ \frac{1}{\sqrt{2}} \end{bmatrix}$$

The same result is obtained if, rather than choosing an arbitrary value for x_{11} and normalizing, one uses the expression

$$\sqrt{x_{11}^2 + x_{21}^2 + x_{31}^2} = 1$$

together with two out of the three equations in (3.8.10). The components of the eigenvector are then its direction cosines. In a similar fashion one obtains for $\lambda_2 = 1$

$$\{\bar{x}_2\} = \begin{bmatrix} \frac{1}{\sqrt{3}} \\ \frac{1}{\sqrt{3}} \\ \frac{1}{\sqrt{3}} \end{bmatrix}$$

and for $\lambda_3 = -1$

$$\{x_3\} = \begin{bmatrix} \frac{1}{\sqrt{2}} \\ -\frac{1}{\sqrt{2}} \\ 0 \end{bmatrix}$$

The angle between the eigenvectors can be obtained by taking their scalar products. Thus

$$\cos\theta_{12} = \frac{2}{\sqrt{6}}, \text{ so } \theta_{12} = 35.26^{\underline{0}}$$

$$\cos\theta_{23} = 0, \text{ so } \theta_{23} = 90^{\underline{0}}$$

$$\cos\theta_{31} = \frac{1}{2}, \text{ so } \theta_{31} = 60^{\underline{0}}$$

In this particular example we had three distinct real roots to our cubic equation in λ and three independent eigenvectors. However, if two of the roots are equal we may or may not have independent eigenvectors. In this text we will encounter nothing but distinct eigenvalues and eigenvectors when the matrix of the transformation is not symmetric. When the matrix is symmetric the eigenvalues are all real and one can always find three real invariant directions, which will be shown in section 3.10 to be principal too.

(b) Consider now the matrix [a]. Its inverse is

$$[a]^{-1} = \begin{bmatrix} -0.5 & 0.5 & 1.0 \\ 2.0 & 1.0 & -2.0 \\ 0.5 & 0.5 & 0.0 \end{bmatrix}$$

A similarity transformation of a matrix [b], where

$$[b] = \begin{bmatrix} 2 & -1 & 0 \\ 9 & 4 & 6 \\ -8 & 0 & -3 \end{bmatrix},$$

yields a matrix [c] such that

$$[c]=\begin{bmatrix}-0.5 & 0.5 & 1.0\\ 2.0 & 1.0 & -2.0\\ 0.5 & 0.5 & 0.0\end{bmatrix}\begin{bmatrix}2 & -1 & 0\\ 9 & 4 & 6\\ -8 & 0 & -3\end{bmatrix}\begin{bmatrix}-2 & -1 & 4\\ 2 & 1 & -2\\ -1 & -1 & 3\end{bmatrix}$$

$$[c]=\begin{bmatrix}14.0 & 7.0 & -23.0\\ -66.0 & -39.0 & 148.0\\ -11.0 & -7.0 & 28.0\end{bmatrix}$$

As expected, the determinant of [b] = -3 and the determinant of [c] is also –3; the trace of [b] is 3 and the trace of [c] is 3; and the eigenvalues of [b] and [c] are both equal to 1.0, -1.0 and 3.0.

(c) Consider also the similarity transformation of the matrix [a] through the use of its modal matrix [M] and its inverse $[M]^{-1}$

$$\begin{bmatrix}-1 & -1 & 2\\ 1 & 1 & -1\\ 1 & 0 & -1\end{bmatrix}\begin{bmatrix}-2 & -1 & 4\\ 2 & 1 & -2\\ -1 & -1 & 3\end{bmatrix}\begin{bmatrix}1 & 1 & 1\\ 0 & 1 & -1\\ 1 & 1 & 0\end{bmatrix}=\begin{bmatrix}2 & 0 & 0\\ 0 & 1 & 0\\ 0 & 0 & -1\end{bmatrix}$$

This is the operation that diagonalizes the matrix [a] and gives the spectral matrix.

The invariants of the transformation are given by Eqs. (3.7.1) to 3.7.3) as

$$\mathrm{I}_1=\lambda_1+\lambda_2+\lambda_3=2$$
$$\mathrm{I}_2=\lambda_1\lambda_2+\lambda_2\lambda_3+\lambda_3\lambda_1=-1$$
$$\mathrm{I}_3=\lambda_1\lambda_2\lambda_3=-2$$

Notice that any combination of invariants is still an invariant of the transformation.

Thus

$$\frac{1}{2}a_{ij}a_{ji}=\frac{1}{2}I_1^2-I_2=3$$

and

$$\frac{1}{3}a_{ij}a_{jk}a_{ki}=\frac{1}{3}I_1^3-I_1I_2+I_3=\frac{8}{3}$$

are also invariant.

d) In section 3.3 it was mentioned that every linear transformation has three principal directions and correspondingly three principal planes. Those directions are normal to each other and often different from the invariant ones which we just found. To find those principal directions we use the methods of analytic geometry:

The equation of the characteristic ellipsoid (see Section 3.3), corresponding to the linear transformation whose matrix is [a], is

$$(-2x_1 - x_2 + 4x_3)^2 + (2x_1 + x_2 - 2x_3)^2 + (-x_1 - x_2 + 3x_3)^2 = R^2$$

or

$$9x_1^2 + 3x_2^2 + 29_3^2 + 10x_1x_2 - 18x_2x_3 - 30x_1x_3 = R^2.$$

The general equation of a quadric centered at the origin is

$$ax_1^2 + bx_2^2 + cx_3^2 + 2fx_2x_3 + 2gx_3x_1 + 2hx_1x_2 + d = 0$$

In this case a=9, b=3, c=29, f=-9, g=-15, h=5. This discriminating cubic is

$$\begin{vmatrix} 9-\beta & 5 & -15 \\ 5 & 3.-\beta & -9 \\ -15 & -9 & 29-\beta \end{vmatrix} = 0$$

The roots of this equation and the corresponding direction cosines of the principal directions are (using an eigenvalue software):

$$\beta_1 = 39.8997 \qquad \beta_2 = 1.000 \qquad \beta_3 = 0.1002$$

$$\begin{Bmatrix} -0.4555 \\ -0.2687 \\ 0.8487 \end{Bmatrix} \qquad \begin{Bmatrix} 0.8451 \\ 0.1690 \\ 0.5071 \end{Bmatrix} \qquad \begin{Bmatrix} -0.2797 \\ 0.9483 \\ 0.1501 \end{Bmatrix}$$

The equation of the characteristic ellipsoid in the principal system of axes (Canonical form) is:

$$39.8997(x_1')^2 + (x_2')^2 + 0.1(x_3')^2 = R^2$$

This example shows that, in a general linear transformation (not symmetric), the invariant directions are different from the principal directions. The former do not rotate and are not orthogonal, and the latter rotate but remain orthogonal. As is shown in section 3.10, the two sets of directions coincide when the matrix of the transformation is symmetric.

[At the end of Section 3.9]

It is also instructive to look at what an antisymmetric transformation does to points located on the reference axes. Consider, for example, points whose coordinates are A(1,0,0), B(0,1,0) and C(0,0,1). Then, for point A, (Fig. 3.17a)

$$\begin{bmatrix} u_1 \\ u_2 \\ u_3 \end{bmatrix} = \begin{bmatrix} 0 & -a_{12} & a_{13} \\ a_{12} & 0 & -a_{23} \\ -a_{13} & a_{23} & 0 \end{bmatrix} \begin{bmatrix} 1 \\ 0 \\ 0 \end{bmatrix} = \begin{bmatrix} 0 \\ a_{12} \\ -a_{13} \end{bmatrix} \text{; and it moves to A}'.$$

For point B,

$$\begin{bmatrix} u_1 \\ u_2 \\ u_3 \end{bmatrix} = \begin{bmatrix} 0 & -a_{12} & a_{13} \\ a_{12} & 0 & -a_{23} \\ -a_{13} & a_{23} & 0 \end{bmatrix} \begin{bmatrix} 0 \\ 1 \\ 0 \end{bmatrix} = \begin{bmatrix} -a_{12} \\ 0 \\ a_{23} \end{bmatrix} \text{; and it moves to B}'.$$

For point C,

$$\begin{bmatrix} u_1 \\ u_2 \\ u_3 \end{bmatrix} = \begin{bmatrix} 0 & -a_{12} & a_{13} \\ a_{12} & 0 & -a_{23} \\ -a_{13} & a_{23} & 0 \end{bmatrix} \begin{bmatrix} 0 \\ 0 \\ 1 \end{bmatrix} = \begin{bmatrix} a_{13} \\ -a_{23} \\ 0 \end{bmatrix} \text{; and it moves to C}'.$$

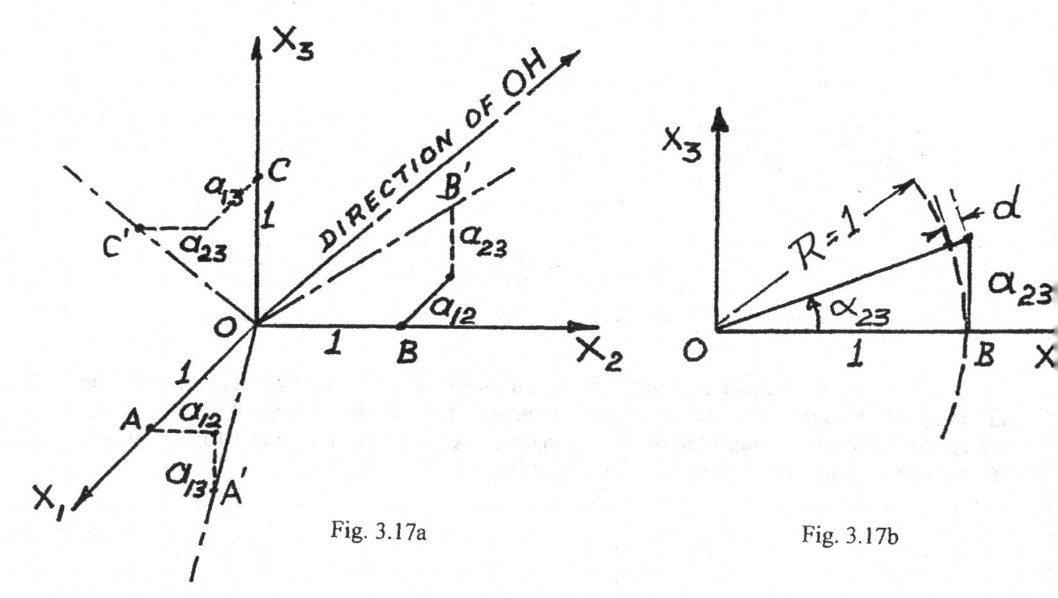

Fig. 3.17a

Fig. 3.17b

As can be seen on Fig. 3.17a, if we imagine a unit cube with sides OA, OB and OC, a_{23} represents a rigid rotation around the OX_1 axis, a_{13} represents a rigid rotation around the OX_2 axis and a_{12} represents a rigid rotation around the OX_3 axis. The rotation vector $\overline{OH}$ is the sum of three rotation vectors $\overline{a_{23}}$ around OX_1, $\overline{a_{13}}$ around OX_2 and $\overline{a_{12}}$ around OX_3. Notice that the word rotation here is used within the limitations imposed by the smallness of a_{23}, a_{13} and a_{12} with respect to 1 and the possibility of ignoring radial dilations. This is best seen in Fig. 3.17b, in two dimensions, where the angle of rotation α_{23} is small enough so that,

$\tan\alpha_{23} = a_{23} \approx \alpha_{23}$ in radians, and the radial dilation d is negligible.

The three direction ratios of OA', OB' and OC' are:

$$OA' \quad OB' \quad OC'$$

$$\begin{Bmatrix} 1 \\ a_{12} \\ -a_{13} \end{Bmatrix}, \quad \begin{Bmatrix} -a_{12} \\ 1 \\ a_{23} \end{Bmatrix}, \quad \begin{Bmatrix} a_{13} \\ -a_{23} \\ 1 \end{Bmatrix}$$

Notice that those three directions are not normal to each other as indicated by their scalar product; unless the products $a_{13}a_{23}, a_{12}a_{13}$ and $a_{12}a_{23}$ are negligible.

As an example consider the antisymmetric transformation where the displacements are given by:

$$\begin{bmatrix} u_1 \\ u_2 \\ u_3 \end{bmatrix} = \begin{bmatrix} 0 & -1 & -0.5 \\ 1 & 0 & -0.5 \\ 0.5 & 0.5 & 0 \end{bmatrix} \begin{bmatrix} x_1 \\ x_2 \\ x_3 \end{bmatrix} \qquad (3.9.13)$$

The first step is to rewrite the matrix so it has the same appearance as that in (3.9.1):

$$\begin{bmatrix} u_1 \\ u_2 \\ u_3 \end{bmatrix} = \begin{bmatrix} 0 & -1 & +(-0.5) \\ 1 & 0 & -0.5 \\ -(-0.5) & 0.5 & 0 \end{bmatrix} \begin{bmatrix} x_1 \\ x_2 \\ x_3 \end{bmatrix} \qquad (3.9.14)$$

The transformation will cause all the points such as M to rotate around a line OH whose direction ratios are (0.5, - 0.5, 1); then to radially expand from this line. The unit radial displacement ε_r is given by (see Eq. 3.9.11):

$$\varepsilon_r = 0.58$$

and the angle of rotation α by (see Eq. 3.9.9):

$\tan\alpha = 1.22$

The direction of OH is invariant and also principal. Eq. (3.9.3) shows that it is invariant. To show that it is also a principal direction we can choose any line normal to OH and show that the two remain normal to each other after they are subjected to the transformation (3.9.13). A line through O normal to OH could have, for example (-0.5, 0.5, 0.5) as direction ratios. The displacements of a point M whose coordinates are (-0.5, 0.5, 0.5) are given by:

$$MM^* = \begin{bmatrix} u_1 \\ u_2 \\ u_3 \end{bmatrix} = \begin{bmatrix} 0 & -1 & -0.5 \\ 1 & 0 & -0.5 \\ 0.5 & 0.5 & 0 \end{bmatrix} \begin{bmatrix} -0.5 \\ 0.5 \\ 0.5 \end{bmatrix} = \begin{bmatrix} -0,75 \\ -0.75 \\ 0 \end{bmatrix},$$

so that

$$OM^* = \begin{bmatrix} -1.25 \\ -.25 \\ 0.5 \end{bmatrix}.$$

The dot product $\overline{OH} \bullet \overline{OM}^* = 0$ showing that $\overline{OH}$ is also a principal direction.

[At the end of Section 3.11]

As another example consider the linear transformation given by:

$$\begin{bmatrix} \xi_1 \\ \xi_2 \\ \xi_3 \end{bmatrix} = \begin{bmatrix} 2 & 2 & 0 \\ 2 & 2 & 0 \\ 0 & 0 & 1 \end{bmatrix} \begin{bmatrix} x_1 \\ x_2 \\ x_3 \end{bmatrix}. \tag{3.11.18}$$

A cursory examination of the matrix indicates that OX_3 is a principal direction. This matrix is symmetric and the invariant directions for this system are also principal. Here the characteristic equation is

$$\begin{vmatrix} 2-\lambda & 2 & 0 \\ 2 & 2-\lambda & 0 \\ 0 & 0 & 1-\lambda \end{vmatrix} = \lambda(\lambda-4)(\lambda-1) = 0$$

and the eigenvalues and eigenvectors are

$$\lambda_1 = 4.0 \qquad \lambda_2 = 1.0 \qquad \lambda_3 = 0.0$$

$$\begin{Bmatrix} 0.70711 \\ 0.70711 \\ 0.00000 \end{Bmatrix}, \begin{Bmatrix} 0.00000 \\ 0.00000 \\ 1.00000 \end{Bmatrix}, \begin{Bmatrix} -0.70711 \\ 0.70711 \\ 0.00000 \end{Bmatrix} \tag{3.11.19}$$

Let us obtain the principal directions using the methods of analytic geometry previously presented in the example at the end of Section 3.8. The equation of the characteristic ellipsoid is

$$8x_1^2 + 8x_2^2 + x_3^2 + 16x_1x_2 = R^2$$

Here a = 8, b = 8, c = 1, h = 8, f = g = 0.

The discriminating cubic is

$$\begin{vmatrix} 8-\beta & 8 & 0 \\ 8 & 8-\beta & 0 \\ 0 & 0 & 1-\beta \end{vmatrix} = 0$$

The roots of this equation and the corresponding direction cosines of the principal directions are

$$\beta_1 = 16.00 \qquad \beta_2 = 1.00 \qquad \beta_3 = 0.00$$

$$\begin{Bmatrix} 0.70711 \\ 0.70711 \\ 0.00000 \end{Bmatrix}, \begin{Bmatrix} 0.00000 \\ 0.00000 \\ 1.00000 \end{Bmatrix}, \begin{Bmatrix} -0.70711 \\ 0.70711 \\ 0.00000 \end{Bmatrix}.$$

Notice that while one obtains the same principal directions, the eigenvalues and the roots of the discriminating cubic are different.

[At the end of Section 3.12]

As another example consider the quadratic form

$$A(x,x) = 2.5x_1^2 + 2.5x_2^2 + 17.5x_3^2 + 6x_1x_2 - 12x_1x_3 - 12x_2x_3. \tag{3.12.13}$$

Eq. (3.12.11) gives:

$$\begin{bmatrix}\xi_1\\ \xi_2\\ \xi_3\end{bmatrix}=\begin{bmatrix}2.5 & 3.0 & -6.0\\ 3.0 & 2.5 & -6.0\\ -6.0 & -6.0 & 17.5\end{bmatrix}\begin{bmatrix}x_1\\ x_2\\ x_3\end{bmatrix}. \tag{3.12.14}$$

The eigenvalues and eigenvectors are obtained as shown in Section 3.8. They are:

$$\lambda_1 = 21.89230 \qquad \lambda_2 = 1.10770 \qquad \lambda_3 = -0.50000$$

$$\begin{Bmatrix}-.32506\\ -.32506\\ .88807\end{Bmatrix}, \quad \begin{Bmatrix}.62796\\ .62796\\ .45971\end{Bmatrix}, \quad \begin{Bmatrix}-.70711\\ .70711\\ .00000\end{Bmatrix} \tag{3.12.15}$$

The canonical form is:

$$A(x',x') = 21.89231(x_1')^2 + 1.10770(x_2')^2 - 0.5(x_3')^2 \tag{3.12.16}$$

[At the end of Section 3.13]

To use Mohr's circles in three dimensions one must be in a principal system of axes. To find the normal and tangential unit displacements of a point with specific coordinates one must first find those coordinates with respect to the principal system. Consider for example the linear transformation for which the unit displacements are given by

$$\begin{bmatrix}u_1\\ u_2\\ u_3\end{bmatrix}=\begin{bmatrix}3 & 1 & 1\\ 1 & 0 & 2\\ 1 & 2 & 0\end{bmatrix}\begin{bmatrix}x_1\\ x_2\\ x_3\end{bmatrix} \tag{3.13.15}$$

It is required to draw Mohr's circles and show on this diagram the image of a point P whose coordinates are (0.3, 0.8, 0.52); OP = 1. First one must find the eigenvalues and eigenvectors of (3.13.15). They are

$$\lambda_1 = 4.0 \qquad \lambda_2 = 1.0 \qquad \lambda_3 = -2.0$$

$$\begin{Bmatrix}0.81650\\ 0.40825\\ 0.40825\end{Bmatrix}, \quad \begin{Bmatrix}-0.57735\\ 0.57735\\ 0.57735\end{Bmatrix}, \quad \begin{Bmatrix}0.00000\\ -0.70711\\ 0.70711\end{Bmatrix}$$

The coordinates of P are then determined in the system of principal axes according to Eq. (3.4.14) or (3.5.3):

$$\begin{bmatrix} x_1' \\ x_2' \\ x_3' \end{bmatrix} = \begin{bmatrix} 0.81650 & 0.40825 & 0.40825 \\ -0.57735 & 0.57735 & 0.57735 \\ 0.00000 & -0.70711 & 0.70711 \end{bmatrix} \begin{bmatrix} 0.3 \\ 0.8 \\ 0.52 \end{bmatrix} = \begin{bmatrix} 0.78384 \\ 0.58920 \\ -0.19799 \end{bmatrix}$$

Referring to Fig. 3.26 and Eqs. (3.13.9) and (3.13.14) we get:

$R = 1.72$ and $\beta = 36.9°$.

Mohr's diagram yields for the point P, n = 2.72 and t = 1.7 (Fig. 3.26c). Those two values can, of course, be obtained with a higher accuracy using Eqs. (3.13.2) to (3.13.5).

The expressions for n and t in Eqs. (3.13.2) to (3.13.5) are referred to the principal axes; but they can, of course, be written within the framework of a general system of axes for a transformation whose displacements are given by $u_i = a_{ij}x_j$. The x_j''s are direction cosines since the points to be transformed fall on a unit sphere. The dot product giving n is nothing but $u_i x_i = a_{ij}x_i x_j$ and t is given by $(u_i u_i - n^2)^{1/2}$. One notices that the expression for n is that of the quadratic form of (3.12.4). Now, using Eq. (3.13.15) we get,

$$n = a_{ij}x_i x_j = 3(0.3)^2 + 2(0.3 \times 0.8) + 2(0.3 \times 0.52) + 4(0.8 \times 0.52) = 2.726$$

and

$$t = (10.334 - 7.431)^{1/2} = 1.70.$$

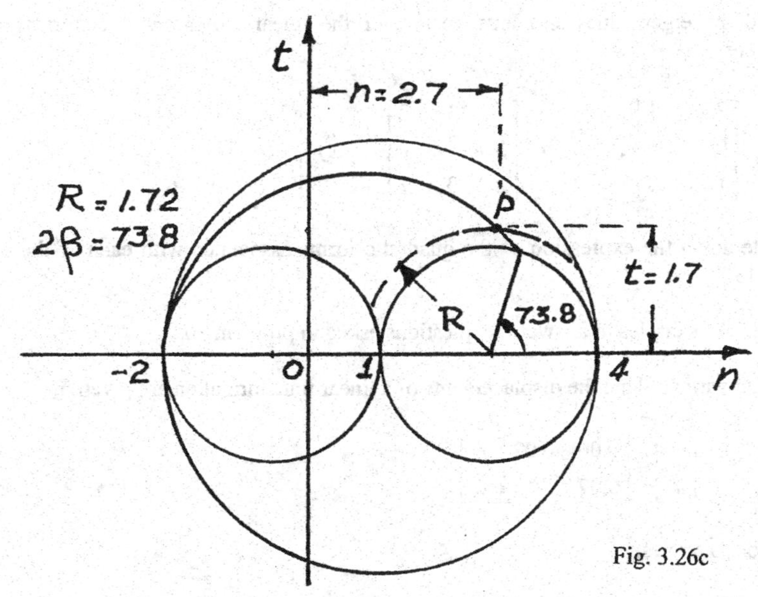

Fig. 3.26c

[Additional Problems]

11. Find the invariant directions and the angles between them for the linear transformation whose matrix is

$$\begin{bmatrix} 7 & 4 & 3 \\ -3 & -2 & -5 \\ -6 & -4 & 0 \end{bmatrix}$$

Verify Eq. (3.6.9).

12. Given a linear transformation whose matrix is

$$k\begin{bmatrix} 6 & 1 & 0 \\ 0 & 4 & 1 \\ 1 & 0 & 8 \end{bmatrix}:$$

a) Write down its symmetric and antisymmetric components.

b) Find the eigenvalues and eigenvectors of its symmetric part.

c) What is the value of the angle of rotation α caused by the antisymmetric part, and what should be the value of the constant k for this angle to be held to one degree?

13. Find the eigenvalues and eigenvectors of the linear symmetric transformations whose matrices are

$$\text{a) } \begin{bmatrix} 3 & 0 & 1 \\ 0 & 2 & 0 \\ 1 & 0 & 3 \end{bmatrix}, \qquad \text{b) } \begin{bmatrix} 2 & 3 & 2 \\ 3 & -12 & 0 \\ 2 & 0 & 1 \end{bmatrix}, \qquad \text{c) } \begin{bmatrix} 11 & -6 & 2 \\ -6 & 10 & -4 \\ 2 & -4 & 6 \end{bmatrix}$$

14. Write down the expression of the quadratic forms associated with each of the matrices in problem 13.

15. Use Mohr's circles to answer the questions asked in problem 10.

16. In an invariant plane the displacements of a linear transformation are given by

$$\begin{bmatrix} u_1 \\ u_2 \end{bmatrix} = \begin{bmatrix} 0.06 & 0.02 \\ 0.02 & 0 \end{bmatrix} \begin{bmatrix} x_1 \\ x_2 \end{bmatrix}$$

Using Mohr's circles:

a) Find the principal directions and the principal unit displacements and show them on a graph.

b) Show on a graph the shape that a unit square, based on the original system of axes OX_1, OX_2 takes after deformation.

c) Show on a graph the shape that a unit square, based on the principal system of axes, O1,O2, takes after deformation.

d) What are the normal and tangential displacements of a point P such that OP is inclined by +30° (counter clockwise) on OX_1 in the original system? Show those displacements on a graph. OP has a length of 2.

e) What would be the displacements of P if the inclination of OP was +30° with the principal direction O1? Show those displacements on a graph. OP has a length of 2.

17. The displacements of a linear transformation are given by the matrix

$$10^{-2}\begin{bmatrix} 4 & 1 & 0 \\ 1 & 0 & 0 \\ 0 & 0 & 3 \end{bmatrix}$$

Use Mohr's circles to find the principal directions and the principal unit displacements. Find the normal and tangential displacement of a point P where OP has a length of 3 and direction cosines $\left(\frac{1}{\sqrt{2}}, \frac{1}{\sqrt{2}}, 0\right)$.

18. What are the direction cosines ℓ_{ij} of a system of axes OX'_1, OX'_2, OX'_3 obtained through a positive rotation of 45° around the OX_1 axis followed by another positive rotation of 45° around the resulting OX'_2 axis. Find the components of the linear transformation matrix

$$\begin{bmatrix} 2.0 & 2.0 & 0.0 \\ 2.0 & 2.0 & 0.0 \\ 0.0 & 0.0 & 1.0 \end{bmatrix}$$

in this new system. Show that the invariants of the two matrices are the same.

CHAPTER 4

[At the end of Section 4.6]

As an example let us assume that the matrix of the state of strain is given by:

$$\varepsilon_{ij} = \begin{bmatrix} 2.5 & 3.0 & -6.0 \\ 3.0 & 2.5 & -6.0 \\ -6.0 & -6.0 & 17.5 \end{bmatrix} 10^{-2} \qquad (4.6.12)$$

This matrix is similar to the one used in Eq. (3.12.14). Eq. (4.4.8) gives:

$$\varepsilon_{MN} = \left[2.5\ell_1^2 + 2.5\ell_2^2 + 17.5\ell_3^2 + 6\ell_1\ell_2 - 12\ell_1\ell_3 - 12\ell_2\ell_3\right]10^{-2} \qquad (4.6.13)$$

which is similar to Eq. (3.12.13). Using (4.6.6) we get $\varepsilon_1 = .218923, \varepsilon_2 = .011077$ and $\varepsilon_3 = -0.005$. The direction cosines of the principal axes are given by (3.12.15). In terms of those principal axes,

$$\varepsilon_{MN} = .218923(\ell_1')^2 + .011077(\ell_2')^2 - 0.005(\ell_3')^2 \qquad (4.6.14)$$

where $\ell_1', \ell_2', \ell_3'$ are the direction cosines of MN with respect to the principal axes. The three invariants are given by:

$$I_1 = 22.5\times10^{-2}, I_2 = 12.75\times10^{-4}, I_3 = 12.125\times10^{-6}$$

[After Eq. (4.9.8)]

Mohr's representation in three dimensions was covered in detail in Sec. 3.13 and the numerical example given at its end can be duplicated here with Eq. (3.13.15) being rewritten as:

$$\begin{bmatrix} du_1 \\ du_2 \\ du_3 \end{bmatrix} = 10^{-3} \begin{bmatrix} 3 & 1 & 1 \\ 1 & 0 & 2 \\ 1 & 2 & 0 \end{bmatrix} \begin{bmatrix} \ell_1 \\ \ell_2 \\ \ell_3 \end{bmatrix}$$

We now have $e_1 = 4\times10^{-3}$, $e_2 = 1\times10^{-3}$ and $e_3 = -2\times10^{-3}$. On Fig. 3.26c we would have $e_n = 2.7\times10^{-3}$ and $e_t = 1.7\times10^{-3}$.

[At the bottom of page 87]

7. *Linear strain in two dimensions.*

Under "Linear strain in two dimensions," the reader is referred to Sec. 3.16 and asked to replace all the a_{ij}'s by e_{ij}'s. The following is essentially a repetition of Sec. 3.16 written in terms of the linear strains e_{ij}. We now have:

7. *Linear strain in two dimensions-plane strain.* Let π be a principal plane through a point M and OX_1, OX_2 be the reference axes (Fig. 4.4a). The unit element MN in this plane is transformed to M*N* in the same principal (also invariant) plane, with the components of the relative displacement being given by:

$$\begin{bmatrix} du_1 \\ du_2 \end{bmatrix} = \begin{bmatrix} e_{11} & e_{12} \\ e_{21} & e_{22} \end{bmatrix} \begin{bmatrix} \ell_1 \\ \ell_2 \end{bmatrix}, \tag{4.9.23}$$

where ℓ_1 and ℓ_2 have replaced dx_1 and dx_2 since MN has a unit length. The corresponding normal and tangential strains of MN are

$$\begin{aligned}(e_{MN})_n &= e_{11}\cos^2\theta + e_{22}\sin^2\theta + 2e_{12}\sin\theta\cos\theta \\ &= \frac{e_{11}+e_{22}}{2} + \frac{e_{11}-e_{22}}{2}\cos 2\theta + e_{12}\sin 2\theta\end{aligned} \tag{4.9.24}$$

$$(e_{MN})_t = -\frac{e_{11}-e_{22}}{2}\sin 2\theta + e_{12}\cos 2\theta . \tag{4.9.25}$$

If the reference axes are rotated an angle θ (Fig. 4.4b) around OX_3, the e_{ij}'s in Eq. (4.9.23) become:

$$\begin{aligned}e'_{11} &= e_{11}\cos^2\theta + e_{22}\sin^2\theta + 2e_{12}\sin\theta\cos\theta \\ &= \frac{e_{11}+e_{22}}{2} + \frac{e_{11}-e_{22}}{2}\cos 2\theta + e_{12}\sin 2\theta\end{aligned} \tag{4.9.26}$$

$$\begin{aligned}e'_{22} &= e_{11}\sin^2\theta + e_{22}\cos^2\theta - 2e_{12}\sin\theta\cos\theta \\ &= \frac{e_{11}+e_{22}}{2} - \frac{e_{11}-e_{22}}{2}\cos 2\theta - e_{12}\sin 2\theta\end{aligned} \tag{4.9.27}$$

$$e'_{12} = -\frac{e_{11}-e_{22}}{2}\sin 2\theta + e_{12}\cos 2\theta \tag{4.9.28}$$

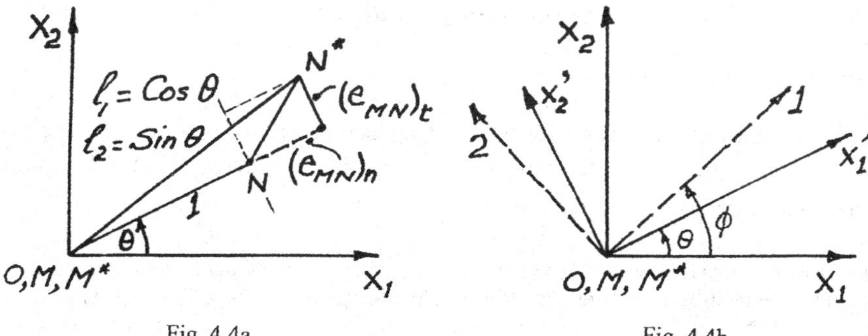

Fig. 4.4a

Fig. 4.4b

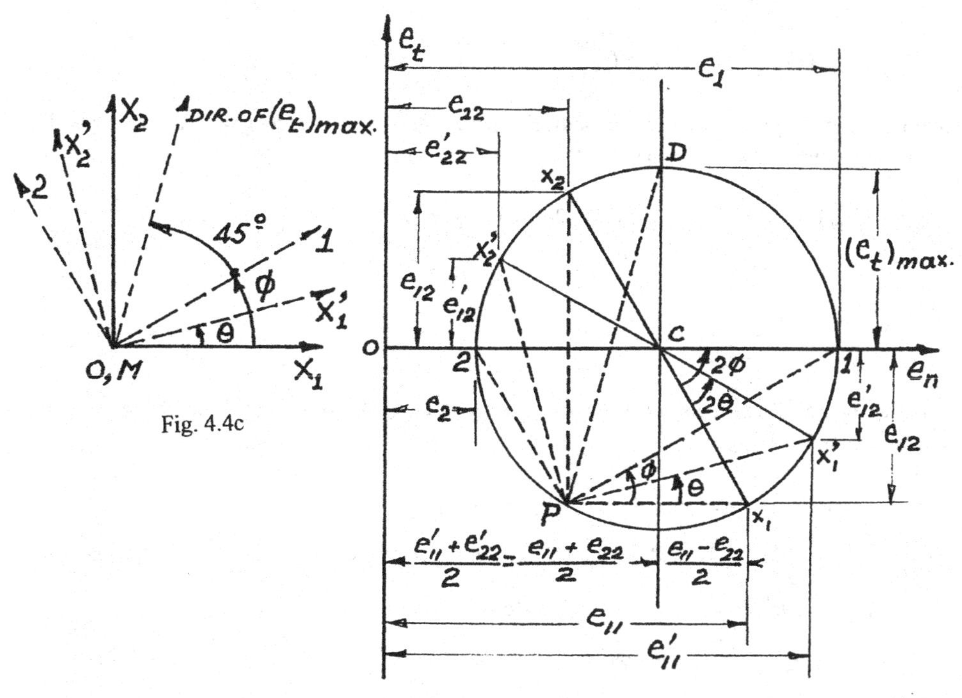

Fig. 4.4c

Fig. 4.4d

Note that (4.9.24) and (4.9.26) are identical. The same is true for (4.9.25) and (4.9.28).

The eigenvalue problem in the principal plane π yields two principal directions, O1 and O2, given by

$$\tan 2\phi = \frac{2e_{12}}{e_{11} - e_{22}} \tag{4.9.29}$$

and two principal strains given by:

$$\begin{matrix} e_1 \\ e_2 \end{matrix} = \frac{e_{11} + e_{22}}{2} \pm \sqrt{\left(\frac{e_{11} - e_{22}}{2}\right)^2 + e_{12}^2} \tag{4.9.30}$$

The representation on Mohr's diagram is extremely useful, not only in giving the normal and shearing strains corresponding to a unit element at a specified angle to the reference axes, but also in giving the components of the strain matrix of Eq. (4.9.23) in a new rotated system of axes; as well as the magnitude and directions of the principal strains e_1 and e_2. While ample explanations were given in Section 3.16 as to sign convention and procedure, it is worthwhile repeating some, using the notation of strain and a more detailed wording. Let us assume that we are given the components of the state of strain in a principal plane (plane strain) at a point M. Relative to a system of axes OX_1, OX_2, those components are e_{11}, e_{22} and e_{12}. It is required (Fig. 4.4c):

1. To find the components of this plane state of strain in a system of axes OX_1', OX_2' defined by the angle θ;
2. To find the magnitude and directions of the principal strains;
3. To find the normal and shear strains associated with an element making an angle θ with OX_1.

The convention for the normal strains is that they are plotted positive to the right of the origin on the Oe_n axis (Fig. 4.4d). The convention for the shear strains is as follows: If the shear strain e_{12} is positive, the point representing the more clockwise of the two axes enclosing the first quadrant (here OX_1) is plotted at a distance e_{12} "below" the Oe_n axis and the point representing the counterclockwise axis (here OX_2) is plotted at a distance e_{12} "above" the Oe_n axis.[1] In Fig. 4.4d, e_{12} is positive. Since in the real body OX_1' makes an angle θ counterclockwise with OX_1, the angle 2θ is taken in the same direction with Cx_1 in Fig. 4.4d. Point x_1' gives e_{11}' and point x_2' gives e_{22}'. e_{12}' can be obtained from either of the two points and is positive (according to the previous convention). Therefore, in the system of coordinates OX_1', OX_2', the right angle between two elements parallel to these axes will become an acute angle after transformation. The direction corresponding to the major principal strain makes an angle ϕ

[1] See S.H. Crandall and N. C. Dahl "An Introduction to the Mechanics of Solids," 1st edition, McGraw Hill, 1959.

with OX_1 and the direction corresponding to the maximum shear strain makes an angle (ϕ+45°) with OX_1.

For an element making an angle θ with OX_1 the value of e_n and e_t is the same as e'_{11} and e'_{12} specified by the angle 2θ. Whenever one is asked to obtain normal and shearing strains corresponding to a certain direction, it is advisable to consider at the same time a direction normal to it and deduce whether the right angle becomes acute or obtuse after transformation. This, allows one to find the direction in which the element moves.

Now, directions on the actual Fig. 4.4c can be exhibited on Mohr's circle through the determination of a point P called the "pole" (Fig. 4.4d). The pole is determined by the line through x_1 parallel to OX_1 or by the line through x_2 parallel to OX_2. Joining the pole to any other point on the circle gives the direction of the element whose normal and shear strains are given by this point. This is clearly shown by Figs. 4.4c and 4.4d: Px_1 is parallel to OX_1 and since the angle $x_1Cx'_1$ is equal to 2θ then Px'_1 makes an angle θ with Px_1 and is parallel to OX'_1. P1 and P2 give the actual directions of the major and minor principal strains. The angle ϕ between the OX_1 direction and the major principal strain is given by

$$\tan 2\phi = \frac{e_{12}}{(e_{11} - e_{22})/2}.$$

Example

A sheet of metal is deformed uniformly in its own plane so that the strain components at all its points related to a set of axes OX_1 and OX_2 are

$$e_{11} = \text{-}500\text{x}10^{-6}$$

$$e_{22} = 2000\text{x}10^{-6}$$

$$e_{12} = 1000\text{x}10^{-6}$$

Find the strain components associated with a set of axes OX'_1 and OX'_2 inclined at an angle of 45° clockwise to the OX_1, OX_2 set, as shown in figure (4.4e). Also find the principal strains and the direction of the principal axes

It is always instructive to begin by writing down the transformation in the form

$$\begin{bmatrix} du_1 \\ du_2 \end{bmatrix} = 10^{-6} \begin{bmatrix} -500 & 1000 \\ 1000 & 2000 \end{bmatrix} \begin{bmatrix} \ell_1 \\ \ell_2 \end{bmatrix}$$

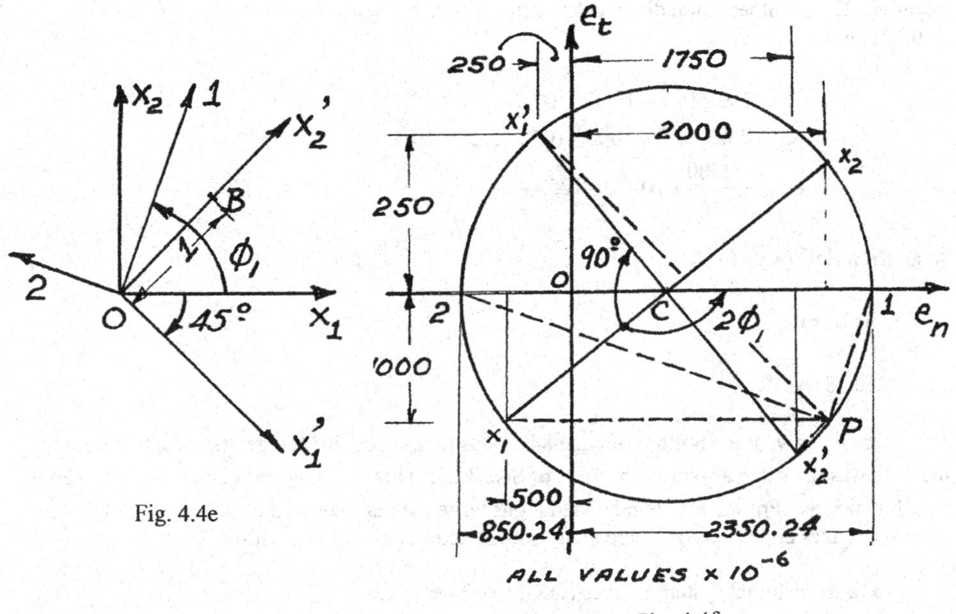

Fig. 4.4e

Fig. 4.4f

and refer to the system of axes OX_1 and OX_2 in Fig. 4.4e. Fig. 4.4f shows Mohr's circle for the given data. Point x_1' lies at a relative angular position twice that existing in the actual body, i.e., at a position 90° clockwise from x_1 on Mohr's circle. The pole P is obtained by drawing from x_1 a line parallel to OX_1. P1 and P2 give the major and minor principal directions. The major principal direction makes an angle ϕ_1 counterclockwise with OX_1. This angle can be directly measured on Mohr's circle or computed from:

$$tan 2\phi_1 = \frac{2e_{12}}{e_{11} - e_{22}} = \frac{2 \times 1000}{-500 - 2000} = -\frac{4}{5}$$

$$\phi_1 = 70.67^{\circ}$$

and $$\phi_2 = \frac{\pi}{2} + 70.67 = 160.67^{\circ}$$

The strain components associated with the systems of axes (OX_1', OX_2') and (O1, O2) can be measured on Mohr's circle. According to our sign convention OX_1' is the more clockwise of the two new axes and because of the location of x_1' on Mohr's circle the sign of e_{12}' will be negative. On the other hand, direct calculation of e'_{11}, e'_{22} and e'_{12} from equation (4.9.26) to (4.9.28) gives

$$e'_{11} = 750\times10^{-6} - 1000\times10^{-6} = -250\times10^{-6}$$
$$e'_{22} = 750\times10^{-6} + 1000\times10^{-6} = 1750\times10^{-6}$$
$$e'_{12} = -\frac{2500}{2}\times10^{-6} = -1250\times10^{-6}.$$

Also, from Eqs, (4.9.30)

$e_1 = 2350.78\times10^{-6}$

$e_2 = -850.78\times10^{-6}$

In Sec. 4.9, at the bottom of page 84, it was mentioned that the geometrical meaning of the e_{ij}'s was the same as that of the a_{ij}'s of Sec. 3.15. This meaning can be very clearly seen in two dimensions. For the problem at hand let us write the matrices of the state of strain in three systems of axes and show the shape a unit square takes after transformation:

1. For the original system of axes (OX_1, OX_2) we have:

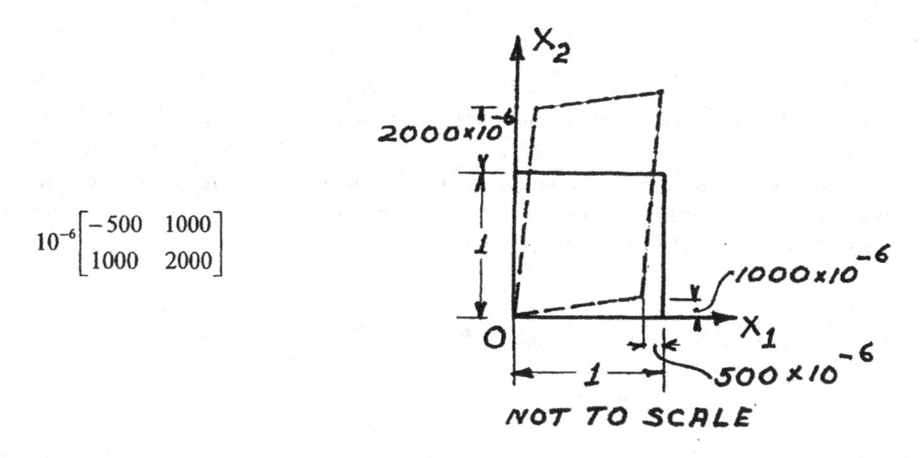

$$10^{-6}\begin{bmatrix} -500 & 1000 \\ 1000 & 2000 \end{bmatrix}$$

e_{12} being positive the right angle becomes acute.

2. For a system of axes (OX'_1, OX'_2) with OB falling along OX'_1 we have:

$$10^{-6}\begin{bmatrix} -250 & -1250 \\ -1250 & 1750 \end{bmatrix}$$

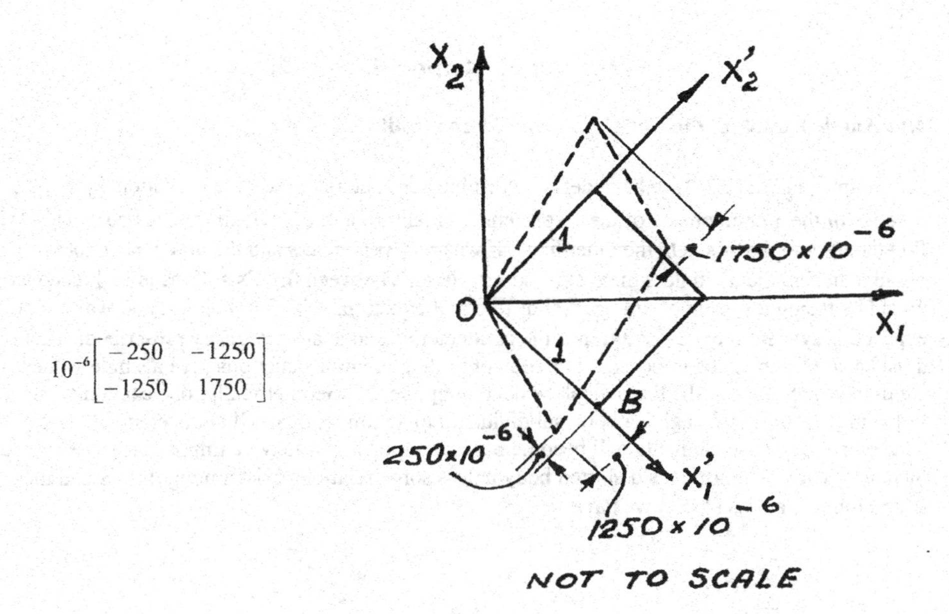

Notice how the use of two directions at right angle, namely OX_1' and OX_2', and the knowledge that this angle becomes obtuse, helps to ascertain the direction in which B moves.

3. For a system of axes (O1,O2) we have:

$$10^{-6}\begin{bmatrix} 2350.78 & 0 \\ 0 & -850.78 \end{bmatrix}$$

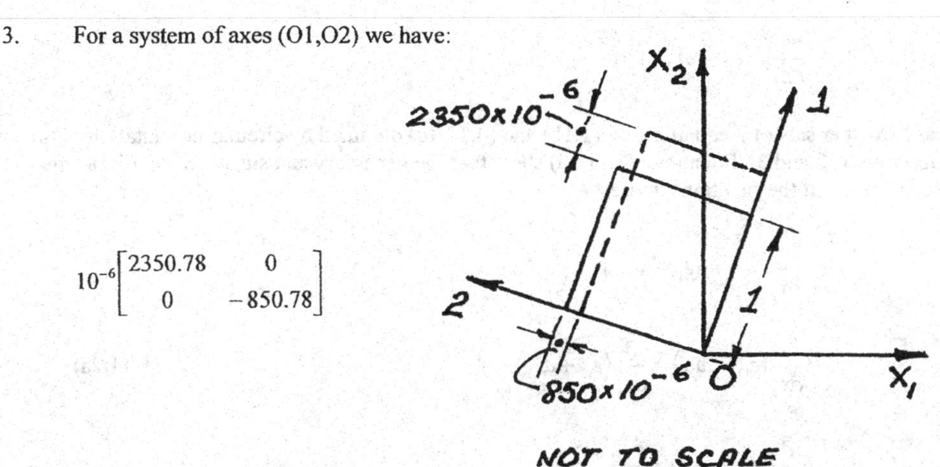

The principal axes are invariant and the right angle remains unchanged.

Finally, if the elements along axes are different from 1, the normal and shearing strains must be multiplied by the length to obtain the magnitude of the relative normal and tangential displacements.

[At the end of Chapter 4, page 91]

4.11 Another Look at the Conditions for Compatibility

In equation (1.21) we have defined the nine parameters of the transformation e_{ij} and ω_{ij} in terms of the 3 components of the displacement u_1 and u_2 and u_3. Given the u's one can always find the e's and the ω's. On the other hand, if we are given the e's and the ω's, it will not be possible to find the u's unless there exist some relations between the e's and the ω's. Indeed we would be facing a system of nine partial differential equations with 3 unknowns and we cannot expect this system to have a solution if the parameters e and ω are arbitrarily prescribed. There must be conditions to be imposed on them in order that the nine equations give a single valued continuous solution for the three displacement components. Geometrically, one can make the following reasoning. Imagine a body subdivided into a number of small cubic elements. After transformation, these elements will become parallelepiped and it may be impossible to rearrange them to become a continuous distorted body unless some relations exist among the e's and the ω's. From equations (1.2.1) we have

$$\frac{\partial u_1}{\partial x_1} = e_{11}$$

$$\frac{\partial u_1}{\partial x_2} = e_{12} - \omega_{21} \qquad (4.11.1a)$$

$$\frac{\partial u_1}{\partial x_3} = e_{13} + \omega_{13}$$

and two other sets of 3 equations (4.11.1b) and (4.11.1c) obtained by circular permutation of the subscripts 1, 2 and 3. Equations (4.11.1a) show that the necessary and sufficient conditions for the existence of the functions u_1 are:

$$\frac{\partial e_{11}}{\partial x_2} = \frac{\partial}{\partial x_1}(e_{12} - \omega_{21})$$

$$\frac{\partial}{\partial x_3}(e_{12} - \omega_{21}) = \frac{\partial}{\partial x_2}(e_{13} + \omega_{13}) \qquad (4.11.2a)$$

$$\frac{\partial}{\partial x_1}(e_{13} + \omega_{13}) = \frac{\partial e_{11}}{\partial x_3}$$

Similarly, the necessary and sufficient conditions for the existence of the functions u_2 and u_3 are given by two sets of 3 equations (4.11.2b) and (4.11.2c). These 2 sets can be directly obtained by circular permutation of (4.11.2a).

$$\frac{\partial e_{22}}{\partial x_3} = \frac{\partial}{\partial x_2}(e_{23} - \omega_{32})$$

$$\frac{\partial}{\partial x_1}(e_{23} - \omega_{32}) = \frac{\partial}{\partial x_3}(e_{21} + \omega_{21}) \quad (4.11.2b)$$

$$\frac{\partial}{\partial x_2}(e_{21} + \omega_{21}) = \frac{\partial e_{22}}{\partial x_1}$$

and

$$\frac{\partial e_{33}}{\partial x_1} = \frac{\partial}{\partial x_3}(e_{31} - \omega_{13})$$

$$\frac{\partial}{\partial x_2}(e_{31} - \omega_{13}) = \frac{\partial}{\partial x_1}(e_{32} + \omega_{32}) \quad (4.11.2c)$$

$$\frac{\partial}{\partial x_3}(e_{32} + \omega_{32}) = \frac{\partial e_{33}}{\partial x_2}$$

In all, we have 9 equations of compatibility to insure the existence of u_1, u_2 and u_3. In these equations $e_{ij} = e_{ji}$ and $\omega_{ij} = -\omega_{ji}$.

Now, as demonstrated in Sections 3.15 and 3.9, and restated in 4.9 the e_{ij}'s characterize the change in shape of a body subjected to a linear transformation while the ω_{ij}'s express a rigid rotation (provided the radial dilation is neglected). Thus, while the e_{ij}'s are related to the internal stresses at each point, the ω_{ij}'s are not. It is therefore of interest to find compatibility relations in terms of the e_{ij}'s alone. For this, the three functions ω_{21}, ω_{32} and ω_{13} must be eliminated from the nine equations (4.11.2a,b,c). For this we add the second equation of (4.11.2a) to the second equation of (4.11.2b) and subtract from the result the second equation of (4.11.2c) to obtain

$$\frac{\partial \omega_{21}}{\partial x_3} = \frac{\partial e_{32}}{\partial x_1} - \frac{\partial e_{31}}{\partial x_2} \quad (4.11.3a)$$

From the first equation of (4.11.2a) we have

$$\frac{\partial \omega_{21}}{\partial x_1} = \frac{\partial e_{12}}{\partial x_1} - \frac{\partial e_{11}}{\partial x_2} \quad (4.11.3b)$$

and from the third equation of (4.11.2b) we have

$$\frac{\partial \omega_{21}}{\partial x_2} = \frac{\partial e_{22}}{\partial x_1} - \frac{\partial e_{21}}{\partial x_2} \tag{4.11.3c}$$

The three equations (4.11.3a,b,c) show that the necessary and sufficient conditions for the existence of the function ω_{21} are:

$$\begin{aligned}
&\frac{\partial}{\partial x_1}\left[\frac{\partial e_{32}}{\partial x_1} - \frac{\partial e_{31}}{\partial x_2}\right] = \frac{\partial}{\partial x_3}\left[\frac{\partial e_{12}}{\partial x_1} - \frac{\partial e_{11}}{\partial x_2}\right] \\
&\frac{\partial}{\partial x_2}\left[\frac{\partial e_{12}}{\partial x_1} - \frac{\partial e_{11}}{\partial x_2}\right] = \frac{\partial}{\partial x_1}\left[\frac{\partial e_{22}}{\partial x_1} - \frac{\partial e_{21}}{\partial x_2}\right] \\
&\frac{\partial}{\partial x_3}\left[\frac{\partial e_{22}}{\partial x_1} - \frac{\partial e_{21}}{\partial x_2}\right] = \frac{\partial}{\partial x_2}\left[\frac{\partial e_{32}}{\partial x_1} - \frac{\partial e_{31}}{\partial x_2}\right]
\end{aligned} \tag{4.11.4}$$

These three equations can be written

$$\begin{aligned}
&\frac{\partial^2 e_{11}}{\partial x_2 \partial x_3} = \frac{\partial}{\partial x_1}\left[-\frac{\partial e_{23}}{\partial x_1} + \frac{\partial e_{31}}{\partial x_2} + \frac{\partial e_{12}}{\partial x_3}\right] \\
&2\frac{\partial^2 e_{12}}{\partial x_1 \partial x_2} = \frac{\partial^2 e_{11}}{\partial x_2^2} + \frac{\partial^2 e_{22}}{\partial x_1^2} \\
&\frac{\partial^2 e_{22}}{\partial x_3 \partial x_1} = \frac{\partial}{\partial x_2}\left[-\frac{\partial e_{31}}{\partial x_2} + \frac{\partial e_{12}}{\partial x_3} + \frac{\partial e_{23}}{\partial x_1}\right]
\end{aligned} \tag{4.11.5}$$

Similarly we can deduce 2 other sets of three equations necessary to insure the existence of ω_{13} and ω_{32}. However, these two sets can directly be obtained by circular permutation of (4.11.5). This gives a total of nine equations of which three will be found to be repeated twice. Indeed, we can see at a glance that one circular permutation of the first equation of (4.11.5) gives the third equation. We are thus reduced to six compatibility equations, the two first of (4.11.5) and two other sets of two equations obtained by circular permutation. Those equations are the ones given in (4.10.14) on page 91.

4.12 Example (use an eigenvalue software)

Let us look at the overall picture of the state of strain at a point M using the transformation:

$$\begin{bmatrix} d\xi_1 \\ d\xi_2 \\ d\xi_3 \end{bmatrix} = \begin{bmatrix} 2 & -2 & 3 \\ 1 & 1 & 1 \\ 1 & 3 & -1 \end{bmatrix} \begin{bmatrix} dx_1 \\ dx_2 \\ dx_3 \end{bmatrix} \tag{4.12.1}$$

In real problems the matrix has diagonal elements barely larger than 1 and the off diagonal elements are also quite small compared to 1. The values above have been chosen strictly for demonstration purposes and to magnify the differences among the various cases at hand.

The relative displacements are given by:

$$\begin{bmatrix} du_1 \\ du_2 \\ du_3 \end{bmatrix} = \begin{bmatrix} 1 & -2 & 3 \\ 1 & 0 & 1 \\ 1 & 3 & -2 \end{bmatrix} \begin{bmatrix} dx_1 \\ dx_2 \\ dx_3 \end{bmatrix} \tag{4.12.2}$$

a) <u>The invariant directions</u> of (4.12.1) can be found following the same procedures presented in Section 3.6 and the example at the end of Section 3.8. They are:

$$\begin{matrix} \lambda_1 = 3 & \lambda_2 = 1 & \lambda_3 = -2 \\ \begin{Bmatrix} 0.57735 \\ 0.57735 \\ 0.57735 \end{Bmatrix}, & \begin{Bmatrix} 0.57735 \\ -0.57735 \\ -0.57735 \end{Bmatrix}, & \begin{Bmatrix} 0.61685 \\ 0.05608 \\ -0.78508 \end{Bmatrix} \end{matrix} \tag{4.12.3}$$

The angles between those directions can be obtained by taking the scalar products of the eigenvectors given above. Thus,

$$\begin{aligned} \cos\theta_{12} &= -0.33333 \rightarrow \theta_{12} = 109.47^\circ \\ \cos\theta_{13} &= -0.06470 \rightarrow \theta_{13} = 93.71^\circ \\ \cos\theta_{23} &= 0.77702 \rightarrow \theta_{23} = 39.01^\circ \end{aligned} \tag{4.12.4}$$

The invariant directions do not rotate under the linear transformation and they are not orthogonal unless the matrix is symmetric; in which case they are also principal.

b) <u>The principal directions</u> of (4.12.1) can be found using the methods of analytic geometry presented at the end of Section 3.3. The equation of the characteristic ellipsoid is:

$$(2x_1 - 2x_2 + 3x_3)^2 + (x_1 + x_2 + x_3)^2 + (x_1 + 3x_2 - x_3)^2 = R^2$$

or

$$6x_1^2 + 14x_2^2 + 11x_3^2 + 12x_1x_3 - 16x_2x_3 = R^2 \tag{4.12.5}$$

The discriminating cubic is

$$\begin{vmatrix} 6-\beta & 0 & 6 \\ 0 & 14-\beta & -8 \\ 6 & -8 & 11-\beta \end{vmatrix} = 0$$

The roots of this equation and the principal directions are given by β's and their corresponding direction cosines:

$$\beta_3=21.65730,\quad \beta_2=9.16126,\quad \beta_1=0.18144$$

$$\begin{Bmatrix} 0.25614 \\ -0.69831 \\ 0.66840 \end{Bmatrix}, \begin{Bmatrix} 0.70076 \\ 0.61043 \\ 0.36921 \end{Bmatrix}, \begin{Bmatrix} 0.66583 \\ -0.37382 \\ -0.64570 \end{Bmatrix} \tag{4.12.6}$$

The principal directions remain orthogonal under the linear transformation but they rotate; unless the matrix is symmetric in which case they are also invariant.

c) The matrix of the state of strain ε_{ij} can be obtained from the displacement gradient matrix of (4.12.2) and Eqs. (4.3.3):

$$\varepsilon_{ij} = \begin{bmatrix} 2.5 & 0.0 & 3.0 \\ 0.0 & 6.5 & -4.0 \\ 3.0 & -4.0 & 5.0 \end{bmatrix}. \tag{4.12.7}$$

Note that the components of this matrix were "created" by Eqs. (4.3.3) and nowhere does this matrix appear in a linear transformation similar to (4.12.1). Those components transform according to Eq. (4.5.8), and there exists at M three directions that remain at right angle to one another when the state of strain (4.12.7) is activated there. They are the three principal directions of this state of strain. As explained in Section 4.6 those directions can be obtained from the solution of an eigenvalue problem (see Eq. 4.6.6) to give in our case:

$$\varepsilon_1 = 10.32865 \quad \varepsilon_2 = 4.08063 \quad \varepsilon_3 = -0.40928$$

$$\begin{Bmatrix} 0.25614 \\ -0.69831 \\ 0.66840 \end{Bmatrix}, \begin{Bmatrix} 0.70076 \\ 0.61043 \\ 0.36921 \end{Bmatrix}, \begin{Bmatrix} 0.66583 \\ -0.37382 \\ -0.64570 \end{Bmatrix} \tag{4.12.8}$$

Notice that the direction cosines in (4.12.8) are the same as the direction cosines in (4.12.6). As stated at the beginning of Section 4.6 the principal directions of the state of strain are the principal directions of the linear transformation. They rotate but remain orthogonal. This is obvious since the ω_{ij}'s are part of the ε_{ij}'s [see Eq. (4.3.3)].

d) <u>The rotationless matrix of the state of strain</u> ε^*_{ij} which is obtained from (4.12.2) and (4.3.3), and setting the ω_{ij}'s $= 0$ is:

$$\varepsilon^*_{ij} = \begin{bmatrix} 3.625 & 1.25 & 0.5 \\ 1.25 & 2.125 & -0.5 \\ 0.5 & -0.5 & 4.0 \end{bmatrix}.$$

The principal directions of this rotationless state of strain are again obtained using Eqs. (4.6.6):

$$\varepsilon^*_1 = 4.45193, \quad \varepsilon^*_2 = 4.05025, \quad \varepsilon^*_3 = 1.24782$$

$$\begin{Bmatrix} 0.78920 \\ 0.31004 \\ 0.53013 \end{Bmatrix}, \quad \begin{Bmatrix} -0.36898 \\ -0.45067 \\ 0.81286 \end{Bmatrix}, \quad \begin{Bmatrix} -0.49093 \\ 0.83712 \\ 0.24127 \end{Bmatrix} \tag{4.12.9}$$

Notice that here the ε^*_{ij} contain e_{ij}'s as well as their squares and products. Since the ω_{ij}'s have been set equal to zero those principal directions will not rotate when ε^*_{ij} is activated.

e) <u>The symmetric part of (4.12.2) is</u>

$$\begin{bmatrix} du_1 \\ du_2 \\ du_3 \end{bmatrix} = \begin{bmatrix} 1 & -0.5 & 2 \\ -0.5 & 0 & 2 \\ 2 & 2 & -2 \end{bmatrix} \begin{bmatrix} dx_1 \\ dx_2 \\ dx_3 \end{bmatrix} \tag{4.12.10}$$

The coefficients of this matrix are the linear strains e_{ij}. The principal directions and the principal linear strains e_i's are

$$e_1 = 2.14704 \quad e_2 = 0.86966 \quad e_3 = -4.01670$$

$$\begin{Bmatrix} 0.78920 \\ 0.31004 \\ 0.53013 \end{Bmatrix}, \quad \begin{Bmatrix} -0.49093 \\ 0.83712 \\ 0.24127 \end{Bmatrix}, \quad \begin{Bmatrix} -0.36898 \\ -0.45067 \\ 0.81286 \end{Bmatrix} \tag{4.12.11}$$

Here the matrix being symmetric the principal directions are also invariant. Those directions are the same as the ones in (4.12.9) even though the matrices are different. Indeed the ε^*_{ij}'s are different from the e_{ij}'s but in both cases we have no rotation. The discussion at the beginning of Section 4.9 is numerically exhibited in this example.

f) The antisymmetric part of (4.12.2) is

$$\begin{bmatrix} du_1 \\ du_2 \\ du_3 \end{bmatrix} = \begin{bmatrix} 0 & -1.5 & 1 \\ 1.5 & 0 & -1 \\ -1 & 1 & 0 \end{bmatrix} \begin{bmatrix} dx_1 \\ dx_2 \\ dx_3 \end{bmatrix} \tag{4.12.12}$$

As explained in Section 3.9, this antisymmetric matrix causes all the points in the vicinity of M to rotate around an axis through M whose direction ratios are (1, 1, 1.5) and direction cosines are (0.48507, 0.48507, 0.727606). This direction is both principal and invariant. The radial dilation is given by

$$\varepsilon_r = \sqrt{1+1+1+2.25} - 1 = 1.29,$$

and the rotation by

$$\tan\alpha = \sqrt{1+1+2.25} = 2.06$$

Recall that by definition an antisymmetric transformation is one whose hodograph is expressed by an antisymmetric matrix. So the antisymmetric transformation whose hodograph is (4.12.12) is

$$\begin{bmatrix} d\xi_1 \\ d\xi_2 \\ d\xi_3 \end{bmatrix} = \begin{bmatrix} 1 & -1.5 & 1 \\ 1.5 & 1 & -1 \\ -1 & 1 & 1 \end{bmatrix} \begin{bmatrix} dx_1 \\ dx_2 \\ dx_3 \end{bmatrix} \tag{4.12.13}$$

Like any other linear transformation, (4.12.13) has both invariant directions and principal directions. The search for the invariant directions yields one real direction, namely (0.48507, 0.48507, 0.727606) and two imaginary ones. For the principal directions one has to use analytic geometry since the matrix is not symmetric. The equation of the characteristic ellipsoid is:

$$4.25x_1^2 + 4.25x_2^2 + 3x_3^2 - 2x_1x_2 - 3x_1x_3 - 3x_2x_3 = R^2$$

The discriminating cubic is

$$\begin{vmatrix} 4.25-\beta & -1.0 & -1.5 \\ -1.0 & 4.25-\beta & -1.5 \\ -1.5 & -1.5 & 3-\beta \end{vmatrix} = 0$$

The roots of this equation and the corresponding direction cosines are:

$$\beta_1 = 5.25 \qquad \beta_2 = 5.25 \qquad \beta_3 = 1.00$$

$$\begin{Bmatrix} 0.77861 \\ -0.61835 \\ -0.10684 \end{Bmatrix}, \begin{Bmatrix} 0.39809 \\ 0.61835 \\ -0.67762 \end{Bmatrix}, \begin{Bmatrix} 0.48507 \\ 0.48507 \\ 0.72761 \end{Bmatrix}$$

Note that $\beta_3 = 1.0$ provides a direction that is both principal and, as previously shown, invariant. The two other directions are in a plane normal to the axis of rotation and correspond to β_1 and β_2. As a matter of fact any two directions in this plane, at right angle to each other could be chosen as the two other principal directions of the antisymmetric transformation; two principal axes of the characteristic ellipsoid being equal. This is the second of the three cases mentioned in Section 3.3. The ellipsoid is of revolution.

[Additional Problems]

13. In a solid subjected to transformation, the components of the displacements are given by:

$$u_1 = c_1 x_1 x_2 x_3,\ u_2 = c_2 x_1 x_2 x_3,\ u_3 = c_3 x_1 x_2 x_3 .$$

It is know that the displacements of a point E (1.5,1.0,2.0) are such that its coordinates after transformation are (1.62, 1.03, 1.88). $c_1, c_2,$ and c_3 are constants.

a) Determine the value of $c_1, c_2,$ and c_3.

b) Determine the components of the matrix of the state of strain ε_{ij} at E.

c) Determine the components of the linear strain e_{ij} at E.

d) Determine the components of the rotation ω_{ij} and the radial dilation ε_r at E.

14. Given the following displacement field:

$$u_1 = 10^{-3}(3x_1^2 x_2 + 6)$$
$$u_2 = 10^{-3}(x_2^2 + 6x_1 x_3)$$
$$u_3 = 10^{-3}(6x_3^2 + 12x_2 x_3 + 60)$$

a) Find the components $\frac{\partial u_i}{\partial x_j}$ of the displacement gradient matrix at a point A (1,0,2).

Write down the matrix.

b) Write down the matrix of the state of strain e_{ij} and that of the rotation ω_{ij} at the same point A.

c) What is the magnitude of the angle of rotation α and the radial dilation ε_r at A. Show on a sketch the direction of the axis of rotation.

15. The relative displacements at a point M for a small element MN are given by

$$\begin{bmatrix} du_1 \\ du_2 \\ du_3 \end{bmatrix} = \begin{bmatrix} 2 & 1 & 1 \\ -1 & 2 & 1 \\ 1 & -1 & 2 \end{bmatrix} \begin{bmatrix} dx_1 \\ dx_2 \\ dx_3 \end{bmatrix}$$

a) Find the principal directions of the strain tensor ε_{ij}.

b) Find the principal directions of the linear strain tensor e_{ij}.

c) Find the principal directions of the strain tensor ε_{ij} when the ω_{ij}'s are neglected.

d) Show that there is one direction common to all the above and that it is also an invariant direction in the transformation.

16. The state of linear strain e_{ij} at a point M is given by:

$$e_{ij} = 10^{-3} \begin{bmatrix} 2 & 2 & 0 \\ 2 & 2 & 0 \\ 0 & 0 & 1 \end{bmatrix}$$

a) Draw Mohr's circles corresponding to this state of strain.

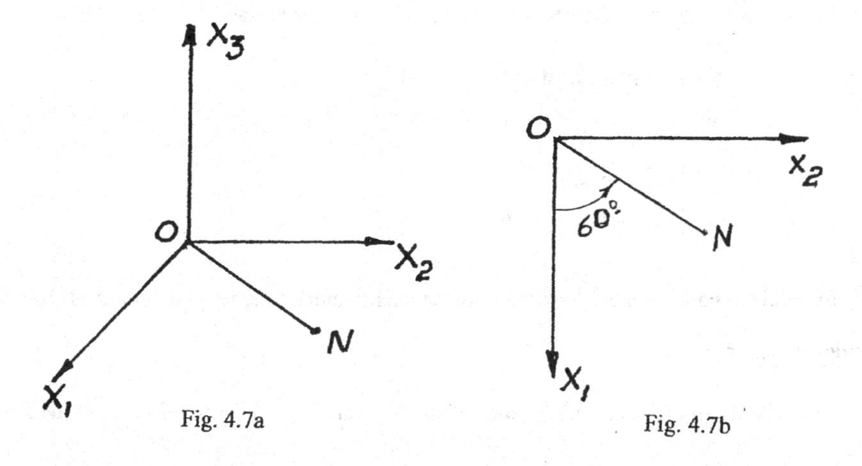

Fig. 4.7a

Fig. 4.7b

b) Using Mohr's circles find the normal and tangential or shearing strains corresponding to an element MN (Fig. 4.7a) whose direction cosines are $\left(\frac{1}{2}, \frac{\sqrt{3}}{2}, 0\right)$. Show on a sketch $(e_{MN})_n$ and $(e_{MN})_t$. Does N move towards the OX_1 axis or the OX_2 axis in Fig. 4.7b.

c) Using the same Mohr circles find the magnitude and direction of the principal linear strains. Show those directions on Fig. 4.7b.

17. Three strain gages are used to measure the normal strain e_n in three directions at 60 degrees to each other, at a point M of a thin plate (Fig. 4.8):

$$(e_n)_{MA} = 1{,}224 \times 10^{-6}$$
$$(e_n)_{MB} = -66 \times 10^{-6}$$
$$(e_n)_{MC} = 442 \times 10^{-6}$$

Find the magnitude and directions of the principal linear strains e_1 and e_2. Show the principal directions on Fig. 4.8

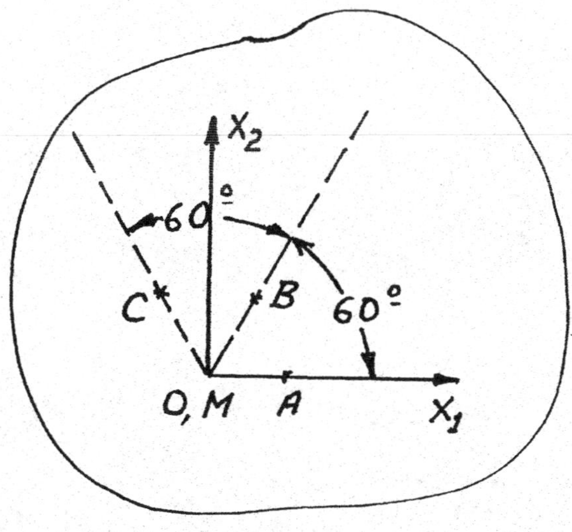

Fig. 4.8

18. If the strain gages in Fig. 4.8 are at 45 degrees with each other and

$$(e_n)_{MA} = 2\times10^{-3}$$
$$(e_n)_{MB} = 1.5\times10^{-5}$$
$$(e_n)_{MC} = 0.5\times10^{-3}$$

a) Write down the matrix of the two dimensional state of strain at point M.

b) Draw Mohr's circle for this state of linear strain and find the magnitudes and directions of the principal strains. Show on a sketch those directions with respect to the OX_1, OX_2 system.

c) Use Mohr's circle to find the matrix of the two dimensional state of strain in a system of axes making 45 degrees counterclockwise with the original system OX_1, OX_2.

CHAPTER 7

[At the end of Section 7.3]

Cauchy's three equations (7.3.8) are very often referred to in many of the following chapters specially when one is looking at the conditions on the boundaries of a solid. For example, given components of the stress tensor at a point such as A on the surface of a solid subjected to outside stresses (Fig. 7.6a), Eq. (7.3.8) gives the components of the stress vector acting on the plane P tangent to the boundary at A. If those components do not coincide with what is being applied from the outside, then the stress tensor that is given is incorrect or does not correspond to the problem at hand.

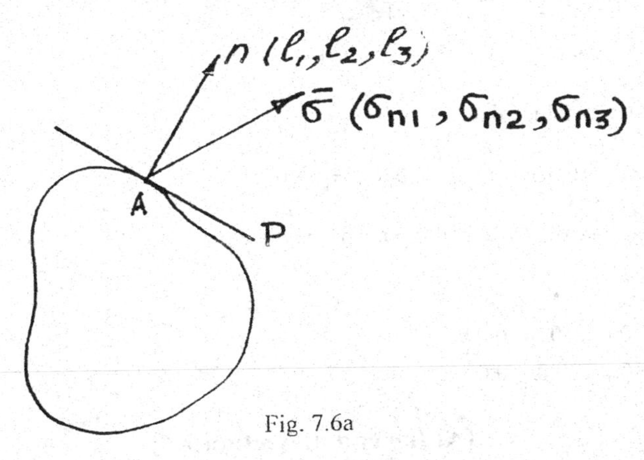

Fig. 7.6a

As an example consider a thin plate subjected on its edges to the stress vectors shown in Fig. 7.6b. What are the components of the state of stress assuming that they are uniformly distributed across the plate? Assume that the components related to the OX_3 direction are equal to zero.

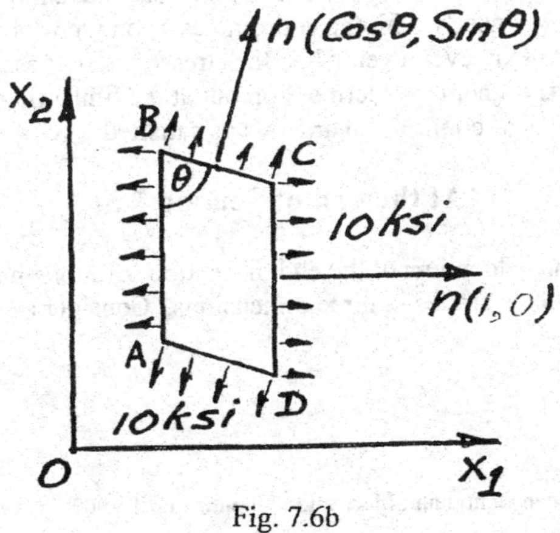

Fig. 7.6b

Here we know the conditions on the boundaries. On AB and CD we have stress vectors of 10ksi acting in the horizontal direction. The direction cosines of the normal to CD are (1, 0). On AD and BC the stress vectors act along the normals to those faces and the direction cosines of the normal to BC are (cosθ, sinθ). Applying Eq. (7.3.8) in two dimensions to CD and BC we get:

$$\begin{bmatrix} 10.0 \\ 0 \end{bmatrix} = \begin{bmatrix} \sigma_{11} & \sigma_{12} \\ \sigma_{12} & \sigma_{22} \end{bmatrix} \begin{bmatrix} 1 \\ 0 \end{bmatrix}$$

and

$$\begin{bmatrix} 10.0\cos\theta \\ 10.0\sin\theta \end{bmatrix} = \begin{bmatrix} \sigma_{11} & \sigma_{12} \\ \sigma_{12} & \sigma_{22} \end{bmatrix} \begin{bmatrix} \cos\theta \\ \sin\theta \end{bmatrix}.$$

From the above we conclude that:

$$\sigma_{11} = 10.0ksi,\ \sigma_{22} = 10.0ksi,\ \sigma_{12} = 0$$

The matrix of the state of stress at all points of the plate is:

$$\begin{bmatrix} 10.0 & 0.0 & 0.0 \\ 0.0 & 10.0 & 0.0 \\ 0.0 & 0.0 & 0.0 \end{bmatrix}$$

[At the end of Section 7.4]

The equations of equilibrium in Section 7.4 can all be deduced by considering an elementary parallelepiped, summing the forces acting on the faces and taking the moments of those forces around the axes of reference. This is the approach taken in many texts[2]. The stresses are assumed to vary linearly with the coordinates as one moves from one face of the parallelepiped to the other. However, even when the stresses are not assumed to vary linearly one gets the same results as higher order terms drop out at the limit when the size of the element tends to zero. The equations of equilibrium are always satisfied.

[At the end of Section 7.5]

As was stated in this addendum at the end of Section 3.13, one has to be in a principal system of axes to draw Mohr's circles in three dimensions. Consider as an example the stress tensor

[2] See Y.C. Fund "A First Course in Continuum Mechanics," Prentice Hall, 1969.

$$\sigma_{ij} = \begin{bmatrix} 2.75 & 1.06066 & 1.75 \\ 1.06066 & 4.5 & 1.06066 \\ 1.75 & 1.06066 & 2.75 \end{bmatrix} ksi$$

at a point O of a solid subjected to a system of surface forces and couples. It is required to find and show on Mohr's circles the normal and tangential (or shear) components of the stress vector acting on a plane whose normal has direction cosines (-0.5, 0.70711, 0.5). The solution of the eigenvalue problem [see Eq. (7.5.3)] yields the following principal stresses and their corresponding principal directions:

$$\sigma_1 = 6.0ksi \qquad \sigma_2 = 3ksi \qquad \sigma_3 = 1ksi$$

$$\begin{Bmatrix} 0.50000 \\ 0.70711 \\ 0.50000 \end{Bmatrix}, \quad \begin{Bmatrix} -0.50000 \\ 0.70711 \\ -0.50000 \end{Bmatrix}, \quad \begin{Bmatrix} -0.70711 \\ 0.00000 \\ 0.70711 \end{Bmatrix}$$

In the principal system of coordinates the direction cosines of the normal to the plane in question are given by Eq. (3.5.3):

$$\begin{bmatrix} \ell_1 \\ \ell_2 \\ \ell_3 \end{bmatrix} = \begin{bmatrix} 0.50000 & 0.70711 & 0.50000 \\ -0.50000 & 0.70711 & -0.50000 \\ -0.70711 & 0.00000 & 0.70711 \end{bmatrix} \begin{bmatrix} -0.5 \\ 0.70711 \\ 0.5 \end{bmatrix} = \begin{bmatrix} 0.5 \\ 0.5 \\ 0.70711 \end{bmatrix}$$

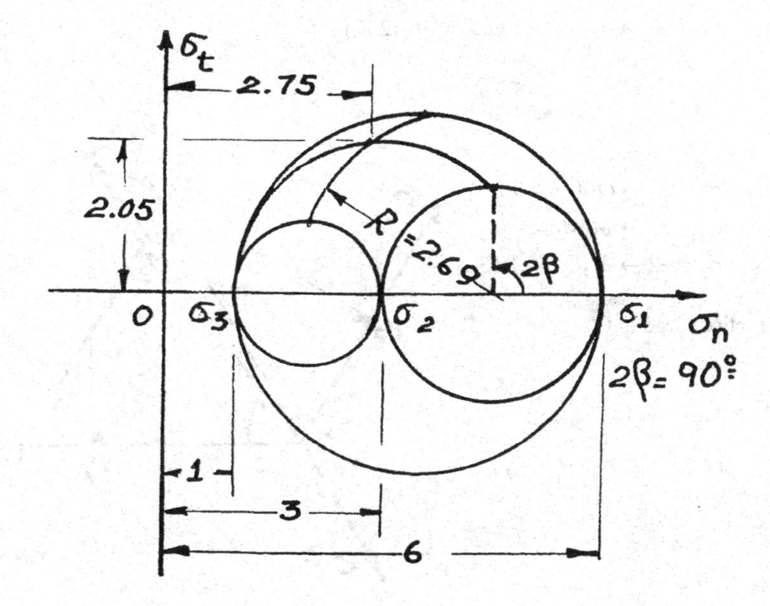

Fig. 7.10c

The value of σ_n can be obtained either from Eq. (7.3.10) in the original system of axes or from (7.5.9) in the principal system of axes. The same can be said for σ_t which can be obtained from either (7.3.11) or (7.5.10). Thus,

$$\sigma_n = 2.75ksi\,,\quad \sigma_t = 2.046ksi$$

Mohr's circles are drawn as shown on Fig. 7.10c. The point p giving the values of σ_n and σ_t is located using Eqs. (3.13.9) and (3.13.14); the value of h being 0.70711. Therefore,

$$R = 2.69, \qquad \beta = 45^\circ$$

[At the end of Section 7.9]

Another way at arriving at the same point C′ is to multiply each of the three principal stresses by $\sqrt{\frac{2}{3}}$ and, starting from the origin, draw lines parallel to the O1, O2, O3 axes in Fig. 7.16. As an example, consider the state of stress $\sigma_1 = 6ksi$, $\sigma_2 = 5ksi$ and $\sigma_3 = -2ksi$. We have (Fig. 7.16a):

$$\left.\begin{array}{l} \sigma_1 = 6 \;\rightarrow S_1 = 3 \;\rightarrow OH_1 = 3.6742 \\ \sigma_2 = 5 \;\rightarrow S_2 = 2 \;\rightarrow OH_2 = 2.4495 \\ \sigma_3 = -2 \rightarrow S_3 = -5 \rightarrow OH_3 = -6.1237 \end{array}\right\} OC' = 6.1644ksi$$

and

$$\begin{array}{l} \sigma_1 = 6 \;\rightarrow OM_1 = 4.90 \\ \sigma_2 = 5 \;\rightarrow M_1M_2 = 4.08 \\ \sigma_3 = -2 \rightarrow M_2C' = -1.63 \end{array}$$

Both constructions lead to C′

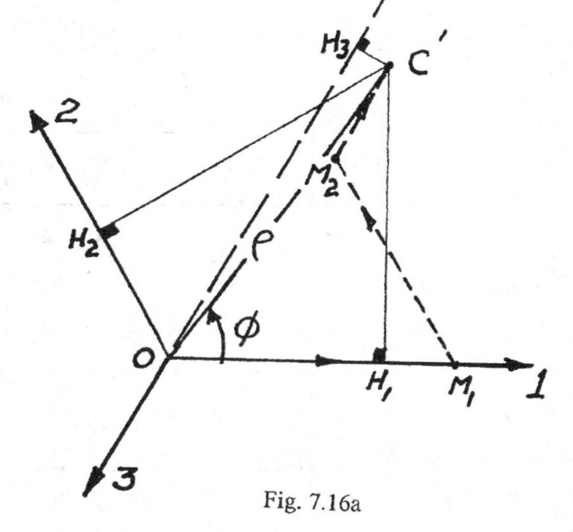

Fig. 7.16a

[At the end of Section 7.11]

A situation, which often occurs, is one in which two or more states of stress at a point have to be added to one another. This may occur if a solid is subjected to two different loading conditions and measurements are made with reference to two different systems of coordinates. In such a case the various components of the stress tensors can be algebraically added if, and only if, they are referred to the same system of axes. Mohr's representation is a convenient way to get to the answer as well as to obtain principal stresses and principal directions for the combined states. Let us assume that the two states of stress shown in Figs. 7.21a and 7.21b are to be added. First, Mohr's circle is used to rotate the system of axes of Fig. 7.21b by 30°clockwise as shown in Fig. 7.21c. This results in Fig. 7.21d. The stresses shown on Fig. 7.21d can now be added to those of Fig. 7.21a to give Fig. 7.21e. A new Mohr circle can be used for this resultant state of stress to find the principal stresses and their direction. This is done in Figs. 7.21f and 7.21g. The combined state of stress is:

$$\sigma_{11} = 19.33ksi\,,\ \sigma_{22} = 5.67ksi\,,\ \sigma_{12} = 5.502ksi$$

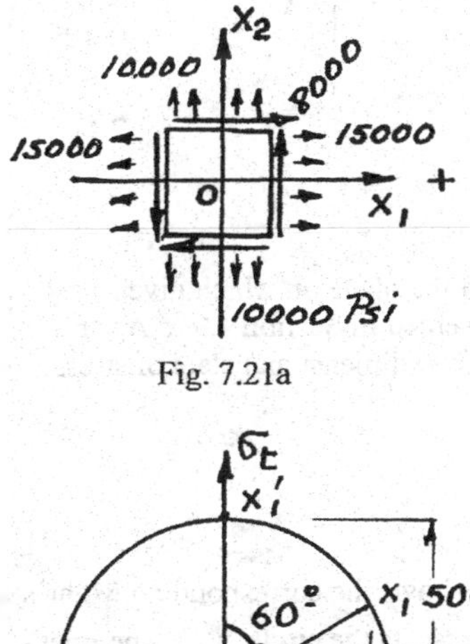

Fig. 7.21a

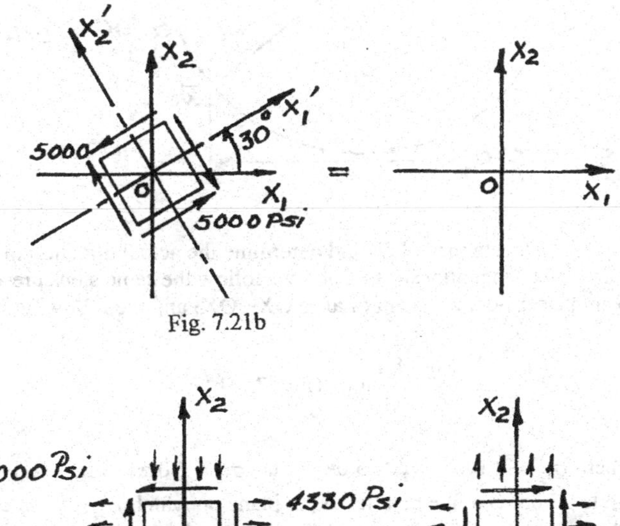

Fig. 7.21b

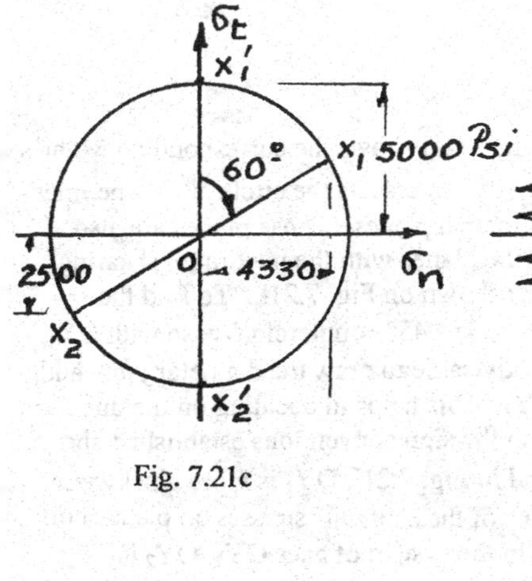

Fig. 7.21c

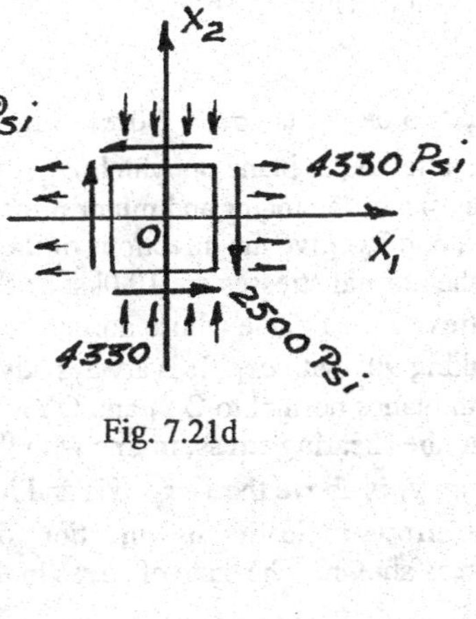

Fig. 7.21d

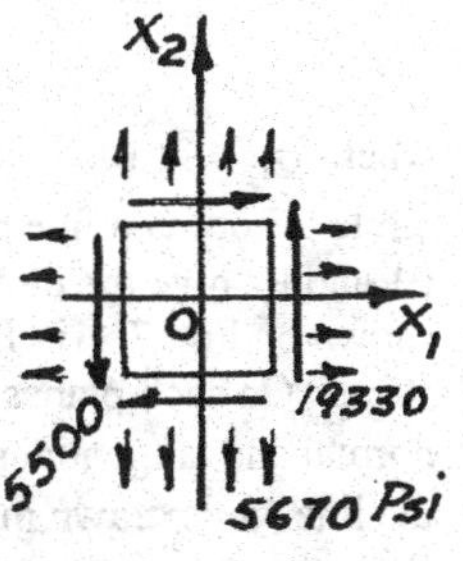

Fig. 7.21e

The principal stresses are

$$\sigma_1 = 21.50ksi, \ \sigma_2 = 3.75ksi$$

making an angle of 19.5° with OX_1

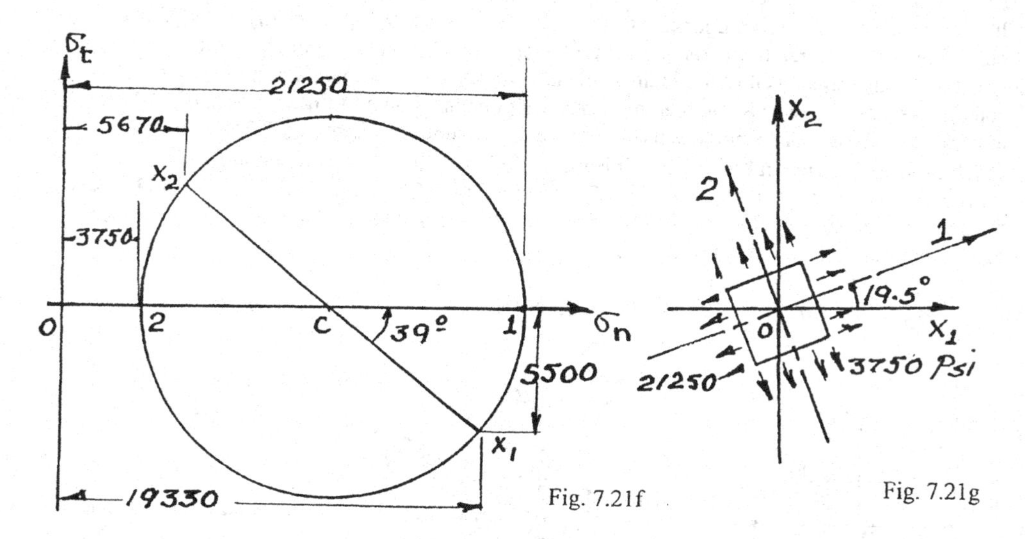

Fig. 7.21f

Fig. 7.21g

To locate the pole P and determine the actual direction of the plane on which given normal and tangential stresses act, we follow the same steps presented in Section 4.97. As an example consider a system of axes OX_1, OX_2 and the following two dimensional state of stress:

$$\begin{bmatrix} 11.0 & 4.0 \\ 4.0 & 5.0 \end{bmatrix} ksi, \text{ (Fig. 7.21h)},$$

where $\sigma_{11} = 11.0ksi$, $\sigma_{22} = 5.0ksi$, and $\sigma_{12} = 4.0ksi$. Fig. 7.21i shows the corresponding Mohr's circle. From x_1 a line parallel to the plane on which σ_{11} acts intersects the circle at P. The lines P1 and P2 give the directions of the major and minor principal planes. Those planes are also shown on Fig. 7.21j. Ps_1 and Ps_2 give the directions of the planes with the maximum shearing stress. On such planes the normal stresses are 10.0ksi as shown on Fig. 7.21k. To find the normal and tangential stresses on a plane whose normal makes 45° counterclockwise with OX_1, the line cy_1 is drawn making 90° with cx_1. It is always advisable to draw the diameter y_1cy_2 and show the stresses on both planes normal to OY_1 and OY_2. This helps in deciding on the direction of the arrow representing the shearing stress, in line with the sign conventions established above. On Mohr's circle the point y_1 is above the $O\sigma_n$ axis and in Fig. 7.21l, OY_1 is more clockwise than OY_2. Therefore the arrows indicating the directions of the shearing stresses on planes normal to OY_1 and OY_2 must be as shown. The state of stress in the system of axes OY_1, OY_2 is

$$\begin{bmatrix} 12.0 & -3.0 \\ -3.0 & 4.0 \end{bmatrix} ksi$$

with $\sigma_{11} = 12.0ksi$, $\sigma_{22} = 4.0ksi$, and $\sigma_{12} = -3.0ksi$.

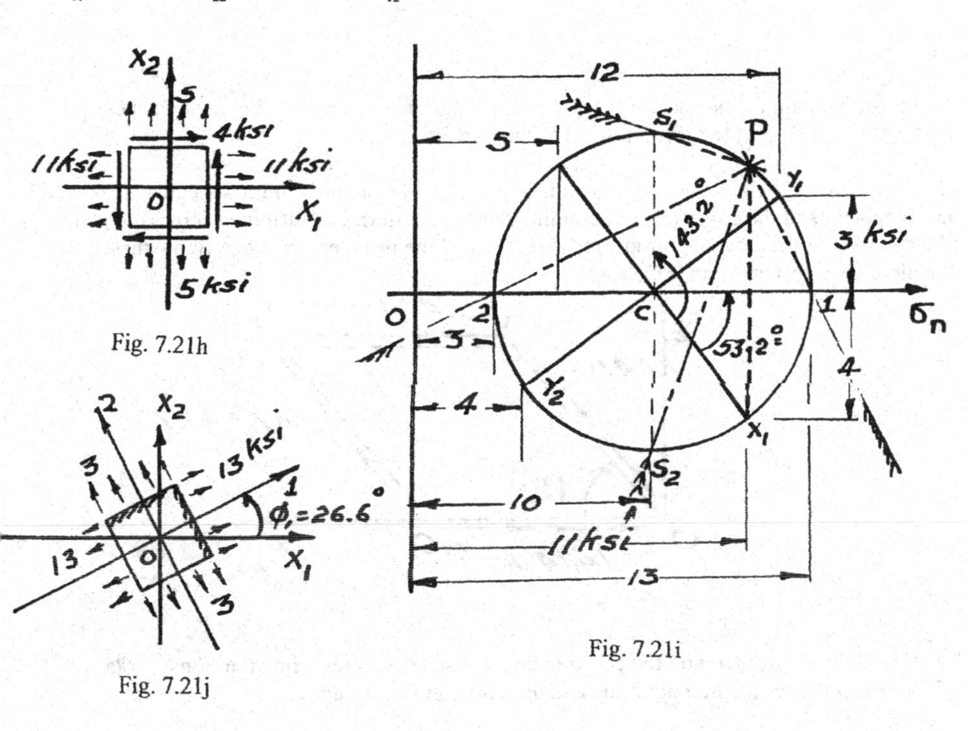

Fig. 7.21h

Fig. 7.21i

Fig. 7.21j

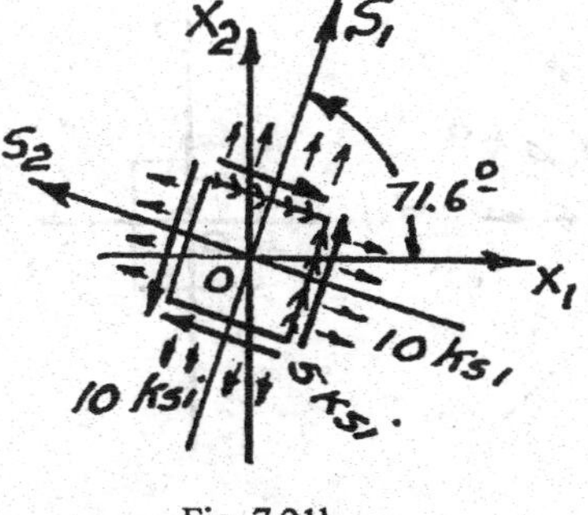

Fig. 7.21k

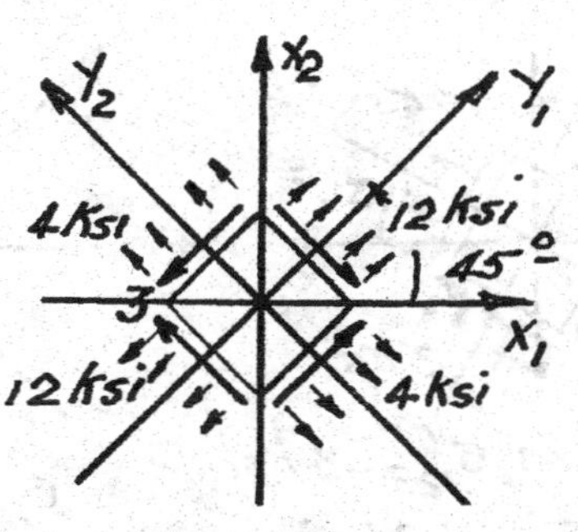

Fig. 7.21l

[Additional Problems]

11. Repeat problem 6 with the stress tensor:

$$\sigma_{ij} = \begin{bmatrix} 30 & 10 & 10 \\ 10 & 0 & 20 \\ 10 & 20 & 0 \end{bmatrix} ksi$$

But in part b obtain the normal and tangential stresses on a plane with direction cosines (0.47507, 0.82285, 0.31180) with respect to the principal system.

12. A very thin plate is loaded as shown in Fig. 7.27. The parallelogram has equal sides. Find the components of the stress tensor assuming they are uniformly distributed across the plate. Use Mohr's circle to find the magnitude and directions of the principal stresses. Show these directions on a system of axes OX_1, OX_2.

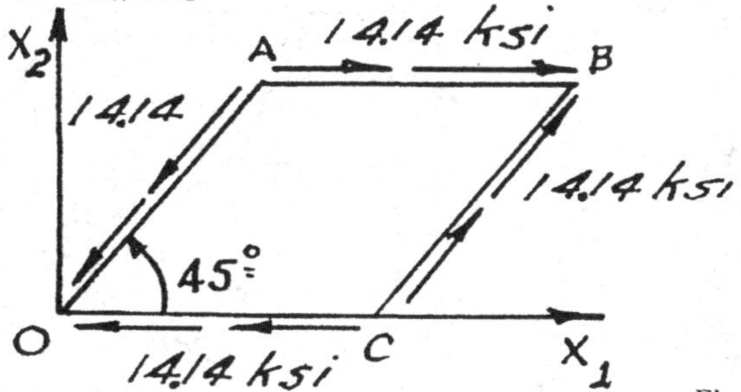

Fig. 7.27

13. Use Mohr's circles to add the two dimensional states of stress shown in Figs. 7.28a and 7.28b. Show the results in magnitude and direction on Fig. 7.28c.

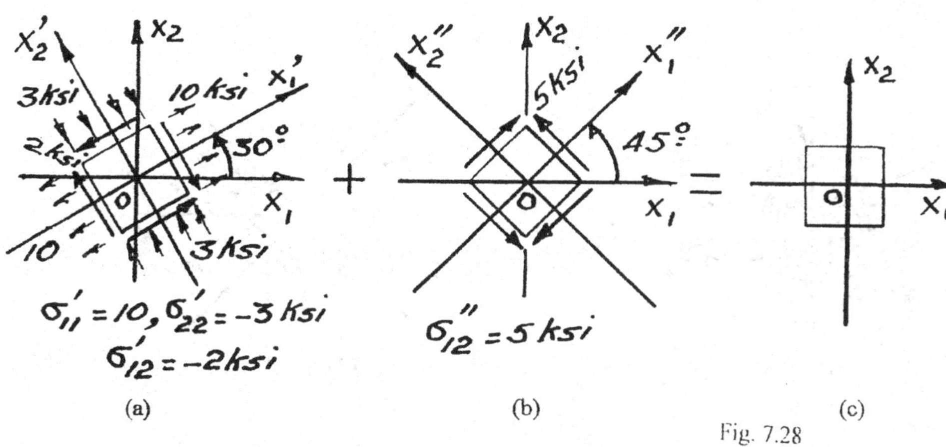

Fig. 7.28

14. The state of stress at a point is given by:

$$\sigma_{ij} = \begin{bmatrix} 4.0 & 1.0 & 0.0 \\ 1.0 & 0.0 & 0.0 \\ 0.0 & 0.0 & 3.0 \end{bmatrix} ksi$$

Find its components in a system whose direction cosines are $\left(\frac{\sqrt{3}}{2}, \frac{1}{2}, 0\right)$, $\left(-\frac{1}{2}, \frac{\sqrt{3}}{2}, 0\right)$, (0,0,1).

15. Repeat problem 14 with the state of stress

$$\sigma_{ij} = \begin{bmatrix} 6.0 & 2.0 & 0.0 \\ 2.0 & 0 & 0 \\ 0 & 0 & 3.0 \end{bmatrix} ksi$$

Use Mohr's circles to obtain the magnitude and directions of the principal stresses. Show your results on a sketch.

16. For the stress tensor

$$\sigma_{ij} = \begin{bmatrix} 3.0 & 1.0 & 1.0 \\ 1.0 & 0.0 & 2.0 \\ 1.0 & 2.0 & 0.0 \end{bmatrix}$$

a) Determine the principal stresses.

b) Find the principal components of the deviator stress tensor.

c) Show the magnitude and phase angle of the deviator on the Π plane using the two methods given in Sec. 7.9.

17. In the 3 cases of Problem 9,

a) Find the components of the state of stress in a system of axes rotated 30° counterclockwise with the OX_1 axis.

b) Locate the pole and find the normal and shearing stress on planes whose normals are inclined 45° clockwise with OX_1.

c) Show on graphs similar to those on Figs. 7.21h to 7.21l the directions of the stresses and the planes on which they act.

CHAPTER 8

[At the end of Section 8.4]

At this stage it is appropriate to discuss briefly the concept of isotropy. Tensors were introduced in Chapter 5; and the states of strain and stress, as well as the elastic relations between them were studied in subsequent sections. It was mentioned that tensors (except for a couple of exceptions) describe a physical state or a physical phenomenon and that this state remains unchanged in a change of the frame of reference in which the phenomenon is described. The states of strain and stress were referred to as strain and stress tensors and, through linear elasticity, were related one to the other by means of the compliance or the stiffness tensor:

$$e_{kl} = S_{klmn}\sigma_{mn}\text{, and } \sigma_{kl} = C_{klmn}e_{mn}. \tag{8.4.27}$$

Now, we have to distinguish between isotropic states and isotropic relations or laws.

An isotropic state is represented by an isotropic tensor, in other words a tensor whose components remain the same under any orthogonal transformation of coordinates. We are mostly interested in rotations of coordinate axes, in other words in proper orthogonal transformations of coordinates. For example, what we referred to as spherical states of strain and stress [Eqs. (4.9.10) and (7.5.19)] are nothing but isotropic tensors of strain and stress. If the rotation of axes expressed by Eq. (7.10.11) is applied to the isotropic stress tensor

$$\sigma_{ij} = \begin{bmatrix} \sigma_m & 0 & 0 \\ 0 & \sigma_m & 0 \\ 0 & 0 & \sigma_m \end{bmatrix}$$

we would get

$$\sigma'_{ij} = \begin{bmatrix} \sigma_m & 0 & 0 \\ 0 & \sigma_m & 0 \\ 0 & 0 & \sigma_m \end{bmatrix}.$$

The above illustrates the fact that an isotropic tensor of the second rank must be diagonal and can be written as $\alpha[1] = \alpha\delta_{ij}$, where α is a scalar; in our case equal to σ_m. It can also be shown that isotropic tensors of rank three are scalar multiples of ε_{ijk}[3].

Now, a physical law is a relation among various tensors such as $F(\sigma_{ij}, e_{ij}) = 0$. The relation is isotropic if it is still valid after a rotation of coordinate axes. If (8.4.27) is isotropic, then

[3] See Y.C. Fang "A First Course in Continuum Mechanics," Prentice Hall, 1977.

$$\sigma'_{kl} = C'_{klmn} e'_{mn} \tag{8.4.28}$$

If the material is isotropic we must have

$$\sigma'_{kl} = C_{klmn} e'_{mn} \tag{8.4.29}$$

regardless of the system of reference axes used (see p. 198). Comparing (8.4.27) and (8.4.29) we deduce that the tensor of rank four C_{klmn} must be isotropic. The same can be said for S_{klmn}.

Now, any isotropic tensor of rank 4, C_{ijkl} has the form:

$\lambda\delta_{ij}\delta_{kl} + \mu(\delta_{ik}\delta_{jl} + \delta_{il}\delta_{jk}) + \nu(\delta_{ik}\delta_{jl} - \delta_{il}\delta_{jk})$ where λ, μ and ν are scalar quantities. Also, if C_{ijkl} has symmetry properties,

$$C_{ijkl} = C_{jikl} \text{ and } C_{ijkl} = C_{ijlk}$$

which is the case in linear elasticity, then,

$$C_{ijkl} = \lambda\delta_{ij}\delta_{kl} + \mu(\delta_{ik}\delta_{jl} + \delta_{il}\delta_{jk}) \tag{8.4.30}$$

The proof for the above is elementary and can be found in Fang's treatise.

If we now take this expression for C_{ijkl} and substitute it in Hooke's law we get

$$\sigma_{ij} = C_{ijkl} e_{kl} = \lambda e_{kk}\delta_{ij} + 2\mu e_{ij} \tag{8.4.31}$$

This is the most general form of the relation between stress and strain for an isotropic elastic solid for which the stresses are linear functions of the strains. There are only two material constants λ and μ. Eq. (8.4.31) is the same as (8.5.2) and (8.5.3).

Finally, we recall that in Section 8.2 it was shown that elasticity (in our case Hyperelasticity) meant the existence of a strain energy density function U whose derivatives with respect to the strain components gives the corresponding stress components. Isotropy can be expressed by stating that U only depends on the strain invariants, i.e.

$$U(e_{ij}) = U(I_1, I_2, I_3)$$

Since the invariants retain their form and values under all rotations of coordinates, the same applies to

$$\sigma_{ij} = \frac{\partial U}{\partial e_{ij}}.$$

Notice that the invariants I_1, I_2 and I_3 could be as given by Eqs. (4.9.3 to 4.9.5) or combinations as indicated at the end of Section 5.6.

[Additional Problems]

15. A thin flat plate has a circle drawn over it. The equation of the circle is

$$x_1^2 + x_2^2 = 9.$$

Subsequently it is loaded on its edges such that $\sigma_{11} = 20.0ksi, \sigma_{22} = -14.0ksi$ and $\sigma_{21} = 16.0ksi$. The state of stress is assumed to be uniform in the plate. The circle becomes an ellipse. What are the magnitudes and directions of the principal axes of the ellipse? E = 30x10^6 psi, ν = .25.

16. Figure 8.10 shows a tooth in a plate in a state of plane stress. The faces of the tooth CA and BA are free from force and the plate is loaded as shown. Assuming that the stress components are finite and continuous throughout, prove that the state of stress is equal to zero at the apex A.

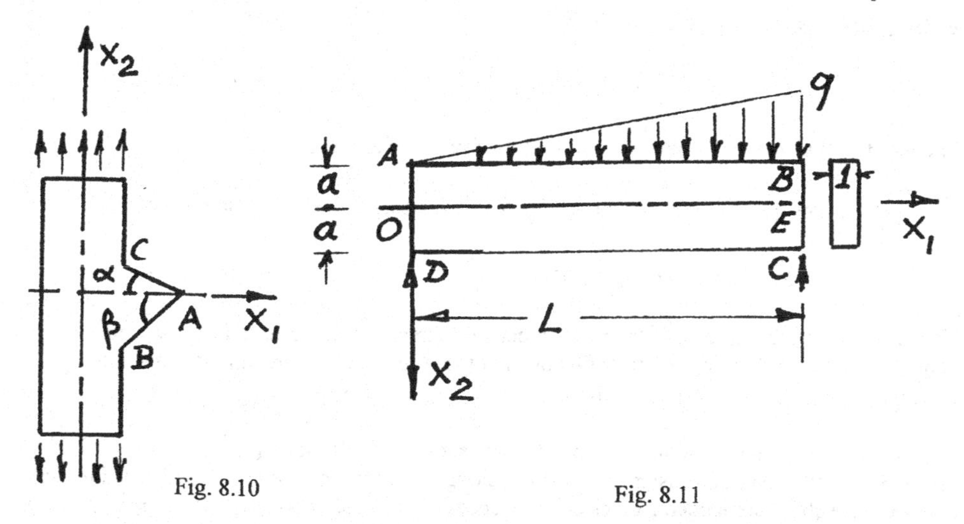

Fig. 8.10

Fig. 8.11

17. Assuming the material in problems 18 and 19 of Chapter 4 is aluminum with $\nu = 0.33$ and $E = 30 \times 10^6\, psi$. Find the magnitude and directions of the principal stresses in both cases.

18. The simple beam of length L shown in Fig. 8.11 is loaded such that the vertical stress on AB varies linearly from 0 to q. Write down the stress boundary conditions on the four faces AB, BC, CD and DA. Where would you need to invoke St. Venant's principle? The beam is in a state of plane stress.

19. The semi-infinite plate in Fig. 8.12 is subjected to a uniform shearing force per unit area S. Using cylindrical coordinates write down the stress boundary conditions along AO and OB.

20. Repeat problem 19 with the loading shown in Fig. 8.13. The vertical pressure to the left of OX_2 varies linearly, is equal to P at a distance d from O and extends indefinitely.

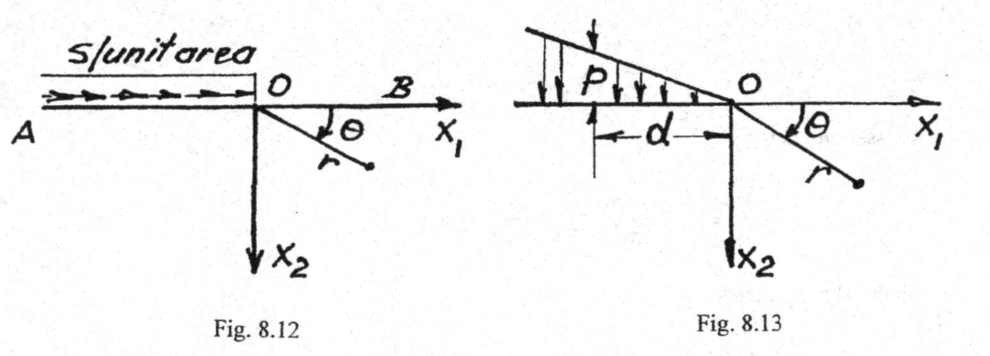

Fig. 8.12

Fig. 8.13

CHAPTER 9

[Following Eq. (9.4.11)]

Thus, using Eq. (9.4.11), and in the absence of body forces.

$$u_1 = \frac{1}{2G}\frac{\partial\phi}{\partial x_1},\ u_2 = \frac{1}{2G}\frac{\partial\phi}{\partial x_2},\ u_3 = \frac{1}{2G}\frac{\partial\phi}{\partial x_3}.$$

The stresses are then given by

$$\sigma_{11} = \frac{\partial^2\phi}{\partial x_1^2},\ \sigma_{22} = \frac{\partial^2\phi}{\partial x_2^2},\ \sigma_{12} = \frac{\partial^2\phi}{\partial x_1 \partial x_2},\ \text{etc.}$$

In cylindrical coordinates,

$$u_r = \frac{1}{2G}\frac{\partial\phi}{\partial r},\ u_\theta = \frac{1}{2Gr}\frac{\partial\phi}{\partial\theta},\ u_z = \frac{1}{2G}\frac{\partial\phi}{\partial z},$$

and the stresses are given by

$$\sigma_{rr} = \frac{\partial^2\phi}{\partial r^2},\ \sigma_{\theta\theta} = \frac{1}{r}\frac{\partial\phi}{\partial r} + \frac{1}{r^2}\frac{\partial^2\phi}{\partial\theta^2},\ \sigma_{zz} = \frac{\partial^2\phi}{\partial z^2}$$

$$\sigma_{r\theta} = \frac{\partial}{\partial r}\left(\frac{1}{r}\frac{\partial\phi}{\partial\theta}\right),\ \sigma_{\theta z} = \frac{1}{r}\frac{\partial^2\phi}{\partial\theta\partial z},\ \sigma_{rz} = \frac{\partial^2\phi}{\partial r\partial z}$$

[Additional Problems]

11. Show that the Lame' strain potential

$$\phi = c\theta$$

where c is a constant provides the solution to the problem of a circular disk subjected to a uniformly distributed shearing stress τ on the circumference. A physical situation corresponding to this problem is shown in Fig. 9.8. Find the value of u_θ, $\sigma_{r\theta}$ and c in terms of M_t at a distance r from the origin. G is the shear modulus.

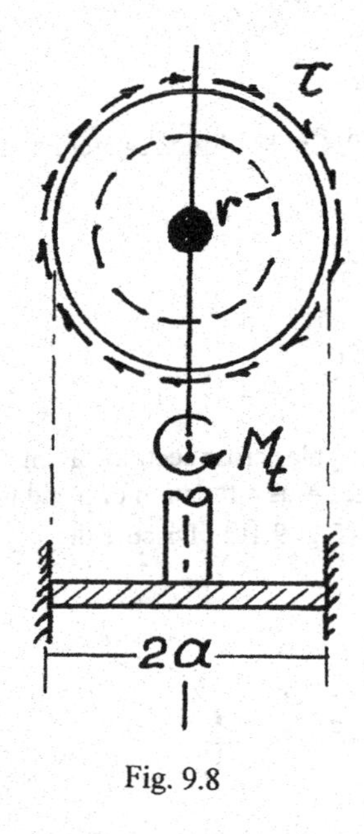

Fig. 9.8

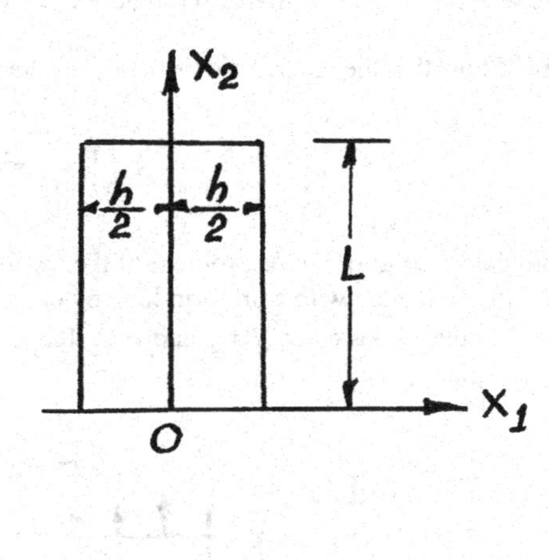

Fig. 9.9

12. Investigate the following stress distribution as a possible solution for an elasticity plane stress problem with no body forces.

$$\sigma_{11} = px_1^3x_2 - 2ax_1x_2 + bx_2$$
$$\sigma_{22} = px_1x_2^3 - 2px_1^3x_2$$
$$\sigma_{12} = -1.5px_1^2x_2^2 + ax_2^2 + 0.5px_1^4 + c$$

where p, a, b, and c are constants. Then,

a) Find the Airy stress function ϕ corresponding to this stress distribution.

b) If this state of stress acts in the thin plate shown in Fig. 9.9 find the constants a, b and c such that there is no shearing stress on the edges $x_1 = \pm\frac{h}{2}$ and no normal stress on the edge $x_1 = -\frac{h}{2}$.

c) Determine the stress distribution on the boundaries of the plate.

13. Show that the Airy stress function

$$\phi = A\left[\frac{1}{2}x_2^2\ln(x_1^2 + x_2^2) + x_1x_2\tan^{-1}\frac{x_2}{x_1} - x_2^2\right]$$

provides the solution to the problem shown in Fig. 8.12 (probl. 19, Ch. 8). Find A in terms of S. (Use cylindrical coordinates and ignore the body forces).

14. Show that the stress field provided by the Airy stress function

$$\phi = A\left[x_1^2\left(-x_2^3 - \frac{3}{2}hx_2^2\right) - \frac{1}{10}x_2^2\left(-2x_2^3 - 5hx_2^2 - 4h^2x_2 - h^3\right)\right]$$

provides a solution to the problem of the cantilever beam of rectangular cross section having unit width and loaded with a uniform load p per unit length. Determine A as a function of p and h as well as the values of σ_{11}, σ_{22} and σ_{12}. Ignore the body forces (Fig. 9.10). Discuss the conditions at $x_1 = 0$.

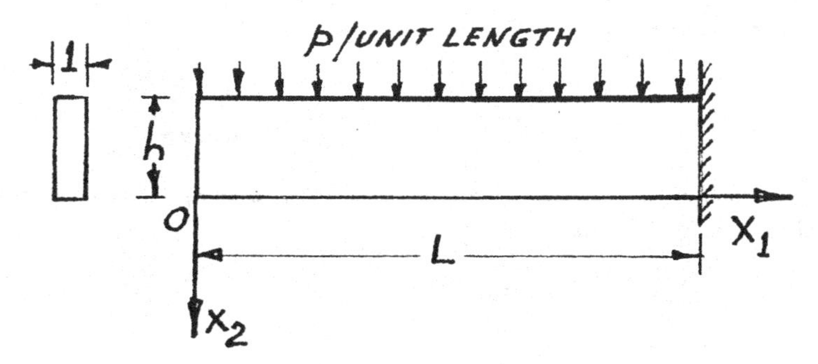

Fig. 9.10

15. For the triangular built in plate of unit thickness, (Fig. 9.11) carrying a normal pressure p per unit length, the following Airy stress function is known to provide the solution:

$$\phi = A\left[r^2(\alpha - \theta) + r^2 \sin\theta \cos\theta - r^2 \cos^2\theta \tan\alpha\right]$$

a) Write the expression of the stresses in cylindrical coordinates and find the constant A so that the boundary conditions are satisfied. Ignore the body forces.

b) What are the principal stresses along OB?

c) For $\alpha = 20^\circ$ find the expression of the normal stress at C and B and compare those values with the ones obtained from the elementary theory of beams.

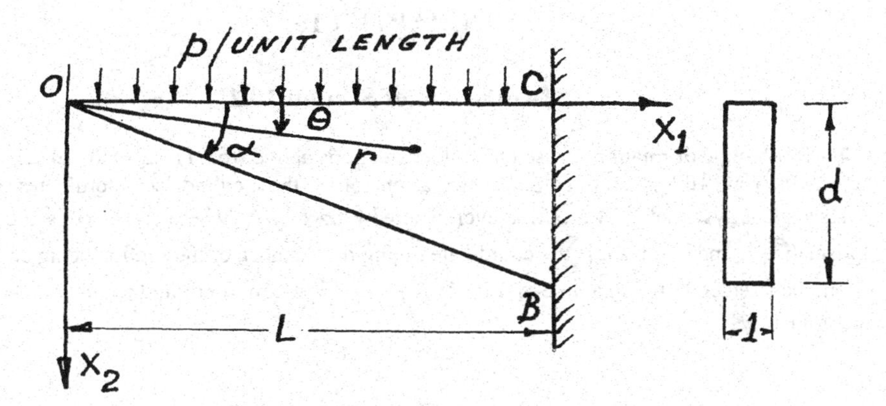

Fig. 9.11

16. Show that the Airy stress function

$$\phi = \frac{cp}{d}\left[\left(\frac{1}{3}x_1^3 + x_1 x_2^2\right)\tan^{-1}\frac{x_2}{x_1} + \frac{1}{3}x_2^3 \ln\left(x_1^2 + x_2^2\right) - \frac{1}{3}x_1^2 x_2\right]$$

satisfies the conditions on the edge $x_2 = 0$ of the plate in Problem 20, of Chapter 8. What is the value of the constant c? (Hint: The use of cylindrical coordinates may simplify the calculations. Ignore the body forces.)

CHAPTER 10

[At the end of Section 10.2]

Note: The angle or rotation θ used in all the equations above from (10.2.1) to (10.2.13) and in Figs. 10.1b and 10.5 is not to be interpreted as one of the three cylindrical coordinates (r, θ, z). This angle is systematically replaced everywhere by αz or αx_3; where α is a constant of the prismatic bar and, z or x_3, is the coordinate giving the distance of the section being considered from the fixed end. For example, in Eq. (10.2.3) u_θ is a function of r and z and not the coordinate θ.

[At the end of Section 10.9]

As an example consider a steel bar whose section is shown if Fig. 10.27a. the bar is subjected at its ends to a torque of 740 lb. ft. The shear modulus G is 12 x 10^6psi. Find the value of the angle of twist per unit length α and that of the maximum shearing stress in the flange AB.

Now, the total torsional rigidity is the sum of the individual rigidities of the flanges and web, each given by Eq. (10.9.15). Thus

$$GJ = G\frac{2(3\times 0.4^3)+(6\times 0.25^3)+2(3\times 0.25^3)}{3} = 0.1905G$$

The angle α is given by Eq. (10.9.14) as

$$\alpha = \frac{740\times 12}{0.1905G} = 0.0038845 \text{ rad}$$

The maximum shearing stress is given by Eq. (10.9.22) and occurs in AB and EF:

$$(\sigma_t)_{AB} = G\alpha\times 0.4 = 18{,}645.67\,psi$$

Notice that each component of the channel carries a part of the applied torque proportional to its own torsional rigidity

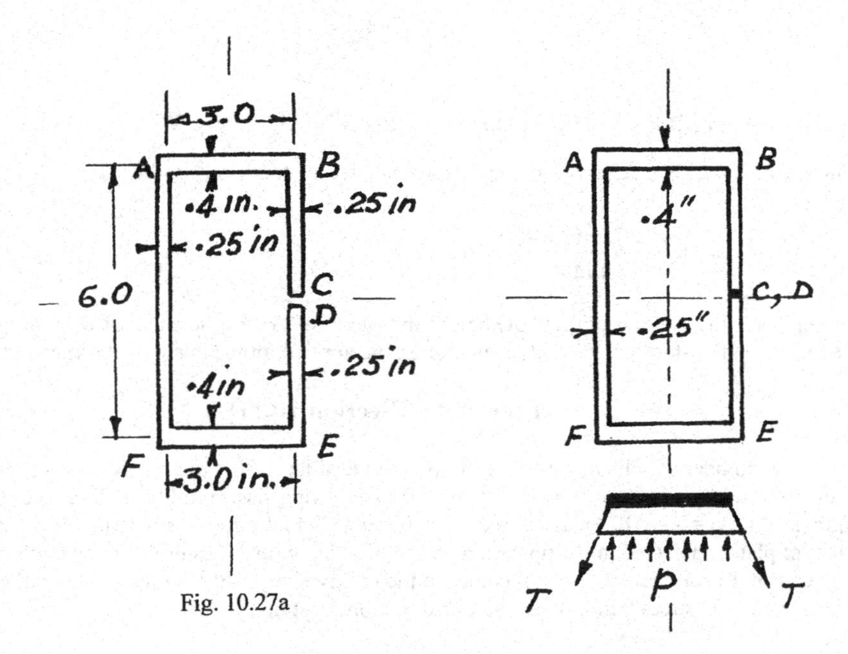

Fig. 10.27a

Fig. 10.27b

[At the end of Section 10.10]

As an exercise let us assume that the gap between C and D in Fig. 10.27a is welded, and let us compare the torsional rigidity to the one prior to welding, as well as calculate the torque which would give the same stress in AB, namely 18,645 psi.

We can, of course use Eqs. (10.10.12) to (10.10.17), but let us start from the equilibrium of a plate of area A = 18 in^2 subjected to an upward pressure p and a membrane with a tension T pulling it down (Fig. 10.27b). The equilibrium equation of the plate is

$$p \times 18 = \frac{2T \times 6h}{0.25} + \frac{2T \times 3h}{0.4}$$

Replacing $\frac{P}{T}$ by $2G\alpha$ we get

$$\alpha = \frac{7h}{4G}.$$

The twisting moment is twice the volume under the plate so that,

$$M_{33} = 2 \times 18h\,.$$

Now the stress in AB is given by the slope $\frac{h}{t}$ so that, $\frac{h}{t} = 18,646 psi$. With t = 0.4 in. we get h = 7458 and $M_{33} = 22,375$ ft. lb. Knowing α one can get the torsional rigidity as

$$JG = \frac{22,375 \times 12 \times 4G}{7 \times 7458} = 20.57G$$

This compares to a rigidity of 0.1905G before the welding. As for the value of α, it drops from 3.885 x 10^{-3} rad. to 1.088 x 10^{-3} rad. even though the applied torque is about 30 times larger.

[At the end of Section 10.11]

The number of cells is not limited to two. Indeed let us find the stresses and the angle of rotation of one end with respect to the other of the steel wing shown in Fig. 10.31a, and whose length is 10 ft (G = 12 x 10^{-6} psi). The wing is subjected to a torque of 300,000 in. lb. The pattern of plates subjected to the upward pressure p is shown in the figure. Notice that if the assumed pattern is incorrect (when it comes to the relative height of the plates), a negative sign in the value of the stress will indicate that its direction must be reversed.

The assumed directions of the shearing stresses in the various elements are shown on Fig. 10.31a.

We have three equations of equilibrium for the plates and one stating that the twisting moment is equal to twice the volume under the plates.

For plate 1:

$$15.71T \frac{h_1}{0.025} - 10T \frac{(h_2 - h_1)}{0.05} = 39.27p$$

For plate 2:

$$24T \frac{h_2}{0.03} + 10T \frac{(h_2 - h_1)}{0.05} + 10T \frac{(h_2 - h_3)}{0.04} = 120p$$

For plate 3:

$$41.23T \frac{h_3}{0.025} - 10T \frac{(h_2 - h_3)}{0.04} = 100p$$

and

$$2(39.27 h_1 + 120 h_2 + 100 h_3) = 300,000$$

Replacing $\frac{p}{T}$ by $2G\alpha$, those four equations become:

$$828.4\,h_1 - 200\,h_2 + 0.0\,h_3 - 942{,}480{,}000\,\alpha = 0$$

$$-200\,h_1 + 1250\,h_2 - 250\,h_3 - 2{,}880{,}000{,}000\,\alpha = 0$$

$$0.0\,h_1 - 250\,h_2 + 1899.20\,h_3 - 2{,}400{,}000{,}000\,\alpha = 0$$

$$78.54\,h_1 + 240\,h_2 + 200\,h_3 - 0.0\,\alpha = 300{,}000$$

The solution of this system of linear equations gives: $h_1 = 469.9443$, $h_2 = 746.1146$, $h_3 = 420.1154$, and $\alpha = 0.0002547$ rad/in. Since the shearing stresses are equal to the slope of the membrane:

$$\sigma_{t1} = \frac{h_1}{0.025} = 18{,}798\,psi$$

$$\sigma_{t2} = \frac{h_2}{0.03} = 24{,}870\,psi$$

$$\sigma_{t3} = \frac{h_3}{0.025} = 16{,}805\,psi$$

$$\sigma_{t4} = \frac{h_2 - h_1}{0.05} = 5{,}523\,psi$$

$$\sigma_{t5} = \frac{h_2 - h_3}{0.04} = 8{,}150\,psi$$

The angle of rotation of one end with respect to the other is 0.0002547 x 10 x 12 = 0.030564 rad. or 1.75°.

A hydrodynamic analogy helps in checking the direction of the arrows in Fig. 10.31 a. At junctions such as B, C, D and E there must be continuity in the shear flow; meaning that the flow into the junction must equal to the flow out of the junction. Thus, at C for example, $(\sigma_{t1} \times 0.025 + \sigma_{t4} \times 0.05)$ must be equal to $\sigma_{t2} \times 0.03$; which it is. For point E, $\sigma_{t3} \times 0.025 = (\sigma_{t5} \times 0.04 + \sigma_{t2} \times 0.03)$.

It is instructive to study the changes that occur in the torsional rigidity and the stress distribution if two cuts are made at G and F. In this case, the wing becomes a box with four fins at its corners. The torsional rigidity is the sum of the torsional rigidities of the box and the fins, namely

$$JG = \left[\frac{4 \times 120^2}{\frac{2 \times 12}{0.03} + \frac{10}{0.04} + \frac{10}{0.05}} + \left(\frac{15.71 \times (0.025)^3}{3} + \frac{41.23 \times (0.025)^3}{3} \right) \right] G$$

$$JG = \left[\underbrace{46.08}_{box} + \underbrace{0.0002965}_{fins} \right] G$$

The rigidity of the fins is negligible and, for all practical purposes they carry no torque and suffer no stress. On the other hand the stress in CD and BE reach the value of Eq. (10.10.12)

$$(\sigma_t)_{CD} = \frac{300000}{2 \times 120 \times 0.03} = 41{,}666.67\, psi$$

which is nearly double of what it was prior to the cuts.

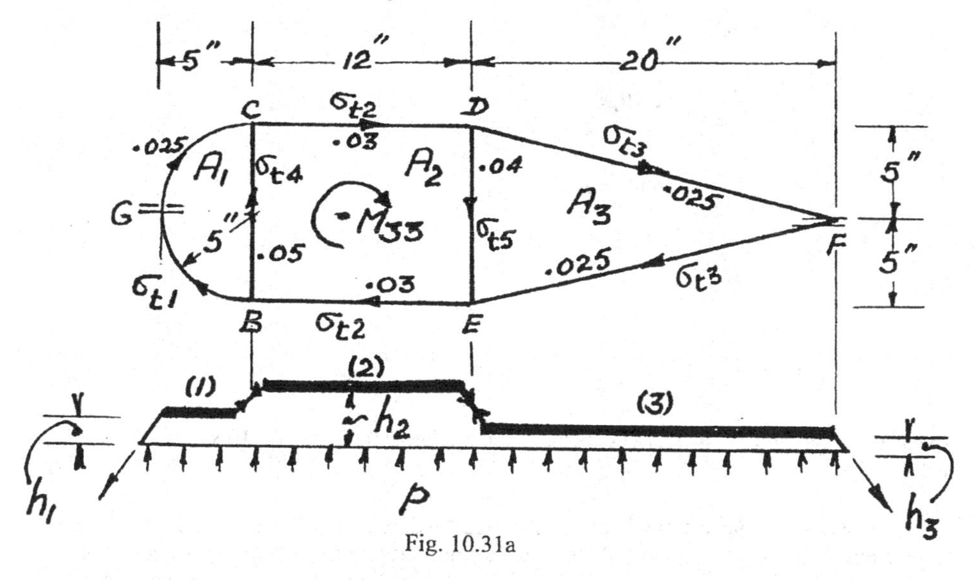

Fig. 10.31a

[Additional Problems]

11. A heat exchanger in the shape of a hollow tube with longitudinal fins (Fig. 10.44) is subjected to a twisting moment of 20,000 in.lb. Find the percentage of the twisting moment that is taken by the fins as well as the shearing stresses in tube and fins.

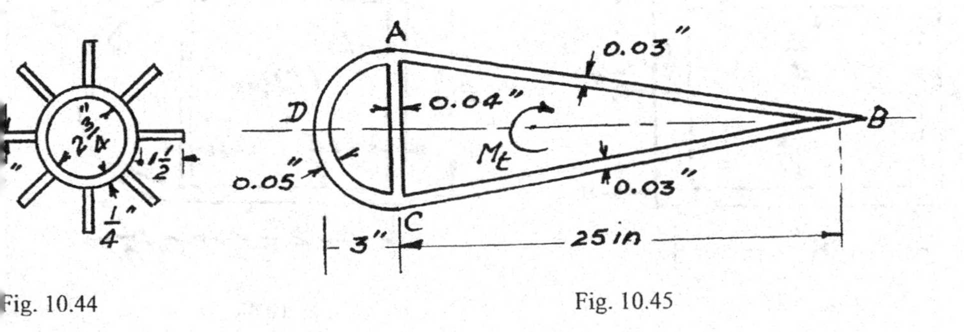

Fig. 10.44

Fig. 10.45

2. An airplane elevator (Fig. 10.45) made of an aluminum alloy is 5 ft. long and is subjected to torque of 200 ft.lb. Calculate the shearing stresses in the walls as well as the angle by which the end rotates with respect to the other (G = 3.8 x 10^6psi). Show the direction of the shearing stresses and check the shear flow at junctions A and C.

3. Five solid steel rods each 1 in. in diameter are built into a solid wall at one end and into a rigid support at the other (Fig. 10.46). Ignoring gravity forces determine the couple P x d that is required to cause an angular rotation of 0.1 degree of the rigid support. What portion of the total stiffness is furnished by the central rod?

Note: E = 3 x 10^7 psi, G = 1.15 x 10^7 psi and the force R at the end of a fixed end beam whose end is displaced by an amount u is $R = \frac{12uEI}{L^3}$

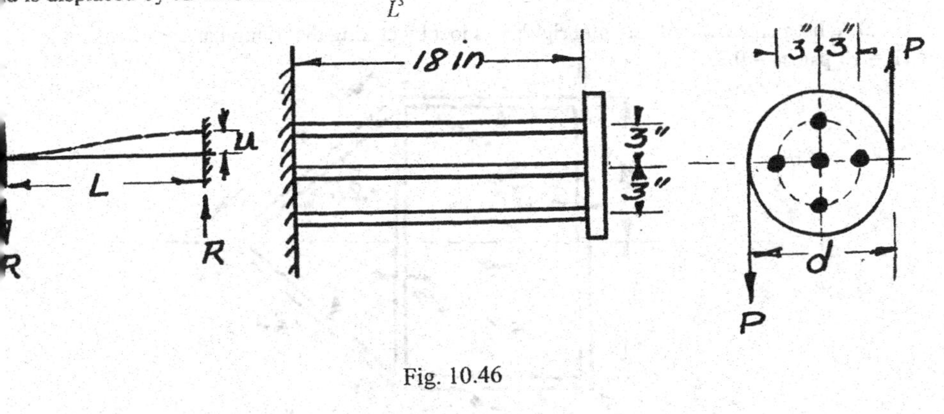

Fig. 10.46

4. The cross section shown in Fig. 10.47 is subjected to a torque of 100,000 in.lb. Find the shearing stress in each wall, the angle of twist per unit length α and the torsional rigidity GJ. G = 12 x 10^6psi.

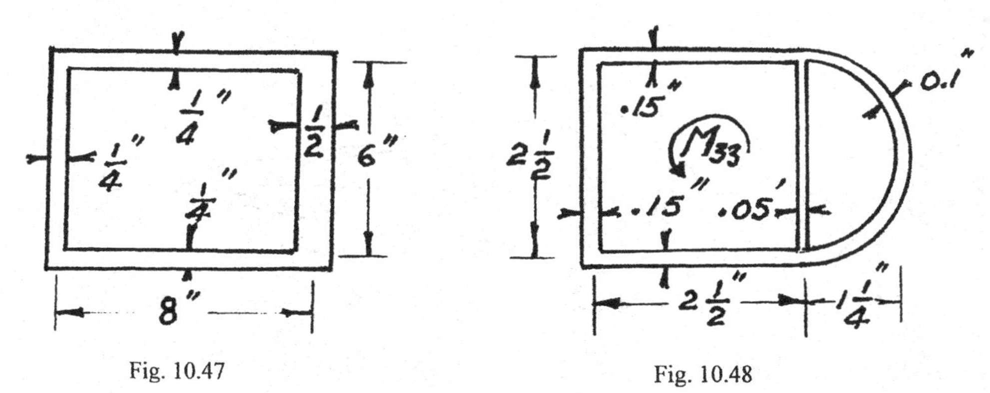

Fig. 10.47 Fig. 10.48

15. A hollow thin wall torsion member has two compartments with cross sectional dimensions as indicated in Fig. 10.48. The aluminum alloy of which this member is made has a shear modulus G = 3.77 x 10^6psi. Determine the torque and the unit angle of twist if the maximum shearing stress at locations away from stress concentration is 5800 psi. Show the directions of the shearing stress in the walls.

16. The duralumin thin wall torsion member shown in Fig. 10.49 is 10 ft. long and is subjected to a torque of 10,000 ft. lb.

a) Determine the magnitude and direction of the shearing stresses in the walls and the angle by which one end rotates with respect to the other.

b) What percentage of the total rigidity is lost by removing the internal partition BD?

c) What percentage of the total rigidity is lost by cutting the triangular compartment at C? (E = 10^7psi, ν = 0.3)

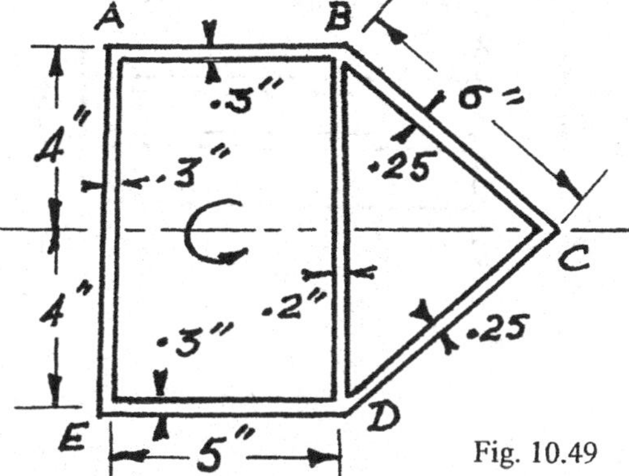

Fig. 10.49

17. The tube shown in Fig. 10.50 is subjected to a twisting moment of 1,000,000 in.lb. Find the magnitude and direction of the shearing stresses as well as the angle of twist per unit length α. G = 12 x 10^6psi.

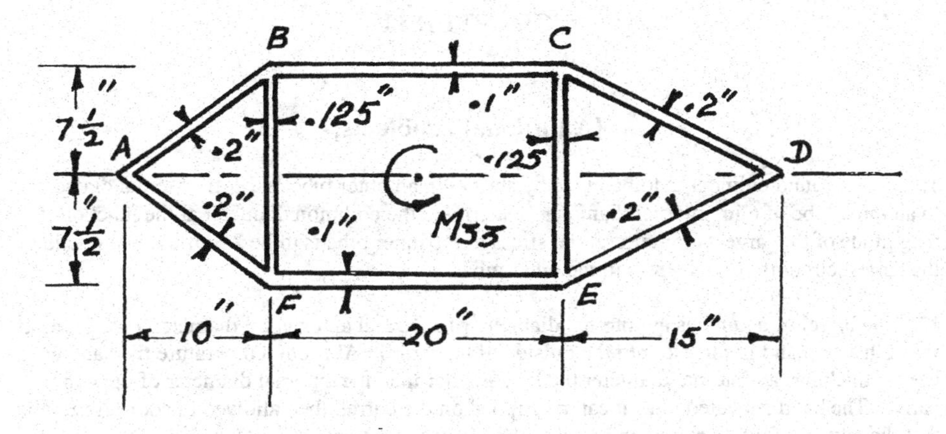

Fig. 10.50

CHAPTER 11

[Additional Problems]

16. A compound cylinder is to be made by shrinking an outer tube of 12 in. external diameter on to an inner tube of 6 in. internal diameter. Determine the common diameter at the junction if the magnitude of the largest circumferential stress in the inner tube is to be 2/3 of the magnitude of the largest circumferential stress in the outer tube.

17. The barrel of a gun has an internal diameter of 3 in. and an external diameter 9 in. A high strength steel band is shrunk onto the outside of the barrel. At room temperature the band which is 1 in. thick has an internal diameter 0.03 in. smaller than the external diameter of the gun barrel. The band is heated until it can be slipped on the barrel, then allowed to cool. Assuming that the gun is infinitely rigid compared to the band, determine:

a) The external pressure on the gun barrel due to the band.

b) The temperature to which the band must be heated in order to place it on the barrel.

c) The value of both σ_{rr} and $\sigma_{\theta\theta}$ on both sides of the interface when the gun is subjected to an internal pressure of 100,000 psi during firing.
$E = 30\times10^{6}\,psi,\ \nu = 0.3,\ \alpha = 6\times10^{-6}\,/^{\circ}F$.

18. An aluminum cylinder $(E_a = 10\times10^{6}, \nu_a = 0.33)$ with an outer radius of 6 in. is to be shrunk fitted over a steel cylinder $(E_s = 30\times10^{6}, \nu_s = 0.29)$ with an outer radius of 4.0098 in. and an inner radius of 2 in. Determine the interface pressure and the maximum stress in the aluminum cylinder.

19. A disk with radii b = 6 in. and a = 1.00 in. is press fitted on a shaft of radius = 1.003 in. Both are made of steel. Determine the stress distribution in the disk at 500 rpm and the speed at which the interface pressure goes to zero $(E_s = 30\times10^{6}, \nu_s = 0.29, \gamma = 0.28 lb/in^{3})$.

20. A bronze ring of 16 in. outside diameter is shrunk around a steel shaft of 8 in. diameter. At room temperature the shrink allowance is 0.004 in. on the radius. Calculate:

a) The temperature above room temperature to which the entire assembly must be raised in order to loosen the press fit.

b) The angular velocity in R.P.M. at room temperature, which will loosen the fit.

The constants are:
For steel: $E = 30 \times 10^6\, psi, \nu = 0.3, \alpha = 6.67 \times 10^{-6} /^\circ F, \gamma = 0.28 lb/in^3$
For bronze: $E = 15 \times 10^6\, psi, \nu = 0.3, \alpha = 10 \times 10^{-6} /^\circ F, \gamma = 0.33 lb/in^3$

The expression for the displacement u_r is given by Eq. (11.5.21) for the ring and by the equation in problem 12 for the shaft.

21. A steel tube 10 ft.long with 3 in. internal and 9 in. external diameter respectively and with fixed ends is to be subjected to an internal hydraulic pressure. Assuming that the tube remains elastic, calculate the amount of water that must be forced into it to raise the internal pressure to 30000psi. $E = 30 \times 10^6\, psi, \nu = 0.3$ and the bulk modulus of water K is 3 x 10^5psi.

22. A steel cylinder has an inside diameter of 5 in. and an outside diameter of 8 in. Another cylinder with an inside diameter of 7.995 in. and an outside diameter of 10 in. is heated and shrunk over it. An internal pressure of 25000 psi is then applied in the inner cylinder. Find the value of $\sigma_{\theta\theta}$ for the inner cylinder at r = 2.5 in. and r = 4.0 in; and the value of $\sigma_{\theta\theta}$ for the outer cylinder at r = 4 in. and r = 5 in. $E = 30 \times 10^6\, psi, \nu = 0.3$.

23. A long steel cylinder just fits around an inner long copper cylinder at room temperature. The ends are free. For the steel the outer diameter is 36 in. and the inner one is 24 in.; $E = 30 \times 10^6\, psi, \nu = 0.3$ and $\alpha = 6.5 \times 10^{-6} /^\circ F$. For the copper the outer diameter is 24 in. and the inner one is 18 in; $E = 13 \times 10^6\, psi, \nu = 0.3$ and $\alpha = 9.3 \times 10^{-6} /^\circ F$. The temperature is raised $150^\circ F$:

a) Find the pressure at the interface.

b) Find the value of the maximum tensile stress created in the steel cylinder and the maximum compressive stress created in the copper cylinder.

24. A short thick cylindrical sleeve with an outer radius of 2 in. is shrunk on a short solid shaft whose radius of 1 in. If the radial pressure p between the two is 10,000 psi, what should the inner radius of the sleeve be before the shrink fit operation? If the combination has a unit length, what is the expression of the strain energy density at a radius r of the sleeve? What is the value of the total strain energy in the sleeve? $E = 30 \times 10^6\, psi, \nu = 0.3$.

CHAPTER 15

[At the end of Section 15.10]

If one is looking for the displacement δ along a force F, Eq. (15.10.4) is written as

$$\delta = \frac{\partial U_t}{\partial F}.$$

If one is looking for the angle of rotation β at a point where the applied moment is M_{13} Eq. (15.10.4) is written as

$$\beta = \frac{\partial U_t}{\partial M_{13}}$$

[Additional Problems]

12. Fig. 15.28 shows a 1/2 in. diameter steel rod with one end fixed into a horizontal table. The remainder is bent into the form of $\frac{3}{4}$ of a circle and the free end is constrained by guides to move in a vertical direction. If the mean radius to which the rod is bent is 6 in., determine the vertical deflection of the free end when a load of 50 lbs is gradually applied there. Consider the effect of bending only. $E = 30 \times 10^6\ psi$.

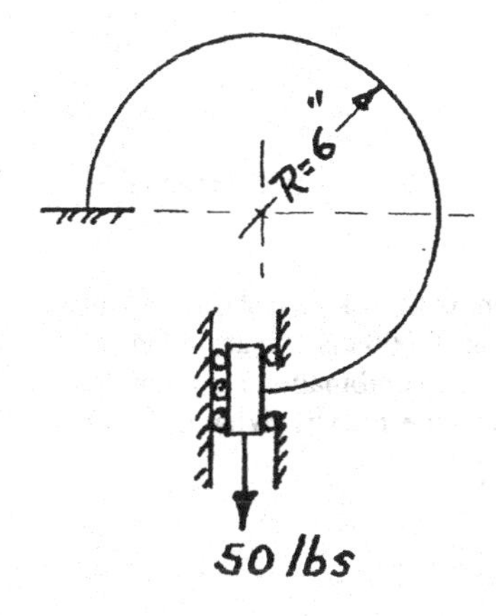

Fig. 15.28

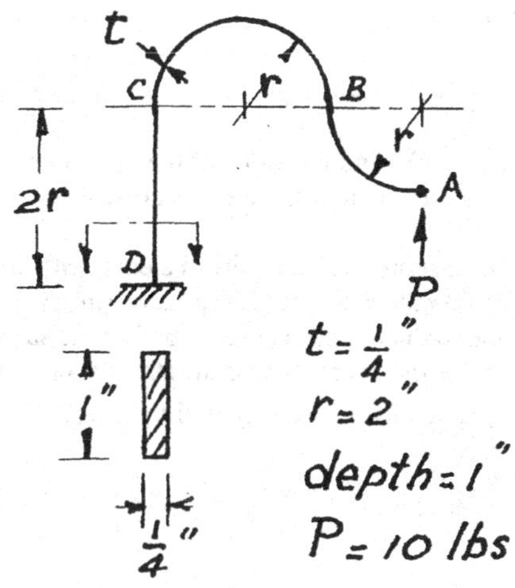

Fig. 15.29

Using any energy theorem and considering the bending terms alone, obtain the expression the horizontal and vertical displacements of point A in terms of P, r and EI (Fig. 15.29). 'en the dimensions shown on the figure, what are those displacements? $E = 30 \times 10^6\ psi$

The thick circular curved beam shown in Fig. 15.30 has a 1 in. square cross section and a an radius of 2.5 in. The beam is made of steel for which $E = 30 \times 10^6\ psi$ and $\nu = 0.29$. If P = lbs.,

a) Use the elementary theory of beams and an energy method to compute the vertical placement of A.

b) What is the error in the displacement if the energy terms containing the normal and aring forces were neglected?

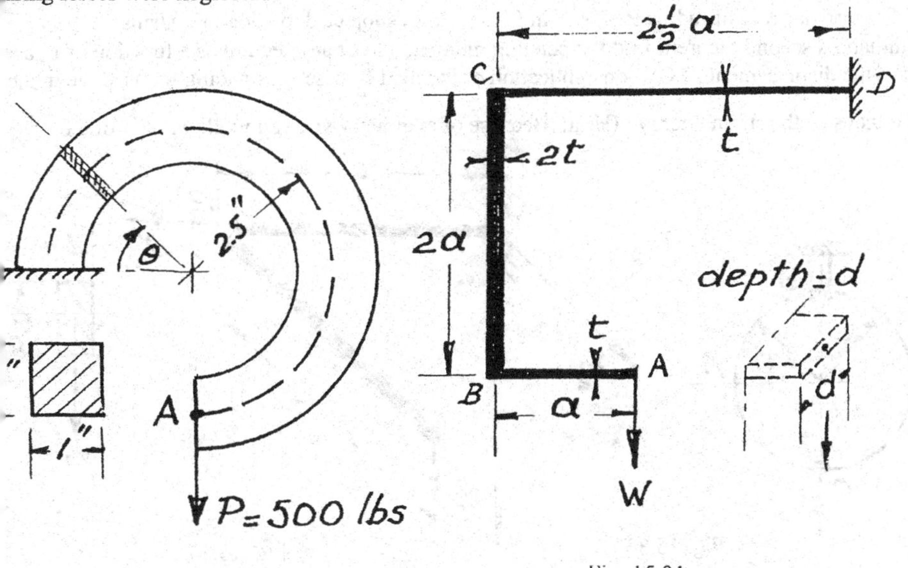

Fig. 15.30

Fig. 15.31

Obtain an expression for the horizontal displacement of a point A of the bent cantilever in . 15.31. Use bending terms only in the energy expressions. If a = 2 in., $t = \frac{1}{4}$ in., and $= 30 \times 10^6\ psi$, what is this displacement for W = 5 lbs? The depth d is 1 in.

When the load P is applied at B in the system shown in Fig. 15.32, the carriage moves and ults in a reaction and a moment at C. Using Castigliano's second theorem three times, find s reaction and moment as well as the deflection at B. Consider only the effects of the bending ns and give your answers in terms of P, R and EI when needed

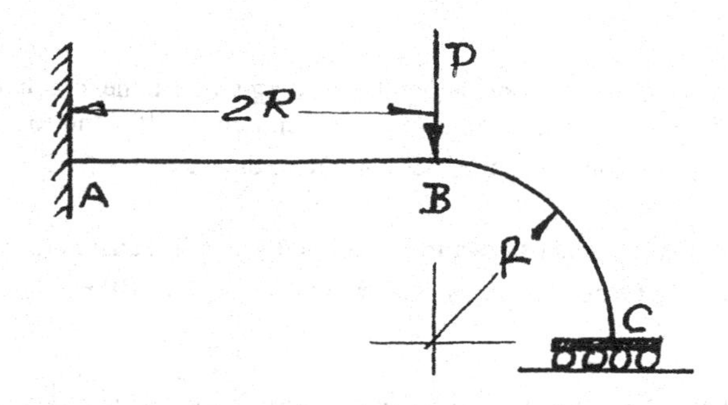

Fig. 15.32

17. The circular ring of radius R shown in Fig. 15.33 is subjected to a load P. Using Castigliano's second theorem find the bending moment M_{13} at any section as a function of θ as well as the displacement of C upon application of the load P. Use the bending terms alone in the expressions of the strain energy. (Hint: Because of symmetry you can work with $\frac{1}{4}$ of the ring).

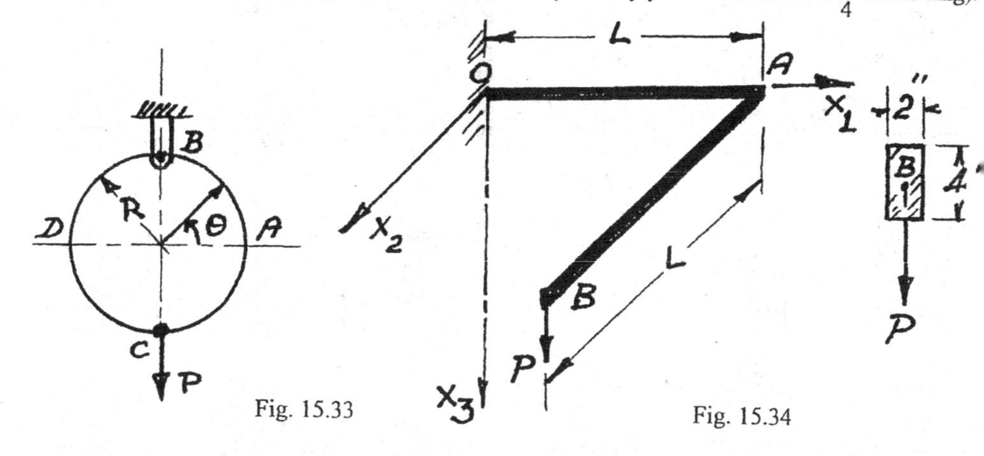

Fig. 15.33

Fig. 15.34

18. The right angle beam OAB shown in Fig. 15.34 is in the horizontal plane OX_1, OX_2 and is fixed at O. The load P is hanging vertically at B. P and L are known quantities; so are the properties of the material (E and G) and the cross section's properties (A, I, C_2, C_3 and J). Using all the terms of the energy equation (i.e. bending, shear and twist):

a) Write down the expression of the vertical displacement of B. Which is the dominant term?

b) What is the value of this displacement given that the cross section is rectangular (2 in x 4 in), P = 100 lbs, L = 5 ft, $E = 30\times10^6\ psi$ and $G = 12\times10^6\ psi$. A, I, C_2, C_3 and J can all be computed or obtained from tables.

CHAPTER 16

[Additional Problems]

8. A pinned strut AB of constant section carries an axial load P and terminal couples M_A and M_B both deflecting the strut in the same direction. Obtain a formula for the maximum stress in the strut.

9. A simple supported beam is loaded along an axis at a distance e from the centroid of the section by a load P. (Fig. 16.17). At the same time a uniformly distributed lateral load of intensity q per unit length acts so as to produce the maximum total bending effect. Obtain a formula for the maximum bending moment and deduce the value of the maxim deflection δ.

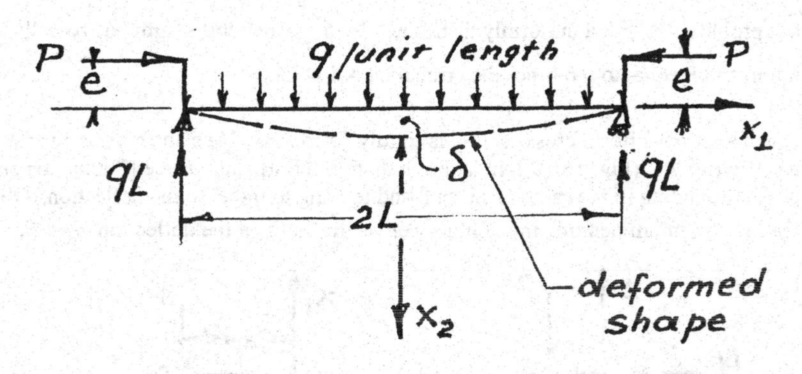

Fig. 16.17

10. An initially straight column of uniform cross section and of length L is fixed at the base and, in its unloaded condition, slopes at an angle α with the vertical. If a vertical load P is applied at the top (Fig. 16.18):

a) What is the maximum bending moment in the column?

b) If α is 2 degrees and P is 0.9 times the critical buckling load for a vertical column, obtain the ratio of the deflection at the top to the length of the column (i.e. $\frac{\delta}{L}$).

11. Repeat problem 10 with an added force $\frac{P}{5}$ applied horizontally at A, and P equal to 0.6 times the critical buckling load.

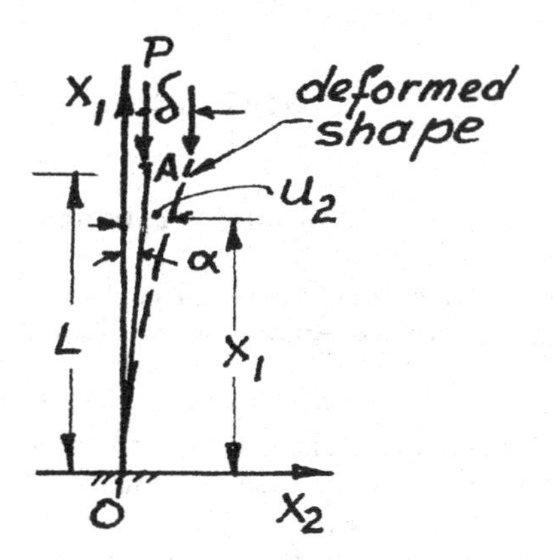

Fig. 16.18

12. Repeat problem 10 with a uniformly distributed load $\frac{P}{10}$ per unit length horizontally applied to the column and P equal to 0.6 times the critical buckling load.

13. A vertical strut of uniform cross section is rigidly fixed at the base and carries a vertical load P with an eccentricity *e* (Fig. 16.19). In addition there is a horizontal force H at the top acting so as to produce bending in the same plane as P and tending to increase the deflection. Obtain a formula for the maximum bending moment as well as the value of the deflection δ at $x_1 = L$.

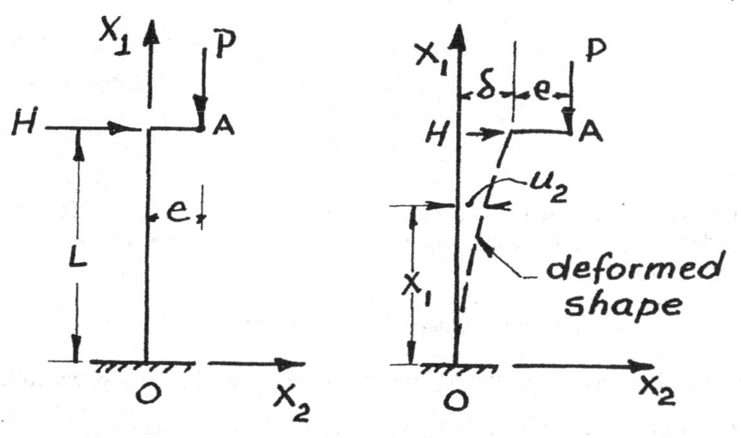

Fig. 16.19

14. A long slender strut of length L is fixed at one end B and pin-jointed at the other end. It carries a known axial load P and a known couple M_0 at the pinned end (Fig. 16.20). Its flexural rigidity is EI_3. What is the value of the fixed end moment M_B at B and what is the value of the force Q which is needed to have the pin move along a horizontal line AB.

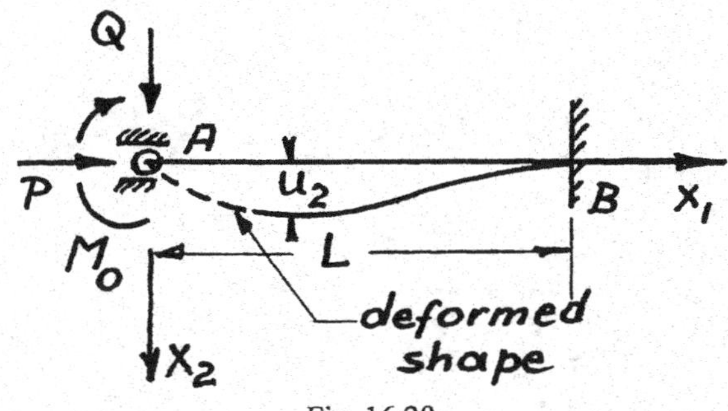

Fig. 16.20

5. The steel strut shown in Fig. 16.21 is to be subjected to an axial load P. Before the load is
pplied the strut is found to be slightly curved so that its center line has the shape

$$u_{20} = \delta_0 \cos\frac{\Pi x_1}{L} \text{ (Fig. 16.21a):}$$

hen the axial load is applied, an additional deformation takes place, so that, $u_2 = u_{20} + u_{21}$.

a) Find the expression for the ordinate of the deflection curve after the load P is applied.

b) Show that the ratio $\frac{\delta}{\delta_0}$ is given by $\frac{\delta}{\delta_0} = \frac{P_{cr}}{P_{cr} - P}$

ere P_{cr} is buckling load for a strut hinged at both ends.

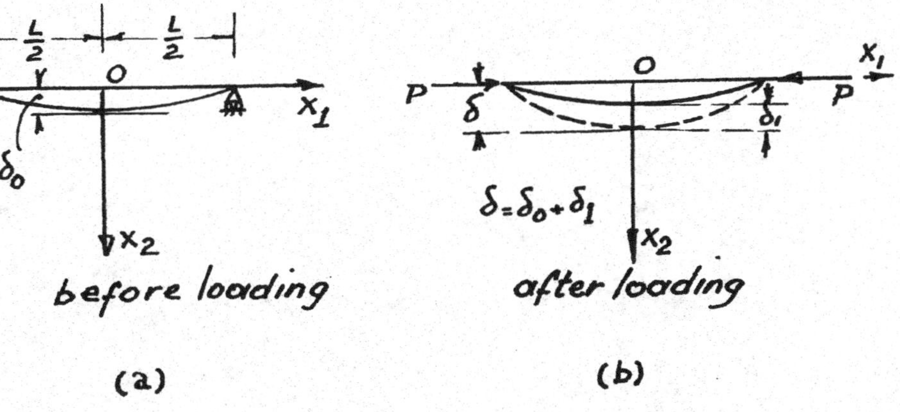

Fig. 16.21

Index